Advances in Science, Technology & Innovation

IEREK Interdisciplinary Series for Sustainable Development

Advances in Science, Technology & Innovation (ASTI) is a series of peer-reviewed books based on important emerging research that redefines the current disciplinary boundaries in science, technology and innovation (STI) in order to develop integrated concepts for sustainable development. It not only discusses the progress made towards securing more resources, allocating smarter solutions, and rebalancing the relationship between nature and people, but also provides in-depth insights from comprehensive research that addresses the **17 sustainable development goals (SDGs)** as set out by the UN for 2030.

The series draws on the best research papers from various IEREK and other international conferences to promote the creation and development of viable solutions for a **sustainable future and a positive societal** transformation with the help of integrated and innovative science-based approaches. Including interdisciplinary contributions, it presents innovative approaches and highlights how they can best support both economic and sustainable development, through better use of data, more effective institutions, and global, local and individual action, for the welfare of all societies.

The series particularly features conceptual and empirical contributions from various interrelated fields of science, technology and innovation, with an emphasis on digital transformation, that focus on providing practical solutions to **ensure food, water and energy security to achieve the SDGs**. It also presents new case studies offering concrete examples of how to resolve sustainable urbanization and environmental issues in different regions of the world.

The series is intended for professionals in research and teaching, consultancies and industry, and government and international organizations. Published in collaboration with IEREK, the Springer ASTI series will acquaint readers with essential new studies in STI for sustainable development.

ASTI series has now been accepted for Scopus (September 2020). All content published in this series will start appearing on the Scopus site in early 2021.

More information about this series at http://www.springer.com/series/15883

Inamuddin · Rajender Boddula ·
Mohd Imran Ahamed · Anish Khan
Editors

Advances in Green Synthesis

Avenues and Sustainability

Springer

Editors
Inamuddin
Department of Applied Chemistry,
Zakir Husain College of Engineering
and Technology, Faculty of Engineering
and Technology
Aligarh Muslim University
Aligarh, India

Mohd Imran Ahamed
Department of Chemistry,
Faculty of Science
Aligarh Muslim University
Aligarh, India

Rajender Boddula
CAS Key Laboratory of Nanosystems
and Hierarchical Fabrication
Beijing, China

Anish Khan
Department of Chemistry,
Faculty of Science
King Abdulaziz University
Jeddah, Saudi Arabia

ISSN 2522-8714 ISSN 2522-8722 (electronic)
Advances in Science, Technology & Innovation
IEREK Interdisciplinary Series for Sustainable Development
ISBN 978-3-030-67883-8 ISBN 978-3-030-67884-5 (eBook)
https://doi.org/10.1007/978-3-030-67884-5

Yurchanka Siarhei/shutterstock

This Springer imprint is published by the registered company Springer Nature Switzerland AG
The registered company address is: Gewerbestrasse 11, 6330 Cham, Switzerland

Contents

Biomass-Derived Polyurethanes for Sustainable Future

Felipe M. de Souza, Pawan K. Kahol, and Ram K. Gupta

Abstract

This book chapter starts with an overview of the ever-growing consumption of polyurethanes along with the main challenges faced by researchers and industry demonstrating how they fit into the current scenario. The main routes to synthesize polyols are discussed in detail, showing the versatility of polyurethane chemistry. Also how green approaches can be introduced to turn process sustainable, profitable, and many times demonstrating to be more effective than petrochemical-based. This chapter also focuses on biomaterials that are currently being used in industry such as corn, soybean, rapeseed, and castor oil as well as bio-renewable sources in ongoing research like terpenes, terpenoids, and lignin. Their chemical modification paths were discussed in detail, to give a proper understanding of how to synthesize them and how their properties can be matched with their petrochemical counterparts. The main procedures covered for the green synthesis include epoxidation, ring-opening, hydroformylation, hydrogenation, ozonolysis, thiol-ene, and transesterification.

Keywords

Bio-polyol • Polyurethanes • Green chemistry • Vegetable oil • Terpenes • Lignin

F. M. de Souza · R. K. Gupta (✉)
Kansas Polymer Research Center, Pittsburg State University, Pittsburg, KS 66762, USA
e-mail: rgupta@pittstate.edu

F. M. de Souza · R. K. Gupta
Department of Chemistry, Pittsburg State University, Pittsburg, KS 66762, USA

P. K. Kahol
Department of Physics, Pittsburg State University, Pittsburg, KS 66762, USA

1 Introduction

The polyurethanes are a vital class of polymers due to their vast range of applications and properties. They are as classified as rigid, flexible, elastomer, thermoplastic, waterborne, adhesive, coating, sealant, binders, and more which transformed the modern industry by implementing low-cost starting materials with high effectiveness. The polyurethanes improved human quality of life in so many aspects that it would be nearly impossible to detach them from daily life nowadays (Szycher 1999). The polyurethane industry established its share in the market back in 1937, thanks to the research of Professor Dr. Otto Bayer and colleagues (Seymour and Seymour 1989). Since then polyurethane industry thrived reaching currently a $69.2 billion net worth market in 2019, which projects a promising future with an expected growth of around 5.0–5.6% per year by 2025 (Bhatnagar 2018). This increase comes from the many applications of polyurethanes, for example, furniture industries, where a large quantity of polyurethanes are used for making chairs, mattresses, and sofas. The flexible structure of some foams allows them to be soft yet resistant to creep maintaining their shape for long period; a common example is the memory foam that adapts according to the body to provide a proper rest, conceding its use in health centers to relieve pressure sores (Collier 1996). The automotive industry also takes advantage of polyurethanes to manufacture seats, armrests, headrests, and interior components due to both lightweight and strong mechanical properties, which makes them fit precisely for this use, providing comfort as well as fuel efficiency because of overall weight decrease (Hăloiu and Iosif 2013). The footwear industry uses it for similar reasons, to add comfort, durability, and resistance to abrasion (Sacchetti et al. 1991). Due to their low heat transfer properties, polyurethanes are usable as thermal insulators in buildings, which besides providing a pleasant environment also contributes to saving energy due to less use of heaters and air conditioning, decreasing emissions of

Inamuddin et al. (eds.), *Advances in Green Synthesis*, Advances in Science, Technology & Innovation,
https://doi.org/10.1007/978-3-030-67884-5_1

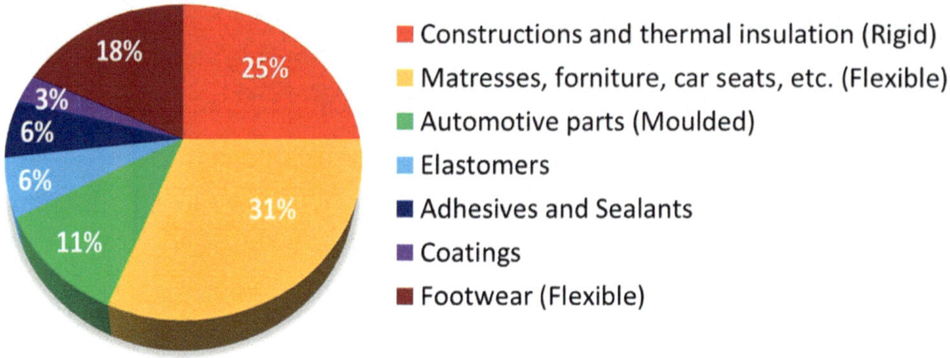

Fig. 1 Global usage of polyurethanes

CO_2 and greenhouse gases. For the same motive, they are widely used in freezers, being responsible for an increase in up to 60% of refrigeration efficiency after their implementation (Somarathna et al. 2018). Another field of their application is on coatings, bonding agents, and sealants due to their stability against corrosion, temperature amplitude, and radiation, enabling them to connect or cover many surfaces like rubber, wood, and even glass (Akindoyo et al. 2016). The classifications and applications of polyurethanes worldwide can be observed in Fig. 1.

1.1 Chemicals for Preparation of Polyurethanes

The polyurethanes are synthesized through a reaction between a hydroxyl (–OH) and isocyanate (–N=C=O) groups that leads to the urethane linkage [–HN–C(O)–O–]. To form a polyurethane, both starting materials must have at least bifunctionality as described in Fig. 2 (Ottenbrite 2001). This reaction can be performed by using a vast number of starting materials, granting many possible combinations that provide a broad set of properties for the final products, which is why polyurethanes are so versatile and continue to grow every year (Charlon et al. 2014). The main reagents required to make polyurethanes are polyols, isocyanates, catalysts, surfactants, and blowing agents.

The polyols are compounds with two or more –OH groups. There are two larger groups of polyols namely polyether and polyester, which represent more than 80% of oligo-polyols manufactured together (Ionescu 2006). The main starting materials for the synthesis of polyether-based polyols are given in Fig. 3. Because their structure has tense bond angles (Bayer's tension), they can be polymerized through ring-opening polymerization, which can be initiated by a starter polyol, which is a compound that contains typically 2 or 3 hydroxyl groups in its structure as described in Fig. 4.

Polyesters are another large group used as polyols that can be synthesized through an esterification reaction mostly between dicarboxylic acids and di or polyols. A general reaction describing esterification is shown in Fig. 5. Other groups of polyols used for the synthesis of polyurethanes are the polycarbonates, polyacrylics, and polybutadiene diols. Their general structures are provided in Fig. 6. The polyols' main chain size is an important factor that dictates the properties of the final polyurethane in a way that high molecular weight polyols (2000–10,000 Da) may yield more flexible polyurethanes while shorter molecular weights or smaller polyol molecules yield rigid ones (Ionescu 2006).

The isocyanates are the other main half component for the synthesis of polyurethanes. Most of the isocyanates used for the synthesis of polyurethanes are bifunctionality in nature and thus referred to diisocyanates. The molecular structure of isocyanates can vary from a more rigid to flexible which can influence the properties of the final polyurethanes. For example, with a proper combination of polyols, toluene diisocyanate (TDI) provides polyurethanes with the most rigid structure while hexamethylene diisocyanate (HDMI)

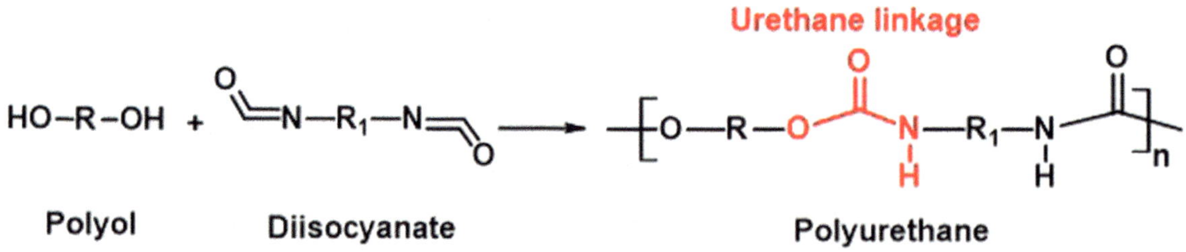

Fig. 2 General reaction between a polyol and diisocyanate to form a polyurethane

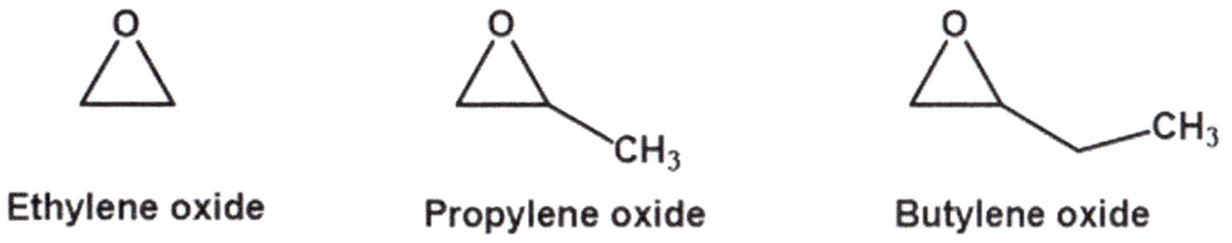

Fig. 3 Main raw materials for polyether polyols synthesis

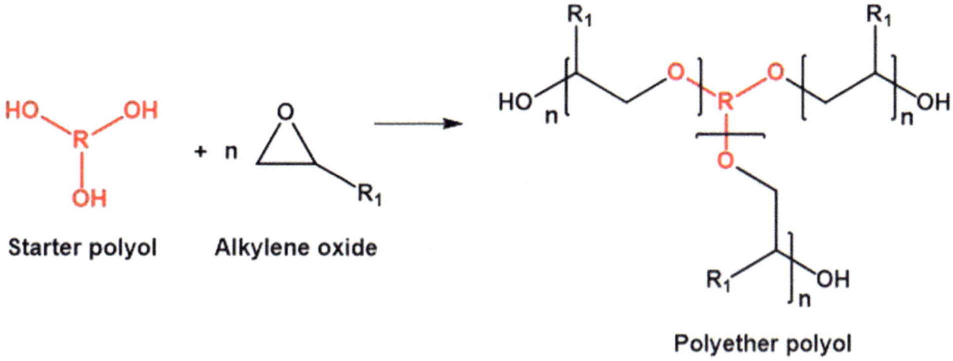

Fig. 4 General ring-opening polymerization for polyether polyols

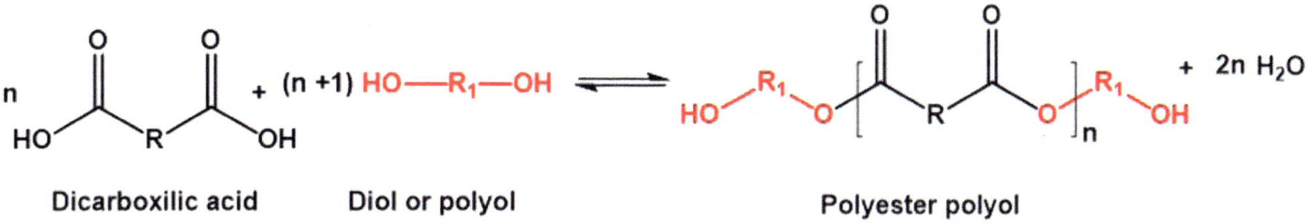

Fig. 5 General polyesterification reaction for the synthesis of polyester polyols

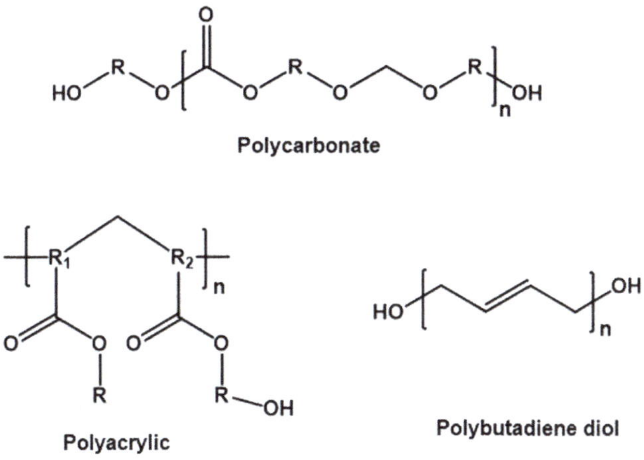

Fig. 6 General structure of polycarbonate, polyacrylic, and polybutadiene based polyols

produces flexible polyurethanes. Based on their properties, they are applicable in thermal insulators (rigid), production of mattresses (flexible), soles of shoes (elastomer), etc.

Figure 7 shows the name and structure of the most commonly used diisocyanates. Despite the essential role of isocyanates in the synthesis of polyurethanes, the isocyanates raise concern regarding their toxicity and environmental issues. There is a growing need to find alternative synthetic routes for polyurethanes that do not demand or at least diminish the use of isocyanates. Recently, non-isocyanate-based polyurethanes are being researched. For example, the reaction between carbonated vegetable oil with aminosiloxanes yielded a non-isocyanate compound able to react with hydroxyl groups from natural-based materials such as lignin, which demonstrated a promising alternative route (Lee and Deng 2015).

The catalysts for polyurethanes are important to accelerate the foaming process, which is crucial for the industry. Mainly two types of catalysts are used, even simultaneously, during the formulation of a foam. They can be amines or/and metal complex-based compounds based on zinc, bismuth tin, and lead. Some of the most common amine-based catalysts are displayed in Fig. 8. Likewise the isocyanates, catalysts

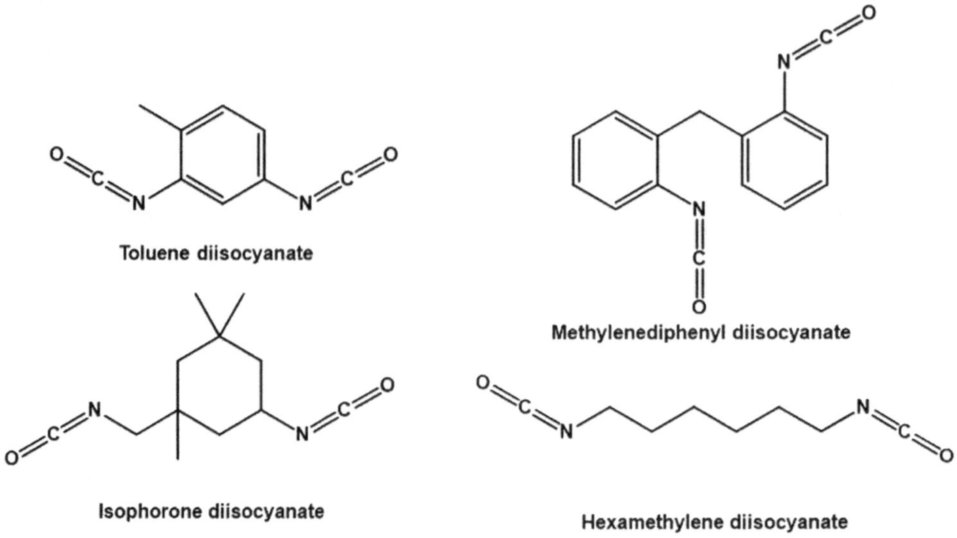

Fig. 7 Name and structure of most common diisocyanates used for making polyurethanes

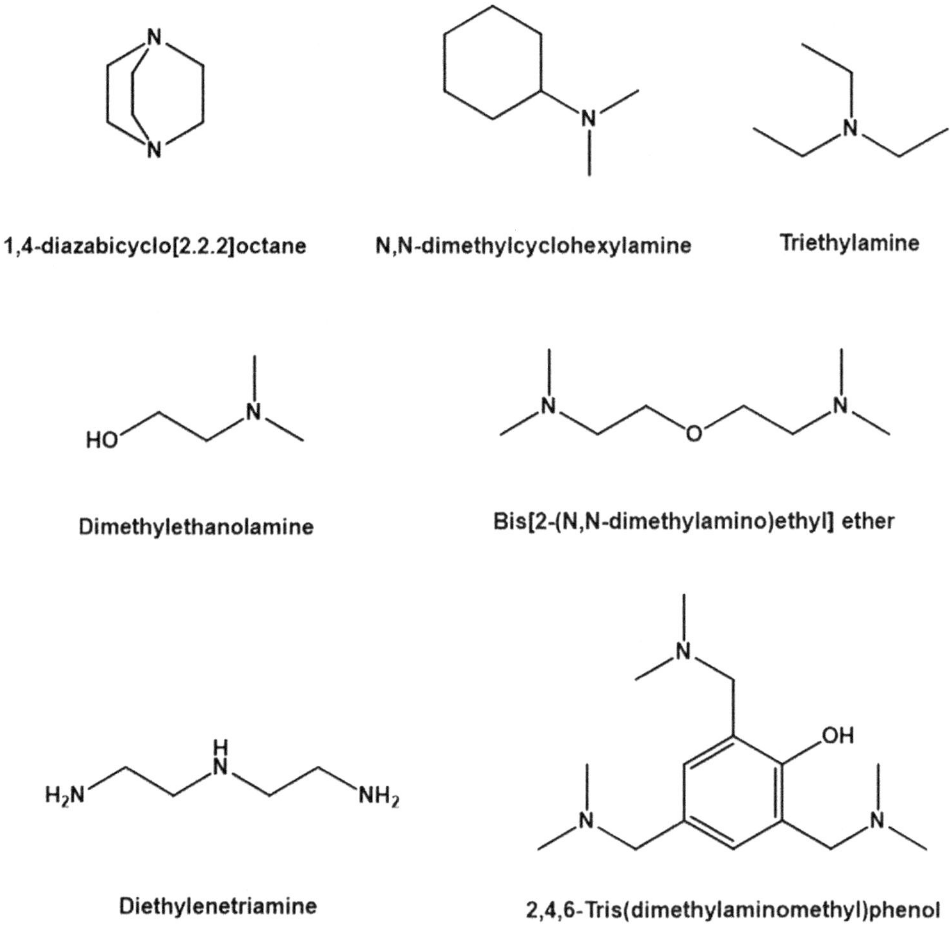

Fig. 8 Name and structure of common amine-based catalysts

for polyurethanes are essential components, especially for large scale applications, but also present both health and environmental concerns regarding the lead and tin-based, which again drive researchers to develop new approaches that are less harmful and eco-friendly.

The surfactants in polyurethanes play the role of forming an emulsion to allow proper mixing of the components, which is important during the foaming process to control and stabilize cell size as well as the structure of the foam to avoid flaws. When used in non-foaming procedures, surfactants are used to prevent inner bubbles to obtain a uniform surface. Generally, surfactants can be cationic or anionic, the first one used for better emulsification action while the second to provide stability against corrosion (Akindoyo et al. 2016). Similar to surfactants, the blowing agents also have the role of controlling the cellular size because of the formation of bubbles within the foam. Blowing agents also minimize the cost due to an increase in the volume of the foams, making them less dense, which requires less material to cover a larger area. However, there is an optimum amount of blowing agent that can be added to the formulation; otherwise, it can compromise the foam's mechanical properties. There are two types of blowing agents, chemical and physical. Water is a chemical blowing agent as water reacts with isocyanate during the foaming process to produce CO_2 gas. Water is a preferred blowing agent for many industrial foaming processes due to its effectiveness, low-cost, and environmentally friendly nature. Physical blowing agents are mainly gases that are physically injected during the foaming process. The most commonly used physical blowing agents were chlorofluorocarbon (CFC)-based compounds, although they are effective but raised many environmental concerns as they cause ascend to the stratosphere and deplete the ozone layer (Kumaran and Bomberg 1990). The use of chlorofluorocarbon-based compounds was restricted by the Montreal Protocol back in 1989.

1.2 Importance of Green Chemicals and Synthesis Methods

The worldwide concern about environmental issues came mostly after the escalation of global warming due to high emission levels of greenhouses gases, which led to international protocols that pushed the development of new materials obtained from renewable sources. It became a general concern as the production of plastics increased from 5 to 300 million tons per year from 1950 to 2019 (Dissanayake and Sinha 2013). Very few countries are recycling plastics, for example, the USA recycled about 8.4% of the produced plastics while Europe recycled about 40% in 2017 (Zhao et al. 2008; Kale et al. 2007). Notably, most nations are slowly taking care of the environment, but there are some issues to overcome such as contamination and technical constraint that limits more recycling, emission of greenhouse gases during the production of virgin plastics, and contamination of the oceans that are estimated to be around 100 million tons. These factors are compelling to develop sustainable routes for the production of commercial polymers (Zhao et al. 2008; Chidambarampadmavathy et al. 2017; Dodbiba and Fujita 2004).

Green synthesis involving new methods and chemicals derived from plant-based sources offers several advantages of being environmentally friendly and cost-effective compared to oil-derived sources. Plant-based chemicals can be produced annually while oil-based chemicals take several years to reproduce. Another important feature of bio-based materials is the possibility of turning them into biodegradable, which not only gives them a proper final destination but also implements a more effective recycling step to the process compared to petrochemical-based materials (Akindoyo et al. 2016; Llevot et al. 2016). This scenario provides a stable schedule for long-time production and pricing since the harvesting of these renewable sources no longer requires petroleum quarries that are spread randomly around the globe, which leads to less friction between nations and decrease the dependence in petro-based materials. Petrochemicals are widely used in many areas such as automotive, households, aircraft, and clothing, which create a large demand for the starting materials; hence, the use of alternative sources decreases the dependence of non-renewable raw materials by providing reasonable and even cheaper paths.

1.3 Characteristics of Biomaterials for Polyurethanes

Many biomass-derived compounds can be used as raw materials for the synthesis of polyols, mostly because of unsaturations that can be converted into hydroxyls groups in a variety of ways. Usually, primary or secondary hydroxyl groups are often desired in the bio-derived compounds as they are more reactive with isocyanates (Kolb et al. 2001; Ramanujam et al. 2019; Ionescu and Petrović 2010). The biomaterials suitable for polyols cover a broad range of structures that grant distinct properties for their derived polyurethanes. Few examples of suitable biomaterials and their sources are given in Table 1 (Blasbalg et al. 2011; Satoh et al. 2006; Sissener et al. 2018; Belgacem and Gandini 2008).

2 Bio-Oils as a Renewable Resource for Polyurethanes

The depletion and price instability of petroleum-based resources and growing environmental concerns created a need to find sustainable starting materials that provide similar properties with low production cost compared to petrochemical-derived products. To attend this requirement, the scientific community made great efforts to create alternative routes for the synthesis of polyurethanes using renewable sources. From this line of work, bio-polyols were synthesized using various methods. For example, soybean oil-based polyurethanes provide a rigid polyurethane with

Table 1 Suitable compounds for polyols and its main characteristics and sources

Name	C=C bonds	Carbons	Main sources
Oleic	1	18	Olive, canola, rapeseed, refined tall, sunflower and fish
Linoleic	2	18	Corn, cottonseed, soybean, refined tall and sunflower
α-linolenic	3	18	Linseed, refined tall, rapeseed and canola
γ-linolenic	3	18	Linseed, refined tall, rapeseed and canola
Arachidonic	4	20	Eggs, beef, pork, and fish
Erucic	1	22	Fish, mustard seed, and rapeseed
Docosapentaenoic acid (DPA)	5	22	Poultry, finfish, and shellfish
Docosahexaenoic acid (DHA)	6	22	Finfish, poultry, and shellfish
Limonene	2	10	Orange peel, lemon, lime, and other citric fruits
Carvone	2	10	Caraway, spearmint, and dill
α-phellandrene	2	10	Turmeric, dill, angelica root and eucalyptus dive
Myrcene	3	10	Mango, hops, and lemongrass

mechanical, insulating, and thermal properties that are similar to petrochemical-based polyurethanes (Mu et al. 2012). In another approach, transesterification with glycerol catalyzed by triethanolamine was used to synthesize castor oil-based rigid polyurethane (Petrović et al. 2008). Another interesting example was the conversion of rapeseed oil into polyol through microwave synthesis that yielded a flexible polyurethane (Dworakowska et al. 2012).

The starting point to consider the use of a bio-oil is the presence of double bonds, which are the versatile reactive sites that can be functionalized in many ways. Conversion of a double bond to –OH group is one of the highly studied and developed processes as it provides a reaction site for the isocyanate to form urethane linkage (Ramanujam et al. 2019; Zhang et al. 2018; Elbers et al. 2017). Some common examples of bio-oils that have double bonds in their structures can be converted into polyols are soybean, corn, canola, limonene, carvone, and palm oil (Ramanujam et al. 2019; Javni et al. 2003; de Souza et al. 2020). Some natural oils can be directly used as polyols to make polyurethanes due to the presence of hydroxyl groups in their structure. Castor and Lesquerella oils are among the few natural oils which possess hydroxyl groups in their structure and therefore could be used for the preparation of polyurethanes. Such oils are currently implemented in the industrial production of polyurethanes as well as in research and development.

Initially, the castor oil extracted from the castor plant, *Ricinus communis*, had no commercial use due to its unsuitability for diet as it contains a deadly protein named ricin (Challoner and McCarron 1990). Also, the castor plant spreads easily to other commercial crops making it even more undesirable for formers to grow. However, after seeing its commercial value for the production of polyurethanes and increasing demand for renewable resources for industrial applications, there has been increased interest in harvesting castor plants mostly in India (Haynes et al. 2001; Milliano et al. 2010; Mittal et al. 1991). The increased demand for castor oil for industrial applications is due to the uniqueness of the oil. Castor oil contains almost 90% of a single triglyceride from glycerin and ricinoleic acid, which contains a secondary hydroxyl group as well as unsaturation, while most of the other oils possess a mixture as described in Fig. 9 (Yeadon et al. 1959; Babb 2012; Ehrlich et al. 1959). As noted, castor oil contains three hydroxyl groups in its structure which allow castor oil to use polyols without any chemical modification for the polyurethane industry. Also, the presence of unsaturation opens many possibilities for chemical modifications, which can provide a wide range of properties (Hansen 1972; Moore and Norton 1953; Müller 2014). Currently, castor oil is widely used for flexible polyurethane foams, rigid foams, elastomers, and lubricants due to facile growth, extraction, low-cost, and improved properties (Szycher 1999; Hablot et al. 2008; Javni et al. 1998).

Lesquerella oil, extracted from *Lesquerella fendleri*, also presents a chemical structure similar to castor oil. Lesquerella oil contains a double bond at C_{11} and –OH group at C_{14} in a backbone of 21 carbons. *L. fendleri* prefers slightly dry and alkaline environments, therefore mostly found in southwestern regions in the USA (Hayes and Kleiman 1992; Dierig et al. 1996; Puppala and Fowler 1999). Although the chemical structure of Lesquerella oil is suitable to use for the polyurethane industry, the low quantity of extractable oil from the plant (a maximum of 25% by weight) restricts its industrial applications. Despite the low industrial demand of *L. fendleri* for the polyurethane industry, it can be used for food purposes as its seed is non-toxic and contains a high protein of around 35% (Babb 2012; Dierig et al. 1996;

Fig. 9 The chemical structure of castor oil

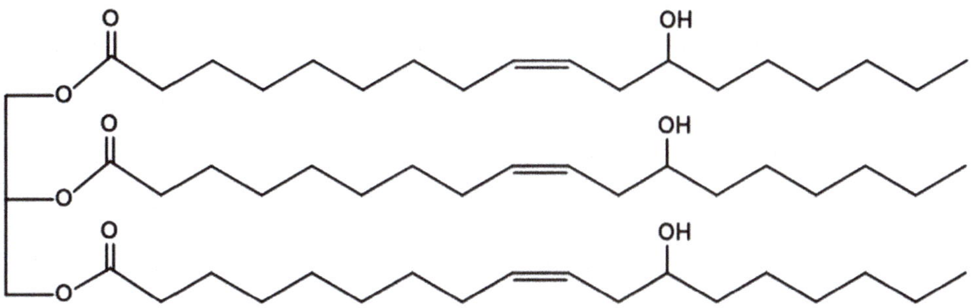

Puppala and Fowler 1999). Nevertheless, Lesquerella-based polyurethanes were developed for coating applications, due to its proper chemical structure (Sharmin et al. 2015).

There are many versatile procedures to convert natural oils into bio-polyols, such as epoxidation and ring-opening, hydroformylation and hydrogenation, ozonolysis, thiol-ene, and transesterification. The following sections discuss the various methods to convert bio-oils into polyols.

2.1 Epoxidation and Ring-Opening Reactions

The epoxidation reaction consists of the insertion of an oxygen atom into a double bond to create an oxirane ring, which is a very reactive group due to the tension of the bonds in a short cycle of three atoms, famously known as 'Bayer's tension.' This characteristic allows the epoxy groups to be used as raw materials to produce epoxy resins, functionalized alcohols, alkanolamines, glycols, and polymers such as polyethers, polyesters, and polyurethanes (Cai et al. 2008). There are four main ways to synthesize epoxy compounds from olefins. The first method is epoxidation using percarboxylic acids; being *m*-chloroperoxybenzoic acid (MCPBA) the most common, its mechanism can be described in Fig. 10 (Rios et al. 2005; Warwel and Klaas 1995; Bohnet 2003). The second method is epoxidation through inorganic or organic peroxides, such as transition metal-based, nitrile hydrogen, and alkaline peroxides (Finn and Sharpless 1985). This method is considered a clean method due to low cost and high yield and widely used in industries (Goud et al. 2006). The third method uses hypohalous acids of general formula HOX and respective salts mostly to form epoxies in olefins that present electron-withdrawing neighbor groups (Bohnet 2003). The fourth and most effective way is epoxidation through molecular oxygen. This method does not require a catalyst and is considered as the cleanest method. However, it is not applied to the industry because it demands too specific equipment and extensive safety measures due to the risk of explosions (Cai et al. 2008; Bohnet 2003).

The synthesis of epoxies is applied to a variety of bio-oils derived from corn, soybean, sunflower, linseed, safflower, etc. The follow-up procedure usually consists of opening the oxirane ring to turn it into a secondary hydroxyl group that can react with diisocyanates to make polyurethane foams. This reaction can be carried out in most of the epoxidized bio-oils by performing reflux of methanol or other alcohols under mild conditions such as temperatures between 35 and 45 °C, pressure around 0.1–0.2 MPa and catalyst, at which the largest used one is tetrafluorboricacid (HBF_4) a strong Lewis acid (Ionescu et al. 2007). A general ring-opening reaction is described in Fig. 11 (Ionescu et al. 2007; Dai et al. 2009). Some possibilities reported by authors involve ring-opening reactions for epoxidized bio-oils with larger aliphatic alcohols that provided a change in properties causing the resulting polyurethanes to lose some of its mechanical properties due to increase of dangling groups but increase environment stability because of higher hydrophobicity (Zhang et al. 2015). Also, synthetic routes have been performed to introduce primary hydroxyl groups due to their higher reactivity toward isocyanates by performing a ring-opening reaction with ethylene oxide, mentioned as ethoxylation (Ionescu et al. 2007).

Fig. 10 Epoxy formation mechanism reaction for a general olefin with MCPBA

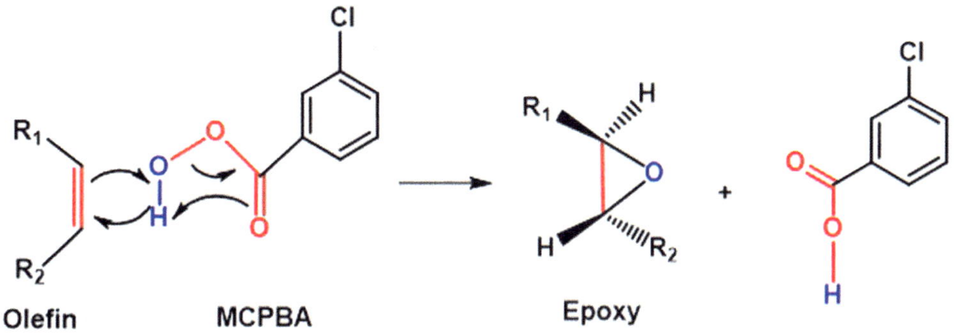

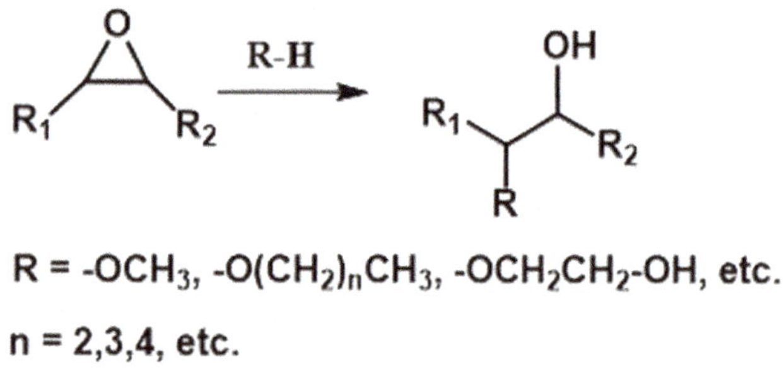

Fig. 11 General ring-opening reaction of epoxides

2.2 Hydroformation and Hydrogenation Reactions

Hydroformylation involves the conversion of a double bond into an aldehyde through a reaction with a gaseous mixture of CO and H_2 which requires a high temperature and pressure to shift equilibria toward the product formation and catalyst such as rhodium or cobalt phosphine complexes. The obtained aldehyde can be reduced to alcohol through hydrogenation, which is a reaction with hydrogen under pressure and catalysts such as Pt, Raney Ni. General reactions of both procedures were given in Fig. 12. These technologies were implemented around 1950 and caused a great impact in the chemical industry because they allowed the conversion of many unsaturated organic compounds such as precursors for fragrances, detergents, drugs and currently used to synthesize polyols (Pino and Botteghi 2003; Ojima et al. 2000; Franke et al. 2012).

Hydroformation and hydrogenation reactions were successfully used to converted soybean oil into polyol (Guo et al. 2002). It was observed that catalysts have a significant role in these reactions and affect the properties of the polyurethanes. For example, rigid polyurethanes were obtained when the synthesized polyol was catalyzed by rhodium-based compounds, while rubbery polyurethanes resulted when cobalt-based catalysts were used. The polyols synthesized through this method present primary hydroxyl groups, which are more reactive with isocyanates. The –OH group in the middle of the chain causes a part of the backbone to act as a pendant group, leading to a plasticizing effect. Therefore, such reactions and modifications can be used to obtain flexible polyurethanes (Guo et al. 2002; de Souza et al. 2012). Besides, hydroformylation and hydrogenation are green and clean methods that are used in a broad range of applications, allowing the chemical introduction of primary hydroxyl groups within the backbone of an unsaturated compound. Despite the requirements of high temperature, pressure, and expensive catalyst, it is a widely used technology in the industry (Frankel and Thomas 1972; Frankel 1973; Frankel 1976).

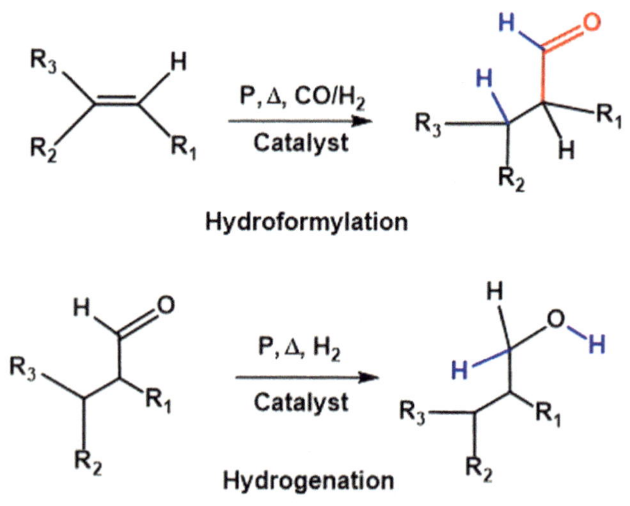

Fig. 12 General hydroformylation and hydrogenation reaction for the conversion of double bonds into hydroxyl groups

2.3 Ozonolysis

The unsaturation within the vegetable oils' backbone is the main source for different reactions that allow their conversion into polyols. Ozonolysis is another method to take advantage of unsaturation in the vegetable oils to convert them into polyols. This method differs from other mentioned methods as it causes cleavage of both π and σ bonds within the carbon chain leading to the formation of two new compounds. If the carbon attacked by ozone is primary, carboxylic acid and an aldehyde or a terminal primary alcohol (depending on condition) can be formed. If the carbon attacked by ozone is secondary, a ketone will be formed instead of an aldehyde as described in Fig. 13a, b. The aldehyde and ketone can be reduced to primary and secondary alcohol respectively in subsequent reactions as described in Fig. 13c, d. For most known cases, fatty acids derived from vegetable oils have linear chains, which lead to the formation of aldehyde as a product of ozonolysis. These aldehydes can be reduced to an terminal primary alcohol yielding the general structure of a polyol shown in Fig. 14 (Petrović et al. 2005). Primary alcohol at the end of the chain allows the whole backbone to be part of polyurethane's structure. Differently from a secondary hydroxyl group in the middle of the chain, which after the reaction with isocyanate would leave a hanging group that could act as plasticizers, hence decreasing the mechanical properties. This effect explains why rigid polyurethanes can be obtained through this method, as it has been successfully employed to synthesize polyols derived from vegetable oils, such as corn, canola, and castor oil (Petrović et al. 2005; Narine et al. 2007a, b). Ozonolysis leads to side products such as carboxylic acids which are impurities since their reaction with

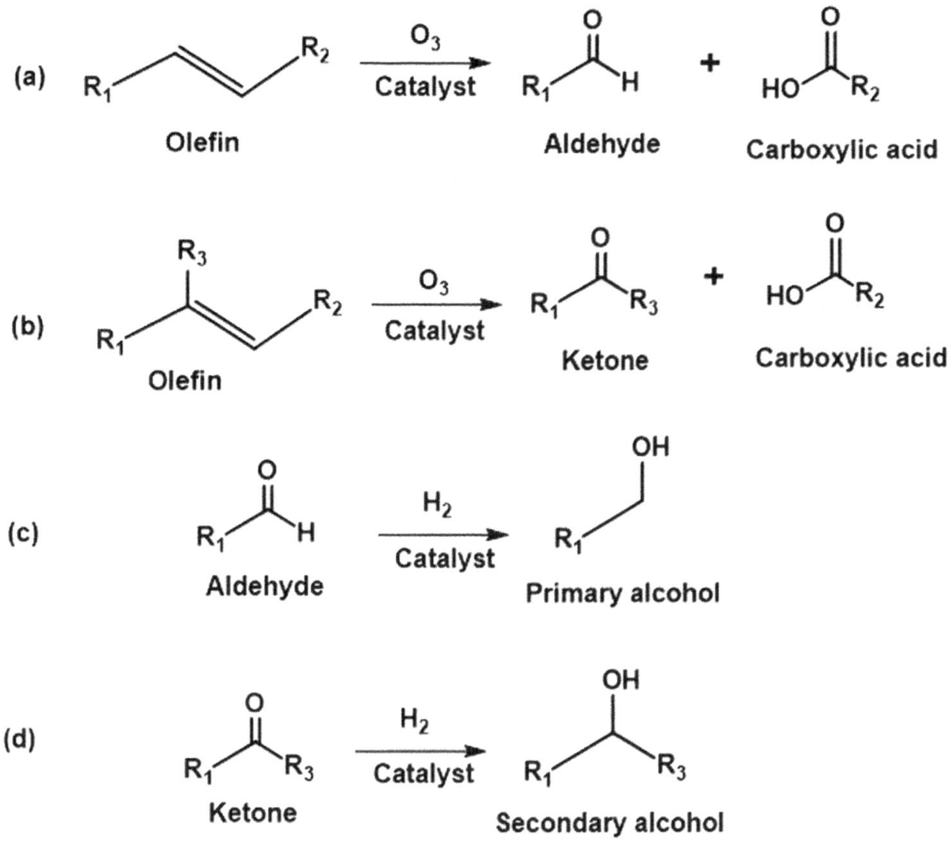

Fig. 13 General ozonolysis of in an olefin **a** not substituted, **b** substituted followed by reduction to a **c** primary alcohol, and **d** secondary alcohol

diisocyanate does not yield a polyurethane Fig. 13. This makes a purification process necessary for this method. Another factor that reinforces this need is that previous studies showed that a vegetable oil converted into polyol through ozonolysis may have a smaller hydroxyl number than other methods. It happens because the chain is cleaved at the double bond, which allows the formation of only one terminal hydroxyl group per chain of a triglyceride, despite the number of unsaturations that previously existed as demonstrated in Fig. 14 (Zlatanić et al. 2002; Frollini et al. 2016). The procedure can be performed in many ways, for example, by bubbling ozone into the reaction mixture along with catalysts such as $CaCO_3$, pyridines, or NaOH under low temperature, followed by wash and purification of the polyol under vacuum or low pressure (de Souza et al. 2012; Petrović et al. 2005; Tran et al. 2005).

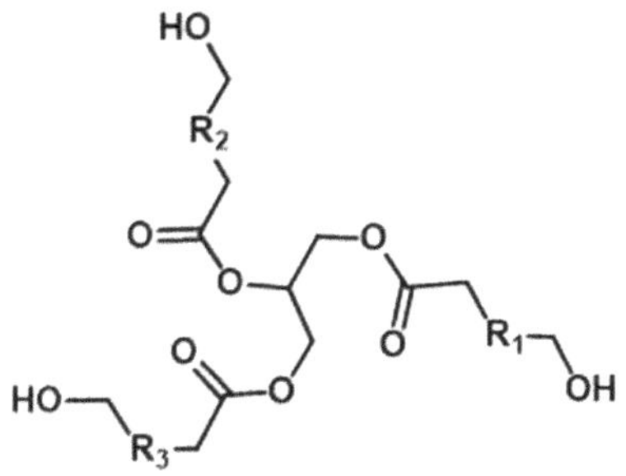

Fig. 14 General structure of a vegetable-derived polyol from ozonolysis

2.4 Thiol-Ene Reaction

The thiol-ene reaction was first reported around 1905 and quickly became a very useful tool for many chemicals, biochemical, and industrial procedures to fabricate coatings, electronic adhesives, shape-memory foams for medicinal applications and nanoengineered materials (Posner 1905; Pappas 1985; Killops et al. 2008; Nair et al. 2010; Morgan et al. 1977). The searching for sustainable routes to produce renewable materials made both industry and researchers land into the thiol-ene chemistry that grew and still growing into one of the most beneficial methods. Thiol-ene reactions are green, low-cost, high yield, and facile procedures that do not require solvents and high temperatures (Desroches et al. 2011; Griesbaum 1970; Lluch et al. 2010). This method can be performed by using a photoinitiator in the presence of ultraviolet light, which prompted the addition of a mercaptan (S–H) into a carbon–carbon double bond (C=C) through the formation of a thiyl radical that attach to the double bond followed by abstraction of hydrogen radical. It could also lead to side reactions such as disulfide or hydrogen

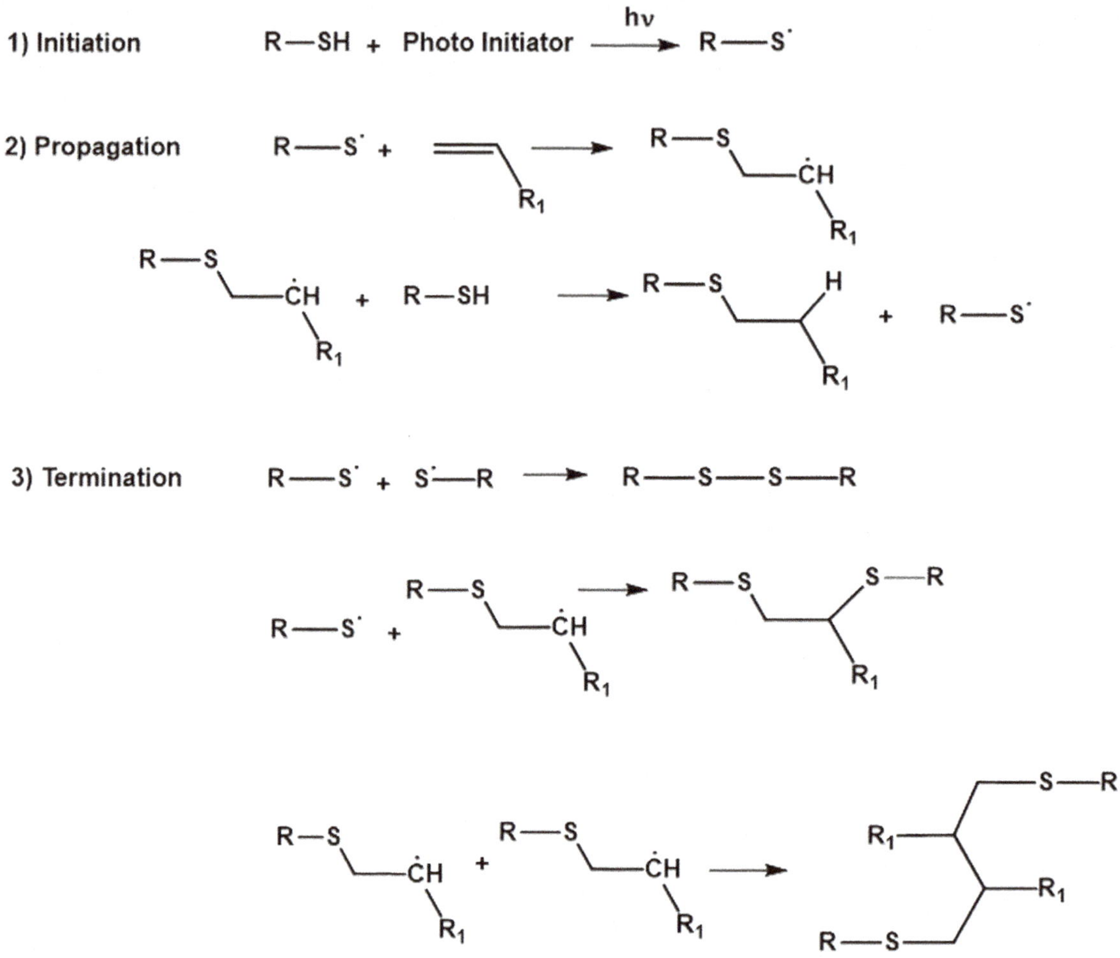

Fig. 15 General radical mechanism for a thiol-ene reaction. Adapted with permission (Hoyle et al. 2004). Copyright (2004) Wiley Periodicals, Inc.

recombination and formation of dimers as given in Fig. 15 (Morgan et al. 1977; Desroches et al. 2011; Hoyle et al. 2004). It can also be performed through ionic mechanisms when clays and metal sulfides are used as catalysts (Hagberg et al. 2007). The versatility of the thiol-ene reaction allows the conversion of olefins using a mercaptan with virtually any end group. Therefore, many bio-polyols can be synthesized using bio-renewable oils such as corn (Ramanujam et al. 2019), soybean (Yang et al. 2017), castor (Ionescu et al. 2016), lignin (Liu and Chung 2017), limonene (Zhang et al. 2018), carvone (de Souza et al. 2020), and many others by functionalization of their unsaturation with mercapto compounds containing a hydroxyl group at the end of their structure, allowing a facile and green route to obtain bio-polyols. Some kinetics studies made on thiol-ene reaction showed that the higher electron density in the double bonds is more the reactivity toward the thiyl radical. This phenomenon is favorable for bio-oils since the majority of them present linear and aliphatic chains that increase the reactivity of the double bonds through the inductive effect of the backbone (Morgan et al. 1977). Another factor is that the thiol-ene reaction yields a primary hydroxyl group into the chain as demonstrated in Fig. 16, which is more reactive with isocyanate (Hoyle et al. 2004).

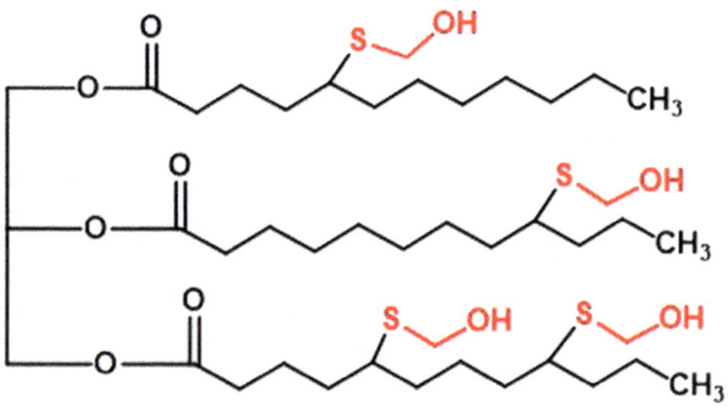

Fig. 16 General structure of a thiol-ene bio-derived polyol

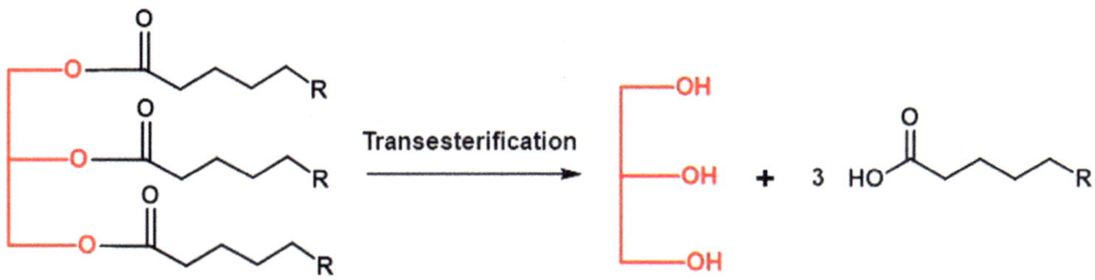

Fig. 17 General transesterification reaction

2.5 Transesterification Reaction

As previously mentioned, most of the oils derived from plants are found in their triglyceride form, which requires functionalization through many methods to convert them into useful bio-based polyols for polyurethane industries. The transesterification reaction is a method that relies on the breaking of one or two of the ester bonds in the triglyceride to yield one or two hydroxyl groups, hence increasing the functionality and allowing proper reaction with diisocyanates to make polyurethanes. The general transesterification reaction can be described in Fig. 17. As noted, this method provides a primary hydroxyl group that besides increasing reactivity with isocyanate also does not leave any pendant groups which generally increase mechanical properties. The reaction can be performed under mild temperature and catalyzed through alkaline media such as sodium, potassium, or aluminum hydroxide beside others (Arniza et al. 2015). Usually, transesterification is performed in conjunction with other technologies such as epoxidation and ring-opening, as described in Fig. 18, since there are remaining unsaturated bonds from the other chains that can be further functionalized with hydroxyl groups (Arniza et al. 2015; Ji et al. 2013; Kamil et al. 2011).

3 Terpenes as Green Starting Chemicals for Polyurethanes

As demonstrated so far, many vegetable oils can be used in industry as an alternative for non-renewable sources. Among non-renewable sources for chemicals, terpenes are a large group of bio-derived materials mostly originated from plants as well as from insects, fungi, and aqueous microorganisms (Wilbon et al. 2013). In plants, the terpenes are secondary metabolic products that can function as a defense mechanism to expel predators or to attract some insects to induce pollination.

Dr. Otto Wallach, the Nobel Prize winner in Chemistry in 1910 for his research on alicyclic compounds, was the scientist that discovered the structural pattern of terpenes and created the 'isoprene rule' which stated that terpenes are mostly substances with a 2-methyl-1,4-butadiene backbone that derived from condensation reactions (Clark 1999). Later in 1950, Leopold Ruzicka, also a Nobel Prize winner in Chemistry in 1939, proposed the 'biogenic isoprene rule' that stated besides condensation, terpenes can also be obtained through cyclization and rearrangement from precursors such as geranyl pyrophosphate, which can form

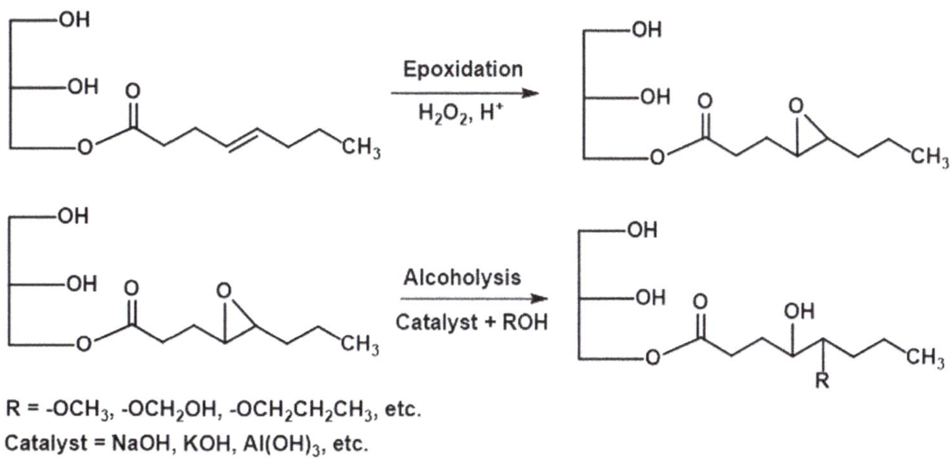

Fig. 18 Follow-up procedure for further functionalization of general monoglyceride through epoxidation and ring-opening

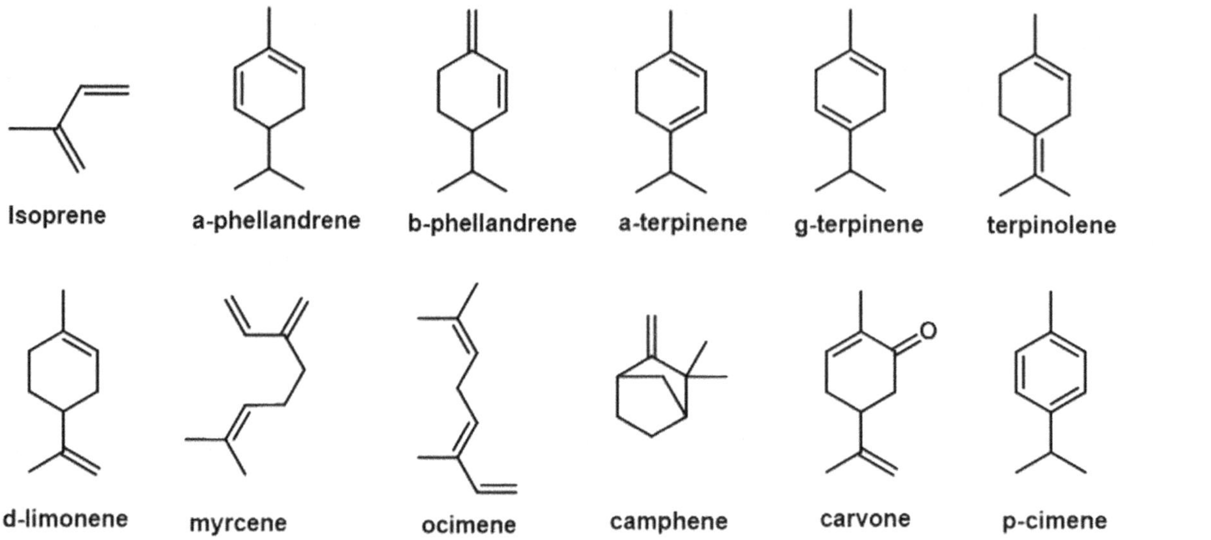

Fig. 19 Chemical structure of some terpenes

different structures of terpenes Ruzicka 1953). The chemical structure of a few terpenes is shown in Fig. 19.

As noted, terpenes present a diverse number of structures that derive from similar carbon skeleton structures, yielding many isomers (Ruzicka 1953). They also present oxygenated derivatives named terpenoids that bear functions as carboxylic acids, aldehydes, ketones, and alcohols. This broad range of materials find applications as insecticides, repellents, cosmetics, medicine, polymers, and many others (Wilbon et al. 2013; Carvalho and Fonseca 2006; Ameh 2014; Garin 1976; Silvestre and Gandini 2008; Beatson 2011).

The terpenes had early and fast passed applications in polymer and materials science. But they got noticed mostly after the work of Hermann Staudinger, the 'father of polymer science,' due to his work in the 'macromolecular hypothesis' that awarded him the Nobel Prize in 1953 Furukawa 1998). The base of his work was isoprene which had a great impact in the industry for the production of latex that has many applications such as gloves, balloons, and swim caps. The polymerization reactions of the isoprene are described in Fig. 20.

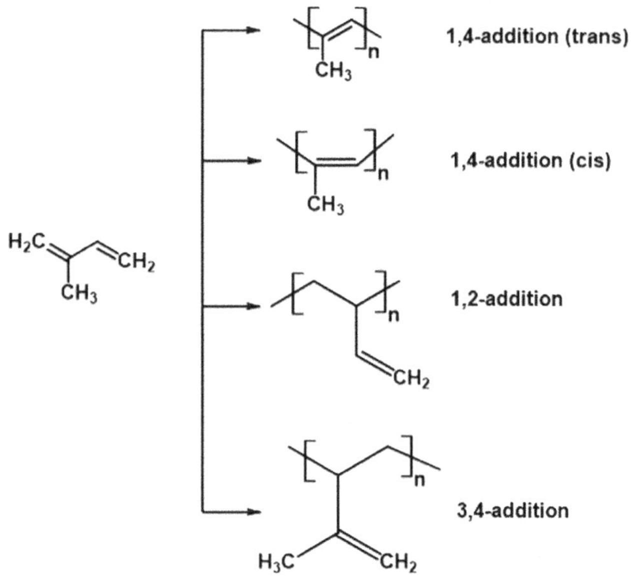

Fig. 20 Polymerization reactions of isoprene

The 1,4-addition polyisoprene received special attention because the double bond in the middle of the main chain allows crosslink bonds. Later, Goodyear developed a process named 'vulcanization.' It consisted of a mixture of a copolymer of polyisoprene and polystyrene with orthorhombic sulfur (S_6) that breaks the double bonds creating a chemical linkage between two polymeric chains. This process enhanced the mechanical properties, which made possible the application in tires for vehicles (Obrecht et al. 2011).

Among terpenes, limonene is a low-cost and versatile chemical widely used in the preparation of polyurethanes. Hauenstein and his group used an interesting green approach using limonene oxide and carbon dioxide as monomers to synthesize a bio-derived polymer named poly(limonene carbonate) shown in Fig. 21 (Hauenstein et al. 2016). This synthesis is one of the rare cases in which carbon dioxide, a greenhouse gas, is implemented into a polymeric chain along with a bio-derived material. These starting materials make this process both profitable and environmentally friendly.

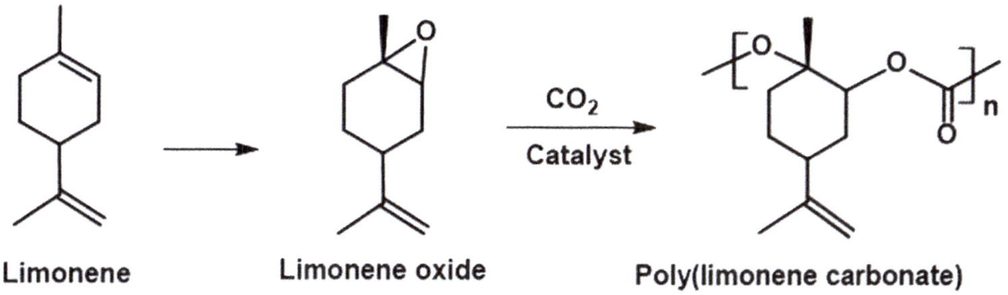

Fig. 21 Polymerization reaction of poly(limonene carbonate). Adapted from Hauenstein et al. (2016)

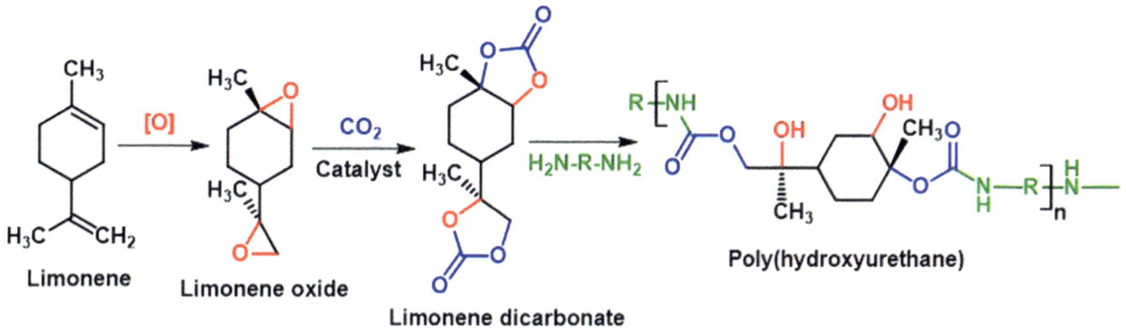

Fig. 22 Synthesis of a limonene oxide-based polyurethane through isocyanate-free polymerization. Adapted with permission from Bähr et al. (2012). Copyright (2012) The Royal Society of Chemistry

In theory, this process can be applied to most bio-derived materials that present at least one double bond that can be epoxidized. Also, the polymerization procedures do not require large use of energy or expensive catalysts (Hauenstein et al. 2016; Park et al. 2013). Even 100% bio-derived and isocyanate-free routes for polyurethane were obtained from the work of Bahr et al. (2012) as described in Fig. 22.

Terpenes can be converted to polyols in several ways as previously described. Among those, thiol-ene is a convenient method that uses UV-light, either from an electric source or sunlight, mild conditions such as room temperature and solvent-free procedures, hence making it a facile and environmentally friendly method (de Souza et al. 2020; Gupta et al. 2014; Ranaweera et al. 2017). Many authors reported terpene or terpenoid-based polyols for polyurethane that yielded satisfactory properties compared to petro-based polyurethanes (Zhang et al. 2018; Elbers et al. 2017; de Souza et al. 2020). For example, Gupta and his research group used a thiol-ene reaction to synthesize limonene-based polyol for polyurethanes (Elbers et al. 2017; Ranaweera et al. 2017; Gupta et al. 2015). A derivative of limonene, limonene dimercaptan, was also used for the preparation of polyol as shown in Fig. 23. The polyurethane foams prepared using these limonene-based polyols showed thermal stability up to 250 °C with regular shape cells and uniform cell size distribution. The highest compressive strength of 195 kPa was observed for the foams from limonene-based polyols.

Terpenes are also used as starting materials for different chemicals. For example, Gupta and his team used limonene as a starting material for the synthesis of Mannich polyol (Gupta et al. 2015). Mannich polyols are amino-based aromatic compounds and polyurethane foams prepared using aromatic polyols could provide fire resistance, high thermal, and superior mechanical properties. The Mannich polyol was synthesized in three steps. In the first step, Friedel–Crafts alkylation of phenol with limonene was carried out in the presence of HBF_4 as a catalyst (Fig. 24). In the second step, the Mannich base was synthesized by the reaction of oxazolidine with phenol alkylated limonene. In the third step, the Mannich base was propoxylated to synthesize Mannich polyol. The polyurethane foams prepared using limonene base Mannich polyol showed thermal stability up to 250 °C with a high glass transition temperature of about ~200 °C. The high reactivity of the limonene-based Mannich polyol makes it very suitable to use for ‘spray’ polyurethane foams.

α-Phellandrene is a monoterpene which contains two endocyclic carbon double bonds in its structure. α-phellandrene-based polyol was synthesized using thiol-ene chemistry for the preparation of rigid polyurethane foams (Elbers et al. 2017). α-Phellandrene was reacted with 2-mercaptoethanol and α-thioglycerol to synthesize two polyols with different hydroxyl functionalities as shown in Fig. 25. The prepared polyurethane foams showed high closed-cell content (over 90%) with apparent density in the

Fig. 23 Synthesis of polyols based on limonene by using thiol-ene 'click' chemistry. Adapted with permission from Gupta et al. (2014). Copyright (2014) Springer Nature

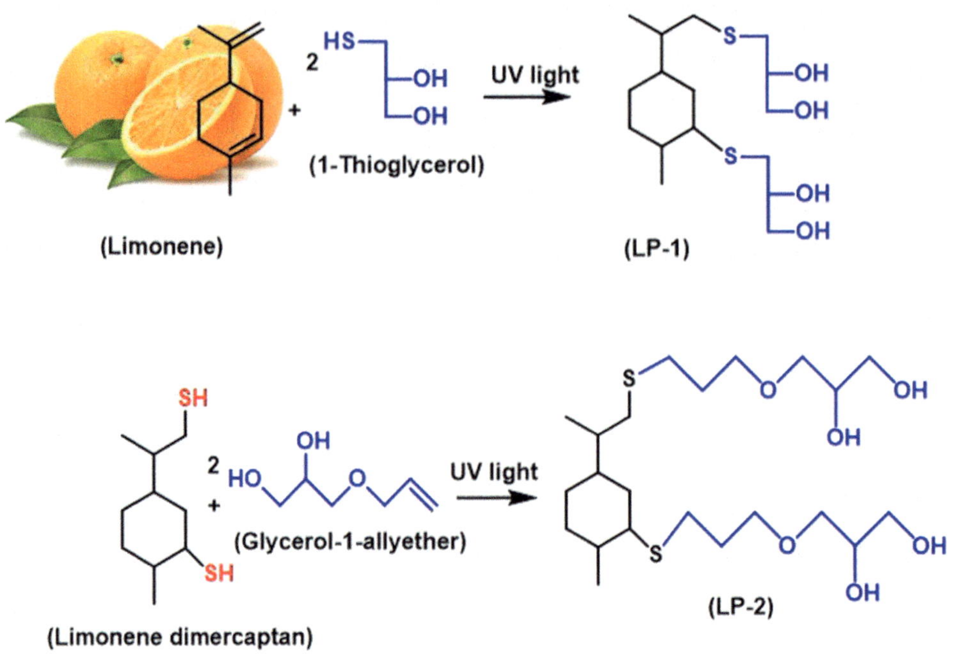

Fig. 24 Possible structures of phenol alkylated with limonene. Adapted with permission from Gupta et al. (2015). Copyright (2015) Springer Nature

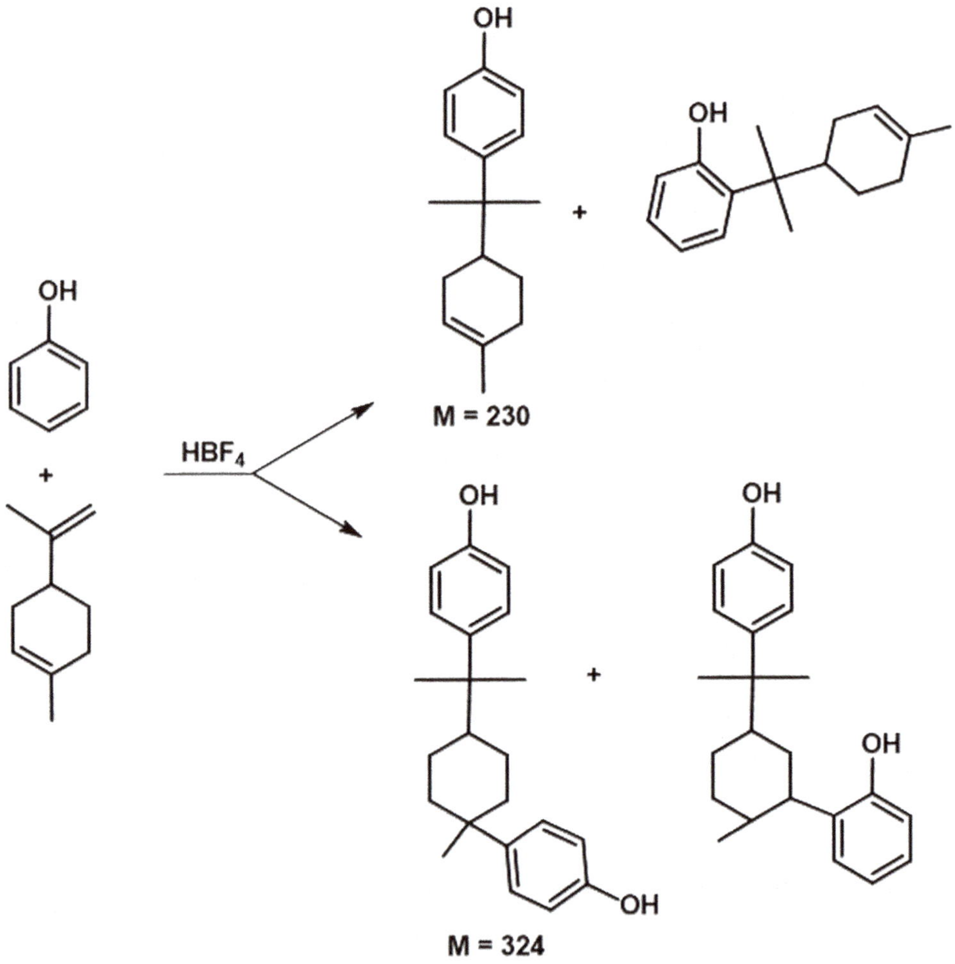

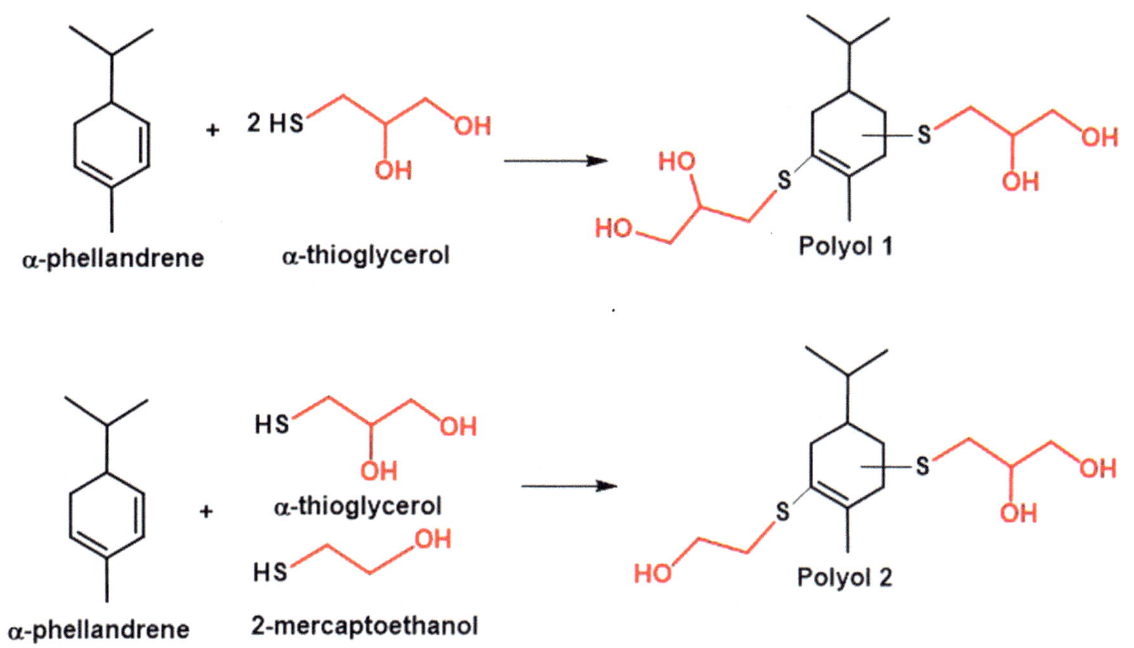

Fig. 25 Synthesis of polyols based on α-phellandrene by thiol-ene 'click' chemistry. Adapted from Elbers et al. (2017)

range of 28–39 kg/m^3. It was observed that the polyol synthesized using α-phellandrene and α-thioglycerol provided higher compressive strength of 220 kPa which could be due to the higher hydroxyl functionalities of the polyol compared to polyol synthesized using α-phellandrene and 2-mercaptoethanol.

4 Lignin for Green Polymers

Lignin is a raw material found in the cellular wall of the plants, mostly in gymnosperm and angiosperms. It plays the role of giving physical support, strength, protection against external threats (animals, insects, microorganisms, etc.) and forms a vascular system to transport water, due to its hydrophobic nature. After cellulose, it is the largest component of a plant (Belgacem and Gandini 2008). Cellulose is a biopolymer composed of defined repeating units of sugar, named as polysaccharides. Lignin, however, is different because it presents a random tridimensional network structure composed of different proportions of coumaryl, coniferyl, and sinapyl alcohols as described in Fig. 26. These compounds undergo biosynthetic routes that provide three segments noted as *p*-hydroxyphenyl, guaiacyl, and syringil. The different ratios of these segments within the plant give different types of lignin (Laurichesse and Avérous 2014).

Lignin is largely produced by plants reaching almost 1 to 3 billion tons per year (Gellerstedt and Henriksson 2008). The abundance of this resource makes it relevant to find more industrial implementations to obtain sustainable, environmentally friendly, and low-cost materials. Some examples of the use of lignin are in animal feed, surfactants, dyes, additives, and others; however, they represent around only 2% of the usage of lignin (Lora 2008; Holladay et al. 2007; Lora and Glasser 2002). The largest consumption of lignin comes from the pulp and paper industry, which use almost 50 million tons per year (Ragauskas et al. 2006). The paper industry uses several ways to process lignin such as soda process, kraft and sulfite pulping that consists of cleaving the ether bonds with the attack of strong alkali or acid compounds to fragment lignin into lower molecular weight products to be used for the production of paper with better quality (Saake and Lehnen 2012; Alén 2000; Gierer 1980). Other examples are organosolv and steam explosion lignin. Organosolv lignin process consists of organic solvents, such as methanol, ethanol, methanoic, and ethanoic acid used to separate lignin from woody components. It is efficient but is also harmful to the environment (Pan et al. 2007). The steam explosion of lignin consists of high pressure and temperature during a short time to promote the cleavage of ether bonds. It is a promising and green approach that yields lignin-based compounds with similar characteristics of those from organosolv (Li et al. 2007). The scientific community has been finding new routes to functionalize lignin to give new applications. One of the reasons to implement its use, besides the previously mentioned, is due to the large presence of aromatic groups in its structure that enables it to replace some of the petrochemical-based materials. Two large areas had emerged for this studies; one is catalytic cleavage and the other is polymeric modification.

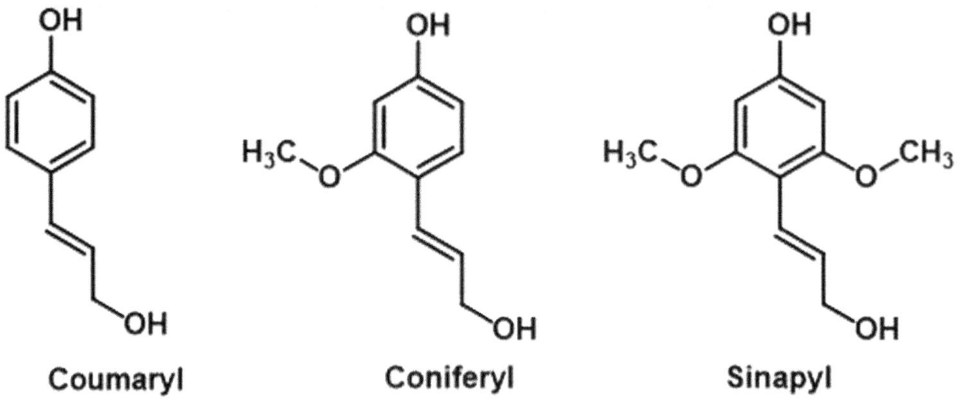

Fig. 26 Main alcohol unit components of lignin. Adapted with permission from Zakzeski et al. (2010). Copyright (2010) American Chemical Society

Catalytic cleavage is, in theory, similar to the industrial application, in which the main objective is to break the ether bonds in lignin's structure to obtain fine chemicals and fuels at the end of the process, and it is presenting ongoing progress using some models of lignin (Zakzeski et al. 2010; Sergeev and Hartwig 2011). Polymeric modifications are divided into physical blending or chemical bonding with another polymer. In many reports, chemical bonding demonstrated better enhancement of properties if compared with blended ones (Kai et al. 2016, 2017). Some examples are copolymers of lignin which presented good antioxidants and in other case self-healing properties making it potentially suitable for medical applications (Kai et al. 2015, 2017). Following the ideas of the functionalization of lignin, some other methods such as thiol-ene were also adopted. Liu and Chung (2017) were able to functionalize lignin using a facile thiol-ene reaction through photoinitiators and UV-light to introduce terminal double bonds that could be further used to implement polymeric chains or several other functional groups, as described in Fig. 27. It shows the versatility of chemistry that can be applied to lignin allowing potential industry applications.

Normally, lignin-based polyurethanes are tough and brittle materials. To reduce brittleness, some studies were performed such as a chemical introduction of longer chains that act as plasticizers which caused an improvement in elasticity and glass transition temperature. A green approach for that manner used cardanol oil, which presents a backbone of 15 carbons along with a phenolic group that acts as a reactive site to be introduced into lignin's structure (Tan 1996). Other approaches developed a lignin-based polyol for polyurethane foam by performing direct oxypropylation under alkaline media into lignin powder. It extended the chain by introducing different ether groups that turned into a processable low-viscosity liquid polyol (Tan 1996). In comparison with a commercial polyurethane foam based on sucrose and glycerol, the pure lignin-based presented higher mechanical properties, which can clear a path for industry applications. To perform the oxypropylation lignin and propylene oxide were placed under high temperature and pressure and catalyzed by alkaline media as described in Fig. 28.

Lignin has been also used as a filler in polyurethanes. The use of lignin as a filer offers the advantage of improved mechanical strength and enhanced bio-content in the polyurethanes (Feldman and Lacasse 1994). Vanillin, an extract of lignocellulosic, was used as a chain extender in the synthesis of polyurethanes (Gang et al. 2017). Vanillin, as a chain extender, does not only increase the bio-content in the polyurethane but also enhanced Young's modulus of 128% and strain of 147% compared to control polyurethane.

Chemical modification of lignin can provide polyurethanes foams with specific applications. For example, polyurethane foams with highly resilient properties were

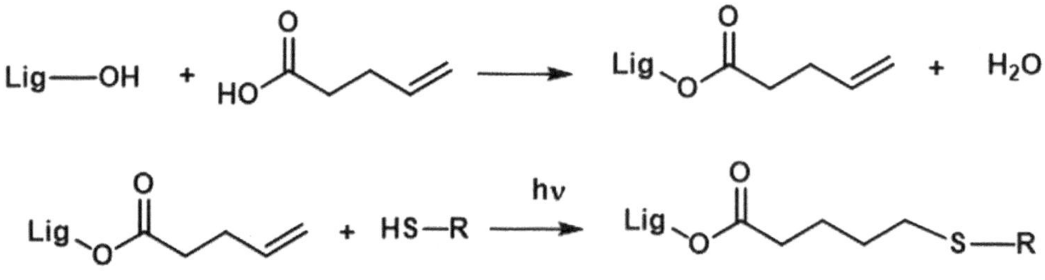

Fig. 27 Functionalization of lignin through esterification followed by a thiol-ene reaction. Adapted with permission from Liu and Chung (2017). Copyright (2017) American Chemical Society

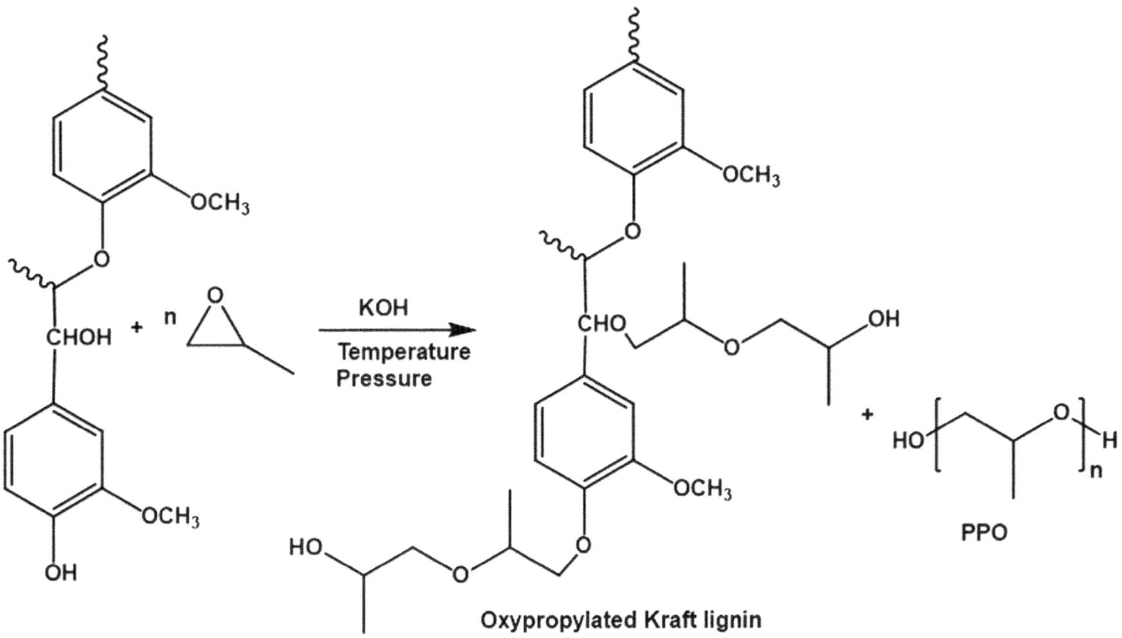

Fig. 28 Reaction involved in lignin oxypropylation. Adapted with permission from Li and Ragauskas (2012). Copyright (2012) Taylor & Francis Group, LLC

prepared using polyethylene glycol-grafted lignin (Wang et al. 2019). Figure 29 shows the chemical modification of alkali lignin with polyethylene glycol (PEG 2000). The polyurethane foams prepared using polyethylene glycol-grafted lignin showed high elastic recovery (>93%) along with improved compression strength.

As discussed so far, lignin presents a broad range of potential applications due to its versatility. The introduction of this renewable material into the industry helps the environment and decrease overall cost. The polyurethane industry, however, is highly dependent on isocyanate, which is a petrochemical-based material that brings health concerns due to its toxicity. To decrease its use, Lee et al. obtained a non-isocyanate route to produce polyurethane elastomers based on both carbonated soybean oil and lignin (Lee and Deng 2015). The introduction of lignin provided an increase in mechanical properties, which shows an advantageous use of this bio-renewable material. The synthesis of this green elastomer was performed in a four-step reaction. First, soybean oil was epoxidized. In the second step, it was carbonated with CO_2, while in the third step, an aminoalkyl siloxane coupling agent was added to the carbonated group yielding the urethane linkage. In the last step, lignin was implemented into the chain through the reaction between its phenolic groups with the siloxane groups. The reaction can be described in Fig. 30.

5 Conclusion

The ongoing research is determined to find new alternative routes that are low-cost, environmentally friendly, and sustainable. It has been proven to be a challenging task; however, the crescent consumption of polyurethanes around the globe pushes both industry and scientific community to create new approaches that are sustainable and profitable. Despite the increase of petrochemicals over the years, there was also an increase in the consumption of bio-based materials. Polyurethanes as extremely versatile polymers cover a broad range of applications as well as materials that can be used for its synthesis, opening many paths for new sustainable and green approaches.

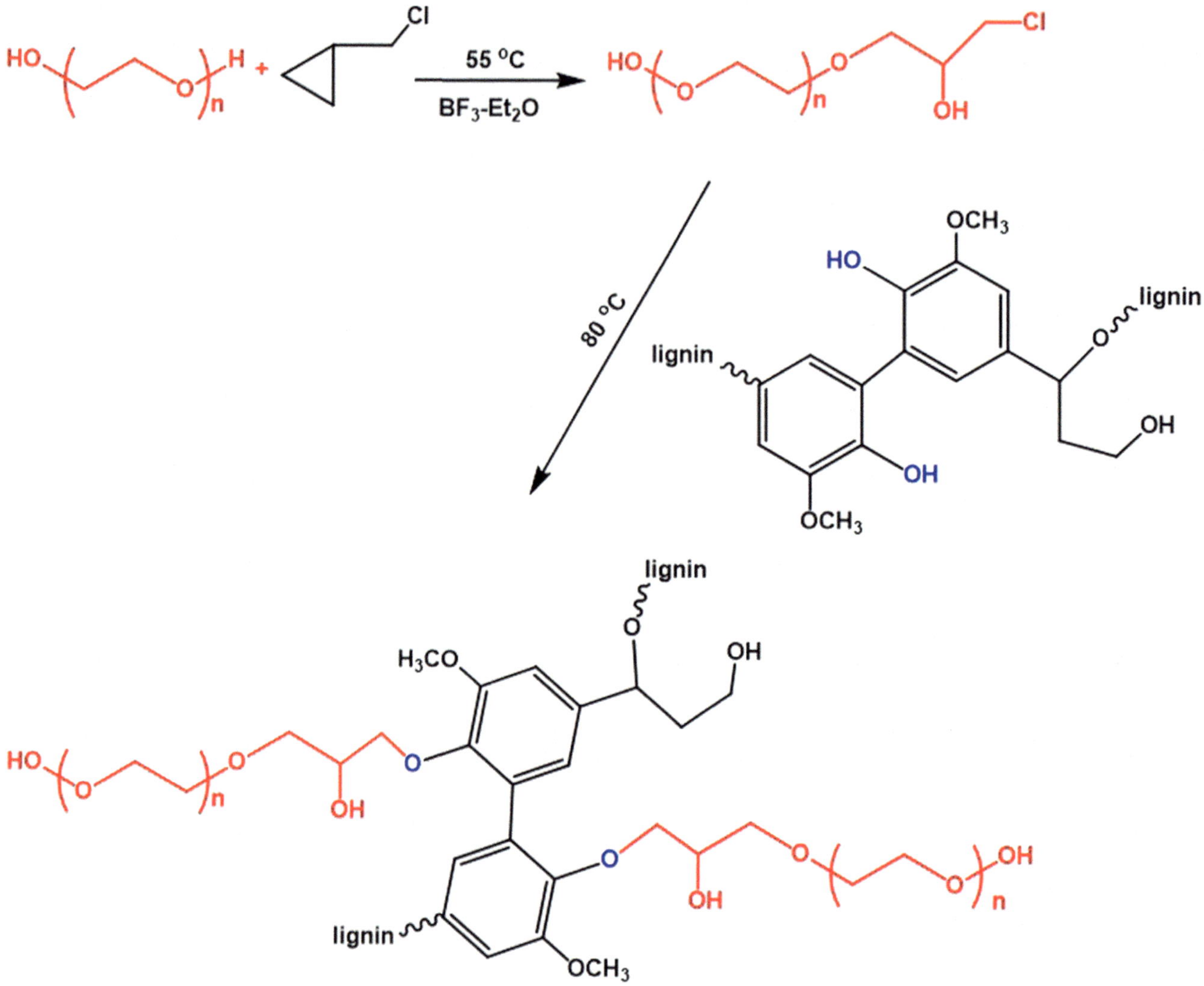

Fig. 29 Modification of alkali lignin with PEG2000. Adapted with permission (Wang et al. 2019). Copyright (2018) American Chemical Society

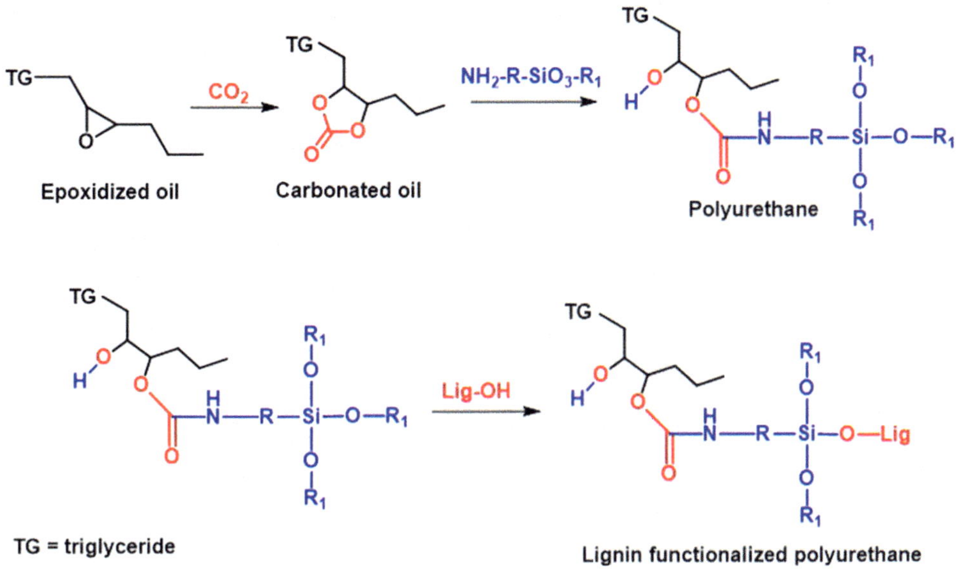

Fig. 30 Isocyanate-free route for oil-lignin-based polyurethane. Adapted with permission from Lee and Deng (2015). Copyright (2015) Elsevier

References

Akindoyo JO, Beg MDH, Ghazali S, Islam MR, Jeyaratnam N, Yuvaraj AR (2016) Polyurethane types, synthesis and applications-a review. RSC Adv 6:114453–114482. https://doi.org/10.1039/c6ra14525f

Alén R (2000) Basic chemistry of wood delignification. Forest products chemistry. Papermaking science and technology. Fapet Oy, Helsinki, Finland, pp 59–104

Ameh S (2014) Diversity, utility, analytical methods and use implications of aroma-active compounds from select angiosperm families. Eur J Med Plants 4:1046–1086. https://doi.org/10.9734/ejmp/2014/10779

Arniza MZ, Hoong SS, Idris Z, Yeong SK, Hassan HA, Din AK, Choo YM (2015) Synthesis of transesterified palm olein-based polyol and rigid polyurethanes from this polyol. J Am Oil Chem Soc 92:243–255. https://doi.org/10.1007/s11746-015-2592-9

Babb DA (2012) Polyurethanes from renewable resources. Adv Polym Sci 245:315–360. https://doi.org/10.1007/12-2011-130

Bähr M, Bitto A, Mülhaupt R (2012) Cyclic limonene dicarbonate as a new monomer for non-isocyanate oligo- and polyurethanes (NIPU) based upon terpenes. Green Chem 14:1447–1454. https://doi.org/10.1039/c2gc35099h

Beatson RP (2011) Chemicals from extractives. In: Zhu J, Zhang XP (eds) Sustainable production of fuels, chemicals, and fibers from forest biomass. American Chemical Society, pp 279–297

Belgacem MN, Gandini A (2008) Monomers, polymers and composites from renewable resources, 1st edn. Elsevier Ltd., Amsterdam

Bhatnagar A (2018) Marketing strategies of multinational chemical companies in emerging markets with special reference to polyurethane industry in India. Dayalbagh Educational Institute, India

Blasbalg TL, Hibbeln JR, Ramsden CE, Majchrzak SF, Rawlings RR (2011) Changes in consumption of omega-3 and omega-6 fatty acids in the United States during the 20th century. Am J Clin Nutr 93:950–962. https://doi.org/10.3945/ajcn.110.006643

Bohnet M (2003) Ullmann's encyclopedia of industrial chemistry, 3rd edn. Springer, Berlin

Cai C, Dai H, Chen R, Su C, Xu X, Zhang S, Yang L (2008) Studies on the kinetics of in situ epoxidation of vegetable oils. Eur J Lipid Sci Technol 110:341–346. https://doi.org/10.1002/ejlt.200700104

Challoner KR, McCarron MM (1990) Castor bean intoxication. Ann Emerg Med 19:1177–1183. https://doi.org/10.1016/S0196-0644(05)81525-2

Charlon M, Heinrich B, Matter Y, Couzigné E, Donnio B, Avérous L (2014) Synthesis, structure and properties of fully biobased thermoplastic polyurethanes, obtained from a diisocyanate based on modified dimer fatty acids, and different renewable diols. Eur Polym J 61:197–205. https://doi.org/10.1016/j.eurpolymj.2014.10.012

Chidambarampadmavathy K, Karthikeyan OP, Heimann K (2017) Sustainable bio-plastic production through landfill methane recycling. Renew Sustain Energy Rev 71:555–562

Clark JH (1999) Green chemistry: challenges and opportunities. Green Chem 1:1–8. https://doi.org/10.1039/a807961g

Collier ME (1996) Pressure-reducing mattresses. J Wound Care 5:207–211. https://doi.org/10.12968/jowc.1996.5.5.207

Dai H, Yang L, Lin B, Wang C, Shi G (2009) Synthesis and characterization of the different soy-based polyols by ring opening of epoxidized soybean oil with methanol, 1,2-ethanediol and 1,2-propanediol. J Am Oil Chem Soc 86:261–267. https://doi.org/10.1007/s11746-008-1342-7

De Carvalho CCCR, Da Fonseca MMR (2006) Carvone: why and how should one bother to produce this terpene. Food Chem 95:413–422. https://doi.org/10.1016/j.foodchem.2005.01.003

De Milliano JW, Woolnough A, Reeves A, Shepherd D (2010) Ecologically significant invasive species: a monitoring framework for natural resource management groups in Western Australia. Department of Agriculture and Food, Western Australia, Perth

De Souza VHR, Silva SA, Ramos LP, Zawadzki SF (2012) Synthesis and characterization of polyols derived from corn oil by epoxidation and ozonolysis. J Am Oil Chem Soc 89:1723–1731. https://doi.org/10.1007/s11746-012-2063-5

de Souza FM, Choi J, Bhoyate S, Kahol PK, Gupta RK (2020) Expendable graphite as an efficient flame-retardant for novel partial bio-based rigid polyurethane foams C 1–13. https://doi.org/10.3390/c6020027

Desroches M, Caillol S, Lapinte V, Auvergne R, Boutevin B (2011) Synthesis of biobased polyols by thiol-ene coupling from vegetable oils. Macromolecules 44:2489–2500. https://doi.org/10.1021/ma102884w

Dierig DA, Coffelt TA, Nakayama FS, Thompson AE (1996) Lesquerella and vernonia: oilseeds for arid lands. Progress in new crops. ASHS Press, Alexandria, VA, pp 347–354

Dissanayake G, Sinha P (2013) Sustainable waste management strategies in the fashion industry sector. Int J Environ Sustain 8:77–90. https://doi.org/10.18848/2325-1077/cgp/v08i01/55036

Dodbiba G, Fujita T (2004) Progress in separating plastic materials for recycling. Phys Sep Sci Eng 13:165–182. https://doi.org/10.1080/14786470412331326350

Dworakowska S, Bogdal D, Prociak A (2012) Microwave-assisted synthesis of polyols from rapeseed oil and properties of flexible polyurethane foams. Polymers (Basel) 4:1462–1477. https://doi.org/10.3390/polym4031462

Ehrlich A, Smith MK, Patton TC (1959) Castor polyols for urethane foams. J Am Oil Chem Soc 36:149–154. https://doi.org/10.1007/BF02636962

Elbers N, Ranaweera CK, Ionescu M, Wan X, Kahol PK, Gupta RK (2017) Synthesis of novel biobased polyol via thiol-ene chemistry for rigid polyurethane foams. J Renew Mater 5:74–83. https://doi.org/10.7569/jrm.2017.634137

Feldman D, Lacasse MA (1994) Polymer–filler interaction in polyurethane kraft lignin polyblends. J Appl Polym Sci 51:701–709. https://doi.org/10.1002/app.1994.070510416

Finn MG, Sharpless KB (1985) On the mechanism of asymmetric epoxidation with titanium-tartrate catalysts. In: Asymmetric synthesis. De Gruyter, pp 247–308

Franke R, Selent D, Börner A (2012) Applied hydroformylation. Chem Rev 112:5675–5732. https://doi.org/10.1021/cr3001803

Frankel EN (1973) Selective hydroformulation of unsaturaed fatty acid esters. In: Annals of the New York Academy of sciences. Wiley Online Library, pp 79–93

Frankel EN (1976) Catalytic hydroformylation of unsaturated fatty derivatives with cobalt carbonyl. J Am Oil Chem Soc 53:138–141. https://doi.org/10.1007/BF02586351

Frankel EN, Thomas FL (1972) Selective hydroformylation of polyunsaturated fats with a rhodium-triphenylphosphine catalyst. J Am Oil Chem Soc 49:10–14. https://doi.org/10.1007/BF02545129

Frollini E, Rodrigues BVM, Da Silva CG, Castro DO, Ramires EC, De Oliveira F, Santos RPO (2016) Polymeric materials from renewable resources. In: AIP conference proceedings. AIP Publishing

Furukawa Y (1998) Inventing Polymer science: staudinger, carothers, and the emergence of macromolecular chemistry. University of Pennsylvania Press

Gang H, Lee D, Choi KY, Kim HN, Ryu H, Lee DS, Kim BG (2017) Development of high performance polyurethane elastomers using vanillin-based green polyol chain extender originating from lignocellulosic biomass. ACS Sustain Chem Eng 5:4582–4588. https://doi.org/10.1021/acssuschemeng.6b02960

Garin DL (1976) Steam distillation of essential oils—carvone from caraway. J Chem Educ 53:105. https://doi.org/10.1021/ed053p105

Gellerstedt G, Henriksson G (2008) Lignins: major sources, structure and properties. In: Monomers, polymers and composites from renewable resources. Elsevier, pp 201–224

Gierer J (1980) Chemical aspects of kraft pulping. Wood Sci Technol 14:241–266. https://doi.org/10.1007/BF00383453

Goud VV, Patwardhan AV, Pradhan NC (2006) Studies on the epoxidation of mahua oil (Madhumica indica) by hydrogen peroxide. Bioresour Technol 97:1365–1371. https://doi.org/10.1016/j.biortech.2005.07.004

Griesbaum K (1970) Problems and possibilities of the free-radical addition of thiols to unsaturated compounds. Angew Chemie Int Ed English 9:273–287. https://doi.org/10.1002/anie.197002731

Guo A, Demydov D, Zhang W, Petrovic ZS (2002) Polyols and polyurethanes from hydroformylation of soybean oil. J Polym Environ 10:49–52. https://doi.org/10.1023/A:1021022123733

Gupta RK, Ionescu M, Radojcic D, Wan X, Petrovic ZS (2014) Novel renewable polyols based on limonene for rigid polyurethane foams. J Polym Environ 22:304–309. https://doi.org/10.1007/s10924-014-0641-3

Gupta RK, Ionescu M, Wan X, Radojcic D, Petrovic ZS (2015) Synthesis of a novel limonene based mannich polyol for rigid polyurethane foams. J Polym Environ 23:261–268. https://doi.org/10.1007/s10924-015-0717-8

Hablot E, Zheng D, Bouquey M, Avérous L (2008) Polyurethanes based on castor oil: Kinetics, chemical, mechanical and thermal properties. Macromol Mater Eng 293:922–929. https://doi.org/10.1002/mame.200800185

Hagberg EC, Malkoch M, Ling Y, Hawker CJ, Carter KR (2007) Effects of modulus and surface chemistry of thiol-ene photopolymers in nanoimprinting. Nano Lett 7:233–237. https://doi.org/10.1021/nl061217f

Hăloiu A, Iosif D (2013) Bio-source composite materials used in automotive industry. Sci Bull Automot Ser 24:57–61

Hansen E (1972) Grease compositions containing magnesium salts of unsaturated fatty acids as rust inhibitors

Hauenstein O, Reiter M, Agarwal S, Rieger B, Greiner A (2016) Bio-based polycarbonate from limonene oxide and CO_2 with high molecular weight, excellent thermal resistance, hardness and transparency. Green Chem 18:760–770. https://doi.org/10.1039/c5gc01694k

Hayes DG, Kleiman R (1992) Recovery of hydroxy fatty acids from lesquerella oil with lipases. J Am Oil Chem Soc 69:982–985. https://doi.org/10.1007/BF02541062

Haynes J, McLaughlin J, Vasquez L, Hunsberger A (2001) Low-maintenance landscape plants for South Florida. Environ Hortic Dept, Florida Coop Ext Serv Univ Florida-IFAS Publ ENH854, Florida, USA, pp 1–49

Holladay JE, White JF, Bozell JJ, Johnson D (2007) Top value-added chemicals from biomass-Volume II—results of screening for potential candidates from biorefinery lignin. Pacific Northwest National Lab, Richland, WA (United States)

Hoyle CE, Lee TY, Roper T (2004) Thiol-enes: chemistry of the past with promise for the future. J Polym Sci Part a Polym Chem 42:5301–5338. https://doi.org/10.1002/pola.20366

Ionescu M (2006) Chemistry and technology of polyols for polyurethanes. Polimeri 26:218–218

Ionescu M, Petrović ZS (2010) High functionality polyether polyols based on polyglycerol. J Cell Plast 46:223–237. https://doi.org/10.1177/0021955X09355887

Ionescu M, Petrović ZS, Wan X (2007) Ethoxylated soybean polyols for polyurethanes. J Polym Environ 15:237–243. https://doi.org/10.1007/s10924-007-0065-4

Ionescu M, Radojčić D, Wan X, Shrestha ML, Petrović ZS, Upshaw TA (2016) Highly functional polyols from castor oil for rigid polyurethanes. Eur Polym J 84:736–749. https://doi.org/10.1016/j.eurpolymj.2016.06.006

Javni I, Petrovic ZS, Waddon A (1998) Polyurethanes Expo'98. Dallas, TX, Sept 17–20

Javni I, Zhang W, Petrović ZS (2003) Effect of different isocyanates on the properties of soy-based polyurethanes. J Appl Polym Sci 88:2912–2916. https://doi.org/10.1002/app.11966

Ji D, Fang Z, Wan ZD, Chen HC, He W, Li XL, Guo K (2013) Rigid polyurethane foam based on modified soybean oil. Adv Mater Res 724–725:1681–1684. https://doi.org/10.4028/www.scientific.net/AMR.724-725.1681

Kai D, Low ZW, Liow SS, Abdul Karim A, Ye H, Jin G, Li K, Loh XJ (2015) Development of lignin supramolecular hydrogels with mechanically responsive and self-healing properties. ACS Sustain Chem Eng 3:2160–2169. https://doi.org/10.1021/acssuschemeng.5b00405

Kai D, Ren W, Tian L, Chee PL, Liu Y, Ramakrishna S, Loh XJ (2016) Engineering Poly(lactide)-lignin nanofibers with antioxidant activity for biomedical application. ACS Sustain Chem Eng 4:5268–5276. https://doi.org/10.1021/acssuschemeng.6b00478

Kai D, Zhang K, Jiang L, Wong HZ, Li Z, Zhang Z, Loh XJ (2017) Sustainable and antioxidant lignin-polyester copolymers and nanofibers for potential healthcare applications. ACS Sustain Chem Eng 5:6016–6025. https://doi.org/10.1021/acssuschemeng.7b00850

Kale G, Kijchavengkul T, Auras R, Rubino M, Selke SE, Singh SP (2007) Compostability of bioplastic packaging materials: an overview. Macromol Biosci 7:255–277. https://doi.org/10.1002/mabi.200600168

Kamil RNM, Yusup S, Rashid U (2011) Optimization of polyol ester production by transesterification of Jatropha-based methyl ester with trimethylolpropane using Taguchi design of experiment. Fuel 90:2343–2345. https://doi.org/10.1016/j.fuel.2011.02.018

Killops KL, Campos LM, Hawker CJ (2008) Robust, efficient, and orthogonal synthesis of dendrimers via thiol-ene "click" chemistry. J Am Chem Soc 130:5062–5064. https://doi.org/10.1021/ja8006325

Kolb HC, Finn MG, Sharpless KB (2001) Click chemistry: diverse chemical function from a few good reactions. Angew Chemie Int Ed 40:2004–2021. https://doi.org/10.1002/1521-3773(20010601)40:11%3c2004::AID-ANIE2004%3e3.0.CO;2-5

Kumaran MK, Bomberg MT (1990) Thermal performance of sprayed polyurethane foam insulation with alternative blowing agents. J Therm Insul 14:43–57. https://doi.org/10.1177/109719639001400105

Laurichesse S, Avérous L (2014) Chemical modification of lignins: towards biobased polymers. Prog Polym Sci 39:1266–1290. https://doi.org/10.1016/j.progpolymsci.2013.11.004

Lee A, Deng Y (2015) Green polyurethane from lignin and soybean oil through non-isocyanate reactions. Eur Polym J 63:67–73. https://doi.org/10.1016/j.eurpolymj.2014.11.023

Li Y, Ragauskas AJ (2012) Kraft lignin-based rigid polyurethane foam. J Wood Chem Technol 32:210–224. https://doi.org/10.1080/02773813.2011.652795

Li J, Henriksson G, Gellerstedt G (2007) Lignin depolymerization/repolymerization and its critical role for delignification of aspen wood by steam explosion. Bioresour Technol 98:3061–3068. https://doi.org/10.1016/j.biortech.2006.10.018

Liu H, Chung H (2017) Visible-light induced thiol-ene reaction on natural lignin. ACS Sustain Chem Eng 5:9160–9168. https://doi.org/10.1021/acssuschemeng.7b02065

Llevot A, Dannecker PK, von Czapiewski M, Over LC, Söyler Z, Meier MAR (2016) Renewability is not enough: recent advances in the sustainable synthesis of biomass-derived monomers and polymers. Chem A Eur J 22:11510–11521. https://doi.org/10.1002/chem.201602068

Lluch C, Ronda JC, Galiá M, Lligadas G, Cádiz V (2010) Rapid approach to biobased telechelics through two one-pot thiol-ene click reactions. Biomacromol 11:1646–1653. https://doi.org/10.1021/bm100290n

Lora J (2008) Industrial commercial lignins: sources, properties and applications. In: Monomers, polymers and composites from renewable resources. Elsevier, pp 201–224

Lora JH, Glasser WG (2002) Recent industrial applications of lignin: a sustainable alternative to nonrenewable materials. J Polym Environ 10:39–48. https://doi.org/10.1023/A:1021070006895

Mittal JP, Dhawan KC, Thyagraj CR (1991) Energy scenario of castor crop under dryland agriculture of Andhra Pradesh. Energy Convers Manag 32:425–430. https://doi.org/10.1016/0196-8904(91)90003-2

Moore RJ, Norton WJ (1953) Manufacture of grease compositions

Morgan CR, Magnotta F, Ketley AD (1977) Thiol/ene photocurable polymers. J Polym Sci Polym Chem Ed 15:627–645. https://doi.org/10.1002/pol.1977.170150311

Mu Y, Wan X, Han Z, Peng Y, Zhong S (2012) Rigid polyurethane foams based on activated soybean meal. J Appl Polym Sci 124:4331–4338. https://doi.org/10.1002/app.35612

Müller R (2014) Production of lubricants. Encyclopedia of lubricants and lubrication. Google Patents, USA, pp 1428–1450

Nair DP, Cramer NB, Scott TF, Bowman CN, Shandas R (2010) Photopolymerized thiol-ene systems as shape memory polymers. Polymer (Guildf) 51:4383–4389. https://doi.org/10.1016/j.polymer.2010.07.027

Narine SS, Kong X, Bouzidi L, Sporns P (2007a) Physical properties of polyurethanes produced from polyols from seed oils: II. Foams. J Am Oil Chem Soc 84:65–72. https://doi.org/10.1007/s11746-006-1008-2

Narine SS, Kong X, Bouzidi L, Sporns P (2007b) Physical properties of polyurethanes produced from polyols from seed oils: I. Elastomers. J Am Oil Chem Soc 84:55–63. https://doi.org/10.1007/s11746-006-1006-4

Obrecht W, Lambert J, Happ M, Oppenheimer-Stix C, Dunn J, Krüger R (2011) Rubber, 4. Emulsion rubbers. In: Ullmann's encyclopedia of industrial chemistry. Wiley Online Library, Verlag

Ojima I, Tsai C, Tzamarioudaki M, Bonafoux D (2000) The hydroformylation reaction. In: Organic reactions. Wiley Online Library, pp 1–354

Ottenbrite RM (2001) Functional monomers and polymers. IEEE Electr Insul Mag 17:67. https://doi.org/10.1109/MEI.2001.917546

Pan X, Xie D, Kang K-Y, Yoon S-L, Saddler JN (2007) Effect of organosolv ethanol pretreatment variables on physical characteristics of hybrid poplar substrates. Applied biochemistry and biotecnology. Humana Press, Jinju, pp 367–377

Pappas SP (1985) UV curing by radical, cationic and concurrent radical-cationic polymerization. Radiat Phys Chem 25:633–641. https://doi.org/10.1016/0146-5724(85)90143-8

Park HJ, Ryu CY, Crivello JV (2013) Photoinitiated cationic polymerization of limonene 1,2-oxide and α-pinene oxide. J Polym Sci Part a Polym Chem 51:109–117. https://doi.org/10.1002/pola.26280

Petrović ZS, Zhang W, Javni I (2005) Structure and properties of polyurethanes prepared from triglyceride polyols by ozonolysis. Biomacromol 6:713–719. https://doi.org/10.1021/bm049451s

Petrović ZS, Cvetković I, Hong DP, Wan X, Zhang W, Abraham T, Malsam J (2008) Polyester polyols and polyurethanes from ricinoleic acid. J Appl Polym Sci 108:1184–1190. https://doi.org/10.1002/app.27783

Pino P, Botteghi C (2003) Aldehydes from olefins: cyclohexanecarboxaldehyde. In: Organic syntheses. Wiley Online Library, pp 11–11

Posner T (1905) Beiträge zur Kenntniss der ungesättigten Verbindungen. II. Ueber die Addition von Mercaptanen an ungesättigte Kohlenwasserstoffe. Berichte Der Dtsch Chem Gesellschaft 38:646–657. https://doi.org/10.1002/cber.190503801106

Puppala N, Fowler JL (1999) Growth analysis of lesquerella in response to moisture stress. ASHS Press, Alexandria, VA

Ragauskas AJ, Williams CK, Davison BH, Britovsek G, Cairney J, Eckert CA, Frederick WJ, Hallett JP, Leak DJ, Liotta CL, Mielenz JR, Murphy R, Templer R, Tschaplinski T (2006) The path forward for biofuels and biomaterials. Science (80-)311:484–489. https://doi.org/10.1126/science.1114736

Ramanujam S, Zequine C, Bhoyate S, Neria B, Kahol P, Gupta R (2019) Novel biobased polyol using corn oil for highly flame-retardant polyurethane foams. C 5:13. https://doi.org/10.3390/c5010013

Ranaweera CK, Ionescu M, Bilic N, Wan X, Kahol PK, Gupta RK (2017) Biobased polyols using thiol-ene chemistry for rigid polyurethane foams with enhanced flame-retardant properties. J Renew Mater 5:1–12. https://doi.org/10.7569/jrm.2017.634105

Rios LA, Weckes P, Schuster H, Hoelderich WF (2005) Mesoporous and amorphous Ti-silicas on the epoxidation of vegetable oils. J Catal 232:19–26. https://doi.org/10.1016/j.jcat.2005.02.011

Ruzicka L (1953) The isoprene rule and the biogenesis of terpenic compounds. Experientia 9:357–367. https://doi.org/10.1007/BF02167631

Saake B, Lehnen R (2012) Lignin. In: Ullmann's encyclopedia of industrial chemistry. Wiley Online Library

Sacchetti G, Mussini S, Maccari B (1991) New cfc free microcellular polyether polyurethane for footwear. J Cell Plast 27:88–89. https://doi.org/10.1177/0021955X91027001115

Satoh K, Sugiyama H, Kamigaito M (2006) Biomass-derived heat-resistant alicyclic hydrocarbon polymers: Poly(terpenes) and their hydrogenated derivatives. Green Chem 8:878–888. https://doi.org/10.1039/b607789g

Sergeev AG, Hartwig JF (2011) Selective, nickel-catalyzed hydrogenolysis of aryl ethers. Science (80-)332:439–443. https://doi.org/10.1126/science.1200437

Seymour RB (1989) Pioneers in polymer science. In: Seymour RB (ed) Pioneers in polymer science. Springer, Netherlands, Dordrecht, p 272

Sharmin E, Zafar F, Akram D, Alam M, Ahmad S (2015) Recent advances in vegetable oils based environment friendly coatings: a review. Ind Crops Prod 76:215–229. https://doi.org/10.1016/j.indcrop.2015.06.022

Silvestre AJD, Gandini A (2008) Terpenes: major sources, properties and applications. In: Monomers, polymers and composites from renewable resources. Elsevier, pp 17–38

Sissener NH, Ørnsrud R, Sanden M, Frøyland L, Remø S, Lundebye AK (2018) Erucic acid (22:1n–9) in fish feed, farmed, and wild fish and seafood products. Nutrients 10:1–12. https://doi.org/10.3390/nu10101443

Somarathna HMCC, Raman SN, Mohotti D, Mutalib AA, Badri KH (2018) The use of polyurethane for structural and infrastructural engineering applications: a state-of-the-art review. Constr Build Mater 190:995–1014

Szycher M (1999) Szycher's handbook of polyurethanes, 1st edn. CRC Press, New York

Tan TTM (1996) Cardanol-lignin-based polyurethanes. Polym Int 41:13–16. https://doi.org/10.1002/(sici)1097-0126(199609)41:1%3c13::aid-pi557%3e3.0.co;2-8

Tran P, Graiver D, Narayan R (2005) Ozone-mediated polyol synthesis from soybean oil. J Am Oil Chem Soc 82:653–659. https://doi.org/10.1007/s11746-005-1124-z

Wang S, Liu W, Yang D, Qiu X (2019) Highly resilient lignin-containing polyurethane foam. Ind Eng Chem Res 58:496–504. https://doi.org/10.1021/acs.iecr.8b05072

Warwel S, Klaas MR (1995) Chemo-enzymatic epoxidation of unsaturated carboxylic acids. J Mol Catal B Enzym 1:29–35. https://doi.org/10.1016/1381-1177(95)00004-6

Wilbon PA, Chu F, Tang C (2013) Progress in renewable polymers from natural terpenes, terpenoids, and rosin. Macromol Rapid Commun 34:8–37. https://doi.org/10.1002/marc.201200513

Yang Z, Feng Y, Liang H, Yang Z, Yuan T, Luo Y, Li P, Zhang C (2017) A solvent-free and scalable method to prepare soybean-oil-based polyols by thiol-ene photo-click reaction and biobased polyurethanes therefrom. ACS Sustain Chem Eng 5:7365–7373. https://doi.org/10.1021/acssuschemeng.7b01672

Yeadon DA, McSherry WF, Goldblatt LA (1959) Preparations and properties of castor oil urethane foams. J Am Oil Chem Soc 36:16–20. https://doi.org/10.1007/BF02540259

Zakzeski J, Bruijnincx PCA, Jongerius AL, Weckhuysen BM (2010) The catalytic valorization of lignin for the production of renewable chemicals. Chem Rev 110:3552–3599. https://doi.org/10.1021/cr900354u

Zhang C, Madbouly SA, Kessler MR (2015) Biobased polyurethanes prepared from different vegetable oils. ACS Appl Mater Interfaces 7:1226–1233. https://doi.org/10.1021/am5071333

Zhang C, Bhoyate S, Ionescu M, Kahol PK, Gupta RK (2018) Highly flame retardant and bio-based rigid polyurethane foams derived from orange peel oil. Polym Eng Sci 58:2078–2087. https://doi.org/10.1002/pen.24819

Zhao R, Torley P, Halley PJ (2008) Emerging biodegradable materials: starch- and protein-based bio-nanocomposites. J Mater Sci 43:3058–3071. https://doi.org/10.1007/s10853-007-2434-8

Zlatanić A, Petrović ZS, Dušek K (2002) Structure and properties of triolein-based polyurethane networks. Biomacromol 3:1048–1056. https://doi.org/10.1021/bm020046f

Mechanochemistry: A Power Tool for Green Synthesis

Demet Ozer

Abstract

Mechanochemistry has gained significant interest as a powerful, more sustainable, timesaving, environmentally friendly, and more economical synthesis method to prepare new functional materials. This method depends on the chemical and physicochemical transformations through mechanical force forming by grinding and milling. This study is a systematic review of the history, principles, mechanisms, and kinetics of mechanochemistry. The effects of mechanochemical synthesis parameters (milling types, materials, size, time, temperature, atmosphere, revolution speed, frequency, ball/powder weight ratio, filling ratio, process control agents) were detailed explained. The current researches about the mechanochemical synthesis of co-crystals, inorganic materials, metal–organic frameworks, porous organic materials, and polymers, their respective characteristics, challenges, and future improvements were briefly discussed.

Keywords

Mechanochemistry • Mechanochemical synthesis • Grinding • Milling • History • Principles • Mechanisms • Kinetics • Synthesis parameters

1 Introduction

Green synthesis has excellent advantages compared with the conventional methods such as safe to handle, the environmentally friendly, eliminates wastes, reduces derivatives, relatively inexpensive, energy-efficient, has renewable feedstock, and biodegradable final products (Gonzalez-Moragas et al. 2015; Pérez-Venegas and Juaristi 2020). In recent years, green synthesis methods have widely applied to prepare nanomaterials and nanocomposites (Mondal et al. 2020), catalysis (Gómez-López et al. 2020), metal–organic frameworks (Kumar et al. 2020a), graphene quantum dots (Kumar et al. 2020b), etc., for important applications such as environmental, medical, pharmaceutical, drug delivery, sensing, bio-imaging, energy storage, and cancer therapy (Emami and Shayanfar 2020; Vaid et al. 2020).

Among the green synthesis methods, **mechanochemistry** has already been extensively applied to prepare novel materials in different industries as a clean, safe, time, and energy-efficient synthesis approach (James et al. 2012). Especially, the decrement of the large-scale use of a volatile organic solvent has increased the usage of mechanochemistry as a green and eco-friendly synthesis method compared with other solvent-based methods (Anastas and Tundo 2000). Today, mechanochemistry is accepted as a splendid method for green chemistry due to several advantages such as overall simplicity, consumption of minimum energy, low-cost, less use or lack of hazardous solvents, recycling, purification, and like these (Giannakoudakis et al. 2020). The mechanochemistry is based on mechanical energy that has formed from various types of grinding and milling (Palazon et al. 2019; Mursalat et al. 2019). Hand grinding, ball milling, mixer/shaker milling, and planetary milling have mostly used types (Sopicka-Lizer 2010), and in recent years, liquid (LAG), ion-liquid (ILAG), and polymer-assisted grinding (POLAG) have developed to diversify the efficiency of the process (Hasa et al. 2015). The milling and grinding parameters (milling types, materials, size, time, temperature and atmosphere, rotation speed, frequency, ball/powder weight ratio, filling ratio, process control agents) have directly affected the aim and direction of the studies (Palaniandy and Jamil 2009). The mechanical grinding and milling of solids not only produce new materials but also reduce to particles size, produce new surfaces,

D. Ozer (✉)
Department of Chemistry, Hacettepe University, 06590 Ankara, Turkey
e-mail: demetbaykan@hacettepe.edu.tr

Inamuddin et al. (eds.), *Advances in Green Synthesis*, Advances in Science, Technology & Innovation,
https://doi.org/10.1007/978-3-030-67884-5_2

and increase the surface area. During grinding and milling, the dislocations and defects were occurred in the crystalline structure and new phases were formed in polymorphic materials. Different types of chemical reactions were observed like decomposition, oxidation–reduction, complex formation, etc. (Fernandez-Bertran 1999).

Mechanochemistry has been carried out the systems depending upon covalent (Beyer and Clausen-Schaumann 2005), coordination (Braga et al. 2006), and supramolecular bonds (Braga et al. 2007). The mechanochemical synthesis can be described as a chemical transformation through the absorption of mechanical energy (McNaught and Wilkinson 1997) and widely used to prepare valuable products from alloys to organic compounds (Boldyrev and Tkáčová 2000) and resulted in many exciting developments in green chemistry, supramolecular chemistry, pharmaceutical chemistry, organic synthesis, catalysis, inorganic chemistry, metal–organic frameworks, and organometallics (James and Friščić 2013). The most used application areas were shown in Fig. 1 and some recent examples and their application areas were given in Table 1.

2 History of Mechanochemistry

The mechanochemistry has been applied dates back to prehistoric times to ensure living conditions such as ceramics for potteries and metals for submunitions via mechanical treatments. The first inorganic materials, pigments, and drugs were produced through grinding using the basics of mechanochemistry (Boldyreva 2013). Spring and Lea published the first systematic study at the end of the nineteenth century (Takacs 2013). In 1966, mechanical activation was developed as a powder preparation method that permitted the fabrication of homogeneous products from elemental powder mixtures (Suryanarayana 2001). After that new investigations about minerals, inorganic compounds, and polymers have been found using mechanochemistry (Baláž et al. 2005; Fox 1975). Mechanical activation helps to extend solid solubility limits and create new phases with fine dispersion and diminution grain sizes. Different phase formations like amorphous, crystalline, and quasicrystalline have been possible at low temperatures and scalable process (Suryanarayana 2001).

The nomenclature of mechanochemistry got first used in the scientific literature in 1919 (Ostwald 1919). The IUPAC defined the mechanochemical synthesis as: "a chemical reaction that is induced by the direct absorption of mechanical energy" (Rightmire and Hanusa 2016). Mechanochemistry is engaged in mechanochemical transformations like compression, shear, or friction. The mechanical actions such as grinding, milling, shearing, sliding, or plastic deformation start the reactions (Kaupp 2009) and their effects are changeable according to released heat. Explaining, utilizing, and progressing this method is a significant topic from history to today for mechanochemistry (Takacs 2013).

3 Principles of Mechanochemistry

In the mechanochemical synthesis, chemical reactions are excited by mechanical energy. It is principally supplied from hand grinding and mechanical milling without the need of bulk solvent. Hand **grinding** (manual grinding) is generally applied by a mortar and pestle (Takacs 2007) and it is the simplest way for the generation of mechanical energy (Cinčić et al. 2012). For mechanical **milling**, a shaker, ball, or mill have generally used at various frequencies from 5 to 60 Hz (Stolle et al. 2014). Common grinding and milling apparatus were shown in Fig. 2. The disadvantages of hand grinding are open and sensitive to environmental factors, while mechanical milling presents an enclosed solvent-free medium. The other advantages of milling do not require any physical effort and greater power allowed further systematic researches. The mechanical energy is obtained from grinding and milling that effects an enlarged crystalline solid like heating, decreasing particle size, increasing surface area, fabricating new interfaces and crystal defects, removal of passivating layers, amorphization, and the formation of metastable polymorphs (Suryanarayana 2001).

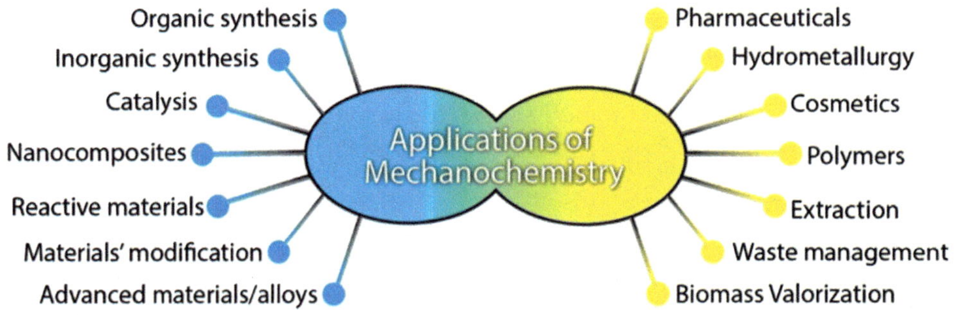

Fig. 1 Applications of mechanochemistry (Giannakoudakis et al. 2020). Copyright 2019 Springer Nature

Table 1 Some recent examples prepared by mechanochemical synthesis and their application areas

Example	Synthesis method	Application area	References
Ti-bearing blast furnace slag (TS)	Mechanochemical activation	Reduction of nitric oxide (NO) (removed 80.5% at 350 °C)	Hou et al. (2020)
Nanofiber-cage $LiFePO_4/C$	Ultrasound-assisted wet media milling	Electrification of the automobile industry	Li et al. (2020)
Calcium diglyceroxide	Semi-continuous mechanochemical process	Biodiesel production (90% yield) as a heterogeneous catalyst	Malpartida et al. (2020)
Nanocrystalline nickel disulfide, NiS_2	Ball milling	Adsorption of methylene blue (MB) dye (94% adsorbed)	Ulbrich et al. (2020)
5-(4-hydroxy-3-methoxybenzylidene) barbituric acid	The solvent-free mechanochemical reaction in planetary ball mill	The Knoevenagel condensation	Burmeister et al. (2020)
Fe/FeO_x@citrate colloids	Oscillatory mill and coprecipitation	Biomedical and environmental remediation applications	Medina et al. (2020)
The Diels–Alder mechanophore and its corresponding polymers	Tunable flow-induced polymer mechanochemistry	Mechanophore design	Willis-Fox et al. (2020)
Carbamoyl isatins and benzamides	LAG mechanochemistry	C–N coupling of amides and isocyanates	Dayaker et al. (2020)
Phenylhydrazono-*N*-methylene fluorescein (PHMF)	Mortar and pestle	Detecting cyanide ions as a PHMF's paper-strip sensor	Rathod et al. (2020)
A pharmaceutical co-crystal of ambrisentan	Mechanochemical grinding	Drugs in the pharmaceutical industry	Haneef and Chadha (2020)
Polycrystalline targets of complex sulfide semiconductors, $CuInS_2$	Mechanosynthesis	Film deposition in physical vacuum techniques	Delmonte et al. (2020)
High-value nano-lead sulfide	A combined vacuum calcination and two-step mechanochemical reaction	The recycling of spent lead-acid batteries	Liu et al. (2020)

For the development of functional materials with mechanochemical synthesis, a new approaches have been tried. One of them is liquid-assisted grinding (LAG) (kneading) which is an effective method for polymorph control (Trask et al. 2005) and increases co-crystallization rate (Nguyen et al. 2007). A small amount of liquid is used and the ratio of liquid volume to reactant weight (η) is 0–2 μL mg^{-1}. It gives an opportunity to effectively optimize milling parameters (Friščić et al. 2009a) and is widely used in pharmaceutical and organic chemistry (Colacino et al. 2019a). In the ion-liquid-assisted grinding (ILAG) method, small amounts of salt added to promote the mechanochemical construction of materials (Friščić et al. 2010). Polymer-assisted grinding (POLAG) improves the reaction rate and diversity during mechanochemical co-crystallization (Hasa et al. 2015). To eliminate the disadvantageous of LAG, a small amount of polymer additive can be added and the replacement of small-molecule liquids with macromolecules can control the particle size. In an example, various polyethylene glycols with different molecular weight (200–10,000), viscosities, and melting points were utilized, and POLAG has ensured better control of powder particle size without solvate formation (Hasa et al. 2015).

3.1 Mechanisms and Kinetics of Mechanochemistry

In 1967, Thiessen offered the magma-plasma theory to explain mechanochemistry (Thiessen et al. 1967). In this theory, due to mechanical energy, many excitation states occurred at different relaxation times (Baláž 2008). After that, several approaches and models were used to explain the mechanism and kinetics of mechanochemistry (Tumanov et al. 2011; Urakaev and Boldyrev 2000; Urakaev 2010). One of them is eutectic melting and the submerged eutectic temperatures below room temperature enable the formation of co-crystal (Chadwick et al. 2007)). The dissolution in a small amount of liquid added on co-grinding (Friščić et al. 2009b), a gas–solid reaction resulting from the sublimation (Mikhailenko et al. 2004), and direct solid-state interdiffusion of the components (Kuroda et al. 2004) are some of the other methods using mechanochemistry. The major problem of these methods is that the sample was treated with impermanent mechanical pulses, and the time differences were found between total and actual time for mechanical treatment (Friščić et al. 2013). Kaupp explained the mechanism under mechanochemical milling through three basic steps. First is the diffusion of reactants via mobile phase

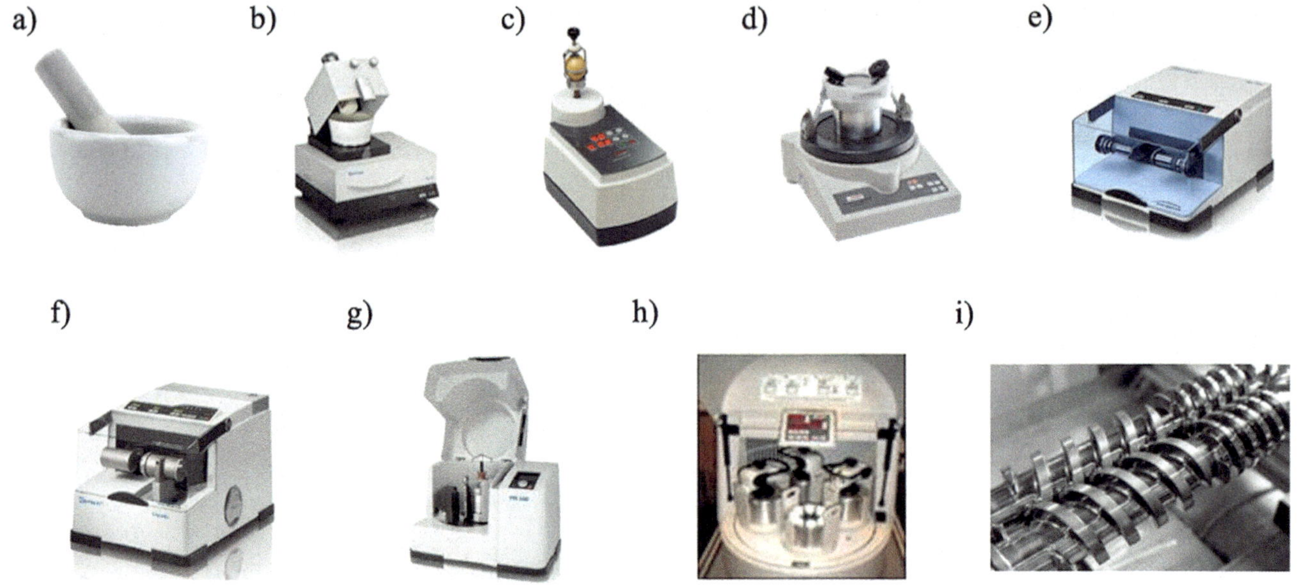

Fig. 2 Most used grinding and milling types of equipment. **a** Mortar and pestle. **b** Automatic mortar (*R*). **c** Vertical vibrational mini-mill (*F*). **d** Vibratory micro-mill (*F*). **e** Vibrational ball mill (*R*) and **f** with temperature control (*C*). **g** Planetary ball mill (*R*). **h** Multisample mill (*A*). (i) Twin-screw for continuous mechanochemical synthesis. *A* = Automaxion, *R* = Retsch, *F* = Fritsch, *C* = Cryomill. Reprinted with permission from Pérez-Venegas and Juaristi (2020). Copyright (2020) ACS

resulting a chemical reaction. Second is the nucleation and growth of the product phase, and the last is the product separation to expose fresh reactant surface (Kaupp 2003). Currently, the mechanical activation and alloying are among the emerging technologies to produce high-value products. The mechanical treatment has come in an alternative technique for green chemistry with or without solvent (Mucsi 2019).

3.2 Effects of Reaction Parameters

The milling types (shaker, platenary, ball mill, and attritor), the milling materials (vessels, jars, balls, bowls, and vials), the milling size, the ball/powder weight ratio, the filling of a vial, balls numbers, milling time, temperature and atmosphere, rotation speed, frequency and process control agents are the important milling parameters to optimize reaction conditions. All of them directly affect the kinetic energy of the balls and how mechanochemical reactions perform (Howard et al. 2018).

Different **mill types** such as shaker mills (Takacs and McHenry 2006), platenary mills (Brun et al. 1993), attritors (Rydin et al. 1993), and uni-ball mills were studied in the literature. Their capacities and speeds differ from each other. They check the process by varying the milling temperature and minimizing contaminations. The most used milling types are shaker and planetary mills. The shaker mills are applied to small samples, while planetary mills are used on the industrial scale for sample processing and colloidal grinding. The planetary mill is an effective method with good reproducibility and reliability (Burmeister and Kwade 2013). In shaker mills, rapid side-to-side motions were shown in a reaction vessel. The planetary mills have pots that are connected to disk and rotate around the central axis. The large strike energies of milling balls inside the pots that increase the efficiency of grinding were formed through the high rotational speed (Janot and Guérard 2005). In general, the **milling materials** (vessels, jars, balls, bowls, and vials) (Fig. 3) are tungsten carbide, stainless steel, tempered steel, zirconia, corundum, agate, silicon nitride, chromium steel, Cr–Ni steel, and plastic polyamide (Di and Bakker 1991). Some of them are naturally occurring porous minerals. The stainless steel (7.5 g/mL) is widely applied in short milling and zirconia (5.6 g/mL) is generally selected in long milling to avoid metal-contaminated (Štefanić et al. 2013). The higher density can cause higher energy transfer, and in conclusion, the product yield can be increased.

The **milling size** directly changes the surface properties of the synthesized materials. The large (or high density) balls can supply high collision energy and increase the surface activity that causes form the thermodynamically stable product. The small balls increase the frictional action that helps to form amorphous and metastable phases and they are generally selected when if more efficient mixing is required. The **ball-to-powder weight ratio** has been a crucial parameter that correlates with kinetics. This ratio can be changed from low values as 1:1 to high values 220:1

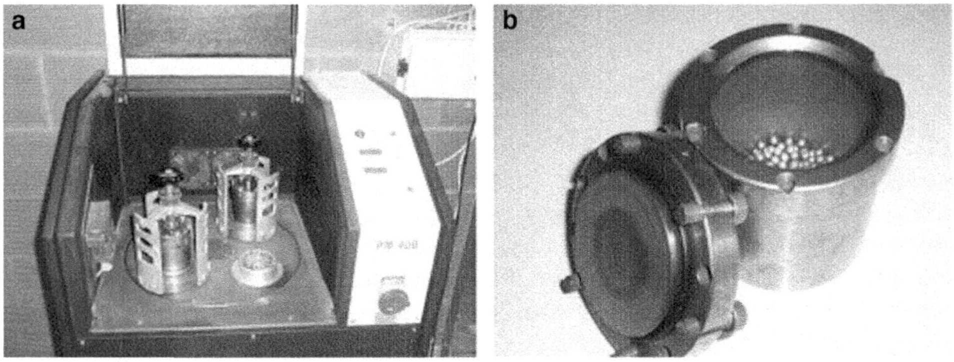

Fig. 3 **a** Planetary mill, **b** stainless steel vial (Janot and Guérard 2005). Copyright (2005) with permission from Elsevier

(Kis-Varga 1996). At high ratio, the number of collisions increases, and more energy transfers, more heat originates, and so mechanical activation takes place faster. By changing these parameters, novel materials that have desired physical, chemical, and optical properties have been produced. **The filling ratio of the vial** is another significant parameter due to the need for enough space for moving freely in the container. When the amounts of balls and powders have been small, the production rate has also been small. If the amounts are large, not enough gaps find to move and the energy affects less. For these reasons, 50% of the vial space must be empty. The **milling time** has increased the yield of reaction because the collision numbers and the total amount of transferred energy have also increased. In some cases, the product can be decomposed due to the increment of contamination and the formation of undesirable phases with long milling time (Suryanarayana 1995). The milling time has also changed the morphology and mechanical properties of products (Liu et al. 2012).

The revolution speed must be minimized to obtain the desired product with high yield and quality. The results of more powerful mixing and the increasing number of collisions, the revolution speed increase, and the energy input increase, higher yields, and better substrate conversion can be obtained. High speeds may cause by-product formation or the increasing temperature may cause the decomposition of metastable phases (Kaloshkin et al. 1997). The temperature increase can be useful if diffusion is required for powder alloying (Maurice and Courtney 1990). The lower revolution speed causes less energy consumption, lower cost, and more energy-efficiency.

The **milling temperature** is also an important parameter for mechanochemical synthesis. To decrease the temperature, liquid nitrogen can be used on the milling container. For increasing temperature, electrical heating can be used in the milling vial. Temperature changes directly affect the solid solubility levels and phase formation (Hong et al. 1994). **Rotation/oscillation frequency** affects the yield, selectivity, and similar properties of the reaction according to the kinetics of the reactions. If the frequency increases, side reactions occur and this negatively affects the yield (Stolle et al. 2011). The **milling atmosphere** is another important parameter. The inert atmospheres (helium and argon) are widely used to reduce contaminations. Nitrogen atmospheres can be used to form nitrides (Miki et al. 1992). Hydrogen atmospheres can be applied for hydride formation (Chen and Williams 1996). If the powders are reactive, both oxides and nitrides can be produced in the air atmosphere.

The lubricants and surfactants can be utilized as a **process control agent** to minimize the effects of cold welding between powders and prevent agglomeration. When the surfactants adsorb on the particle surfaces interfere with cold welding, the surface tension decrease. Benzene, C-wax, dodecane, paraffin, stearic acid, sodium chloride, boric acid, borax, alumina, etc., have widely used as the process control agents and during the milling process, they decompose and interact with powder particles and produce the compounds. The process control agents directly influence the final phase, the solid solubility and contamination levels (Gaffet et al. 1993), and the glass-forming range (Ivison et al. 1992).

4 Mechanochemical Synthesis of Materials

4.1 Mechanochemical Synthesis of Co-crystals

Due to the green and clean process, mechanochemical synthesis is utilized to investigate co-crystal in the pharmaceutical industry (Vishweshwar et al. 2005), organic chemistry (Friščić and MacGillivray 2005), and electronic materials (Sokolov et al. 2006). Co-crystals are described by US-FDA as "crystalline materials composed of two or more different molecules, typically active pharmaceutical ingredient (API) and co-crystal formers (coformers), in the same crystal lattice" (U.S. Department of Health and Human Services Food and Drug Administration 2018). Co-crystals consist of at least two neutral organic molecules and connect with intermolecular interactions (Tröbs and Emmerling 2014). Co-crystallization has been used to improve properties like dissolution rate, solubility, thermal stability, etc. (Aitipamula et al. 2012; Braga et al. 2011). The first mechanochemical synthesis of co-crystal was prepared by Toda and coworkers in 1987 (Toda et al. 1987). The Kuroda and coworkers were used simple co-grinding to form a new crystal depend on racemic bis-β-naphthol, benzoquinone, and anthracene with novel coloristic properties (Kuroda et al. 2002). Direct mixing is not enough to prepare new crystals (Lu and Rohani 2009), but it helps to obtain precursor, hydrogen bonds formation (Etter 1991), and determines polymorph interconversion (Ojala and Etter 1992). The construction of ternary and higher-order co-crystals still has challenges without solvent (Friščić et al. 2006). Crystals are generally soluble in waters or organic solvents. Liquid-assisted grinding mechanochemistry has been used for co-crystals in various industrial and technological processes (Braga et al. 2013). The kinetics of co-crystal changed through adding a little solvent. Cyclohexane-1,3cis,5cis-tricarboxylic acid, and hexamethylenetetramine mixed in a ball mill with small amounts of solvent for 20 min. According to solvent type, the obtained co-crystals were changed (Shan et al. 2002). When caffeine and citric acid did not form a co-crystal through neat grinding, liquid-assisted grinding with water or organic solvents gave the pharmaceutical solid into a 25 mL stainless steel grinding jar for 20 min (Karki et al. 2007).

Mechanochemistry is also given a chance to prepare multicomponent co-crystals. Thakuria and coworkers prepared drug-drug and drug–nutraceutical multicomponent solids. After neat grinding (NG), the conversion was incomplete, and the amorphous phase formed. After liquid-assisted grinding (LAG), coamorphous olanzapine-nateglinide, and crystalline salts/salt hydrates of olanzapine with the remaining coformers formed (Sarmah et al. 2020). Roex and coworkers synthesized nine new multicomponent crystalline materials using a series of three triazole, 7-chloroquinoline antimalarials, and two carboxylic acid coformers. They showed that coformers can change some physicochemical aspects of the drug molecules (Clements et al. 2019).

4.2 Mechanochemistry in Inorganic Synthesis

Mechanical activation is widely used for inorganic synthesis and ceramic technology to increase solid solubility and reactivity at low temperature. Besides, for the preparation of complex materials, ball milling has been used as the first step of the process. The high-energy milling has been applied to alter microstructure and to activate materials.

The **metal oxides**, mixed metal oxide, and supported metal oxides were successfully synthesized using mechanochemical synthesis as efficient catalysts for various catalytic systems (Ralphs et al. 2013). Kamolphop and coworkers have achieved low-temperature hydrocarbon selective catalytic reduction of NO_x without hydrogen and solvent using Ag/Al_2O_3 catalyst prepared in a ball mill for the first time. The silver precursor (nitrate, oxide, or powder) was mixed with the alumina support through ball milling for 1 h and then calcined at 550 °C for 2 h. The catalytic efficiency was compared with the wet-impregnation method. The ball milling prepared catalyst shows higher activity for the reduction of NO_x than the Ag/Al_2O_3 catalyst prepared via the wet-impregnation method due to the formation of the defects on the alumina surface. The best catalyst prepared from silver oxide indicated a 50% NO_x conversion at 240 °C and 99% at 302 °C (Kamolphop et al. 2011). Pardeshi and Patil prepared zinc oxide using two steps solvent-free mechanochemical method and examined the effects of morphology on the photocatalytic activity. The catalysts were synthesized through milling using zinc oxide and oxalic acid in the agate. Calcination temperatures were changed to control the morphology of the crystallites. The growth rate of ZnO changes with the calcination temperature range. The catalytic efficiency was investigated in the photocatalytic activity of ZnO in oxidative photocatalytic degradation of resorcinol in water under the irradiation. With increasing calcination temperature, the catalytic activity decreases due to increasing particle size. When the zinc oxide calcined from 400 to 550 °C, the same crystallite growth rate and maximum photocatalytic degradation of resorcinol were obtained (Pardeshi and Patil 2009). The mechanochemical synthesis of titanium dioxide has been prepared using different titanium source. During synthesis, the milling duration, the ball milling power, and the atmosphere are important parameters. In general, ball milling has been used to reduce particle size and obtain metastable polymorphs. However a small amount TiO_2 was obtained due to difficulties in controlling the temperature (Giannakoudakis et al. 2020). The high-pressure modification of TiO_2 with an α-PbO_2-type structure is formed through the grinding of anatase in a planetary ball mill with stainless steel vial. Rutile has not been formed directly ball milling of anatase. Transient phases appear at the beginning and disappear after long ball milling time. Three hours later, the rutile was only found (Begin-Colin et al. 1994). In the atmosphere, effects were studied using air, nitrogen, and ammonia in planetary ball mill with 300 rpm rotation. When Ti powder has been milled in N_2 or NH_3 atmospheres, the resultants have been TiN. In the air atmosphere, in open vials, titanium powder turned to titanium oxide and titanium oxynitride while only titanium oxynitride formed in closed vials (Lu et al. 2004). The kinetics and mechanisms of titanium dioxide phase transformation were studied by Colin and coworkers in 2000 and the powder/ball weight ratio is the other important parameter for titanium dioxide transformation (Begin-Colin et al. 2000). As an example of mixed oxide, high-density lead zirconate titanate (PZT) has been successfully synthesized by using a novel mechanochemical method from the low-cost, widely available oxides without calcination an intermediate temperature. According to milling time, particle sizes and crystallinity of the products were changed (Xue et al. 1999). Strontium manganite ($SrMnO_3$) and strontium-doped lanthanum manganite ($La_{0.7}Sr_{0.3}MnO_3$, LSM3) have been prepared by grinding the constituent oxides of $SrO–MnO_2$ and La_2O_3–$SrO–Mn_2O_3–MnO_2$ using a planetary mill. The reactions have proceeded with escalating grinding time. The product has a strong agglomeration of fine grain nanosize particles (Zhang et al. 2000). $LiMO_2$ (M=Ti, Mn and Fe) has been synthesized using lithium transition metals and metal oxides in a ball mill. The resulting oxides have the rocksalt structure. The transition metal and lithium ions were randomly ordered in the cation sites and after prolonged milling, lithium oxide and oxygen were lost. As α-$LiFeO_2$, cathodes of the obtained materials have poor electrochemical performance in lithium cells (Obrovac et al. 1998).

The efficient and rapid synthesis of various **metal complexes** like mononuclear complexes, coordination clusters, and spacious coordination cages was prepared by the mechanochemical method (Garay et al. 2007). Tsuchimoto and coworkers prepared polymeric vanadyl salen compound using Schiff base reproduced from 5-nitrosalicylaldehyde and diamines through mechanochemical reaction. The monomeric form of the obtained product is green, and the polymeric form of the obtained product is orange. Through grinding with sixteen zirconia balls (10 mm diameter in a zirconia-lined vessel, the orange form turned to green form Tsuchimoto et al. (2000). Orita and coworkers synthesized tetraplatinum square through a solvent-free mechanochemical method using bipyridyl and $[Pt(NO_3)_2(en)]$ (en = ethylene diamine) in only 10 min and with only a few percents of by-products and 76% yield (Orita et al. 2002). The same reaction was obtained with the solvent at 100 °C for 4 weeks. That shows the time and energy efficiency of the synthesis method (Fig. 4) (Garay et al. 2007). Bowmaker and coworkers used solvent-mediated mechanochemical synthesis for the formation of metal complexes using silver halides and ethylene thiourea. For the AgI, no reaction occurred between the dry reactants, whereas the reaction was complete with a small amount of solvent in a snap (Bowmaker et al. 2008).

Mechanochemistry has recently been used in inorganic coordination chemistry. The first report about **organometallic** synthesis appeared in the early 1990s and the preparation of different cyclopentadienyl and metal-carborane complexes were explained (Rightmire and Hanusa 2016). In previous studies, grinding only activated the reaction and no products were obtained without heat annealing (Petrova et al. 2002). In an example, mechanochemical synthesis of iron chloride with cyclopentadienides of alkaline metals was studied to form ferrocene, and milling time increased the yield of ferrocene formation (Makhaev et al. 1999). Multimetallic clusters are also prepared through mechanochemical synthesis. The reactions of $Au(CuCPh)PPh_3$ and Ag(OTf) lead to the high nuclearity species $Ag_{12}Au_{10}(CuCPh)_{17}(OTf)_5(PPh_3)_3$. The same products are formed whether the reactions are conducted in acetone or with ball milling (Blanco et al. 2012). The first example of mechanochemical synthesis of organometallic pincer complexes was prepared by Aleksanyan and coworkers in 2017. The synthesis of an organometallic Pd^{II} pincer complex has been synthesized through C-H bond activation of the bis(thiocarbamate) ligand with $PdCl_2(NCPh)_2$ under mechanochemical conditions both grinding in mortar for 1.5 h and in ball mill for 115 s at gram scale (Aleksanyan et al. 2017).

Nanocrystalline **chalcogenides** display the novel physical and chemical properties that differ from their bulk equivalents and are applicable in several fields (Baláz et al. 2020). Chalcogenides have gained a great variety of properties through mechanochemical treatment (Baláž et al. 2017). The antimony and bismuth sulfide are semiconductors with interesting thermoelectric properties. The mechanochemical synthesis of Sb_2S_3 and Bi_2S_3 nanoparticles using high-energy milling was prepared at ambient temperature in a planetary laboratory mill. Individual nanoparticles sound in generate nanoparticle agglomerates during milling. The particle sizes are 30 nm and 24 nm for Sb_2S_3 and Bi_2S_3, respectively (Dutková et al. 2013). Besides metal sulfides, metal selenides are used as semiconductors in materials science. They have successfully been synthesized through a simple, fast, and less-consumptive mechanochemical method in a planetary and vibratory mill using the selected metal and selenium powders as starting materials (Achimovičová et al. 2012; Kristl et al. 2016; Gotor et al. 2013). Chalcogenide nanocomposites were successfully prepared mechanochemical method. The $Cu_2ZnSnSe_4$ (CZTSe) solar cells and Cu_2ZnSnS_4 (CZTS) nanocrystals were prepared by a simple, environmentally friendly, and scalable mechanochemical method (Park et al. 2014). The effects of the milling time were observed and according to the XRD analyses (Fig. 5a, b), after 3 h, the CZTS nanocrystals were formed. The characteristic Raman scatterings (289, 335, and 364 cm^{-1}) can be identified to CZTS (Fig. 5c) and no impurity was found (Fernandes et al. 2011). The milling time was also decreased the particle sizes of the obtained nanocrystals.

4.3 Mechanochemistry in Organic Synthesis

For organic chemistry, ball milling has been an efficient method for C–C bond formation (Rodriguez et al. 2007), aldol condensation (Raston and Scott 2000), asymmetric

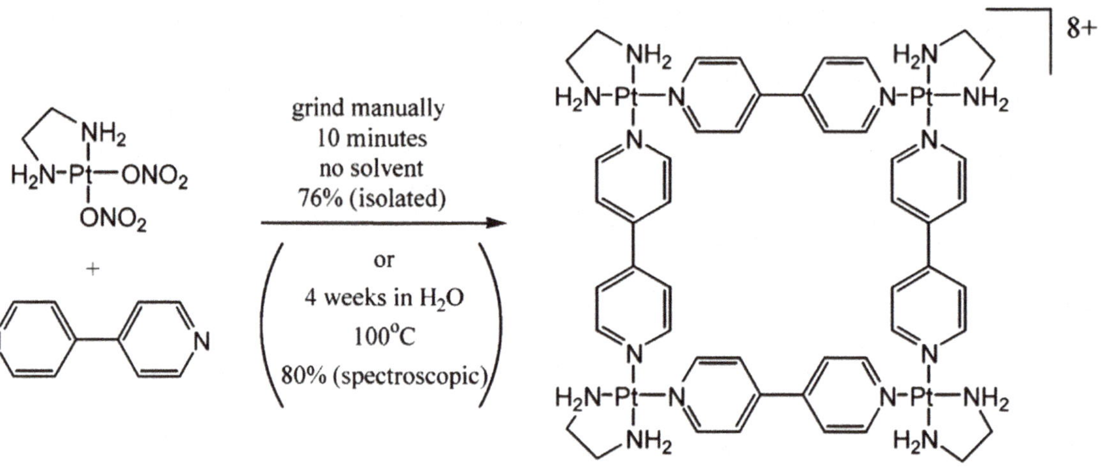

Fig. 4 Synthesis of a molecular square without solvent Garay et al. 2007). Reproduced by permission of RSC

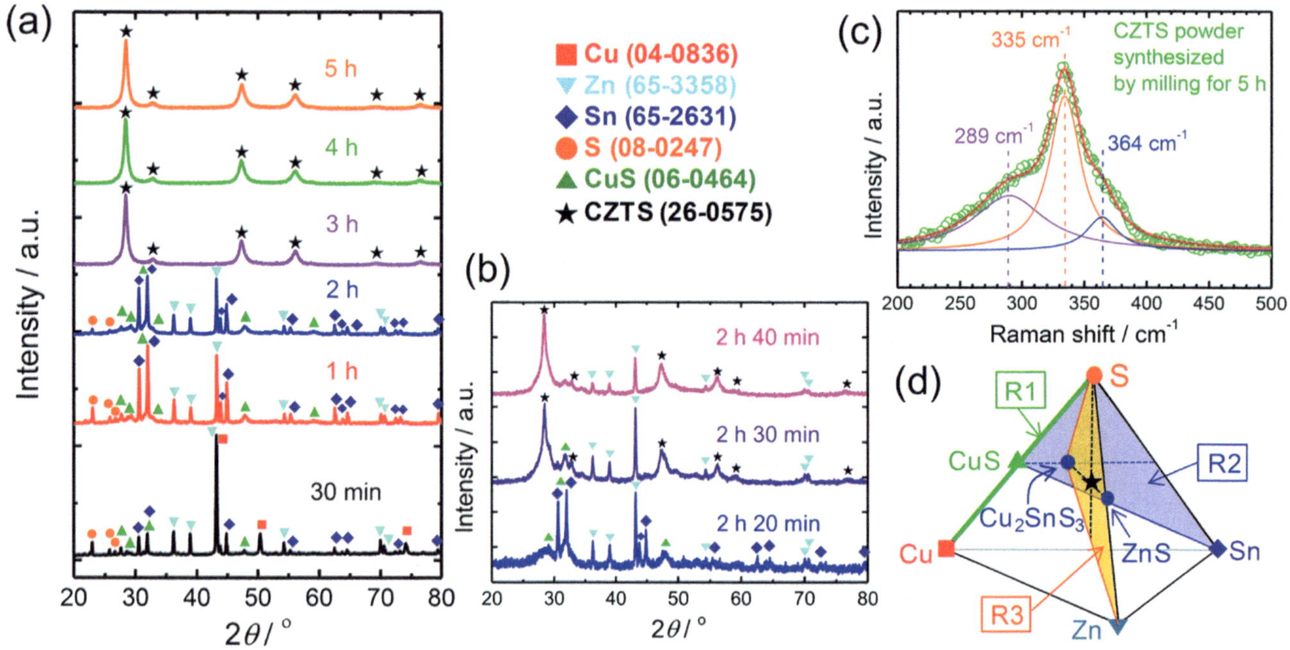

Fig. 5 **a**, **b** XRD patterns of the precursors milled at various milling times. **c** Raman spectrum CZTS synthesized for 5 h. **d** The quaternary phase diagram (Park et al. 2014). Reproduced by permission of RSC

organic synthesis (Egorov et al. 2020), amine condensation (Kaupp 2006), syntheses of heterocycles, enantioselective synthesis (Avila-Ortiz et al. 2019), Baylis–Hillman reactions (Mack and Shumba 2007), fullerene modifications (Komatsu 2005) under solvent-free conditions (Cave et al. 2001). For organic synthesis, mechanochemistry was widely used through mixer mill and planetary mill under solvent-free conditions (Wang 2013). Stolle and coworkers prepared a review about ball milling in organic syntheses like C–C bond formation, preparation of heterocycles and fullerenes, redox reactions, etc. (Stolle et al. 2011). Axelsson and coworkers were prepared the Suzuki reaction under mechanochemical and solvent-free conditions for the first time (Nielsen et al. 2000). They published the coupling of phenylboronic acid with aryl bromides using $Pd(PPh_3)_4$ as a catalyst, and K_2CO_3 as a base. After 30–60 min milling, the reaction was completed with excellent yields. Schneider and coworkers have studied milling conditions on the Suzuki–Miyaura reaction. The effects of milling parameters (the revolutions per minute, milling time, the material, size, and the number of milling balls and beakers) on the reaction yield are investigated. The order of the parameters on the formation of 4-acetylbiphenyl is revolution > milling time > milling ball size > balls numbers > grinding

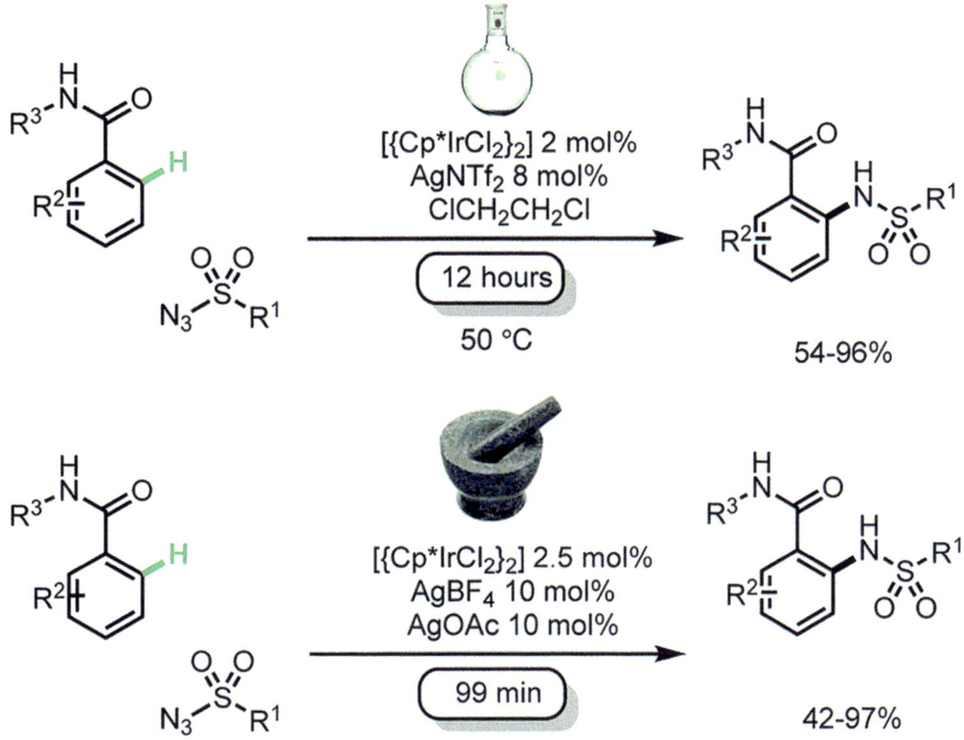

Fig. 6 Synthesis conditions of C–H amidation of benzamides (Howard et al. 2018). Published by RSC

material (Schneider et al. 2009). The catalyst also affected in organic mechanochemical synthesis. Hermann and coworkers were prepared the amidation of benzamides with sulfonyl azides using mechanochemical synthesis (Hermann et al. 2016) and synthesis conditions were shown in Fig. 6 (Howard et al. 2018). The active cationic Ir(III) catalyst was prepared in mixer mill using (Cp * $IrCl_2)_2$ and $AgNTf_2$. The amidated products were achieved high yield and short reaction times (99 min) compared with solution-based methods (12 h) as published by Chang and coworkers (Lee et al. 2013).

The fullerenes are important materials because of their potential applications in fuel cells (Coro et al. 2016), molecular devices (Lai et al. 2014), optoelectronics (Cravino and Sariciftci 2002), biomedical science (Partha and Conyers 2009), and nanotechnology (Rapoport et al. 2005). The fullerene synthesis via mechanochemistry has some difficulties due to solubility problems. Fullerenes are slightly soluble in less polar solvents like CS_2, toluene, chlorobenzene, and insoluble in polar solvents such as methanol, DMF, and water. Consequently, the solvent-free synthesis of fullerenes using mechanochemistry has several advantages such as without harmful organic solvent usage, fast reaction, high yield, and selectivity compared with the solution-based method. The serendipitous found of dumbbell-shaped C_{120} (Wang et al. 1997) and the several fullerene dimers and trimers are made possible to use the mechanochemical synthesis. The abundant fullerene molecules nearby can trap the reaction intermediates to produce the dimeric (Murata et al. 2001) and trimeric (Kunitake et al. 2002) forms by assuming the excessively high concentration of C_{60}/C_{70}.

Mechanochemical organic synthesis is also important for the pharmaceutical industry to produce drugs, active pharmaceutical ingredients, and commodity chemicals (Colacino et al. 2019b). This has also been named "medicinal mechanochemistry" (Tan et al. 2016) that is a clean and simple synthesis approach without solvent with highly reducing the environmental impact for the fabrication of biomolecules. As an example, using mechanochemical blitz synthesis both in a vibrating and planetary mill, the antiepileptic drug phenytoin was produced in moderate yield (10–44%). With modified read synthesis, it was obtained in 86% yield by mechanochemistry (Fig. 7). The mechanochemical synthesis was successfully applied without solvent, reagents excess, and purification. As a result of these, mechanochemistry was used as an eco-friendly, cheap, time, and energy-efficient method to design the antiepileptic drug phenytoin in good yield. In future studies, mechanochemistry can be an alternative eco-friendly synthesis method to Biltz synthesis for the pharmaceutically interesting drug on a large scale (Konnert et al. 2014).

4.4 Mechanochemistry in Metal–Organic Frameworks (MOFs)

Metal–organic frameworks have been important functional materials due to various properties like designable structure,

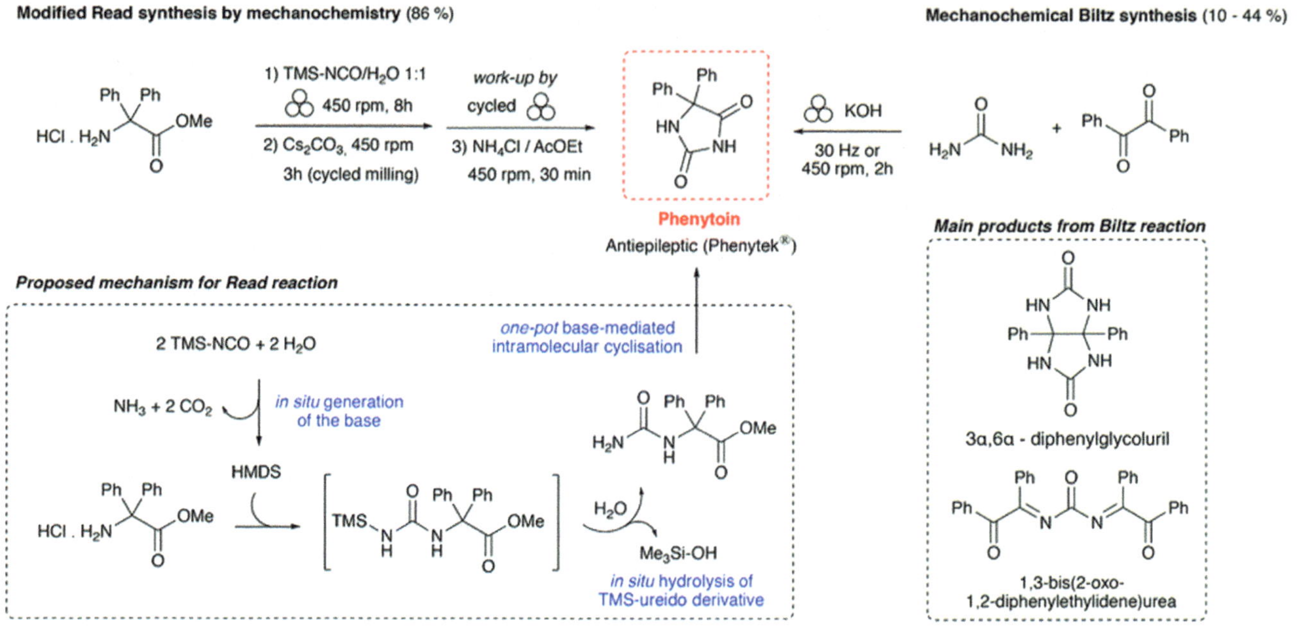

Fig. 7 Mechanochemical syntheses of the antiepileptic drug phenytoin (Colacino et al. 2019b). Copyright 2019 by RSC

controllable morphology, high porosity, and surface area, surface functionality, optical, electrical, and magnetic properties (Ozer 2020). In recent years, mechanochemical synthesis has been an effective synthesis approach for metal–organic frameworks due to simple, fast, and efficient reactions with short reaction times, quantitative conversion, and lack solvent (Do and Friščić 2017). While soluble metal sources have been needed for the solution method, poorly soluble sulfates, oxide, and carbonates have been used as reactants resulting in a cleaner, more atom-efficient processes that prevent the production of external bases and mineral acids as by-products (Adams et al. 2008).

In recent years, liquid-assisted grinding (LAG) was widely used to produce coordination polymers. Water and organic solvents can affect the formation of intermediates as kinetic products and different polymorphs were formed according to liquid types (Strobridge et al. 2010). The liquid type is both facilitating the reaction and acts as a structure-directing and space-filling agent. Using different types and amounts of grinding liquids, from zinc oxide and fumaric acid, different products were formed and shown in Fig. 8. With four equivalent water, the zigzag 1D polymer was formed while linear 1D polymer was formed with three equivalent water. Using organic solvent (ethanol or methanol), anhydrous 3D coordination polymer with tetrahedral zinc center. When adding water to organic solvent, due to the formation of hydrogen bonds, the 2D polymer with octahedral zinc center was formed (Friščić and Fábián 2009).

With changing milling time, MOF-74 was directly prepared using ball milling from a metal oxide (ZnO), 2,5-dihydroxy terephthalic acid, and without bulk solvent through liquid-assisted grinding in poly(methyl)methacrylate jar (14 mL) with a single stainless steel ball. With increasing milling time, the reaction mixture changed according to colors. The waiting product was formed at 40 min and after 45 min, no traces of ZnO were visible in the p-XRD. Collected pale-yellow microcrystalline product was formed without any by-product as shown in Fig. 9 (Julien et al. 2016).

In previous studies, the solvothermal method has generally been used for MOF synthesis, but the remaining solvents in the pores can cause problems in different applications. To show the synthesis method effects on the structural and surface properties, the intensively studied metal–organic framework HKUST-1 was prepared three different synthesis approaches including mechanochemical, electrochemical, and solution method. The crystallite sizes were found as 50 nm for the electrochemical method, 55 nm for the mechanochemical method, and 31 nm for the solution method according to XRD analyses. BET surface analysis was given in Table 2 and as to the synthesis method, surface areas of the samples were changed. With the activation of mechanochemical synthesis materials, solvents (acetic acid) were removed and the highest surface area was obtained compared with the other methods (Klimakow et al. 2010).

4.5 Mechanochemistry in Porous Organic Materials (POMs)

Porous organic materials (POMs) have gained great attention due to their potential applications as gas adsorption and storage, separation, heterogeneous catalysis, etc. (Das et al.

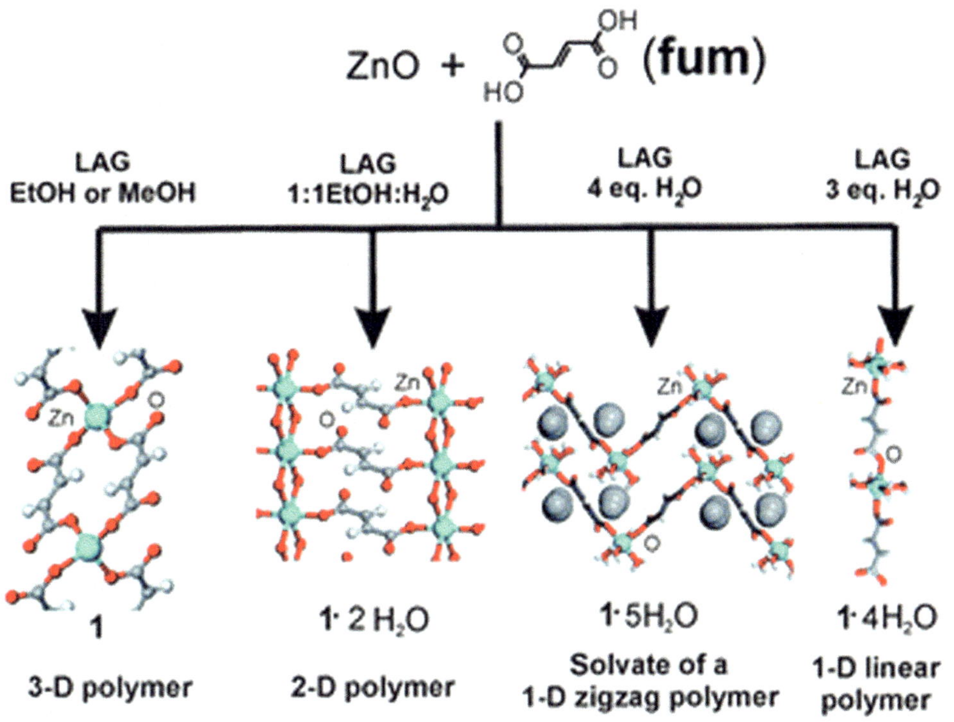

Fig. 8 Effects of grinding liquid types and amounts (Friščić and Fábián 2009). Copyright 2009 by RSC

2017). They are comprised of only light elements that have accessible functionality, high stability in the air, and chemical robustness to acids and bases (Zhang and Dai 2017). PIM-1 was prepared through ball or manually grinding in a mortar using different milling times and the obtained products were shown in Fig. 10. Zhang and coworkers were prepared the polycondensation of 5,5,6,6-tetrahydroxy-3,3,3,3-tetramethyl-1,1′-spirobisindane and tetrafluoroterephthalonitrile toward K_2CO_3 using stainless steel reactor along with twelve stainless steel ball bearings (Zhang et al. 2015). Color changes from white through green to yellow during polymerization. Compared with the solvent-based method (72 h), ball milling synthesis was complete only 15 min with a 98% yield without solvent. As a result of mechanochemical synthesis, the PIM-1 has a high molecular weight with high solubility in dimethylformamide and tetrahydrofuran due to the physical effects of the mechanical action. The ball milling prevents rapid chain growth and chain termination (Zhang and Dai 2017).

4.6 Mechanochemical Synthesis of Polymers

In the beginning, mechanochemistry has been applied to break polymer chains or form short chains with low molecular weights for polymer synthesis. The mechanochemical parameters like energy input, dilution of the monomers, etc., are important to investigate the potential usage of this method for polymers. As an example, in 2014, poly(phenylene) vinylenes that are an important electroluminescent material, were synthesized by a mechanochemical Gilch reaction as a rapid, simple, and solvent-free synthetic route in a Retsch mixer mill 400 at 30 Hz for 30 min. Effect of milling time and frequency, base strength, solid-state dilution, and size of milling balls were investigated and found that the polymerization through ball milling is a rapid and efficient process with up to 40 kDa average molecular weight and 70% yield with the optimized condition (Ravnsbæk and Swager 2014). For mechanochemical polymerization of styrene, wet-grinding of quartz was applied in the styrene using a vibrating ball mill with a laboratory scale and the polymerization was connected with the total surface area of the ground quartz (Hasegawa et al. 2001). Mechanochemistry is also efficient for surface modification of polymers, as shown by Fiss and coworkers who used P_4O_{10} to conduct phosphorylation of synthetic polymers and cellulose, leading to new flame-retardants (Fiss et al. 2019). The Moore and coworker utilized a combined method with milling and followed by aging to produce high molecular weight chitosan biopolymers of unprecedented length. For deacetylation, mechanochemical synthesis is a versatile and applicable method with 98% yield and remarkably high molecular weights (Nardo et al. 2019).

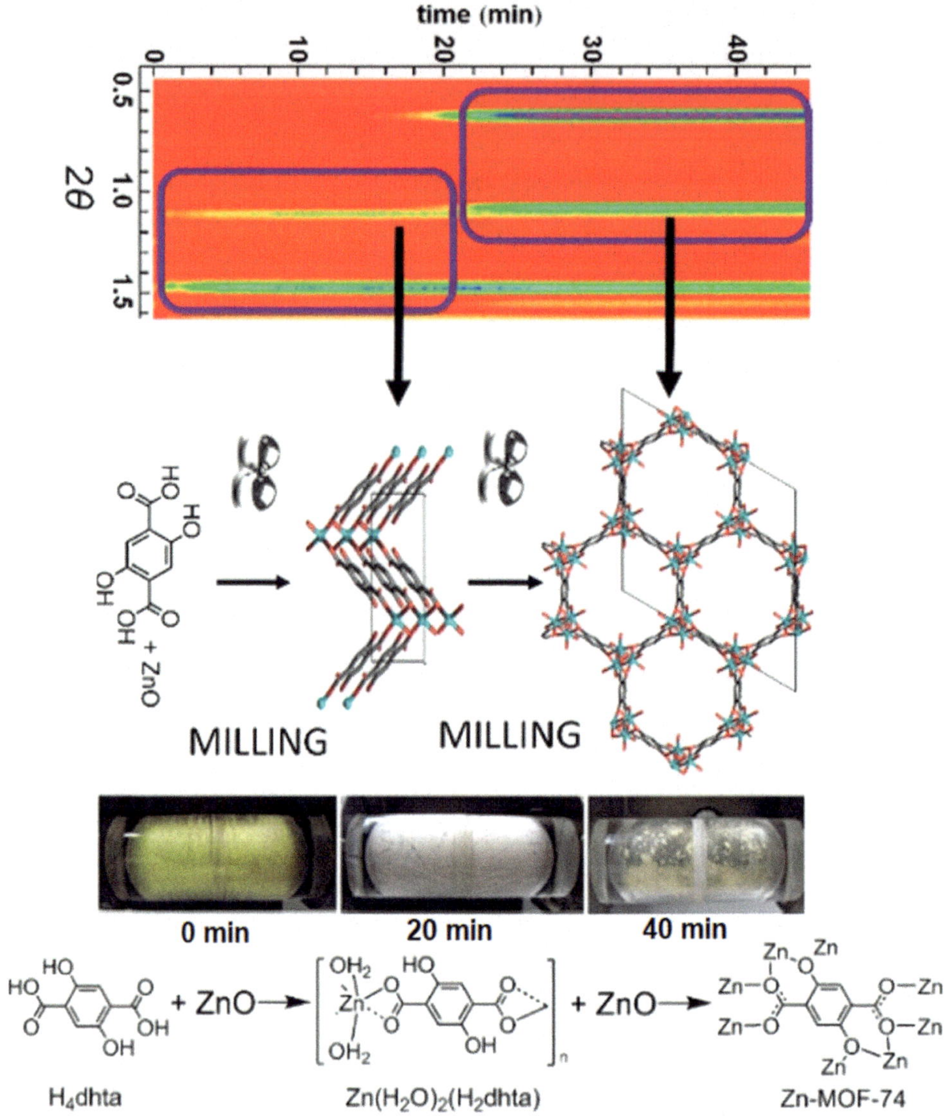

Fig. 9 Mechanochemical synthesis of MOF-74 (Julien et al. 2016). Copyright (2016) ACS

Table 2 Surface areas of HKUST-1 synthesized by different methods

Synthesis method	BET surface area (m^2/g)	p/p_o range for BET fit
Mechanochemistry	758	0.004–0.029
Electrochemistry	1836	0.001–0.011
Solution	1184	0.001–0.011
Mechanochemistry and activation	1713	0.005–0.029

Reprinted with permission from Klimakow et al. (2010). Copyright (2010) ACS

Fig. 10 PIM-1 grinding different times and solving different solvents (dimethylformamide and tetrahydrofuran) (Zhang et al. 2015). Copyright 2015 RSC

5 Conclusions

In summary, mechanochemical synthesis is a green, more powerful, more sustainable, timesaving, environmentally friendly, and more economical preparative method for the synthesis of new and effective functional materials. According to synthesis parameters (milling types, materials, size, time, temperature and atmosphere, revolution speed, frequency, ball/powder weight ratio, filling ratio, process control agents), the chemical, thermal, surface, and other properties of the materials can be improved. To reduce the solvent usage, to prevent the solubility problem, to protect the environment, mechanochemistry has been widely used in different research areas and industries up to now and it will have great importance for future studies.

References

Achimovičová M, Gotor FJ, Real C, Daneu N (2012) Mechanochemical synthesis and characterization of nanocrystalline BiSe, Bi_2Se_3 semiconductors. J Mater Sci Mater Electron 23(10):1844–1850

Adams CJ, Kurawa MA, Lusi M, Orpen AG (2008) Solid state synthesis of coordination compounds from basic metal salts. CrystEngComm 10(12):1790–1795

Aitipamula S, Banerjee R, Bansal AK, Biradha K, Cheney ML, Choudhury AR, Desiraju GR, Dikundwar AG, Dubey R, Duggirala N (2012) Polymorphs, salts, and cocrystals: what's in a name? Cryst Growth Des 12(5):2147–2152

Aleksanyan DV, Churusova SG, Aysin RR, Klemenkova ZS, Nelyubina YV, Kozlov VA (2017) The first example of mechanochemical synthesis of organometallic pincer complexes. Inorg Chem Commun 76:33–35

Anastas PT, Tundo P (2000) Green chemistry: challenging perspectives. Oxford University Press, Oxford

Avila-Ortiz CG, Pérez-Venegas M, Vargas-Caporali J, Juaristi E (2019) Recent applications of mechanochemistry in enantioselective synthesis. Tetrahedron Lett 60(27):1749–1757

Baláž P (2008) Mechanochemistry in nanoscience and minerals engineering. Springer, Berlin

Baláž P, Aláčová A, Achimovičová M, Ficeriova J, Godočíková E (2005) Mechanochemistry in hydrometallurgy of sulphide minerals. Hydrometallurgy 77(1–2):9–17

Baláž P, Baláž M, Achimovičová M, Bujňáková Z, Dutková E (2017) Chalcogenide mechanochemistry in materials science: insight into synthesis and applications (a review). J Mater Sci 52(20):11851–11890

Baláz M, Achimovicová M, Baláz P, Dutková E, Fabián M, Kovácová M, Bujnáková ZL, Tóthová E (2020) Mechanochemistry as a versatile and scalable tool for nanomaterials synthesis: Recent achievements in Košice, Slovakia. Curr Opin Green Sustain Chem 24:7–13

Begin-Colin S, Caer GL, Mocellin A, Zandona M (1994) Polymorphic transformations of titania induced by ball milling. Philos Mag Lett 69(1):1–7

Begin-Colin S, Girot T, Le Caër G, Mocellin A (2000) Kinetics and mechanisms of phase transformations induced by ball-milling in anatase TiO_2. J Solid State Chem 149(1):41–48

Beyer MK, Clausen-Schaumann H (2005) Mechanochemistry: the mechanical activation of covalent bonds. Chem Rev 105(8):2921–2948

Blanco MC, Cámara J, Gimeno MC, Laguna A, James SL, Lagunas MC, Villacampa MD (2012) Synthesis of gold-silver luminescent honeycomb aggregates by both solvent-based and solvent-free methods. Angew Chem Int Ed 51(39):9777–9779

Boldyrev V, Tkáčová K (2000) Mechanochemistry of solids: past, present, and prospects. J Mater Synth Process 8(3–4):121–132

Boldyreva E (2013) Mechanochemistry of inorganic and organic systems: what is similar, what is different? Chem Soc Rev 42 (18):7719–7738

Bowmaker GA, Chaichit N, Pakawatchai C, Skelton BW, White AH (2008) Solvent-assisted mechanochemical synthesis of metal complexes. Dalton Trans 22:2926–2928

Braga D, Curzi M, Johansson A, Polito M, Rubini K, Grepioni F (2006) Simple and quantitative mechanochemical preparation of a porous crystalline material based on a 1D coordination network for uptake of small molecules. Angew Chem Int Ed 45(1):142–146

Braga D, Giaffreda S, Curzi M, Maini L, Polito M, Grepioni F (2007) Mechanical mixing of molecular crystals: a green route to co-crystals and coordination networks. J Therm Anal Calorim 90 (1):115–123

Braga D, d'Agostino S, Dichiarante E, Maini L, Grepioni F (2011) Dealing with crystal forms (the kingdom of serendip?). Chem Asian J 6(9):2214–2223
Braga D, Maini L, Grepioni F (2013) Mechanochemical preparation of co-crystals. Chem Soc Rev 42(18):7638–7648
Burmeister CF, Kwade A (2013) Process engineering with planetary ball mills. Chem Soc Rev 42(18):7660–7667
Burmeister CF, Schmidt R, Jacob K, Breitung S, Stolle A, Kwade A (2020) Effect of stressing conditions on mechanochemical Knoevenagel synthesis. Chem Eng J 396:124578
Cave GW, Raston CL, Scott JL (2001) Recent advances in solventless organic reactions: towards benign synthesis with remarkable versatility. Chem Commun 21:2159–2169
Chadwick K, Davey R, Cross W (2007) How does grinding produce co-crystals? Insights from the case of benzophenone and diphenylamine. CrystEngComm 9(9):732–734
Chen Y, Williams JR (1996) Hydriding reactions induced by ball milling. In: Materials science forum. Trans Tech Publications, pp 881–888
Cinčić D, Brekalo I, Kaitner B (2012) Effect of atmosphere on solid-state amine–aldehyde condensations: gas-phase catalysts for solid-state transformations. Chem Commun 48(95):11683–11685
Clements M, Blackie M, de Kock C, Lawrence N, Smith P, Tl R (2019) Investigation into the structures and properties of multicomponent crystals formed from a series of 7-chloroquinolines and aromatic acids. Cryst Growth Des 19(3):1540–1549
Colacino E, Porcheddu A, Charnay C, Delogu F (2019b) From enabling technologies to medicinal mechanochemistry: an eco-friendly access to hydantoin-based active pharmaceutical ingredients. React Chem Eng 4(7):1179–1188
Colacino E, Dayaker G, Morère A, Friščić T (2019) Introducing students to mechanochemistry via environmentally friendly organic synthesis using a solvent-free mechanochemical preparation of the antidiabetic drug tolbutamide. J Chem Educ 96(4):766–771
Coro J, Suárez M, Silva LS, Eguiluz KI, Salazar-Banda GR (2016) Fullerene applications in fuel cells: a review. Int J Hydrogen Energy 41(40):17944–17959
Cravino A, Sariciftci NS (2002) Double-cable polymers for fullerene based organic optoelectronic applications. J Mater Chem 12 (7):1931–1943
Das S, Heasman P, Ben T, Qiu S (2017) Porous organic materials: strategic design and structure–function correlation. Chem Rev 117 (3):1515–1563
Dayaker G, Tan D, Biggins N, Shelam A, Do J-L, Katsenis AD, Friscic T (2020) Catalytic room-temperature C–N coupling of amides and isocyanates using mechanochemistry. ChemSusChem 13(11):2966–2972
Delmonte D, Manfredi R, Calestani D, Mezzadri F, Righi L, Mazzer M, Pattini F, Rampino S, Spaggiari G, Gilioli E (2020) An affordable method to produce $CuInS_2$ 'mechano-targets' for film deposition. Semicond Sci Technol 35(4):045026
Di Nardo T, Hadad C, Van Nhien AN, Moores A (2019) Synthesis of high molecular weight chitosan from chitin by mechanochemistry and aging. Green Chem 21(12):3276–3285
Di L, Bakker H (1991) Phase transformation of the compound V3Ga induced by mechanical grinding. J Phys: Condens Matter 3 (20):3427
Do J-L, Friščić T (2017) Mechanochemistry: a force of synthesis. ACS Central Sci 3(1):13–19
Dutková E, Takacs L, Sayagués MJ, Baláž P, Kováč J, Šatka A (2013) Mechanochemical synthesis of Sb_2S_3 and Bi_2S_3 nanoparticles. Chem Eng Sci 85:25–29
Egorov IN, Santra S, Kopchuk DS, Kovalev IS, Zyryanov GV, Majee A, Ranu B, Rusinov VL, Chupakhin ON (2020) Ball-milling: an efficient and green approach for asymmetric organic synthesis. Green Chem 22(2):302–315
Emami S, Shayanfar A (2020) Deep eutectic solvents for pharmaceutical formulation and drug delivery applications. Pharm Dev Technol 25:1–18
Etter MC (1991) Hydrogen bonds as design elements in organic chemistry. J Phys Chem 95(12):4601–4610
Fernandes P, Salomé P, Da Cunha A (2011) Study of polycrystalline Cu_2ZnSnS_4 films by Raman scattering. J Alloy Compd 509 (28):7600–7606
Fernandez-Bertran JF (1999) Mechanochemistry: an overview. Pure Appl Chem 71(4):581–586
Fiss BG, Hatherly L, Stein RS, Friščić T, Moores A (2019) Mechanochemical phosphorylation of polymers and synthesis of flame-retardant cellulose nanocrystals. ACS Sustain Chem Eng 7 (8):7951–7959
Fox P (1975) Mechanically initiated chemical reactions in solids. J Mater Sci 10(2):340–360
Friščić T, Fábián L (2009) Mechanochemical conversion of a metal oxide into coordination polymers and porous frameworks using liquid-assisted grinding (LAG). CrystEngComm 11(5):743–745
Friščić T, MacGillivray LR (2005) Reversing the code of a template-directed solid-state synthesis: a bipyridine template that directs a single-crystal-to-single-crystal [2 + 2] photodimerisation of a dicarboxylic acid. Chem Commun 46:5748–5750
Friščić T, Trask AV, Jones W, Motherwell WS (2006) Screening for inclusion compounds and systematic construction of three-component solids by liquid-assisted grinding. Angew Chem Int Ed 45(45):7546–7550
Friščić T, Childs SL, Rizvi SA, Jones W (2009a) The role of solvent in mechanochemical and sonochemical cocrystal formation: a solubility-based approach for predicting cocrystallisation outcome. CrystEngComm 11(3):418–426
Friščić T, Meštrović E, Škalec Šamec D, Kaitner B, Fabian L (2009b) One-pot mechanosynthesis with three levels of molecular self-assembly: coordination bonds, hydrogen bonds and host–guest inclusion. Chem Eur J 15(46):12644–12652
Friščić T, Reid DG, Halasz I, Stein RS, Dinnebier RE, Duer MJ (2010) Ion-and liquid-assisted grinding: improved mechanochemical synthesis of metal–organic frameworks reveals salt inclusion and anion templating. Angew Chem Int Ed 49(4):712–715
Friščić T, Halasz I, Beldon PJ, Belenguer AM, Adams F, Kimber SA, Honkimäki V, Dinnebier RE (2013) Real-time and in situ monitoring of mechanochemical milling reactions. Nat Chem 5(1):66
Gaffet E, Harmelin M, Faudot F (1993) Far-from-equilibrium phase transition induced by mechanical alloying in the Cu–Fe system. J Alloy Compd 194(1):23–30
Garay AL, Pichon A, James SL (2007) Solvent-free synthesis of metal complexes. Chem Soc Rev 36(6):846–855
Giannakoudakis DA, Chatel G, Colmenares JC (2020) Mechanochemical forces as a synthetic tool for zero-and one-dimensional titanium oxide-based nano-photocatalysts. Top Curr Chem 378(1):2
Gómez-López P, Puente-Santiago A, Castro-Beltrán A, do Nacimiento LAS, Balu AM, Luque R, Alvarado-Beltrán CG (2020) Nanomaterials and catalysis for green chemistry. Curr Opin Green Sustain Chem 24:48–55
Gonzalez-Moragas L, Yu S-M, Murillo-Cremaes N, Laromaine A, Roig A (2015) Scale-up synthesis of iron oxide nanoparticles by microwave-assisted thermal decomposition. Chem Eng J 281:87–95
Gotor F, Achimovicova M, Real C, Balaz P (2013) Influence of the milling parameters on the mechanical work intensity in planetary mills. Powder Technol 233:1–7
Haneef J, Chadha R (2020) Sustainable synthesis of ambrisentan–syringic acid cocrystal: employing mechanochemistry in the

development of novel pharmaceutical solid form. CrystEngComm 22(14):2507–2516

Hasa D, Schneider Rauber G, Voinovich D, Jones W (2015) Cocrystal formation through mechanochemistry: from neat and liquid-assisted grinding to polymer-assisted grinding. Angew Chem Int Ed 54 (25):7371–7375

Hasegawa M, Kimata M, Kobayashi SI (2001) Mechanochemical polymerization of styrene initiated by the grinding of quartz. J Appl Polym Sci 82(11):2849–2855

Hermann GN, Becker P, Bolm C (2016) Mechanochemical iridium (III)-catalyzed C−H bond amidation of benzamides with sulfonyl azides under solvent-free conditions in a ball mill. Angew Chem Int Ed 55(11):3781–3784

Hong L, Bansal C, Fultz B (1994) Steady state grain size and thermal stability of nanophase Ni_3Fe and Fe_3X (X=Si, Zn, Sn) synthesized by ball milling at elevated temperatures. Nanostruct Mater 4 (8):949–956

Hou H, Zhou J, Ji M, Yue Y, Qian G, Zhang J (2020) Mechanochemical activation of titanium slag for effective selective catalytic reduction of nitric oxide. Sci Total Env 743:140733

Howard JL, Cao Q, Browne DL (2018) Mechanochemistry as an emerging tool for molecular synthesis: what can it offer? Chem Sci 9(12):3080–3094

Ivison P, Soletta I, Cowlam N, Cocco G, Enzo S, Battezzati L (1992) The effect of absorbed hydrogen on the amorphization of CuTi alloys. J Phys Condens Matter 4(23):5239

James SL, Friščić T (2013) Mechanochemistry. Chem Soc Rev 42 (18):7494–7496

James SL, Adams CJ, Bolm C, Braga D, Collier P, Friščić T, Grepioni F, Harris KD, Hyett G, Jones W (2012) Mechanochemistry: opportunities for new and cleaner synthesis. Chem Soc Rev 41 (1):413–447

Janot R, Guérard D (2005) Ball-milling in liquid media: applications to the preparation of anodic materials for lithium-ion batteries. Prog Mater Sci 50(1):1–92

Julien PA, Užarević K, Katsenis AD, Kimber SA, Wang T, Farha OK, Zhang Y, Casaban J, Germann LS, Etter M (2016) In situ monitoring and mechanism of the mechanochemical formation of a microporous MOF-74 framework. J Am Chem Soc 138 (9):2929-2932

Kaloshkin S, Tomilin I, Andrianov G, Baldokhin U, Shelekhov E (1997) Phase transformations and hyperfine interactions in mechanically alloyed Fe–Cu solid solutions. In: Materials science forum. Trans Tech Publications, pp 565–570

Kamolphop U, Taylor SF, Breen JP, Burch R, Delgado JJ, Chansai S, Hardacre C, Hengrasmee S, James SL (2011) Low-temperature selective catalytic reduction (SCR) of NO_x with *n*-octane using solvent-free mechanochemically prepared Ag/Al_2O_3 catalysts. ACS Catal 1(10):1257–1262

Karki S, Friščić T, Jones W, Motherwell WS (2007) Screening for pharmaceutical cocrystal hydrates via neat and liquid-assisted grinding. Mol Pharm 4(3):347–354

Kaupp G (2003) Solid-state molecular syntheses: complete reactions without auxiliaries based on the new solid-state mechanism. CrystEngComm 5(23):117–133

Kaupp G (2006) Waste-free large-scale syntheses without auxiliaries for sustainable production omitting purifying workup. CrystEngComm 8(11):794–804

Kaupp G (2009) Mechanochemistry: the varied applications of mechanical bond-breaking. CrystEngComm 11(3):388–403

Kis-Varga M, Beke DL (1984) Phase transitions in Cu–Sb systems induced by ball milling. In: Materials science forum, 1996. Trans Tech Publications, Aedermannsdorf, Switzerland, pp 465–470

Klimakow M, Klobes P, Thünemann AF, Rademann K, Emmerling F (2010) Mechanochemical synthesis of metal–organic frameworks: a fast and facile approach toward quantitative yields and high specific surface areas. Chem Mater 22(18):5216–5221

Komatsu K (2005) kThe mechanochemical solid-state reaction of fullerenes. In: Organic solid state reactions. Springer, pp 185–206

Konnert L, Reneaud B, de Figueiredo RM, Campagne J-M, Fdr L, Martinez J, Colacino E (2014) Mechanochemical preparation of hydantoins from amino esters: application to the synthesis of the antiepileptic drug phenytoin. J Org Chem 79(21):10132–10142

Kristl M, Gyergyek S, Srt N, Ban I (2016) Mechanochemical route for the preparation of nanosized aluminum and gallium sulfide and selenide. Mater Manuf Process 31(12):1608–1612

Kumar S, Jain S, Nehra M, Dilbaghi N, Marrazza G, Kim K-H (2020a) Green synthesis of metal–organic frameworks: a state-of-the-art review of potential environmental and medical applications. Coord Chem Rev 420:213407

Kumar YR, Deshmukh K, Sadasivuni KK, Pasha SK (2020b) Graphene quantum dot based materials for sensing, bio-imaging and energy storage applications: a review. RSC Adv 10(40):23861–23898

Kunitake M, Uemura S, Ito O, Fujiwara K, Murata Y, Komatsu K (2002) Structural analysis of C_{60} trimers by direct observation with scanning tunneling microscopy. Angew Chem Int Ed 41(6):969–972

Kuroda R, Imai Y, Tajima N (2002) Generation of a co-crystal phase with novel coloristic properties via solid state grinding procedures. Chem Commun 23:2848–2849

Kuroda R, Higashiguchi K, Hasebe S, Imai Y (2004) Crystal to crystal transformation in the solid state. CrystEngComm 6(76):464–468

Lai Y-Y, Cheng Y-J, Hsu C-S (2014) Applications of functional fullerene materials in polymer solar cells. Energy Environ Sci 7 (6):1866–1883

Le Brun P, Froyen L, Delaey L (1993) The modelling of the mechanical alloying process in a planetary ball mill: comparison between theory and in-situ observations. Mater Sci Eng A 161(1):75–82

Lee D, Kim Y, Chang S (2013) Iridium-catalyzed direct arene C–H bond amidation with sulfonyl-and aryl azides. J Org Chem 78 (21):11102–11109

Li H, Cabañas-Gac F, Hadidi L, Bilodeau-Calame M, Abid A, Mameri K, Rigamonti MG, Rousselot S, Mickael D, Patience GS (2020) Ultrasound assisted wet media milling synthesis of nanofiber-cage $LiFePO_4/C$. Ultrason Sonochem 68:105177

Liu Z, Xu S, Xiao B, Xue P, Wang W, Ma Z (2012) Effect of ball-milling time on mechanical properties of carbon nanotubes reinforced aluminum matrix composites. Compos A Appl Sci Manuf 43(12):2161–2168

Liu K, Tan Q, Liu L, Li J (2020) From lead paste to high-value nanolead sulfide products: a new application of mechanochemistry in the recycling of spent lead-acid batteries. ACS Sustain Chem Eng 8(9):3547–3552

Lu J, Rohani S (2009) Preparation and characterization of theophylline −nicotinamide cocrystal. Org Process Res Dev 13(6):1269–1275

Lu C, Zhang J, Li Z (2004) Structural evolution of titanium powder during ball milling in different atmospheres. J Alloy Compd 381(1–2):278–283

Mack J, Shumba M (2007) Rate enhancement of the Morita–Baylis–Hillman reaction through mechanochemistry. Green Chem 9 (4):328–330

Makhaev V, Borisov A, Petrova L (1999) Solid-state mechanochemical synthesis of ferrocene. J Organomet Chem 590(2):222–226

Malpartida I, Maireles-Torres P, Vereda C, Rodríguez-Maroto JM, Halloumi S, Lair V, Thiel J, Lacoste F (2020) Semi-continuous mechanochemical process for biodiesel production under heterogeneous catalysis using calcium diglyceroxide. Renew Energy 159:117–126

Maurice DR, Courtney T (1990) The physics of mechanical alloying: a first report. Metall Trans A 21(1):289–303

McNaught AD, Wilkinson A (1997) Compendium of chemical terminology, vol 1669. Blackwell Science, Oxford

Medina GM, van Raap MF, Coral D, Muraca D, Sánchez F (2020) Synthesis of highly stable Fe/FeO$_x$@ citrate colloids with strong magnetic response by mechanochemistry and coprecipitation for biomedical and environmental applications. J Magn Magn Mater 508:166759

Mikhailenko MA, Shakhtshneider TP, Boldyrev VV (2004) On the mechanism of mechanochemical synthesis of phthalylsulphathiazole. J Mater Sci 39(16–17):5435–5439

Miki M, Yamasaki T, Ogino Y (1992) Preparation of nanocrystalline NbN and (Nb, Al) N powders by mechanical alloying under nitrogen atmosphere. Mater Trans JIM 33(9):839–844

Mondal P, Anweshan A, Purkait MK (2020) Green synthesis and environmental application of Iron-based nanomaterials and nanocomposite: a review. Chemosphere 259:127509

Mucsi G (2019) A review on mechanical activation and mechanical alloying in stirred media mill. Chem Eng Res Des 148:460-474

Murata Y, Kato N, Komatsu K (2001) The reaction of fullerene C_{60} with phthalazine: the mechanochemical solid-state reaction yielding a new C_{60} dimer versus the liquid-phase reaction affording an open-cage fullerene. J Org Chem 66(22):7235–7239

Mursalat M, Hastings DL, Schoenitz M, Dreizin EL (2019) Microspheres with diverse material compositions can be prepared by mechanical milling. Adv Eng Mater 22(3):1901204

Nguyen KL, Friščić T, Day GM, Gladden LF, Jones W (2007) Terahertz time-domain spectroscopy and the quantitative monitoring of mechanochemical cocrystal formation. Nat Mater 6(3):206–209

Nielsen SF, Peters D, Axelsson O (2000) The Suzuki reaction under solvent-free conditions. Synth Commun 30(19):3501–3509

Obrovac M, Mao O, Dahn J (1998) Structure and electrochemistry of $LiMO_2$ (M=Ti, Mn, Fe Co, Ni) prepared by mechanochemical synthesis. Solid State Ionics 112(1–2):9–19

Ojala WH, Etter MC (1992) Polymorphism in anthranilic acid: a reexamination of the phase transitions. J Am Chem Soc 114 (26):10288–10293

Orita A, Jiang L, Nakano T, Ma N, Otera J (2002) Solventless reaction dramatically accelerates supramolecular self-assembly. Chem Commun 13:1362–1363

Ostwald W (1919) Die chemische Literatur und die Organisation der Wissenschaft, vol 1. Akad. Verlag, Gesel

Ozer D (2020) Fabrication and functionalization strategies of MOFs and their derived materials "MOF architecture". In: Applications of metal–organic frameworks and their derived materials, pp 63–100

Palaniandy S, Jamil NH (2009) Influence of milling conditions on the mechanochemical synthesis of $CaTiO_3$ nanoparticles. J Alloy Compd 476(1–2):894–902

Palazon F, El Ajjouri Y, Bolink HJ (2019) Making by grinding: mechanochemistry boosts the development of halide perovskites and other multinary metal halides. Adv Energy Mater 10 (13):1902499

Pardeshi S, Patil A (2009) Effect of morphology and crystallite size on solar photocatalytic activity of zinc oxide synthesized by solution free mechanochemical method. J Mol Catal A Chem 308(1–2):32–40

Park B-I, Hwang Y, Lee SY, Lee J-S, Park J-K, Jeong J, Kim JY, Kim B, Cho S-H, Lee D-K (2014) Solvent-free synthesis of Cu_2 ZnSnS 4 nanocrystals: a facile, green, up-scalable route for low cost photovoltaic cells. Nanoscale 6(20):11703–11711

Partha R, Conyers JL (2009) Biomedical applications of functionalized fullerene-based nanomaterials. Int J Nanomed 4:261

Pérez-Venegas M, Juaristi E (2020) Mechanochemical and mechanoenzymatic synthesis of pharmacologically active compounds: a green perspective. ACS Sustain Chem Eng 8(24):8881–8893

Petrova L, Borisov A, Makhaev V (2002) Solid-phase synthesis of zinc (II) b-diketonates upon mechanical activation. Russ J Inorg Chem 47(12):1827–1832

Ralphs K, Hardacre C, James SL (2013) Application of heterogeneous catalysts prepared by mechanochemical synthesis. Chem Soc Rev 42(18):7701–7718

Rapoport L, Fleischer N, Tenne R (2005) Applications of WS_2 (MoS_2) inorganic nanotubes and fullerene-like nanoparticles for solid lubrication and for structural nanocomposites. J Mater Chem 15 (18):1782–1788

Raston CL, Scott JL (2000) Chemoselective, solvent-free aldol condensation reaction. Green Chem 2(2):49–52

Rathod RV, Mondal D, Bera S (2020) Mechanochemical synthesis of fluorescein-based receptor for CN-ion detection in aqueous solution and cigarette smoke residue. Anal Bioanal Chem 412(13):3177–3186

Ravnsbæk JB, Swager TM (2014) Mechanochemical synthesis of poly (phenylene vinylenes). ACS Macro Lett 3(4):305–309

Rightmire NR, Hanusa TP (2016) Advances in organometallic synthesis with mechanochemical methods. Dalton Trans 45 (6):2352–2362

Rodriguez B, Bruckmann A, Rantanen T, Bolm C (2007) Solvent-free carbon-carbon bond formations in ball mills. Adv Synth Catal 349 (14–15):2213–2233

Rydin R, Maurice D, Courtney T (1993) Milling dynamics: part I. Attritor dynamics: results of a cinematographic study. Metall Trans A 24(1):175–185

Sarmah KK, Nath N, Rao DR, Thakuria R (2020) Mechanochemical synthesis of drug–drug and drug–nutraceutical multicomponent solids of olanzapine. CrystEngComm 22(6):1120–1130

Schneider F, Stolle A, Ondruschka B, Hopf H (2009) The Suzuki −Miyaura reaction under mechanochemical conditions. Org Process Res Dev 13(1):44–48

Shan N, Toda F, Jones W (2002) Mechanochemistry and co-crystal formation: effect of solvent on reaction kinetics. Chem Commun 20:2372–2373

Sokolov AN, Friščić T, MacGillivray LR (2006) Enforced face-to-face stacking of organic semiconductor building blocks within hydrogen-bonded molecular cocrystals. J Am Chem Soc 128 (9):2806–2807

Sopicka-Lizer M (2010) High-energy ball milling: mechanochemical processing of nanopowders. Elsevier

Štefanić G, Krehula S, Štefanić I (2013) The high impact of a milling atmosphere on steel contamination. Chem Commun 49(81):9245–9247

Stolle A, Szuppa T, Leonhardt SE, Ondruschka B (2011) Ball milling in organic synthesis: solutions and challenges. Chem Soc Rev 40 (5):2317–2329

Stolle A, Schmidt R, Jacob K (2014) Scale-up of organic reactions in ball mills: process intensification with regard to energy efficiency and economy of scale. Faraday Discuss 170:267–286

Strobridge FC, Judaš N, Friščić T (2010) A stepwise mechanism and the role of water in the liquid-assisted grinding synthesis of metal–organic materials. CrystEngComm 12(8):2409–2418

Suryanarayana C (1995) Does a disordered γ-TiAl phase exist in mechanically alloyed TiAl powders? Intermetallics 3(2):153–160

Suryanarayana C (2001) Mechanical alloying and milling. Prog Mater Sci 46(1–2):1–184

Takacs L (2007) The mechanochemical reduction of AgCl with metals. J Therm Anal Calorim 90(1):81–84

Takacs L (2013) The historical development of mechanochemistry. Chem Soc Rev 42(18):7649–7659

Takacs L, McHenry J (2006) Temperature of the milling balls in shaker and planetary mills. J Mater Sci 41(16):5246–5249

Tan D, Loots L, Friščić T (2016) Towards medicinal mechanochemistry: evolution of milling from pharmaceutical solid form screening to the synthesis of active pharmaceutical ingredients (APIs). Chem Commun 52(50):7760–7781

Thiessen PA, Meyer K, Heinicke G (1967) Grundlagen der Tribochemie: mit 24 Tab. im Text. Akad.-Verlag

Toda F, Tanaka K, Sekikawa A (1987) Host–guest complex formation by a solid–solid reaction. J Chem Soc Chem Commun 4:279–280

Trask AV, Shan N, Motherwell WS, Jones W, Feng S, Tan RB, Carpenter KJ (2005) Selective polymorph transformation via solvent-drop grinding. Chem Commun 7:880–882

Tröbs L, Emmerling F (2014) Mechanochemical synthesis and characterisation of cocrystals and metal organic compounds. Faraday Discuss 170:109–119

Tsuchimoto M, Hoshina G, Yoshioka N, Inoue H, Nakajima K, Kamishima M, Kojima M, Ohba S (2000) Mechanochemical reaction of polymeric oxovanadium (IV) complexes with Schiff base ligands derived from 5-nitrosalicylaldehyde and diamines. J Solid State Chem 153(1):9–15

Tumanov IA, Achkasov AF, Boldyreva EV, Boldyrev VV (2011) Following the products of mechanochemical synthesis step by step. CrystEngComm 13(7):2213–2216

U.S. Department of Health and Human Services Food and Drug Administration (2018) Regulatory classification of pharmaceutical co-crystals: guidance for industry. Center for Drug Evaluation and Research (CDER), Silver Spring, US

Ulbrich K, Nishida E, Souza B, Campos C (2020) NiS_2–NiS nanocrystalline composite synthesized by mechanochemistry and its performance for methylene blue dye adsorption. Mater Chem Phys 252:123226

Urakaev F (2010) Mechanism and kinetics of mechanochemical processes. In: High-energy ball milling. Elsevier, pp 9–44

Urakaev FK, Boldyrev V (2000) Mechanism and kinetics of mechanochemical processes in comminuting devices: 1. Theory. Powder Technol 107(1–2):93–107

Vaid P, Raizada P, Saini AK, Saini RV (2020) Biogenic silver, gold and copper nanoparticles—a sustainable green chemistry approach for cancer therapy. Sustain Chem Pharm 16:100247

Vishweshwar P, McMahon JA, Peterson ML, Hickey MB, Shattock TR, Zaworotko MJ (2005) Crystal engineering of pharmaceutical co-crystals from polymorphic active pharmaceutical ingredients. Chem Commun 36:4601–4603

Wang G-W (2013) Mechanochemical organic synthesis. Chem Soc Rev 42(18):7668–7700

Wang G-W, Komatsu K, Murata Y, Shiro M (1997) Synthesis and X-ray structure of dumb-bell-shaped C_{120}. Nature 387(6633):583–586

Willis-Fox N, Rognin E, Baumann C, Aljohani TA, Göstl R, Daly R (2020) Going with the flow: tunable flow-induced polymer mechanochemistry. Adv Funct Mater 30(27):2002372

Xue J, Wan D, Lee SE, Wang J (1999) Mechanochemical synthesis of lead zirconate titanate from mixed oxides. J Am Ceram Soc 82 (7):1687–1692

Zhang P, Dai S (2017) Mechanochemical synthesis of porous organic materials. J Mater Chem A 5(31):16118–16127

Zhang Q, Nakagawa T, Saito F (2000) Mechanochemical synthesis of $La_{0.7}Sr_{0.3}MnO_3$ by grinding constituent oxides. J Alloys Compd 308(1–2):121–125

Zhang P, Jiang X, Wan S, Dai S (2015) Advancing polymers of intrinsic microporosity by mechanochemistry. J Mater Chem A 3 (13):6739–6741

Future Trends in Green Synthesis

Suman Chowdhury, Atanu Rakshit, Animesh Acharjee, and Bidyut Saha

Abstract

Green approach is an advance technique of synthetic chemistry. It opens up a new vista of modified version of known synthetic reactions in a newer way with associated enlarged potentiality accompanied by sustainability. Revolutionary measures against solvent-based reaction have been invited in green routes. Organic solvents have been substituted by non-organic media in organic synthetic reactions to overcome the hazard of volatile organic solvents and to preserve the greenness of the reactions, following green principles, supported by green matrices. This chapter illustrates the successive green revolutionary measures in solvent-based organic chemistry followed by solvent-free synthetic routes and their future trends in a nut shell.

Keywords

Sustainable chemistry • Biocatalysis • Biphasic reaction • Solvent-free synthesis

1 Introduction

The synthetic organic chemistry was generally discovered in 1828 by Wöhler while synthesising urea, a natural product, from ammonium isocyanate (Dunn et al. 2010a). Since then, time to time, many advancements in the synthetic procedures have been attained and in the present times, the welfare of human race is inconceivable without abundant synthetic organic products obtained industrially. Quality of our survival does intensely depend upon various pharmaceutical products, foods and so many other products obtained synthetically. But in most cases, the reagents or their sub-products are detrimental to health and also the processes are not sustainable on the ground of pollution, energy consumption and economy. The environmental protection agency (EPA) of USA reported that more or less 100 billion tons of wastes come out of industries every year. The wastes, cost around US$5 billion, have terrifying environmental effect which affect human health extremely. Ozone layer depletion is a typical example of environmental threat which is caused by chlorofluorocarbons (CFCs) come from volatile hazardous organic solvents (Mikami 2005). Hence, some greener eco-friendly or sustainable procedures were really welcome from past few decades. At the beginning of the 1990s, Environmental Protection Agency formulated the definition of green Chemistry or sustainable chemistry as the "Design of chemical products and processes to reduce or eliminate the use and generation of hazardous substances" (Anastas and Eghbali 2010). Green Chemistry being economically profitable aims to reduce hazards across all the life-cycle stages. Synthetic procedures in association with greener approaches can truly termed as green synthesis. The designing concept is of prime important feature of Green Chemistry. Designing requires novelty, planning and systematic approaches. The twelve Principles of Green Chemistry (Fig. 1), coined by Paul Anastas and John Warner in 1998, are guiding framework to reach the goal of sustainability and to reduce adverse consequences by careful planning of chemical synthesis and molecular design (Anastas and Eghbali 2010). In this chapter, some major and prominent applications so far of green synthesis and associated future trends have been focussed.

2 Green Chemistry Metrics

Scientists were put to a challenge to be deeply concerned about the environmental impact of chemical processes for the evolution of green chemistry. The evolution necessitated

S. Chowdhury · A. Rakshit · A. Acharjee · B. Saha (✉)
Homogeneous Catalysis Laboratory, Department of Chemistry, The University of Burdwan, Burdwan, WB 713104, India
e-mail: bsaha@chem.buruniv.ac.in

A. Acharjee
Department of Chemistry, Hooghly Mohsin College, Chinsurah, WB, India

Inamuddin et al. (eds.), *Advances in Green Synthesis*, Advances in Science, Technology & Innovation,
https://doi.org/10.1007/978-3-030-67884-5_3

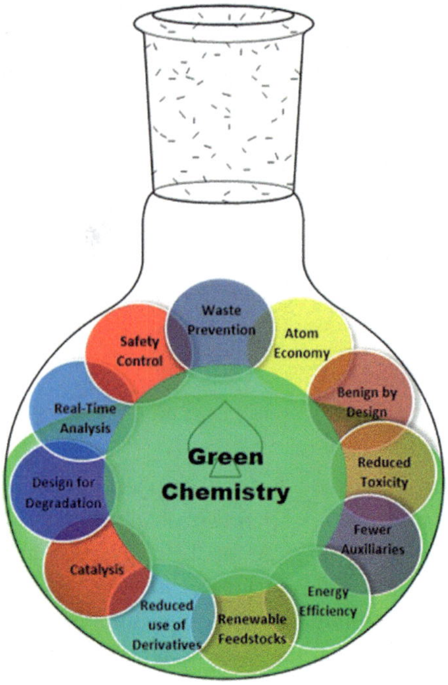

Fig. 1 Principles of green chemistry

practical changes in industrial and academic processes significantly, and to measure the "greenness" caused by the practical changes, few valid and reliable approaches were essential. Determining greenness does mean not only to measure the waste quantities but also some additional but meaningful factors are to be considered. So, green metrics are that kind of means which can measure how greener a chemical process is. The most common and useful metrics are atom economy (AE), environmental factor (E factor), process mass intensity (PMI) and reaction mass efficiency (RME).

2.1 Atom Economy (AE)

AE, proposed by Trost in 1991, is very widely known measures of efficient chemical processes and is computed from Eq. (1) (Dunn 2012)

$$\text{Atom Economy} = \frac{\text{Molecular mass of desired product}}{\text{Molecular mass of all products}} \times 100\% \tag{1}$$

A higher value of AE clearly indicates more greenness of any process. The atom economy can be easily understood with the help of Baylis–Hillman reaction where the product is impregnated with all the atoms present in the reactant molecules and there is no side product. Hence, this reaction has 100% AE value (Scheme 1).

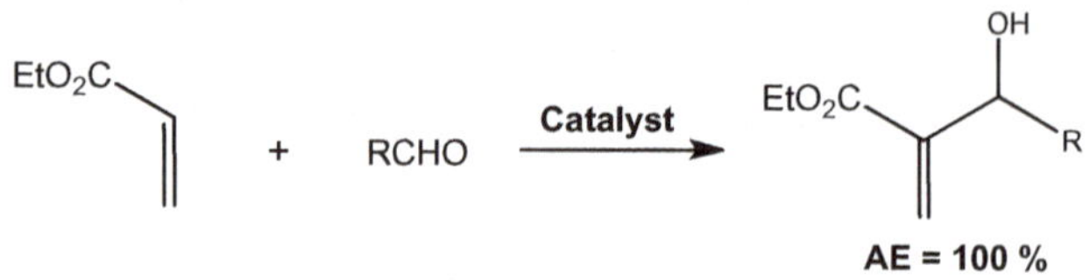

Scheme 1 Completely atomically economic Baylis–Hillman reaction

On the other side, for Wittig reaction having three by-products, atom economy is only 18.5% (Scheme 2).

Another elegant example of contrasting AE is production of ibuprofen, a non-steroidal anti-inflammatory drug, in two different routes via the common intermediate, *p*-isobutylacetophenone. The classical route (developed by the Boots Pure Drug Company) with six successive steps has comparatively low AE for sizeable amount inorganic salt formation whereas the other route (developed by the Boots Hoechst- Celanese company) produces it only in three catalytic steps with high AE (Dunn et al. 2010a).

AE is very simple to understand but it possess some demerits. It is calculated only from the reaction scheme, but the yield percentage, even the stoichiometry, is not taken into consideration.

2.2 Environmental Factor (E Factor)

This environmental factor, coined by Sheldon in 1992, is formulated as kilograms of waste per kilogram of product which measures waste products formed in a synthesis of fine chemicals, pharmaceuticals or their intermediates (Table 1) (Dunn 2012). An illustrative example is production of a pharmaceutical intermediate phloroglucinol, a reprographic chemical also, with $Cr_2(SO_4)_3$, NH_4Cl, $FeCl_2$ and $KHSO_4$ as wastes (by-products) from 2,4,6-trinitrotoluene (TNT) in nineteenth century. Calculation shows copious amount of waste is formed in this synthesis with E factor of ca. 20. Ranges of E factor values, differ according to different chemical branches and industries, have been given in tabulated form which are equally applicable for any product, even in the production of laptops or mobiles. A higher E factor indicates more waste with more environmental trace and ideality is attained when the value is zero. This metric includes the chemical yield, reagents, solvent losses, process aids and even fuel. In the computation of E factor in aqueous waste stream, only the inorganic salts and the organic compounds in the water are considered but water is generally removed from consideration. Otherwise, water makes high value of E factor making meaningful comparisons of processes difficult (Dunn et al. 2010a; Dunn 2012) (Scheme 3).

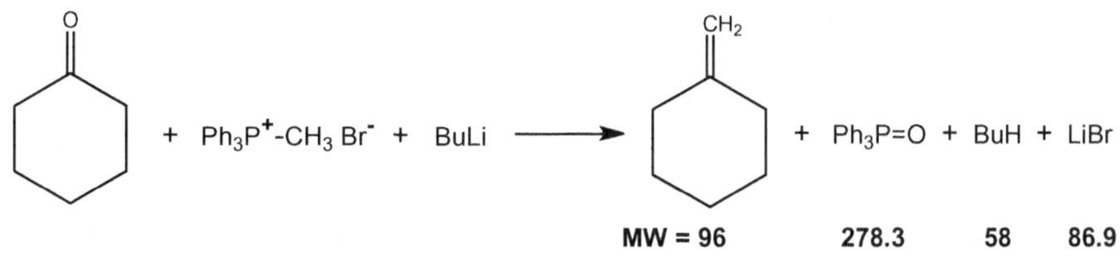

Scheme 2 Wittig reaction with poor AE value

Table 1 E factors in different chemical industry segments (Dunn 2012)

Sector	Volume/tonnes per year	E factor (kg waste per kg product)
Oil refining	10^6–10^8	<0.1
Bulk chemicals	10^4–10^6	<1–5
Fine chemicals	10^2–10^4	5 to >50
Pharmaceutical	10–10^3	25 to >100

2.3 Process Mass Intensity (PMI)

PMI, quite similar to E factor, is computed as the ratio of mass of all the materials in any process and mass of the desired product (Eq. 2). The ideal E factor value, zero, is supposed to be a better reflection of the target of zero waste than the ideal PMI of 1. In this case also water can be included or excluded as per requirements.

$$\begin{aligned}\text{Process Mass Intensity} &= \frac{\text{Mass of all materials used to make the product}}{\text{Mass of product}}\\ &= \text{E Factor} + 1\end{aligned} \tag{2}$$

Process chemists choose PMI as the key metric in pharmaceutical industry for the measurement of sustainability of any synthetic process. Besides that, it is very easy to generate, measure, communicate data and can estimate greenness very fast. This mass-based metric focuses on process input rather than output, and targets can be set to measure the process sustainability quantitatively (Monteith et al. 2020).

2.4 Reaction Mass Efficiency (RME)

In 2001, the company GlaxoSmithKline (GSK) invented RME as a practical metric for explaining the greenness of a process. RME is computed as the percentage of mass of product with respect to the sum of the masses of all corresponding reactants in the balanced chemical equation (Eq. 3).

$$\text{Reaction Mass Efficiency} = \frac{\text{Mass of product}}{\text{Mass osf all reactants}} \times 100 \tag{3}$$

Perhaps, RME is one of the most helpful metrics for researchers in knowing the actual greenness of any processes currently because it considers yield, stoichiometry and AE. Solvents, catalysts, acids or bases for the neutralisation of by-products cannot be included in the calculation but, any solvent acting as reactant can be used (only the portion of mass responsible for the product formation) (Dunn et al. 2010b; Dicks and Hent 2015).

3 Application of Green Concept in Synthesis

Actually, it is more important to apply green concepts in practical purposes to check their viability than only to be written theoretically on pages. There are no scarcity of such practical examples from many past decades where the green principles really have shown their profound positive impacts.

3.1 Solvent-Based Organic Synthesis

It is undeniable that organic synthesis is one of the biggest field wherein the green principles have provided its deepest footprints effectively. It was from the very beginning, organic synthesis was incurred with hazardous organic solvents and the lion's share of the reaction mass is possessed by solvents of which 70% is burnt up for retrieving heat of the reaction. Therefore, solvents are probably the most essential and key portion of green chemistry (Anastas and Eghbali 2010). Waste production, hazardousness, catalysis, inflammability, etc., and many important features of synthetic reactions can be controlled by choosing effective solvent only, even an effective solvent can reduce multistep reactions responsible for their separation, multiple

Scheme 3 Phloroglucinol preparation from TNT

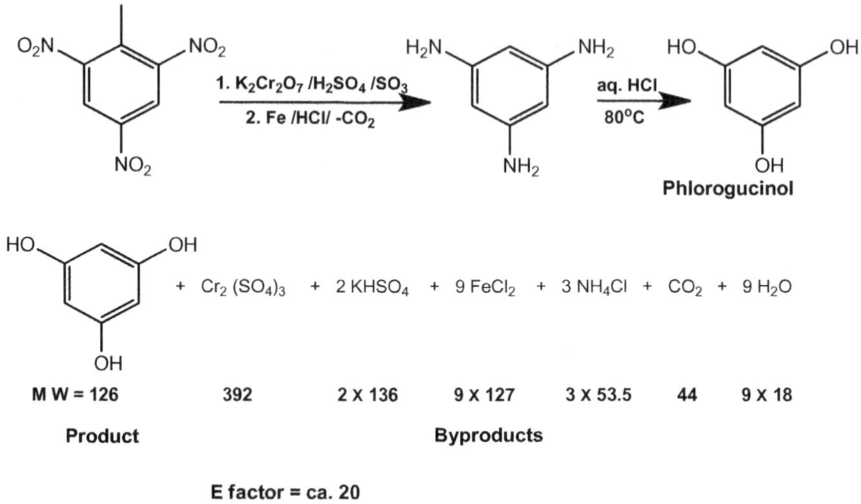

by-products, their costly separation and huge energy consumption also. So, revolutionary measures have been adopted to gain sustainability, conforming to the green matrices, by choosing effective greener solvents replacing the organic solvents.

3.2 Aqueous Medium

There is no doubt that water is the greenest solvent of all time for its abundance, cheap availability, non-inflammability, nontoxicity and non-volatility making it fully eco-friendly. Organic synthesis in aqueous medium sounds weird to hear but, really it has come into existence probably after the work of Breslow in 1980. It was a Diels–Alder (D.A.) reaction (cycloaddition) of butenone with cyclopentadiene undergone in water with 740 folds faster rate than in isooctane and increased selectivity for endo product (Rideout and Breslow 1980). As organic compounds are insoluble in water, it was considered to be a contaminant and was taken as a poor solvent for organic syntheses from the inception. But in particular, polarity, hydrogen bonding, hydrophobic effect and trans-phase interactions arising from the structural uniqueness and associated physicochemical properties of water impart influences in the reaction route profoundly with much reactivity and selectivity. The findings in aqueous solvent are justified by the hydrophobic effect (Fig. 2) (Breslow 1991), as the work of Breslow was also tested in protic solvents like ethanol and methanol which resulted same as obtained in case of hydrocarbons. The hydrophobic effect arises from the repulsion of hydrophobic organic reagents with water molecules. This effect inhibits the surface of hydrophobic aggregation of reagents to come in contact with water molecules. Now, to keep the hydrogen bonding among water molecules unaffected, water molecules wrap around the hydrophobic aggregates increasing the pressure inside the wrapper and this internal pressure enhance the reaction rate to make activation volume negative. Though, some reaction rates may be facilitated also due to interfacial interaction between the organic molecules and free hydroxyl groups of water (Simon and Li 2012).

Cycloaddition-type Diels-Alder (D.A.) reaction, revealed by Engberts et al., of 3-aryl-1-(2-pyridyl)2-propen-1-ones in aqueous solvent is an excellent example of our present discussion. The said reaction showed 287 times greater rate in aqueous medium than in acetonitrile (Scheme 4) (Li 2005). It is worthy to note that the same reaction in presence of water, Lewis acid and micellar catalyst [2.4mM Cu $(OSO_3C_{12}H_{25})_2$] was reported with 1,800,000 fold rate enhancement than in acetonitrile.

Pirrung tested the influence of water in the multicomponent transformations, for example, Ugi and Passerini reactions thinking over the negative activation volumes. According to their report, the reaction of 3-methylbut-2-enoic acid, 3-methylbutanal and 2-isocyano-2-methylpropane in water medium (Scheme 5) resulted in the desired product with 100% yield only within 3.5 h whereas in dichloromethane and dimethylformamide, the yields were 50% after 18 h and 15% after 24 h, respectively. They also reported that no product was found in methanol (Pirrung and Sarma 2003, 2005).

In an industrial-level synthesis of 1-substituted-4-cyano-1,2,3-triazoles from 2-chloroacrylonitrile and organic azides, Novartis took water as solvent (Scheme 6) (Chanda and Fokin 2009). He compared the result of his experiment with the results of the same reaction in organic solvents and he reported that 98% yield was obtained in water which is very high in comparison to the others (46% in n-heptane, 51% in toluene, 78% in dimethylformamide and 40% in ethanol). In water medium, unwanted side reactions were minimised to improve the yield. Generally, hydrogen

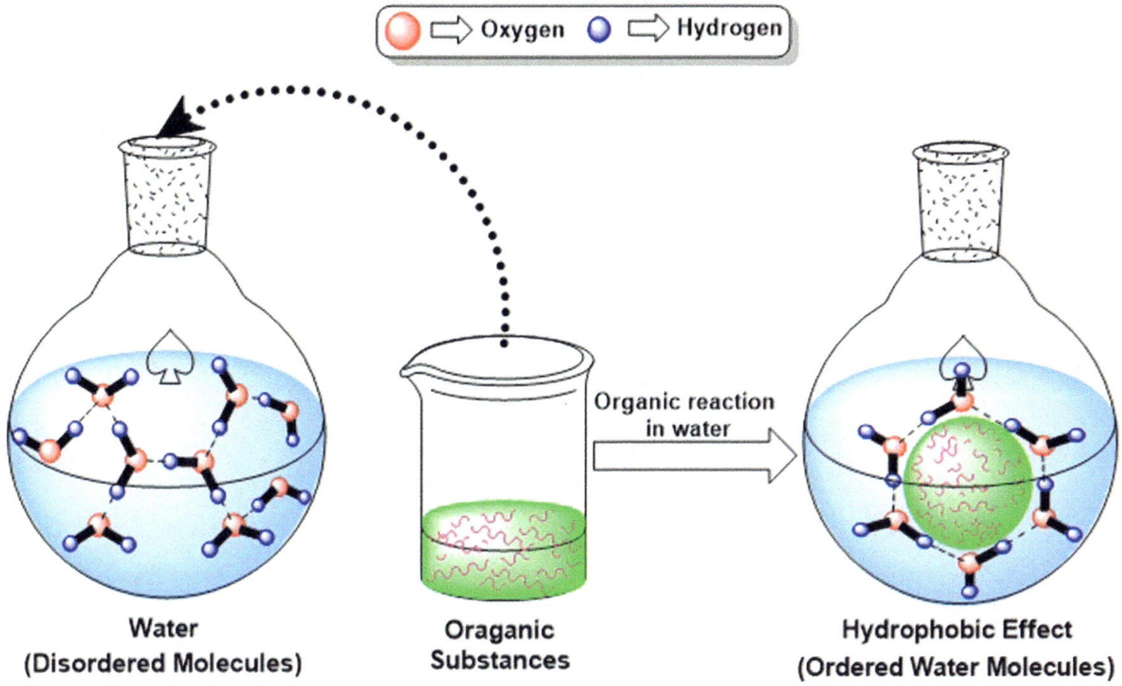

Fig. 2 Hydrophbic effect in water solvent

Scheme 4 D.A. reaction in water solvent

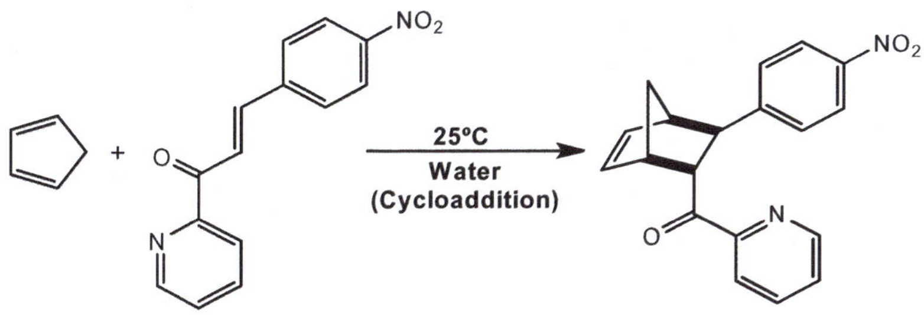

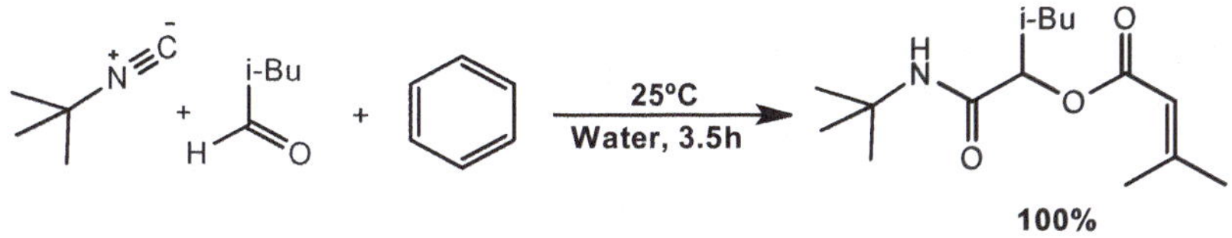

Scheme 5 Passerini reaction in aqueous medium

chloride, a by-product, is produced in the 1,3-dipolar cycloaddition after aromatisation and the medium becomes acidic. Now, 2-chloroacrylonitrile gets polymerised both in acidic and basic media, but in present case, the produced hydrogen chloride goes to aqueous phase and 2-chloroacrylonitrile goes to organic phase and the reaction proceeds easily without any hindrance of olefinic poly-merisation yielding high percentage of product.

A nice reaction was reported by Aziz et al. where selectivity of water increased the sustainability decreasing environment impact. The reaction was synthesis of β-aminoalcohols at room temperature, with high regio- and stereoselectivity, from the reaction of epoxides and amines in high yields in water (Scheme 7) (Azizi and Saidi 2005). In every cases, the yields percentage in water were >90%, but in ethanol, it was only 50% and in dichloromethane, ace-tonitrile and di-ethylether, no product was found.

Hayashi et al. reported an asymmetric Mannich reaction of some ketones with dimethoxyacetaldehyde, in an aqueous solution, and p-anisidine in presence of an excellent organocatalyst (Scheme 8) (Hayashi et al. 2007). The organocatalyst was prepared by the Hayashi group using

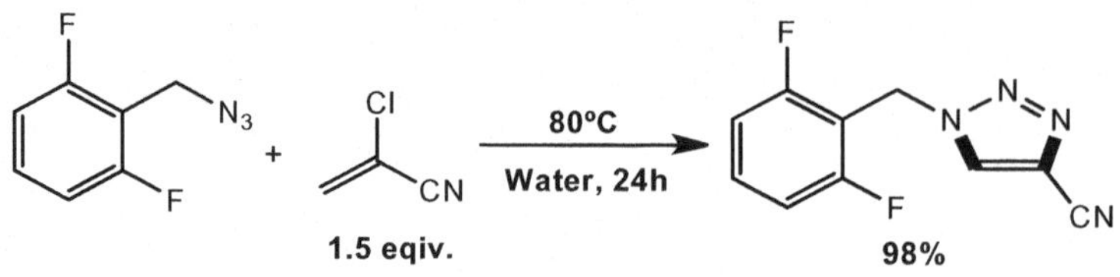

Scheme 6 Triazole synthesis from azides and acrylonitrile in aqueous medium

Scheme 7 β-aminoalcohols synthesis from epoxides in water

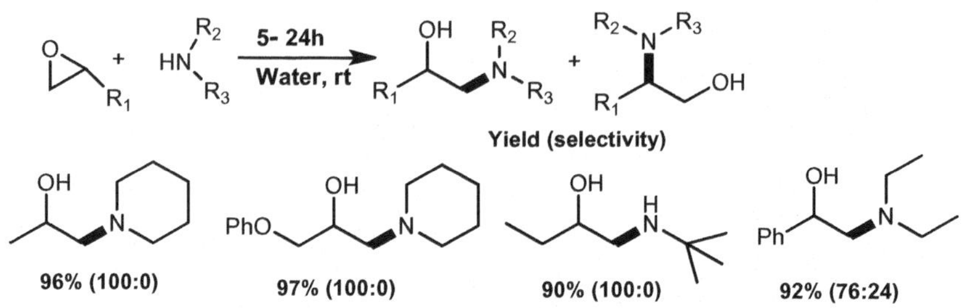

siloxy and tetrazole functional groups in presence of pyrrolidine scaffold. This reaction gave high yield on using only the aqueous solution of dimethoxyacetaldehyde (60% solution in water) and required no additional water to gain the benefit of high yield or selectivity (yields = 78% with 95% ee). Use of aqueous solution of the aldehyde also made the work up process easier and the product extraction from the crude mixture was possible by chromatography through a silica gel column.

Mizoroki–Heck reaction an important organometallic reaction which was repeated by Firouzabadi et al. in water using insoluble Pd(0)L_2 complex (Scheme 9). The Pd catalyst, synthesised using 2-aminophenyldiphenylphosphinite and palladium acetate, was found to catalyse the reaction of aromatic halides and different alkenes in presence of water and air at two different temperatures 80 and 95 °C (Firouzabadi et al. 2009). After the completion of the reaction, the catalyst can be used again about six more times to get again 79–83% yields on an average.

Lemaire et al. used water soluble ammonium derivative of BINAP analogue for ruthenium-catalysed asymmetric hydrogenation of ethyl acetoacetate (Scheme 10) (Shaughnessy 2009). The reaction was brought about in aqueous phase efficiently giving 100% chemical yield with 97% enantiomeric excess. The most promising fact is that the catalyst can be reused about 9 times more with quite same activity.

Gruttadauria and groups reported an asymmetric aldol reaction of cycohexanone and *p*-nitrobenzaldehyde in water medium catalysed by a substituted proline derivative at room temperature with excellent selectivity and chemical yields (99%) of the desired product (Scheme 11) (Giacalone et al. 2008). They developed the 4-substituted prolines organocatalyst particularly with n-propylpyrene-1-carboxylate as a substituent which shows high selectivity at very low concentration even in presence of tap water instead of distilled water and the said catalyst can be repeatedly used with same activity. In organic solvents, the reaction was found to give very poor yields.

The versatile function of water in organic syntheses came into knowledge again after the work of Auge′ and groups. They perfomed Claisen rearrangement of 6-b-glycosylallyl vinylethers and concomitant reduction of the produced aldehyde using sodium borohydride in water (Scheme 12) (Xu and Queneau 2014). The reaction was completed with full conversion to the desired product within one hour only at 80 °C and no additional protection or deprotection steps were needed making the reaction away from green

Scheme 8 Organocatalyzed asymmetric Mannich reaction in aqueous medium

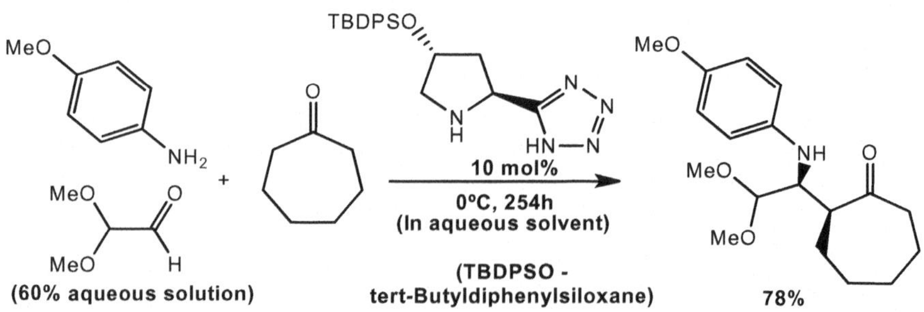

Scheme 9 Mizoroki–Heck reaction in aqueous medium using insoluble Pd–phosphinite complex catalyst

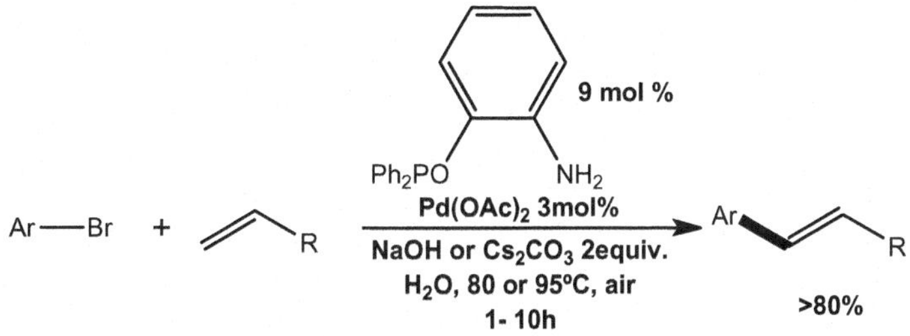

Scheme 10 Keto-ester hydrogenation asymmetrically in aqueous phase

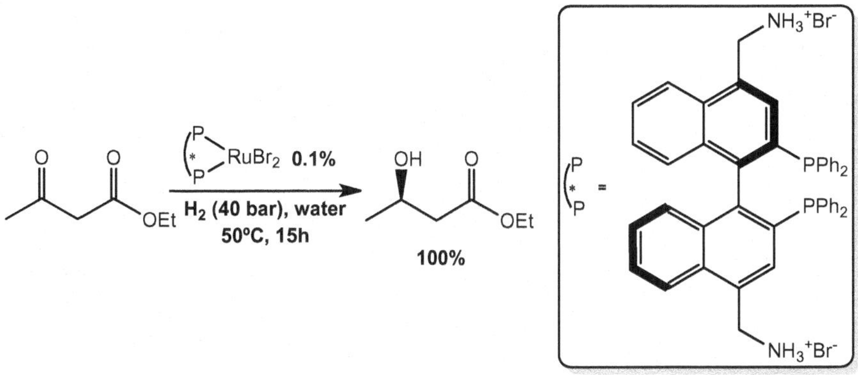

Scheme 11 Proline derivative-catalysed asymmetric aldol reaction in water medium

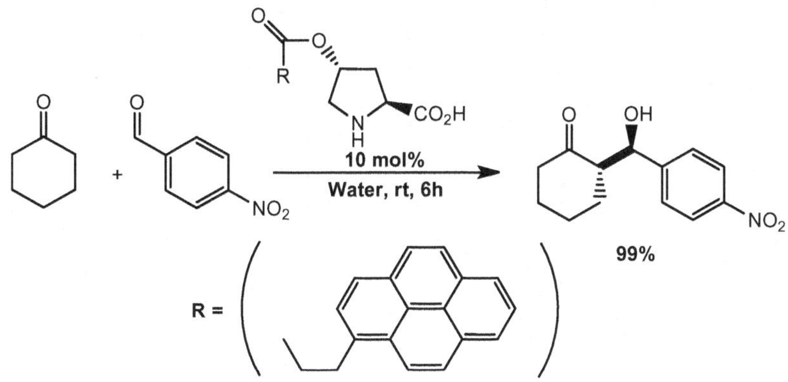

chemistry. It is relevant to mention here that in toluene, the reaction took 13 days to be completed from the substrate.

Such many reported synthetic reactions in water medium have not been mentioned here, which really opened up new avenues in the field of green synthesis (Vilotijevic and Jamison 2007; Ohnmacht et al. 2008; Rogozinska et al. 2011; Wang et al. 2003; Hu and Manetsch 2010; Mamidyala and Finn 2010).

Biocatalysis was the best way of green synthesis for all time and to start any discussion about this the name "enzyme" must be spelt, in fact enzyme is the prime ingredient that authorises the chemistry as green. History is the witness that searching for suitable enzymes from nature for organic syntheses is in vogue from old ages. Upto 1980s, enzymatic reactions were performed in aqueous medium (Mikami 2005). But, enzymes being organic molecule hardly soluble

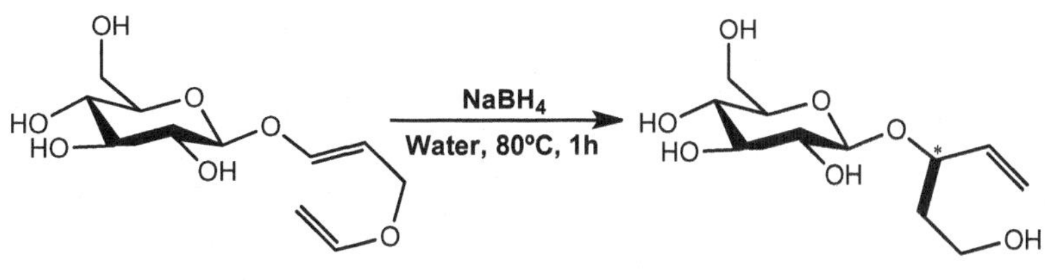

Scheme 12 Claisen rearrangement in aqueous medium

in water to be potential to conduct reactions with no limitations. Non-aqueous solvents are more welcome in this purpose. In life processes, also liposomes, the lipid bilayers assist enzymatic reactions to happen in aqueous medium. A smallest extent of water is essential to maintain the active conformation of enzymes, which varies with the enzyme, but proteins have no activity in fully dehydrated condition. Therefore, the presence of water is critical for any kind of biocatalytic reaction. Use of biocatalytic reaction in aqueous medium is very common in food industries (Marques et al. 2012). But, biphasic medium, i.e. aqueous medium coupled with another non-aqueous medium, is promising for biocatalytic reactions. Some discussions in this regard have been provided in the later sections.

3.2.1 Micellar Media

Micelles, the nano-scale assemblies, (Fig. 3a) formed by aggregation of amphiphilic surfactants (cationic, anionic, neutral or Gemini type) are in equilibrium thermodynamically, where monomers of surfactant swiftly exchange among aggregates with an average lifetime of order 10^{-3}–10^{-2} s. One thing should be mentioned here that some liquid polymers are also included in non-ionic or neutral surfactants (Chowdhury et al. 2019a; Dwars et al. 2005). Surfactants, above critical micelle concentration (CMC), exhibit good catalytic activities in aqueous medium (Dwars et al. 2005). Micellar catalysis is one of the simplest methods, even economic, in the field of catalytic reactions. It is also abundantly used in detergency for attaining cleansing property. Therefore, surfactants are termed not only as soapy version of homogeneous catalysis but also as nano-reactors having distinct features. Surfactants are versatile molecules and very useful for many inconceivable purposes with some magical features (Chowdhury et al. 2019a; Dwars et al. 2005; Sar et al. 2019). It being amphiphilic in nature helps solubilising hydrophobic organic substrates in aqueous medium. Moreover, surfactants make sure contacts between substrates of different polarity controlling mutual heat transfer and favouring their interaction that accounts for the final transformation making the whole process environment friendly with less waste production and less greenhouse emissions as well as making E factor close to zero. Micellar assembly encompasses the organic substrates and thus increases the local concentration of the substrate favouring compartmentalisation (Fig. 3b). Now, the concentration of surfactants can be increased in aqueous medium to increase the no of micelles – even to generate micro-emulsions which encompass more substrates. Hydrophobic effect of surfactants stabilises the organic substrates in the micellar cavity making the overall concentration of the substrate in micellar media greater than that in organic solvent and enhance reactivity as well with distinctive chemo-, regio- and stereo-selectivities. Water insoluble-charged metal catalysts are also inherently soluble in oppositely charged micellar media for coulombic interactions and reside on the peripheral surface as the second coordination sphere (Fig. 3b). Thus, they can easily come close to the apolar substrates to catalyse the reaction easily (Sorella et al. 2015). There are so many examples of successful organic syntheses performed in aqueous micellar media (Tomasek and Schatz 2013; Ohara et al. 2014; Lipshutz et al. 2011; Isley et al. 2014a, b; Linstadt et al. 2014).

The synthesis of quinoxalines from 1, 2-diamines (aromatic, hetero-aromatic or aliphatic) and 1,2-dicarbonyl compounds (aromatic or aliphatic) at room temperature, a dehydration reaction, using about fifty different surfactants is a good example of micellar catalytic reaction, reported by Chakraborti and co-workers (Scheme 13) (Kumar et al. 2013a). The micellar reactions produced more yield and were faster than the reaction with no surfactants. The neutral surfactants, for example, tween—40 among the all surfactants displayed better yield production and their catalytic activity follows the sequence: non-ionic surfactants > anionic surfactants > Brønsted acid surfactants > cationic surfactants. The high local concentration of the nonpolar substrates within the micellar cavity facilitates the water expulsion.

Another traditional reaction named Friedel-Craft acylation of aromatic rings which also gave very high yield in aqueous medium in presence of cationic surfactants, for example, cetyltrimethylammonium bromide (CTAB) or cetyltrimethylammonium chloride (CTAC) at room temperature. The reaction was carried out between acetyl chloride and 1-halo-2-methoxynaphthalenes (Scheme 14). Some other substrates, for example, 2-methoxynaphthalene, anisole, 2-methoxypyridine, and 2-methoxypyrimidine were also used under the same condition (Rajendar Reddy et al. 2013). In this process, sluggishness of the reaction, severity of reaction condition, metal triflates, mineral acids and production of wastes can be reduced by this method.

Hydration of alkyne is a very popular C–heteroatom bond forming reaction which needs presence of transition metals as catalyst to increase the electrophilicity of the alkyne. But, in presence of surfactants, the reaction proceeded very nicely in aqueous medium at 140 °C without using any transition metal with effective hydration of aromatic alkynes terminally and internally (Scheme 15). Terminal alkenes were also found to be hydrated in this process (Nairoukh et al. 2013).

Oxidation of alcohols in aqueous acidic medium by Cr (VI) is a very common reaction and it is very slow to move to the product. The rate of oxidation of 2-propanol got higher in presence of trace amount of Ru(III) and also in SDS (sodium dodecyl sulphate) micellar medium separately at 30 °C. But in presence of both Ru(III) and SDS micellar media, the rate became around 8 times higher than that of the uncatalysed propanol oxidation in absence of surfactant at

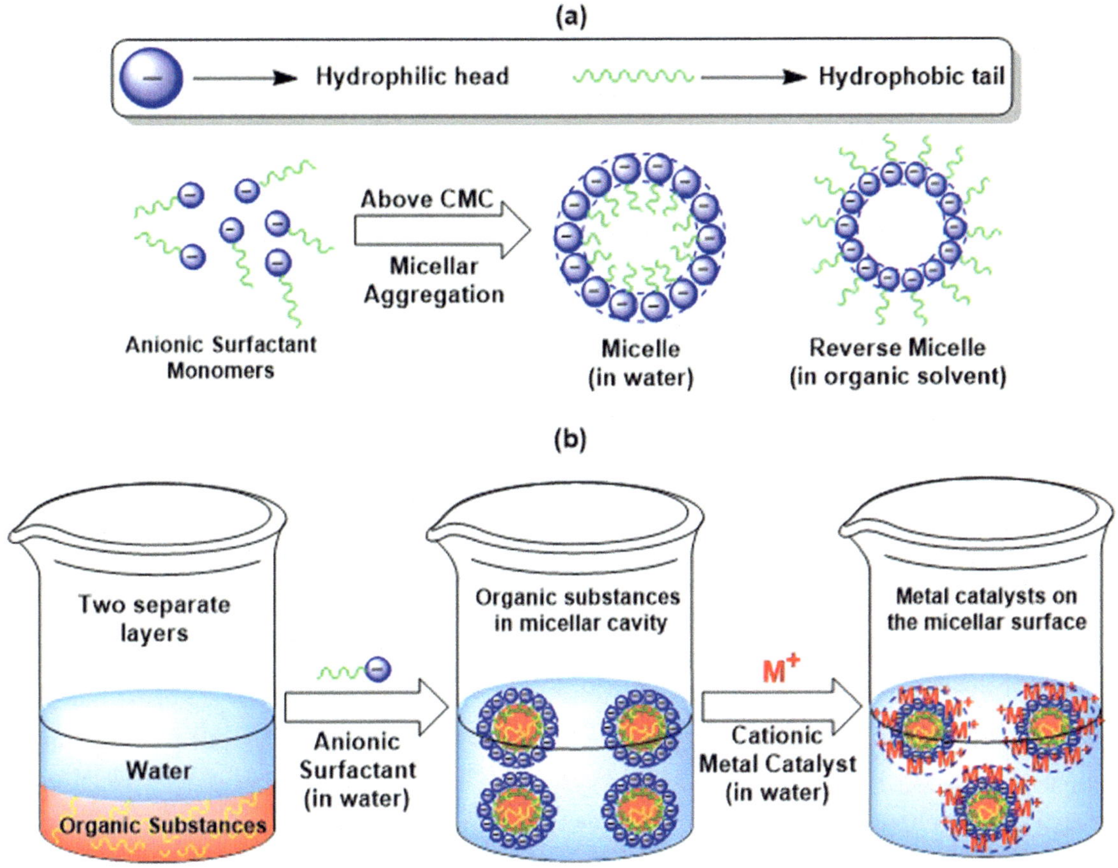

Fig. 3 (a) Micellar aggregation. (b) Compartmentalisation of organic compounds by micellar formation in aqueous medium

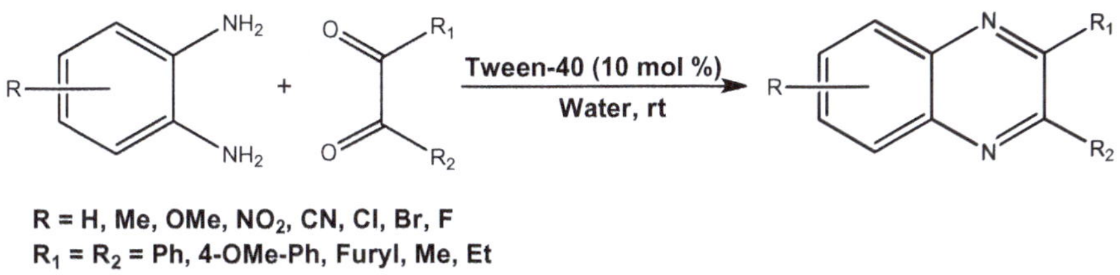

Scheme 13 Neutral surfactant-assisted synthesis of quinoxalines in aqueous medium

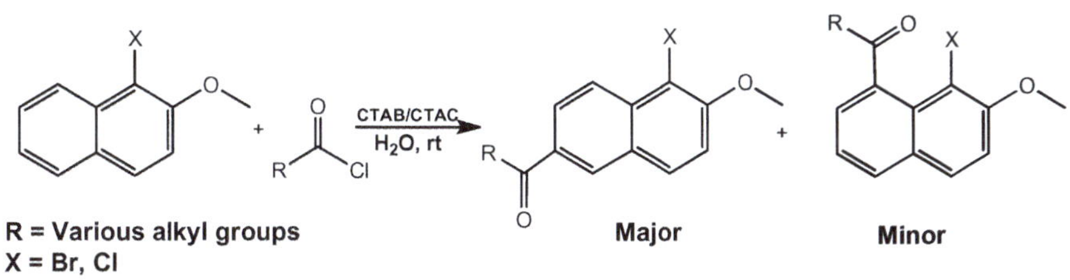

Scheme 14 Cationic surfactants assisted Friedel-Craft acylation in aqueous medium

the same temperature (Scheme 16) (Chowdhury et al. 2019b). The micellar reaction is truly an enhancement of the oxidation reaction.

Another oxidation of an alicyclic alcohol in acidic aqueous medium, cyclohexanol, was expedited repeating the reaction in micellar media. In this, oxidation reaction Ce(IV)

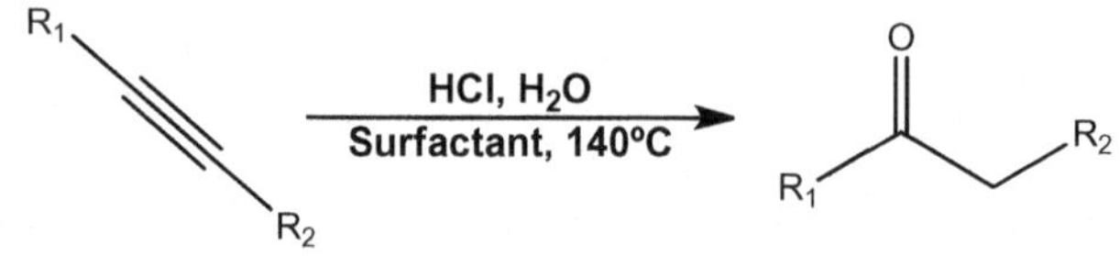

Scheme 15 Surfactant-assisted hydration of alkyne

was used as an oxidant which oxidised cyclohexanol in around 6 h at 28 °C having no catalyst but in presence of small quantity of an anionic surfactant, SDS, the reaction completed within only 21 min though in presence of CPC (*N*-cetylpyridinium chloride), a cationic surfactant, the reaction rate slowed down (Scheme 17) (Chowdhury et al. 2020a).

Oxidation of long-chain fatty aldehydes in aqueous medium is not at all very easier. Octanal, such a long-chain aldehyde was successfully oxidised to corresponding fatty acid, caprylic acid, by Cr(VI) in acidic aqueous medium with the help of surface phenomena in micellar media (Scheme 18) (Chowdhury et al. 2020b). The reaction was reported with good solubilising power and catalytic activity of surfactants. In the reaction, SDS and TX-100 both elevated the rate of the reaction considerably with their increasing concentration and CPC does the opposite. The micellar catalytic reaction was also reported with some rate enhancing effect by some promoters (Phen, PA and Bpy) which are actually hetero-aromatic bases in nature. The micellar-promoted reactions display more enhancement where the combination of SDS and Bpy was reported as the best for such observation.

Micellar media is very useful in Diels-Elder reaction, with a high atom economy, in the preparation of 4-amidyl-2-methyl-1, 2, 3, 4-tetrahydroquinolines, a pharmacologically relevant molecule (Scheme 19) (Merchán Arenas et al. 2013). In this case, using SDS in acidified water a very high yield was produced diastereospecifically and increasing the concentration of SDS above critical micelle concentration the yield percentage was improved more effectively.

Ruthenium-catalysed ring closing metathesis using, for example, first-generation Grubbs, Hoveyda–Grubbs and Zhan catalysts, is a very popular reaction. Such reactions responded the catalytic behaviour of surfactants effectively. In the ring closing reaction of *N,N*-diallyl tosylamine Gemini

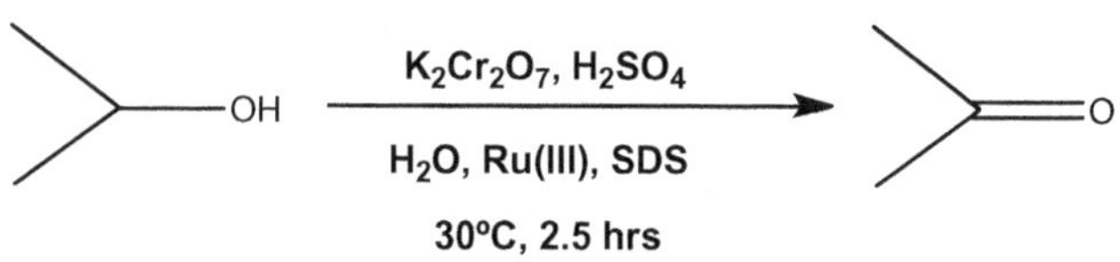

Scheme 16 Ruthenium-catalysed oxidation of alcohols in micellar medium

bis-cationic surfactants (various common surfactants linked with a spacer) showed improvement of catalytic efficiency of metal catalysts in aqueous medium (Scheme 20) (Laville et al. 2012). By NMR investigation, it was examined that only the substrate molecule gets dissolved in the medium of reaction, whereas the metal catalysts and the products remained undissolved.

A very impressive contribution of micellar reaction was announced by Ismail and groups in case of nitration of aromatic compounds using nitric acid. The researchers reported that in SDS micellar medium, nitration of aromatic compounds steadily gave only para-isomer with strong regioselectivity and higher atom economy even at room temperature only within 30 min. Not only that, but also the regiolselectivity was equally valid for other aromatic electron rich (like aniline) and electron poor substrates (having nitro, carboxylic acid groups, etc.) as well and no protection required in that case (Scheme 21) (Dey et al. 2013). High yield, less tedious work up, no poly nitration and usages of organic solvent make the process highly sustainable with very small E factor. The substrate molecule gets dissolved in the micellar medium of SDS and the micelles facilitate the approach of substrate to the cationic nitronium ion.

Enantioselective asymmetric hydrogenation of itaconic acid and dimethyl itaconate like important benchmark substrates with chiral Rh(I) catalysts in methanol medium is a very important reaction. The reactions were performed in SDS and TX-100 media successfully with high turnover-numbers (up to 1000) though the selectivity was quite similar to the previous and the rates were rather lesser than that in methanol (Scheme 22) (Schwarze et al. 2011). The micellar media provide suitable membrane for the recycling of the Rh(I) catalyst.

Suzuki coupling of aryl bromides and aryl chlorides is a very well-known coupling reaction in presence of ligand-free Pd catalyst which was repeated by Li and co-workers in guanidinium ionic liquids containing six lipophilic alkyl chains and reported with high yield. Here, the guanidium ionic liquids acted as micellar media was ensured by TEM analysis. Actually, the micelles of those ionic liquids catalyse the reaction by stabilising the generation of elemental Pd-nanoparticles ($\sim$5.1 nm). The length of the lipophilic chain of the guanidium molecule has strong impact upon the reaction efficiency. In presence of hexaethylguanidinium bromide, the reaction between 4-bromoanisole and phenylboronic acid produced 30% more product than the usual and in presence of hexa-dodecyl-guanidinium bromide, almost 100% product was obtained (Scheme 23) (Lin et al. 2011). Here also, the micellar media provided an additional benefit of recycling the elemental Pd catalyst without losing its significant activity.

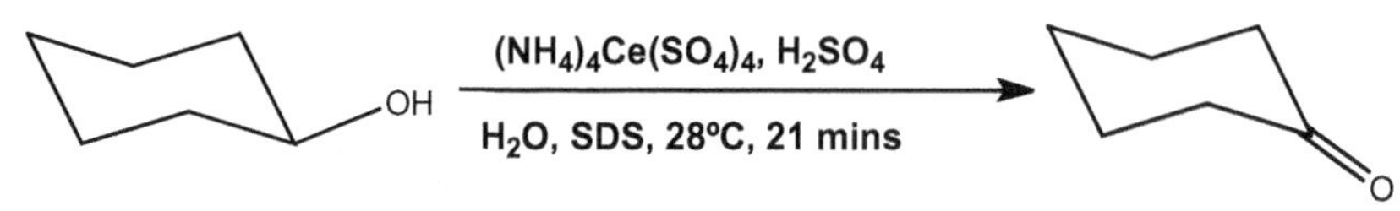

Scheme 17 Micellar-promoted oxidation of alicyclic alcohols in water

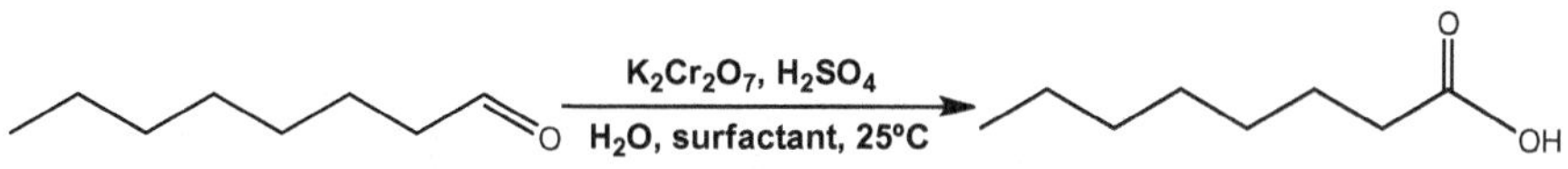

Scheme 18 Micellar-assisted oxidation of long-chain fatty aldehydes in aqueous medium

Scheme 19 Diels-Elder reaction in micellar medium

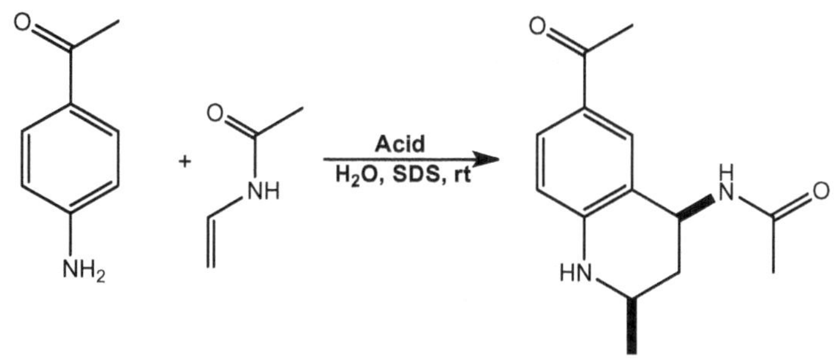

Scheme 20 Ruthenium metathesis in presence of Gemini surfactants

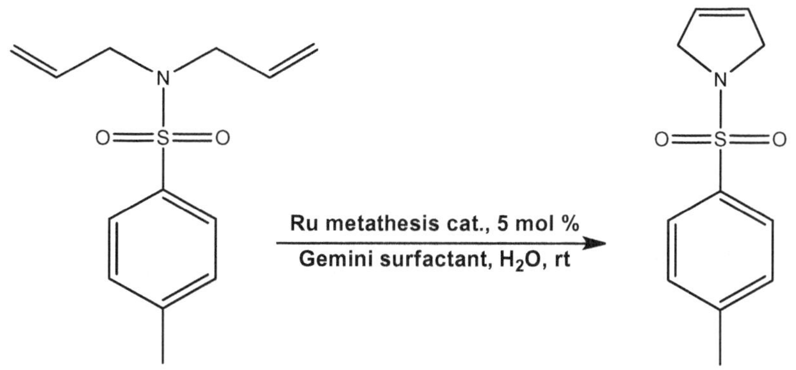

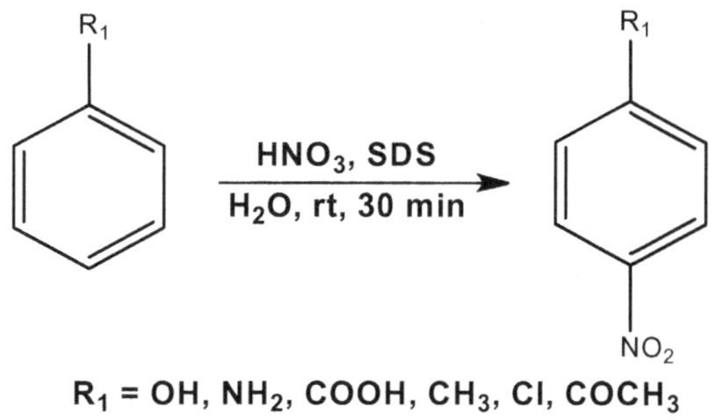

Scheme 21 Nitration of aromatic compounds in SDS micellar medium

Multicomponent reactions (MCRs) are good expansion of chemist's toolbox of sustainable organic synthesis. In this type of reactions, at least three reactants combine in a single pot to make a product having most of the atoms (preferably all) of the reactants. High atom economy, mild reaction condition, efficiency, great convergence and concerted steps economy have made this methodology a transcendence. Some very well-known examples of MCRs are Strecker, Biginelli, Mannich, Passerini, Ugi, Groebke–Blackburn–Bienaymé, Orru, etc., synthetic reactions (Cioc et al. 2014). In all cases, the minimum atom economy is 80% and waste production is also very minimal. But, use of benign solvents was not in vogue beforehand. It comes into play comprehensively in this century (Touré and Hall 2009). Use of micellar medium as solvent has really provided good consequences.

In three-component one-pot synthesis of spirooxindole derivatives, very important for the synthesis of pharmaceuticals and natural products, weakly basic sodium stearate was proved to be very efficient as micellar medium. The micellar aggregations of weakly basic sodium stearates

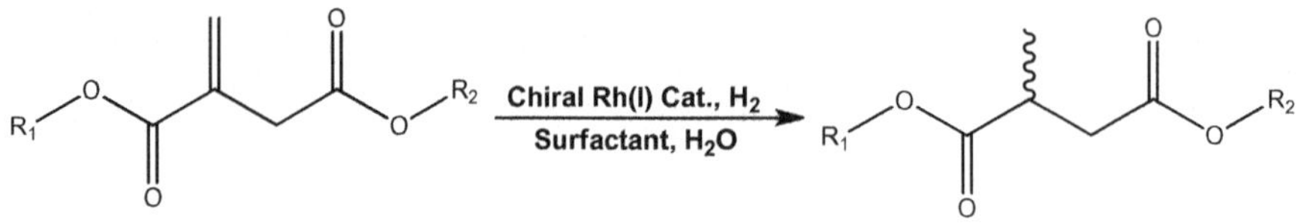

Scheme 22 Enantioselective asymmetric hydrogenation in presence of surfactants

Scheme 23 Influence of guanidinium ionic liquids in Suzuki coupling

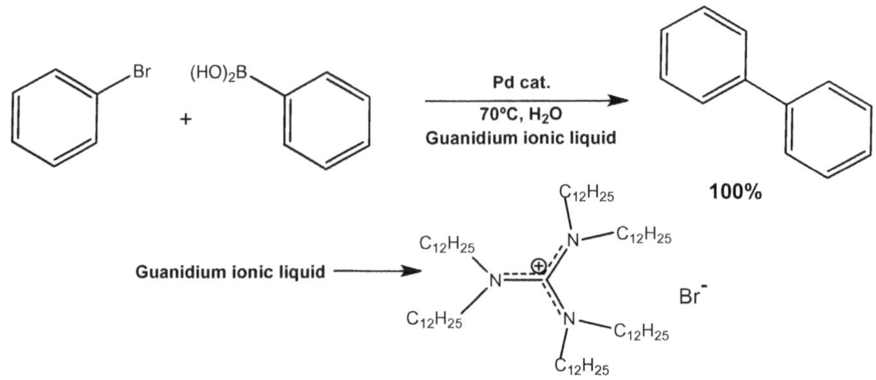

dissolved the substrate molecule forming colloidal dispersion and then stimulated the deprotonation of malononitrile which reacted with isatin and next with a 1,3-dicarbonyl compound to give 91–97% yield of the product (Scheme 24) (Wang et al. 2010). It is very relevant to mention here that use of SDS, SDBS (sodium dodecylbenzenesulfonate) or any other weak bases, for example, acetate gave different amount of yields in each cases.

Another example of multicomponent reaction is one-pot synthesis of 3-amino alkylated indoles in aqueous micellar medium via a Mannich-type reaction of secondary amines, aldehydes and indoles. In presence of any Bronsted or Lewis acid, the reaction produced mostly bis-indole derivative whereas in micellar medium of SDS, it produced 3-amino alkylated indoles in a smooth way with high selectivity and very high yields (Scheme 25) (Kumar et al. 2013b). The hydrophobic moiety of the anionic surfactant primarily promotes the dehydration and cationic iminium formation and the cationic additive reacts with indole for product formation. In this procedure, about 25 different products were synthesised easily within only 2–14 h with high yields (78–94%).

Cu(I)-mediated Sonogashira-cyclisation of three-components, i.e. *o*-halo-*N*-phenylbenzamides (I and Br), phenylacetylene and indoles in one-pot is a very interesting regioselective reaction which were tested using different micellar media (Scheme 26) (Sarkar et al. 2013). Neutral surfactants, PTS (polyoxyethanyl-α-tocopheryl sebacate) in particular, and a base, triethylamine, gave the best result with 76% at 80 °C. The ligand, 2,20-(1E,10E)-(1R,2R)-cyclohexane-1,2-diylbis(azan1-yl-1-lidene)bis(methan-1-yl-1-ylidene)diphenol has a vital role in this reaction for the activation of Cu(I) metal centre.

Micellar enzymology is a very important physicochemical line of research in biocatalysis. Enzymes can be of two types: (1) hydrophilic or lyophobic and (2) hydrophobic or lyophilic. The hydrophilic and hydrophobic moieties impregnated surfactants can solubilise enzymes above CMC forming supramolecular assemblies named micelles or vesicles or o/w micro-emulsions, in aqueous medium, and

Scheme 24 Three-component one-pot synthesis in sodium stearate micellar medium

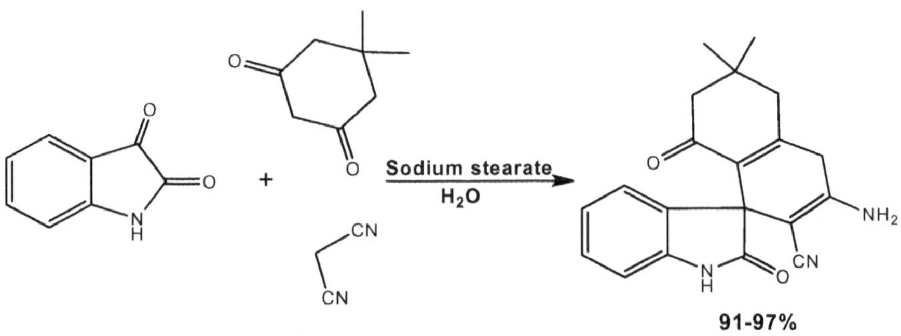

Scheme 25 Multicomponent reaction in aqueous micellar medium

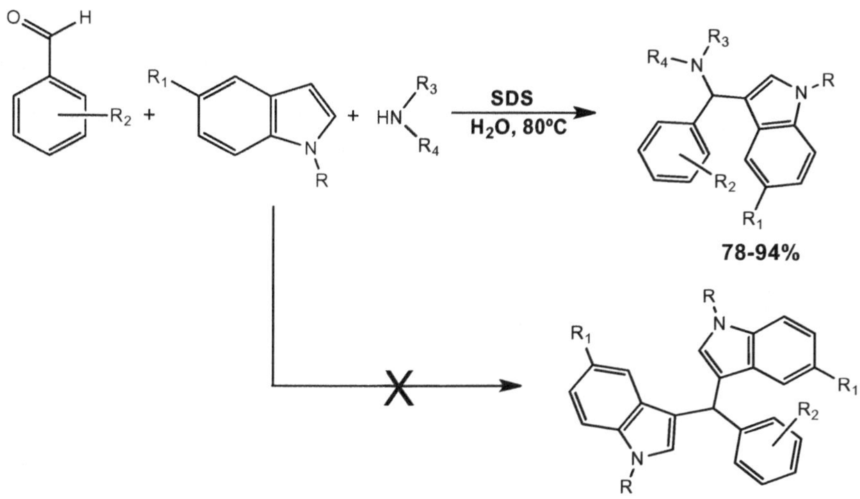

Scheme 26 Sonogashira-cyclisation of three-components in micellar media

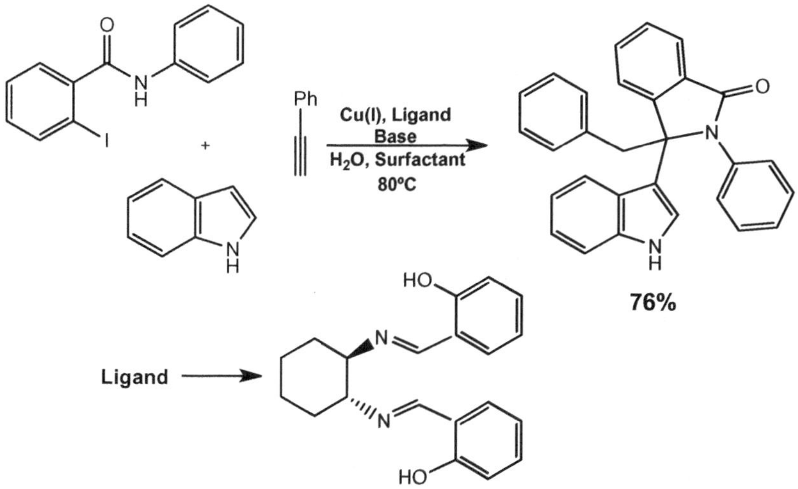

reverse micelles or reverse vesicles or w/o micro-emulsions, in organic solvents. In most cases, enzymes were found to be more active in micellar medium (normal micelle or reverse micelle) than in any pure solvent (water or organic solvent) (Biasutti et al. 2008; Adlercreutz 2013). Stability of enzymes under non-aqueous conditions is very crucial, which depends upon (1) the character of the enzymes, (2) their state of mobility, (3) the presence or absence of additives, for example, sugars, (4) on the water content, and (5) sort of non-aqueous solvent (Adlercreutz 2013; Cao and Matsuda 2016). In micellar media, all the factors are maintained very carefully as per requirements. So, for in vitro biocatalytic reaction, micellar enzymology has worldwide applications (Biasutti et al. 2008; Adlercreutz 2013).

In 2004, Falcone and groups studied spectrophotometrically the kinetics of a successful enzymatic hydrolysis of 2-naphthyl acetate by α-chymotrypsin (α-CT), a hydrophilic and globular enzyme, in reverse micelles constructed by sodium bis(2-ethylhexyl)sulfosuccinate (AOT), water and n-heptane [water/AOT/n-heptane] (Scheme 27) (Darío Falcone et al. 2004). In this reverse micellar reaction mixture, addition of glycerine (GY) [water and a GY-water (38% v/v) mixture], a non-aqueous polar solvent, exhibited significant rate enhancement at pH ≈8.7 at about 25 °C. The half-life of the enzymatic reaction in reverse micellar medium was obtained 0.5 min, which is much larger than that of homogeneous non-enzymatic reaction (about 420 min). In both, the reverse micellar hydrolysis quantitative yields of the product were obtained. The rate enhancement, on addition of water soluble GY to the reverse micellar medium, is attributed to the increased order of micellar matrix structure causing decrease of the mobile nature of the enzyme molecule and, hence, increase of its stability. In 2006, a biomimetic route was reported where haemoglobin was used in polymerisation of aniline in SDS, CTAB and TX-100 micellar media. Good yields were obtained at pH = 2.0 (Hu et al. 2006). Shome et al. presented a very relevant aspect of micellar enzymology. They upheld that non-ionic surfactants

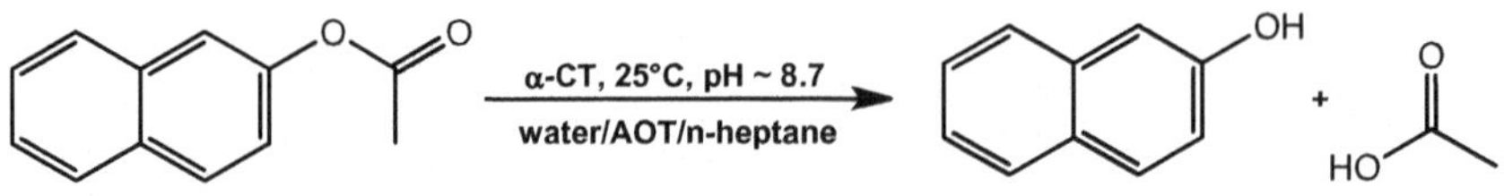

Scheme 27 Hydrolysis of 2-naphthyl acetate in presence of enzyme in reverse micelles

can enhance the catalytic efficiency of surface-active enzymes, for example, lipase, in cationic reverse micellar media (Shome et al. 2007). They used CTAB as cationic surfactant and displayed up to 200% rate enhancement in presence of non-ionic surfactants, for example, Brij-30, Brij-92, Tween-20, and Tween80. The rate enhancement was mainly ascribed to the attenuated positive charge density at the interface of cationic W/O microemulsions in presence of the non-ionic surfactants. Similar trend of rate acceleration was also inspected in case of peroxidase activity (Shome et al. 2007).

3.2.2 Different Non-Aqueous Media

Some different non-aqueous media are also in good books of modern researchers to replace hazardous organic solvents in organic syntheses. Ionic liquids, fluorous media and supercritical fluids are those promising non-aqueous solvents which have been used extensively in this regard.

Ionic Liquids

Ionic liquids [ILs] are a kind of salts which are basically liquid unlike solid salts and form by the ionic interaction of mismatched cations and anions. Some structures of common ILs have been produced for convenience (Fig. 4) (Mikami 2005). Amphiphilic nature of ILs has been revealed by many investigations and ILs with long-chain imidazolium- and pyridinium moieties behave like cationic surfactants in aqueous solutions (Pei et al. 2018). ILs have some very good qualities to be chosen as a green solvent. Firstly, the liquids are salts; they cannot vapourise to pollute atmosphere and a simple distillation is enough for product separation. Secondly, different organic, organometallic, inorganic and polymeric compounds are highly soluble in the media including catalysts, so organic syntheses can be carried out in high concentration of the reagents. Even, it can dissolve gaseous compound better than organic solvents. Thirdly, ionic liquids are stable over a wide range of temperatures (−96 to 300 °C) and kinetics can be controlled in these media better than in water and organic solvents. Fourthly, many organic solvents are immiscible with ionic liquids for which they can be used in biphasic catalytic reactions and product can be extracted easily from the catalyst. At the fifth, though ionic liquids are of high price, large number derivatives of the ionic liquids can be synthesised easily at low cost and they are very tunable solvent (Mikami 2005; Pârvulescu and Hardacre 2007; Hallett and Welton 2011).

In ionic liquid epoxidation of enones was done efficiently in a basic aqueous solution of hydrogen peroxide requiring no metal catalyst. In this case, hydrophobic [bmim][PF_6] and hydrophilic [bmim][BF_4](bmim$^+$=1-butyl-3-methylimidazolium cation) both were used as the medium producing quantitative yields of the desired epoxide in a very short time (Scheme 28). The same strategy was used for the epoxidation of chromones and flavonoids also (Bortolini et al. 2002; Bernini et al. 2004).

The Suzuki–Miyaura coupling with good yield was reported in a mixture of water and ammonium salt or pyrrolidinium salt of BF_4^- with ligand-free palladium catalyst (Scheme 29) (Calò et al. 2005). The salts are of high melting point (above 120 °C) but in water, they melt easily and form a biphasic mixture at lower temperature (∼50 °C and 80 °C, respectively). In this reaction, the salts were separated and purified though the recycling was not so good.

After few years, an interesting example of Swern oxidation of a secondary alcohol in sulphide containing ionic liquids was reported. Here, non-volatile sulphur-containing compounds were grafted to the scaffold of imidazolium ionic liquids (n = 2, 3 and 6) and oxidised to corresponding sulfoxides by periodic acid. These sulfoxides actually oxidised the secondary alcohol to ketones and the sulphur-containing compound was regenerated easily in the process (Scheme 30) (He and Chan 2006). The advantage over normal Swern oxidation is that the sulphides or sulphur-containing compounds used or generated in this whole process were non-volatile and odourless, so this process is quite eco-friendly. There are many more reactions as such in the same field.

Microwave (MW)-assisted heterocyclic compounds synthesis of ionic liquids is an advancement in the field of green chemistry. MW acts as high-frequency electric field and it can heat charged material (cations or anions) to synthesise ionic liquids very fast reducing the reaction time to great extent. There are many examples of MW-assisted reactions reported so far. A couple of important examples of them have been given below. (Pathak et al. 2016; Palou 2010).

A nice example of MW-assisted pyrazole derivatives synthesis at room temperature was reported by Raghuvanshi et al. The reaction of phthalhydrazide, aromatic aldehydes and malononitrile was carried out in [bmim][OH] medium at 45 °C under the irradiation of high-frequency electric field to synthesise 1H-pyrazolo[1,2-b]phthalazine-5,10-dione as sole product (Scheme 31) (Singh et al. 2011).

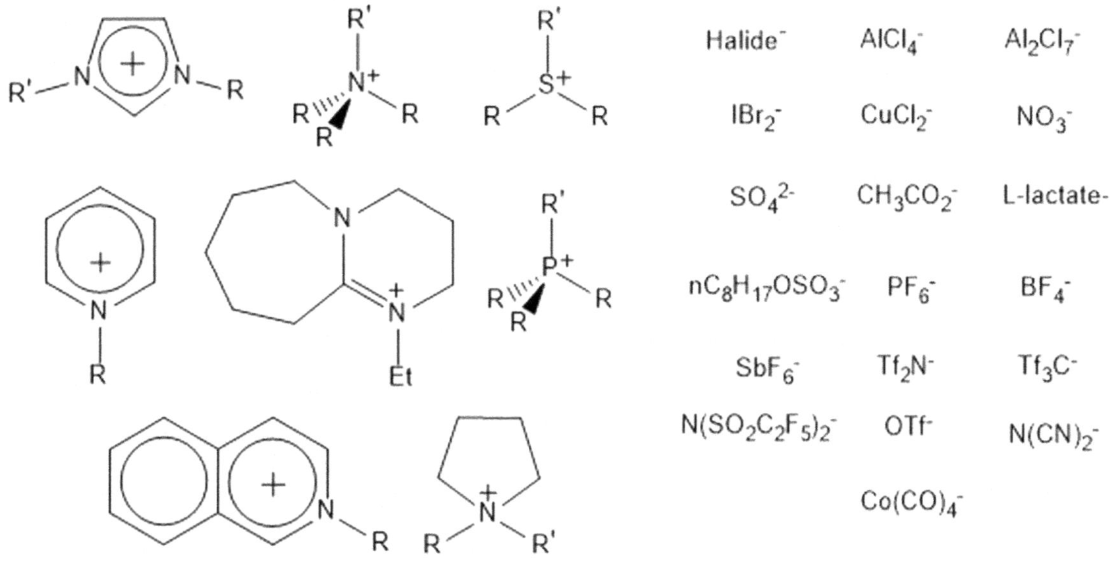

Fig. 4 Cationic and anionic entities of some common ionic liquids

Scheme 28 Epoxidation of enones in IL medium

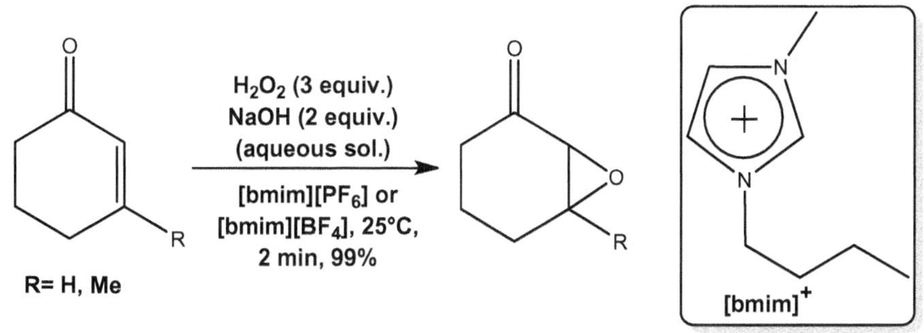

Scheme 29 Suzuki–Miyaura coupling in IL/water biphasic mixture

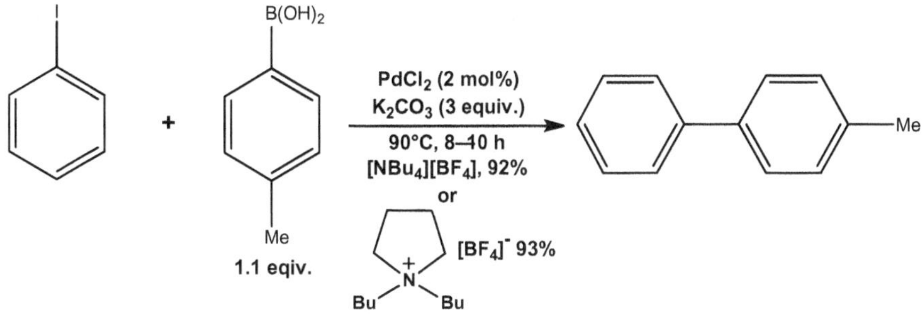

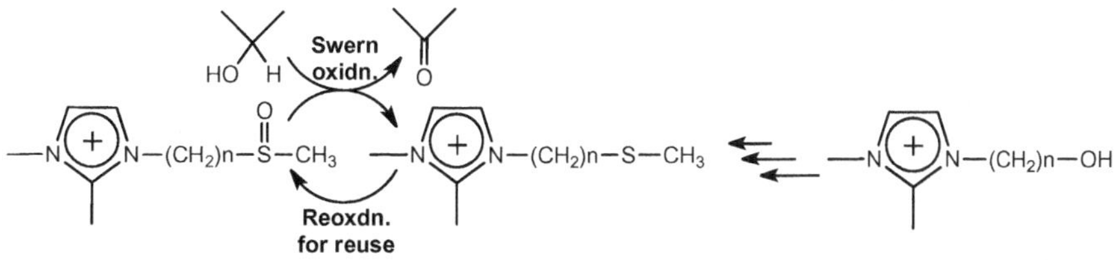

Scheme 30 Swern oxidation of a secondary alcohol in sulphide containing IL

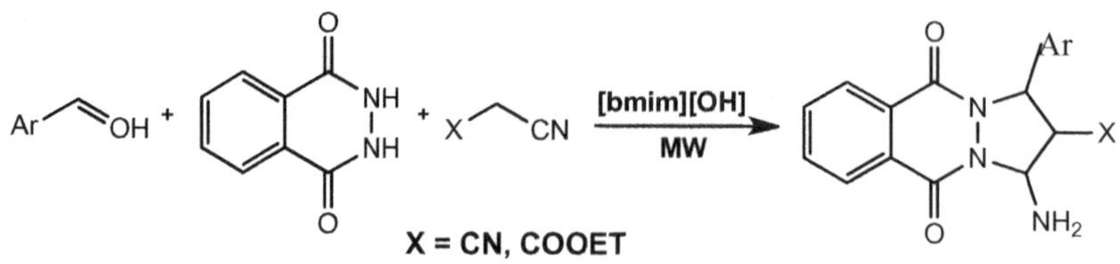

Scheme 31 MW-assisted pyrazole derivatives synthesis in IL medium

Few years ago, a derivative of xanthene [1,8-dioxo-octahydroxanthene] was synthesised very efficiently from 5,5-dimethyl-1,3cyclohexanedione (dimedone) and benzaldehyde in [cmmim][BF_4](cmmim$^+$=1-carboxylmethyl-3-methylimidazolium cation) IL medium by Dadhania and co-workers, which was irradiated by MW (Scheme 32) (Singh and Savoy 2020). The reaction procedure provided reusability of catalyst, quantitative yields in a very little time and work up was also very easier.

Biocatalysis in ionic liquids is a very relevant and ardent topic of green chemistry which cannot be omitted in ILs by any chance. Enzymes being an organic molecule have poor solubility in aqueous medium. So, some non-aqueous media were solicited and ionic liquids are really that kind of media. Beside that the charged functional groups of enzymes may interact with the counter ions of ionic liquids which may enhance catalytic behaviour of enzymes (Sheldon et al. 2002; Sheldon and Pereira 2017).

Lipases (hydrolases) are the most extensively used biocatalytic enzymes because it can catalyse almost all kind of organic synthetic reactions. There are many updated instances of lipase-mediated synthesis of biodiesel using ILs as reaction media (Scheme 33) (Itoh 2017). It is really an interesting method of production of energy, one of the most important needs in the world, by a sustainable way and so, we accepted this example among the numerous. Moreover, organic solvent-free separation of hydrophobic diesel from the reaction medium has been perceived. There are many such recent lipase-catalysed reactions in ionic liquids media. Not only lipase but also many such enzymes like proteases, cellulase, epoxide hydrolase, horseradish peroxidase (HRP), alcohol dehydrogenase, cytochrome C, lyase, etc., have profound positive impact in organic syntheses in a very eco-friendly process (Li et al. 2015; Abe et al. 2012; Ou et al. 2016; Mohammadyazdani et al. 2016; Dong et al. 2016; Daneshjoo et al. 2011; Rodrigues et al. 2014). But we have no opportunity to discuss them here in detail.

Fluorous Media

Perfluorocarbons (PFCs) (Fig. 5) (Mikami 2005), the fluorinated hydrocarbon analogues, can dissolve gases (due to low surface tension) and they can be separated easily from organic hydrocarbon solvents. Only for these qualities, they were considered to give trial as solvent in biphasic organic synthetic reactions of fluids (gas and liquid) and sometimes for purification of products and catalysts. They were first used by Zhu in 1993 as solvent in organic reaction. Time to time, many more interesting features of PFCs were revealed, which assists them to be greener solvent (Mikami 2005). In fact, fluorous biphasic systems (FBS), first interpreted by Horváth and Rábai, are widely exploited in many reactions (Sheldon 2005). The characteristic features of fluorous solvents (Mikami 2005; Sheldon 2005; Hobbs and Thomas 2007) are

1. At room temperature, the fluorous and organic phases are not miscible. But, on increasing temperature, miscibility and homogeneity of the two phases are achieved (Fig. 6). On lowering temperature, the phases separate again. So, the catalyst (organometallic catalyst) can be separated from the product only by temperature variation, which provides a basis for performing biphasic catalysis reaction using fluorous solvents.
2. In catalytic reactions, fluorous catalyst can be extracted as well as recycled directly from the reaction mixture and organic product can be extracted using organic solvent.
3. Fluorous solvents are very inactive towards oxidant, radicals, nucleophiles and electrophiles almost unlike the common organic solvents.
4. They are less polar, non-protic solvents and not strongly Lewis acidic or basic.
5. They are not too viscous to inhibit the movement of reagents.
6. They are very friendly to environment for being less volatile, non-toxic and having zero potential to deplete ozone layer. Their greenhouse potential is also very low.

There are numerous examples of organic synthetic reactions, where fluorous media have been discovered as a great solvent and catalyst as well (Goto et al. 2005; Akiyama and Mori 2015; Yan et al. 2010; Brittain et al. 2005; Trindade et al. 2009; Zhang 2009).

Acetalisation of benzaldehyde with 1,3-propanediol and Mukaiyama aldol reaction of an enol silyl ether with benzaldehyde in neat organic medium in presence of fluorous super Brønsted acid catalyst are interesting examples of biphasic organic reaction (Scheme 34) (Akiyama 2007). In this process, the recovery of the fluorous catalyst was very

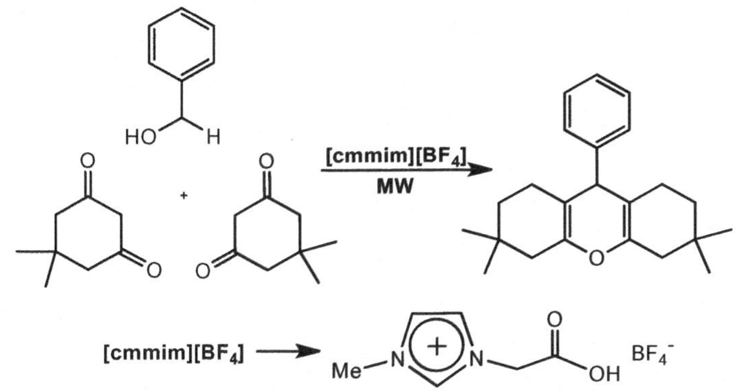

Scheme 32 MW-irradiated MCR in IL for the synthesis of xanthene derivative

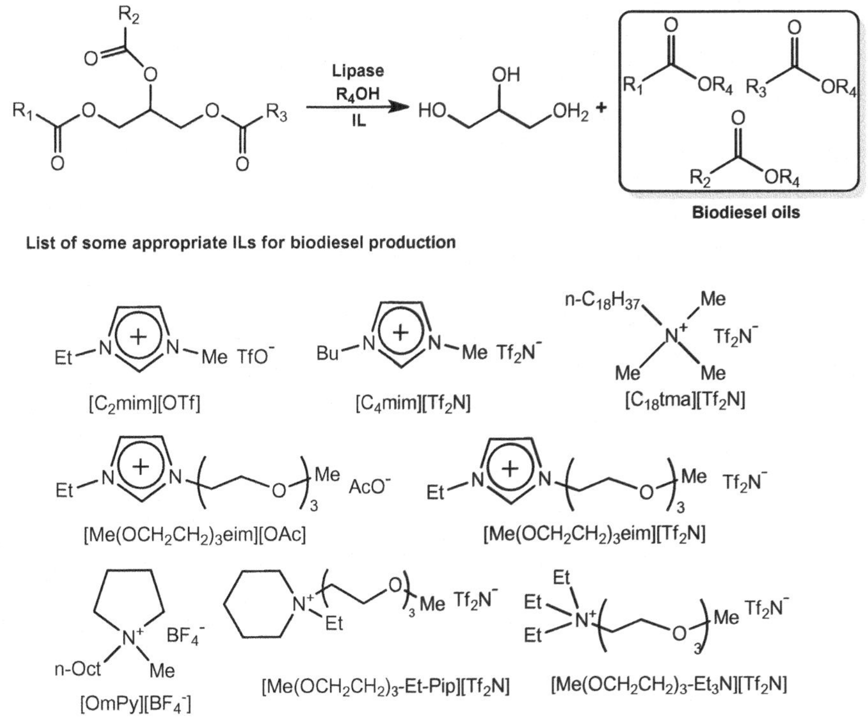

Scheme 33 Lipase-mediated synthesis of biodiesel in ILs Media

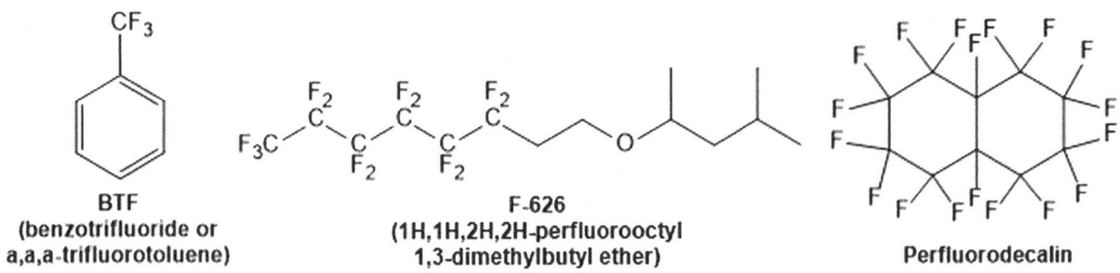

Fig. 5 Common perfluorocarbons (PFCs)

easier, it only needed precipitation by decreasing the temperature of the reaction mixture to ambient temperature.

Mizoroki–Heck arylation reaction of acrylic acid in fluorous medium was reported with both catalytic as well as good solvent behaviour of fluorous medium. A fluorous Pd catalyst was used as catalyst and F-626 was adopted as reaction medium (Scheme 35) (Molnár 2011). The catalyst was synthesised in situ using a fluorous IL and palladium acetate. The products precipitated in the reaction as the solubility of arylated carboxylic acids is very low in F-626.

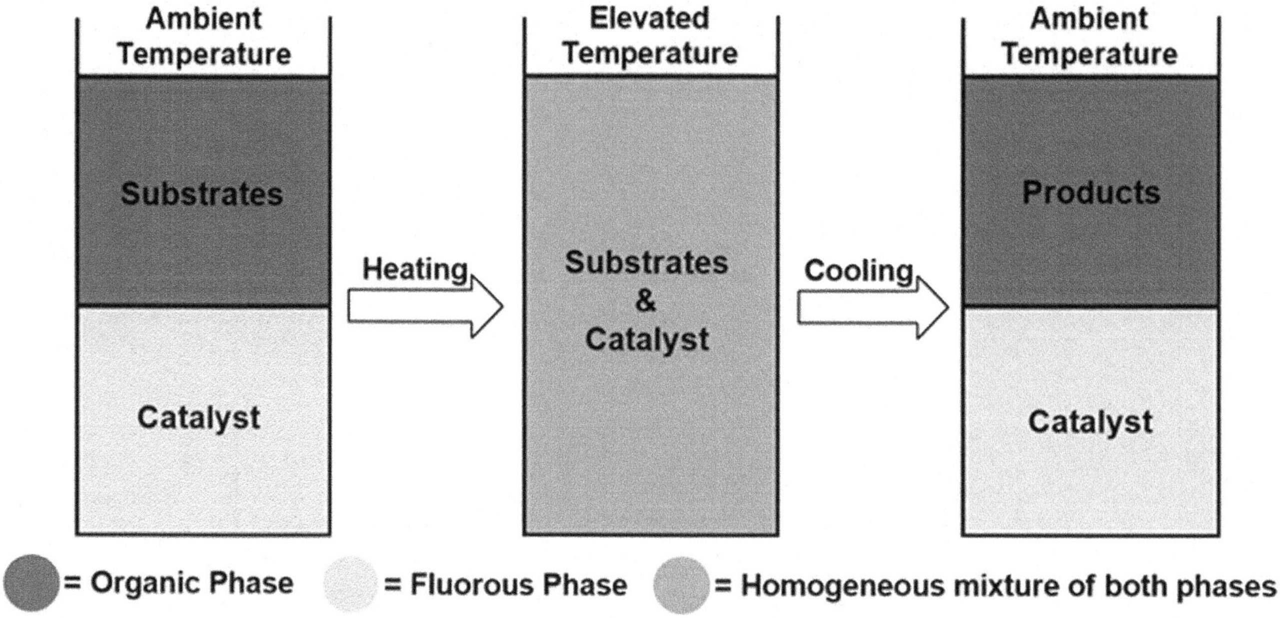

Fig. 6 Miscibility of fluorous media with temperature variation

Scheme 34 Acetalisation reaction and Mukaiyama aldol reaction of benzaldehyde in organic/fluorous biphasic media

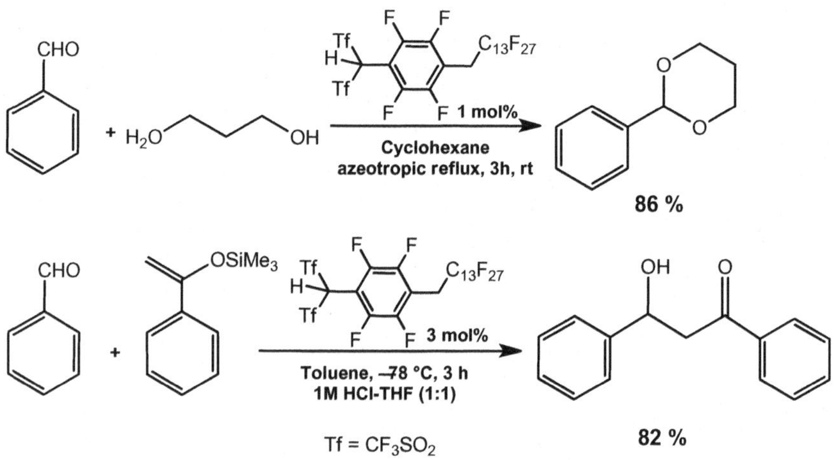

After the filtration of product and amine salts, recycling of the fluorous Pd catalyst was possible for some more times.

The cross-coupling reactions, Mizoroki–Heck and the Sonogashira reactions, also gave good response in organic/ fluorous biphasic medium (Scheme 36) (Matsubara et al. 2014). These reactions moved forward steadily producing good yields of the desired product. The Pd catalyst was recyclable and reusable in presence of the F-DMF in the reaction medium.

A novel benzylation reagent, 2-benzyloxy-1methylpyridinium triflate, was reported by Dudley et al. in BTF solvent with higher yields (>95%) than in organic solvents like toluene and 1,2-dichloroethane (Scheme 37) (Sowmiah et al. 2018). Later, some studies of the reaction

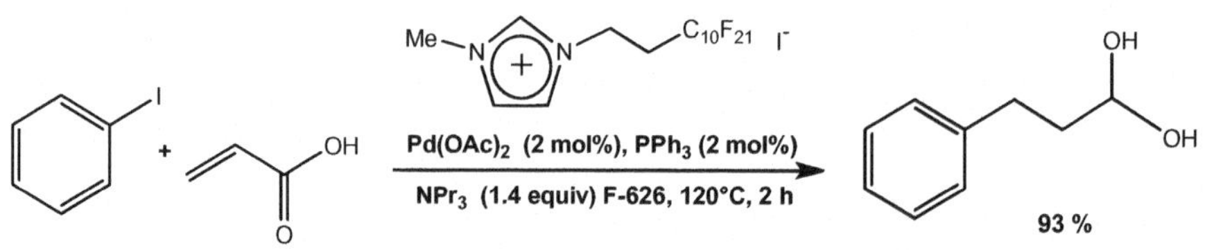

Scheme 35 Fuorous Pd-catalysed Mizoroki–Heck arylation reaction in fluorous medium

Scheme 36 Mizoroki–Heck and the Sonogashira reactions in organic/fluorous biphasic medium

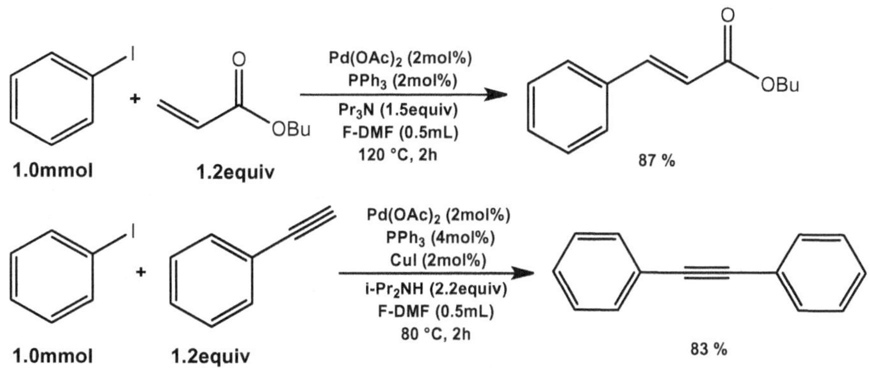

confirmed the observations. Apart from this, the benzylation also was successful by preparation of Dudley's reagent in situ (Scheme 38) (Yang and Dudley 2010).

Isomeric mixtures of methyl perfluorobutyl ether ($C_4F_9OCH_3$) are organic hybrid ether solvents, commercially known as Novec 7100. In a report, Novec-7100 was successfully discovered as a co-solvent with quite good yields, where *N*-fluoro-2,4,6-trimethylpyridinium tetrafluoroborate was used for electrophilic fluorination of aryl Grignard reagents (Scheme 39) (Petrone et al. 2016).

Fluorous solvent-based biocatalytic reaction is quite a recent topic in research area. But not much work has been reported yet thereof, for partial insolubility of enzymes in fluorous solvents. But, "like dissolves like" so highly fluorinated enzymes are soluble in fluorous solvents and catalysts can be fluorinated by incorporating fluorous ponytails therein (Hobbs and Thomas 2007; Sheldon and Woodley 2018). This strategy allows fluorous solvents are excellent to be used in multiphasic (biphasic or triphasic) biocatalytic reactions coupled with organic solvent. Highly fluorinated enzymes are readily soluble in fluorous media and organic substrate is soluble in organic solvent. The immiscible fluorous solution and organic solution can be made miscible by raising temperature and after successful reaction, the products can be obtained in the organic medium only by cooling the reaction mixture. Even the enzyme, in the fluorous medium, can be recycled and reused without losing its activity (Lozano 2010; Zhang and Cai 2008).

Enantioselective esterification of rac-2-methylpentanoic acid with a fluorinated alcohol catalysed by candida rugosa lipase (CRL) is a good example of heterogeneous-type biocatalytic reaction in FBS (perfluorohexane and hexane), reported by Beier and O'Hagan (Scheme 40) (Beier and O'Hagan 2002). The acidic substrate is soluble in hexane, the alcohol and product (S-fluorinated ester) are soluble in fluorous solvent. The reaction is termed as heterogeneous due to insolubility of the CRL in either solvents. In the reaction, only the (S)-2-methylpentanoic acid was selectively converted to corresponding (S)-fluorinated ester and unreacted (R)-2-methylpentanoic acid was obtained in hexane. The catalyst was separated by filtration after the reaction is over. A real homogeneous FBS reaction, where the enzyme is soluble in flourous solvent, is very rare because of trouble and difficulty in separation of enzyme from products within the same fluorous solvent. If the substrates get into organic solvent and only fluorinated enzyme gets into fluorous solvent, the case will be more facilitating. But, more research is expected in this field.

Supercritical Fluid

Any substance above its critical point is termed as supercritical fluid (SCF). The critical point indicates the maximum temperature (critical temperature, T_c) and pressure (critical pressure, P_c) at which both phases (vapour and liquid) of the substance exist in equilibrium (Fig. 7). Carbon dioxide has the critical pressure 73.8 bar and temperature 31.1 °C. Water is also a supercritical fluid for reaction chemistry (T_c = 374.0 °C, P_c = 220.6 bar) but its high critical point values limits its application appreciably as a solvent. SCFs are essential because of its dramatic changes in physical properties at small changes in pressure and temperature, especially around critical points (Mikami 2005; Boyère et al. 2014).

At critical point, the density of gas phase intersects and equals to the density of the liquid phase of any substance. So, there is no discrimination between the two phases at that point (Fig, 8), which leads to an end of the boiling curve (Fig. 7). The density of SCF is 100 times higher than that of the gas and lower than the half density of the traditional liquid (Mikami 2005; Hobbs and Thomas 2007).

The viscosity and diffusivity both depend upon temperature and pressure. SCF has ten-fold lower viscosity and higher diffusion rate in comparison to the liquid, which is very helpful to make a reaction faster.

Most of the SCFs exert increase in dielectric constants and thermal conductivities steeply with increase in pressure near critical points.

SCFs having no surface tension, like gases, can spread rapidly in the whole reaction volume and left no option other than being miscible with gas compound.

Scheme 37 Benzylation in BTF solvent

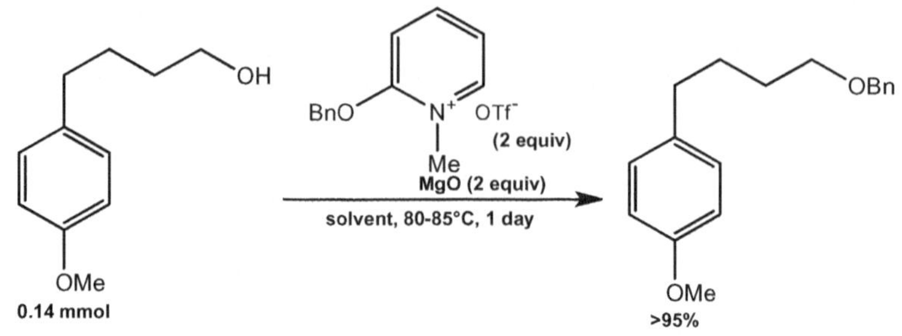

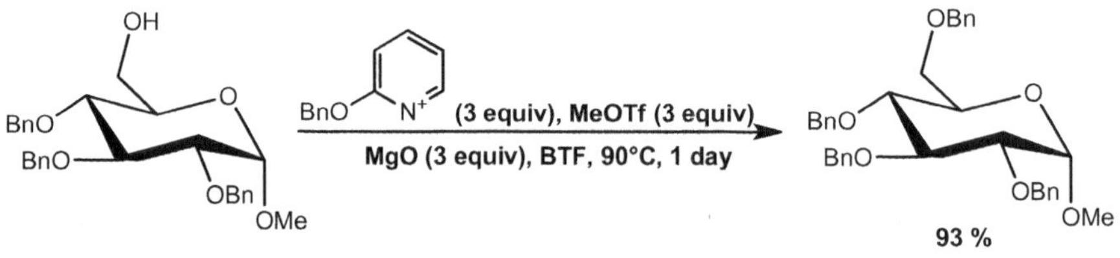

Scheme 38 Benzylation in BTF solvent by synthesis of reagents in situ

Scheme 39 Electrophilic fluorinations of aryl Grignard reagents in Novec-7100

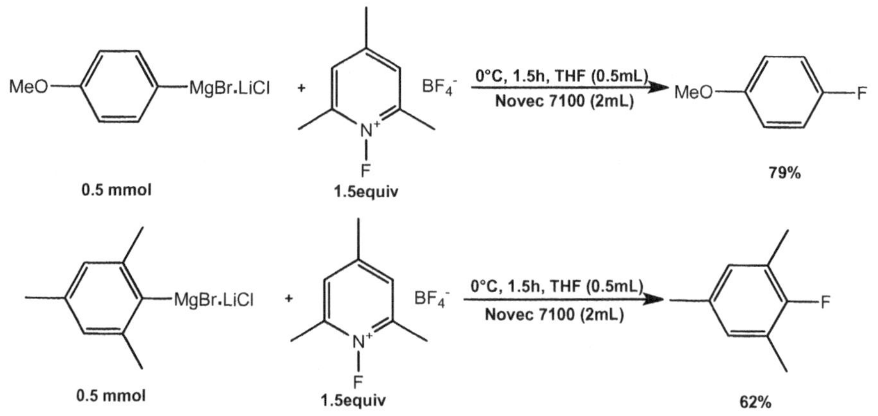

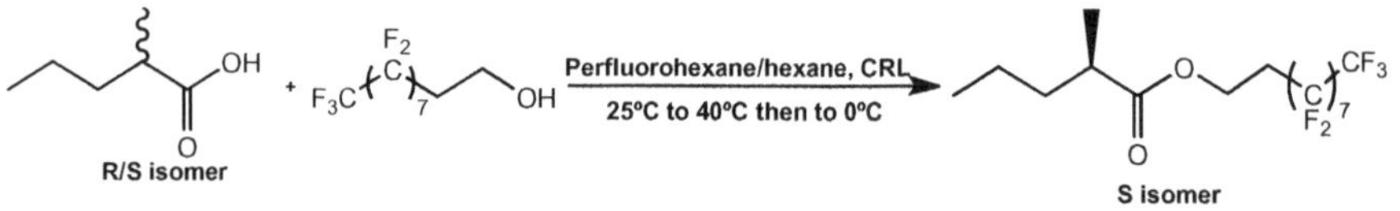

Scheme 40 CRL-catalysed enantioselective esterification of rac-2-methylpentanoic acid in FBS

SCFs can dissolve liquid and solid compounds also in its highly pressurised and liquid-like dense condition. So, the solubility of the compounds can be changed by varying pressure and products can also be separated by precipitation according to our intention. So, SCFs are very powerful solvent in organic synthesis and exceedingly important to the bulk chemical industry. An extra benefit of using SCFs as reaction medium is that, by changing the pressure of the reaction medium, the phase number (single phase to bi-phase or vice versa) of reaction can be controlled easily (Hobbs and Thomas 2007).

For these salient features, SCFs are immensely useful as catalysts and solvents in homogeneous or heterogeneous organic reactions for many times (Machida et al. 2011; Firin et al. 2013; Pavlovič et al. 2013; Onwudili and Williams 2006; Housaindokht and Monhemi 2013; Eckert et al. 1996). Among the SCFs, supercritical carbon dioxide ($scCO_2$) is a very widely used ideal green reaction medium for its unique, well understood physical and chemical properties as well as its nontoxicity to the nature and to human beings. It is an asphyxiant at high concentrations, but precautions can be taken suitably to minimise the risk. It produces no hazardous

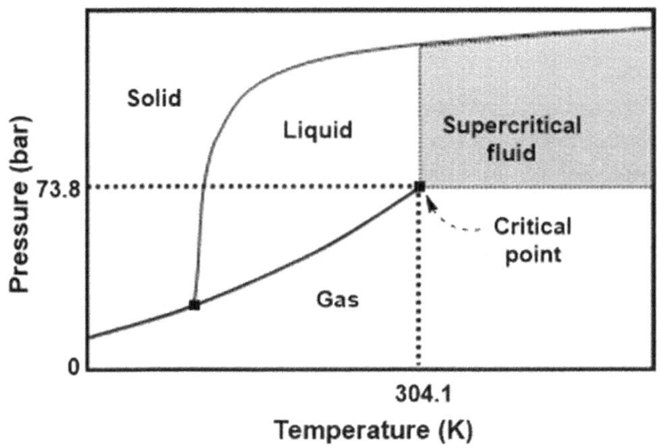

Fig. 7 Phase diagram for a SCF

waste or effluent like other organic solvents, because at atmospheric pressure, CO_2 is in gaseous state and easily can escape from reaction medium without any severe treatment. Non-polar solutes are more soluble in CO_2 than polar solutes, but with larger molecular quadrupole, the case is quite opposite for $scCO_2$. However, the solubility in $scCO_2$ can be tuned by change in density, addition of co-solvent and modification of solutes (Hobbs and Thomas 2007). These unique features of $scCO_2$ have smoothen many uneven circumstances of organic synthesis intensely and we have mentioned here some interesting examples.

Metal catalysts are necessary for hydrogenation and nanoparticles of the metal catalysts are far more active in this purpose. Such a case was reported where colloidal Pd-nanoparticles supported with polymers were discovered very efficient in catalysis of hydrogenation of 4-methoxycinnamic acid to 4-methoxyhydrocinnamic acid in $scCO_2$ at a 10 bar hydrogen-pressure at 56 °C (Scheme 41) (Astruc 2007). More than 99% of the reactant was successfully converted to the product in 20 s in this process with high turnover frequency. The nanoparticles of Pd were stabilised in the micro-emulsion of water-$scCO_2$ medium and uniform dispersion of the nanoparticles there caused fast movement of the reaction to the product. Similar examples are also there, efficient hydrogenation of naphthalene and benzene using nanoparticles of Rhodium catalyst in $scCO_2$ (Ohde et al. 2004).

Many coupling reactions have been processed in $scCO_2$ so far. Stille coupling is one of the remarkable and exemplary of them using $Pd_2(dba)_3$ (dba—dibenzalacetone) and P $[3,5\text{-}(CF_3)_2C_6H_3]_3$. In this reaction, the product, vinyl-coupled-phenyl, was obtained quantitatively (>99%) using iodobenzene and vinyl(tributyl)tin as reagent in the supercritical solvent which is much higher than that obtained in conventional method using triphenylphosphine as ligand (Scheme 42) (Morita et al. 1998). The responsibility of this achievement has been attributed to the greater solubility of Pd complex in $scCO_2$, which was also supported by another Pd-catalysed popular coupling reaction, Heck reaction, using $PPh(C_6F_5)_2$—another fluorinated ligand (Fujita et al. 2002).

Base (DABCO—1,4-diazabicyclo[2.2.2]octane)-catalysed Morita–Baylis–Hillman reaction of *p*-nitrobenzaldehyde with an electrophilic alkene in $scCO_2$ was found with higher rate than in organic solvents. The pressure required in the experiment was as low as 80–100 bar and integrated dehydration of the product led to symmetrical ether (Scheme 43) (Rose et al. 2002). Unsymmetrical ethers was also generated using other alcohol at the etherification step.

In the field of polymer synthesis, the contribution of $scCO_2$ is just indispensable. For high mass transfer, diffusivity, low viscosity and tunable solvation property, $scCO_2$ has captured a unique possession in this field. In this medium, polymers can be synthesised smoothly via both step-growth and chain-growth polymerisation. It is very

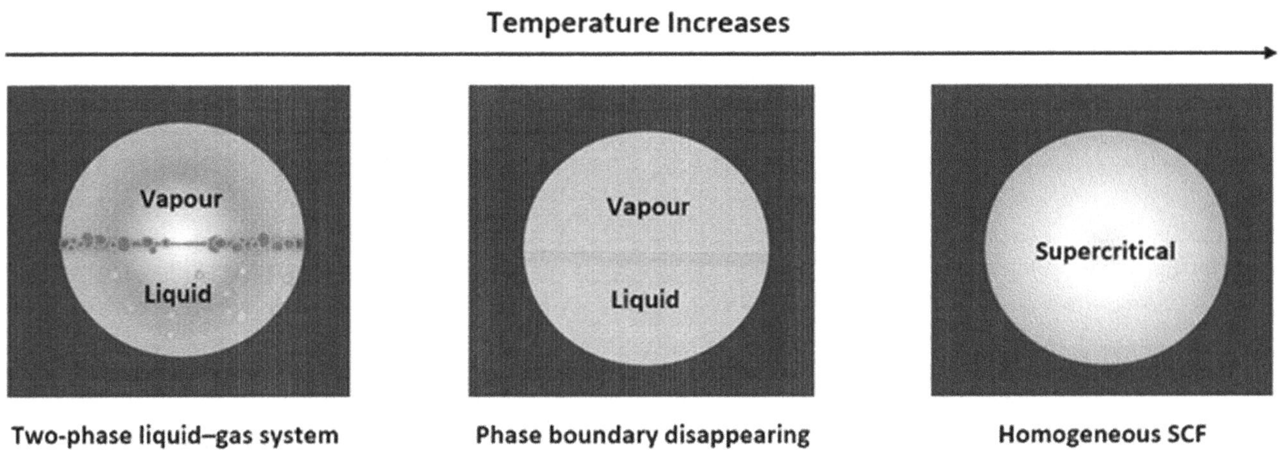

Fig. 8 Change in physical state of biphasic system towards supercritical temperature

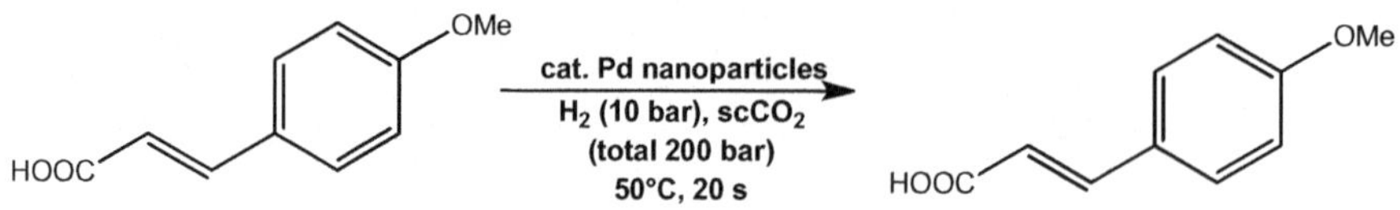

Scheme 41 Nano-Pd-catalysed hydrogenation of 4-methoxycinnamic acid in $scCO_2$

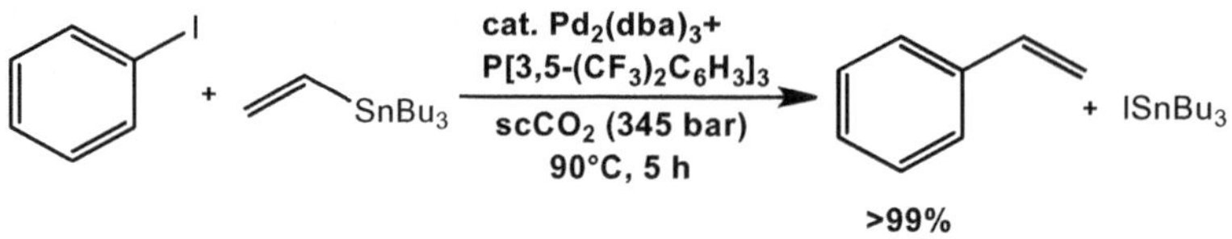

Scheme 42 Efficient Stille coupling in $scCO_2$

Scheme 43 DABCO-catalysed Morita–Baylis–Hillman reaction in $scCO_2$

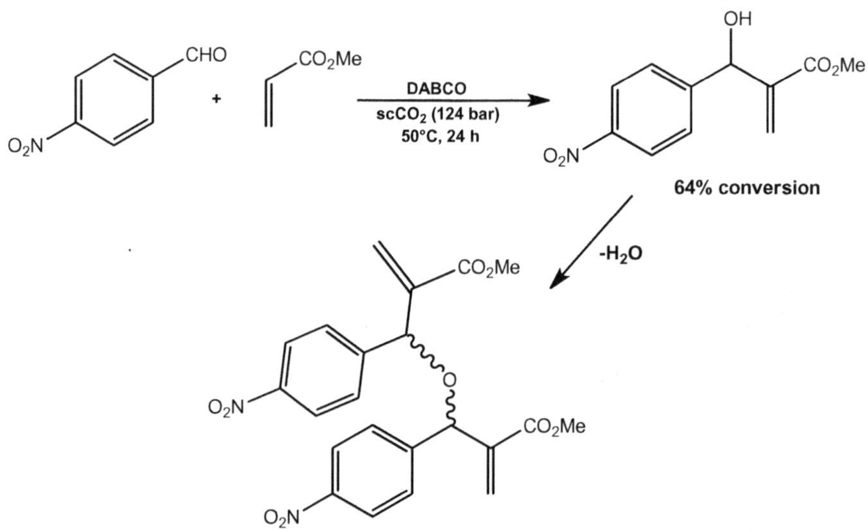

notable that after the synthesis, the solid product can be obtained only by simple depressurisation (Boyère et al. 2014).

The offering of $scCO_2$ in polymerisation reaction was realised towards the end of the last century. Till now, many reactions have been processed in the medium including RAFT (Reversible addition fragmentation chain transfer) polymerisation (Birkin et al. 2011, 2013; Zong et al. 2008). Many polyesters, polycarbonates, polyamides (nylons), polyurethanes, polyureas, poly(ether carbonate)s and poly (ether ester)s were easily synthesised in the supercritical fluid medium (Fig. 9) (Boyère et al. 2014). An important achievement by using this non-aqueous solvent, $scCO_2$, is the first homogeneous and heterogeneous synthesis of polymeric fluorinated polymers (e.g. polymerisation of 1,1-dihydrofluorooctyl acrylate) (Scheme 44) (Du et al. 2009). With the help of biotechnology, a novel strategy of bio-fuel production has been revealed on treatment of cellulosic materials, the most abundant renewable biopolymer, in this unique $scCO_2$ medium (Medina-Gonzalez et al. 2012). This is really a fascinating and unprecedented fact in the field of green chemistry. Metal-catalysed polymeric syntheses are also not an exception of improvement in same reaction medium (Islam et al. 2014a, b; Akbarinezhad 2014; Guironnet et al. 2009).

In the mid-1980s, the enzymatic activities were developed in $scCO_2$ and there are some merits of enzymatic reactions in this medium unlike water. The biocatalytic substances are insoluble in this medium, so easily separable from it after the reaction is over. As the medium possesses high diffusivity, the $scCO_2$/solid reaction will be more beneficial than in other liquid solvents. Moreover, the separation of the product is also easier than in aqueous solvent. Apart from these, the tunable physical properties of this solvent should not be forgotten ever in any adverse situation. One relevant thing is to be mentioned here that, a minimum volume of water is always present in the medium to prepare buffer solution (Mikami 2005; Matsuda 2013).

In this purpose, lipases have been used widely as biocatalytic substance (Dias et al. 2018). Recently, in a

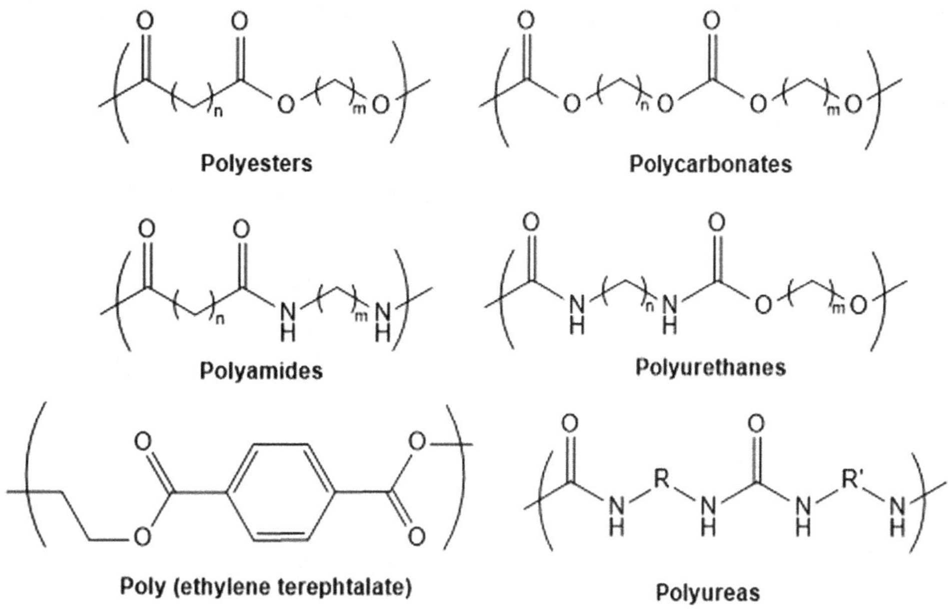

Fig. 9 Some polymers synthesized in the supercritical fluid medium

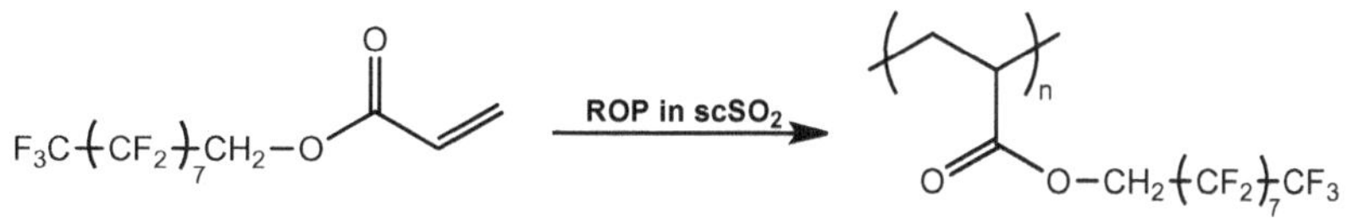

Scheme 44 Synthesis of polymeric fluorinated polymers by ROP in $scCO_2$

synthesis of a biodegradable star polycaprolactone by ring opening polymerisation (ROP), Novozym 435 (Lipase B from Candida Antarctica fixed on a solid support) was utilised as biocatalyst in $scCO_2$ effectively. In this reaction, monomeric ε-caprolactone (ε-CL) and D-sorbitol were used to synthesise the polymer star polycaprolactone. In presence of $scCO_2$, the conversion to the product was found very fast (96% conversion in 7 h) at 60 °C. The same reaction was also carried out in presence of conventional metal catalyst, tin(II) 2-ethylhexanoate [Sn(Oct)$_2$], and 96% conversion took place in that case too but after 56 h at 95 °C (Scheme 45) (Baheti et al. 2018).

Mixture of $scCO_2$ and IL is a highly admiring biphasic system for effective organic synthesis. ILs and $scCO_2$ both have tunable physical properties. ILs, polar solvent of no vapour pressure, can solubilise organic, inorganic and organometallic compounds. It can be miscible with $scCO_2$ at high temperature and so ILs are more compatible with $scCO_2$ in a two-phase system than any other solvents. Isolation of products from ILs can also be undergone by using $scCO_2$ smoothly without requiring any organic solvent. Metal-catalysed reactions get additional benefit of getting metal complexes dissolved in ILs to enhance the reaction rate (Pârvulescu and Hardacre 2007; Hallett and Welton 2011; Horng et al. 2016; Jessop et al. 2003).

There are striking evidences of effective biocatalytic organic synthesis also in the biphasic system (Mikami 2005; Sheldon and Pereira 2017). A fine instance is transesterification of racemic secondary alcohols in presence of lipase in $scCO_2$/[bmim][NTf$_2$] biphasic medium (Scheme 46) (Reetz et al. 2002; Sheldon 2008). The benefits of this reaction unlike the traditional methods were associated with easy product/enzyme, product/solvent separations and effortless recycling of the biocatalyst with no considerable loss of activity at 40 °C. A similar transesterification reaction was reported in the same year in presence of same enzyme in $scCO_2$/[bmim][NTf$_2$] and $scCO_2$/[bmim][PF$_6$] medium with equal benefits (Scheme 47) (Laszlo and Compton 2002).

Solvent-Free Synthesis

In chemistry, solvent is an essential part whenever any reaction is supposed to carry out. It makes reaction medium and takes care of transporting heat throughout the medium to cause the reaction happen effectively. It takes care of the solubilisation of solutes and rates and equilibrium status of reactions considerably. But, the solvent, the largest chemical species in any reaction, is also responsible for eco-unfriendliness of organic reactions in many ways. In following the guidelines of green chemistry, few hazardless and benign solvents have been discussed so far. But, it was

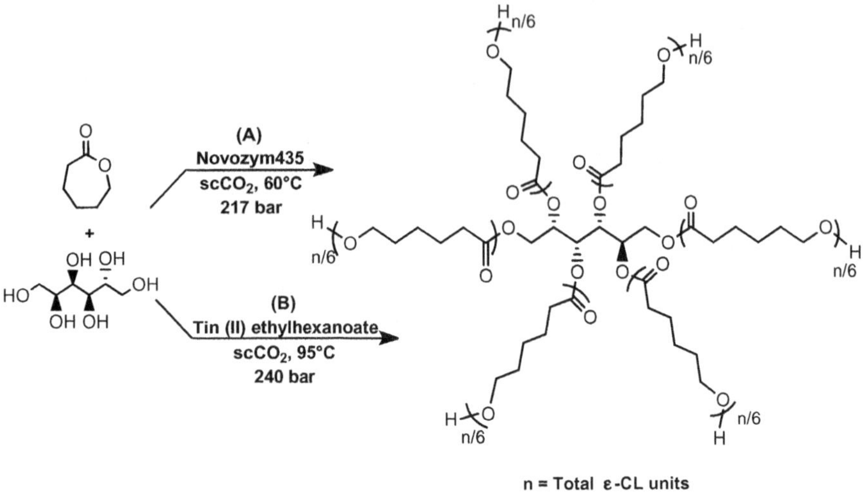

Scheme 45 Synthesis of star polycaprolactone by ROP in $scCO_2$ catalysed by (A) Novozym 435 (B) $Sn(Oct)_2$

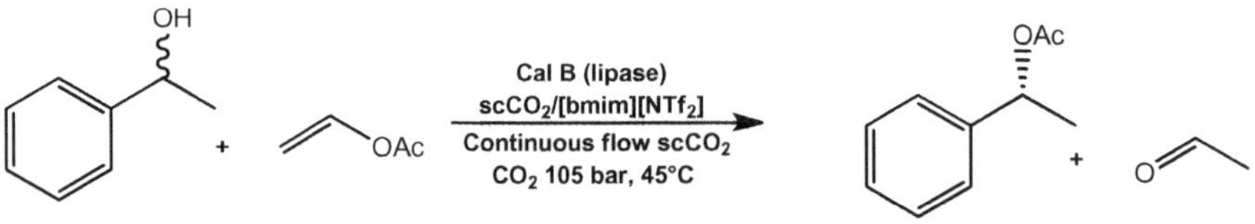

Scheme 46 Lipase-catalysed transesterification of racemic secondary alcohols in $scCO_2$/[bmim][NTf_2]

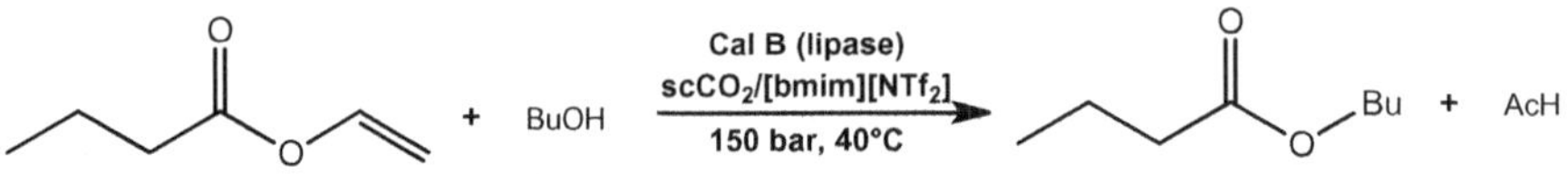

Scheme 47 Lipase-catalysed transesterification of butanol in $scCO_2$/[bmim][NTf_2]

said that "the best solvent is no solvent" to overcome the maximum unsustainability caused by solvents. So, solvent-free synthesis is supposed to be the best possible choice hitherto. It sounds weird to hear a reaction without solvent but it happened in many common and important reactions, for example, condensation of carbonyl compounds, cycloadditions, alkylations, aromatic substitutions, additions of amines, water, and alcohols, cyclisations, eliminations, rearrangements, C–C coupling, cascade reactions and catalysed reactions (Toda and Tanaka 2000).

There are many positive outcomes of solvent-free syntheses over the solvent-based reactions: (1) there is no question of any collection, purification, and recycling from reaction medium; (2) high yields are obtained (3) products are pure enough and sometimes no chromatographic purification, even recrystallisations, is necessary; (4) reactions are sometimes extraordinarily faster than conventional organic solvent-based reactions; (5) often no modern instrument and their setting up is required; (6) mostly energy requirement is also not too high; (7) preformed salts and metal-metalloid complexes are not in use often; (8) there is no need of protection-deprotection of functional groups; (i) batch experiments at low cost is possible more eco-friendly than conventional solvent-based methods. In association with these advantages some disadvantages, for example, formation of "hot spots" and the possibility of runaway reactions, are also present, which can be overcome technically (Cave et al. 2001; Dunk and Jachuck 2000).

The understanding of molecular movements and contacts between reactants of solvent-free synthetic reactions where reagents are solids at reaction temperature, unlike the miscible combinations of solid–liquid or liquid–liquid reagents, were quite complex. But, later, it was comprehended and demonstrated (Martins et al. 2009; Tanaka 2009). Actually, in case of solid–solid combination, minimum requirements are to be hosted to process the reaction. In solvent-free solid–solid reaction, four mechanisms are employed for activation (Fig. 10)—(1) mechanochemistry (grinding); (2) microwave irradiation (MW); (3) ultrasound irradiation (US); and (4) conventional thermal heating. In mechanochemistry, the solid solute substances are grinded physically by using mortar and pestle or by ball milling. In MW irradiation, microwave is employed to the solutes and the solutes absorb the waves to convert into thermal energy, which propagates through reaction mixture by dielectrical heating. MW irradiation is the most effective solvent-free solid–solid reaction process and in this internal heating process reaction rate increases, yield percentage increases significantly in comparison to the traditional heating method. In US technique, powerful ultrasound is used to irradiate the reagents till the product raising the reaction temperature of range 4900–5200 K by cavitation (Hobbs and Thomas 2007; Martins et al. 2009; Tanaka 2009). In the conventional thermal heating method, the reaction mixture is heated using magnetic stirring and oil bath. In this connection, it is worthy to mention that this conventional thermal heating is quite inefficient and sometimes unable to produce any yield (Martins et al. 2009; Tanaka 2009).

In last few decades, numerous reactions have been performed in solvent-free conditions (Chandna et al. 2013; Hernández and Juaristi 2010; Kanagaraj and Pitchumani 2010; Hasaninejed et al. 2012; Kumar and Sharma 2017; Li et al. 2012; Panja and Saha 2013; Singh and Chowdhury 2012; Wen et al. 2012; Zhanga and Wang 2012). Atom economy of these reactions are very high. Many organic molecules can very easily be synthesised by this method with less trouble.

Michael addition is an atom economic reaction. In 2011, a simple catalyst-free as well as solvent-free synthesis of nitroamines and nitrosulfides by following Michael addition were reported by Choudhary et al. In the effective experiment, high yields (>99%) were produced simply by grinding the mixtures of amines and thiols individually with nitroolefins (Scheme 48) (Choudhary and Peddinti 2011). The process is so effective that it ended within few minutes only at room temperature and requiring no purification of product.

In the same year, another solvent-free synthesis of poly-substituted quinolines under microwave irradiation was reported, in which a polyethylene glycol supported catalyst (PEG-6000) was used effectively. The speciality of the polymeric catalyst is its bio-degradability and modified by sulphuric acid, presenting as PEG-OSO_3H. In this Friedländer-type synthesis, poly-substituted quinolines and 4-aminoquinolines were synthesised under the said condition swiftly by condensing 2-aminoarylketones or anthranilonitrile with carbonyl compounds, unlike the conventional method (Scheme 49) (Hasaninejad et al. 2011). The

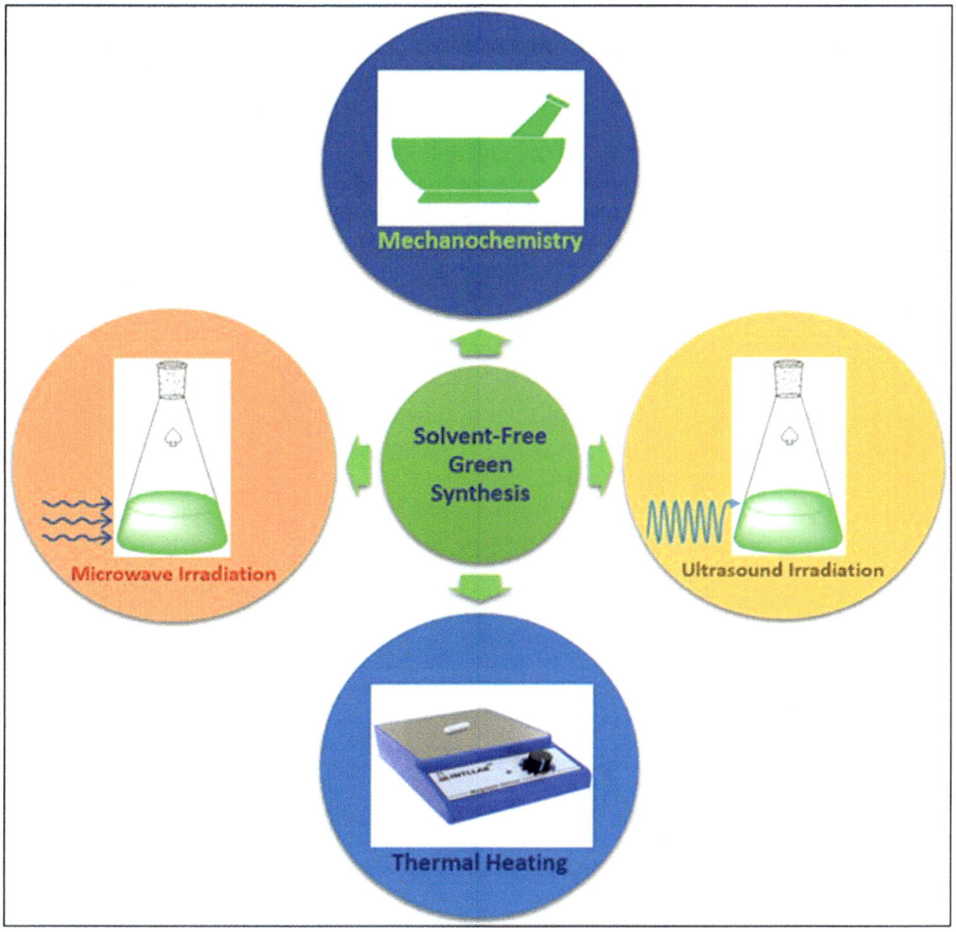

Fig. 10 Mechanisms of solid-solid solvent synthesis reactions

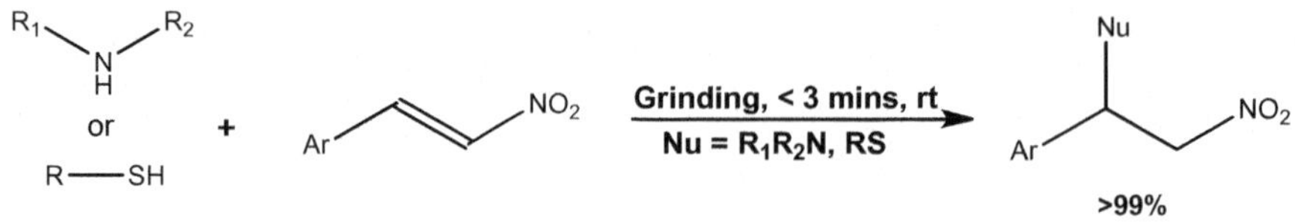

Scheme 48 Mechanochemical Michael addition of nitroamines and nitrosulfides in solvent-free pathway

maximum yields (80–95%) were obtained within 12 min, when 3 mol% of PEG-OSO_3H and 600 W of microwave irradiation was used at 130 °C in solvent-free condition. After the reaction was over, the catalyst could be recycled adding water in the medium followed by evaporation and the water-insoluble product was separated and recrystallised using ethanol. The recycled catalyst can be reused ten times without undergoing any loss of activation.

In 2012, Huskić et al. reported appealing mechanochemical solvent-free synthesis of nitrosobenzenes by oxidation of different para derivatives of aniline using potassium hydrogenperoxosulphate, also known as Oxone®, ($K_2SO_5.K_2SO_4.KHSO_4$) and featuring solvent-free purification of the product only by sublimation under reduced pressure ($\sim$0.1 mbar) (Scheme 50) (Huskić et al. 2012). From good to excellent yields (50–85%) were gained in the study based upon the para substituent of anilines. The para-iodoaniline yielded maximum. The yield % was found to be enhanced while adding equimolar extent of basic $NaHCO_3$ to the reaction vessel.

In 2013, another outstanding nano-S-catalysed solvent-free MCR was performed by Das and co-workers to synthesise 1 Amidoalkyl-2naphthols using mortar and pestle. In the preparation reaction, aldehyde derivatives, naphthol and amides were grinded in mortar using the synthesised S_8 nanoparticles and quantitative yields were produced in a while at 50 °C, unlike the solvent-based method using the conventional catalysts (Scheme 51) (Das et al. 2013). The nano-sized S_8 catalyst could be used another five times with no loss of its substantial activity. The determined green metrics supported the intensive greenness of this process.

Another efficient example of solvent-free heterocyclic compounds synthesis in room temperature catalysed by recyclable, in situ prepared, ZnO-NPs by ball-milling strategy was reported by Sharma and co-workers. Different

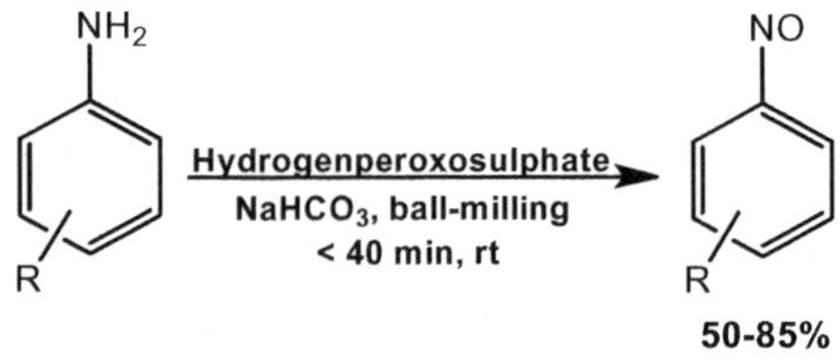

Scheme 50 Solvent-free synthesis of nitrosobenzenes by ball-milling process

derivatives of benzothiazole, benzimidazole and benzoxazole were synthesised with quantitative yields (> 85%) in 30 min in this experiment which are environment friendly with low E-factor. Various aldehydes were considered to react with 2-aminothiophenol, *o*-Phenylenediamine and 2-aminophenol to obtain benzothiazole, benzimidazole and benzoxazole, respectively (Scheme 52) (Sharma et al. 2014). But the substituents in aldehydes were not found to show any significant effect on the percentage of yields. It is noteworthy that turnover number (TON), turnover frequency (TOF) increased linearly with the amount of NPs but upto up to 0.5 mol%. Maximum yields were obtained at rpm 600 of ball-mill.

Solvent-free organic synthesis of heterocyclic compounds using renewable chemicals is a very exciting fact and demanding also nowadays. Such an interesting example of solvent-free and metal-free short time (15–40 min) synthesis of 1,4-dihydropyridines in presence of slight amount chitosan nanoparticles (NPs) was reported few years ago. The reaction was got going with Hantzsch synthesis using aldehyde, ethylacetoacetate, ammonium acetate and chitosan NPs in a round bottom flask fitted with a magnetic stirrer at 80 °C (Scheme 53) (Safari et al. 2015). The chitosan NPs were prepared by gelation of chitosan using heptamolybdate anions and then it was made suitable to use by drying with dry CO_2 for 30 min. In this solvent-free synthesis, the rate

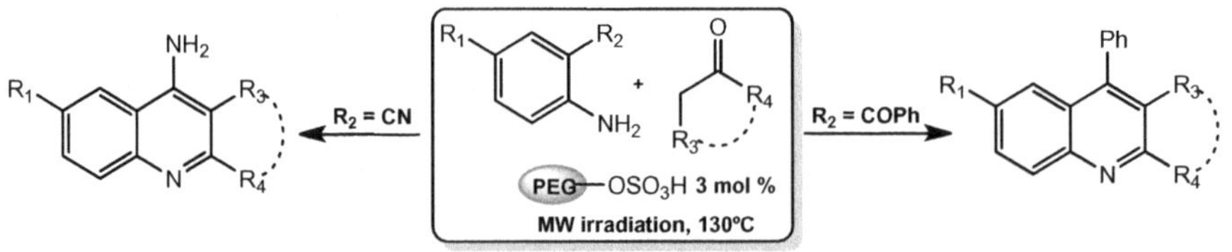

Scheme 49 MW-irradiated solvent-free synthesis of poly-substituted quinolines

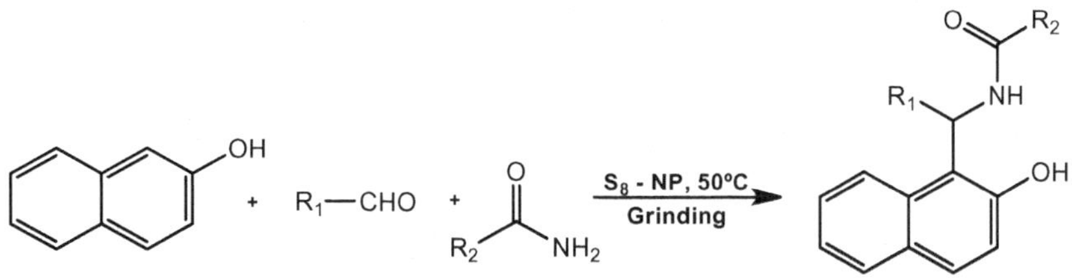

Scheme 51 Mechanochemical solvent-free MCR catalysed by nano-S

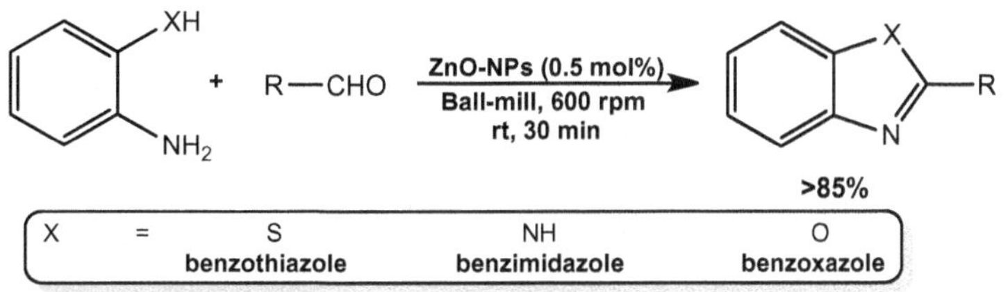

Scheme 52 ZnO-NPs catalysed solvent-free synthesis of heterocyclic compounds by ball milling

and yields (~ 90%) both were higher than that of solvent-based synthesis.

Paal-Knorr synthesis is a very popular organic reaction for heterocyclic compounds synthesis. Cho and co-workers reported this reaction for synthesising underivatised pyrroles by reaction of hexan-2,5-diones with aqueous NH_4OH in catalyst and solvent-free condition assisted by MW irradiation (80 °C) and US technique (room temperature) (Scheme 54) (Cho et al. 2015). In conventional method, about 120 h is required to get quantitative product but in this present method, the same was obtained in less than 1 h. MW irradiation was discovered more effective than US strategy in this process. Similarly, efficient solvent-free synthesis of N-substituted pyrroles was also reported using hexan-2,5-diones and amines by the same group with more yields, even faster than the previous, but without using MW irradiation or US strategy.

There are significant drawbacks of mechanochemistry—lack of temperature control and scalability—for which twin screw extrusion (TSE) was introduced. The melting temperatures of reactants are very crucial in reactions especially in solvent-free synthesis. TSE involves moving of the substances through a narrow restricted area by the rotation of a pair of intermeshing, modular screws to get the desired condition. Already, it has extensive use in food, polymer and pharmaceutical industries, including reactive extrusion of polymers (Crawford et al. 2017). In 2017, Crawford and co-workers studied solvent-free Knoevenagel condensation, a benchmark reaction, of barbituric acid with aldehydes (vanillin, veratraldehyde and 5-bromovanillin) by TSE (Scheme 55). In conventional solvent-based method, these reactions take around 50 days to reach equilibrium. But, in this ball-milling process, the reaction-completion times were not more than 90 min at 25 Hz. TSE takes care of the reaction temperature, and in these reactions, maximum yields were produced at 160 °C at 55 rpm screw speed. The same group also reported efficient solvent-free Aldol condensation and Michael addition using same apparatus (Crawford et al. 2017).

Use of no solvent is the best way in the green synthesis. In case of sustainable biocatalytic reactions also, an extra benefit can be achieved under solvent-free condition. If one or more of the reagents in biocatalytic reactions is a liquid then mixing of substrate and enzyme is as simple as reactions in non-aqueous solvents. But, for solid substrates and solid catalysts, then, the reaction will not be accessed smoothly. The lack of miscibility of substrate and enzyme will limit the rate and yield of the reaction. There are two approaches in this solvent-free biocatalytic reactions—“heterogeneous eutectic” and “solid-to-solid” reactions.

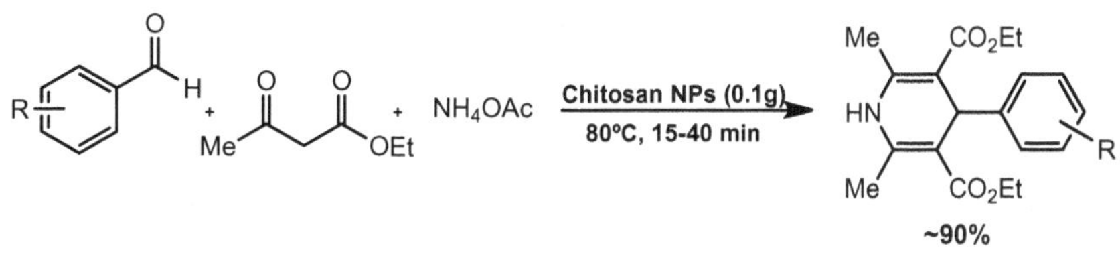

Scheme 53 Solvent-free Hantzsch synthesis by thermal heating catalysed by chitosan-NPs

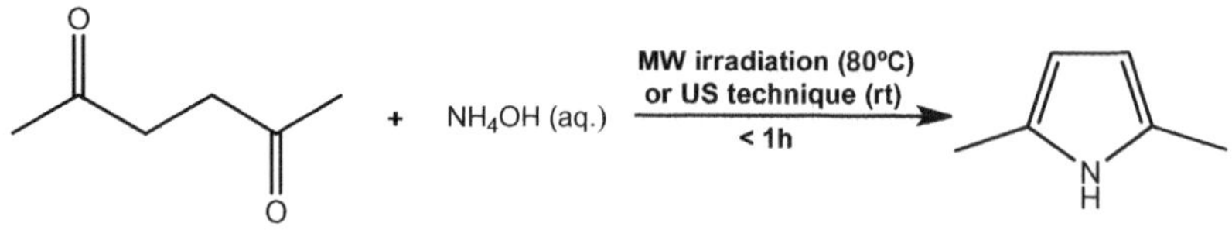

Scheme 54 Solvent-free Paal-Knorr synthesis by MW irradiation or US technique

Scheme 55 Solvent-free Knoevenagel condensation using TSE technique

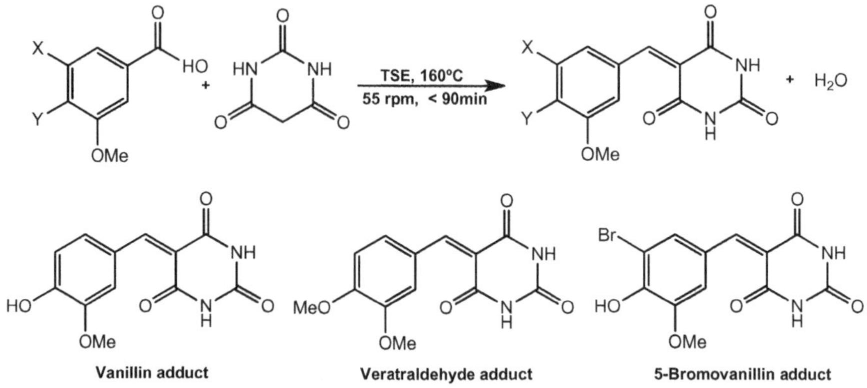

A minimum extent liquid phase is essential in both the approaches, so that enzyme can conduct the reaction. In heterogeneous eutectic approach, enzyme gets compatible with the low melting (in general, <60 °C) eutectic mixture of substrates (or in combination with very little adjuvant, for example, water or organic compound). In the solid-to-solid reaction, as the reactants are mostly in the solid state, an added liquid phase is required for the reaction to occur. Both the approaches are used extensively in organic synthesis (Hobbs and Thomas 2007). To provide enzymes enough resistance to influential conditions, for example, pH or temperature enzyme immobilisation is very effective, especially for commercial uses. Enzyme immobilisation involves modification of enzyme by binding to a solid support (carrier), entrapment and cross-linking (Sheldon and Pereira 2017). Though, immobilisation of enzymes are not fully free of disadvantages. Enzymes have already been used in many synthesis reactions in immobilised condition (Hobbs and Thomas 2007).

In 2012, a solvent-free interesterification of acrylate derivatives was catalysed by a lipase enzyme in a synthetic process. Actually, allyl and dichloropropyl acrylates were synthesised in the process from allyl and dichloropropyl dodecanoates using whole cells (fungal resting cells) supported lipase-based commercial biocatalysts (e.g. CALB—Candida antarctica lipase B) at temperature below 50 °C under solvent-free condition (Scheme 56) (Varón et al. 2012). Quantitative yields of chlorohydrin acrylates were produced in this green process which was a maiden achievement and more advantageous than conventional methods. There are examples where lipases have been also used towards feasible and scalable solvent-free enzymatic polycondensations (Pellisa et al. 2015).

Few years ago, Liu et al. proposed solvent-free enantioselective aldol reactions of isatin derivatives with cyclic ketones catalysed by Nuclease P1 extracting from Penicillium citrinum (Scheme 57). The effective reaction was discovered to be enantioselective in association with high yielding (up to 95%) at mild condition. In conventional solvent-based methods, poor yields were produced under same reaction conditions. Harnessing Nuclease in this biocatalytic organic synthesis widens its applicability in pharmaceutical industry also (Liu et al. 2014).

4 Future Trends

According to the Nobel Laureate Professor Ryoji Noyori "Green Chemistry is not just a mere catch phrase; it is the key to the survival of mankind." (Noyori 2010) In addition, the long discussion about organic synthesis with greener approaches, the dependency of chemical reaction upon solvent is understood in essence. Therefore, discovering more appropriate solvent or solvent mixture from the hidden treasure of chemistry is a prime task to climb the summit of transcendence of green chemistry in the forthcoming future.

The hydrophobic effect and trans-phase activity of water ameliorated organic synthetic path with a new version. It provided mild reaction condition, good reactivity, selectivity, easy separation and recycling of catalysts. But, uncommon reactivity and reverse selectivity of water unlike organic solvents limited its further use in synthesis. Even, its

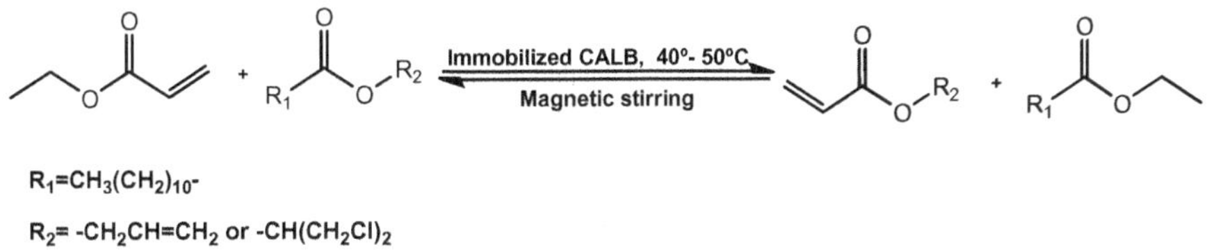

Scheme 56 Lypase-catalysed solvent-free interesterification of acrylate derivatives by thermal by magnetic stirring

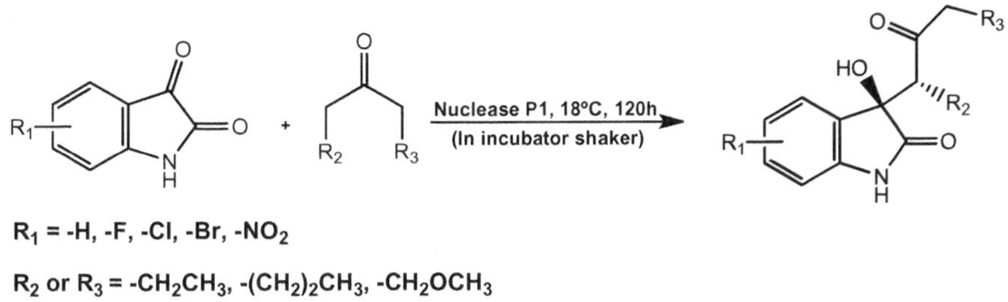

Scheme 57 Nuclease P1-catalysed solvent-free enantioselective aldol reactions of isatin derivatives

nontoxicity could not make it completely green solvent. On the contrary, the supramolecular aspects of micellar aggregation have drawn a great attention in the field of green synthesis (Acharjee et al. 2019a, b, 2020; Rakshit et al. 2020). Contribution of surfactants to almost all classes of chemical transformations, as medium as well as catalyst of reaction, is just undeniable. Sophisticate optimisation of chemical nature, concentration and molar ratio of surfactants can greatly influence the desired outcome of reaction. Unique combination of catalyst, substrate and surfactant can result some incredible happenings. Recently, preparation of molecular capsule by new surfactants and their photoactive encapsulation property has gained potential applications in many technological fields, in catalysis and drug delivery (Moscoso and Ballester 2017; Beaudoin et al. 2016). In future, further research is necessary for some more multiple applications, understanding and development of photocontrolled molecular capsules. In heterogeneous catalysis, stabilisation of nanoparticles with controlled shape and its fate in micellar media is attracting researchers for more research in near future (Yin et al. 2015; Xi et al. 2018). Another very important research is to discover suitable reversible bio-surfactants for upgraded biocatalytic organic synthesis to mimic green metabolic reactions in aqueous medium in living organisms.

ILs are getting more attention, as non-aqueous media, than fluorous media. The tunable property of ILs by selecting appropriate ion-pairs has made them so unique. Enzymes are very stable in some ILs for which they are suitable for enzymatic reactions. Many organic compounds are immiscible in ILs unlike fluorous solvents. Therefore, ILs are very satisfactorily acceptable in biphasic reaction. Whereas, in case of FBS, volatile organic solvents are to be consumed for organic product separation. Hence, ILs are supposed to be getting wider appreciation in near future in green synthesis including industrial purpose than fluorous media. The high price of fluorous solvent is also a limitation of using it. The coupling of $scCO_2$ with ILs is a great combination in harnessing green synthesis. Some liquid polymers, for example, poly(ethyleneglycol) (PEG), poly (propyleneglycol) (PPG) or poly(tetrahydrofuran) (PTHF), being biodegradable and widely tunable may be alternative to slightly biodegradable ILs (Lozano 2010). CO_2 in its supercritical state with tunable solvent property and nontoxicity has occupied a remarkable position as a green solvent. But, in enzymatic reaction, the carbamate formation and pH reduction due to carbonic acid formation have created problems (Lozano 2010). So, some scope of research is there to find way-out to overcome the problems. Synthesis of solid nanocomposites, assembly of metal-organic framework and amphiphiles may solve many challenging problems hereafter in $scCO_2$/IL biphasic medium (Zhang et al. 2014, 2016; Ivanova et al. 2015).

In the wake of this, one thing should not be forgotten that solvent-free synthesis is more successful than solvent-based system. In biocatalytic reactions also no exception happened. But, in following solvent-based biocatalysis in living organisms, finding some better green solvents is of utmost importance and more exquisite research is necessary to unfold this fact.

References

Abe Y, Yagi Y, Hayase S, Kawatsura M, Itoh T (2012) Ionic liquid engineering for lipase-mediated optical resolution of secondary alcohols: design of ionic liquids applicable to ionic liquid coated-lipase catalyzed reaction. Ind Eng Chem Res 51:9952–9958

Acharjee A, Rakshit A, Chowdhury S, Malik S, Barman MK, Ali MA, Saha B (2019a) Micellar catalysed and heteroaromatic base promoted rate enhancement of oxidation of an alicyclic alcohol in aqueous medium. J. Mol. Liq. 277:360–371. https://doi.org/10.1016/j.molliq.2018.12.082

Acharjee A, Rakshit A, Chowdhury S, Datta I, Barman MK, Ali MA, Saha B (2019b) Micellar catalysed oxidation of hydrophobic fatty alcohol in aqueous medium. J Mol Liq 293:111475. https://doi.org/10.1016/j.molliq.2019.111475

Acharjee A, Rakshit A, Chowdhury S, Ali MA, Singh B, Saha B (2020) Mixed anionic-nonionic micelle catalysed oxidation of aliphatic alcohol in aqueous medium. J Mol Liq 303:112655. https://doi.org/10.1016/j.molliq.2020.112655

Adlercreutz P (2013) Immobilisation and application of lipases in organic media. Chem Soc Rev 42:6406–6436. https://doi.org/10.1039/C3CS35446F

Akbarinezhad E (2014) Synthesis of conductive polyaniline–graphite nanocomposite in supercritical CO_2 and its application in zinc-rich epoxy primer. J Supercrit Fluids 94:8–16. https://doi.org/10.1016/j.supflu.2014.06.018

Akiyama T (2007) Stronger Brønsted acids. Chem Rev 107(12):5744–5758. https://doi.org/10.1021/cr068374j

Akiyama T, Mori K (2015) Stronger Brønsted acids: recent progress. Chem Rev 115(17):9277–9306. https://doi.org/10.1021/acs.chemrev.5b00041

Anastas P, Eghbali N (2010) Green chemistry: principles and practice. Chem Soc Rev 39:301–312. https://doi.org/10.1039/b918763b

Astruc D (2007) Palladium nanoparticles as efficient green homogeneous and heterogeneous carbon−carbon coupling precatalysts: a unifying view. Inorg Chem 46(6):1884–1894. https://doi.org/10.1021/ic062183h

Azizi N, Saidi MR (2005) Highly chemoselective addition of amines to epoxides in water. Org Lett 7:3649

Baheti P, Gimello O, Bouilhac C, Desmazes PL, Howdle SM (2018) Sustainable synthesis and precise characterisation of bio-based star polycaprolactone synthesised with a metal catalyst and with lipase. Polym Chem 9:5594. https://doi.org/10.1039/c8py01266k

Beaudoin D, Rominger F, Mastalerz M (2016) Chirality-assisted synthesis of a very large octameric hydrogenbonded capsule. Angew Chem Int Ed 55:1–6. https://doi.org/10.1002/anie.201609073

Beier P, O'Hagan D (2002) Enantiomeric partitioning using fluorous biphase methodology for lipase-mediated (trans)esterifications. Chem Commun 1680–1681

Bernini R, Mincione E, Coratti A, Fabrizi G, Battistuzzi G (2004) Epoxidation of chromones and flavonoids in ionic liquids. Tetrahedron 60:967.

Biasutti MA, Abuin EB, Silber JJ, Correa NM, Lissi EA (2008) Kinetics of reactions catalyzed by enzymes in solutions of surfactants. Adv Colloid Interface Sci 136:1–24. https://doi.org/10.1016/j.cis.2007.07.001

Birkin NA, Arrowsmith NJ, Park EJ, Richez AP, Howdle SM (2011) Synthesis and application of new CO_2-soluble vinyl pivalate hydrocarbon stabilisers via RAFT polymerisation. Polym Chem 2:1293. https://doi.org/10.1039/C1PY00062D

Birkin NA, Wildig OJ, Howdle SM (2013) Effects of poly(vinyl pivalate)-based stabiliser architecture on CO_2-solubility and stabilising ability in dispersion polymerisation of N-vinyl pyrrolidone. Polym Chem 4:3791. https://doi.org/10.1039/c3py00275f

Bortolini O, Conte V, Chiappe C, Fantin G, Fogagnolo M, Maietti S (2002) Epoxidation of electrophilic alkenes in ionic liquids. Green Chem 4:94

Boyère C, Jérôme C, Debuigne A (2014) Input of supercritical carbon dioxide to polymer synthesis: an overview. Eur Polym J 61:45–63. https://doi.org/10.1016/j.eurpolymj.2014.07.019

Breslow R (1991) Hydrophobic effects on simple organic reactions in water. Acc Chem Res 24:159

Brittain SM, Ficarro SB, Brock A, Peters EC (2005) Enrichment and analysis of peptide subsets using fluorous affinity tags and mass spectrometry. Nat Biotechnol 23:463–468. https://doi.org/10.1038/nbt1076

Calò V, Nacci A, Monopoli A, Montingelli F (2005) Pd Nanoparticles as efficient catalysts for suzuki and stille coupling reactions of aryl halides in ionic liquids. J Org Chem 70(15):6040–6044. https://doi.org/10.1021/jo050801q

Cao C, Matsuda T (2016) Chapter 3—Biocatalysis in organic solvents, supercritical fluids and ionic liquids, organic synthesis using biocatalysis, pp 67–97. https://doi.org/10.1016/B978-0-12-411518-7.00003-2

Cave GWV, Raston CL, Scott JL (2001) Recent advances in solventless organic reactions: towards benign synthesis with remarkable versatility. Chem Commun 21:2159

Chanda A, Fokin VV (2009) Organic synthesis "On Water." Chem Rev 109:725–748. https://doi.org/10.1021/cr800448q

Chandna N, Chandak N, Kumar P, Kapoor JK, Sharma PK (2013) Metal- and solvent-free synthesis of N-sulfonylformamidines. Green Chem 15:2294. https://doi.org/10.1039/c3gc40797g

Cho H, Madden R, Nisanci B, Török B (2015) The Paal-Knorr reaction revisited a catalyst and solvent-free synthesis of underivatized and N-substituted pyrroles. Green Chem 17:1088–1099. https://doi.org/10.1039/C4GC01523A

Choudhary G, Peddinti RK (2011) An expeditious, highly efficient, catalyst-free and solvent-free synthesis of nitroamines and nitrosulfides by Michael addition. Green Chem 13:276. https://doi.org/10.1039/c0gc00830c

Chowdhury S, Rakshit A, Acharjee A, Saha B (2019a) Novel amphiphiles and their applications for different purposes with special emphasis on polymeric surfactants. Chem Sel 4:1–19. https://doi.org/10.1002/slct.201901160

Chowdhury S, Rakshit A, Acharjee A, Ghosh A, Mahali K, Saha B (2019b) Ru(III) catalysed oxidation of 2-propanol by Cr(VI) in micellar media. J Mol Liq 290:111247. https://doi.org/10.1016/j.molliq.2019.111247

Chowdhury S, Rakshit A, Acharjee A, Mahali K, Saha B (2020b) Surface phenomenon in micellar media: an excellent controlling factor for oxidation of fatty aldehyde in aqueous medium. J Mol Liq 310:113224. https://doi.org/10.1016/j.molliq.2020.113224

Chowdhury S, Rakshit A, Acharjee A, Ghosh A, Mahali K, Saha B (2020a) Trivalent ruthenium and iridium salt: excellent homogeneous catalysts for alicyclic alcohol oxidation in micellar media. Tenside Surfactant 57:298–309

Cioc RC, Ruijter E, Orru RVA (2014) Multicomponent reactions: advanced tools for sustainable organic synthesis. Green Chem 16:2958. https://doi.org/10.1039/c4gc00013g

Crawford DE, Miskimmin CKG, Albadarin AB, Walker G, James SL (2017) Organic synthesis by Twin Screw Extrusion (TSE): continuous, scalable and solvent-free. Green Chem 19:1507–1518. https://doi.org/10.1039/C6GC03413F

Daneshjoo S, Akbari N, Sepahi AA, Ranjbar B, Khavarinejad R-A, Khajeh K (2011) Imidazolium chloride-based ionic liquid-assisted improvement of lipase activity in organic solvents. Eng Life Sci 11:259–263

Darío Falcone R, Alicia Biasutti M, Mariano Correa N, Silber JJ, Lissi E, Abuin E (2004) Effect of the addition of a nonaqueous polar solvent (glycerol) on enzymatic catalysis in reverse micelles.

Hydrolysis of 2-naphthyl acetate by r-chymotrypsin. Langmuir 20:5732–5737. https://doi.org/10.1021/la036243x
Das VK, Borah M, Thakur AJ (2013) Piper-betle-shaped nano-S-catalyzed synthesis of 1-amidoalkyl-2naphthols under solvent-free reaction condition: a greener, "Nanoparticle-Catalyzed Organic Synthesis Enhancement" approach. J Org Chem 78 (7):3361–3366. https://doi.org/10.1021/jo302682k
Dey J, Saha M, Kumar Pal A, Ismail K (2013) Regioselective nitration of aromatic compounds in an aqueous sodium dodecylsulfate and nitric acid medium. RSC Adv (2013) 3:18609
Dias ALB, Santos PD, Martínez J (2018) Supercritical CO_2 technology applied to the production of flavor ester compounds through lipase-catalyzed reaction: a review. J. CO_2 Util 23:159–178. https://doi.org/10.1016/j.jcou.2017.11.011
Dicks AP, Hent A (2015) Atom economy and reaction mass efficiency. In: Green chemistry metrics, springer briefs in molecular science. Springer, Cham, pp 17–44. https://doi.org/10.1007/978-3-319-10500-0_2
Dong X, Fan Y, Zhang H, Zhong Y, Yang Y, Miao J, Hua S (2016) Inhibitory Effects of ionic liquids on the lactic dehydrogenase activity. Int J Biol Macromol 86:155–161
Du L, Kelly JY, Roberts GW, De Simone JM (2009) Fluoropolymer synthesis in supercritical carbon dioxide. J Supercrit Fluids 47 (3):447–457
Dunk B, Jachuck R (2000) A novel reactor for UV irradiated reactions. Green Chem 2:G13. https://doi.org/10.1039/B000158I
Dunn PJ, Wells AS, Williams MT (2010a) Introduction to green chemistry, organic synthesis and pharmaceuticals. In: Green chemistry in the pharmaceutical industry, Wiley, pp 1–18
Dunn PJ, Wells AS, Williams MT (2010b) Green chemistry metrics, in green chemistry in the pharmaceutical industry. Wiley, 21–46
Dunn (2012) The importance of green chemistry in process research and development. Chem Soc Rev 41:1452–1461. https://doi.org/10.1039/c1cs15041c
Dwars T, Paetzold E, Oehme G (2005) Reactions in micellar systems. Angew. Chem Int Ed 44:7174–7199. https://doi.org/10.1002/anie.200501365
Eckert CA, Knutson BL, Debenedetti PG (1996) Supercritical fluids as solvents for chemical and materials processing. Nature 383:313–318
Firin ŸZ, Demirkol O, Akbaflar D, Giray ES (2013) Clean and efficient synthesis of flavanone in sub-critical water. J Supercrit Fluids 217–220. https://doi.org/10.1016/j.supflu.2013.05.014
Firouzabadi H, Iranpoor N, Gholinejad M (2009) 2-Aminophenyl diphenylphosphinite as a new ligand for heterogeneous palladium-catalyzed Heck–Mizoroki reactions in water in the absence of any organic co-solvent. Tetrahedron 65:7079
Fujita S, Yuzawa K, Bhanage BM, Ikushima Y, Arai MJ (2002) Palladium-catalyzed heck coupling reactions using different fluorinated phosphine ligands in compressed carbon dioxide and conventional organic solvents. Mol Catal A 180:35
Giacalone F, Gruttadauria M, Meo PL, Riela S, Noto R (2008) New simple hydrophobic proline derivatives as highly active and stereoselective catalysts for the direct asymmetric aldol reaction in aqueous medium. Adv Synth Catal 350:2747
Goto K, Miura T, Mizuno M (2005) Synthesis of peptides and oligosaccharides by using a recyclable fluorous tag. Tetrahedron Lett 46:8293–8297. https://doi.org/10.1016/j.tetlet.2005.09.160
Guironnet D, Friedberger T, Mecking S (2009) Ethylene polymerization in supercritical carbon dioxide with binuclear nickel(II) catalysts. Dalton Trans 8929–893. https://doi.org/10.1039/b912883b
Hallett JP, Welton T (2011) Room-temperature ionic liquids: solvents for synthesis and catalysis. 2. Chem Rev 111:3508–3576. https://doi.org/10.1021/cr1003248
Hasaninejad A, Zare A, Shekouhya M, Rada JA (2011) Sulfuric acid-modified PEG-6000 (PEG-OSO_3H): an efficient, biodegradableandreusablepolymericcatalystforthesolvent-freesynthesis of poly-substituted quinolines under microwave irradiation. Green Chem 13:958. https://doi.org/10.1039/c0gc00953a
Hasaninejed A, Kazerooni MR, Zare A (2012) Solvent-free, one-pot, four-component synthesis of 2H-indazolo[2,1-b]phthalazine-triones using sulfuric acid-modified PEG-6000 as a green recyclable and biodegradable polymeric catalyst. Catal Today 196:148–155. https://doi.org/10.1016/j.cattod.2012.05.026
Hayashi Y, Urushima T, Aratake S, Okano T, Obi K (2007) Organic solvent-free, enantio- and diastereoselective, direct mannich reaction in the presence of water. Org Lett 10:21
He X, Chan TH (2006) New non-volatile and odorless organosulfur compounds anchored on ionic liquids. Recyclable reagents for swern oxidation. Tetrahedron 62:3389–3394
Hernández JG, Juaristi E (2010) Green synthesis of r, β- and β, β-dipeptides under solvent-free conditions. J Org Chem 75:7107–7111. https://doi.org/10.1021/jo101159a
Hobbs HR, Thomas NR (2007) Biocatalysis in supercritical fluids, in fluorous solvents, and under solvent-free conditions. Chem Rev 107:2786–2820. https://doi.org/10.1021/cr0683820
Horng RS, Lee SK, Hsu WT, Wu TA (2016) Pressure influences on CO_2 reaction with cyclohexylamine using ionic liquid as reaction medium and catalyst. J Chin Inst Eng 39(7):876–881. https://doi.org/10.1080/02533839.2016.1203734
Housaindokht MR, Monhemi H (2013) The open lid conformation of the lipase is explored in the compressed gas: new insights from molecular dynamic simulation. J Mol Catal B Enzym 87:135–138
Hu X, Manetsch R (2010) Kinetic target-guided synthesis. Chem Soc Rev 39:1316
Hu X, Shu XS, Li XW, Liu S-G, Zhang Y-Y, Zou G-L (2006) Hemoglobin-biocatalyzed synthesis of conducting polyaniline in micellar solutions. Enzyme Microb Technol 38:675–682. https://doi.org/10.1016/j.enzmictec.2005.08.006
Huskić I, Halasz I, Friščić T, Vančik H (2012) Mechanosynthesis of nitrosobenzenes: a proof-of-principle study in combining solvent-free synthesis with solvent-free separations. Green Chem 14:1597. https://doi.org/10.1039/c2gc35410a
Islam MT, Haldorai Y, Nguyen VH, Islam MN, Ra CS, Shim JJ (2014) Controlled radical polymerization of vinyl acetate in supercritical CO_2 catalyzed by CuBr/terpyridine. Korean J Chem Eng 31 (6):1088–1094. https://doi.org/10.1007/s11814-014-0031-5
Islam MN, Haldorai Y, Nguyen VH, Shim JJ (2014) Synthesis of poly (vinyl pivalate) by atom transfer radical polymerization in supercritical carbon dioxide. Eur Polym J 61:93–104. https://doi.org/10.1016/j.eurpolymj.2014.09.003
Isley NA, Linstadt RTH, Slack ED, Lipshutz BH (2014) Copper-catalyzed hydrophosphinations of styrenes in water at room temperature. Dalton Trans 43:13196
Isley NA, Dobarco S, Lipshutz BH (2014) Installation of protected ammonia equivalents onto aromatic & heteroaromatic rings in water enabled by micellar catalysis. Green Chem 16:1480
Itoh T (2017) Ionic liquids as tool to improve enzymatic organic synthesis. Chem Rev 117:10567–10607. https://doi.org/10.1021/acs.chemrev.7b00158
Ivanova M, Kareth S, Spielberg ET, Mudring AV, Petermann M (2015) Silica ionogels synthesized with imidazolium based ionic liquids in presence of supercritical CO_2. J Supercrit Fluids 105:60–65. https://doi.org/10.1016/j.supflu.2015.01.014
Jessop PG, Stanley RR, Brown RA, Eckert CA, Liotta CL, Ngo TT, Pollet P (2003) Neoteric solvents for asymmetric hydrogenation: supercritical fluids, ionic liquids, and expanded ionic liquids. Green Chem 5:123–128. https://doi.org/10.1039/b211894g

Kanagaraj K, Pitchumani K (2010) Solvent-free multicomponent synthesis of pyranopyrazoles: per-6-amino-β-cyclodextrin as a remarkable catalyst and host. Tetrahedron Lett 51:3312–3316. https://doi.org/10.1016/j.tetlet.2010.04.087

Kumar A, Sharma S (2017) A grinding-induced catalyst- and solvent-free synthesis of highly functionalized 1,4-dihydropyridines via a domino multicomponent reaction. Green Chem 2011:13. https://doi.org/10.1039/c1gc15223h

Kumar D, Seth K, Kommi DN, Bhagat S, Chakraborti AK (2013) Surfactant micelles as microreactors for the synthesis of quinoxalines in water: scope and limitations of surfactant catalysis. RSC Adv 3:15157

Kumar A, Kumar Gupta M, Kumar M, Saxena D (2013) Micelle promoted multicomponent synthesis of 3-amino alkylated indoles via a Mannich-type reaction in water. RSC Adv 3:1673

Laszlo JA, Compton DL (2002) Adv Chem Ser 818:387

Laville L, Charnay C, Lamaty F, Martinez J, Colacino E (2012) Chem Eur J 18:760

Li CJ (2005) Organic reactions in aqueous media with a focus on carbon−carbon bond formations: a decade update. Chem Rev 105:3095–3166. https://doi.org/10.1021/cr030009u

Li M, Cao H, Wang Y, Lv XL, Wen LR (2012) One-pot multicomponent cascade reaction of N,S-ketene acetal: solvent-free synthesis of imidazo[1,2-a]thiochromeno[3,2-e]pyridines. Org Lett 14(13). https://doi.org/10.1021/ol301441v

Li X, Zhang C, Li S, Huang H, Hu Y (2015) Improving catalytic performance of candida rugosa lipase by chemical modification with polyethylene glycol functional ionic liquids. Ind Eng Chem Res 54:8072−8079

Lin L, Li Y, Zhang S, Li S (2011) Enhancing activity of suzuki reactions in water by using guanidinium ionic liquid stabilized palladium micelle catalyst. Synlett 1779

Linstadt RTH, Peterson CA, Lippincott DJ, Jette CI, Lipshutz BH (2014) Stereoselective silylcupration of conjugated alkynes in water at room temperature. Angew Chem Int Ed 53:4159

Lipshutz BH, Ghorai S, Wen W, Leong T, Taft BR (2011) Manipulating micellar environments for enhancing transition metal-catalyzed cross-couplings in water at room temperature. J Org Chem 76:5061.

Liu ZQ, Xiang ZW, Shen Z, Wu Q, Lin XF (2014) Enzymatic enantioselective aldol reactions of isatin derivatives with cyclic ketones under solvent-free conditions. Biochimie 101:156–160. https://doi.org/10.1016/j.biochi.2014.01.006

Lozano P (2010) Enzymes in neoteric solvents: from one-phase to multiphase systems. Green Chem 12:555–569. https://doi.org/10.1039/b919088k

Machida H, Takesue M, Smith RL Jr (2011) Green chemical processes with supercritical fluids: properties, materials, separations and energy. J Supercrit Fluids 60:2–15. https://doi.org/10.1016/j.supflu.2011.04.016

Mamidyala SK, Finn MG (2010) In situ click chemistry: probing the binding landscapes of biological molecules. Chem Soc Rev 39:1252

Marques MPC, Lourenço NMT, Fernandes P, de Carvalho CCCR (2012) Green solvents for biocatalysis. In: Mohammad A, Inamuddin (eds) Green solvents I. Springer, Dordrecht. https://doi.org/10.1007/978-94-007-1712-1_3

Martins MAP, Frizzo CP, Moreira DN, Buriol L, Machado P (2009) Solvent-free heterocyclic synthesis. Chem. Rev. 109:4140–4182. https://doi.org/10.1021/cr9001098

Matsubara H, Maegawa T, Kita Y, Yokoji T, Nomoto A (2014) Synthesis and properties of fluorous benzoquinones and their application in deprotection of silyl ethers. Org Biomol Chem 12:5442–5447. https://doi.org/10.1039/C4OB00783B

Matsuda T (2013) Recent progress in biocatalysis using supercritical carbon dioxide. J Biosci Bioeng 115:233–241. https://doi.org/10.1016/j.jbiosc.2012.10.002

Medina-Gonzalez Y, Camy S, Condoret J-S (2012) Cellulosic materials as biopolymers and supercritical CO_2 as a green process: chemistry and applications. Int J Sust Eng 5:47–65. https://doi.org/10.1080/19397038.2011.613488

Merchán Arenas DR, Martínez Bonilla CA, Kouznetsov VV (2013) Org Biomol Chem 11:3655.

Mikami K (2005) Green reaction media in organic synthesis. Blackwell Publishing, Oxford, UK

Mohammadyazdani N, Reza Bozorgmehr M, MomenHeravi M (2016) Conformation changes and diffusion of a-amylase in 1hexyl-3-methylimidazolium chloride ionic liquid: a molecular dynamics simulation perspective. J Mol Liq 221:463–468

Molnár Á (2011) Efficient, selective, and recyclable palladium catalysts in carbon−carbon coupling reactions. Chem Rev 111(3):2251–2320. https://doi.org/10.1021/cr100355b

Monteith ER, Mampuys P, Summerton L, Clark JH, Maes BUW, Robert McElroy C (2020) Why we might be misusing process mass intensity (PMI) and a methodology to apply it effectively as a discovery level metric. Green Chem 22:123. https://doi.org/10.1039/c9gc01537j

Morita, DK, Pesiri DR, David SA, Glaze WH, Tumas W (1998) Palladium-catalyzed cross-coupling reactions in supercritical carbon dioxide. Chem Commun 1397

Moscoso AD, Ballester P (2017) Light-responsive molecular containers. Chem Commun 53:4635. https://doi.org/10.1039/c7cc01568b

Nairoukh Z, Avnir D, Blum J (2013) Acid-catalyzed hydration of alkynes in aqueous microemulsions. Chemsuschem 6:430–432

Noyori R (2010) Insight: Green chemistry: the key to our future. Tetrahedron 66:1028

Ohara M, Hara Y, Ohnuki T, Nakamura S (2014) Direct enantioselective three-component synthesis of optically active propargylamines in water. Chem Eur J 20:8848–8851.

Ohde H, Ohde M, Wai CM (2004) Swelled plastics in supercritical CO_2 as media for stabilization of metal nanoparticles and for catalytic hydrogenation. Chem Commun 930–931. https://doi.org/10.1039/B311522D

Ohnmacht SA, Mamone P, Culshaw AJ Greaney MF (2008) Direct arylations on water: synthesis of 2,5-disubstituted oxazoles balsoxin and texaline. Chem Commun 1241

Onwudili JA, Williams PT (2006) Flameless supercritical water incineration of polycyclic aromatic hydrocarbons. Int J Renew Energy Res 30:523–533

Ou G, He B, Halling P (2016) Ionization basis for activation of enzymes soluble in ionic liquids. Biochim Biophys Acta Gen Subjects 1860:1404−1408

Palou RM (2010) Microwave-assisted synthesis using ionic liquids. Mol Divers 14:3–25. https://doi.org/10.1007/s11030-009-9159-3

Panja SK, Saha S (2013) Recyclable, magnetic ionic liquid bmim [FeCl4]catalyzed, multicomponent, solvent-free, green synthesis of quinazolines. RSC Adv 3:14495–14500. https://doi.org/10.1039/c3ra42039f

Pârvulescu VI, Hardacre C (2007) Catalysis in ionic liquids. Chem Rev 107:2615–2665. https://doi.org/10.1021/cr050948h

Pathak AK, Ameta C, Ameta R, Punjabi PB (2016) Microwave-assisted organic synthesis in ionic liquids. J Heterocycl Chem 53:1697–1705. https://doi.org/10.1002/jhet.2515

Pavlovič I, Knez Ž, Škerget M (2013) Hydrothermal reactions of agricultural and food processing wastes in sub- and supercritical water: a review of fundamentals, mechanisms, and state of research. J Agric Food Chem 61:8003–8025

Pei Y, Hao L, Ru J, Zhao Y, Wang H, Bai G, Wang J (2018) The self-assembly of ionic liquids surfactants in ethanolammonium nitrate ionic liquid. J Mol Liq 254:130–136. https://doi.org/10.1016/j.molliq.2018.01.095

Pellisa A, Corici L, Sinigoia L, D'Amelioc N, Fattor D, Ferrario V, Ebert C, Gardossi L (2015) Towards feasible and scalable solvent-free enzymatic polycondensations: integrating robust biocatalysts with thin film reactions. Green Chem 17:1756–1766. https://doi.org/10.1039/C4GC02289K

Petrone DA, Ye J, Lautens M (2016) Modern transition-metal-catalyzed carbon-halogen bond formation. Chem Rev 116(14):8003–8104. https://doi.org/10.1021/acs.chemrev.6b00089

Pirrung MC, Sarma KD (2003) Multicomponent reactions are accelerated in water. J Am Chem Soc 126:444

Pirrung MC, Sarma KD (2005) Aqueous medium effects on multi-component reactions. Tetrahedron 61:11456

Rajendar Reddy K, Rajanna KC, Uppalaiah U (2013) Environmentally benign contemporary Friedel–Crafts acylation of 1-halo-2-methoxynaphthalenes and its related compounds under conventional and nonconventional conditions. Tetrahedron Lett 54:3431

Rakshit A, Chowdhury S, Acharjee A, Datta I, Dome K, Biswas S, Bhattacharyya SS, Saha B (2020) Hetero-aromatic N-base-promoted oxidation of 4-chlorobenzyl alcohol by Cr(VI) in micellar media. Res Chem Intermediate 46:2559–2578. https://doi.org/10.1007/s11164-020-04106-x

Reetz MT, Wiesenhöfer W, Franciò G, Leitner W (2002) Biocatalysis in ionic liquids: batchwise and continuous flow processes using supercritical carbon dioxide as the mobile phase. Chem Commun 992–993. https://doi.org/10.1039/B202322A

Rideout DC, Breslow R (1980) Hydrophobic acceleration of Diels-Alder reactions. J Am Chem Soc 102:7816

Rodrigues JV, Ruivo D, Rodríguez A, Deive FJ, Esperanca JMSS, Marrucho IM, Gomes CM, Rebelo LPN (2014) Structural−functional evaluation of ionic liquid libraries for the design of co-solvents in lipase-catalysed reactions. Green Chem 16:4520–4523

Rogozinska M, Adamkiewicz A, Mlynarski J (2011) Efficient "on water" organocatalytic protocol for the synthesis of optically pure warfarinanticoagulant. Green Chem 13:1155

Rose PM, Clifford AA, Rayner CM (2002) The Baylis–Hillman reaction in supercritical carbon dioxide: enhanced reaction rates, unprecedented ether formation, and a novel phase-dependent 3-component coupling. Chem Commun 968–969. https://doi.org/10.1039/B111347J

Safari J, Azizi F, Sadeghi M (2015) Chitosan nanoparticles as a green and renewable catalyst in the synthesis of 1,4-dihydropyridine under solvent-free conditions. NewJ Chem 39:1905. https://doi.org/10.1039/c4nj01730g

Sar P, Ghosh A, Scarso A et al (2019) Surfactant for better tomorrow: applied aspect of surfactant aggregates from laboratory to industry. Res Chem Intermed 45:6021–6041. https://doi.org/10.1007/s11164-019-04017-6

Sarkar S, Pal R, Kumar Sen A (2013) Efficient synthesis of 3-benzyl-3-(indol-3-yl)-2-phenyl-2,3-dihydroisoindolinone derivatives via a simple and convenient MCR in aqueous micellar system. Tetrahedron Lett 54:4273

Schwarze M, Milano-Brusco JS, Strempel V, Hamerla T, Wille S, Fischer C, Baumann W, Arlt W, Schomäcker R (2011) RSC Adv 1:474

Sharma H, Singh N, Jang DO (2014) A ball-milling strategy for the synthesis of benzothiazole, benzimidazole and benzoxazole derivatives under solvent-free conditions. Green Chem 16:4922–4930. https://doi.org/10.1039/C4GC01142B

Shaughnessy KH (2009) Hydrophilic ligands and their application in aqueous-phase metal-catalyzed reactions. Chem Rev 109:643–710. https://doi.org/10.1021/cr800403r

Sheldon RA (2005) Green solvents for sustainable organic synthesis: state of the art. Green Chem 7:267–278. https://doi.org/10.1039/b418069k

Sheldon RA (2008) E factors, green chemistry and catalysis: an odyssey. Chem Commun 3352–3365. https://doi.org/10.1039/B803584A

Sheldon RA, Pereira PC (2017) Biocatalysis engineering: the big picture. Chem Soc Rev 46:2678–2691. https://doi.org/10.1039/C6CS00854B

Sheldon RA, Woodley JM (2018) Role of biocatalysis in sustainable chemistry. Chem Rev 118(2):801–838. https://doi.org/10.1021/acs.chemrev.7b00203

Sheldon RA, Lau RM, Sorgedrager MJ, Rantwijk FV, Seddon KR (2002) Biocatalysis in ionic liquids. Green Chem 4:147–151. https://doi.org/10.1039/B110008B

Shome A, Roy S, Das PK (2007) Nonionic surfactants: a key to enhance the enzyme activity at cationic reverse micellar interface. Langmuir 23:4130–4136. https://doi.org/10.1021/la062804j

Simon MO, Li CJ (2012) Green chemistry oriented organic synthesis in water. Chem Soc Rev 41:1415–1427. https://doi.org/10.1039/c1cs15222j

Singh MS, Chowdhury S (2012) Recent developments in solvent-free multicomponent reactions: a perfect synergy for eco-compatible organic synthesis. RSC Adv 2:4547–4592. https://doi.org/10.1039/c2ra01056a

Singh SK, Savoy AW (2020) Ionic liquids synthesis and applications: an overview. J Mol Liq 297:112038. https://doi.org/10.1016/j.molliq.2019.112038

Singh D, Krishna R, Singh N (2011) A highly efficient green synthesis of 1H-pyrazolo[1,2-b]phthalazine-5,10-dione derivatives and their photophysical studies. Tetrahed Lett 52:5702–5705. https://doi.org/10.1016/j.tetlet.2011.08.111

Sorella GL, Strukul G, Scarso A (2015) Recent advances in catalysis in micellar media. Green Chem 17:644–683

Sowmiah S, Esperança JMSS, Rebelo LPN, Afonso CAM (2018) Pyridinium salts: from synthesis to reactivity and applications. Org Chem Front 5:453–493. https://doi.org/10.1039/C7QO00836H

Tanaka K (2009) Solvent free organic synthesis. Wiley, Weinheim, Germany

Toda F, Tanaka K (2000) Solvent-free organic synthesis. Chem Rev 100:1025

Tomasek J, Schatz J (2013) Olefin metathesis in aqueous media. Green Chem 15:2317

Touré BB, Hall DG (2009) Natural product synthesis using multicomponent reaction strategies. Chem Rev 109(9):4439–4486. https://doi.org/10.1021/cr800296p

Trindade AF, Gois PMP, Afonso CAM (2009) Recyclable stereoselective catalysts. Chem Rev 109(2):418–514. https://doi.org/10.1021/cr800200t

Varón EY, Joli JE, Torres M, Sala N, Villorbina G, Méndez JJ, Garayoa RC (2012) Solvent-free biocatalytic interesterification of acrylate derivatives. Catal Today 196:86–90. https://doi.org/10.1016/j.cattod.2012.02.055

Vilotijevic I, Jamison TF (2007) Epoxide-opening cascades promoted by water. Science 317:1189

Wang Q, Chan TR, Hilgraf R, Fokin VV, Sharpless KB, Finn MG (2003) Bioconjugation by copper(I)-catalyzed azide-alkyne [3 + 2] cycloaddition. J Am Chem Soc 125:3192

Wang LM, Jiao N, Qiu J, Yu J-J, Liu J-Q, Guo F-L, Liu Y (2010) Sodium stearate-catalyzed multicomponent reactions for efficient synthesis of spirooxindoles in aqueous micellar media. Tetrahedron 66:339

Wen LR, Li ZR, Li M, Cao H (2012) Solvent-free and efficient synthesis of imidazo[1,2-a]pyridine derivatives via a one-pot three-component reaction. Green Chem 14:707. https://doi.org/10.1039/c2gc16388h

Xi W, Phan HT, Haes AJ (2018) How to accurately predict solution-phase gold nanostar stability. Anal Bioanal Chem 410:6113–6123. https://doi.org/10.1007/s00216-018-1115-6

Xu R, Queneau Y (2014) How the polarity of carbohydrates can be used in chemistry. In: Rauter AR, Queneau YQ, Lindhorst TK (eds) Carbohydrate, chemical, royal society of chemistry, vol 40, pp 31–50. https://doi.org/10.1039/9781849739986-00031

Yan N, Xiao C, Kou Y (2010) Transition metal nanoparticle catalysis in green solvents. Coord Chem Rev 254:1179–1218. https://doi.org/10.1016/j.ccr.2010.02.015

Yang J, Dudley GB (2010) Pyridine-directed organolithium addition to an enol ether. Adv Syn Cat 352:3438–3442. https://doi.org/10.1002/adsc.201000495

Yin Y, Yu S, Shen M, Liu J, Jiang G (2015) Fate and transport of silver nanoparticles in the environment. In: Liu J, Jiang G (eds) Silver nanoparticles in the environment. Springer, Berlin, Heidelberg

Zhang W (2009) Fluorous linker-facilitated chemical synthesis. Chem Rev 109(2):749–795. https://doi.org/10.1021/cr800412s

Zhang W, Cai C (2008) New chemical and biological applications of fluorous technologies. Chem Commun 5686–5694. https://doi.org/10.1039/B812433G

Zhang J, Peng L, Han B (2014) Amphiphile self-assemblies in supercritical CO_2 and ionic liquids. Soft Matter 10:5861–5868. https://doi.org/10.1039/C4SM00890A

Zhang B, Zhang J, Han B (2016) Assembling metal–organic frameworks in ionic liquids and supercritical CO_2. 11:2610–2619. https://doi.org/10.1002/asia.201600323

Zhanga X, Wang L (2012) TBHP/I_2-promoted oxidative coupling of acetophenones with amines at room temperature under metal-free and solvent-free conditions for the synthesis of α-ketoamides. https://doi.org/10.1039/c2gc35489f

Zong M, Thurecht KJ, Howdle SM (2008) Dispersion polymerisation in supercritical CO_2 using macro-RAFT agents. Chem Commun 5942–5944. https://doi.org/10.1039/b812827h

Plant-Mediated Green Synthesis of Nanoparticles

Hira Munir, Muhammad Bilal, Sikandar I. Mulla, Hassnain Abbas Khan, and Hafiz M. N. Iqbal

Abstract

Nanoparticles are an inspiring group of nanostructured materials with broad-spectrum applications in different fields such as catalysis, antimicrobial treatment, drug delivery, nanomedicine, environmental remediation, electronics, and chemical sensors. Nevertheless, the techniques used for preparation are environmentally unfriendly. Aiming to promote the greener synthesis of nanoparticles, this chapter spotlights plant-mediated eco-friendly and sustainable development of nanoparticles. Naturally occurring plant extracts are enriched with a plethora of various biologically active biomolecules and secondary metabolites, including alkaloids, terpenoids, flavonoids, enzymes, and phenolic substances. These bioactive compounds can catalyze the reduction of metal ions into biogenic nanoparticles in an eco-sustainable single-step biosynthetic process. Additionally, the utilization of plant extracts and their derived compounds circumvents the necessity for capping and stabilizing agents and generates bioactive size and shape-dependent green nanoparticles. Herein, we have made an effort that describes the synthesis of a wide range of metal-based nanoparticles (platinum, gold, zinc oxide, silver, and titanium dioxide nanoparticles) by using plant extract as a green synthesis matrix. In addition, different parts of plants that have widely been utilized for the biosynthesis of these NPs with several sizes and shapes by biological methodologies are briefly described. In conclusion, the greener synthesis approaches are safer and easier to exploit the massive preparation of nanostructured particles.

Keywords

Nanotechnology • Metal nanoparticles • Green chemistry • Plant extract • Secondary metabolites

1 Introduction

Nanotechnology may be described as the modification of matter by different physical or chemical tactics for the formation of substances with particular applications (Herlekar et al. 2014). It can also be defined as the microscopic-sized particle that has at least, one dimension much lesser than one hundred nanometers in size (Thakkar et al. 2010). Nanoparticles (NPs) possess many interesting applications and multi-functional properties in diverse fields which includes nutrition energy and medicine (Abbasi et al. 2016; Ghorbanpour and Fahimirad 2017) because of their significant surface-to-volume ratio and abundance of surface atoms. Other considerable advantageous features of plant-based nanoparticles are shown in Fig. 1. The biogenic synthesis of monodispersed nanoparticles with particular sizes and shapes was an undertaking in biomaterial science. Also, it has garnered prodigious interest within the industry of pharmacology for the cure of viral and bacterial infections (Song and Kim 2009). The biological synthesis strategies are prospective alternative as compared to other classical

H. Munir (✉)
Department of Biochemistry and Biotechnology, University of Gujrat, Gujrat, Pakistan
e-mail: hira.munir@uog.edu.pk

M. Bilal
School of Life Science and Food Engineering, Huaiyin Institute of Technology, Huaian, 223003, China

S. I. Mulla
Department of Biochemistry, School of Applied Sciences, REVA University, Bangalore, 560064, India

H. Abbas Khan
Clean Combustion Research Centre, Division of Physical Sciences and Engineering, King Abdullah University of Science and Technology (KAUST), 6900 Thuwal, 23955, Saudi Arabia

H. M. N. Iqbal
Tecnologico de Monterrey, School of Engineering and Sciences, Campus Monterrey, Ave. Eugenio Garza Sada 2501, 64849 Monterrey, N.L., CP, Mexico

Inamuddin et al. (eds.), *Advances in Green Synthesis*, Advances in Science, Technology & Innovation,
https://doi.org/10.1007/978-3-030-67884-5_4

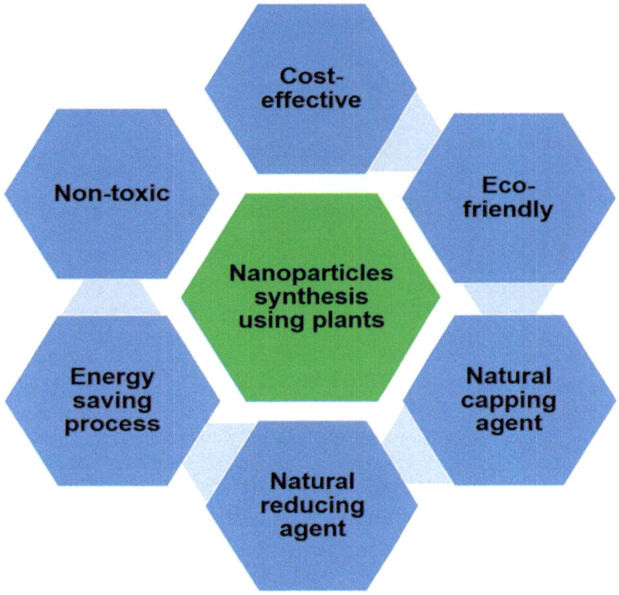

Fig. 1 Advantageous features of plant-based biosynthesis of nanoparticles

methods of synthesis due to the usage of extra-biological compounds. The easy accessibility and rich biodiversity of natural material had been enormously studied for the biosynthesis of nanomaterials (NMs) (Monda et al. 2011). Recently, the green synthesis of nanosized tubes, particles, wires, and flowers was described. These biosynthesized NMs exhibit great prospective in many diverse fields like diagnosis, treatment, commercial product manufacturing, and improvement surgical nanodevices (Bar et al. 2009). Nanomedicines had made a great influence in the healthcare sector by curing numerous chronic ailments. Therefore, green biosynthesis of NPs is taken into consideration as the building blocks of the coming near generations by controlling various health issues (Cruz et al. 2010).

Nowadays, nanotechnology has attracted great interest as a promptly growing scientific field. Manipulation, characterization, and fabrication of NPs are the foremost aims observed in this new technology (Ghorbanpour and Fahimirad 2017; Ahmed et al. 2016; Beyene et al. 2017). Plant crude extracts of plants consist of unique secondary metabolites like phenolic acid, terpenoids, flavonoids, and alkaloids, wherein these metabolites are known to be the reason for ions reduction during the biosynthesis of NPs from different metals (Aromal and Philip 2012). These plant metabolites are frequently used for redox reactions for the preparation of eco-friendly NPs. Many earlier reviews are revealing that these synthesized nanoparticles are efficiently controlling the apoptosis, genotoxicity, and oxidative stress-associated changes (Kim et al. 2007). Moreover, NPs possess a wider range of applications in plant sciences and the agriculture industry. For example, with the help of bioprocessing technology, nanoparticles can convert agricultural wastes and food into energy and many other useful products.

2 Methods for Metallic Nanoparticle Biosynthesis

Numerous procedures are utilized for the formation of different nanoparticles like chemical, biological, physical as well as enzymatic. Chemical protocols are utilized to prepare nanoparticles by way of the sol–gel, electrodeposition, and vapor deposition (Oliveira et al. 2005), moist chemical, and co-precipitation techniques (Gan et al. 2012), hydrolysis, catalytic route, Langmuir–Blodgett approach and soft chemical technique (Pileni 1997). Likewise, physical techniques that are being used include ball milling, plasma arcing, thermal evaporate, ultra-thin films, spray pyrolysis, lithographic techniques, pulsed laser desorption, layer by using layer growth, molecular beam epistaxis, sputter deposition, and diffusion flame (Joerger et al. 2000). Chemical and physical techniques are using stabilizing agents, highly concentrated reductants, and high radiations that are toxic and unsafe for humans and the environment. Therefore, the biosynthesis of NPs is a single-step bio-reduction procedure that uses lesser energy for the synthesis of eco-friendly nanoparticles (Sathishkumar et al. 2009). Plant extracts, enzymes, fungi, bacteria, and microalgae are different resources that are being used for the biosynthesis of nanoparticles (Iravani (2011)). Table 1 depicts a recent list of biosynthesis of metallic nanoparticles from various plant sources.

3 Green Biosynthesis of Metallic NPs

The procedures for acquiring NPs through different naturally occurring compounds like sugars, plant extracts, vitamins, micro-organisms, and biodegradable polymers might be considered as appealing for nanotechnology. This biological synthesis has led to the fabrication of a confined variety of inorganic NPs (particularly, metallic NPs, though numerous salts and metal oxides also are described). Among the material noted above, primarily, plant-based substances are to be the greatest source and are appropriate for the large-scale formation of NPs (Iravani 2011). For the biological synthesis of metallic NPs, different parts of plants like root, stem, leaf, seed, and latex are being utilized. Polyphenols that are present in plants, for example, in wine, tea, red grape pomace, and winery waste are supposed to be the active and important agent required for the synthesis. Green biosynthesis of nanoparticles offers advancement as compared to other strategies as it is cost-efficient,

Table 1 Synthesis of metallic nanoparticles from various plant sources

Name of plant	Part used	Nanoparticles	Size (nm)	Morphology	Functional groups/Bioactive compounds	Applications	References
Musa paradisiaca	Peel	Gold	100	–	Amine, hydroxyl and carboxyl groups	–	[32]
Tea	Leaves	Gold	20	–		–	[33]
Pelargonium graveolens	Leaves	Gold	–	–	Terpenoids	–	[35]
Azadirachta indica	Leaves	Gold	–	–		–	[36]
lemongrass	Leaves	Gold	–	Spherical		–	[37]
Diospyros kaki	Leaves	Gold	5–300	Spherical and hexagonal		–	[39]
Mentha arvensis	Leaves	Gold	39 ± 15	Spherical and hexagonal			[40]
Pear	Fruit	Gold	200–500	Hexagonal and triangular	Organic acids, proteins, peptides, amino acids, and saccharides	–	[41]
Garcinia combogia	Fruit extract	Gold	–	Spherical shapes		–	[42]
Pistacia integerrima	(leaf galls)	Gold	20–200		Carboxylic acid and hydroxyl groups of polyphenols	Antimicrobial and antinociceptive activities	[43]
Diopyros kaki	Leaves	Platinum	2–12				[58]
Cacumen platycladi	Leaves	Platinum	2.4 ± 0.8		Flavonoids and reducing sugars		[53]
Ocimum sanctum	Leaves	Platinum	23 nm		Gallic acid, ascorbic acid, proteins, and terpenoids		[52]
Terminalia chebula	Fruit		<4	Nearly spherical	Polyphenols		[59]
Tea		Platinum	30–60		Polyphenol		[60]
Pelargonium graveolens	Leaves	Silver	16–40				[78]
Cinnamomum camphora	Leaves	Silver	55–80	spherical or triangular	Water-soluble heterocyclic and polyol components		[79]
Podophyllum hexandrum	Leaves	Silver	12–40	spherical shaped			[82]
Alternanthera dentate	Leaves	Silver	50–100	spherical shaped		Antibacterial	[84]
Boerhaavia diffusa	Whole plant	Silver	25	spherical		Antibacterial	[86]
Taraxacum officinale	Leaves	Silver	15	Spherical	Flavonoids, terpenoids, and triterpenes	Role in disease management	[83]
Sesuvium portulacastrum	Callus	Silver	5–20	spherical			[87]
Tribulus terrestris	Fruit	Silver	16–28	spherical		Antibacterial	[88]
Cocous nucifera	Inflorescence	Silver	22	spherical		Antimicrobial	[89]
Abutilon indicum	Leaves	Silver	7–17	Spherical			[90]
Ziziphoratenuior	Leaves	Silver	8–40	Spherical	Carbonyl, hydroxyl, amine		[91]

(continued)

Table 1 (continued)

Name of plant	Part used	Nanoparticles	Size (nm)	Morphology	Functional groups/Bioactive compounds	Applications	References
Chenopodium album	Leaves	Silver	10–30	Spherical			[93]
Azadirachta indica	Leaves	Silver	10–35	Spherical			[94]
Rosa canina	flesh	Zinc oxide		Spherical	Polar groups, a carbonyl group	Antibacterial, non-toxic and antioxidant	[109]
Aloe barbadensis	Leaves	Zinc oxide	25–40	Spherical			[115]
Nyctanthes arbortristis	Leaves	Titanium	100–150				[130]
Annona squamosa	Fruit peel	Titanium	23 ± 2	Spherical			[132]
Solanum trilobatum	Leaves	Titanium				Pediculocidal and larvicidal activities	[134]
Catharanthus roseus	Leaves	Titanium	25–110			Larvicidal and adulticidal potential	[133]

comparatively reproducible, simple, and it results in more stable products (Kalaiarasi et al. 2010). Generally, biological material offers an environmentally-friendly and green chemical protocol for the production of valuable materials as the biomaterial-based routes reduce the usage of toxic chemicals (Parsons et al. 2007). Figure 2 shows a schematic illustration of the biosynthesis of metallic nanoparticles using various parts of the plant. Few different types of synthesis for metallic NPs are as follows.

3.1 Gold Nanoparticles

Gold (Au) nanoparticles have been stated as an extensive research tool in different fields of agriculture, medicine, and health center due to their stability and biocompatibility (Sheoran and Kaur 2018). Au is considered as one of the extraordinary metals having a melting and boiling point of 1064 and 2808 °C, respectively. Numerous properties of Au, like its incapability to react with oxygen and water and its exceptional conductive properties, have made it very beneficial for mankind. During the 5th millennium BC, gold extraction began close to Bulgaria and is assumed that "soluble" Au come to be seen approximately the 4th or fifth centuries BC in China. The spectacular statue of Touthankamon that was built in this time is proof. Earlier, gold was known with various terms, like drinkable and soluble Au, and Graham (Graham 1861) coined the term "colloid." The colloids of Au and the stunning ruby red color have attracted people for lots of eras (Marie-Christine and Didier 2004; Hough et al. 2011). Gold was widely utilized for medicinal, cosmetic, and ornamental purposes (Savage 1975; Kunckels 1976). Drinkable Au was also used for the cure of several health problems like arthritis, heart diseases, tumors, dysentery, epilepsy, and venereal disease (Anil et al. 2013).

Gold nanoparticles (AuNPs) fascinated some researchers inside the area of plant-based biosynthesis due to their exceptional applications and properties in biomedical, nanodevices, nonlinear optics, catalysis, and nanodevices (Huo and Worden 2007). Gold nanoparticles offer promising scaffolds for gene and drug delivery (Siddiqi and Husen 2017). These nanoparticles possess many useful features such as monodispersity, tunable core size, tunning, and transport of delivery processes and the large surface-to-volume ratio (Han et al. 2007). The AuNPs can be formed by green synthesis (Aromal and Philip 2012; Shankar et al. 2004a); however, the number of gold synthesis reports is considerably lesser as compared to AgNPs. The size of AuNPs varies within a range of 20–300 nm. For example, gold nanoparticles (100 nm) have been prepared from the peel extract of banana (*Musa paradisiaca*) by using simple, eco-friendly, and non-toxic material (Bankar et al. 2010). The crushed, boiled, acetone-precipitated, and dried powder of banana peel was used for the reduction of chloroauric acid. In this study, the enlargement of NPs into microwire and microcubes networks to the periphery of the banana sample was seen. The contribution of amine, hydroxyl, and

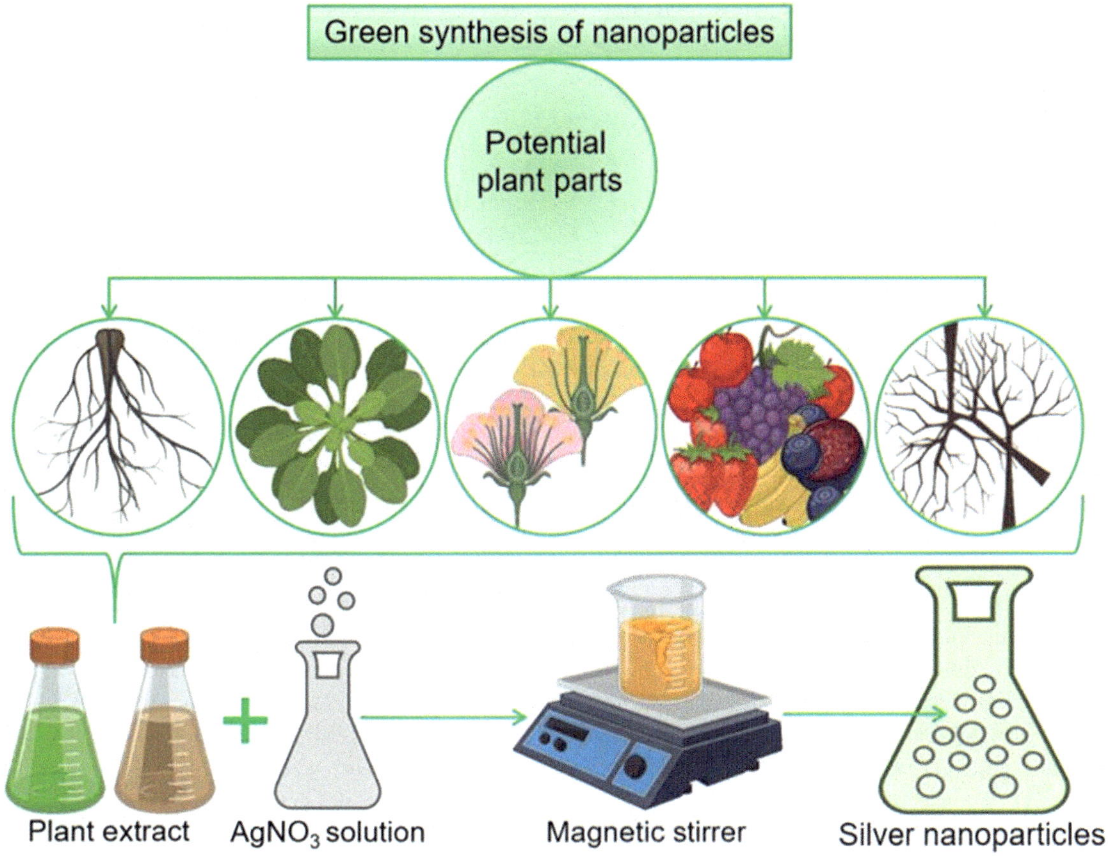

Fig. 2 Schematic illustration of biosynthesis of metallic nanoparticles from various parts of plant

carboxyl groups was observed during the synthesis. Using tea extracts, nanocomposites of gold nanoparticles can be prepared. The extract was made in the solution of 1-methyl-2-pyrrolidinone during the process of nanoparticles (20 nm) formation (Afzal et al. 2009). The AuNPs formation in the polyaniline matrix was confirmed through TEM. In another research by Wu and Chen (Wu and Chen 2007), a facile and green route was reported, by mixing rice wine, soda, and Au (III) at pH 6.5 at a temperature of 25–55 °C without the use of any protective agent. No precipitation occurred, and the resultant solution was stable after a few months.

Shankar et al. (2003a) described the formation of AuNPs by using the leaves of *Pelargonium graveolens*. It was stated that present terpenoids in geranium leaves act as reducing and capping material for a quick reduction of chloroaurate ions toward the AuNPs of different sizes. Later on, these research groups determined the formation of gold nanoparticles by using different plants like *Azadirachta indica* (Shankar et al. 2004b) and lemongrass (Shankar et al. 2005). Prism-, trapezoid-, rod-, and sphere-shaped gold nanoparticles were formed from the black tea leaf extract. The tea phenols and flavonoids are known to be the reason for efficient reduction caused by the extract (Begum et al. 2009). Song et al. (2009) stated that leaf extracts of *Diospyros kaki* and *Magnolia kobus* successfully synthesize the extracellular gold nanoparticles within a range of 5–300 nm having pentagonal, triangular, spherical, and hexagonal shapes at the temperature of 95 °C within few minutes. Babu et al. (2010) cited that leaves of *Mentha arvensis* (ethanolic extract) formed AuNPs of spherical and hexagonal shapes having approximately a size of 39 ± 15 nm. Extract of pear fruit biosynthesized hexagonal and triangular nanoparticles at room temperature of 200–500 nm (Ghodake et al. 2010). The reason for the synthesis of AuNPs is the presence of organic acids, proteins, peptides, amino acids, and saccharides in pear fruit extract. Rajan et al. (2014) used *Garcinia cambogia* (fruit extract) for the biosynthesis of AuNPs of anisotropic and spherical shapes. It was concluded that the shape of nanoparticles relies on the reaction temperature and quantity of extract. In 2015, Islam et al. (2015) and his research group stated that *Pistacia integerrima* (leaf galls) reduces ions into the gold nanoparticles. Carboxylic acid and hydroxyl groups of polyphenols were known to be the reason for the process of reduction. These polyphenols capped the NPs and making them more stable in varied pH solution and at high temperatures. These synthesized Au nanoparticles showed great potential in antimicrobial, enzyme inhibition, muscle relaxant, and antinociceptive activities. The mechanism of antimicrobial activity of nanoparticles is shown in Fig. 3.

3.2 Platinum Nanoparticles

Noble metallic nanoparticles like silver, gold, platinum, and palladium possess a wider range of different applications, such as material science, medicine, pharmaceuticals, and chemistry (Karikalan et al. 2016; Sahin et al. 2017). Among them, platinum has few specific properties like good resistance toward chemical attacks and corrosion, and high surface area. For the synthesis of platinum nanoparticles (PtNPs), till now, numerous methods have been used such as sol–gel route, vapor deposition, and chemical precipitation. However, these protocols have few limitations such as usage of toxic and unsafe chemical, high cost and energy requirements, multistep for preparation. To overcome these limitations, recently, different plant-based syntheses of these metallic nanoparticles, especially platinum nanoparticles, have gained great attention because of their simple usage, eco-friendly, and non-toxic nature.

Platinum (Pt) is one of the expensive, rare, and high-density metals. Platinum nanoparticles are mostly used in the form of suspension or colloid. The antioxidant ability of PtNPs is one of the reasons for their extensive research (Siddiqi and Husen 2016). The main application areas of PtNPs are cancer therapy, polymer membranes, catalytic converters, plastics, textiles, and nanofibers. PtNPs are being extensively utilized as catalysts and in different biomedical applications (Akhtar et al. 2013). In comparison with silver, gold, and platinum, NPs are notably limited. The reaction between an aqueous solution of Pt and plant extracts leads to the following mechanism:

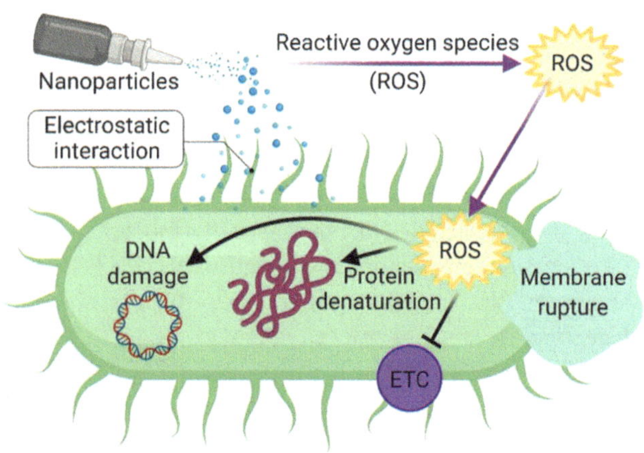

Fig. 3 Mechanism of antimicrobial activity of nanoparticles

$$H_2Pt^{+}6Cl^{-} \cdot 6H_2O + \text{Plant molecule} \rightarrow Pt^{0}\text{nano - particles}$$

Specifically, the interest in PtNPs is because of their unique structural, catalytic, and optical properties making them a promising nanoparticle in catalysis and biomedical applications (Chen and Holt-Hindle 2010; Pedone et al. 2017). Biological approaches for the biosynthesis of PtNPs with the help of plant extracts have not been extensively used. According to the literature survey, the plants that have been used for the green synthesis for PtNPs are as follows: *Diospyros kaki* (Song et al. 2010), *Ocimum sanctum* (Soundarrajan et al. 2012), *Anacardium occidentale* (Sheny et al. 2013), *Cacumen platycladi* (Zheng et al. 2013), *Bidens tripartitus* (Dobrucka 2015), *Punica granatum* (Dauthal and Mukhopadhyay 2015), *Cochlospermum gossypium* (Vinod et al. 2011), *Azadirachta indica,* and *Quercus glauca* (Thirumuran et al. 2016). Jae et al. (2010) prepared PtNPs (2–12 nm) from a solution of $H_2PtCl_6{}^{\cdot}6H_2O$, where leaf extracts of *Diopyros kaki* were used as the agent for reduction of ions. At a temperature of 95 °C with an extract concentration of more than 10%, around 90% of platinum ions were successfully converted into PtNPs. Zheng et al. (2013) biologically synthesize PtNPs (2.4 ± 0.8 nm) by using an extract of *Cacumen platycladi* at 90 °C having an extract of 70% and reaction time of 25 h. During this reaction, the flavonoids and reducing sugars performed an important part in the reduction of the platinum ion as compared to proteins. Likewise, Soundarrajan et al. (2012) used the leaves of *O. sanctum* that acts as the reducing agent during the biosynthesis of PtNPs of size 23 nm, whereas the solution of $H2PtCl_6{\bullet}6H_2O$ was used. Different compounds of plants such as gallic acid, ascorbic acid, proteins, and terpenoids acted as agents for reduction during the formation of PtNPs. Kumar et al. (2013) isolated polyphenols from the extract of *Terminalia chebula* that causes the one single-step synthesis of platinum nanoparticles. They demonstrated that the reduction of platinum (IV) to platinum (0) was due to the presence of polyphenols in the *T. chebula* extracts.

Platinum nanoparticles, which were synthesized from tea polyphenol, act as both capping and reducing agents. These PtNPs were crystalline in nature, having a particle size of 30–60 nm, with a structure of face-centered cubic. Transmission electron microscopy determines that the NPs were of flower-shaped. The polyphenols present in tea include an amount of different phenolic components that form the complexes with the ions and quickly reduce them into metallic nanoparticles of various sizes and shapes (Nadagouda and Varma 2008; Kim et al. 2008; Porcel et al. 2010). The usage of different extracts from plant extracts for the green biosynthesis of PtNPs relies on the fact that the procedure is easier, faster, reliable, cost-effective, and eco-friendly and forms more stable nanoparticles as compared to other classical methods (Mohanpuria et al. 2008).

3.3 Silver Nanoparticles

Usually, silver (Ag) is used as a catalyst for the oxidation of ethylene to ethylene oxide and methanol to formaldehyde (Nagy and Mestl 1999). Some of the features of Ag are good conductivity, catalytic, antibacterial potential, and chemical stability (Frattini et al. 2005). In the era of the Roman Empire, silver nanoparticles were utilized by the founders of glass, for the creation of the Lycurgus cup (fourth century AD); now, this cup is in British Museum (Nagy and Mestl 1999; Parida et al. 2016). Before the 1980s, silver nanoparticles (AgNPs) were used for isolated supports that were utilized for signals in Raman spectroscopy for practical and scientific concern (Frattini et al. 2005), revealing that AgNPs exhibit a unique organization of high electrical double-layer capacitance (Henglein 1999). Currently, the formation of AgNPs is one of the most actively rising developments in the colloid chemistry and continuously increases in the scientific publications in around the last 20 years.

The mechanism of AgNPs synthesis is due to the occurrence of the polyphenols in the extracts of different plants, which causes the reduction of NPs. This reduction is carried out by the removal of hydrogen due to the OH groups present in polyphenols. The formation of silver nanoparticles is being performed by various biological systems (Baker et al. 2013; Nath and Banerjee 2013; Makarov et al. 2014; Płaza et al. 2014). Numerous plants and their respective parts have been utilized for the formation of silver nanoparticles. This formation involves the reaction of AgN salt with the plant. The presence of a brownish-yellow color confirms the formation of AgNPs.

At present, AgNPs have gained great attention and are being considered as the most capable and useful NP for different biological applications, including biomolecular detection, therapeutics goals, food production and preservation, agricultural purposes, antibacterial agents, drug delivery (Ghorbanpour et al. 2018), bio-labeling, wound healings, sensing, cosmetics, and water purifications (Zhang et al. 2016). These NPs can be formed through protocols (Mittal et al. 2013). Though the method of chemical reduction is a common one for the formation of silver nanoparticles, the usage of precarious and expensive chemicals has diverted the attention of researchers for the search of new and alternative methods (Khan et al. 2015). Some potential health risks were another reason for the search for new methods and had attracted the attention of researchers worldwide (Thakkar et al. 2010; Gade et al. 2011).

Recently, plant-based green biosynthesis of AgNPs is rising into an important subdivision of nanotechnology, as it gained importance and is developed due to its cost-effectiveness and lesser toxicity (Chanel et al. 2017). Silver has been considered and studied very extensively for plant-based synthesis, and it has been known as a more rapid and easier process as compared to the monotonous and time-taking microbial synthetic methods (Akhtar et al. 2013). Shankar et al. (2003b) used leaf extracts of *Pelargonium graveolens* for the extracellular formation of AgNPs. Through mixing AgN solution and plant extract, a quick reduction of Ag ions takes place followed by the synthesis of crystalline and highly stable AgNPs (16–40 nm) in solution. Later on, in 2007, Huang et al. (2007) displayed that spherical or triangular-shaped silver nanoparticles of size 55–80 nm can be prepared from the leaf extract of *Cinnamomum camphora*. He discovered that water-soluble heterocyclic and polyol compounds in the leaves of *C. camphora* are responsible for silver ions reduction. Leela and Vivekanandan (2008) compared the leaf extracts of some different plants such as *Sorghum bicolor, Basella alba, Helianthus annuus, Saccharum officinarum, Zea mays, Oryza sativa and* for the formation of AgNPs and determined that *H. annuus* showed a speedy reduction of silver ions. Similarly, Ahmad et al. (2010) used broth of *Ocimum sanctum* for the biosynthesis of AgNPs and observed highly stable NPs with a size of 5 ± 1.5 nm to 10 ± 2. In a research conducted by Jeyaraj et al. (2013), the biological preparation of silver nanoparticles from the leaf extracts of *Podophyllum hexandrum* was described. The complete reduction of silver ions was completed in 2.5 h at the temperature of 60 °C and pH 4.5 with the formation of spherical-shaped nanoparticles in the range of 12–40 nm. Saratale et al. (2018) biosynthesized spherically and monodispersed silver nanoparticles (15 nm) by using the leaf extract of *Taraxacum officinale*. The presence of flavonoids, terpenoids, and triterpenes is known to be the active compounds present in extract for the synthesis of silver nanoparticles.

By using the aqueous extract of *Alternanthera dentate,* a rapid and green synthesis of AgNPs having spherical shaped with a size of 50–100 nm was prepared. Within 10 min, the silver ions were reduced into silver nanoparticles by the leaf extract. These prepared AgNPs showed antibacterial potential against *Escherichia coli, Klebsiella pneumonia, Enterococcus faecalis,* and *Pseudomonas aeruginosa* (Kumar et al. 2014a). For the formation of AgNPs, *Acorus calamus* was also used, and its antibacterial, anticancer, and antioxidant effects were determined (Nakkala et al. 2014a). In 2014, Nakkala et al. (2014b) and his research members used the extract of *Boerhaavia diffusa* plant as the reducing material for the formation of AgNPs. It was revealed from

TEM and XRD that the prepared NPs possess a particle size of 25 nm, spherical shape with a face-centered cubic structure. The antibacterial potential was determined against *Flavobacterium branchiophilum*, *Aeromonas hydrophila,* and *Pseudomonas fluorescens*; the highest sensitivity was observed toward *F. branchiophilumin* as compared to the two other bacteria. Similarly, Nabikhan et al. (2010) prepared spherical AgNPs (5–20 nm) by using the callus extract of *Sesuvium portulacastrum* L. a salt marsh plant.

The fruit of *Tribulus terrestris* L was dried and reacted with Ag nitrate for the formation of silver nanoparticles. The newly synthesized AgNPs possess a size of 16–28 nm with spherical shape and were used to determine their antibacterial potential against few multidrug-resistant bacteria like *Streptococcus pyogenes, Staphylococcus aureus, Pseudomonas aeruginosa, Escherichia coli,* and *Bacillus subtilis* (Gopinath et al. 2012). Methanolic and ethyl acetate extracts of *Cocos nucifera* were used for the successful synthesis of Ag nanoparticles (22 nm) and exhibited good antimicrobial activity toward different bacterial strains such as *Klebsiella pneumoniae, Bacillus subtilis, Salmonella paratyphi, and Pseudomonas aeruginosa* (Mariselvam et al. 2014). Another spherical-shaped and stable AgNPs were synthesized from *Abutilon indicum* which also possesses good antimicrobial potential against *E. coli, S. typhi, B. substilus,* and *S. aureus* (Ashok et al. 2015). Leaves of *Ziziphora tenuior* were also utilized for the preparation of AgNPs, and the nanoparticles were characterized through different techniques. TEM analysis reveals the spherical shape and size of 8 to 40 nm, and FTIR showed carbonyl, hydroxyl, amine, and some other stabilizing functional groups (Sadeghi and Gholamhoseinpoor 2015). By using the leaves of *Acalypha indica*, a rapid and green protocol for the formation of nanoparticles was reported by Krishnaraj et al. (2010), where a successful formation of nanoparticles was completed in 30 min (Krishnaraj et al. 2010). A weed *Chenopodium album* was reported in a rapid and facile biosynthesis of AgNPs. Its leaf extract was used for the synthesis and results in NPs of the size range of 10–30 nm, and its shape was spherical that was inferred through TEM analysis (Dwivedi and Gopal 2010). Similarly, silver nitrate was reduced by *Azadirachta indica* (leaf extract) that results in the biosynthesis of silver nanoparticles (10–35 nm) with spherical shape (Ramyal and Subapriya 2012).

3.4 Zinc Oxide Nanoparticles

Zinc oxide (ZnO) is an n-type semiconducting metallic oxide. ZnO nanoparticles (ZnONPs) have diverted the attention of researchers in the past few years because of its wider variety of applications in the field of optics, biomedical systems, and electronics (Anbuvannan et al. 2015a; Jamdagni et al. 2018). Different categories of inorganic metallic oxides have been prepared in some latest findings such as CuO, ZnO, and TiO_2. Of all these metallic oxides, ZnONPs have gained much interest as they are not costly to synthesize, safe, and easily prepared (Jayaseelan et al. 2012). ZnO as a metal oxide has been registered as generally recognized as safe (GRAS) by FDA US (Pulit et al. 2016). These nanoparticles possess great semiconducting features due to the large bandgap and high exciton binding energy such as catalytic potential, wound healing, UV filtering, and optics (Elumalai et al. 2015; Mirzaei and Darroudi 2017). ZnONPs have been widely utilized in different cosmetics such as sunscreen lotion, depending on its UV filtering potential (Wodka et al. 2010). It also has extensive use in biomedical fields like anticancer, antifungal, antibacterial, antidiabetic, and drug delivery (Jain et al. 2014; Hameed et al. 2016), while ZnO is being utilized in targeted drug delivery and still has the constraint of cytotoxicity, which has to be solved (Ma et al. 2013). Nanoflower, nanoflake, nanorod, nanowire, and nanobelt are different morphologies of ZnONPs that have been reported (Paulkumar et al. 2014; Rajeshkumar et al. 2014).

Different parts of plants like leaf, fruit, stem, seed, and root have been used for the formation of ZnONPs, as they contain a sufficient quantity of phytochemicals. The usage of natural extracts for biosynthesis is an eco-friendly and cheap process that does not require any intermediate base groups (Heinlaan et al. 2008). Plants produce large-scale synthesis of stable nanoparticles with various sizes and shapes, and it is one of the reasons that plants are the preferred source for the formation of NPs (Qu et al. 2011a). The synthesis of ZnO nanoparticles is achieved by bio-reduction, which involves the reduction of metallic ions into 0 valences metallic NPs. This process is assisted by different phytochemicals such as polyphenolic compounds, amino acids, vitamins, polysaccharides, terpenoids, and alkaloids secreted from several plants (Heinlaan et al. 2008; Qu et al. 2011a).

Jafarirad et al. (2016) and his research team experimented and compared the results of zinc oxide nanoparticles that were synthesized through two different methods—microwave irradiation (MI) and conventional heating (CH) and results proved that MI took lesser time and faster reaction rate for the formation of nanoparticles (Jafarirad et al. 2016). Plants of family Lamiaceae have been widely used like *Plectranthus amboinicus* (Fu and Fu 2015), *Vitex negundo* (Ambika and Sundrarajan 2015), and *Anisochilus carnosus* (Anbuvannan et al. 2015b) which showed NPs of different shapes and sizes like quasi-spherical, rod-shaped, spherical, and hexagonal. The findings demonstrate that an increase in the concentration of the plant extract causes a decrease in the size of nanoparticles (Fu and Fu 2015; Ambika and Sundrarajan 2015; Anbuvannan et al. 2015b). *Azadirachta indica* (leaf extract) of family Meliaceae has been commonly used for the ZnONP biosynthesis (Bhuyan et al. 2015; Madan et al. 2016). Sangeetha et al. (2011) synthesized spherical and highly stable ZnONPs of size 25–40 nm from the leaf extract of *Aloe barbadensis*. It was described that the synthesized NPs were polydispersed and their particle size can be controlled by changing the concentration of extract. Similarly, spherical and highly stable NPs were synthesized using *Parthenium hysterophorus* leaf extracts. Additionally, *Plectranthus amboinicus* (leaf extract) was utilized for the formation of zinc oxide NPs by Vijayakumar et al. (2015). The synthesis of ZnONPs has also been performed by using *Sedum alfredii* (Qu et al. 2011a), *Physalis alkekengi* (Qu et al. 2011b), *Pongamia pinnata* (Sundrarajan et al. 2015), flowers of *Trifolium pratense* (Dobrucka and Długaszewska 2016), and *Cassia Auriculata* (Ramesh et al. 2014; Suresh et al. 2015). Qu et al. (2011b) synthesized crystalline ZnO nanoparticles (72.5 nm) from *Physalis alkekengi, and* it can grow in soils with high levels of Zn and can incorporate zinc in its aerial parts.

3.5 Titanium Dioxide Nanoparticles

Titanium dioxide (TiO_2) is the oxide form of Ti and occurs naturally. It can be obtained from different minerals as brookite, rutile, and anatase. The manufacturing and utilization of TiO_2 only in the USA have stayed is 1100 thousand tons since 1997, and it has been categorized as a possible carcinogen for humans (Group 2B) by International Agency for Research on Cancer (2010). TiO_2 is a main and vital metallic oxide nanoparticle that is being extensively used in industrial photocatalytic processes, printing ink, paints, paper, rubber, sunscreens, cosmetics, air cleaning products, and car materials because of its biological, chemical and physical features (Rajakumar et al. 2013). The nanoparticles of TiO_2 are being utilized in a wide variety of applications such as sunscreens, drug delivery systems, food preparation, and cosmetics (Grand and Tucci 2016; Shi et al. 2013; Robertson et al. 2010).

Due to its bright, white pigment, and high refractive index, titanium dioxide is an ideal material used as a whitening agent in different applications. Discovery of the super-hydrophilicity and photocatalytic properties of TiO_2 has also led to applications in some industries by producing self-cleaning products as well as enhancing sterilization and deodorizing processes. The biomedical applications of nanoparticles synthesized from titanium dioxide have greatly developed in recent few years. Research is being conducted on these NPs to attenuate the effects of chemotherapy by making cancer therapy more efficient and targeted (Wang et al. 2015; You et al. 2016). Additionally, food-grade titanium dioxide nanoparticles (TiO_2NPs) are present in a range of food products like gum, candy, donuts, marshmallows, cookies, and some others. Toothpaste, shaving creams, deodorants, conditioners, shampoos, and sunscreens are some personal care products also containing food-grade TiO_2NPs (Weir et al. 2012).

Sundrarajan and Gowri (2011) synthesized titanium nanoparticles by using titanium isopropoxide solution and leaf extract of *Nyctanthes arbortristis* with a size range of (100–150 nm). Similarly, TiO_2NPs (25–100 nm) were also biosynthesized from *Jatropha curcas* (aqueous extract) (Hudlikar et al. 2012). An enzyme, curcain, and cyclic peptides were recognized as the possible capping and reducing compound in latex of *J. curcas*. Spherical-shaped (23 ± 2 nm) nanoparticles were synthesized from the fruit peel of *Annona squamosa* at room temperature with a time of 6 h (Roopan et al. 2012). At room temperature, *Solanum trilobatum* (leaves) were used for the formation of titanium NPs that were having pediculicidal and larvicidal activities (Rajakumar et al. 2013). Velayutham et al. (2011) use leaf extracts of *Catharanthus roseus* for the synthesis of TiO_2NPs, and the synthesized nanoparticles were having rough shape and size of 25–110 nm. These NPs possess great adulticidal and larvicidal potential against *Bovicola ovis* and *Hippobosca maculata.* Similarly, the synthesis of TiO_2NPs from plant *Eclipta prostrata* was reported by Rajakumar et al. (2012). Santhoshkumar et al. (2014) biosynthesize TiO_2NPs from *Psidium guajava* (aqueous leaf extract), antibacterial, and antioxidant potential of these nanoparticles which was explored. By using 20 g/mL of nanoparticles, the highest zone of inhibition was recorded against *E. coli* and *S. aureus*. In another research conducted by Priyadarshani et al. (2019), TiO_2NPs of *Cissus quadrangularis* having significant bactericidal potential against *Staphylococcus* and *E. coli* were also reported. Kumar et al. (2014b) compared the antibacterial potential of two types of biosynthesized titanium dioxide nanoparticles—one from extracts of *Hibiscus rosa-sinensis* and one chemically synthesized. The author concluded that nanoparticles synthesized from plant extract showed higher activity as compared to chemically synthesized ones.

4 Different Parts Used for the Synthesis of Metallic Nanoparticles

Recently, green plant-based nanotechnology has diverted the attention because of its wide applications in many different fields. The metallic NPs like gold, zinc, silver, platinum, nickel, copper, titanium oxide, and magnetite were biosynthesized by using various plant parts and are being studied extensively. Stem, fruit, root, seed, peel, callus, flower, and gums are different parts that are utilized for the synthesis of nanoparticles with various sizes and shapes by biological methodologies (Chandran et al. 2006).

4.1 Fruit

Fruit bodies of *Tribulus terrestris* and solution of silver nitrate were utilized for the eco-friendly biosynthesis of silver NPs (Gopinath et al. 2012). The presence of different phytochemicals in the plant extracts causes the formation of nanoparticles in only one step of reduction. The shape of synthesized NPs was spherical and exhibits excellent antimicrobial potential against the multidrug-resistant human pathogens. Similar research was performed by Amarnath et al. (2012) in which polyphenol from grapes was utilized for the formation of palladium NPs and these nanoparticles act very efficiently against different bacterial diseases. Fruit extracts of *Rumex hymenosepalus* act as reducing and stabilizing agents in the biosynthesis of silver nanoparticles.

4.2 Stem

The methanolic extract of *Callicarpa maingayi* (stem) was used for the biosynthesis of AgNPs (Shameli et al. 2012). The prepared extract consists of a group aldehyde that mainly involves in the process of reduction of Ag into silver nanoparticles. The functional groups such as amide I and polypeptides are considered as the responsible groups for the capping of ions to metallic NPs. The molecular findings on the synthesis of Ag crystals are complicated and still not completely known. According to some latest studies, AgNPs bind with the proteinaceous outer cell of fungi and bacteria that results in the breakage of lipoproteins of the microbial cell wall. This is followed by the blockage of cell division and leads to cell death. Vanaja et al. (2013) stated the photosynthesis of silver nanoparticles by using extracts of *Cissus quadrangularis*. The functional groups mainly the amine, phenolics, and carboxyl are directly involved in the reduction reaction. The synthesized NPs possess good antibacterial potential toward *Bacillus subtilis* and *Klebsiella planticola* (pathogenic bacteria).

4.3 Seeds

The seed extract of fenugreek contains many naturally occurring bioactive compounds like lignin, saponin, vitamins, and high content of flavonoids. In the presence of reducing agents, this extract of seed acts as a good surfactant in the process of reduction of chloroauric acid for the formation of nanoparticles. Seed extracts of different plants contain some functional groups such as COO^- group, C,C, and C,N. This functional group serves as a surfactant of AuNPs and flavonoids easily stabilize the electrostatic stabilization during the synthesis of AuNPs (Mittal et al. 2013). *Macrotyloma uniflorum* (aqueous extract) increases the rate of reduction of Ag ion in the biosynthesis of AgNPs. Caffeic acid present in the seed extracts is recognized to be the reason for the increase in the reduction rate. Hence, the present caffeic acid completes the reduction reaction of nanoparticle synthesis within a minute.

4.4 Flowers

Petals of rose were utilized for the formation of gold nanoparticles through an eco-friendly method by Noruzi et al. (2011). This extract contains an ample amount of proteins and sugars that are considered to be the main constituents during the reduction of salt tetrachloroaurate into the bulk of gold nanoparticles. Likewise, flowers of *Clitoria ternatea* and *Catharanthus roseus* are also utilized for the biosynthesis of different metallic nanoparticles of desired sizes and shapes. Vankar and Bajpai (2010) also synthesize AuNPs by using the flower extract of *Mirabilis jalapa*.

4.5 Leaves

Extracts of plant leaves also serve as a facilitator in the biosynthesis of different metallic nanoparticles. Leaves of several plants like *Alternanthera sessilis*, *Murraya koenigii*, and *Centella asiatica* have been used for the said purpose. Recently, *Piper nigrum* leaves were utilized for successful biosynthesis of silver nanoparticles through an eco-friendly protocol, as they contain important bioactive products. The silver nanoparticles have an effective role in cancer medicine for the treatment of different dreadful diseases. The presence of piper longminine and longumine in *Piper nigrum* acts as the agent for capping in preparation of silver nanoparticles and can also improve the cytotoxic effects of the tumor cells (Jacob et al. 2012). Another green preparation of AgNPs from *Artemisia nilagirica* (leaves extract) was described by Vijayakumar et al. (2013). These metallic nanoparticles serve as a significant antimicrobial agent. Similarly, AgNPs formed from different plant leaves can control many pathogenic problems in humans.

5 Conclusions

In this chapter, we have summarized information about the green and environmentally responsive production of various metal nanoparticles. The presence of biologically active molecules and secondary metabolites in plant extracts is responsible to reduce metal ions to nanostructures in a rapid single-step synthetic strategy. A large number of plant species, their parts, and derived extract has been effectively employed for the preparation of numerous kinds of

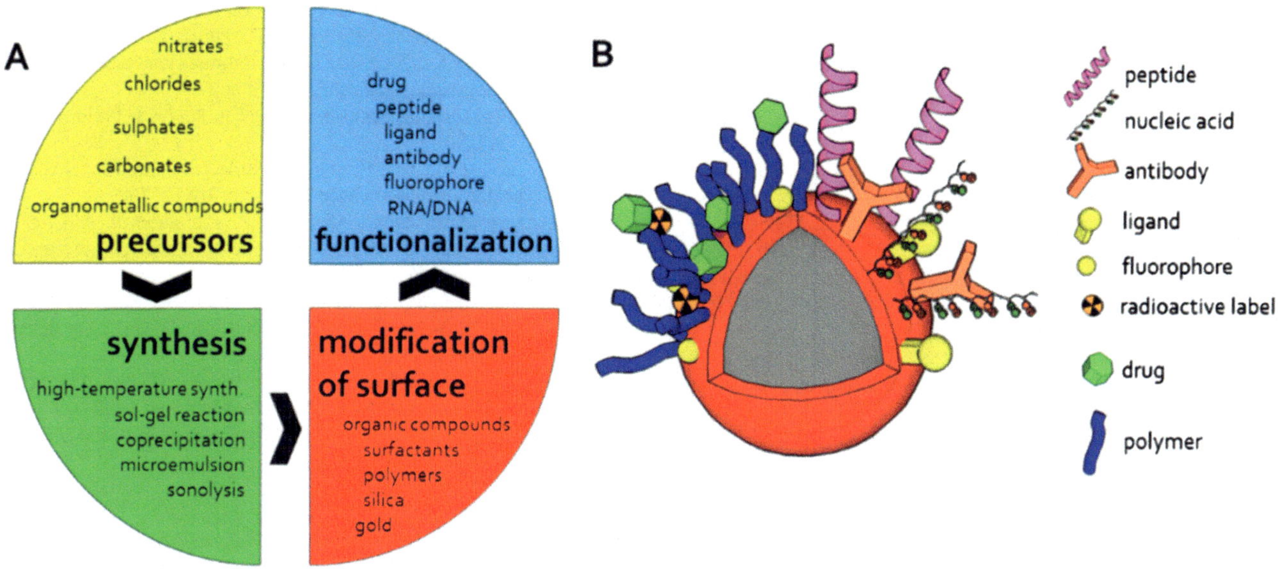

Fig. 4 Scheme of nanoparticles design workflow (A) and possible modification and functionalization of nanoparticles (B). Reprinted from Kudr et al. (2017), anopen access article distributed under the Creative Commons Attribution License. Copyright (2017) the authors. Licensee MDPI, Basel, Switzerland

nanoparticles and other nanostructured materials due to the occurrence of abundant metabolic compounds, including phenols, alkaloids, carbohydrates, terpenoids, and bio-enzymes. It is worth noting that the utilization of plant sources for nanoparticle biosynthesis obviates the requirement for capping and stabilizing agents. Taking into consideration the inescapable applications of nanotechnology and nanoscience in a range of modern everyday life, additional research is necessary to explore the unique physical and chemical attributes of newly synthesized nanostructures. From the future development perspective, Fig. 4 shows the involvement of nanotechnology to further advance the characteristics features and functionalization of nanoparticles (Kudr et al. 2017; Bilal et al. 2019).

References

Abbasi E, Milani M, Fekri Aval S, Kouhi M, Akbarzadeh A, Tayefi Nasrabadi H, Nikasa P, Joo SW, Hanifehpour Y, Nejati-Koshki K, Samiei M (2016) Silver nanoparticles: synthesis methods, bio-applications and properties. Crit Rev Microbiol 42(2):173–180

Afzal AB, Akhtar MJ, Nadeem M, Hassan MM (2009) Investigation of structural and electrical properties of polyaniline/gold nanocomposites. J Phys Chem C 113:17560–17565

Ahmad N, Sharma S, Alam MK, Singh VN, Shamsi SF, Mehta BR, Fatma A (2010) Rapid synthesis of nanoparticles using medicinal particles of basil. Coll Surf B: Bionterf 81:81–86

Ahmed S, Ahmad M, Swami BL, Ikram S (2016) A review on plants extract mediated synthesis of silver nanoparticles for antimicrobial applications: a green expertise. J Adv Res 7:17–28

Akhtar MS, Panwar J, Yun YS (2013) Biogenic synthesis of metallic nanoparticles by plant extracts, *ACS Sustain.* Chem Eng 1:591–602

Amarnath K, Kumar J, Reddy T, Mahesh V, Ayyappan SR, Nellore J (2012) Synthesis and characterization of chitosan and grape polyphenols stabilized palladium nanoparticles and their antibacterial activity. Colloid Surf B 92:254–261

Ambika S, Sundrarajan M (2015) Antibacterial behaviour of *Vitex negundo* extract assisted ZnO nanoparticles against pathogenic bacteria. J Photochem Photo-Biol B Biol 146:52–57

Anbuvannan M, Ramesh M, Viruthagiri G, Shanmugam N, Kannadasan N (2015) *Anisochilus carnosus* leaf extract mediated synthesis of zinc oxide nanoparticles for antibacterial and photocatalytic activities. Mater Sci Semicond Process 39:621–628

Anbuvannan M, Ramesh M, Viruthagiri G, Shanmugam N, Kannadasan N (2015) Synthesis, characterization and photocatalytic activity of ZnO nanoparticles prepared by biological method. Spectrochim Acta A Mol Biomol Spectrosc 143:304–308

Anil K, Xu Z, Xing-Jie L (2013) Gold nanoparticles: emerging paradigm for targeted drug delivery system. Biotechnol Adv 31 (5):593606

Aromal SA, Philip D (2012) Green synthesis of gold nanoparticles using *Trigonella foenum-graecum* and its size dependent catalytic activity. Spectrochim Acta A 97:1–5

Ashok KS, Ravi S, Kathiravan V, Velmurugan S (2015) Synthesis of silver nanoparticles using *A. indicum* leaf extract and their antibacterial activity. Spectrochim Acta Part A: Mol Biomol Spectrosc 134:34–39

Babu PJ, Sharma P, Borthakur BB, Das RK, Nahar P, Bora U (2010) Synthesis of gold nanoparticles using Mentha arvensis leaf extract. Int J Green Nanotechnol Phys Chem 2:62–68

Baker S, Rakshith D, Kavitha KS, Santosh P, Kavitha HU, Rao Y, Sreedharamurthy S (2013) Plants: emerging as nanofactories towards facile route in synthesis of nanoparticles. Bioimpacts 3 (3):111117

Bankar A, Joshi B, Kumar AR, Zinjarde S (2010) Banana peel extract mediated synthesis of gold nanoparticles. Colloids Surf B: Biointerfaces 80:45–50

Bar H, Bhui DK, Sahoo GP, Sarkar P, De SP, Misra A (2009) Green synthesis of silver nanoparticles using latex of *Jatropha curcas.* Colloids Surf A 339:134–139

Begum NA, Mondal S, Basu S, Laskar RA, Mandal D (2009) Biogenic synthesis of Au and Ag nanoparticles using aqueous solutions of black tea leaf extracts. Coll Surf B: Bionterf 71:113–118

Beyene HD, Werkneh AA, Bezabh HK, Ambaye TG (2017) Synthesis paradigm and applications of silver nanoparticles (AgNPs), a review. Sustain Mater Technol 13:18–23

Bhuyan T, Mishra K, Khanuja M, Prasad R, Varma A (2015) Biosynthesis of zinc oxide nanoparticles from *Azadirachta indica* for antibacterial and photocatalytic applications. Mater Sci Semicond Process 32:55–61

Bilal M, Mehmood S, Rasheed T, Iqbal H (2019) Bio-catalysis and biomedical perspectives of magnetic nanoparticles as versatile carriers. Magnetochemistry 5(3):42

Chandran SP, Chaudhary M, Pasricha R, Ahmad A, Sastry M (2006) Synthesis of gold nanotriangles and silver nanoparticles using Aloe vera plant extract. Biotechnol Prog 22:577–583

Chanel TH, Adri H, Muhammad DB, Bambang Y, Fakhili G (2017) Green synthesis of silver nanoparticle and its antibacterial activity. Rasayan J. Chem. https://doi.org/10.7324/rjc.2017.1041875

Chen A, Holt-Hindle P (2010) Platinum-based nanostructured materials: synthesis, properties, and applications. Chem Rev 110:3767–3804

Cruz D, Fale PL, Mourato A, Vaz PD, Serralheiro ML, Lino ARL (2010) Preparation and physicochemical characterization of Ag nanoparticles biosynthesized by *Lippia citriodora* (Lemon Verbena). Colloid Surf B 81:67–73

Dauthal P, Mukhopadhyay M (2015) Biofabrication, characterization, and possible bio-reduction mechanism of platinum nanoparticles mediated by agro-industrial waste and their catalytic activity. J Ind Eng Chem 22:185–191

Dobrucka R (2015) Synthesis and structural characteristic of platinum nanoparticles using herbal *bidens tripartitus* extract. J Inorg Organomet Polym 11:1–7

Dobrucka R, Długaszewska J (2016) Biosynthesis and antibacterial activity of ZnO nanoparticles using *Trifolium pratense* flower extract. Saudi J Biol Sci 23:517–523

Dwivedi AD, Gopal K (2010) Biosynthesis of silver and gold nanoparticles using *Chenopodium album* leaf extract. Physicochem Eng Aspects 369:27–33

Elumalai K, Velmurugan S, Ravi S, Kathiravan V, Ashokkumar S (2015) Green synthesis of zinc oxide nanoparticles using *Moringa oleifera* leaf extract and evaluation of its antimicrobial activity. Spectrochim Acta a Mol Biomol Spectrosc 143:158–164

Frattini A, Pellegri N, de Dand N, Sanctis O (2005) First-principles study of the adsorption of NH3 on Ag surfaces. Mater Chem Phys 94:148152

Fu L, Fu Z (2015) *Plectranthus amboinicus* leaf extract-assisted biosynthesis of ZnO nanoparticles and their photocatalytic activity. Ceram Int 41:2492–2496

Gade A, Rai M, Kulkarni S (2011) *Phoma sorghina*, a phytopathogen mediated synthesis of unique silver rods. Int J Green Nanotechnol 3 (3):153–159

Gan PP, Ng SH, Huang Y, Li SFY (2012) Green synthesis of gold nanoparticles using palm oil mill effluent (POME): a low-cost and eco-friendly viable approach. Bioresour Technol 113:132–135

Ghodake GS, Deshpande NG, Lee YP, Jin ES (2010) Pear fruit extract-assisted room-temperature biosynthesis of gold nanoplates. Coll Surf B: Bionterf 75:584–589

Ghorbanpour M, Fahimirad S (2017) Plant nanobionics a novel approach to overcome the environmental challenges. Medicinal plants and environmental challenges. Springer, Cham, pp 247–257

Ghorbanpour M, Farahani AHK, Hadian J (2018) Potential toxicity of nano-graphene oxide on callus cell of *Plantago major* L. under polyethylene glycol-induced dehydration. Ecotoxicol Environ Saf 148:910–922

Gopinath V, Mubarak AD, Priyadarshini S, Priyadharsshini NM, Thajuddin N, Velusamy P (2012) Biosynthesis of silver nanoparticles from *Tribulus terrestris* and its antimicrobial activity: a novel biological approach. Colloid Surf B 96:69–74

Graham T (1861) Liquid diffusion applied to analysis. Philos Trans R Soc 151:183190

Grand F, Tucci P (2016) Titanium dioxide nanoparticles: a risk for human health? Mini-Rev Med Chem 16(9):762–769

Hameed AS, Karthikeyan C, Ahamed AP, Thajuddin N, Alharbi NS, Al-harbi SA, Ravi G (2016) In vitro antibacterial activity of ZnO and Nd doped ZnO nanoparticles against ESBL producing *Escherichia coli* and *Klebsiella pneumoniae*. Sci Rep 6:24312

Han G, Ghosh P, De M, Rotello VM (2007) Drug and gene delivery using gold nanoparticles. NanoBiotechnology 3:40–45

Heinlaan M, Ivask A, Blinova I, Dubourguier HC, Kahru A (2008) Toxicity of nanosized and bulk ZnO, CuO and TiO2 to bacteria *Vibrio fischeri* and crus- taceans *Daphnia magna* and *Thamnocephalus platyurus*. Chemosphere 71:1308–1316

Henglein A (1999) Formation of colloidal silver nanoparticles: capping action of citrate. J Phys Chem B 103:95339539

Herlekar M, Barve S, Kumar R (2014) Plant-mediated green synthesis of iron nanoparticles. J Nanopart. Article ID 140614. https://doi.org/10.1155/2014/140614

Hough RM, Noble RRP, Reich M (2011) Natural gold nanoparticles. Ore Geol Rev 42(1):5561

Huang J, Li Q, Sun D, Lu Y, Su Y, Yang X (2007) Biosynthesis of silver and gold nanoparticles by novel sundried *Cinnamomum camphora* leaf. Nanotechnology 18:05104–105114

Hudlikar M, Joglekar SS, Dhaygude M, Kodam KM (2012) Green synthesis of TiO_2 nanoparticles by using aqueous extract of *Jatropha curcas* L. latex. Mater Lett 75:196–199

Huo Q, Worden JG (2007) Monofunctional gold nanoparticles: synthesis and applications. J Nanopart Res 9:1013–1025

International Agency for Research on Cancer (2010) Carbon black, titanium dioxide, and talc. IARC monographs on the evaluation of carcinogenic risk to humans, 93, pp 193e275. Retrieved from. https://monographs.iarc.fr/ENG/Monographs/vol93/mono93.pdf. Last accessed 3 Mar 2016

Iravani S (2011) Critical review—green synthesis of metal nanoparticles using plants. Green Chem 13:2638–2650

Islam NU, Jalil K, Shahid M, Muhammad N, Rauf A (2015) *Pistacia integerrima* gall extract mediated green synthesis of gold nanoparticles and their biological activities, Arabian J Chem. https://doi.org/10.1016/j.arabjc.2015.02. 014

Jacob SJP, Finub JS, Narayanan A (2012) Synthesis of silver nanoparticles using *Piper longum* leaf extracts and its cytotoxic activity against Hep-2 cell line. Colloid Surf B 91:212–214

Jae YS, Kwon EY, Kim BS (2010) Biological synthesis of platinum nanoparticles using *Diopyros kaki* leaf extract. Bioprocess Biosyst Eng 33:159–164

Jafarirad S, Mehrabi M, Divband B, Kosari NM (2016) Biofabrication of zinc oxide nanoparticles using fruit extract of *Rosa canina* and their toxic potential against bacteria: a mechanistic approach. Mater Sci Eng C 59:296–302

Jain N, Bhargava A, Panwar J (2014) Enhanced photocatalytic degradation of methylene blue using biologically synthesized "protein-capped" ZnO nanoparticles. Chem Eng J 243:549–555

Jamdagni P, Khatri P, Rana JS (2018) Green synthesis of zinc oxide nanoparticles using flower extract of *Nyctanthes arbortristis* and their antifungal activity. J King Saud Univ Sci 30(2):168–175

Jayaseelan C, Rahuman AA, Kirthi AV, Marimuthu S, Santhoshkumar T, Bagavan A, Gaurav K, Karthik L, Rao KVB (2012) Novel microbial route to synthesize ZnO nanoparticles using *Aeromonas hydrophila* and their activity against pathogenic bacteria and fungi. Spectrochim Acta a Mol Biomol Spectrosc 90:78–84

Jeyaraj M, Rajesh M, Arun R, MubarakAli D, Sathishkumar G, Sivanandhan G, Kapildev G, Manickavasagam M, Thajuddin N, Premkumar K, Thajuddin N, Ganapathi A (2013) An investigation on the cytotoxicity and caspase-mediated apoptotic effect of biologically synthesized silver nanoparticles using *Podophyllum hexandrum* on human cervical carcinoma cells. Coll Surf B: Bionterf 102:708–717

Joerger R, Klaus T, Granqvist CG (2000) Biologically produced silver–carbon composite materials for optically functional thin-film coatings. Adv Mater 12:407–409

Kalaiarasi R, Jayallakshmi N, Perumal V (2010) Phytosynthesis of nanoparticles and its applications. Plant Cell Biotechnol Mol Biol 11:1–16

Karikalan N, Velmurugan M, Chen SM, Karuppiah C (2016) Modern approach to the synthesis of Ni(OH)2 decorated sulfur doped carbon nanoparticles for the nonezymatic glucose sensor. ACS Appl Mater Interfaces 8(34):22545–22553

Khan T, Khan MA, Nadhman A (2015) Synthesis in plants and plant extracts of silver nanoparticles with potent antimicrobial properties: current status and future prospects. Appl Microbiol Biotechnol 99 (23):9923–9934

Kim JS, Kuk E, Yu KN, Jong-Ho K, Park SJ, Lee HJ, Kim SH (2007) Antimicrobial effects of silver nanoparticles. Nanomedicine 3:95–101

Kim EY, Ham SK, Shigenaga MK, Han O (2008) Bioactive dietary polyphenolic compounds reduce nonheme iron transport across human intestinal cell monolayers. J Nutr 138:1647–1651

Krishnaraj C, Jagan EG, Rajasekar S (2010) Synthesis of silver nanoparticles using *Acalypha indica* leaf extracts and its antibacterial activity against water borne pathogens. Colloids Surf B 76:50–56

Kudr J, Haddad Y, Richtera L, Heger Z, Cernak M, Adam V, Zitka O (2017) Magnetic nanoparticles: from design and synthesis to real world applications. Nanomaterials 7(9):243

Kumar KM, Mandal BK, Tammina SK (2013) Green synthesis of nano platinum using naturally occurring polyphenols. RSC Adv 3:4033

Kumar PSM, Francis AP, Devasena T (2014) Biosynthesized and chemically synthesized Titania nanoparticles: comparative analysis of antibacterial activity. J Environ Nanotechnol 3:73–81

Kumar DA, Palanichamy V, Roopan SM (2014) Green synthesis of silver nanoparticles using *Alternanthera dentata* leaf extract at room temperature and their antimicrobial activity. Spectrosc Spectrochim Acta Part. A: Mol Biomol 127:168–171

Kunckels J (1976) Nuetliche Observationes oder Anmerkungen von Auro und Argento Potabili. Schutzens, Hamburg

Leela A, Vivekanandan M (2008) Tapping the unexploited plant resources for the synthesis of silver nanoparticles. Afr J Biotechnol 7:3162–3165

Ma H, Williams PL, Diamond SA (2013) Ecotoxicity of manufactured ZnO nanoparticles—a review. Environ Pollut 172:76–85

Madan HR, Sharma S, Udayabhanu C, Suresh D, Vidya YS, Nagabhushana H, Rajanaik H, Anantharaju KS, Prashantha SC, Maiya PS (2016) Facile green fabrication of nanostructure ZnO plates, bullets, flower, prismatic tip, closed pine cone: their antibacterial, antioxidant, photoluminescent and photocatalytic properties. Spectrochim Acta A Mol Biomol Spectrosc 152:404–416

Makarov VV, Love AJ, Sinitsyna OV, Makarova SS, Yaminsky IV, Taliansky ME, Kalinina NO (2014) "Green" nanotechnologies: synthesis of metal nanoparticles using plants. Acta Naturae 6 (1):3544

Marie-Christine D, Didier A (2004) Gold nanoparticles: assembly, supramolecular chemistry, quantum-size-related properties, and applications toward biology, catalysis, and nanotechnology. Chem Rev 104:293346

Mariselvam R, Ranjitsingh AJA, Nanthini AUR, Kalirajan K, Padmalatha C, Selvakumar MP (2014) Green synthesis of silver nanoparticles from the extract of the inflorescence of *Cocos nucifera* (Family: Arecaceae) for enhanced antibacterial activity. Spectrochim Acta Part A: Mol Biomol Spectrosc 129:537–541

Mirzaei H, Darroudi M (2017) Zinc oxide nanoparticles: biological synthesis and biomedical applications. Ceram Int 43:907–914

Mittal AK, Chisti Y, Banerjee UC (2013) Synthesis of metallic nanoparticles using plant extracts. Biotechnol Adv 31:346–356

Mohanpuria P, Rana NK, Yadav SK (2008) Biosynthesis of nanoparticles: technological concepts and future applications. J Nanopart Res 10:507–517

Monda S, Roy N, Laskar RA, Sk I, Basu S, Mandal D, Begum NA (2011) Biogenic synthesis of Ag, Au and bimetallic Au/Ag alloy nanoparticles using aqueous extract of mahogany (*Swietenia mahogani* JACQ.) leaves. Colloid Surf B 82:497–504

Nabikhan A, Kandasamy K, Raj A, Alikunhi NM (2010) Synthesis of antimicrobial silver nanoparticles by callus and leaf extracts from saltmarsh plant, *Sesuvium portulacastrum* L. Colloids Surf B: Biointerfaces 79:488–493

Nadagouda MN, Varma RS (2008) Green synthesis of silver and palladium nanoparticles at room temperature using coffee and tea extract. Green Chem 10:859–862

Nagy A, Mestl G (1999) High temperature partial oxidation reactions over silver catalysts. Appl Catal A 188:337

Nakkala JR, Mata R, Gupta AK, Sadras SR (2014) Green synthesis and characterization of silver nanoparticles using *Boerhaavia diffusa* plant extract and their antibacterial activity. Indus Crop Prod 52:562–566

Nakkala JR, Mata R, Kumar GA, Sadras SR (2014) Biological activities of green silver nanoparticles synthesized with *Acorous calamus* rhizome extract. Eur J Med Chem 85:784–794

Nath D, Banerjee P (2013) Green nanotechnology—a new hope for medical biology. Environ Toxicol Pharmacol 36(3):9971014

Noruzi M, Zare D, Khoshnevisan K, Davoodi D (2011) Rapid green synthesis of gold nanoparticles using *Rosa hybrida* petal extract at room temperature. Spectrachem Acta Part A 79:1461–1465

Oliveira MM, Ugarte D, Zanchet D, Zarbin AJG (2005) Influence of synthetic parameters on the size, structure, and stability of dodecanethiol-stabilized silver nanoparticles. J Colloid Interface Sci 292:429–435

Parida UK, Das S, Jena PK, Rout N, Bindhani BK (2016) Plant mediated green synthesis of metallic nanoparticles: challenges and opportunities. Fabrication and self-assembly of nanobiomaterials. Elsevier

Parsons JG, Videa JRP, Torresdey JLG (2007) Use of plants in biotechnology: synthesis of metal nanoparticles by inactivated plant tissues, plant extracts, and living plants. Dev Environ Sci 5:463–486

Paulkumar K, Gnanajobitha G, Vanaja M, Rajeshkumar S, Malarkodi C, Pandian K, Annadurai G (2014) *Piper nigrum* leaf and stem assisted green synthesis of silver nanoparticles and evaluation of its antibacterial activity against agricultural plant pathogens. Sci World J 829894

Pedone D, Moglianetti M, Luca ED, Bardi G, Pompa PP (2017) Platinum nanoparticles in nanobiomedicine. Chem Soc Rev 46:4951–5975

Pileni MP (1997) Nanosized particles made in colloidal assemblies. Langmuir 13:3266–3276

Płaza GA, Chojniak J, Banat IM (2014) Biosurfactant mediated biosynthesis of selected metallic nanoparticles. Int J Mol Sci 15 (8):1372013737

Porcel E, Liehn S, Remita H, Usami N, Koayashi K, Furusawa Y, Lesech C, Lacombe S (2010) Platinum nanoparticles: a promising material for future cancer therapy? Nanotechnology 21:085103

Priyadarshini S, Mainal A, Sonsudin F, Yahya R, Abdullah Alyousef AA, Mohammed A (2019) Biosynthesis of TiO_2 nanoparticles and their superior antibacterial effect against human nosocomial bacterial pathogens. Res Chem Intermediates 46:1077–1089

Pulit PJ, Chwastowski J, Kucharski A, Banach M (2016) Applied surface science functionalization of textiles with silver and zinc oxide nanoparticles. Appl Surf Sci 385:543–553

Qu J, Luo C, Hou J (2011) Synthesis of ZnO nanoparticles from Zn-hyperaccumulator (*Sedum alfredii Hance*) plants. Micro Nano Lett 6:174 (b)

Qu J, Yuan X, Wang X, Shao P (2011) Zinc accumulation and synthesis of ZnO nanoparticles using *Physalis alkekengi* L. Environ Pollut 159:1783–1788 (a)

Rajakumar G, Abdul Rahuman A, Priyamvada B, Gopiesh KV, Kishore KD (2012) *Eclipta prostrata* leaf aqueous extract mediated synthesis of titanium dioxide nanoparticles. Mater Lett 68:115–117

Rajakumar G, Rahuman A, Jayaseelan C, Santhoshkumar T, Marimuthu S, Bagavan A, Zahir AA, Kirthi AV, Elango G, Arora P, Karthikeyan R, Manikandan S, Jose S (2013) *Solanum trilobatum* extract-mediated synthesis of titanium dioxide nanoparticles to control *Pediculus humanus capitis*, *Hyalomma anatolicum* and *Anopheles subpictus*. Parasitol Res 113:469–479

Rajan A, Meena KM, Philip D (2014) Shape tailored green synthesis and catalytic properties of gold nanocrystals. Spectrochim Acta Part A 118:793–799

Rajeshkumar S, Malarkodi C, Paulkumar K, Vanaja M, Gnanajobitha G, Annadurai G (2014) Algae mediated green fabrication of silver nanoparticles and examination of its antifungal activity against clinical pathogens. Int J Met 2014:1–8

Ramesh P, Rajendran A, Meenakshisundaram MJ (2014) Green syntheis of zinc oxide nanoparticles using flower extract *Cassia Auriculata*. Nanosci Nanotechnol 2:41–45

Ramyal M, Subapriya MS (2012) Green synthesis of silver nanoparticles. Int J Pharm Med Biol Sci 1(1):54–61

Robertson TA, Sanchez WY, Roberts MS (2010) Are commercially available nanoparticles safe when applied to the skin? J Biomed Nanotechnol 6(5):452–468

Roopan SM, Bharathi A, Prabhakarn A, Abdul Rahuman A, Velayutham K, Rajakumar G, Padmaja RD, Lekshmi M, Madhumitha G (2012) Efficient phytosynthesis and structural characterization of rutile TiO_2 nanoparticles using *Annona squamosa* peel extract. Spectrochim Acta Part A 98:86–90

Sadeghi B, Gholamhoseinpoor F (2015) A study on the stability and green synthesis of silver nanoparticles using *Ziziphora tenuior* (Zt) extract at room temperature. Spectrochim Acta Part A: Mol Biomol Spectrosc 34:310–315

Sahin B, Demir E, Aygun A, Gunduz H, Sen F (2017) Investigation of the effect of pomengranate extract and monodisperse silver nanoparticle combination on MCF-7 cell line. J Biotech 260:79–83

Sangeetha G, Rajeshwari S, Venckatesh R (2011) Green synthesis of zinc oxide nanoparticles by aloe barbadensis miller leaf extract: structure and optical properties. Mater Res Bull 46:2560–2566

Santhoshkumar T, Rahuman AA, Jayaseelan C, Govindasamy R, Marimuthu S, Kirthi AV, Velayutham K, Thomas J, Jayachandran V, Kim SK (2014) Green synthesis of titanium dioxide nanoparticles using *Psidium guajava* extract and its antibacterial and antioxidant properties. Asian Pac J Trop Med 7:968–976

Saratale RG, Benelli G, Kumar G, Kim DS, Saratale GD (2018) Bio-fabrication of silver nanoparticles using the leaf extract of an ancient herbal medicine, dandelion (*Taraxacum Officinale*), evaluation of their antioxidant, anticancer potential, and antimicrobial activity against phytopathogens. Environ Sci Pollut Res 25 (11):10392–10406

Sathishkumar M, Sneha K, Won SW, Cho CW, Kim S, Yun YS (2009) *Cinnamon zeylanicum* bark extract and powder mediated green synthesis of nano-crystalline silver particles and its bactericidal activity. Colloid Surf B 73:332–338

Savage G (1975) Glass and glassware. Octopus Book, London

Shameli K, Ahmad M, Al-Mulla EAJ, Ibrahim NA, Shabanzadeh P, Rustaiyan A, Abdollahi Y (2012) Green biosynthesis of silver nanoparticles using *Callicarpa maingayi* stem bark extraction. Molecules 17:8506–8517

Shankar SS, Ahmad A, Pasricha R, Sastry M (2003a) Bio- reduction of chloroaurate ions by Geranium leaves and its endophytic fungus yields gold nanoparticles of different shapes. J Mater Chem 13:1822–1826

Shankar SS, Ahmad A, Sastry M (2003b) Geranium leaf assisted biosynthesis of silver nanoparticles. Biotechnol Progr 19:1627–1631

Shankar SS, Rai A, Ankamwar B, Singh A, Ahmad A, Sastry M (2004a) Biological synthesis of triangular gold nanoprisms. Nat Mater 3:482–488

Shankar SS, Rai A, Ahmad A, Sastry M (2004b) Rapid synthesis of Au Ag, and bimetallic Au core-Ag shell nanoparticles using Neem (*Azadirachta indica*) leaf broth. J Coll Interf Sci 275:496–502

Shankar SS, Rai A, Ahmad A, Sastry M (2005) Controlling the optical properties of lemongrass extract synthesized gold nanotriangles and potential application in infrared-absorbing optical coatings. Chem Mater 17:566–572

Sheny D, Philip D, Mathew J (2013) Synthesis of platinum nanoparticles using dried *Anacardium occidentale* leaf and its catalytic and thermal applications. Spectrochim Acta A 114:267

Sheoran N, Kaur P (2018) Biosynthesis of nanoparticles using eco-friendly factories and their role in plant pathogenicity: a review. Biotechnol Res Innov 2:63–73

Shi H, Magaye R, Castranova V, Zhao J (2013) Titanium dioxide nanoparticles: a review of current toxicological data. Part Fibre Toxicol 10(15):10–15

Siddiqi KS, Husen A (2016) Green synthesis, characterization and uses of palladium/platinum nanoparticles. Nanoscale Res Lett 2016:11

Siddiqi KS, Husen A (2017) Recent advances in plant-mediated engineered gold nanoparticles and their application in biological system. J Trace Elem Med Biol 40:10–23

Song JY, Kim BS (2009) Biological synthesis of bimetallic Au/Ag nanoparticles using Persimmon (*Diospyros kaki*) leaf extract. Korean J Chem Eng 25:808–811

Song JY, Jang HK, Kim BS (2009) Biological synthesis of gold nanoparticles using *Magnolia kobus* and *Diopyros kaki* leaf extracts. Process Biochem 44:1133–1138

Song JY, Kwon EY, Kim BS (2010) Biological synthesis of platinum nanoparticles using *Diopyros kaki* leaf extract. Bioprocess Biosyst Eng 33:159

Soundarrajan C, Sankari A, Dhandapani P, Maruthamuthu S, Ravichandran S, Sozhan G, Palaniswamy N (2012) Rapid biological synthesis of platinum nanoparticles using *Ocimum sanctum* for water electrolysis applications. Bioprocess Biosyst Eng 35:827

Sundrarajan M, Gowri S (2011) Green synthesis of titanium dioxide nanoparticles by *Nyctanthes arbortristis* leaves extract. Chalcogenide Lett 8:447–451

Sundrarajan M, Ambika S, Bharathi K (2015) Plant-extract mediated synthesis of ZnO nanoparticles using *Pongamia pinnata* and their activity against pathogenic bacteria. Adv Powder Technol 26:1294–1299

Suresh D, Nethravathi PC, Rajanaika H, Nagabhushana H, Sharma SC (2015) Green synthesis of multifunctional zinc oxide (ZnO) nanoparticles using *Cassia fistula* plant extract and their photodegradative, antioxidant and antibacterial activities. Mater Sci Semicond Process 31:446–454

Thakkar KN, Mhatre SS, Parikh RY (2010) Biological synthesis of metallic nanoparticles. Nanomedicine 6(2):257–262

Thirumuran A, Aswitha P, Kiruthika C, Nagarajan S, Christy AN (2016) Green synthesis of platinum nanoparticles using *Azadirachta indica*—an eco-friendly approach. Mater Lett 170:175–178

Vanaja M, Rajeshkumar S, Paulkumar K, Gnanajobitha G, Malarkodi C, Annadurai G (2013) Phytosynthesis and characterization of silver nanoparticles using stem extract of *Coleus aromaticus*. Int J Mater Biomat Appl 3:1–4

Vankar PS, Bajpai D (2010) Preparation of gold nanoparticles from *Mirabilis jalapa* flowers. Indian J Biochem Biophys 47:157–160

Velayutham K, Rahuman AA, Rajakumar G, Santhoshkumar T, Marimuthu S, Jayaseelan C, Bagavan A, Kirthi AV, Kamaraj CZAA, Elango G (2011) Evaluation of *Catharanthus roseus* leaf extract-mediated biosynthesis of titanium dioxide nanoparticles against *Hippobosca maculata* and *Bovicola ovis*. Parasitol Res 111 (6):2329–2337

Vijayakumar M, Priya K, Nancy FT, Noorlidah A, Ahmed ABA (2013) Biosynthesis, characterisation and anti-bacterial effect of plant-mediated silver nanoparticles using *Artemisia nilagirica*. Ind Crops Prod 41:235–240

Vijayakumar S, Vinoj G, Malaikozhundan B, Shanthi S, Vaseeharan B (2015) *Plectranthus amboinicus* leaf extract mediated synthesis of zinc oxide nanoparticles and its control of methicillin resistant *Staphylococcus aureus* biofilm and blood sucking mosquito larvae. Spectrochim Acta—Part A: Mol Biomol Spectrosc 137:886–891

Vinod VTP, Saravanan P, Sreedhar B, Keerthi Devi D, Sashidhar RB (2011) A facile synthesis and characterization of Ag, Au and Pt nanoparticles using a natural hydrocolloid gum kondagogu (*Cochlospermum gossypium*). Colloid Surf B 83:291–298

Wang TY, Jiang HT, Wan L, Zhao QF, Jiang TY, Wang B, Wang SL (2015) Potential application of functional porous TiO_2 nanoparticles in light-controlled drug release and targeted drug delivery. Acta Biomater 13:354–363

Weir A, Westerhoff P, Fabricius L, Hristovski K, von Goetz N (2012) Titanium dioxide nanoparticles in food and personal care products. Environ Sci Technol 46(4):2242–2250

Wodka D, Bielaníska E, Socha RP, Elzbieciak-Wodka M, Gurgul J, Nowak P, Warszynski P, Kumakir I (2010) Photocatalytic activity of titanium dioxide modified by silver nanoparticles, *ACS Appl.* Mater Interfaces 2:1945–1953

Wu CC, Chen DH (2007) A facile and completely green route for synthesizing gold nanoparticles by the use of drink additives. Gold Bull 40:206–212

You DG, Deepagan VG, Um W, Jeon S, Son S, Chang H, Park JH (2016) ROS-generating TiO_2 nanoparticles for non-invasive sonodynamic therapy of cancer. Sci Rep 6:12

Zhang S, Tang Y, Vlahovic B (2016) A review on preparation and applications of silver-containing nanofibers. Nanoscale Res Lett. https://doi.org/10.1186/s11671-016-1286-z

Zheng B, Kong T, Jing X, Odoom-Wubah T, Li X, Sun D, Lu F, Zheng Y, Huang J, Li QJ (2013) Plant-mediated synthesis of platinum nanoparticles and its bioreductive mechanism. Colloid Interf Sci 396:138

Green Synthesis of Hierarchically Structured Metal and Metal Oxide Nanomaterials

Malobi Seth, Hasmat Khan, Susanta Bera, Atanu Naskar, and Sunirmal Jana

Abstract

Scientific and technological innovations are rapidly occurring in today's world. These innovations include development of numerous functional nanomaterials among which hierarchically structured nanomaterials (HSNs) are most important owing to their enormous applications in different fields like biomedical, wastewater management, energy storage, sensing, and so on. Alternative to conventional synthesis methods which can cause harmful effects to human health and environment, several green chemical, physical and biological methods are known for producing different metal and metal oxide-based HSNs. It is to be noted that adoption of these methods is increasing as these methods are environmentally sustainable. Green chemical and physical methods involve use of environmentally benign solvents and reagents. In some cases, these are also energy efficient. Additionally, several biological materials including microorganisms, plants, biomolecules are known to be well accepted as befitting reactants/templates for synthesizing several HSNs. However, these methods further need an in-depth mechanistic realization toward large-scale production for practical applications. This chapter is mainly focused upon an understanding of green chemistry-based methodologies for the synthesis of metal and metal oxide HSNs and their potential applications. Finally, the present challenges and future prospect of the methodologies toward making biocompatible and environmentally sustainable HSNs with useful functional properties for advanced applications are discussed briefly.

M. Seth · H. Khan · S. Bera · A. Naskar · S. Jana (✉)
Specialty Glass Division, CSIR-Central Glass and Ceramic Research Institute, 196 Raja S. C. Mullick Road, Jadavpur, Kolkata, 700032, West Bengal, India
e-mail: sjana@cgcri.res.in

Keywords

Green synthesis • Green chemical, physical and biological methods • Green reagents and templates with green techniques • Metal and metal oxide nanomaterials • Hierarchical nanostructures

1 Introduction

Scientific and technological innovations are rapidly occurring in today's world. More than often, the advancements in research for these innovations are occurring at a cost of multifold enhancement of environmental pollution. Human civilization's immense scientific prowess is causing the worst of natural calamities of this century, making difficult the coexistence of humans and our Mother Nature. After years of exploitation, the scientific community has been alarmed since last two decades and concentrating on minimizing the harmful environmental impact of science toward reducing the consumption of non-renewable resources and approaching new paradigms for prevention of environmental pollution (Gao et al. 2012). On this aspect, green chemistry or sustainable chemistry came into limelight. The subject basically concentrates on designing appropriate synthetic methods toward minimizing the use of harmful chemicals/reactants or eliminating the generation of hazardous side products. The green synthesis of nanomaterials is primarily based upon twelve principles that mainly include the synthesis and use of non-toxic chemicals, utilization of renewable feedstock, designing of energy efficiency and degradable waste products (Kreuder et al. 2017). The important of this green synthesis is to use environmentally benign solvents and reagents; non-toxic reducing and capping agents that can in-turn minimize the generation of toxic derivatives/products. Several nanomaterials can be synthesized by adopting green synthesis methods (Gahlawat and Roy Choudhury 2019; Hulkoti and Taranath 2014). Among

Inamuddin et al. (eds.), *Advances in Green Synthesis*, Advances in Science, Technology & Innovation,
https://doi.org/10.1007/978-3-030-67884-5_5

these nanomaterials, metal and metal oxide-based nanomaterials with hierarchical structures have attracted significant attention for both the aspects of fundamental and technological research (Gahlawat and Roy Choudhury 2019; Yuliarto et al. 2019). In this regard, researchers have already adopted green chemical, physical, and biological methods for synthesizing various metal and metal oxide nanomaterials. Green chemical methods involve the use of environmentally benign solvents and reagents. The physical methods include microwave and light-assisted synthesis, thermal deposition, electrochemical anodization, and so on. In some cases, these methods are found to be energy efficient as well as environmentally sustainable. Biological methods involve the use of biologically active materials that function as reagents and templates. These have advantages of being cost effective, eco-friendly, and biocompatible. Also, these are widely abundant in nature. In this purpose, microorganisms like bacteria, fungi, algae, yeast, and also plant or plant-derived materials are being used. It is noteworthy that in the green synthesis method, the active chemicals/phytochemicals from microorganisms and plants have been used as reducing, structure directing, and capping agents (Hulkoti and Taranath 2014; Mohammadinejad et al. 2016; Ebrahiminezhad et al. 2018). In addition, the use of biotemplates such as cellulose, collagen, eggshell membranes, and butterfly wings as natural templates has also been found in the synthesis of metal and metal oxide nanomaterials (Zan and Wu 2016).

Nowadays, hierarchically structured nanomaterials (HSNs) including metal and metal oxide-based nanomaterials are very interesting in fundamental science and also these are highly useful for potential applications (Trogadas et al. 2016). In this case, upon tactful designing of the reaction conditions in the aforesaid green synthesis methods with appropriate green reagents and templates, metal and metal oxide-based HSNs can be synthesized effectively. It is known that the term *hierarchy* comes from Greek words *hieros* (sacred) and *archein* (rule) that refers to an institutional framework where each and every unit are ranked according to their importance. Also, the structural hierarchy has reined in Mother Nature in umpteen variations of living beings and biological materials from macroscopic to microscopic scales. Moreover, the hierarchy of natural materials arises from a complex reciprocity between surface structure, morphology, and physical as well as chemical properties of the biologically active components. Importantly, synthesized hierarchically structured nanomaterials means higher dimension of a micro- or nanostructures consisting of numerous assemblies of low dimensional nanobuilding blocks arranged in a particularly organized manner with less agglomerated configuration (Lee 2009). The nanobuilding blocks that comprise a hierarchical structure may be the array of nanoparticles creating different shapes including but not limited to 1D nanowires, nanotubes and nanobelts, 2D nanosheets and nanocubes, 3D nanoflowers, superstructures and hollow structures or may be the array of well-aligned meso- and microporous hierarchical structures. Hierarchical metal and metal oxide nanostructures exhibit special properties like large surface area with high surface-to-volume ratio. Sometimes, hierarchical porosity endows high antimicrobial activity, effective photon-harvesting, and efficient charge transfer abilities (Yuliarto et al. 2019; Yang et al. 2017). Such advanced functional properties/special characteristics make them advantageous for wide range of applications including biomedical, photocatalysis, environmental remediation, sensors, optoelectronics, photoelectrochemical energy conversion and storage (Bera et al. 2016a,b,c,2017; Khan et al. 2020; Seth et al. 2020).

In this chapter, we critically discuss the *state-of-the-art* research on the green synthesis of metal and metal oxide-based HSNs, their growth mechanisms and applications in various fields. Further, the present challenges and future prospect of green synthesized hierarchically structured nanomaterials have been briefly discussed.

2 Advantages of Green Synthesis Methods

Typical bottom-up synthesis processes with toxic solvents and hazardous chemical compounds as surfactants/complexing/stabilizing/capping agents that are very tough to decompose and most of them are non-recyclable (Yuliarto et al. 2019) are generally used for obtaining metal and metal oxides. Moreover, the chemical waste produced as side derivatives/products obtained at the end of the synthesis can have detrimental effects on environment, biodiversity, and human health. These harmful wastes are present as (a) toxic gases that lead to air pollution, (b) liquids which cause water pollution, and (c) solids that are disposed on soil increase ground water pollution. Hence, various attempts are found to be taken to mitigate the pollution by minimizing the use of hazardous chemicals by utilizing the synthesis methods of green chemistry. It is worthy to note that the methods, solvents, and chemicals/raw materials used for the synthesis of metal and metal oxide nanomaterials *via* these pathways have to be ecologically sound, energy efficient, recyclable and non-toxic to human health and wildlife as well. Another advantage of these methods is that the microorganisms and plant/plant-derived extracts that are used for the synthesis can mostly be used for multiple purposes in such a way that the active biochemicals present in the extracts can function simultaneously as reducing agents, capping agents, and stabilizing agents as per the need of a particular material synthesis (Hulkoti and Taranath 2014; Mohammadinejad et al. 2016). In this way, the atom economy which is one of the vital principles of green chemistry is also being maintained. Sometimes, the hydrocarbons in the extracts can also act as biofuels for making the final products. These are also cost effective, biodegradable, and largely abundant in nature. Also, the reaction conditions and methods of green chemistry can strategically be used for metal and metal oxide HSNs synthesis. Moreover, the phase structure and morphology as well as the size of the nanomaterials can be controlled by tuning reaction temperature, time, concentration of extracts, and solvents toward the generation of hierarchically structured nanomaterials (Yang et al. 2017). It is also seen that in the biotemplate-mediated synthesis, the creation of hierarchical nanomaterials is particularly efficient as the nanostructures are replicated from the natural micro/nanoscale hierarchical structures of naturally occurring templates. After deposition, the biotemplate is to be burnt off at higher temperatures *via* a suitable calcination/heat treatment process that lead to create hierarchical network and/or arrays of meso- and macroporous structure.

3 Green Synthesis Methods for Hierarchically Structured Metal and Metal Oxide Nanomaterials

Various methods are known today for the green synthesis of HSNs *via* the formation of nanoparticles (NPs). These methods involve the use of energy efficient techniques, non-toxic green solvents and reagents as well as the use of plant extracts or microbes (bacteria, fungi, yeast, and algae) or biotemplates (Khandel et al. 2018). It is now distinct and significant that the green synthesis is the most convenient approach for the synthesis of the nanomaterials toward restoring sustainable environment. It is an authentic and unique way not only because of its non-toxicity to gradually deteriorating environment at present but also, it can produce contamination free nanoparticles with distinct hierarchical morphologies. On the basis of different synthesis methods used for the preparation of metal and metal oxide HSNs, these can mainly be categorized into biological, physical, and chemical methods (discussed details in the next subsections).

3.1 Biological Methods

Application of various kinds of microorganisms (bacteria, fungi, yeast, algae, etc.), plant and plant-derived substances in the synthesis of HSNs have been discussed in this section. These biological components/species are used as reactant and/or templates.

3.1.1 Using Microorganism

Microorganisms as Reactant

Microorganisms (such as bacteria, fungi, and algae) are appropriate candidates for the synthesis of nanomaterials because of their unique ability to reduce metal ions toward the formation with high growth rates (Gahlawat and Roy Choudhury 2019). It is observed that some microbes with a certain defense mechanism are evolved to fight off the toxicity excreted by heavy metals because microbes endure harsh and toxic environment consistently. The defense mechanisms which include intracellular sequestration, efflux pumps, change in metal ion concentration and extracellular precipitation help the bacteria to survive the generated stresses. It is noteworthy that the aforementioned processes are to be utilized for metal and metal oxide-based HSNs synthesis for various applications (to be discussed later).

During synthesis of a specific nanometal, He et al. (2008) observed that the bacteria *Rhodopseudomonas capsulate* extracellularly reduced chloroaurate ions to AuNPs. The proteins in the enzymes secreted by bacteria can reduce the gold ions forming AuNPs which is further capped by the protein molecules developing Au nanowires. Moreover, *Rhizopusoryzae* fungi-mediated *in vitro* synthesis of nanoflower-like stable AuNPs is also reported. In this case, the cell wall of proteins played dominant role in the reduction of gold ions and formation of AuNPs (Kitching et al. 2016). In another example, the nanosheets of TiO_2 architectures had been synthesized by using three kinds of bacteria, namely *Staphylococcus aureus*, *Bacillus subtilis*, and *Escherichia coli* (Zhang et al. 2019a). The phospholipids leaking from bacteria are found to be covering the generated TiO_2 nanoparticles creating steric barrier leading to the formation of nanosheets. Instead of bacteria, yeast mold broth powder has also been used for the synthesis of carbon-doped ZnO with a silver heterostructure at the interfaces (Shen et al. 2015). It is noticed that *Saccharomyces cerevisiae* present in yeast extract can promote secondary ZnO aggregation, amino acids, or other functional groups in the extract that suppressed the growth along the c-axis toward the formation of various ZnO superstructures. Furthermore, dextrose in the broth powder can condense and form an amorphous film of carbon on the ZnO particle surface with hierarchical morphology and helps to introduce silver nanoparticles (Fig. 1). Different metals and metal oxide nanomaterials with their hierarchical morphologies synthesized by using microorganisms as reactant are displayed in Table 1.

Microorganism as Template

Microbial superstructures (like bacteria, fungi, yeast, and algae) are generally used as biotemplate (Table 2) to direct deposition, assembly, and patterning of metal and metal

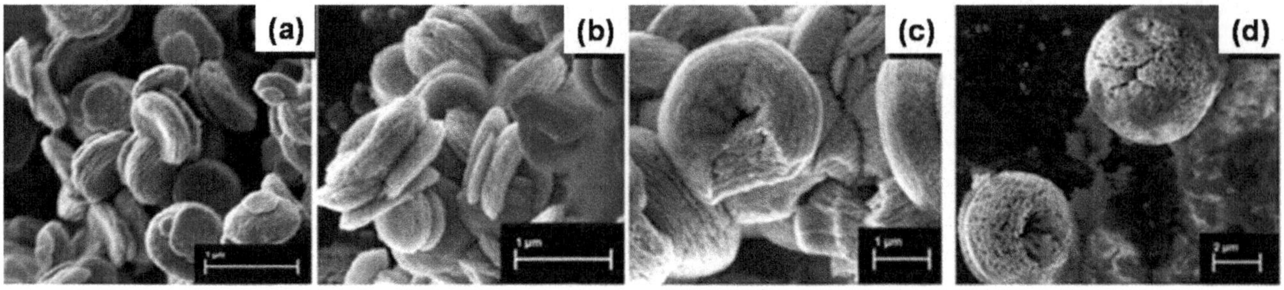

Fig. 1 SEM images of ZnO with hierarchical morphologies synthesized with Zn^{2+}concentration of 12.5 mM (**a**), 25 mM (**b**), 50 mM (**c**), and 100 mM (**d**). Copyrights reserved to the American Chemical Society (Shen et al. 2015)

Table 1 Characteristics of metal/metal oxide HSNs synthesized using microorganisms as reactants

Reactant	Nanomaterial	Morphology/shape	Dimension	Property/application	References
Bacteria					
Rhodopseudomonas capsulata	Au	Nanowires	Diameter, 50–60 nm	Microelectronics, opt oelectronics, nanoscale electronic devices	He et al. (2008)
Staphylococcus aureus, Bacillus subtilis, Escherichia coli	TiO_2	Nanosheets assembled hierarchical architecture	–	Photocatalytic and electrocatalytic applications	Zhang et al. (2019a)
Bacillus subtilis	ZnO	Microsphere assembled by hair-like nanostructure	Diameter, 10–15 nm	Photocatalytic dye degradation	Dhandapani et al. (2020)
Klebsiella pneumoniae, Escherichia coli and Pseudomonas jessinii	AgCl coated with AgNPs	Cubes, flowers	–	–	Müller et al. (2016)
Yeast					
Yeast mold broth	C, Ag modified ZnO	Nanoplatelets, twin-nanodisks, thick microdonuts, microapples wrapped by graphene-like sheets, microspheres, microhamburgers	Nanoplates, 30–50 nm	Photocatalytic activity	Shen et al. (2015)
Fungi					
Rhizopusoryzae	AuNPs	Nanoflowers	24–62 nm	Biomedical applications	Kitching et al. (2016)

oxide-based HSNs by microbes-templating method (Zhou et al. 2007). This method provides a sustainable, economical, and convenient strategy compared to traditional template-directed method. The abundant functional groups like carboxylic, phosphate, amine, etc., are present in microbial cell wall along with enzymes to bind metal ions *via* coordination or electrostatic attraction onto the cell. The steps for surface activation or functionalization are found to be reduced in this method. There are several mechanisms developed by these species to overcome the toxic effects of heavy metals (Hulkoti and Taranath 2014). It is observed that when the cell walls are used as biotemplate, the replica of metal and metal oxide nanostructures is formed. A suitable example is *Spirulina* that had been used as biotemplate for the fabrication of nanosheets assembly of AgNPs (Sun et al. 2019). The intracellular highly ordered texture of *Spirulina* typically contains bioactive components such as nucleoid, layered thylakoid, polyhedral carboxysome, and so on. They can act as supporting base of nanosheets that formed under the spatially repression of the cellular structure. On the other hand, hollow porous ZnO microspheres (Zhou et al. 2007) and hierarchical ZnO nanostructures (Hussein et al. 2009) had been synthesized by using the biotemplates, *Streptococcus thermophiles* and *Bacillus cereus*, respectively. For the growth of hierarchical ZnO nanostructures, *Bacillus cereus* had been used to

Table 2 Characteristics of metal/metal oxide HSNs synthesized using microorganisms as template

Species	Nanomaterial	Morphology/shape	Dimension	Property/application	References
Bacteria					
Staphylococcus aureus, Bacillus subtilis, and Escherichia coli	TiO_2	Nanosheets assembled hierarchical architecture	–	Photocatalytic and electrocatalytic applications	Zhang et al. (2019a)
Streptococcus thermophilus	ZnO	Hollow porous microspheres	Particle size, 20–40 nm; pore size, 2.5–11 nm	–	Zhou et al. (2007)
Bacillus cereus	ZnO	Raspberry (composed of nodules)- and plate-like structures	Nodules, 20–30 nm; plate thickness, ~25 nm	–	Hussein et al. (2009)
Staphylococcus aureus	TiO_2	Hierarchically porous	Macro and mesopores	Photocatalytic oxidation and reduction	He et al. (2014)
Micrococcus lylae	Co_3O_4	Porous, flower-like microspheres	2–10 nm	Electrochemical application	Shim et al. (2013)
Yeast					
Yeast	TiO_2	Hierarchical mesoporous	Pore size, 3–15 nm	Biosensor, fuel cell, and metal–air battery fields	Cui et al. (2009)
Yeast	TiO_2	Lamella	50 nm	Photodegradation of dye	Bao et al. (2012)
Yeast	In_2O_3	Hollow microspheres	~20 nm	Photocatalytic dye degradation	Pan et al. (2018)
Yeast	NiO/C	Hollow microspheres	50 nm	Anode for Li-ion batteries	Tian et al. (2018)
Fungi					
Cladosporium cladosporioides	NiO	Nanostructured microtubules	–	Energy storage applications	Atalay et al. (2016)
Algae					
Spirulina	AgNPs	Nanosheets	Nanogaps, ~4 nm	Antibacterial and surface enhanced Raman scattering (SERS) properties	Sun et al. (2019)
Nannochloropsis oculata	MnO/C	Hierarchically porous	10–100 nm	Catalysis, gas sensing, and energy storage	Xia et al. (2013)
Foraminiferal shells	Co, Ag, Cu, Pt, Au	3D hierarchical structures	–	Filter for water purification, electrocatalysts for ethanol oxidation	Diab et al. (2019)

control intracellularly Zn^{2+}deposition and nucleation/growth of ZnO on or within *B. cereus* cell. This step had been followed by calcination at 500 °C to form raspberry- and plate-like structures, depending upon the organelles where zinc species nucleate and act as template. On the other hand, LPD-modified hierarchically porous TiO_2 nanostructures had been synthesized using *Staphylococcus aureus* as biotemplate (He et al. 2014) (Fig. 2).

3.1.2 Using Plant

In green nanotechnology, plant-mediated synthesis is popularizing among the researchers due to their cost-effectiveness, biodegradability, biocompatibility, and recyclability. Every part of a plant including flower, leaf, fruit, pollen, grain, stem, seed, bark, bran, and peel can be used as green reagents or templates for the biosynthesis of nanomaterials (Ebrahiminezhad et al. 2018). These extracts consist of various biologically active compounds (phytochemicals) such as proteins, amino acids, vitamins, polysaccharides, polyphenols, alkaloids, quinones, organic acids, flavonoids, terpenoids, catechins, and co-enzymes (Mohammadinejad et al. 2016). These compounds can act as reducing agents/capping agents for metal and metal oxide HSNs synthesis. Moreover, the HSNs can be produced by exploiting natural morphology of plant as biotemplate.

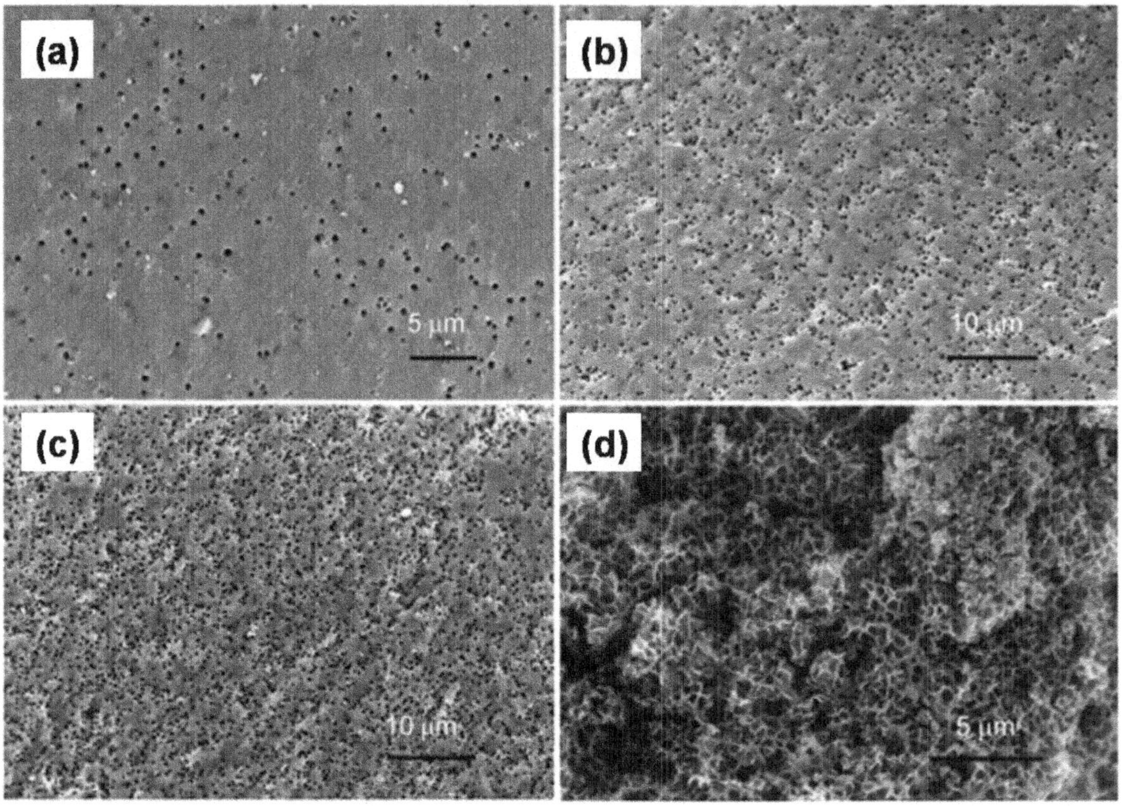

Fig. 2 SEM images of hierarchically porous TiO_2 using **a** 1, **b** 4, **c** 12 and **d** 48 × 10^{11} *S. aureus* as template. Copyrights reserved to the American Chemical Society (He et al. 2014)

Plant as Reactant

Extracellular plant-mediated synthesis of metal and metal oxide nanomaterials/HSNs simply involves the use of plant extract as reactant (Table 3). It is to be noted that the preparation and mixing of extract into aqueous solution of metal ions is to be performed in well-controlled condition. As for example, dendritic silver nanostructure constructed from AgNPs (Fig. 3) had been synthesized by using white grape pomace extract (Carbone et al. 2019). Also, an interesting dendritic nanostructure of Pd was formed in a coffee ring-like fashion while banana peel is used as a reducing agent (Bankar et al. 2010). The functional groups present in the cellulose, pectin, and hemicelluloses—the main constituents in banana peel—are said to be acting as reducing agents. Polyphenols present in the extract had been identified as reducing and capping agents for the synthesis of the different hierarchical nanostructures. In another study, black grape skin extract had been used for the ZnO superstructures/HSNs synthesis (Udayabhanu et al. 2017). Interestingly, with increasing the concentration of the extract, different structures (Mysore pak to canine teeth to hollow pyramid to ornament gem) were formed. The authors also reported the grape extract with polyphenols, flavanoids, tannins, and phytolexins not only can perform as strong reducing or capping agent for Zn^{2+}but also can function as fuel for the combustion of zinc nitrate at 500 °C toward the formation of ZnO superstructures/HSNs. On the other hand, amino acids and sugar in aloe extract had been used as capping agents for encapsulation of Ti^{4+}ions, forming nanorods/nanoflowers of TiO_2. It had been seen that the morphology of the nanomaterial is dependent upon the concentration of aloe extract (Li et al. 2020). It is interesting to note that after adopting hydrothermal process, these morphologies changed into 3D tripyramidal structure.

Plant as Template

The idea of using plant as biotemplate (Table 4) is to introduce naturally occurring well-ordered and hierarchical structures in artificially designed nanomaterials to enhance their properties and functions by virtue of the generated structures (Zan and Wu 2016). Plant-derived biotemplates are very popular because of their availability in large quantity, renewability, and cost-effectiveness. On this aspect, wood is an excellent choice as biotemplate owing to the presence of several species with hierarchically porous macroscopic structures (Liu et al. 2009). In one relevant example, Zn^{2+}ions adsorbed onto the cell walls of wood through capillary adsorption during infiltration of precursor

Table 3 Characteristics of metal/metal oxide HSNs synthesized using plants as reagents

Reagents	Nanomaterial	Morphology/shape	Dimension	Property/application	References
White grape pomace	Ag	Hierarchical dendritic structure	33 ± 6 nm	Electrochemical and antifungal properties	Carbone et al. (2019)
Banana peel	Pd	Coffee ring	50 nm	–	Bankar et al. (2010)
Black grape skin	ZnO	Superstructures	20 nm	Photocatalytic dye degradation, antibacterial, electrochemical sensing of hydrazine	Udayabhanu et al. (2017)
Aloe	TiO_2	3D tripyramidal nanostrcutures, nanorods, nanoflowers	~10 nm	Photocatalytic degradation of antibiotic materials	Li et al. (2020)
Cynodondactylon and *Cyperusrotundus* grass extracts	CuO	Rice spikelet like nanostrcutures, 1D, nanorods, 2D nanoprisms, 3D nanoparticles	16 nm	Antibacterial	Suresh et al. (2020)
Lemon juice, peel	MnO_2	Nanorods	Diameter, ~17 nm	Electrode for LiB	Hashem et al. (2018)
Pepper	Fe_3O_4–Pd	Dendritic Pd, spherical Fe_3O_4	Dendrites diameter, <100 nm; Fe_3O_4 nanoparticles, ~50 nm	Superparamagnetic and photocatalytic properties	Khaghani et al. (2017)
Watermelon rind	ZnO	Nanorods interconnected with small flower formation	Length, 100–200 nm; width, 80–130 nm	Photocatalytic and optoelectronic applications	Singh et al. (2017)
Azadirachta indica	ZnO	Plates, bullets, flower, prismatic tip, closed pine cone	Hexagonal NPs, 10–30 nm	Photocatalytic dye degradation, antibacterial and antioxidant properties	Madan et al. (2015)
Sapindus mukorossi fruit	CuO	Nanowires	~10 nm	Electrochemical sensing of dopamine	Sundar et al. (2018)
Rape pollen grains	SnO_2	Hierarchically porous interconnected network	SnO_2 membrane thickness, ~10 nm; mesopore, 5–20 nm	Gas sensing	Song et al. (2012)

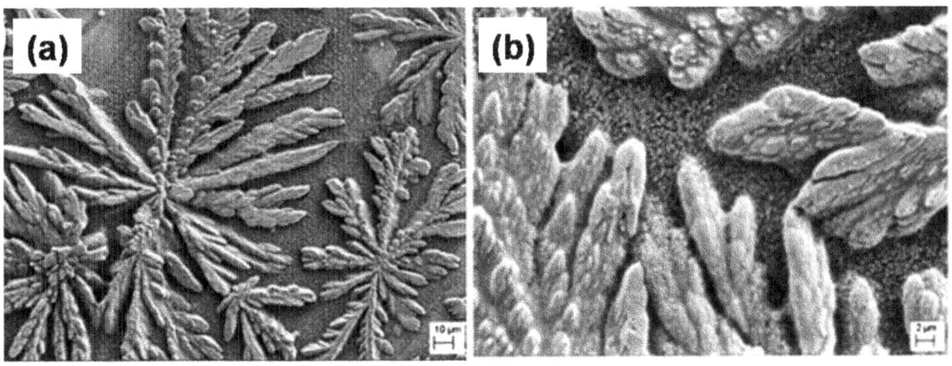

Fig. 3 SEM images (**a, b**) of dendritic Ag from grape pomace. Copyrights reserved to the Elsevier (Carbone et al. 2019)

solution and deposited onto the cell walls homogeneously. After calcination, the wood template burnt off and hierarchically porous ZnO formed by preserving the original template of wood. It is also noted that different nanostructures and nanoflowers of ZnO can be generated using palm olein as biotemplate depending upon the concentration of the template (Ramimoghadam et al. 2013a, b). In this case, initially the self-assembly of palm olein was found to occur followed by the arrangement of zinc acetate over the self-assembly of olein forming a stable inorganic-organic hybrid. In the final step, i.e., after template removal through calcination, the ZnO nanostructures were distinctly formed.

Table 4 Characteristics of metal/metal oxide HSNs synthesized using plants as templates

Template	Nanomaterial	Morphology/shape	Dimension	Property/application	References
Wood	ZnO	Hierarchically porous	Pore size, 25–52 nm	Gas sensing	Liu et al. (2009)
Rice	ZnO	Flower-like, star like structure	40–100 nm	–	Ramimoghadam et al. (2013a)
Palm olein	ZnO	Flowers assembled by nanoplates	~50 nm	–	Ramimoghadam et al. (2013b)
Regenerated cellulose membrane	TiO_2	Nanorods	Length, 45 nm; diameter, 10 nm	Photocatalytic dye degradation	Mohamed et al. (2016)
Green leaves	TiO_2	Porous morph-structures	Layer thickness, ~15 nm; pore size, ~2–10 nm	Photocatalytic dye degradation	Li et al. (2009)
Sunflower Pollen	ZrO_2	Spinous core–shell microspheres	Nanoparticles, ~50–80 nm	Hydrogen storage	Yang et al. (2011)
Collagen fiber	TiO_2–CeO_2	Nanofiber bundles	50–100 nm	Lithium storage	Wei et al. (2019)
Collagen fiber	Ce_x/TiO_2	Mesoporous nanofiber bundles	Thickness, 20–50 nm; pore size, 2–10 nm	Photocatalytic dye degradation	Xiao et al. (2013)
Peltophorum pterocarpum pollen grain	SnO_2	Porous motif	16–25 nm	Potential gas sensing	Fazil et al. (2015)
Basil leave extract	Co_3O_4/C	Porous nanorods assembled by nanospheres, nanocapsules	Nanospheres diameters, 5–15 nm; nanocapsules, 10–20 nm length	Catalysed hydrogen generation	Abu-Zied and Alamry (2019)
Nanocellulose/alginate	CeO_2	3D porous interconnected nanostructures	Pore size, 2–3 nm	–	Moyer et al. (2019)
Cotton	SnO_2	Porous microtubules	Pore size, 23–47 nm	Gas sensing	Ma et al. (2019)
Apple pectin	ZnO	Hollow double-caged peanut built from nanorods	Nanorods, ~100 nm	Photocatalytic dye degradation	Wang et al. (2012)
China rose petal	CeO_2	Porous nanosheets	Thickness, ~7 nm; pores size 2–4 nm	Potential catalyst	Qian et al. (2011)
Grapefruit exocarp	SnO_2	Hierarchically porous	Pore size, 2–10 nm	Formaldehyde gas sensing	Zhang et al. (2016)
Gingko leaves	Co_3O_4	Hierarchically porous structure	Nanoparticles, 30–100; pores, 10–200 nm	Electrochemical biosensing	Han et al. (2015)
Mung bean sprout	$BaCrO_4$	Dendritic superstructures	50–150 μm long side branches	Electronic light device	Yan et al. (2006)
Onion inner coat	$SrCrO_4$	Dumbbell assembled by nanorods	Nanorods width 30–50 nm	–	Chen et al. (2008)
Starch	Bi_2O_3	Nanorods	50 nm	Catalyst for oxidative aromatization	Farzaneh et al. (2017)

In another example, the regenerated cellulose membrane had been used as template for the synthesis of TiO_2 nanorods (Mohamed et al. 2016). This study suggested that the hydrophilic membrane promoted the interaction of hydroxyl groups and Ti^{4+} ions by electrostatic attraction toward the development of TiO_2 nanoparticles that finally made nanorods. It has also been reported that morph-TiO_2 can be synthesized by replication of the structural features of hierarchical and porous template-green leaves (Li et al. 2009).

3.1.3 Using Other Green Templates

As discussed in the previous subsections, a significant attention has been paid upon the synthesis of HSNs using biomaterials as templates. However, the research is not only limited toward synthesizing metal and metal oxide HSNs by using microorganisms or plant and plant-derived parts, rather using different living organisms and their parts with unique microstructures (Table 5). In this respect, Zhang et al. (2018) reported the fabrication of Ag butterfly wing scale arrays using butterfly wing as template. It is well known that butterfly wings have very unique and ordered 3D spatial microstructures that are responsible for brilliant blue coloration and iridescence of their wings (Zhang et al. 2006). In the analysis of chemical composition of wings, the presence of proteins, amino acids are found as major functional groups. These groups can absorb or interact with Zn^{2+} and accelerate to crystallize as ZnO nanoparticles. Upon further calcination and simultaneous pyrolysis of the scales, the formation of ZnO hierarchical structures was observed (Zhang et al. 2006). It is also seen that bioreplication of two types of butterfly wings can led to produce quasi-honeycomb-like, hollow concavities and cross-ribbing structures of TiO_2 deposited over fluorine-doped glass substrate (Zhang et al. 2009).

Eggshell membrane (ESM) is another fascinating material used as a biotemplate for controlling of nucleation, assembling and patterning of unique morphologies of metal and metal oxide HSNs (Zan and Wu 2016). Glycoprotein, a component of shell membrane is made up of $-NH_2$, –OH, –COOH groups that act as structure directing as well as capping agents for the synthesis of metal and metal oxide nanoparticles/HSNs. In aqueous solution of metal oxide precursors, the inorganic material cross-linked and polymerized toward the formation of inorganic/biotemplate complexes which are mesoscopically ordered due to the self-assembling nature of glycoprotein. The organics of ESM biotemplate can be removed through high-temperature calcination, leaving the network-like porous morphology of metal oxide intact (Mallampati and Valiyaveetti 2013). Biomorphic ZnO interwoven microfibers (Dong et al. 2007a), hierarchical mesoporous Mn_3O_4 (Mallampati and Valiyaveetti 2012), and a series of 3D micro/nanocomposite porous structured metal oxides such as CeO_2, Co_3O_4, CuO,

Table 5 Characteristics of metal/metal oxide HSNs synthesized using green templates

Template	Nanomaterial	Morphology/shape	Dimension	Property/application	References
Butterfly wings bioscaffold	AgNPs/graphene	Butterfly wing scale arrays	50–150 nm	Trace chemical detection	Zhang et al. (2018)
Butterfly wings	ZnO	Replica	15 nm	–	Zhang et al. (2006)
Butterfly wings	TiO_2	Quasi-honeycomb-like structure, two-dimensional array shallow concavities structure	50–100 nm	Potential application on dye-sensitized solar cell	Zhang et al. (2009)
Eggshell membrane	CeO_2, Co_3O_4, CuO, NiO and ZnO	Interwoven microporous tubular structures	Nanocrystallites, 20–50 nm	Extraction of nanoparticles from water	Mallampati and Valiyaveetti (2013)
Eggshell membrane	ZnO	Porous interwoven nanofibers	Nanofiber diameter, 200 nm; pore size, ~1 nm	Highly efficient photocatalysts, optical devices	Dong et al. (2007a)
Eggshell membrane	Mn_3O_4	Porous fibrous network	Crystallite size, ~20 nm	Dye adsorption	Mallampati and Valiyaveetti (2012)
Eggshell membrane	SnO_2	Interwoven hollow tubular structure	Tube wall thickness, ~80 nm	–	Dong et al. (2006)
Eggshell membrane	ZnO, Co_3O_4, PdO	Hierarchically porous interwoven nanofibrous structure	Pore size, 20–30 nm for ZnO; 30 nm for Co_3O_4; 80 nm for PdO	–	Dong et al. (2007b)
Eggshell membrane	Co_3O_4	3D hierarchically porous interconnected nanofibers	~50 nm	Non-enzymatic electrochemical detection of glucose	Fan et al. (2016)
Glutamine	ZnO	Nanorods	–	Photocatalytic dye degradation	Alkaim et al. (2016)
Albumen	ZnO	Brush-like morphology assembled by nanorods	90 nm	–	Nouroozi and Farzaneh (2011)

NiO, and ZnO (Dong et al. 2007b) can be produced *via* a hierarchical organization of nanocrystals using ESM as biotemplate.

3.2 Physical and Chemical Methods

3.2.1 Green Techniques

Microwave-assisted process is a convenient approach for heating. It is known as eco-friendly or green technique as in this method, the microwave energy directly interacts with the reaction system as opposed to the traditional heating technique (Lei et al. 2014). This method has several advantages like fast and steady volumetric heating and substantial reduction in synthesis time, causes to develop porous and hierarchically structured metal and metal oxide nanomaterials. As an example, the advantage of microwave-assisted synthesis over conventional hydrothermal heating can be explained by virtue of the growth mechanism of hierarchical CuO@reduced graphene oxide (rGO) (Yin et al. 2019). It is seen that microwave irradiation can promote a heteronuclear nucleation mechanism of monodispersed CuO NPs anchored evenly over rGO nanosheets compared to large particles of CuO randomly distributed on rGO nanosheets under hydrothermal condition. Thermal decomposition of Zn $(NO_3)_2$ at different temperatures led to generate different ZnO superstructures/HSNs like hexagonal pyramids and tulip, bud, apple, dahlia, sunflower, and wheat grains within 500 °C (Udayabhanu et al. 2016). Other than the microwave-assisted synthesis, organic free electrodeposition (Xia et al. 2018; Ji et al. 2019), anodization (Momeni et al. 2016), and light-assisted methods (Das et al. 2017; Hu et al. 2016) had also been reported as efficient green methods for the synthesis of metal oxide HSNs (Table 6).

3.2.2 Green Reagents

Different amino acids (Kang et al. 2013; Gao et al. 2008), metal powders (Zhang et al. 2011), biopolymers (Wang et al. 2016; Zong et al. 2016), salts (Chen et al. 2018) are generally used as green reagents (Table 7) for the green synthesis of different nanomaterials. In this regard, Kang et al. (2013) reported the green synthesis of nanosheets assembled hierarchical silver microspheres in a surfactant/template-free route using different amino acids as structure directing agent and ascorbic acid as reducing agent. Interestingly, amino acids with simple structures (e.g., alanine, glycine) and more complicated structures (e.g., glycine, glutamine, asparagine) would generate different microstructures. Ye et al. (2015) reported the evolution of Pt–Au dendrimer-like HSNs supported on polydopamine-functionalized graphene. In this

Table 6 Characteristics of metal/metal oxide HSNs synthesized using alternate green techniques

Technique	Nanomaterial	Morphology/shape	Dimension	Property/application	References
Microwave	$NiCo_2O_4$	Flower-shaped microsphere consisting of petal-like nanosheets	Nanopetals thickness, ~15 nm; width, 0.1 μm	High performance supercapacitor	Lei et al. (2014)
Microwave	CuO@rGO	Hierarchical nanostructure	4–12 nm	H_2S-sensing	Yin et al. (2019)
Thermal decomposition of precursor	ZnO	Superstructures-hexagonal pyramid, flower, bud, fruit-grain-like structures	–	Photocatalysis of dye, photoluminescence, and electrochemical biosensing	Udayabhanu et al. (2016)
Electrodeposition	Fe_3O_4 @Fe^o	Dendritic	Nanoparticles, 50 nm	Phenol oxidation	Xia et al. (2018)
Organic free electrodeposition	Ag/Cu_2O	Nanopyramids, nanoflakes, nanoplates	Nanopyramids, ~ 311 nm height	SERS applications	Ji et al. (2019)
Electrochemical anodization	CuO	Nanoneedles consist of a bundle of irregular polygonal wires	70–90 nm	Photocatalytic dye degradation	Momeni et al. (2016)
Light assisted method	MnO_2	Nanoflowers assembled by thin intersected nanosheets	Nanosheets thickness, ~4 nm	Photocatalytic degradation of dye	Das et al. (2017)
Light assisted	MnO_2	Desert rose-like 3D hierarchical nanostructures composed of curly and interlaced nanosheets	Nanosheets, ~50 nm	Proposed supercapacitor	Hu et al. (2016)
Microwave	CuO	Hollow cocoon	Thickness, 50 nm	Applications in biosensor, optical	Deng et al. (2011)

(continued)

Table 7 Characteristics of metal/metal oxide HSNs synthesized using alternate green reagents

Reagent	Nanomaterial	Morphology/shape	Dimension	Property/application	References
Polyethylene glycol (PEG)	Au–CuO	Flower-like structure composed of nanosheets	Nanosheets thickness, ~ 30 nm; nanoparticles, ~1.8–12 nm	Catalytic reduction of p-nitrophenol	Gao et al. (2012)
Alanine, glycine, glutamine, asparagine	Ag	Hierarchical Ag microsphere assembles by nanosheets	Microspheres, 2–3 µm, 3–4 µm diameter; nanosheets, 50–150 nm thickness	Sensitive chemical detection and monitoring plasmon-driven reactions	Kang et al. (2013)
Tyrosine	CuO	Hierarchical hollow micro/nanostructure assembled by nanosheets	Nanosheets, 250 nm diameter	Electrode materials for lithium-ion batteries	Gao et al. (2008)
Mg powder	Ag	Dendrites, dendritic flowers and rods	Branches diameter, 40 nm	Potential application in fuel cells, SERS detection	Zhang et al. (2011)
Green reagents	Co_3O_4/C	Hierarchically nanoporous	3–30 nm; pores, 2–4 nm	Supercapacitor	Wang et al. (2016)
Food grade sodium alginate	δ-MnO_2	Nanosheets inter tangled porous flowers	43 nm; micropore, 1–2 nm	Supercapacitor	Zong et al. (2016)
Surfactant, template free	SnO_2	Dahlia-flower like structure	Nanosheets thickness, 10–15 nm; nanoparticles, 20–50 nm	Photocatalytic dye degradation	Chen et al. (2018)
Ascorbic acid	Pt–Au	Dendrimer	26 nm	4-nitrophenol reduction	Ye et al. (2015)
Surfactant free	ZnO	Porous nanoflakes assembled nanostructures	5–40 nm	Solid catalyst	Sinhamahapatra et al. (2012)
D-(+)-glucose powder	Cu_2O–CuO	Hydrangea microspheres assembled by nanosheets	Microspheres diameter, 3–5 µm; nanosheets, 80 nm thickness	–	Yang et al. (2013)
PEG	Au NPs	Hierarchically mesoporous sponge	~12 nm	Potential applications in chemical and biological analysis	Lee et al. (2016)
Green reagents	Ag/WO_{3-x}	Nanowires, nanowires bundles, 3D chestnut-like nanostructures	800 nm nanotips	SERS sensing	Huang et al. (2017)
Green template and precursor	Mn_3O_4, MnO_2	Hierarchical mesoporous microcuboids assembled by nanosheets	Nanosheets thickness, ~70 nm	Anode materials for lithium-ion batteries	Hu et al. (2018)

process, initially, the reduction and self-polymerization of dopamine results in polydopamine-functionalized graphene. Afterward, the amine and catechol groups preferentially attached with $PtCl^{2-}$ and $AuCl^{-}$ by electronic conjugation which finally reduced by ascorbic acid to form the nanocomposite. It is also known that surfactant free synthesis (Chen et al. 2018; Sinhamahapatra et al. 2012) of metal oxide-based HSNs is efficient in regard of atom economized approach of green chemistry.

3.2.3 Green Solvents

A very promising category of green solvents is deep eutectic solvents (DES) because these are cost effective, eco-friendly, non-toxic and convenient for large-scale production compared to traditional ionic liquids (Wang et al. 2018). These green solvents (Table 8) are capable of producing different hierarchical TiO_2 structures (microrods and quasi-crassulapeforata, quasi-peanuts and hierarchical microspheres). Apart from these solvents, glycerol is a widely

Table 8 Characteristics of metal/metal oxide HSNs synthesized using green solvents

Solvent	Nanomaterial	Morphology/shape	Dimension	Property/application	References
DES (deep eutectic solvents)	TiO_2	Microspheres, quasi-microspheres, microrods and quasi-crassulapeforata-like structure constructed by nanodisks	–	Photocatalytic water splitting	Wang et al. (2018)
Green solvent, template free	ZnO	Wool ball structure assembled by nanoflakes	Thickness, 20–25 nm	Photocatalytic dye degradation	Singh et al. (2016)
Template/surfactant free	$ZnFe_2O_4$	Shuttle-shaped mesoporous microrods assembled by 1D nanofiber subunits	Nanofibers, 100–200 nm	Anode for LIBs	Hou et al. (2015)
Green solvent	CuO–ZnO	Flower	ZnO flower, ~554 nm; CuO nanoparticles, ~50–90 nm	Anticorrosion properties of thin film	Beshkar et al. (2017)
Water solvent	Co_3O_4	Nanoflakes assembled nanostructure	Thickness, 2–3 nm	H_2O_2 sensing	Su et al. (2015)
Green solvent	WO_3	Nanowires emerged from the edges of stacked nanoplates	Nanowire diameter, <20 nm	Photochemical water splitting	Nayak et al. (2017)
Green solvent	$CoMn_2O_4$	Porous micro/nanostructures	20–100 nm	Anode for LIB	Li et al. (2017)

Fig. 4 FESEM images of hierarchical self-assembled 3D ZnO superstructures at different magnifications. Copyrights reserved to the American Chemical Society (Singh et al. 2016)

used green solvent. Hierarchical 3D wool ball-like ZnO superstructures (Fig. 4) had been synthesized using urea (additive)-glycerol/water/ethanol assisted hydrothermal method. Glycerol had been used as chelating agent to Zn^{2+} as well as capping agent for regulating the morphology while the presence of different dosages of urea controls the morphology from urchin-like to wool ball-like structure of ZnO (Singh et al. 2016). On the other hand, hierarchical shuttle-shaped mesoporous $ZnFe_2O_4$ microrods assembled by 1D nanorods had been synthesized using the combination of green solvents, i.e., glycerol and water (Hou et al. 2015). In this process, the viscosity of glycerol was more than water. This influenced the diffusion rate of ions in glycerol slower than water that led to higher aggregation rate of the nanorods; eventually fused together to form the shuttle-shaped structures. Similarly, flower-like CuO/ZnO hybrid hierarchical nanostructures had been fabricated on copper substrate in which ethylene glycol was used as reducing agent as well as solvent (Beshkar et al. 2017).

4 Growth Mechanism of Metal and Metal Oxide HSNs

On the basis of the above reported literature as described in the previous sections/subsections of different green synthesis methods/techniques used for the fabrication of metal and metal oxide HSNs, one can classify the methods into

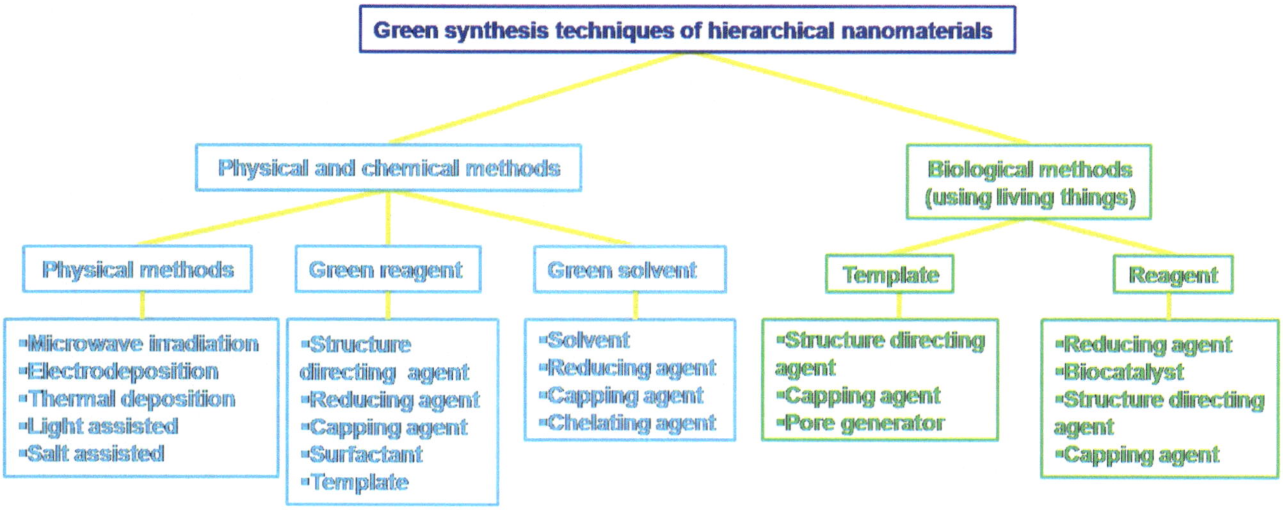

Fig. 5 Different green methods for the synthesis of metal and metal oxide HSNs

(a) physical, (b) chemical, and (c) biological methods (Fig. 5). For understanding the growth mechanism of the nanomaterials, it is essential to enter into the insights of the different methods, i.e., green physical, green chemical and green biological methods. In the literature, there are large numbers of reports available that has already been discussed in this chapter and the explanation on probable mechanistic aspects related to growth of nanoparticles toward the formation of hierarchically structured nanomaterials is also discussed. However, no generalized growth mechanism is yet found. This section describes the mechanistic pathways of formation of metal/metal oxide HSNs synthesized by physical, chemical, and biological methods with submethods also.

4.1 Biological Method

In recent years, the biogenic synthesis process also known as biological synthesis method toward producing metal/metal oxide nanomaterials has attracted substantial attention (Hulkoti and Taranath 2014). In this process, the nanomaterials can be synthesized using microorganisms and plant extracts. The biosynthesis can actually provide nanoparticles with ordered and controlled morphology (shape, size) and physiochemical properties as compared to some chemical and physical methods (Khandel et al. 2018; He et al. 2008). In this respect, there are previous reports on biocompatible and eco-friendly synthesis process based on microorganism (Hulkoti and Taranath 2014). The synthesized products can be used for pharmacological applications. However, one major disadvantage is that the mass production of the nanomaterials using microorganisms is often more expensive because of existing some of the critical handling protocols. On the other hand, the main advantage of plant-based synthesis approaches over conventional methods is being environmentally benign, low-cost, and scalable. In these processes, the use of high temperature, pressure, and toxic chemicals are not necessary. In brief, the microorganism and plant can be used in two ways for the synthesis of nanomaterials—(a) as reagent and (b) as template. In the next subsection, we will further discuss how they can function as reagents and templates.

4.1.1 Biomolecules as Reagents

It is important to mention that a common mechanism for the synthesis of nanoparticles employing microbes, for example, bacteria, fungi, algae, and yeast or biomolecules as reagents has yet not been conceived distinctly. One of the reasons is that the reaction mechanism of a biological reagent with a specific reactant like metal ions leading the formation of nanomaterials does not match with the other. In this respect, the formation of nanomaterials by microbes is known as an outcome of their defense mechanism toward the metal ions. Most of the reported works referred to enzymes, proteins. and lipids as the main biologically active materials that act as reducing, capping, and structure directing agents during the nanomaterials synthesis (Hulkoti and Taranath 2014; Mohammadinejad et al. 2016; Carbone et al. 2019). This is mostly due to of the defense mechanism of the microbes toward reduction of metal ions (He et al. 2008) and their aggregation with catalyzing the reaction medium (Dhandapani et al. 2020). The cell wall of the microorganisms also plays a crucial role in intracellular synthesis of nanoparticles/HSNs. In this process, an electrostatic interaction occurs between the positive charge of metal ions with the negative charge of the cell wall. The proteins that composes the cell wall enzymes reduce metal ions, resulting

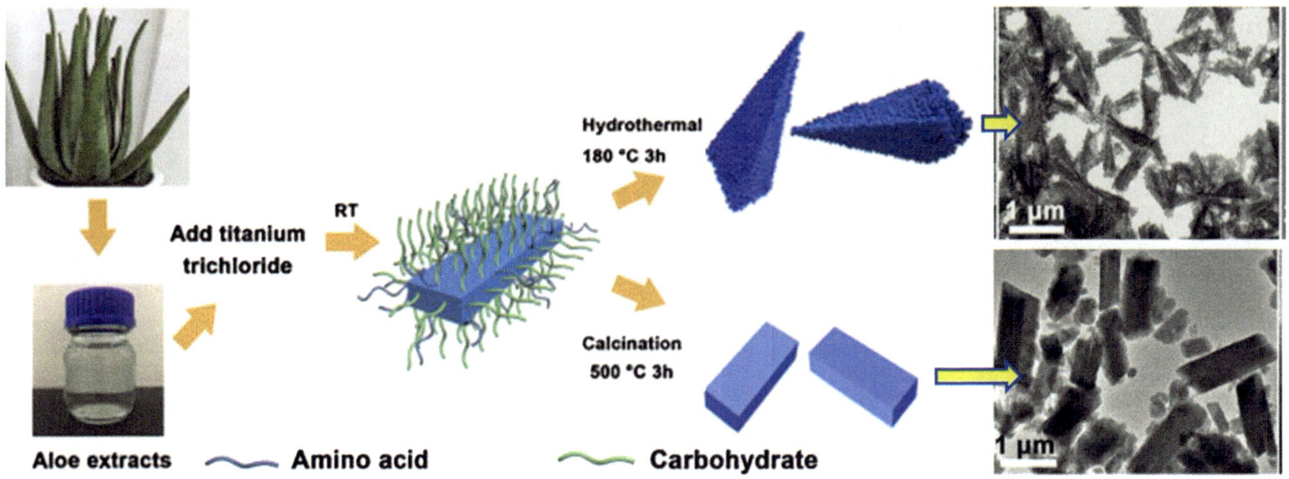

Fig. 6 Scheme for fabrication of tripyramidal and rod-like TiO_2. Copyrights reserved to the Elsevier (Li et al. 2020)

the formation of nanoparticles/nanomaterials with hierarchical structures (Kitching et al. 2016).

Similarly, plants have ability to precisely producing a variety of highly ordered hierarchical nanostructures. It is seen that various water soluble plant metabolites such as polyphenols, flavanoids, tannins, phytolexins, terpenoids, proteins, flavonoids, alkanoids, limonoids, amino acids, cellulose, saponin, pectin and hemicelluloses can act as reducing, capping and complexing agents for the synthesis of metal and metal oxide HSNs (Fig. 6) (Carbone et al. 2019; Bankar et al. 2010; Udayabhanu et al. 2017; Li et al. 2020; Suresh et al. 2020; Khaghani et al. 2017; Singh et al. 2017; Madan et al. 2015; Sundar et al. 2018).

4.1.2 Biomolecules as Templates

In our nature, different microorganisms are present in diverse morphologies such as spheres, rods, spirals, icosahedrons, and so on (Hulkoti and Taranath 2014; Khandel et al. 2018). Sometimes, these are varying size in length starting from nano to mesoscopic scales providing the base for metal ions that can act as low-cost and eco-friendly templates in potential applications of micro- and nanofabrication *via* green synthesis especially for hollow micro/nanoporous metal and metal oxide nanomaterials. So far, bacteria, fungi, yeast and algae have been employed successfully as biotemplates for the synthesis of controllable structures of various nanomaterials. A major advantage of this process is the specific morphology of a microbe. This offers a uniform and tunable biotemplate as base of the metal ions. Functional groups like OH–, CHO–, COO–, etc. present in the microbial cell wall bind the metal ions by electrostatic force of attraction whereas the cell wall acts as nucleation site of the metal ions. These avoid the need of additional surface modifying or templating agent for further ripening, self-assembly, and the growth of metal and/or metal oxide toward the formation of hierarchical nanostructures. After calcination, the decomposition of an organic microbial template as well as escape of CO_2 and H_2O occurs. Thus, one can make hierarchically porous nanostructures and in some cases, it results in the formation of hollow microstructures due to complete decomposition of the template. A brief mechanism of the synthesis of hierarchically ordered mesoporous TiO_2 using yeast (Cui et al. 2009) as biotemplate is displayed in Fig. 7.

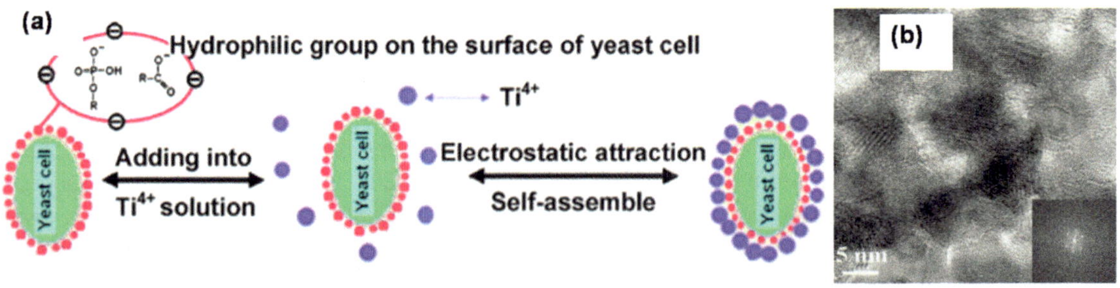

Fig. 7 **a** Illustration of ordered hierarchical mesoporous TiO_2 preparation process and **b** the corresponding HRTEM image of nano TiO_2. Copyrights reserved to the Elsevier (Cui et al. 2009)

Nature has already performed a fine job in creating humongous hierarchy in the structures of the living beings and creatures such as insects, plants and plant-derived products. Particularly, at the micro- and nanometer scales, these naturally abundant structures show such degree of elegance that it overshadows current man-made bioinspired structures as synthesized in conventional manners (Mohammadinejad et al. 2016; Ebrahiminezhad et al. 2018; Zan and Wu 2016). In general, the process consists of two dominant stages—(a) the assembly of the precursor and (b) template removal (Zan and Wu 2016). At first, a biotemplate is dipped into a precursor metal solution, which then diffuses and permeates into the template. In the next, the precursor metal ions self-assemble onto specific sites of the template *via* a molecular recognition process, deposit homogenously over the template by electrostatic attraction between metal ions and the functional groups (carboxylic, hydroxyl, amine, etc.) present in the cell wall template and form a stable organic–inorganic composite/hybrid/complex (Ramimoghadam et al. 2013a; Han et al. 2015; Mallampati and Valiyaveetti 2012). In the following steps, further growth of the precursor through the template occurs in which the template acts as the structure directing agent and the functional groups act as capping agents. When the growth of the nanomaterial extends up to its thickness capacity, the growth is ceased due to lack of space. Afterward, as the template is removed by calcination, porous metal/metal oxide nanomaterials are formed with magnificent hierarchy of the replicated specific template. Wood (Liu et al. 2009) (Fig. 8), pollen (Yang et al. 2011; Fazil et al. 2015), nanofibers (Wei et al. 2019; Xiao et al. 2013; Moyer et al. 2019), leaf (Li et al. 2009; Abu-Zied and Alamry 2019; Han et al. 2015), cotton (Mohamed et al. 2016; Ma et al. 2019), flower petals (Qian et al. 2011), fruit exocarps, inner coats and sprouts (Yan et al. 2006; Chen et al. 2008), rice and starch (Ramimoghadam et al. 2013a; Farzaneh et al. 2017) are some of the plant-derived biotemplates that had been used for the green synthesis for variety of nanomaterials. On the other hand, butterfly wings (Zhang et al. 2018,2006,2009) and eggshell membranes (Mallampati and Valiyaveetti 2013; Dong et al. 2007a,b; Fan et al. 2016), albumen (Nouroozi and Farzaneh 2011), and glutamine (Nouroozi and Farzaneh 2011) are the examples of widely used insect/animal-derived biotemplates.

4.2 Physical and Chemical Methods

The physical methods/techniques that are used in green synthesis of hierarchically structured nanomaterials involve requirement of less energy than the conventional methods. These can also reduce the use of harmful solvents, surfactants, and templates. Moreover, these are environment friendly and often cost effective. Microwave-assisted synthesis method is one of the most energy and time efficient methods that has been used widely for the synthesis of hierarchical nanomaterials (Lei et al. 2014; Tompsett et al. 2006). As the microwave energy increases, the heating rate of the medium and time of the overall reaction are reduced *vis-á-vis* a lower consumption of overall energy is possible. Microwave synthesis allows more uniform heating of the reaction mixture that leads to form homogeneous distribution of nanocrystals. This is expected to be one of the very crucial parameters toward the formation of hierarchical nanostructures.

Thermal decomposition and electrochemical deposition are another green approaches for the synthesis of ultra-pure nanostructured materials without using any fuel or leaving off toxic side products to the environment. As for example, 3D hierarchical nanopyramids and nanosheets array of Ag/Cu_2O and CuO nanoneedles had been fabricated using electrochemical deposition method (Ji et al. 2019; Momeni et al. 2016). In this respect, a reported study suggested that by controlling the reaction time and temperature, diverse superstructures of ZnO can be obtained (Udayabhanu et al. 2016).

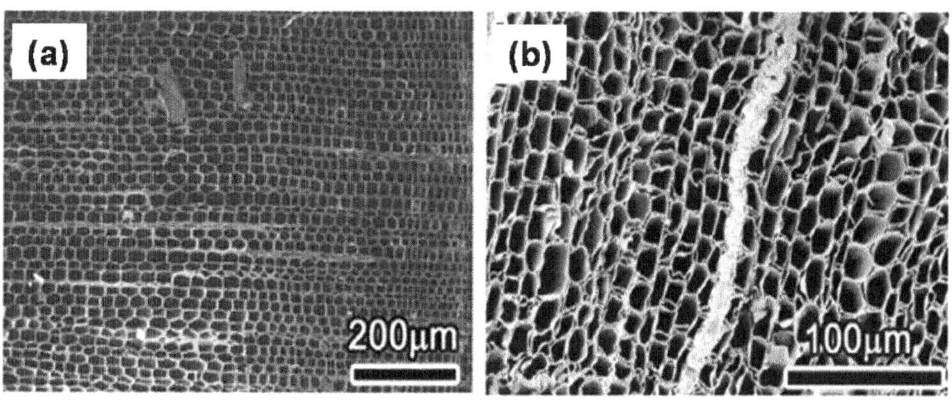

Fig. 8 FESEM images of **a** carbonized wood and **b** prepared ZnO using wood. Copyrights reserved to the Elsevier (Liu et al. 2009)

Using the precursor materials that undergo visible light-assisted decomposition is a unique and eco-friendly method for the synthesis of hierarchical nanostructures of especially metal oxides (Das et al. 2017; Hu et al. 2016). In this method, high temperature, heat or reducing agents are not required and only the presence of light is required for accelerating the reaction toward formation of nanostructures. For example, in a green and sustainable pathway, the sunlight-assisted decomposition phenomenon of $KMnO_4$ gives rise to hierarchical flower-like MnO_2.

Environmentally benign reagents such as amino acid (Gao et al. 2008), metal (Zhang et al. 2011), glucose (Yang et al. 2013), polyethylene glycol (Gao et al. 2012; Lee et al. 2016), and dopamine (Ye et al. 2015) have been used in the green synthesis of HSNs. In this case, the reagents can perform multifunctions such as reducing agents, structure directing agents, and capping agents in a surfactant and template-free synthesis. Thus, atom economy that is one of the principles of green synthesis can also be maintained properly. Although the exact functional mechanism is very difficult to investigate, the organic functional groups like carboxylic and amines groups present in the green regents are known to be responsible for their multifunctionalities. It is noteworthy that biopolymer alginate has generally been used as cross-linker for manganese ions to form hierarchically structured nanoporous metal oxide hybrids (Wang et al. 2016) (Fig. 9) whereas the reducing agent, surfactant and structure directing agents are to be found responsible for creating flower-like nanopetal assembly of δ-MnO_2 (Zong et al. 2016).

Some of the reported studies show that no additional foreign reagents and reducing agents have been used but complex (Sinhamahapatra et al. 2012) and metal oxide framework (Hu et al. 2018) can act as reducing and structure directing/evolving agents for the synthesis of hierarchical metal and metal oxide nanomaterials. Sinhamahapatra et al. (2012) reported the synthesis of hierarchically structured porous ZnO, where the assembled bundles of woolen threads like hydrozincite, $Zn_5(CO_3)_2(OH)_6$ architectures had initially been synthesized and used as the precursor. Upon calcination, the evolution of CO_2 and H_2O took place with the formation of porous nanostructures of ZnO retaining original morphology of hydrozincite (Fig. 10).

Sustainable solvents and additives are other important materials used for the green synthesis of metal and metal oxide HSNs. Water, ethanol, glycerol, ethylene glycol, and urea are some of environmentally benign solvents and additives that are used widely. These solvents function as reducing agents (Beshkar et al. 2017), chelating/capping agents (Singh et al. 2016), and structure directing agents (Singh et al. 2016; Hou et al. 2015; Nayak et al. 2017) in the synthesis process. The combination of hydrogen-bond acceptors (glycine, betaine, or acetylcholine chloride) and donors (urea, oxalic acid, ethylene glycol, or glycerinum) had been used strategically for controlling the morphology of TiO_2 nanostructures where these materials played the role of solvent, template, and inhibitor (Wang et al. 2018).

5 Applications of Hierarchically Structured Metal and Metal Oxide Nanomaterials

In this section, different potential applications of green synthesized metal and metal oxide nanostructures have been discussed. These applications basically rely upon the properties of metal and metal oxides such as morphology and porosity developed during the process of synthesis.

5.1 Biomedical Application

Nanomaterials produced by green synthesis methods have a wide range of applications in biomedical and pharmacological fields for antibacterial, antifungal, and antioxidant applications (Yuliarto et al. 2019). Due to small particle size and high surface area of the nanomaterials, they can easily attach to microbial membranes in a closer proximity

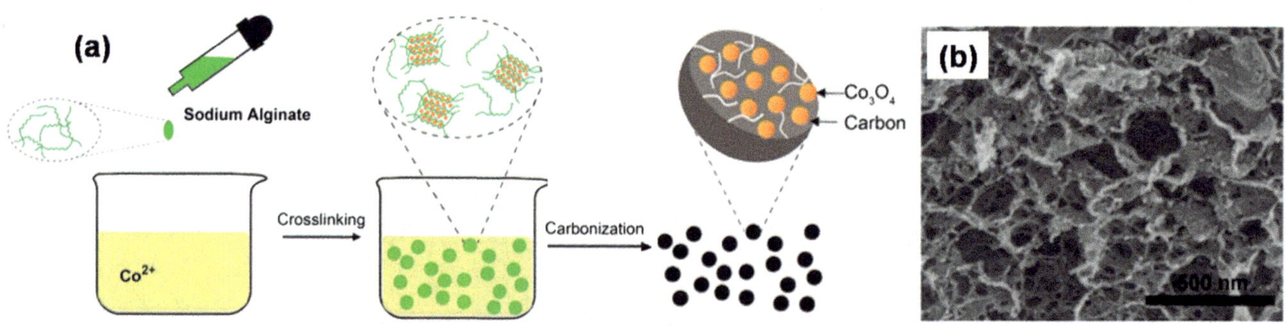

Fig. 9 **a** Schematic of the synthesis process and **b** SEM image of nanoporous Co_3O_4/C hybrids. Copyrights reserved to the American Chemical Society (Wang et al. 2016)

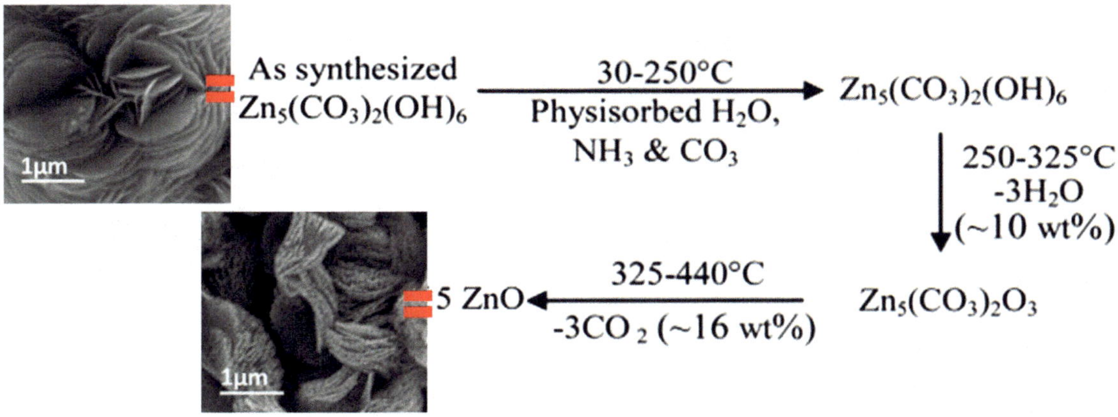

Fig. 10 Schematic for the synthesis of ZnO replications from hydrozincite. Copyrights reserved to the Royal Society of Chemistry (Sinhamahapatra et al. 2012)

compared to their bulk counterparts (Seth et al. 2020). Although, the mechanism of antimicrobial activity has not been understood clearly, the physical damage of the cell wall caused by the nanoparticles had been reported to be responsible for cell death of microbes. Release of metal ions from nanostructures is also said to be playing an important role for the antimicrobial activity. In this respect, *Spirulina* templated Ag nanosheets showed excellent bactericidal activity against *S. aureus*. Additionally, the slow release of Ag^+ ions reduces the chance of adverse impact of ionic silver to our environment (Sun et al. 2019). Moreover, a study revealed that the Ag^+ possess good adaptability. Thus, it can be used in applications such as food safety. Fungicidal activity of dendritic Ag nanoparticles against plant pathogen *F. Graminearum* can practically be used in crop plant protection (Carbone et al. 2019). It has been reported that nanoflowers-shaped Au nanoparticles are hemocompatible and can be used as suitable bioconjugates in therapeutic and biomedical fields (Kitching et al. 2016). It is important to note that the accessibility of Ag^+ ions in the hierarchical dendritic architecture had been reported to be the limiting factor for its fungicidal activity. In another example, ZnO superstructures showed antibacterial activity against both gram positive and negative bacteria (Udayabhanu et al. 2017; Madan et al. 2015). Also, hierarchical CuO nanostructures showed antibacterial activity against gram positive and gram negative bacterial strains (Suresh et al. 2020). The probable factor for killing the bacteria is attributed to generation of reactive oxygen species (superoxide and hydroxide radicals) formed by CuO nanostructures and the affinity of amine as well as carboxylic acid groups present on the cell wall of bacteria may result the formation of Cu^{2+} ions. Moreover, CuO nanoparticles can easily enter into comparatively bigger pores of bacteria causing malfunction in cell enzyme resulting cell death. They also have the potential antioxidant properties, attributed to their small crystal size and hierarchical bullet shaped morphology.

5.2 Environmental Remediation

5.2.1 Wastewater Treatment

To color the final products, various dyes are substantially used in textile, paper, plastic, cosmetic, leather, drug and food processing industries. These dyes are eventually released into water bodies and soil that can severely affect the environment, hamper the ecosystem of water body and cause serious health hazards to human beings (Fang et al. 2019). A considerable amount of research had already been performed for the fabrication of metal/metal oxide nanomaterials as efficient photocatalysts for degradation of the organic dyes under UV/visible light irradiation (Bera et al. 2016c). In the photocatalysis process, toxic organic pollutants photo-degrade into non-toxic by-products through mineralization, without further waste production. In this respect, semiconductor absorbs photon with energy equal or more than the band gap of the semiconductor can generate electrons and holes in the semiconductor system. If the recombination rate of charge carriers is slow, the species will travel to the surface, where the free electrons reduce oxygen and forms peroxides/superoxides and holes that oxidize water and forms OH· (Udayabhanu et al. 2017). These highly reactive and unstable species finally lead to photo-degrade organic dyes. It is also known that biosynthesized nanomaterials exhibit an excellent photocatalytic performance due to high surface-to-volume ratio and the existence of higher number of active sites compared to polycrystalline materials (Fang et al. 2019). Other important factors that affect the photocatalysis performance are crystallinity, porosity, particle size, morphology, particle size distribution and band gap of a photocatalyst.

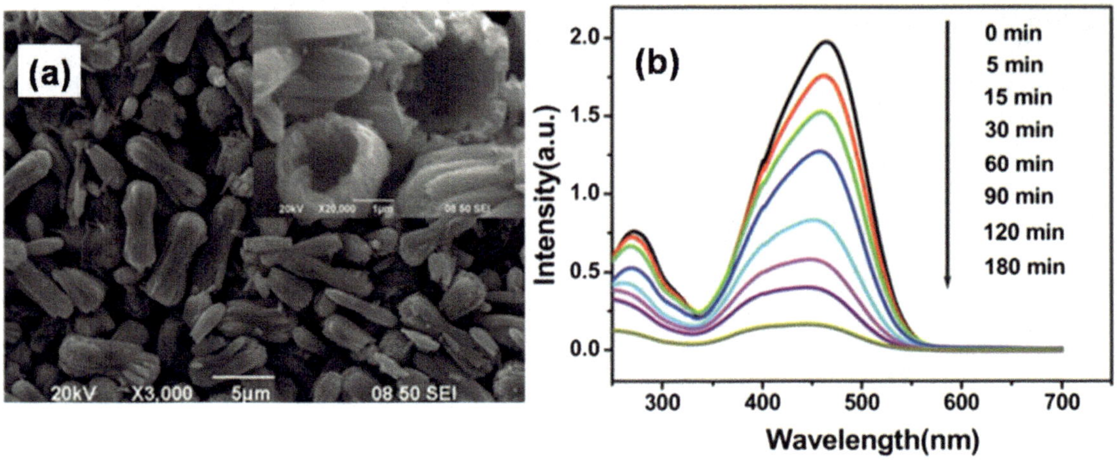

Fig. 11 FESEM image (**a**) and photocatalytic study, (**b**) hollow peanut-like ZnO powder. Copyrights reserved to the Royal Society of Chemistry (Wang et al. 2012).

TiO_2 and ZnO are some of the well-known photocatalysts mostly used in water pollution alleviation *via* photocatalysis as they are cost effective, non-toxic, chemically, and mechanically stable and they can easily form hierarchically structured nanomaterials. In an example, hollow double-caged peanut-like ZnO microstructures had been reported for the photocatalytic degradation of methyl orange under UV irradiation (Wang et al. 2012) (Fig. 11). After 180 min of irradiation, the characteristic peak of the dye was found to be eliminated by ZnO. The large surface area of nanorods assembled peanut structure and commodious interspaces of ZnO microstructure are highly effective for the diffusion and mass transportation of dye molecules and hydroxyl radicals for photocatalytic degradation of dye. In this respect, Li et al. (2009) reported an improvement in photocatalytic activity of leaf templated hierarchical porous morph-TiO_2 than non-templated TiO_2, indicating the contribution of porous and layered nanostructure of morph-TiO_2 toward the catalytic activity.

Adsorption is also an efficient method of dye removal from wastewater. Nanofibrous network of Mn_3O_4 had been used for absorbing a wide range of organic dyes by electrostatic attraction between the surface of Mn_3O_4 and organic compounds in aqueous solution (Mallampati and Valiyaveetti 2012).

5.2.2 Energy Storage

Transition metal oxide-based nanomaterials have been widely used as potential electrode materials (PEMs) for pseudocapacitors (Zhang et al. 2019b), Li-ion batteries (Hashem et al. 2018; Wei et al. 2019; Gao et al. 2008; Hou et al. 2015; Hu et al. 2018; Li et al. 2017), and supercapacitors (Wang et al. 2016; Zong et al. 2016; Sinhamahapatra et al. 2012). On this aspect, flower-like nanostructured $NiCo_2O_4$ that composed of ultrathin nanopetals (thickness ~15 nm) having large specific surface combined with narrow pore size distribution can be considered as one of the significant examples of PEMs (Lei et al. 2014). It is seen that the flower-superstructure composed of mesoporous nanopetals with high surface area is responsible for exhibiting high capacitance and stability of $NiCo_2O_4$, as the nanopetals can provide a large number of electroactive sites for Faraday reaction. Consequently, the microflower can act as an *ion-buffering reservoir* and expedites a quicker permeation process of electrolyte into nanopetal matrix. Zhang *et al.* reported (Zhang et al. 2019b) the synthesis of hierarchical porous MnO@biocarbon (BC) nanocomposite using molted salt assisted method. The highly porous HSN can be used as supercapacitor electrode materials for electrochemical energy storage as well as lithium-ion battery anodes.

5.2.3 Sensing

Gas sensing is analyzed by detection of the variation of conductance of a sensing material that occurs because of the surface reactions between the target gas molecules and the sensing particles (Lee 2009). Obviously, for gas sensing, the surface area with higher hierarchical porosity is required so that the porous network with excellent interconnectivity and mass transportability can supply plentiful connective multiscale channels toward the transportation of gas molecules to directly sensing by metal oxide nanoparticles (Song et al. 2012). In addition, the temperature is a crucial factor affecting the sensitivity of gas sensing performance of metal oxide nanomaterials.

It is also noticed that pollen coats of rape pollen grains have been used to synthesize highly connective hierarchical porous network structure of SnO_2 (Song et al. 2012). Interconnected mesoporous network further can be extended to macro- to nanoscale pores. This makes SnO_2 an excellent candidate for selective sensing of C_2H_5OH, CH_3COCH_3 and

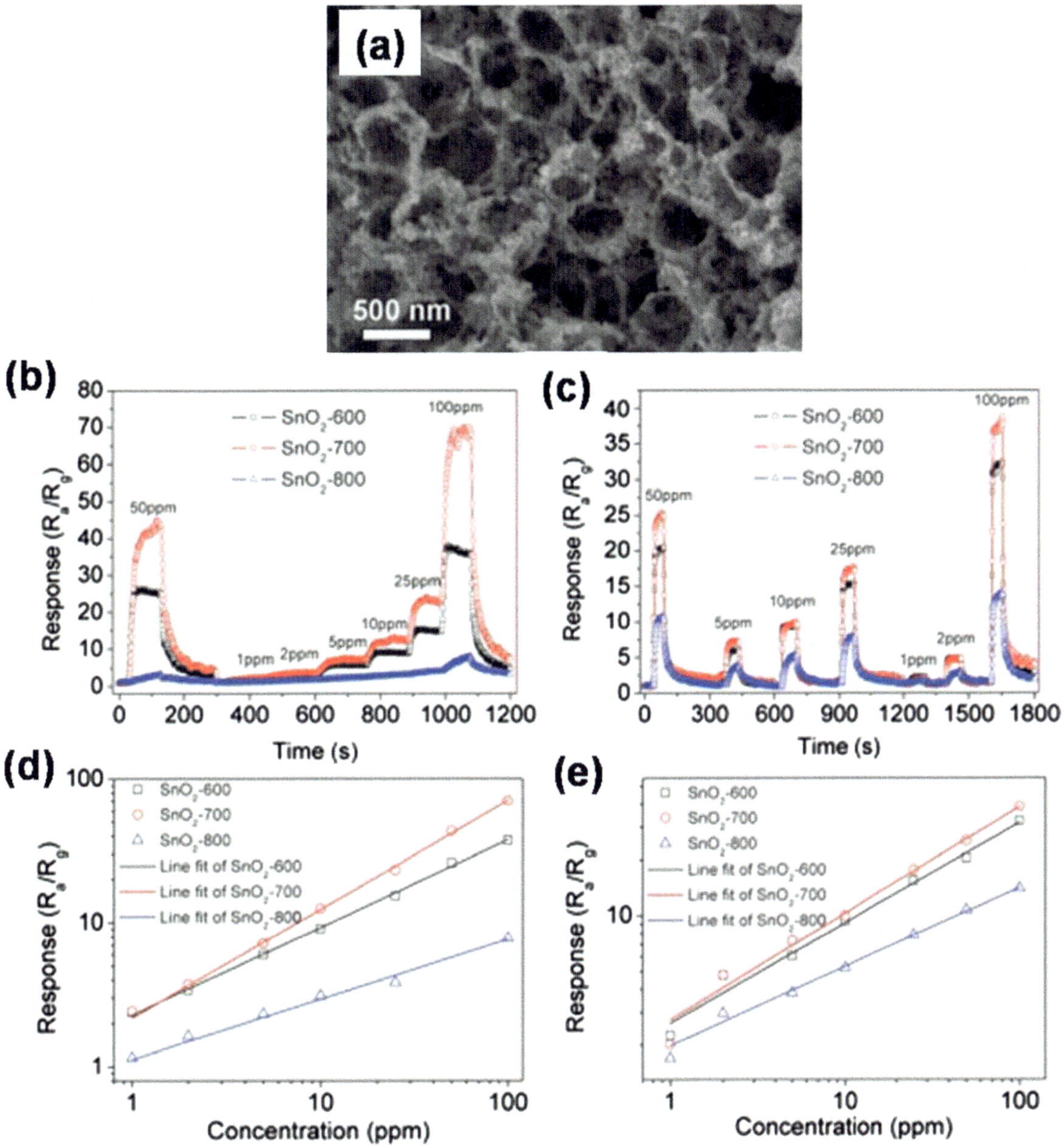

Fig. 12 **a** FESEM image of SnO_2, sintered at 700°C; **b**, **c** real-time response curves and **d**, **e** response dependens on gas concentrations of the sinters; **b**, **d** to C_2H_5OH at 210 °C; **c**, **e** to CH_3COCH_3 at 290 °C. Copyrights reserved to the Royal Society of Chemistry (Song et al. 2012).

Cl_2 (Fig. 12). In another example, wood template-mediated synthesis of hierarchically porous ZnO showed exceptional selective gas sensing for H_2S (Liu et al. 2009). In this case, the hierarchically porous ZnO with higher porosity and surface area provided more surface adsorption positions and reacting areas for oxygen and test gases that helped the target gases to transfer more quickly leading to increase gas response, in comparison to lower gas sensitivity of non-template based ZnO.

Green synthesized 3D chestnut-like structures of Ag decorated WO_{3-x} are found to be used as clean, stable, and recyclable SERS substrate which is able to identify and provide the fingerprint structural information on various analytes even at low concentration (Huang et al. 2017). The material also has exceptional self-cleaning ability.

6 Present Challenges and Future Prospect

Nowadays, hierarchically structured metal and metal oxide nanomaterials have been widely studied for their special surface structures and remarkable applications in various fields including but not limited to pharmaceutical, biomedical, electronics and environmental remediation. These nanomaterials are particularly in demand because of their characteristic features like large surface area with exceptional surface morphology with high porosity. Biological mode has become an important part of nanotechnology where bioagents are used as reagents and templates for the synthesis of hierarchical nanomaterials. However, certain limitations are there in these nanomaterials for their practical production and relevant applications. These must need to be resolved by the scientific community. Firstly, one of the major limitations lies in the lack of complete and in-depth understanding of mechanism of biofabrication of the nanoparticles/nanomaterials. There are several reports available in literature where reasonable hypotheses had been proposed to explain the experimental results and biologically active molecules responsible for biomineralization of metal ions using bioextracts/templates. However, a detailed analysis of the biochemical mechanism is further needed for the development of green HSNs with desired and controlled structure, size, morphology, dispersity, and related properties (Gahlawat and Roy Choudhury 2019; Deng et al. 2011) toward their real applications. Secondly, large-scale production of nanomaterials synthesized by adopting green synthesis methods is mostly obstructed by inability to fully control the structure, size, and morphology of hierarchical nanomaterials along with other concerns toward polydispersity and low yield. The synthesis of nanomaterials at ambient temperature by using natural reagents as reducing/capping/structure directing agents without any toxic additives may make the process for large-scale production in cost-effective and environmental friendly. Also, the large-scale production of monodispersed nanomaterials with narrow size distribution and high yield can be achieved by optimizing the diverse synthesis parameters such as solution pH, temperature and time of reaction, concentration ratio of biomolecules and metal precursors. Interestingly, instead of bacteria, fungi can be used for the production of humongous concentration of nanoparticles as fungi is able to discharge larger amounts of proteins which lead to higher productivity of nanomaterials. In addition, an exploration of microbial diversity of novel, sustainable microorganisms with varying shape may be used as reagent/template for the synthesis of hierarchical nanomaterials. Often, the nanomaterials cannot be effectively separated or regenerated after participating in catalytic/adsorption reactions, hence effective techniques must be established for this purpose. Finally, the synthesis of HSNs using microorganisms is a slow-going process that may take up to several hours or even few days. Green chemical and physical methods as discussed in this chapter can be established as less time-consuming method because they need less time to complete and produce materials with controlled shape and size and higher yield of products. These methods are also easy to handle in comparison to microorganisms which have a high risk of contamination. Another important issue is the biocompatibility and bioavailability of the produced nanomaterials and their environmental sustainability. In this regards, more extensive research and clinical trial are required to study the cytotoxicity and genotoxicity of the synthesized nanomaterials for their practical use in biomedical, pharmaceutical and agricultural fields as well as wastewater purification.

Without causing any further harm to our environment, the green synthesis methods that evolved from the greatness of nanotechnology toward production of large-scale hierarchically structured nanomaterials has a bright future for application in various sectors like health, water, food, and energy. A collaborative research is indeed necessary for understanding the detailed mechanistic aspects, exploration of new biological agents and finding out innovation by cost-effective biological, physical, and chemical green synthesis methods.

Acknowledgements Authors thankfully acknowledge CSIR and DST, Government of India for providing the financial supports to some of the authors (MS, HK) at their doctoral research.

References

Abu-Zied BM, Alamry KA (2019) Green synthesis of 3D hierarchical nanostructured Co_3O_4/carbon catalysts for the application in sodium borohydride hydrolysis. J Alloys Compd 798:820–831

Alkaim AF, Alrobayi EM, Algubili AM, Aljeboree AM (2016) Synthesis, characterization, and photocatalytic activity of sonochemical/hydration–dehydration prepared ZnO rod-like architecture nano/ microstructures assisted by a biotemplate. Environ Technol 38(17):2119–2129

Atalay FE, Asma D, Kaya H, Bingol A, Yaya P (2016) Synthesis of NiO nanostructures using *Cladosporium cladosporioides* fungi for energy storage applications. Nanomater Nanotechnol 6:28

Bankar A, Joshi B, Kumar AR, Zinjarde S (2010) Banana peel extract mediated novel route for the synthesis of palladium nanoparticles. Mater Lett 64(18):1951–1953

Bao S-J, Lei C, Xu M-W, Cai C-J, Jia D-J (2012) Environment-friendly biomimetic synthesis of TiO_2 nanomaterials for photocatalytic application. Nanotechnol 23:205601

Bera S, Naskar A, Pal M, Jana S (2016a) ZnO-graphene-polyaniline nanoflowers: solution synthesis, formation mechanism and electrochemical activity. RSC Adv 6(47):40854–40857

Bera S, Naskar A, Pal M, Jana S (2016b) Low temperature synthesis of graphene hybridized surface defective hierarchical core-shell structured ZnO hollow microspheres with long-term stable and enhanced photoelectrochemical activity. RSC Adv 6(42):36058–36068

Bera S, Pal M, Naskar A, Jana S (2016c) Hierarchically structured ZnO-graphene hollow microspheres towards effective reusable adsorbent for organic pollutant via photodegradation process. J Alloys Compd 669:177–186

Bera S, Pal M, Sarkar S, Jana S (2017) Hierarchically-structured macro with nested mesoporous zinc indium oxide conducting film. ACS Appl Mater Interfaces 9(5):4420–4424

Beshkar F, Khojasteh H, Salavati-Niasari M (2017) Flower-like CuO/ZnO hybrid hierarchical nanostructures grown on copper substrate: glycothermal synthesis, characterization, hydrophobic and anticorrosion properties. Materials 10(7):697

Carbone K, Paliotta M, Micheli L, Mazzuca C, Cacciotti I, Nocente F, Ciampa A, Dell'Abate MT (2019) A completely green approach to the synthesis of dendritic silver nanostructures starting from white grape pomace as a potential nanofactory. Arab J Chem 12(5):597–609

Chen P, Wu Q-S, Ding Y-P, Yuan P-S (2008) Synthesis of $SrCrO_4$ nanostructures by onion inner coat template and their optical properties. Bull Mater Sci 31(4):603–608

Chen X, Chu D, Wang L, Hu W, Yang H, Sun J, Zhu S, Wang G, Tao J, Zhang S (2018) Surfactant-free synthesis of novel hierarchical dahlia-like SnO_2 nanostructures with enhanced visible-light-driven photocatalytic activity. J Alloys Compd 768:517–524

Cui J, He W, Liu H, Liao S, Yue Y (2009) Ordered hierarchical mesoporous anatase TiO_2 from yeast biotemplates. Colloids Surf B 74(1):274–278

Das S, Samanta A, Jana S (2017) Light-assisted synthesis of hierarchical flower-like MnO_2 nanocomposites with solar light induced enhanced photocatalytic activity. ACS Sustain Chem Eng 5 (10):9086–9094

Deng C, Hu H, Zhu W, Han C, Shao G (2011) Green and facile synthesis of hierarchical cocoon shaped CuO hollow architectures. Mater Lett 65(3):575–578

Dhandapani P, Prakash AA, AlSalhi MS, Maruthamuthu S, Devanesan S, Rajasekar A (2020) Ureolytic bacteria mediated synthesis of hairy ZnO nanostructure as photocatalyst for decolorization of dyes. Mater Chem Phys 243:122619

Diab M, Shreteh K, Volokh M, Abramovich S, Abdu U, Mokari T (2019) Calcareous foraminiferal shells as a template for the formation of hierarchal structures of inorganic nanomaterials. ACS Appl Mater Interfaces 11(6):6456–6462

Dong Q, Su H, Zhang D, Zhu N, Guo X (2006) Biotemplate-directed assembly of porous SnO_2 nanoparticles into tubular hierarchical structures. Scr Mater 55(9):799–802

Dong Q, Su H, Song F, Zhang D, Wang N (2007a) Hierarchical metal oxides assembled by nanocrystallites via a simple bio-inspired route. J Am Ceram Soc 90(2):376–380

Dong Q, Su H, Xu J, Zhang D, Wang R (2007b) Synthesis of biomorphic ZnO interwoven microfibers using eggshell membrane as the biotemplate. Mater Lett 61(13):2714–2717

Ebrahiminezhad A, ZareHoseinabadi A, Sarmah AK, Taghizadeh S, Ghasemi Y, Berenjian A (2018) Plant mediated synthesis and applications of iron nanoparticles. Mol Biotechnol 60(2):154–168

Fan S, Zhao M, Ding L, Liang J, Chen J, Li Y, Chen S (2016) Synthesis of 3D hierarchical porous Co_3O_4 film by eggshell membrane for non-enzymatic glucose detection. J Electroanal Chem 775:52–57

Fang X, Wang Y, Wang Z, Jiang Z, Dong M (2019) Microorganism assisted synthesized nanoparticles for catalytic applications. Energies 12(1):190

Farzaneh F, Foruzin LJ, Sharif Z, Rashtizadeh E (2017) Green synthesis and characterization of Bi_2O_3nanorods as catalyst for aromatization of 1,4-dihydropyridines. Science 28(2):113–118

Fazil AA, Bhanu JU, Amutha A, Joicy S, Ponpandian N, Amirthapandian S, Panigrahi BK, Thangadurai P (2015) A facile bio-replicated synthesis of SnO_2 motifs with porous surface by using pollen grains of *Peltophorum pterocarpum* as a template. Micropor Mesopor Mat 212:91–99

Gahlawat G, Roy Choudhury A (2019) A review on the biosynthesis of metal and metal salt nanoparticles by microbes. RSC Adv 9 (23):12944–12967

Gao S, Yang S, Shu J, Zhang S, Li Z, Jiang K (2008) Green fabrication of hierarchical CuO hollow micro/nanostructures and enhanced performance as electrode materials for lithium-ion batteries. J Phys Chem C 112(49):19324–19328

Gao S, Jia X, Li Z, Chen Y (2012) Hierarchical plasmonic-metal/semiconductor micro/nanostructures: green synthesis and application in catalytic reduction of p-nitrophenol. J Nanopart Res 14:748

Han L, Yang D-P, Liu A (2015) Leaf- templated synthesis of 3D hierarchical porous cobalt oxide nanostructure as direct electrochemical biosensing interface with enhanced electrocatalysis. Biosens Bioelectron 63:145–152

Hashem AM, Abuzeid H, Kaus M, Indris S, Ehrenberg H, Mauger A, Julien AM (2018) Green synthesis of nanosized manganese dioxide as positive electrode for lithium-ion batteries using lemon juice and citrus peel. Electrochim Acta 262:74–81

He S, Zhang Y, Guo Z, Gu N (2008) Biological synthesis of gold nanowires using extract of *Rhodopseudomonas capsulate*. Biotechnol Prog 24(2):476–480

He T, Weng Y, Yu P, Liu C, Lu H, Sun Y, Zhang S, Yang X, Liu G (2014) Bio-template mediated in situ phosphate transfer to hierarchically porous $TiO_2$with localized phosphate distribution and enhanced photoactivities. J Phys Chem C 118:4607–4617

Hou L, Hua H, Lian L, Cao H, Zhu S, Yuan C (2015) Green template-free synthesis of hierarchical shuttle-shaped mesoporous $ZnFe_2O_4$ microrods with enhanced lithium storage for advanced Li-ion batteries. Chem Eur J 21(37):13012–13019

Hu X, Shi L, Zhang D, Zhao X, Huang L (2016) Accelerating the decomposition of $KMnO_4$ by photolysis and autocatalysis: a green approach to synthesize layered birnessite-type MnO_2 assembled hierarchical nanostructure. RSC Adv 6(17):14192–14198

Hu X, Lou X, Li C, Yang Q, Chen Q, Hu B (2018) Green and rational design of 3D layer-by- layer MnO_x hierarchically mesoporous microcuboids from MOF templates for high-rate and long- life Li-ion batteries. ACS Appl Mater Interfaces 10(17):14684–14697

Huang J, Ma D, Chen F, Chen D, Bai M, Xu K, Zhao Y (2017) Green in situ synthesis of clean 3D chestnutlike Ag/WO_{3-x} nanostructures for highly efficient, recyclable and sensitive SERS sensing. ACS Appl Mater Interfaces 9(8):7436–7446

Hulkoti NI, Taranath TC (2014) Biosynthesis of nanoparticles using microbes—a review. Colloids Surf B 121:474–483

Hussein MZ, Azmin WHWN, Mustafa M, Yahaya AH (2009) *Bacillus cereus* as a biotemplating agent for the synthesis of zinc oxide with raspberry- and plate-like structures. J Inorg Biochem 103(8):1145–1150

Ji S, Kou S, Wang M, Qiu H, Sun X, Dou J, Yang Z (2019) Two-step synthesis of hierarchical Ag/Cu_2O/ITO substrate for ultrasensitive and recyclable surface-enhanced Raman spectroscopy applications. Appl Surf Sci 489:1002–1009

Kang L, Xu P, Chen D, Zhang B, Du Y, Han X, Li Q, Wang H-L (2013) Amino acid-assisted synthesis of hierarchical silver microspheres for single particle surface-enhanced raman spectroscopy. J Phys Chem C 117(19):10007–10012

Khaghani S, Ghanbari D, Khaghan S (2017) Green synthesis of iron oxide-palladium nanocomposites by pepper extract and its application in removing of colored pollutants from water. J Nanostruct 7 (3):175–182

Khan H, Seth M, Samanta S, Jana S (2020) Nano gold coated hierarchically porous zinc titanium oxide sol-gel based thin film: fabrication and photoelectrochemical activity. J Sol-Gel Sci Technol 94:141–153

Khandel P, Yadaw RK, Soni DK, Kanwar L, Shahi SK (2018) Biogenesis of metal nanoparticles and their pharmacological applications: present status and application prospects. J Nanostruct Chem 8:217–254

Kitching M, Choudhary P, Inguva S, Guo Y, Ramani M, Das SK, Marsili E (2016) Fungal surface protein mediated one- pot synthesis of stable and hemocompatible gold nanoparticles. Enzyme Microb Technol 95:76–84

Kreuder ADV, House-Knight T, Whitford J, Ponnusamy E, Miller P, Jesse N, Rodenborn R, Sayag S, Gebel M, Aped I, Sharfstein I,

Manaster E, Ergaz I, Harris A, Grice LN (2017) A Method for Assessing greener alternatives between chemical products following the 12 principles of green chemistry. ACS Sustainable Chem Eng 5 (4):2927–2935

Lee J-H (2009) Gas sensors using hierarchical and hollow oxide nanostructures: overview. Sensor Actuat B-Chem 140(1):319–336

Lee M-J, Lim S-H, Ha J-M, Choi S-M (2016) Green synthesis of high-purity mesoporous gold sponges using self-assembly of gold nanoparticles induced by thiolated poly(ethyleneglycol). Langmuir 32(23):5937–5945

Lei Y, Li J, Wang Y, Gu L, Chang Y, Yuan H, Xiao D (2014) Rapid microwave-assisted green synthesis of 3D hierarchical flower-shaped $NiCo_2O_4$ microsphere for high-performance supercapacitor. ACS Appl Mater Interfaces 6(3):1773–1780

Li X, Fan F, Zhou H, Chow S-K, Zhang W, Zhang D, Guo Q, Ogawa H (2009) Enhanced light- harvesting and photocatalytic properties in morph-TiO_2 from green- leaf biotemplates. Adv Funct Mater 19 (1):45–56

Li Y, Hou X, Li Y, Ru Q, Wang S, Hu S, Lam K-h (2017) Facile synthesis of hierarchical $CoMn_2O_4$ microspheres with porous and micro-/nanostructural morphology as anode electrodes for lithium-ion batteries. Electron Mater Lett 13(5):427–433

Li Y, Fu Y, Zhu M (2020) Green synthesis of 3D tripyramid TiO_2 architectures with assistance of aloe extracts for highly efficient photocatalytic degradation of antibiotic ciprofloxacin. Appl Catal B 260:118149

Liu Z, Fan T, Zhang D, Gong X, Xu J (2009) Hierarchically porous ZnO with high sensitivity and selectivity to H2S derived from biotemplates. Sens Actuat B-Chem 136(2):499–509

Ma J, Fan H, Ren X, Wang C, Tian H, Dong G, Wang W (2019) A simple absorbent cotton biotemplate to fabricate SnO_2 porous microtubules and their gas-sensing properties for chlorine. ACS Sustain Chem Eng 7(1):147–155

Madan HR, Sharma SC, Udayabhanu SD, Vidya YS, Nagabhushana H, Rajanaik H, Anantharaju KS, Prashantha SC, Maiya PS (2015) Facile green fabrication of nanostructure ZnO plates, bullets, flower, prismatic tip, closed pine cone: their antibacterial, antioxidant, photoluminescent and photocatalytic properties. Spectrochim Acta a 152:404–416

Mallampati R, Valiyaveetti S (2012) Simple and efficient biomimetic synthesis of Mn_3O_4 hierarchical structures and their application in water treatment. J Nanosci Nanotechnol 12(1):618–622

Mallampati R, Valiyaveetti S (2013) Biomimetic metal oxides for the extraction of nanoparticles from water. Nanoscale 5(8):3395–3399

Mohamed MA, Salleh WNW, Jaafar J, Hir ZAM, Rosmi MS, Mutalib MA, Ismail AF, Tanemura M (2016) Regenerated cellulose membrane as bio-template for in-situ growth of visible-light driven C-modified mesoporous titania. Carbohyd Polym 146:166–173

Mohammadinejad R, Karimi S, Iravani S, Varma RS (2016) Plant-derived nanostructures: types and applications. Green Chem 18(1):20–52

Momeni MM, Ghayeb Y, Menati M (2016) Facile and green synthesis of CuO nanoneedles with high photo catalytic activity. J Mater Sci Mater Electron 27(9):9454–9460

Moyer K, Conklin DR, Mukarakate C, Vardon DR, Nimlos MR, Ciesielski PN (2019) Hierarchically structured CeO_2 catalyst particles from nanocellulose/alginate templates for upgrading of fast pyrolysis vapors. Front Chem 7:730

Müller A, Behsnilian D, Walz E, Gräf V, Hogekamp L, Greiner R (2016) Effect of culture medium on the extracellular synthesis of silver nanoparticles using *Klebsiella pneumoniae*, *Escherichia coli* and *Pseudomonas jessinii*. Biocatal Agric Biotechnol 6:106–115

Nayak AK, Sohn Y, Pradhan D (2017) Facile green synthesis of $WO_3{\cdot}H_2O$ nanoplates and WO_3 nanowires with enhanced photo-electrochemical performance. Cryst Growth Des 17(9):4949–4957

Nouroozi F, Farzaneh F (2011) Synthesis and characterization of brush-like ZnO nanorods using albumen as biotemplate. J Braz Chem Soc 22(3):484–488

Pan D, Ge S, Zhang X, Mai X, Lic S, Guo Z (2018) Synthesis and photoelectrocatalytic activity of In_2O_3 hollow microspheres via a bio-template route using yeast templates. Dalton Trans 47:708–715

Qian J, Chen F, Zhao X, Chen Z (2011) China rose petal as biotemplate to produce two-dimensional ceria nanosheets. J Nanopart Res 13:7149–7158

Ramimoghadam D, Hussein MZB, Taufiq-Yap YH (2013a) Hydrothermal synthesis of zinc oxide nanoparticles using rice as soft biotemplate. Chem Cent J 7:136

Ramimoghadam D, Hussein MZB, Taufiq-Yap YH (2013b) Synthesis and characterization of ZnO nanostructures using palm olein as biotemplate. Chem Cent J 7:71

Seth M, Khan H, Jana S (2020) Hierarchically structured alpha-nickel hydroxide based superhydrophobic and antibacterial coating on cellulosic materials for oil-water separation. Mater Chem Phys 249:123030

Shen Z, Liang P, Wang S, Liu L, Liu S (2015) Green synthesis of carbon- and silver-modified hierarchical ZnO with excellent solar light driven photocatalytic performance. ACS Sustain Chem Eng 3 (5):1010–1016

Shim H-W, Lim A-H, Kim J-C, Jang E, Seo S-D, Lee G-H, Kim TD, Kim D-W (2013) Scalable one-pot bacteria-templating synthesis route toward hierarchical, porous- Co_3O_4 superstructures for supercapacitor electrodes. Sci Rep 3:2325

Singh S, Srivastava VC, Lo SL, Mandal TK, Naresh G (2016) Morphology-controlled green approach for synthesizing the hierarchical self assembled 3D porous ZnO superstructure with excellent catalytic activity. Micropor Mesopor Mat 239:296–309

Singh S, Joshi M, Panthari P, Malhotra B, Kharkwal AC, Kharkwal H (2017) Citrulline rich structurally stable zinc oxide nanostructures for superior photo catalytic and optoelectronic applications: a green synthesis approach. Nano Struct Nano Object 11:1–6

Sinhamahapatra A, Giri AK, Pal P, Pahari SK, Bajaj HC, Panda AB (2012) A rapid and green synthetic approach for hierarchically assembled porous ZnO nanoflakes with enhanced catalytic activity. J Mater Chem 22(33):17227–17235

Song F, Su H, Chen J, Moon W-J, Lau WM, Zhang D (2012) 3D hierarchical porous SnO_2 derived from self-assembled biological systems for superior gas sensing application. J Mater Chem 22 (3):1121–1126

Su C-Y, Lan W-J, Chu C-Y, Liu X-J, Kao W-Y, Chen C-H (2015) Photochemical Green synthesis of nanostructured cobalt oxides as hydrogen peroxide redox for bifunctional sensing application. Electrochim Acta 190:588–595

Sun L, Cai J, Sun Y, Zhang D (2019) Three-dimensional assembly of silver nanoparticles spatially confined by cellular structure of *Spirulina*, from nanospheres to nanosheets. Nanotechnology 30 (49):495704

Sundar S, Venkatachalam G, Kwon SJ (2018) Biosynthesis of copper oxide (CuO) nanowires and their use for the electrochemical sensing of dopamine. Nanomaterials 8(10):823

Suresh S, Ilakiya R, Kalaiyan G, Thambidurai S, Kannan P, Prabu KM, Suresh N, Jothilakshmi R, Kumar SK, Kandasamy M (2020) Green synthesis of copper oxide nanostructures using *Cynodondactylon* and *Cyperusrotundus* grass extracts for antibacterial applications. Ceram Int 46:12525–12537

Tian J, Shao Q, Dong X, Zheng J, Pan D, Zhang X, Cao H, Hao L, Liu J, Mai X, Guo Z (2018) Bio-template synthesized NiO/C hollow microspheres with enhanced Li- ion battery electrochemical performance. Electrochim Acta 261:236–245

Tompsett GA, Conner WC, Yngvesson KS (2006) Microwave synthesis of nanoporous materials. Chem Phys Chem 7(2):296–319

Trogadas P, Ramani V, Strasser P, Fuller TF, Coppens M-O (2016) Hierarchically structured nanomaterials for electrochemical energy conversion. Angew Chem 55(1):122–148

Udayabhanu NG, Nagabhushana H, Basavaraj RB, Raghu GK, Suresh D, Rajanaika H, Sharma SC (2016) Green, nonchemical route for the synthesis of ZnO superstructure evaluation of its applications toward photocatalysis, photoluminescence, and biosensing. Cryst Growth Des 16(12):6828–6840

Udayabhanu NG, Nagabhushana H, Suresh D, Anupama C, Raghu GK, Sharma SC (2017) Vitislabruska skin extract assisted green synthesis of ZnO super structures for multifunctional applications. Ceram Int 43(15):11656–11667

Wang A-J, Liao Q-C, Feng J-J, Zhang P-P, Lia A-Q, Wang J-J (2012) Apple pectin-mediated green synthesis of hollow double-caged peanut-like ZnO hierarchical superstructures and photocatalytic applications. Cryst Eng Comm 14(1):256–263

Wang N, Liu Q, Kang D, Gu J, Zhang W, Zhang D (2016) Facile self-cross-linking synthesis of 3D nanoporous Co_3O_4/Carbon hybrid electrode materials for supercapacitors. ACS Appl Mater Interfaces 8:16035–16044

Wang Q, Dong B, Zhao Y, Huang F, Xie J, Cui G, Tang B (2018) Controllable green synthesis of crassulapeforata-like TiO_2 with high photocatalytic activity based on deep eutectic solvent(DES). Chem Engg J 348:811–819

Wei Y, Chen H, Jiang H, Wang B, Liu H, Zhang Y, Wu H (2019) Biotemplate-based engineering of high-temperature stable anatase TiO_2 nanofiber bundles with impregnated CeO_2 nanocrystals for enhanced lithium storage. ACS Sustain Chem Eng 7(8):7823–7832

Xia Y, Xiao Z, Dou X, Huang H, Lu X, Yan R, Gan Y, Zhu W, Tu J, Zhang W, Tao X (2013) Green and facile fabrication of hollow porous MnO/C microspheres from microalgaes for lithium- ion batteries. ACS Nano 7(8):7083–7092

Xia Q, Jiang Z, Li D, Wang J, Yao Z (2018) Green synthesis of a dendritic Fe_3O_4@Fe^ocomposite modified with polar C-groups for fenton-like oxidation of phenol. J Alloys Compd 746:453–461

Xiao G, Huang X, Liao X, Shi B (2013) One-pot facile synthesis of cerium-doped TiO_2 mesoporous nanofibers using collagen fiber as the biotemplate and its application in visible light photocatalysis. J Phys Chem C 117(19):9739–9746

Yan Y, Wu Q-S, Li L, Ding Y-P (2006) Simultaneous synthesis of dendritic superstructural and fractal crystals of $BaCrO_4$ by vegetal bi-templates. Cryst Growth Des 6(3):769–773

Yang X, Song X, Wei Y, Wei W, Hou L, Fan X (2011) Synthesis of spinous ZrO_2 core–shell microspheres with good hydrogen storage properties by the pollen bio-template route. Scripta Mater 64 (12):1065–1068

Yang R, Tang D, Tao T, Ren Y, Zhang X, Xu M, Wang C (2013) One-step green synthesis of hierarchical hydrangea shaped Cu_2O–CuO composite. Mater Lett 113:156–158

Yang X-Y, Chen L-H, Li Y, Rooke JC, Sanchezc C, Su B-L (2017) Hierarchically porous materials: synthesis strategies and structure design. Chem Soc Rev 46(2):481–558

Ye W, Yu J, Zhou Y, Gao D, Wang D, Wang C, Xue D (2015) Green synthesis of Pt–Au dendrimer-like nanoparticles supported on polydopamine-functionalized graphene and their high performance toward 4-nitrophenol reduction. Appl Catal B Environ 181:371–378

Yin L, Wang H, Li L, Li H, Chen D, Zhang R (2019) Microwave-assisted preparation of hierarchical CuO@rGO nanostructures and their enhanced low-temperature H_2S-sensing performance. Appl Surf Sci 476:106–114

Yuliarto B, Septiani NLW, Kaneti YV, Iqbal M, Gumilar G, Kim M, Na J, Wu KC-W, Yamauchi Y (2019) Green synthesis of metal oxide nanostructures using naturally occurring compounds for energy, environmental, and bio-related applications. New J Chem 43(40):15846

Zan G, Wu Q (2016) Biomimetic and bioinspired synthesis of nanomaterials/ nanostructures. Adv Mater 28(11):2099–2147

Zhang W, Zhang D, Fan T, Ding J, Guo Q, Ogawa H (2006) Morphosynthesis of hierarchical ZnO replica using butterfly wing scales as templates. Micropor Mesopor Mat 92(1–3):227–233

Zhang W, Zhang D, Fan T, Gu J, Ding J, Wang H, Guo Q, Ogawa H (2009) Novel photoanode structure templated from butterfly wing scales. Chem Mater 21(1):33–40

Zhang G, Sun S, Banis MN, Li R, Cai M, Sun X (2011) Morphology-controlled green synthesis of single crystalline silver dendrites, dendritic flowers, and rods, and their growth mechanism. Cryst Growth Des 11(6):2493–2499

Zhang C, Wang J, Hu R, Qiao Q, Li X (2016) Synthesis and gas sensing properties of porous hierarchical SnO_2 by grapefruit exocarp biotemplate. Sens Actuat B-Chem 222:1134–1143

Zhang M, Meng J, Wang D, Tang Q, Chen T, Rong S, Liu J, Wu Y (2018) Biomimetic synthesis of hierarchical 3D Ag butterfly wing scale arrays/graphene composites as ultrasensitive SERS substrates for efficient trace chemical detection. J Mater Chem C 6(8):1933

Zhang S, Li H, Wang S, Liu Y, Chen H, Lu Z-X (2019a) Bacteria-assisted synthesis of nanosheet- assembled TiO_2 hierarchical architectures for constructing TiO_2 based composites for photocatalytic and electrocatalytic applications. ACS Appl Mater Interfaces 11(40):37004–37012

Zhang H, Zhang Z, Luo J-D, Qi X-T, Yu J, Cai J-X, Yang Z-Y (2019b) Molten-salt-assisted synthesis of hierarchical porous MnO@biocarbon composites as promising electrode materials for supercapacitors and lithium-ion batteries. Chem Sus Chem 12(1):283–290

Zhou H, Fan T, Zhang D (2007) Hydrothermal synthesis of ZnO hollow spheres using spherobacterium as biotemplates. Micropor Mesopor Mat 100:222–227

Zong L, Wu X, You J, Li M, Li C (2016) Modulating structural hierarchies of manganese oxide in morphology and porosity by marine biopolymer for improved supercapacitors. Electrochim Acta 213:709–716

Bioprivileged Molecules

Shashi Kiran Misra, Devender Pathak, and Kamla Pathak

Abstract

Petrochemical based non-sustainable resources are being utilized for the production of energy and various platform chemicals. However, the fuel-based technology results in hazardous outputs, and hence, sustainable green routes are essential to save our resources and compensate energy requisites. Alternative sustainable green synthesis approach shifts the dependence from fossil fuels to the bio-based feedstock. Lignocellulose (cellulose and hemicellulose), wood, and crop residue, namely sugar cane, herbs and wheat straw, are frequently utilized to generate value-added platform molecules. The chapter portrays the bioprivileged molecules recognized by the US Department of Energy, their contribution to a drop-in replacement, the green route(s) for their preparation and the global market potential.

Keywords

Bioprivileged molecules • Green route preparation • Sustainability • Applications

1 Introduction

Our planet is full of sustainable biomass encompassing renewables and non-renewable assets. Enormous chemicals acquired from biomass or their bioprocessing courses are quite large. Of lately, bioprivileged molecules (biology-derived molecules/intermediates) are being efficiently transformed into miscellaneous innovative chemicals (drop-in replacement and new entities) adorned with advanced properties. The bioprivileged molecules exhibit some key features such as these should be an intermediate chemical molecule procured from biomass and cannot be acquired from the petrochemical route. Also, these molecules can be further transformed into other molecules and require minimum subsequent reactions to produce novel and drop-in replacements (Werpy and Petersen 2004). Bio-based procurement of bioprivileged molecules is certainly advantageous for agriculture, pharmaceutical, petrochemicals/ refineries, nutraceuticals and other consumer provisions (Xiaowei et al. 2019). Many bio-origin polymers (starch, cellulose) and biodegradable plastics such as polyethylene (PE), polyvinylchloride (PVC) and polyethylene terephthalate (PET) are acquired from renewable assets. The annual utilization of approximately 4.2 million tons of bioprivileged molecules recorded in the year 2016 is estimated to increase to about 6.1 million tons by the year 2021 (Brent and Keeling 2017). In the year 2004, a survey was conceded by Gene Petersen and Todd Werpy on the comprehensively employed twelve bioprivileged molecules for their development as innovative molecules and as a drop-in replacement (Bozell and Petersen 2010; Bioprivileged Molecules 2018). Some of them are widely used; some are languished and unloved by researchers depending on their potential. The chapter elaborates the recognized bioprivileged molecules, by the US Department of Energy, their contribution to a drop-in replacement, green synthesis preparation route and their global market potential.

2 Four Carbon 1,4-Diacids

2.1 Succinic Acid

Succinic acid, widely distributed in plants and animal tissues, often known as spirit of amber (Fig. 1) is regarded as a valuable molecule for the industries. Four multinational

S. K. Misra
University Institute of Pharmacy, Chhatrapati Shahujimaharaj University, Kanpur, 208026, India

D. Pathak · K. Pathak (✉)
Faculty of Pharmacy, Uttar Pradesh University of Medical Sciences, Saifai, Etawah, Uttar Pradesh 206130, India

Inamuddin et al. (eds.), *Advances in Green Synthesis*, Advances in Science, Technology & Innovation,
https://doi.org/10.1007/978-3-030-67884-5_6

SUCCINIC ACID

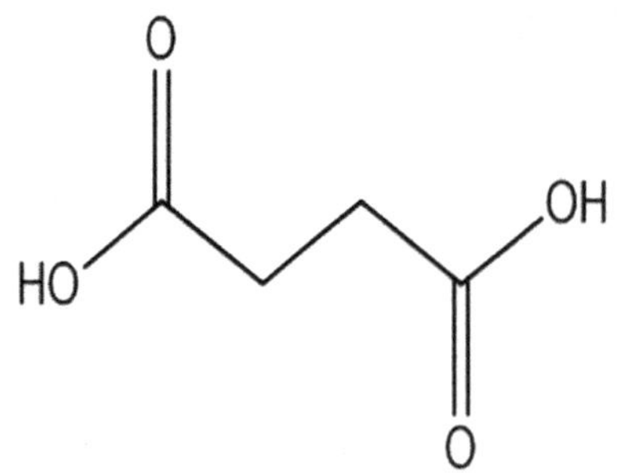

Molecular formula- $C_4H_6O_4$
Molecular weight -118.09 g/mol
IUPAC name- Butanedioic acid
Physical state – White shiny crystals (solid)
Taste & odor- Strong acidic & odorless
Melting point/ boiling point- 188°C/ 235°C
Solubility- 8.32 x 10^4 mg/L
pH/ pKa - 2.7 /4.21 at 25°C

Fig. 1 Physicochemical properties of succinic acid

companies, namely BASF-Purac (Succinity), Myriant, Reverdia (DSM-Roquette) and BioAmber, are engaged in developing newer molecules/valuable replacement products as adipic acid, maleic acid and phthalic anhydride, etc. Myriant is involved in production of pure succinic acid as bio-succinic acid from renewable biomass. Similarly, Reverdia Enterprises, Italy, has been involved in the production of sustainable Biosuccinium that is extensively utilized for the synthesis of polybutylene succinate, phthalate-free plasticizers, polyester polyols and 1,4 butanediol. These molecules are widely used in the production of packaging materials, paints and footwear (Global Succinic Acid Market Analysis Trends 2017).

Conventionally, succinic acid is produced from a petroleum-based maleic anhydride process utilizing a C_4 fraction of naphtha that is expensive and environmentally hazardous. Hence, novel bioprocessing routes/fermentation processes were innovated for the production of succinic acid involving bacterial and fungal cells, enzymes and cofactors that proved safe, economical and eco-friendly (Nhuan et al. 2017). Chiefly, genetically engineered *Escherichia coli* is considered as a viable microbe for the production of succinic acid. Gene deletion of lactose dehydrogenase and pyruvate formate lyase prevents the formation of byproduct and assists the formation of succinic acid (Fig. 2).

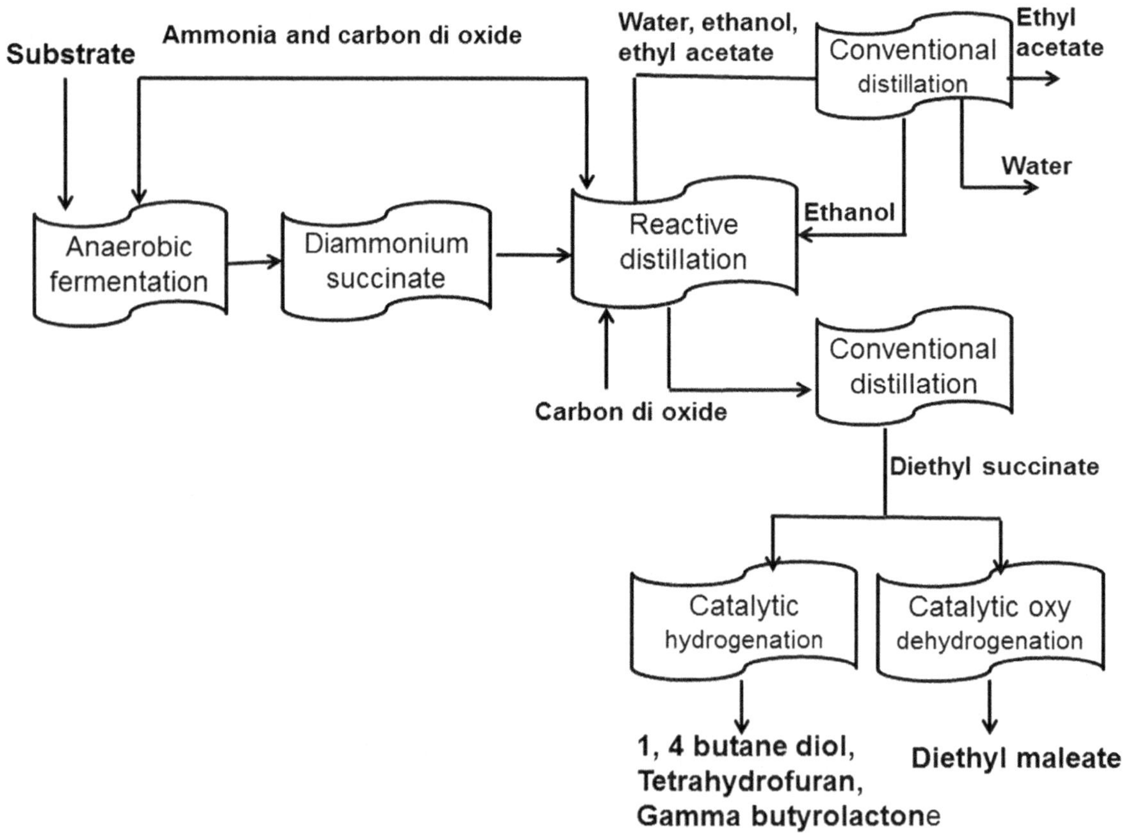

Fig. 2 Schematic illustration of the production of bioprivileged succinic acid

Other microorganisms, namely *Aspergillus niger, Actinobacillus succinogenes, Arthrobactor tumenscens, Salmonella typhimurium, Planococcus eucinatus and Xanthomonas citri,* are also reported for the production succinic acid and its derivatives (tetrahydrofuran, 1,4-diaminobutane, 1,4-butanediol and succindiamide). However, these microorganisms require complex nutrients and culture processing that may be unaffordable commercially.

Succinic acid has huge market potential owing to its involvement in the production of more than thirty commercially significant platform compounds. These compounds find use in pharmaceuticals, nutraceuticals, foods and paints/ink industries. Several precursors, surfactants, antifoams, detergents and ion chelators are frequently produced using bio-based succinic acid (Table 1) due to their nominal processing cost compared to the petroleum-based molecules (Saxena et al. 2017).

Among the platform compounds of succinic acid, the global market of 1,4-butanediol is highest (35%) owing to its involvement in the preparation of plasticizer, food and beverages, polyols, polybutylene succinate and pharmaceutical solvents. Interestingly, 1,4-butanediol is considered as a replacement molecule of maleic anhydride that is further utilized for the synthesis of many other molecules.

2.2 Fumaric Acid

A dicarboxylic acid derivative fumaric acid (Fig. 3) is an indispensable metabolite of several microorganisms. Adorned with multiple functional moieties and procured from renewable biomass, it can be easily transformed into numerous other beneficial substances (Yang 2007). Used as food acidulant since 1946, fumaric acid finds vital applications in the manufacturing of unsaturated polyester resins, plasticizers, lubricating oils, paper resins and for preparation of styrene-butadiene rubber. It is frequently utilized in sundry food products, i.e., bread, corn and wheat pancakes, biscuits, desserts, fruit juices, in beverages and nutraceuticals, etc. owing to its pH lowering, porosity modifier and shelf life amending properties of baked products.

Traditionally, a conventional petroleum-based scheme (n-butane to maleic anhydride route) is utilized for its production owing to cheap production and high yield. But, increased oil prices and pollution-related issues dictated the use eco-friendly renewable feedstock for the production of fumaric acid. Pfizer developed a procedure based on filamentous fungal fermentation of sugar for the preparation of fumaric acid (Goldberg et al. 2006). The scheme requires mild conditions, and is economical, safe without emission of toxic gases and wastes. The production scheme is based on

Table 1 Applications of the platform chemicals developed through succinic acid

Platform chemicals	Applications
1,4-Butanediol	As solvent, in syntheis of plastics, elastic fibers and polyurethanes
Tetrahydrofuran	Polyvinylchloride
Gamma-butyrolactone	Paint strippers, nail polish removers, stain removers and circuit board cleaners
Succinamide	As a key ingredient for synthesis of anti-convulsant therapeutics, i.e., ethosuximide, methsuximide and phensuximide For assay of protein and peptides
Succinonitrile	Vinyl foam
2-Pyrrolidone	Used in inkjet cartridges, key molecule for the synthesis of povidone, cotinone, ethosuximide, doxapram and the racetams, green solvent for water treatment
N-Methyl-2-pyrrolidone	As solvent in petrochemical and plastics industries

FUMARIC ACID

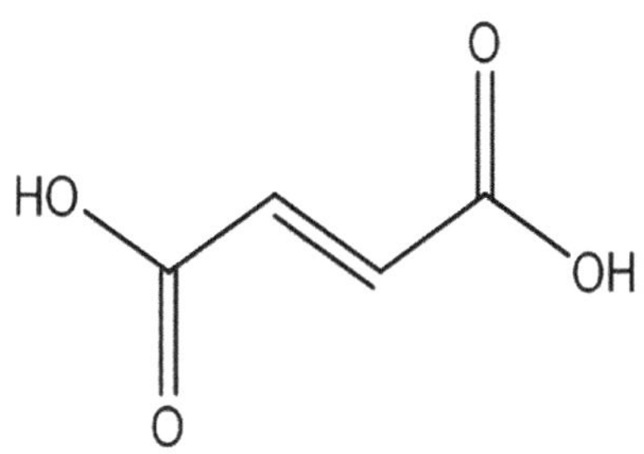

Molecular formula- $C_4H_4O_4$
Molecular weight -116.07 g/ mol
IUPAC name- (E)-butenedioic acid
Physical state – White crystalline powder
Taste & odor- Fruity acidic & odorless
Melting point/ boiling point- 287°C/ 522°C
Solubility- 7 x 10^{-3} mg/L at 25°C
pH / pKa – 3.2 /3.03 at 25°C

Fig. 3 Physicochemical properties of fumaric acid

utilization of the neutralizing agent ($CaCO_3$) that is acidified with H_2SO_4 to attain a pH of 1 followed by extreme heating (160 °C). The precipitated insoluble $CaSO_4$ is removed through filtration, and the crystallized fumaric acid is collected at room temperature (Fig. 4). Generally, *Rhizopus arrhizus* and *Rhizopus oryzae* are the best producers for fumaric acid through aerobic or anaerobic fermentation. The morphology and growth pattern of fungal species challenges the product concentration and production yield, and thus, the process is hard to scale-up commercially (Magnuson et al. 2004). The species of *Mucor, Aspergillus* and *Cirinella* are also reported for the collection of fumaric acid (Jiménez-Quero et al. 2017).

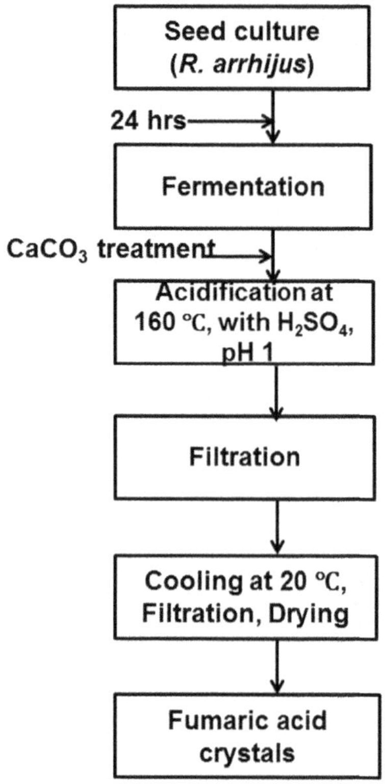

Fig. 4 Schematic illustration of the preparation of fumaric acid through fermentation

The global market of fumaric acid is vast as it is widely used in paper, textile, paint, food and beverage industries (Fig. 5). In pharmaceutical industry, methyl and propyl fumarate esters of fumaric acid are used for the synthesis of dermatological, anti-carcinogenic and anti-inflammatory drugs (Das et al. 2016).

2.3 Malic Acid

Another carbon 1,4-diacid, malic acid, is considered as a key molecule for the production of biodegradable polymers, fabrics and resin products (Fig. 6). Low-value bio-based malate salt is extensively utilized as a flavoring agent in the food industry owing to its acidulated virtue.

Commercially, malic acid is processed by a reductive scheme based on the phosphoenolpyruvate carboxykinase pathway. The process entails carbon dioxide, neutralizing salt and eukaryotic cultures, i.e., *Aspergillus* sp., *Candida* sp., *Penicillium* sp. or prokaryotic *Escherichia coli, Bacillus subtilis*, etc. Another method is the glyoxalate pathway that utilizes acetyl-CoA, water and glyoxylate molecules with *S. cerevisiae* to synthesize malate salt. Both methods are highly reliant on the type of strain and method of processing to

Fig. 5 Applications of fumaric acid

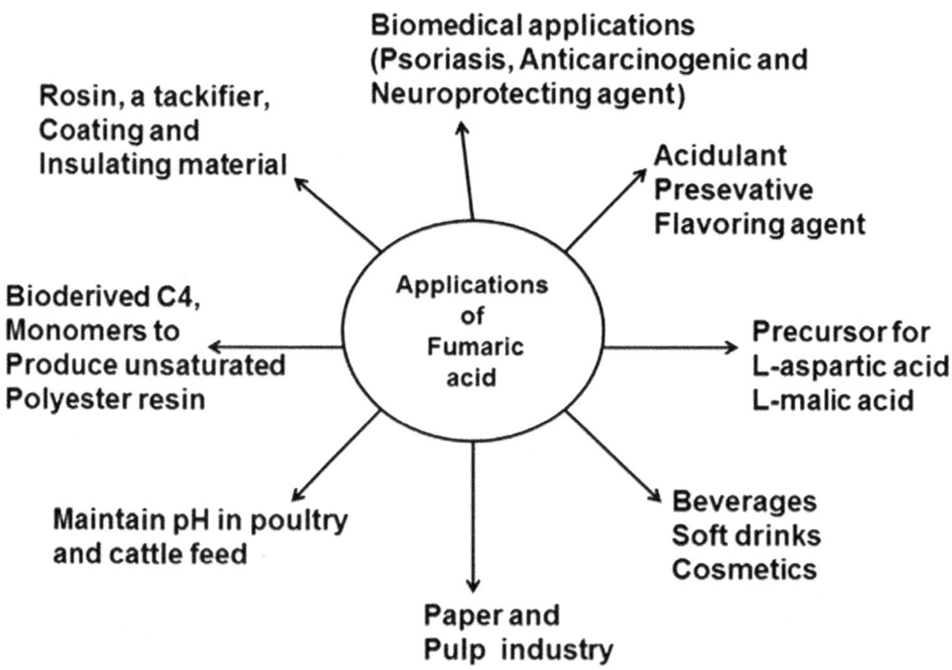

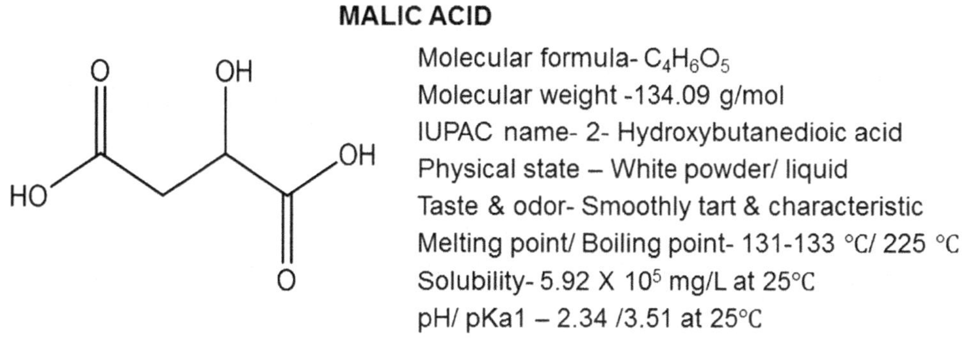

Fig. 6 Physicochemical properties of malic acid

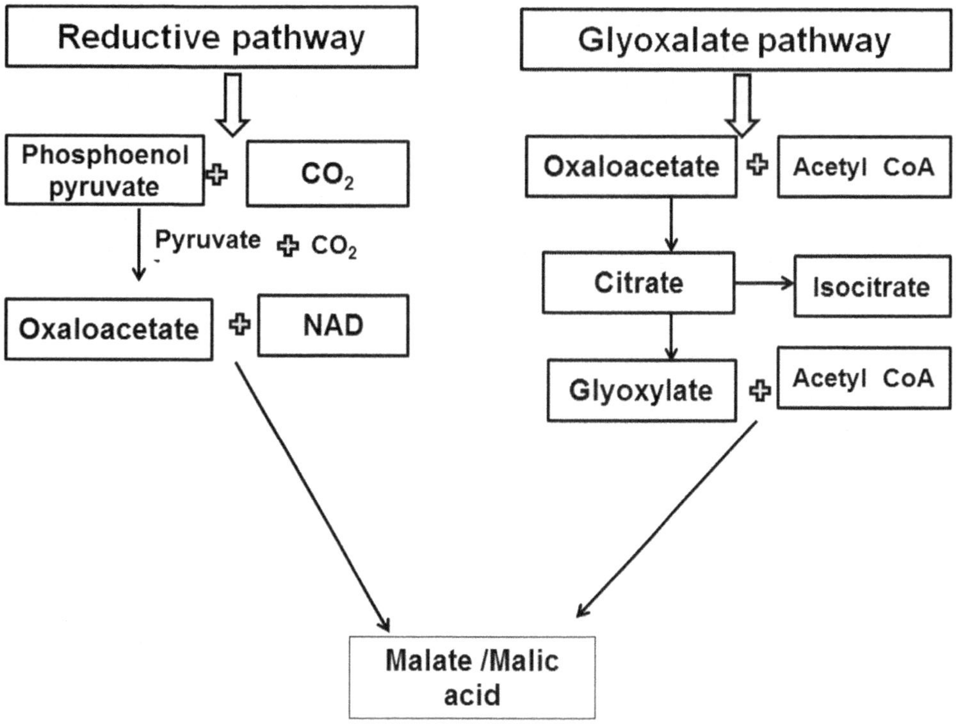

Fig. 7 Schematic representation of malic acid production

obtain cost-effective productivity (Fig. 7). A new method reports utilizing *Ustilago trichophora* biomass for commercial production of malic acid with glycerol at low pH that minimizes the chances of contamination in large-scale production. Moreover, this scheme requires minimum salts ($CaCO_3/Ca(OH)_2$) for processing, thereby pose least environmental hazard (Deng and Zhang 2016).

Like succinic and fumaric acid, malic acid can also be derived into 1,4-butanediol compound that is a favorable platform molecule for the synthesis of other compounds including polymers, resins and plastics. Some fruits, i.e., sour apple, plums, cherries and grapes enriched with malic acid are used for the production of wine globally. A report says that acid deprived wines are flat in taste, and hauling of malic acid makes them acceptable owing to fruity flavor (Saguir et al. 2018). Pharmaceutically, malic acid is extensively utilized as a substitute for citric acid to formulate mouthwashes, dental tablets and effervescent powders. Moreover, its chelating and antioxidant features enable researchers to explore malic acid for delaying oxidation of vegetable oils and enhancement of shelf life. Malic acid synergizes the activity of the antioxidant, butylated hydroxytoluene and thus retards rancidity of oils and fats. Remedially, malic acid has been merged with salicylic acid and benzoic acid to design topical creams for the mitigation of burns, wounds, blisters and ulcers. The literature also envisages oral and parenteral applications of malic acid to combat liver disorders (Brittain 2001).

3 Furan 2,5-Dicarboxylic Acid (FDCA)

A highly stable furan derivative, FDCA is a bioprivileged molecule, adorned with two carboxylic acids in its chemical structure (Fig. 8).

FURAN 2,5- DICARBOXYLIC ACID

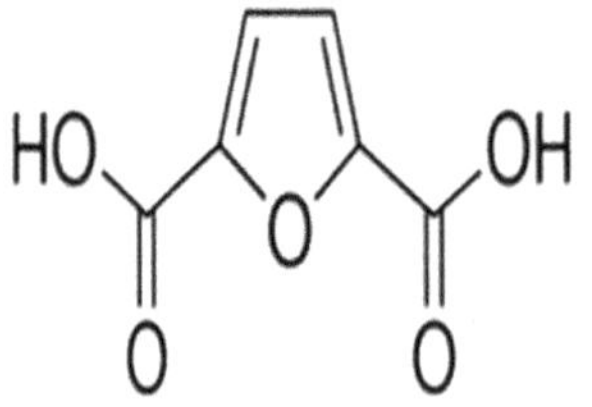

Molecular formula- $C_6H_4O_5$
Molecular weight -156.09 g/mol
IUPAC name- Furan 2, 5- dicarboxylic acid
Physical state – Solid
Melting point/ Boiling point- 342 °C/420°C
Solubility- 1mg/ml at 18°C

Fig. 8 Physicochemical properties of FDCA

Bio-based FDCA obtained from plant-based sugars is transformed into platform molecules that are used for cost-effective development of plastic materials, polyesters and is a suitable alternative for petroleum-based monomer, i.e., terephthalic acid. Recently, commercial production via green synthesis route entailing glucose, fructose and hydroxymethylfurfuraldehyde intermediate has been established.

Several bio-based and microbe oriented schemes are reported for the successful production of FDCA owing to the fact that these are less hazardous and economical with high productivity. Genetically engineered *P. putida* S12, a rod-shaped gram-negative bacterium has been explored for the synthesis of FDCA from hydroxymethylfurfuraldehyde intermediate. The microbially assisted process requires considerable time (3 days) for the conversion of hydroxymethylfurfuraldehyde to FDCA that is not feasible for large-scale production and thus presents a challenge for the sustainability of the green synthesis. An alternate petrochemical route used for the production of furan derivative is convenient but involves harmful compounds that pose risk to the environment.

The production of FDCA from renewable biomass mostly encompasses catalytic, non-catalytic and electrochemical processes. The catalytic route is considered as the potential pathway owing to its high productivity and manageable reaction rate under mild conditions with minimum byproduct generation (Fig. 9). Transition metal oxides and noble metal oxides are frequently utilized for completion of the process to produce FDCA depending upon the cost, availability, recycling and yield.

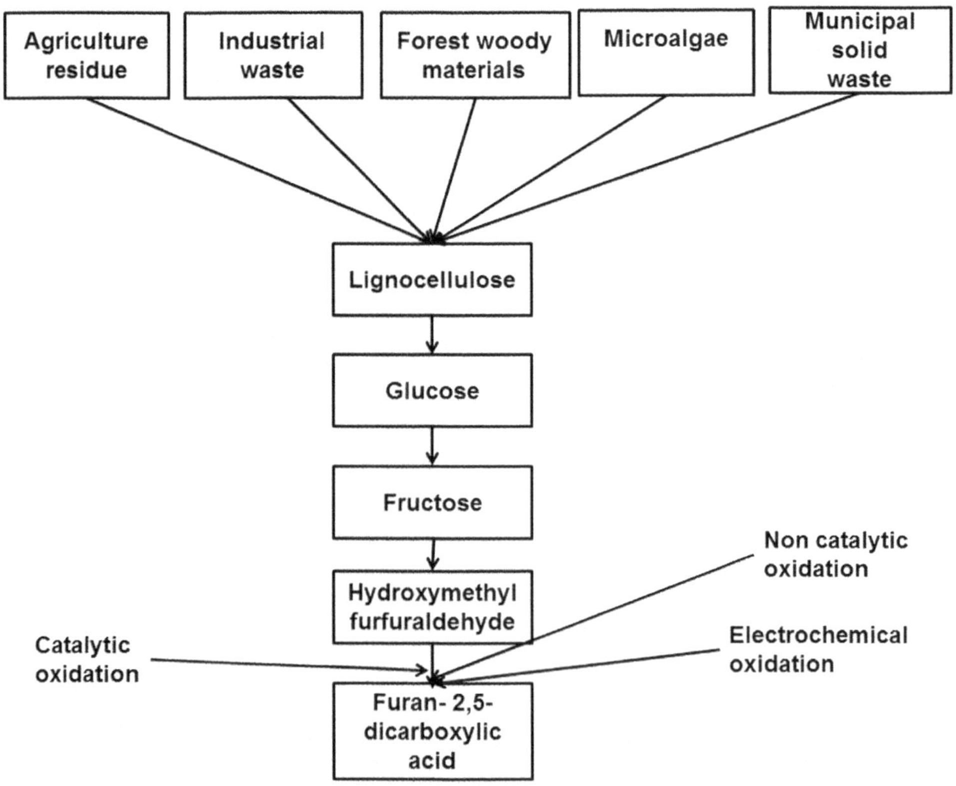

Fig. 9 Production scheme for furan 2,5-dicarboxylic acid from biomass

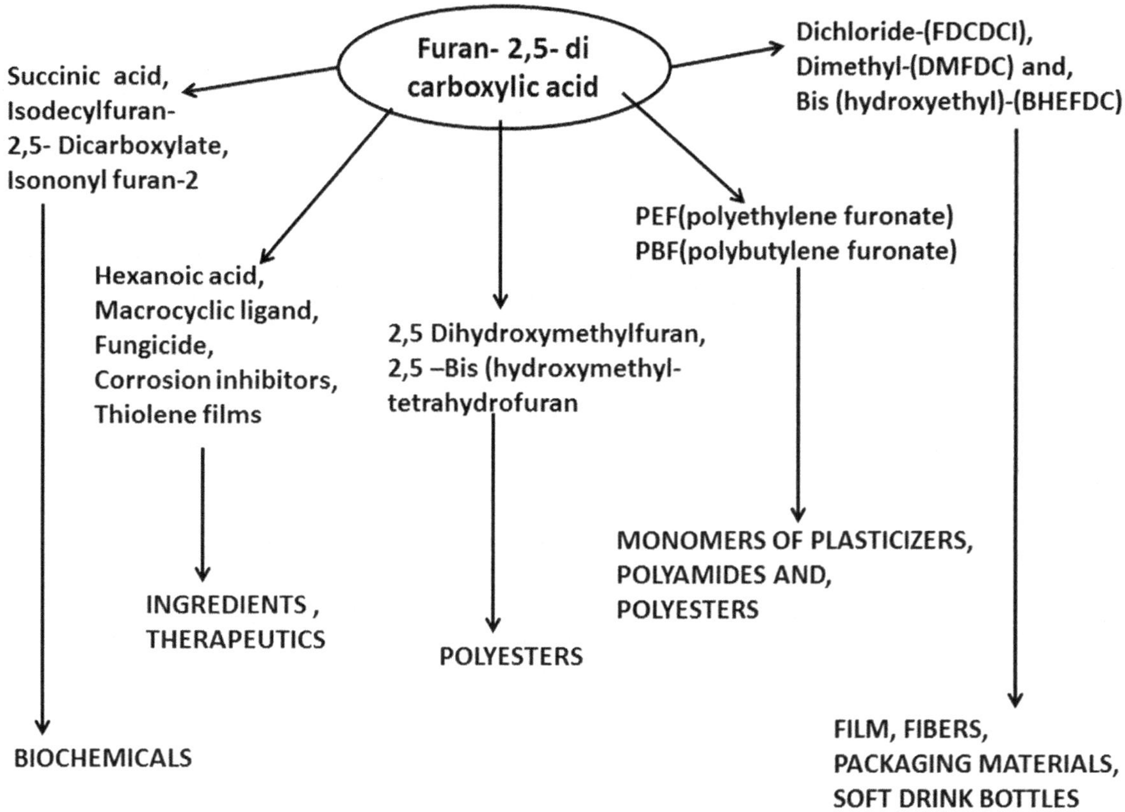

Fig. 10 Derivatives and applications of FDCA

Owing to its stability, renewable feedstock, easy production process and biodegradability, FDCA is a preferred molecule for the manufacturing of polyesters and plastics (Fig. 10). Bio-based FDCA has the potential to replace numerous chemicals that are synthesized from the petroleum route, i.e., adipic acid and terephthalic acid. Avantium technologies, Netherlands, have merged with Cargill Inc. Minnesota, US and developed an economical route to produce FDCA based on chemical catalysis, from easily procurable sugar and starch-based feedstock as raw material (Region and Segment Forecast 2015).

The global market of FDCA was more than 88.2 million USD for the year 2014 and was expected to flourish with an approximately 34% growth rate up to the year 2020. In the pharmaceutical industry, diethyl esters of FDCA are vastly utilized for the preparation of anesthetics. As a chelator, FDCA is being explored for the management of kidney stones (Lewkowski 2001).

4 3-Hydroxypropionic Acid (3-HPA)

3-hydroxypropionic acid is a three carbon optically inactive molecule (Fig. 11), structural isomer of lactic acid and is precursor for the synthesis of acrylic acid and the manufacturing of bioplastics. The presence of two functional groups (hydroxyl and carboxyl) facilitates 3-HPA to transform into a variety of value-added chemicals such as acrylic acid, acrylonitrile, malonic acid, 1,3 propanediol, etc.via chemical route (Werpy and Petersen 2004). These chemicals

Fig. 11 Physicochemical properties of 3-hydroxypropionic acid

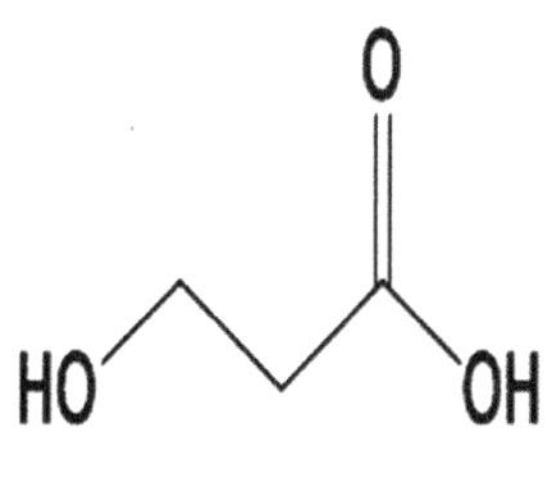

Molecular formula- $C_3H_6O_3$
Molecular weight - 90.08 g/mol
IUPAC name- 3-hydroxypropionic acid
Physical state – Solid
Melting point - < 25 °C
Solubility- Very soluble in water

are 50% cost-effective and lessen greenhouse gas emission by 75% as compared to the petrochemical route (Jung et al. 2014).

Green synthesis route has been opted for efficient production of 3-HPA from biomass encompassing glycerol, glucose and uracil metabolic pathways (Kumar et al. 2013). Both acety-CoA dependent and independent glycerol metabolic processes are widely applicable for the production of 3-HPA. Microorganisms that are acid resistant and capable of synthesizing coenzyme $B_{12,}$ namely *Lactobacillus reuteri, Klebsiella pneumonia, E. coli* and *Corynebacterium glutamicum* are frequently involved in the bioconversion of 3-HPA from glycerol in the bioreactor (Fig. 12).

The selection of microorganisms is based on ease of availability, process suitability, productivity and process cost. Moreover, the strains of microorganisms are highly dependent on the presence of enzymes, by-products and process conditions (Dishisha et al. 2014). Sugarcane, beet sugar and hydrolyzed corn starch are other renewable feedstocks that can be betrothed for the cultivation of microorganisms for the production of 3-HPA. Many physical, chemical and biological pretreatment procedures are proposed for bioconversion from these lignocellulose waste originated sugars.

A valuable molecule, acrylic acid of 3- HPA, has been used for the production of fibers, paints and plastics. Some successful industries like Cargill Inc., Triveni Interchem Private Limited, Lion Apparel, are global developers of 3-HPA (Market Research Report 2019). Table 2 summarizes potential applications of the molecules derived from 3-HPA.

5 Glucaric Acid

Biomass-derived glucaric acid has garnered considerable attention as it can be utilized for the synthesis of bioenergy and several other platform chemicals. Value-added molecule glucaric acid was recognized as bioprivileged molecule by the US Department of Energy in 2004 (Werpy and Petersen 2004). It is also known as saccharic acid. Belonging to a family of oxidized sugars, it is highly functionalized molecule adorned with four chiral carbons (Fig. 13).

Important key intermediates, adipic acid and hyper-branched polyesters are biosynthesized from bio-based

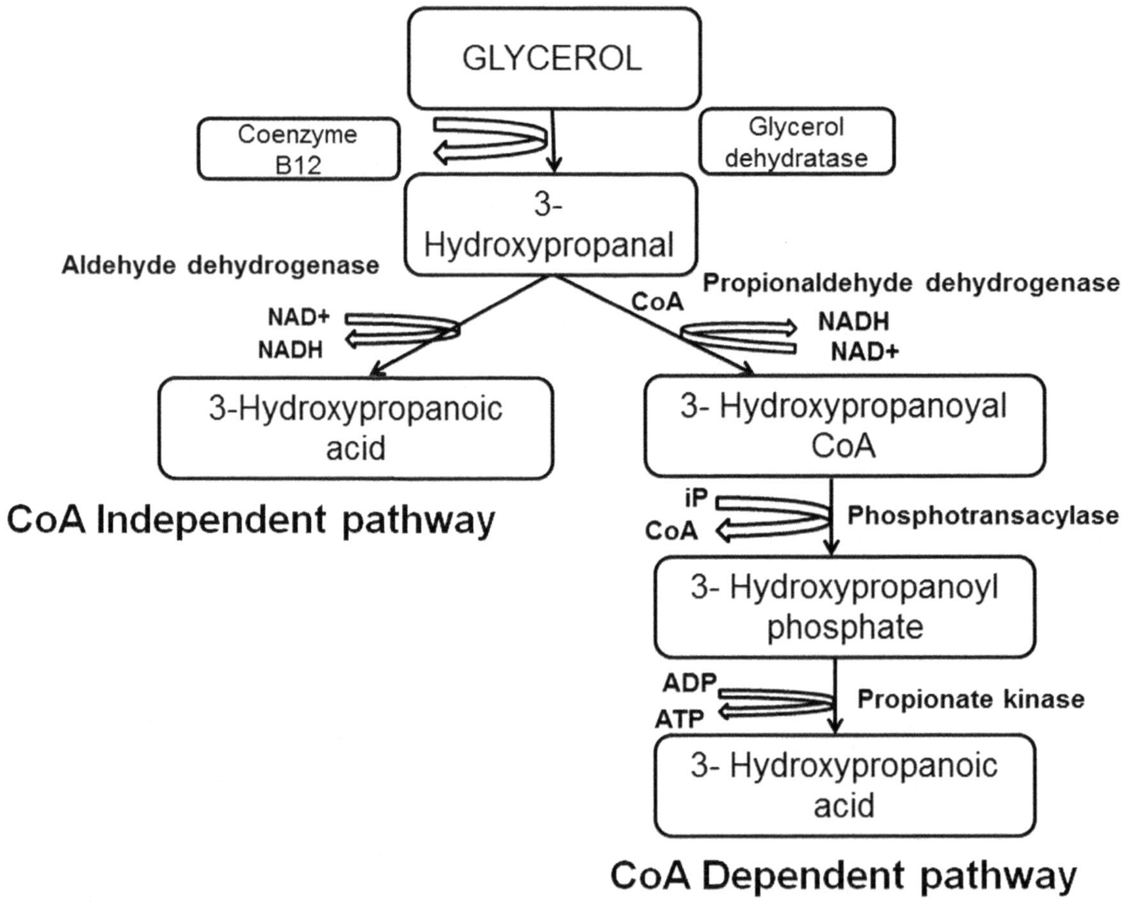

Fig. 12 Processing scheme for 3-hydroxypropionic acid

Table 2 Potential applications of platform molecules derived from 3-HPA

Platform molecules	Potential applications
Acrylic acid	Coating, masonry, adhesive, water treatment chemical
Acrylamide	Paper, dye, plastics, wastewater treatment, thickening agent
Acrylonitrile	acrylic and modacrylic fibers, nitrile rubbers, nitrile barrier resins
1,3-propanediol	Polytrimethylene terephthalate, cosmetics, textile
Malonic acid	Gamma-nonalactone, valproate, cinnamic acid
Methyl acrylate	Amphoteric surfactants, coatings, adhesives, fibers, plastics, textiles, elastomers, thickeners and inks
Propiolactone	Surgical instruments, enzyme, vaccines, tissue grafts and as a sterilant of biological products

GLUCARIC ACID

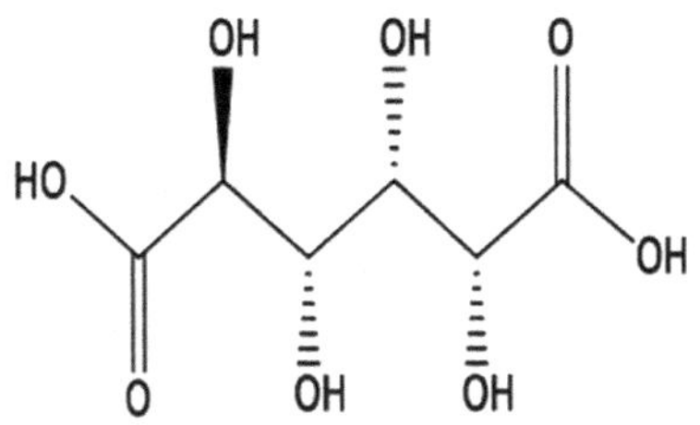

Molecular formula- $C_6H_{10}O_8$
Molecular weight – 210.14 g/ mol
IUPAC name- 2,3,4,5-tetrahydroxy-hexanedioic acid
Physical state – Solid
Melting point – 125.5 °C
Solubility- 63.1 mg/ml

Fig. 13 Physicochemical properties of glucaric acid

glucaric acid (Smith et al. 2012). Furthermore, this molecule is widely used in the formation of phosphate-free detergents, anti-corrosive additives, biodegradable cleaner, adhesive, coating material and various therapeutic molecules (Diamond et al. 2014). Glucaric acid and its derivatives, namely glucaro-γ-lactone, glucaro-δ-lactone, glucarolactone and polyhydroxypolyamides, are utilized in foods and pharmaceuticals.

Commercially, glucaric acid is produced from glucose either by electrochemical, biochemical or chemocatalytic oxidation routes (Bin et al. 2014). Figure 14 summarizes bio-based corn stover biomass-based commercial routes, namely heterogenous catalyst promoted oxidation and homogenous nitric acid oxidation for the production of glucaric acid, (Besson et al. 2014). The former method is a complex process although high yield has been reported in a Rennovian patent (Boussie et al. 2013). The process is accomplished by utilizing platinum on carbon catalyst (Pt/C) and 45% KOH mixed oxidizing mixture. On the other hand, the homogenous nitric acid oxidation is processed under mild atmospheric conditions (40 °C, 2 bars), with glucose and nitric acid as initial precursors. The yield of glucaric acid is comparatively less owing to the simultaneous production of co-products, gluconic acid and tartaric and citric acid (Moon et al. 2009).

The leading industries, Rivertop Renewables and DTI based in Virginia, USA utilize sugar-based feedstock through D-glucose oxidation for efficient production of more than 9 million pounds of sodium glucarate salt per year (Renewables and DTI exceed nameplate capacity of first glucarate production facility 2016). Other units, Johnson Matthey Process Technologies and Rennovia, England follow catalytic aerobic oxidation of glucose for the collection of value-added molecule, glucaric acid (2015).

An exclusive property associated with glucaric acid is its efficiency to modify the tensile strength/ mechanical strength of polymers. Therefore, it is extensively used for altering certain mechanical parameters of polymeric resin by multiple folds. It tends to lessen the melting temperature of polyvinyl alcohol, hence modifies the fiber draw ratio. The literature cites improvement in fiber properties by modulating the strength and moisture resistance by the addition of glucaric acid in lignin (Lu and Ford 2018). Another molecule glucaric acid 1,4-lactone has gained popularity in Asian countries due to its anticancer activity (Walaszek 1990).The D-glucarate exhibited suppression of cell proliferation, inflammation thus improving the body's defense mechanism by removing tumor promoters (Zoltaszek et al. 2008).

6 Glycerol

Glycerol (propane- 1,2,3-triol), another bioprivileged molecule, is frequently derived from biomass and is involved in the development of both bridging chemicals and drop-in

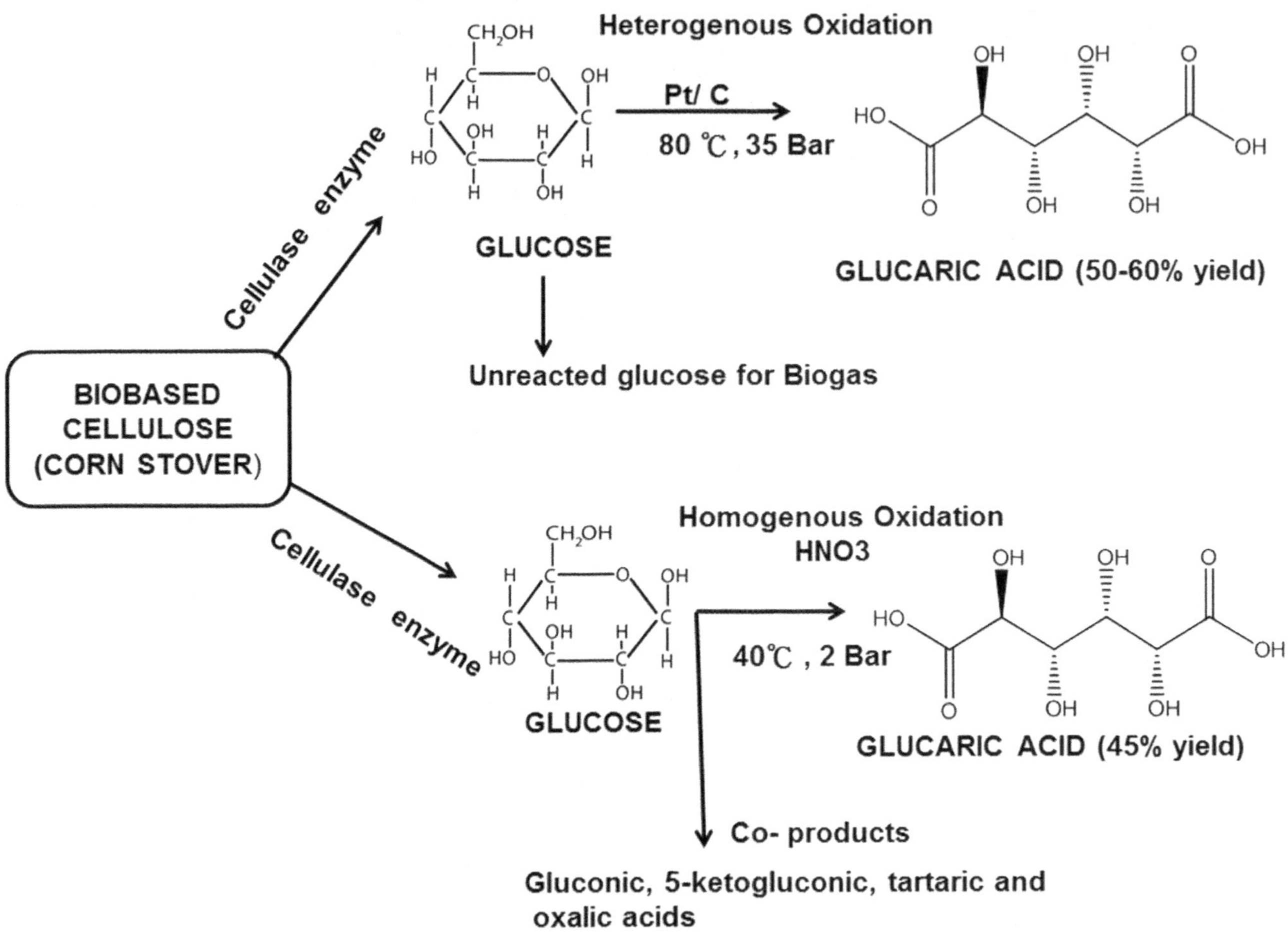

Fig. 14 Different bio-based processes for the development of glucaric acid

Fig. 15 Physicochemical properties of glycerol

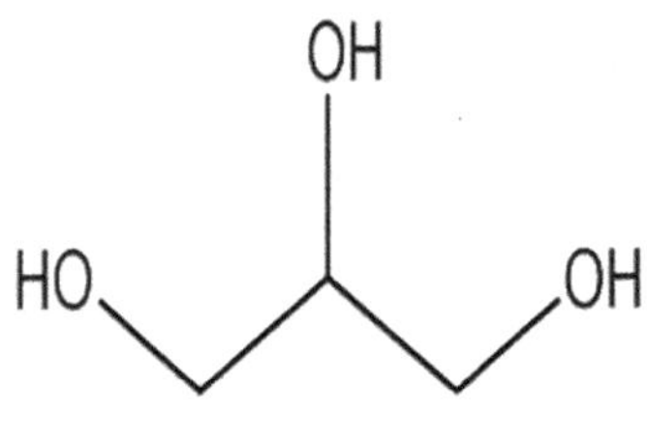

Molecular formula- $C_3H_8O_3$
Molecular weight – 92.09 g/mol
IUPAC name- Propane 1,2,3- triol
Physical state – Colorless to brown color liquid
Boiling point – 290 °C
Solubility- 5.296 X 10^6 mg/L at 25°C
Taste & odor- Sweet warm & mild odor
Log P & viscosity- -1.76 & 954 cp at 25°C

replacements (Fig. 15). The presence of three hydroxyl groups in glycerol accounts for its stability and versatility of the molecule (Kenar 2007).

It finds vital usage in industries by the virtue of its sweetness and has been used as humectant, viscosity modifier, adhesive, plasticizer, icing agent, lubricant and as a cosolvent (Ayoub and Abdullah 2012). It is also a formulation ingredient for the preparation of toothpaste, mouthwash, shaving creams, skincare products, for tablet coating and as levigating agent to reduce particle size. Being an effective moisturizer for the skin, it has a vital position in the cosmeceuticals market. Glycerol is also applied as a non-toxic antifreeze agent owing to the formation of hydrogen bonds with water molecules. In confectionery, it prevents sugar crystallization and acts as a preservative. Cellophane, a packaging material, is prepared after the addition of glycerol in cellulose as it provides plasticity to the film and hence avoids brittleness. Of lately, glycerol has garnered renewed interest owing to the biodiesel production.

Several strategies like transesterification of fats (Fig. 16), saponification of soap, hydrolysis of oil and fermentation of yeast are generally utilized for the commercial development of glycerol (Bagnato et al. 2017).

Glycerol is also produced as a byproduct of biodiesel formation and hence can be termed as green synthesis. Figure 17 summarizes the derivatives of glycerol that are utilized for various industrial applications.

Global crude glycerol produced from biodiesel has augmented from 200,000 tons to 1.224 million tons in the last decade (Yang et al. 2012). One of the reports projects a continuous increment of glycerol market size by 965.8 USD in the timeline of 2019–2023. Several active manufacturers, namely Cargill, Procter and Gamble, Archer Daniels Midland and Emery Oleochemicals are leading producers and marketers of glycerol all over the world. However, increased value drifts in fatty acid/vegetable oils that are core sources for the preparation of glycerol, pose main constraint in its production.

7 Aspartic Acid

Aspartic acid a non-essential amino acid is synthesized from the central metabolic route by the metabolism of carbohydrates. Naturally, aspartic acid is found abundantly in animals (oysters and sausage meat) and vegetables (sugarcane, molasses of sugar beet and sprouting seeds). The presence of both amino and carboxylic acid in its structure (Fig. 18) offers several reactions/processes for the further development of platform molecules that are effectively utilized in nutrition, medicine and physiological functions. Acrylamide, synthesized from aspartic acid via Milliard reaction, is used for the generation of acrylic acid and polyacrylamide used to treat wastewater. Alpha-deamination of aspartic acid results in fumaric and malic acid that are used in the paper and food industry.

Clinically, aspartic acid is used by clinicians for research on depression and immunity. Additionally, it is also used to generate energy from carbohydrates, liver detoxification from drugs and for enhancing resistance to fatigue. Besides serving as a nutritional supplement, aspartic acid is widely used in the agriculture for the development of fertilizers and organic chemicals.

Preparation of bio-based aspartic acid follows one phase process utilizing lignin/glucose-based feedstock mediated through catalytic transformation (Fig. 19). Microbe mediated production of aspartic acid has also been reported using microorganisms like *Pseudomonas fluorescence, E. coli, B.subtilis, Proteus vulgaris* and *Aerobacter aerogenes* (Anderson and Katahira 2016; Lee and Hong 1988).

Fig. 16 Schematic illustration of glycerol preparation

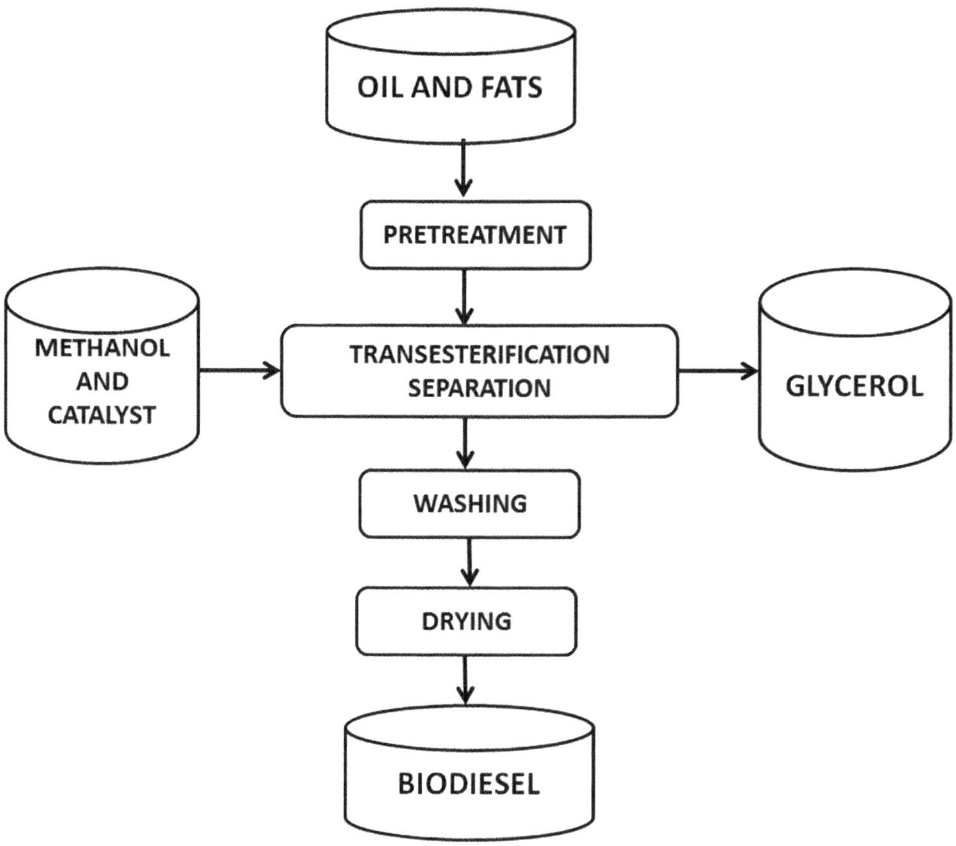

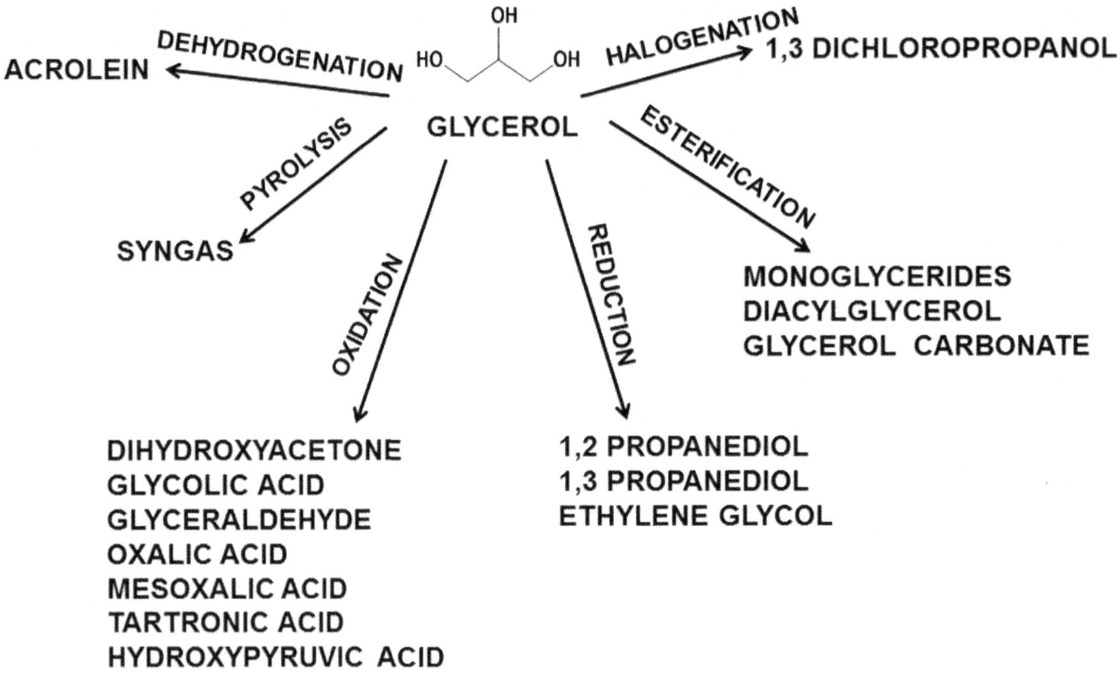

Fig. 17 Glycerol and its transformed derivatives after different reactions

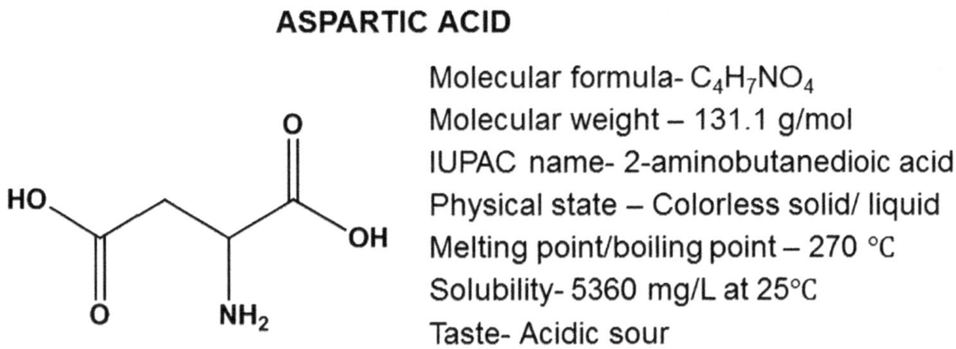

Fig. 18 Physicochemical properties of aspartic acid

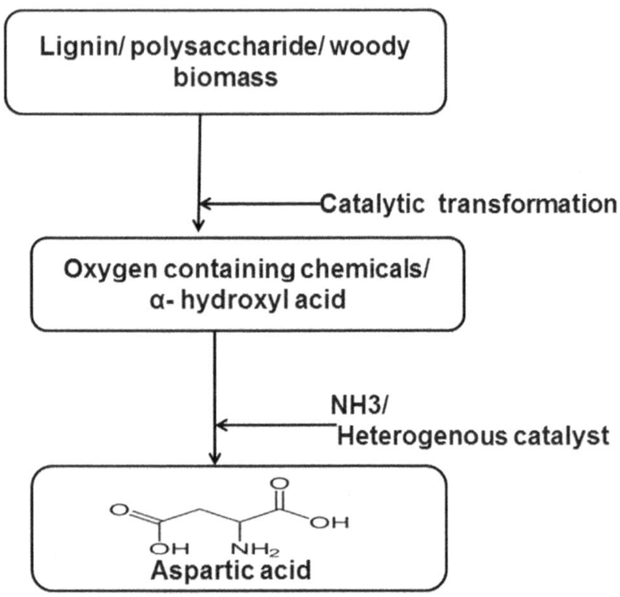

Fig. 19 Green synthetic route of aspartic acid

The literature published in 1965 described a biochemical route for the preparation of aspartic acid utilizing *E.coli* and enzymes. Aspartic acid may also be obtained from the enzymatic conversion of fumaric acid. Although molecules transformed from petrochemicals/fossils feedstocks are easier to process but owing to ecological pollution and emission of carbon dioxide, green chemistry approach is advocated. Thus, fermentation route based on sugar substances is preferred for production of aspartic acid.

Aspartic acid is utilized in food, medicine, nutraceuticals and agriculture sectors. In the year 2012, its global requirement was 35.6 kilo tons that is consistently increasing owing to its utilization for production of aspartame (non-saccharide sweetener) and polyaspartic acid. Augmented applications of polyaspartic acid and its sodium salt in paints, oilfield chemicals and pharmaceuticals affirm an established market of aspartic acid. Increased market growth in medical industries is predicted owing to the escalating demands of magnesium aspartate and calcium aspartate as feed supplements.

8 Itaconic Acid

Naturally occurring itaconic acid (methylene succinic acid) was first reported by Baup in 1836 as a byproduct during citric acid distillation (Pomogailo et al. 2010). It is a value-added molecule owing to its biodegradability and exclusive chemical property due to the presence of two carboxylic groups that offer polymerization and oxidation (Fig. 20). Itaconic acid is more reactive than other bioprivileged molecules, namely fumaric acid and maleic acid (Tomic et al. 2010).

Itaconic acid has been extensively exploited as a key molecule for preparation of plastics, synthetic fibers and paints. Many itaconate metallic salts, diesters and anhydrides are commercially utilized for preparation of *N*-substituted pyrrolidones that are employed in the synthesis of shampoos, herbicides and detergents (Milsom and Meers 1985). It is added to organo-siloxanes (a hardening agent) for the designing contact lens, in napkins/diapers as binders and as biocompatible glass ionomer cement in dentistry (Okabe et al. 2009). In pharmaceuticals, itaconic acid finds versatile applications as antibacterial, wound healing agent, analgesic and for the synthesis of biofuel additives, namely 3-methyl tetrahydrofuran and 2-methylbutanediol (Lucia et al. 2006).

Itaconic acid is a biological metabolite found in numerous strains of *Aspergillus*. Extensive research has been carried out to determine the biosynthetic pathway, its regulation, high yielding strain and the enzymes to catalyze the process of formation of itaconic acid (Klement and Buchs 2013). Commercially, *Aspergillus itaconicus* and *A. terreus* are ideal strains utilized in the fermentation process owing to the high yield and better tolerance to shear/stress. Apart from these, strains of *Ustilago, Candida* and *Pseudomonas* are also used for fermentation (Fig. 21). Submerged solid fermentation including sugar/molasses is typically designed and enzyme catalyzed to process bioconversion of cis-aconitate to itaconate.

Non-toxic bioprivileged molecule, itaconic acid finds global usage in styrene-butadiene rubber, chelant dispersant agents, superabsorbent polymers, methyl methacrylate and latex. Superabsorbent polymers procured from itaconic acid are utilized in the manufacturing of detergents, cosmetics and skin and hygiene care products. Itaconic acid is transformed into various derivatives through several reactions like hydrogenation, amination, reductive amination and carboxylation (Fig. 22).

The annual global production of itaconic acid was approximately 41,000 tons in the year 2011 that increased progressively and is projected to be 50,000 tons by 2020. The worldwide market growth of bio-based itaconic acid is proposed to surge from 83.8 million dollars from the year 2017 to 102.3 million dollars for the year 2022 due to its value-added product: unsaturated polyester resin. Qingdao Langyatai Company, China, is the major producer of itaconic acid with an annual turnover of 20,000 tons. Other countries such as the UK (Itaconix Corporation), India (Alpha Chemika), Japan (Iwata Chemical) and South Korea (Aekyung Petrochemical) are actively engaged in the production of itaconic acid.

9 3-Hydroxybutyrolactone

3-hydroxybutyrolactone finds versatile applications in both, medical and healthcare arena. It is an imperative precursor for various chiral drugs such as atorvastatin calcium, lovastatin, rosuvastatin, linezolid and ezetimibe (Fig. 23). Many pharmaceutical industries such as Pfizer (Lipitor), AstraZeneca (Crestor) and Zybox are involved in the synthesis of 3-hydroxybutyrolactone derived molecules.

The cyclic C4 molecule readily undergoes several chemical transformations for the synthesis of C3 chiral molecules enclosing tetrahydrofuran, amides, epoxides, nitriles and lactones (Hollingsworth and Wang 2000). Apart from statins, this value-added molecule is also utilized for the development of carbapenem (a class of γ-lactam antibiotic), oxazolidinones (antimicrobials) and L-carnitine (nutraceutical). Various conventional chemical routes for the synthesis of 3-hydroxybutyrolactone such as ketoreductase and cofactor recycling, nitrilase catalyzed desymmetrization,

Fig. 20 Physicochemical properties of itaconic acid

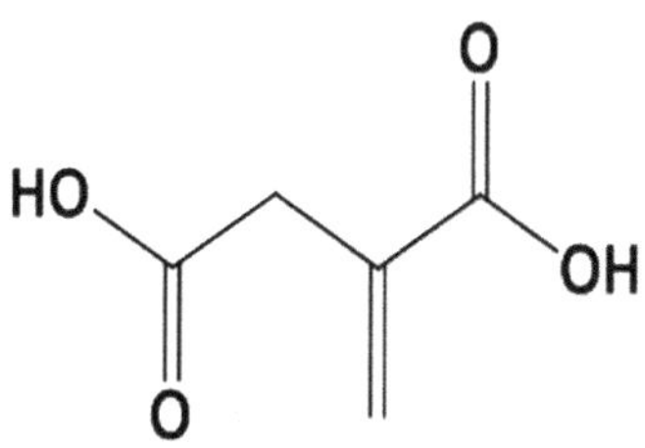

Molecular formula- $C_5H_6O_4$
Molecular weight – 130.1 g/mol
IUPAC name- 2-methylidenebutanedioic acid
Physical state – White solid/ liquid
Melting point/boiling point – 175 °C / 268 °C
Solubility- 76.8 mg/ml at 20°C
Odor - Characteristic
Density- 1.632 g/cm^3

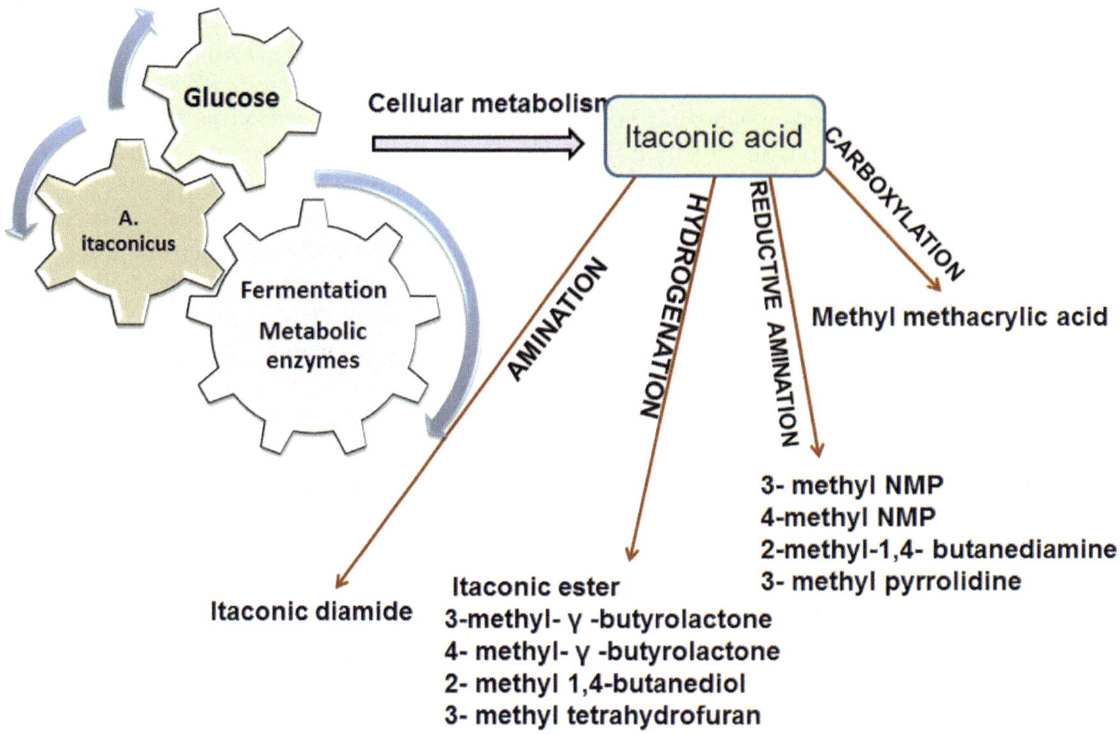

Fig. 21 Production of itaconic acid and its derivatives

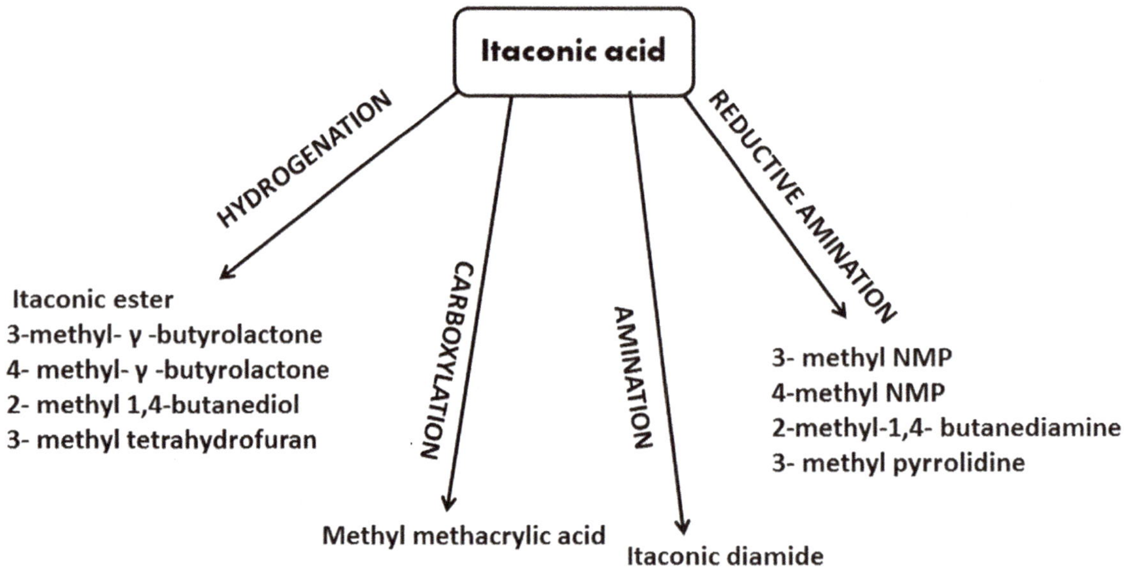

Fig. 22 Derivatives of itaconic acid

aldolase catalyzed condensation and epoxide hydrolase catalyzed hydrolysis are synthetic pathways that are based on non-renewable and non-sustainable sources. These approaches use aggressive processing conditions, catalysts, reagents and tedious purification steps.

Currently, a metabolically engineered technique involving *E. coli* bacteria is being utilized for the production of 3-hydroxybutyrolactone. A renewable and sustainable glucose-based process has been adopted for production of this chiral value-added molecule (Fig. 24).

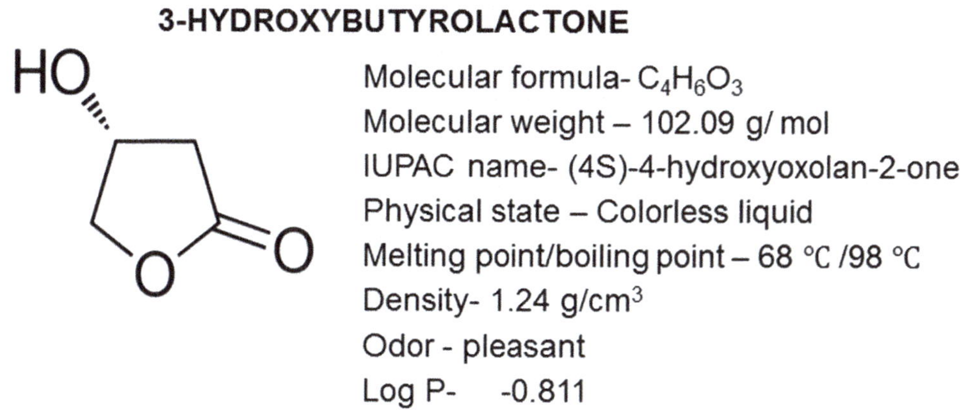

Fig. 23 Physicochemical properties of 3-hydroxybutyrolactone

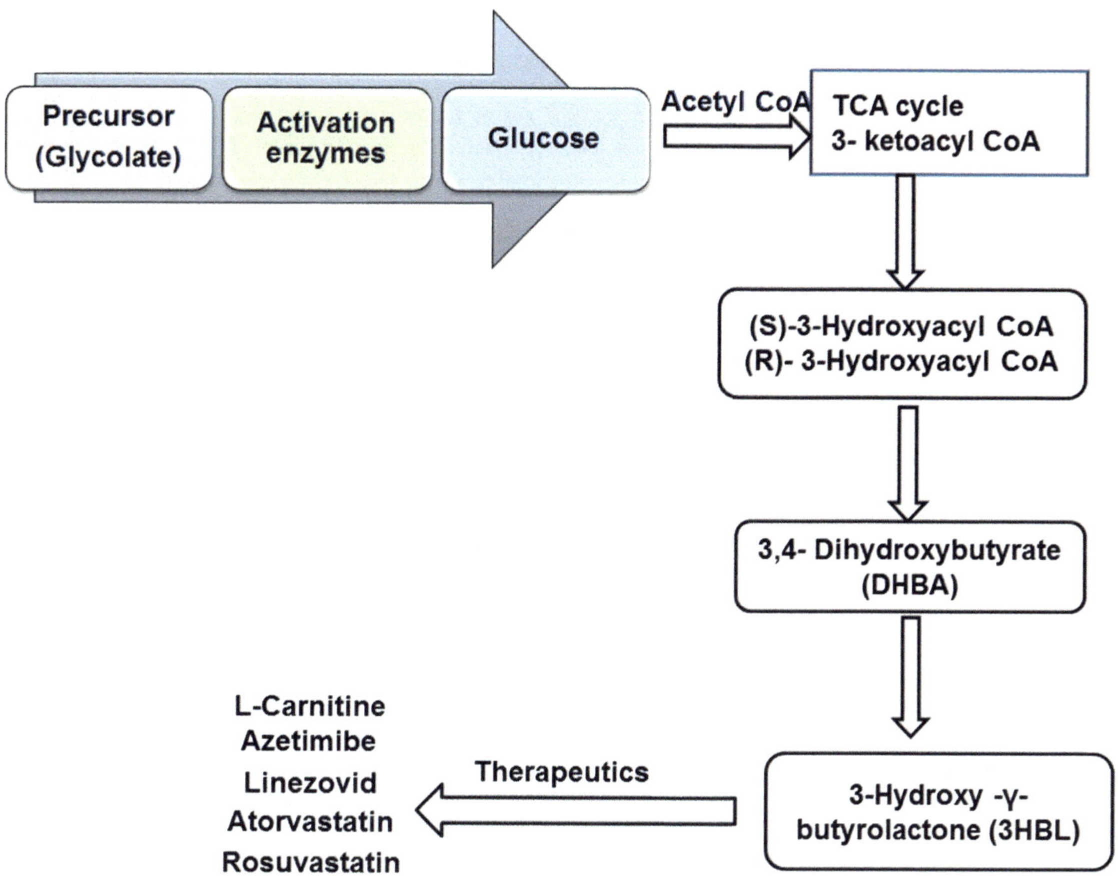

Fig. 24 Bio-based pathway for synthesis of 3-hydroxy-γ-butyrolactone and the therapeutic molecules derived from it

10 Sorbitol

Sorbitol is employed in pharmaceuticals, cosmeceuticals and other industries owing to its favorable properties. The prominent use of sorbitol (Fig. 25) in the pharmaceutical sector is its use in sugar-free liquid preparation, tablet diluent and as a stabilizer for suspensions, vitamins, antacid to prevent crystallization. Table 3 compiles major applications of sorbitol in different domains of industries.

Commercially, sorbitol is synthesized via biosynthetic route through fermentation. Cellulose derivative feedstock is fermented either using yeast (*Saccharomyces cerevisiae*) or fungi (*Candida boidinii*) under anaerobic conditions (Silveira and Jonas 2002). The obtained crude sorbitol is purified either by basic ion exchange resin or by electrodialysis

SORBITOL

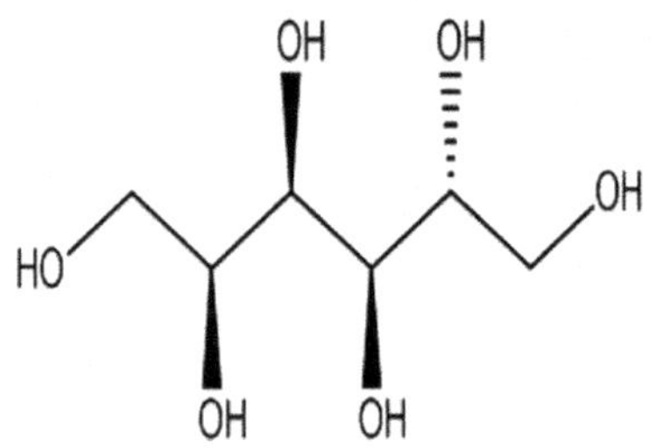

Molecular formula- $C_6H_{14}O_6$
Molecular weight – 182.17 g/mol
IUPAC name- (2S,3R,4R,5R)-Hexane-1,2,3,4,5,6- hexol
Physical state – White solid
Melting point/boiling point – 95 °C /296 °C
Density- 1.489 g/cm^3

Fig. 25 Physicochemical properties of sorbitol

Table 3 A compilation on applications of sorbitol in various sectors

Category	Applications	Reference
Dental products, i.e., toothpaste	Humectant, emulsifier, sweetener, prevents crystallization of abrasives	Deis and Kearsley (2007)
Food, confectionery and beverages	Dietary energy, nutritive sweetener, sugar-free chewing gum	Shwide-Slavin et al. (2012)
Pharmaceuticals	Humectant, diluent, emulsifying agent, stabilizing agent, non-ionic surfactant, a precursor for vitamin C	Sefcovicova et al. (2011)
Platform chemicals production	Ethylene glycol, propylene glycol, glycerol, sorbose	Chen et al. (2013)
Cosmetics	Stable moisturizer, prevent loss of moisturizer, maintain hydration in the skin	Muizzuddin et al. (2013)
Others	Plasticizer, an additive in biocomposite, cross-linking agent	Liu et al. (2013)

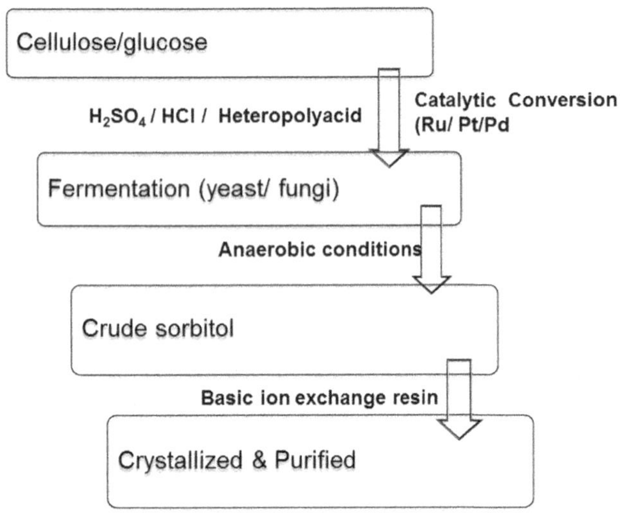

Fig. 26 Green synthesis of sorbitol

(Fig. 26). Although this process is eco-friendly and does not liberate carbon footprints, it is not commercially adaptable compared to the petrochemical route. Process optimization for the selection of high-yielding microbes and operating conditions is required for large-scale production.

Initially, sorbitol production was limited to North America and Europe, but initiation of production of vitamin C through sorbitol encouraged the Asian countries to set up sorbitol production units. China and Indonesia are the top producers and suppliers of sorbitol. The increased demand for low-caloric sweeteners in the United States and European countries has been fueling the global market of sorbitol. The worldwide market size of sorbitol was estimated at approximately 3.79 billion dollars in the year 2017 with a growth rate of 8%. It is a preferred formulative ingredient in medicated chewing gums and dental products as a sugar substitute.

11 Xylitol

A five carbon containing sugar alcohol (Fig. 27), xylitol is naturally found in fruits (grapes, strawberry, banana and yellow plum), vegetables (cauliflower, onion and carrot), seaweeds and mushrooms. However, its extraction from these sources is challenging and tedious. It is commonly used as a natural sweetener in many products meant for diabetics as its metabolism is autonomous and independent of insulin. In medicine and health care, it finds applications in preventing tooth decay, ear infection, as parenteral nutrition and in infusion therapy.

In pharmaceutical sector, it is used in the manufacturing of a variety of pharmaceutical products, namely

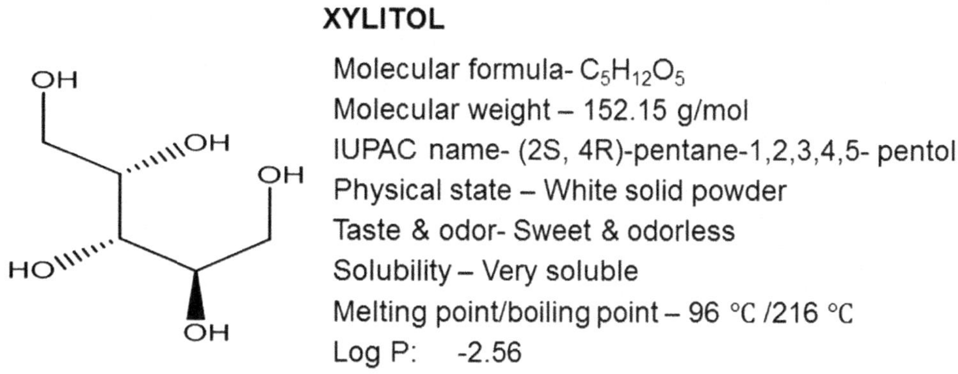

Fig. 27 Physicochemical properties of xylitol

mouthwashes, lozenges, fluoride tablet, tonics, vitamins, coated tablets and syrups (Burt 2006). Compared to other sugar substances, xylitol containing products are preferred in the diabetes care system as it has an insignificant effect on insulin secretion and thus regulates weight management. It checks the development of plaque from bacteria, tooth decay and oral cavity. It facilitates the improvement of dental health by preventing the demineralization of enamel and arouses salivary flow. Its multipurpose functions in oral healthcare products are presented in Fig. 28.

Conventionally, xylitol is synthesized by a chemical route through hydrogenation of D-xylose at elevated temperature in the presence of nickel as catalyst (Ojamo et al. 2009). Exclusive refining steps, i.e., lignocellulose extraction, hydrogenation, crystallization and purification are carried out to get pure xylitol that makes the process quite expensive and non-feasible for large-scale production (Saha 2003). An alternative biotechnological approach to produce xylitol includes microbial fermentation and enzymatic conversion, wherein xylitol is processed from biowaste (corn cob, pulp and paper). These processes are cost-effective and give high yield (Fig. 29). Few bacteria like *Gluconobacter oxydans, Mycobacterium smegmatis* and *Enterobacter liquefaciens* are utilized in the fermenter for collection of xylitol (Suzuki et al. 2002). However, fungus and recombinant yeast are reported as the best producers (approximately 70% yield) among the microorganisms (Chung et al. 2002).

Several companies of the Asia Pacific and Latin America are focusing on the preparation of xylitol through biotechnological route as it is sustainable and eco-friendly. The market size of xylitol was estimated at 737.2 million dollars in the year 2015 that is projected to reach 1.37 billion dollars by the year 2025. The low-calorie molecule xylitol is frequently added as a sugar substitute in diabetic sweetener. The American Academy of Pediatric Dentistry has issued guidelines regarding appropriate uses of xylitol in the pediatric formulations for better health.

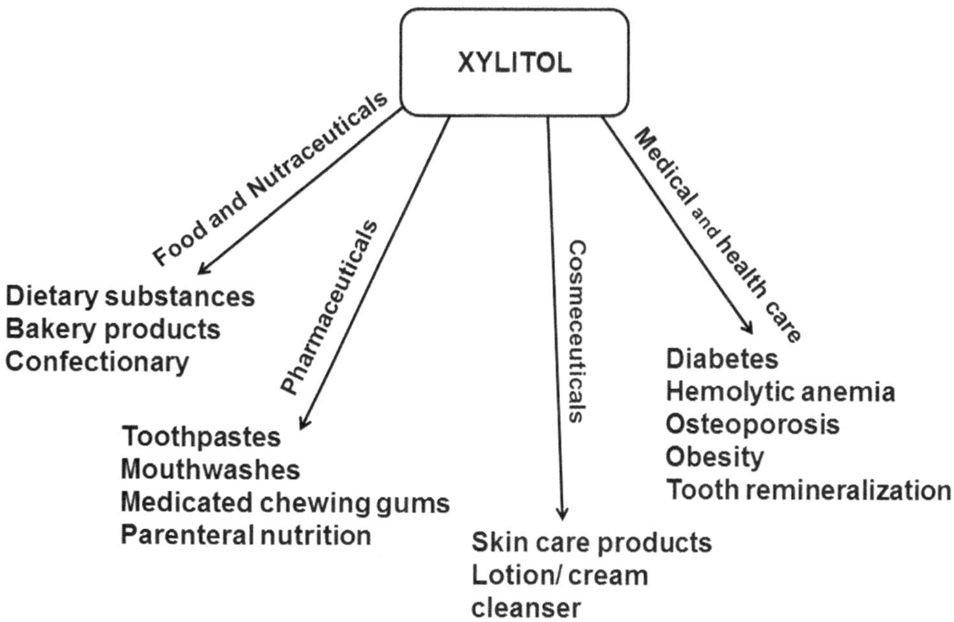

Fig. 28 Comprehensive applications of xylitol

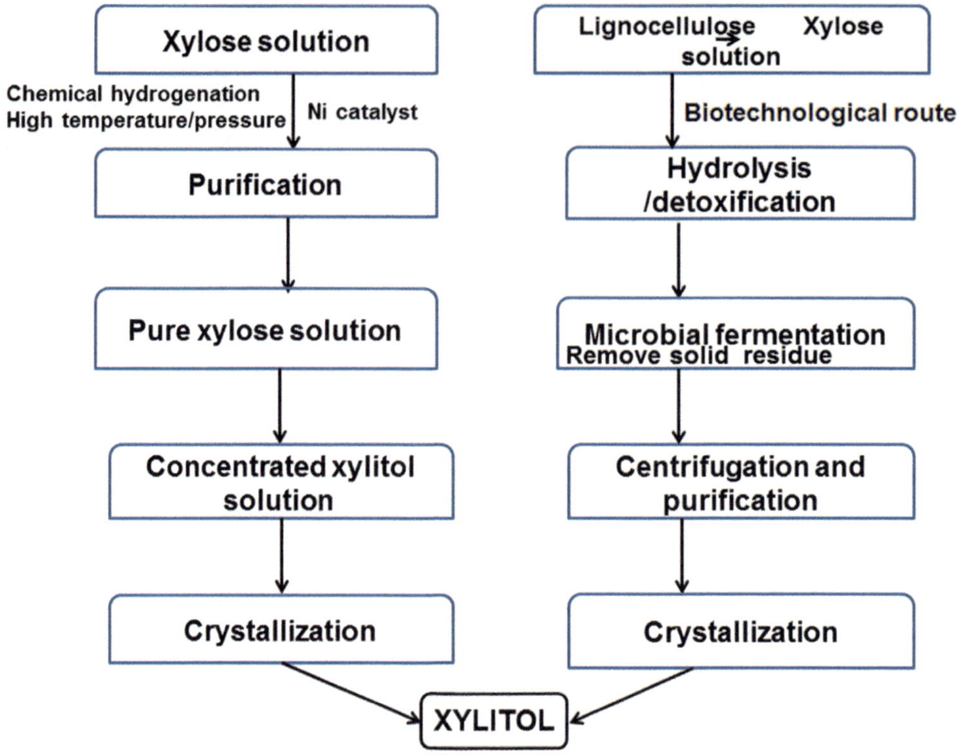

Fig. 29 Comparative chemical and biotechnological routes for xylitol synthesis

12 Glutamic Acid

Glutamic acid is a non-essential amino acid (Fig. 30) and is involved in the biosynthesis of proteins that transmit chemical information to all parts of the body. Glutamic acid was identified by German Chemist K. H. Ritthausen in the year 1866 during the processing of wheat gluten with sulfuric acid (Alger and Teyler 1976). Further, Kikunae Ikeda, a Japanese research scholar collected brown crystals of glutamic acid after evaporation of kombu broth. Thereafter, he patented monosodium glutamate, a salt of glutamic acid; the molecule was extensively explored as a food additive and taste enhancer (Vlakh et al. 2016).

In human body, glutamic acid or its glutamate salt is involved in the biosynthesis of inhibitory neurotransmitter gamma amino butyric acid (GABA) (Reeds et al. 2000). Naturally, it occurs in a variety of foods, i.e., cheese, meat, fish, eggs and soy sauce. The umami tasting glutamic acid is added to beverages and soft drinks to enhance flavor. In cosmeceuticals, it is been used in hair restoration and anti-wrinkle products. L-glutamic acid is often used in pharmaceutical analysis conducted by ion chromatography. A plant growth enhancer auxigro, approved by US Environmental protection agency, contains 30% glutamic acid and is sprayed on vegetables and crops for high productivity (Biorationals: ecological pest management database. Auxigrowp. Attra.ncat.org/https://www.emeraldbio.com).

GLUTAMIC ACID

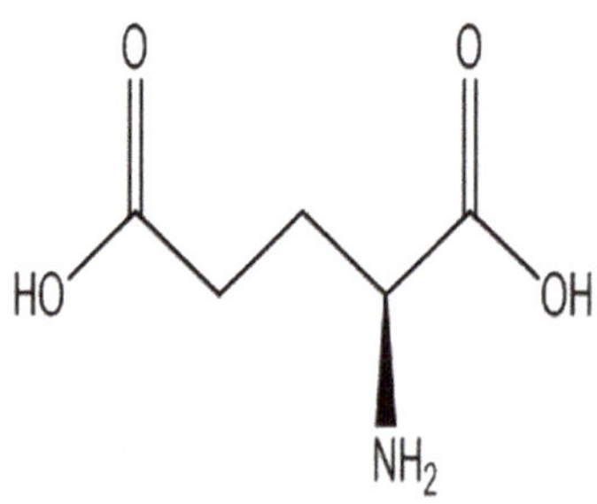

Molecular formula- $C_5H_9NO_4$
Molecular weight – 147.13 g/mol
IUPAC name- (2S)-2-aminopentanedioic acid
Physical state – Solid crystals
Melting point – 224 °C
Density- 1.538g/cm^3
Taste & Odor – Umami sour & odorless
Solubility - Sparingly soluble

Fig. 30 Physicochemical properties of glutamic acid

The production methods of glutamic acid include chemical synthesis, chemo-enzyme route and fermentation methods. Commercially, a biotechnology-based fermentation approach is accepted for large-scale production. In a submerged fermenter, the gram-positive *Corynebacterium glutamicum*, a facultative anaerobic bacteria, carbon and nitrogen sources (glucose and urea) along with biomass (palm/sugarcane/cassava starch wastes) are added together to produce glutamic acid. Ambient conditions such as pH 7.0–8.0, temperature between 30 and 40 °C and stirring for 40–48 h are recommended (Fig. 31). Subsequently, centrifugation, filtration and crystallization are carried out sequentially to obtain pure glutamic acid (Flickinger 2010).

Worldwide, the demand for glutamic acid is anticipated to hike significantly due to its applications in food, poultry and pharmaceutical domains. Monosodium glutamate listed as "Generally recognized as safe" by USFDA is popularly added in preparation of snacks, seasoning blends and noodles. The global market size of this non-essential amino acid was estimated at approximately 2.9 million tons in the year 2014 and is expected to be surge more than 4 million tons to the year 2023 with an annual growth rate of 7.5%. Major companies involved in its preparation are Ajinomoto Co. Inc. (Japan), Sunrise Nutrachem Group (China), Kyowa Hakko Bio Co. (Japan) and Prinova U.S.LLC (North America).

13 Levulinic Acid

Levulinic acid or 4-oxopentanoic acid was listed as bioprivileged molecule in 2004 by the US Department of Energy after screening of 300 molecules. The polar organic keto acid was first identified by a Dutch professor G. J. Mulderin in 1840 during heat treatment of fructose in the presence of hydrochloric acid. In the year 1953, a US company Quaker Oats designed a method for the continuous preparation of levulinic acid that was recognized as a potential platform molecule in the year 1956 (Girisuta 2007). Although it has both keto and carboxylic functional groups (Fig. 32) that undergo various chemical reactions (redox, substitution, esterification and polymerization), the molecule was ignored for quite some time. Of lately, the industrial value of levulenic acid and its derivatives has paved way for their usage (Fachri et al. 2015).

It finds use in different domains of agriculture, pharmaceuticals and cosmetics. The literature highlights the versatile role of this value-added molecule in preparation of resins, animal feed, coating materials, plasticizer and anti-freeze agent and in textiles (Table 4). It is also used in smoke/cigarettes to enhance the delivery of nicotine to neural receptors (Keithly et al. 2005). Its derivative, delta-aminolevulinic acid, is utilized for the eradication of tiny insects from lawns and grain crops owing to its

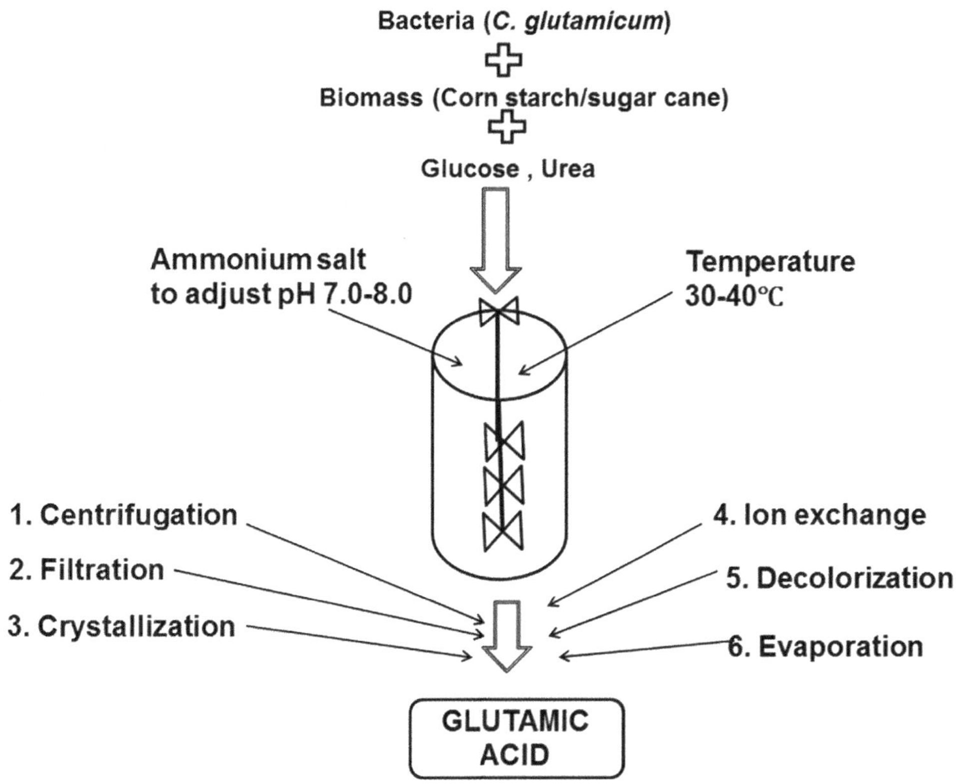

Fig. 31 Green synthesis-based biotechnological route for the preparation of glutamic acid

Fig. 32 Physicochemical properties of levulinic acid

LEVULINIC ACID

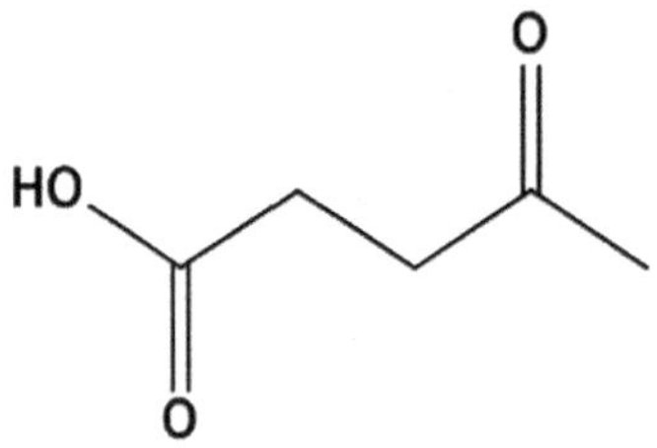

Molecular formula- $C_5H_8O_3$
Molecular weight – 116.17 g/ mol
IUPAC name- 4- Oxopentanoic acid
Physical state – Solid/ liquid
Melting point/ boiling point – 33 °C/ 245.5 °C
Log P/ pka- -0.49/ 4.64 at 18°C

Table 4 Few commercial applications of levulinic acid derivatives (Ji et al. 2015)

Levulinic acid derivatives	Commercial applications
Hydroxyvaloric acid	Fuel additives, paints, resins
Acetyl acrylic acid	Copolymerization
Diphenolic acid	Epoxy resin, paint, adhesive, lubricants
Valerolactone	Biofuel, solvent, fuel additive
Esters of levulinic acid	Fuel additive, plasticizer, fool flavoring
Aminolevulinic acid	Insecticide, herbicide, and chemotherapy
Succinic acid	Solvent, polymer, pesticide
2-methylene valerolactone	Fuel additive, biofuel
1,4-pentanediol	Polymer, fine chemicals, solvent
2-methyltetrahydrofuran	Fuel additive, solvent
Ethyl levulinate	Fragrance and perfumes
Angelica lactone	Fuel additive, solvent

biodegradable herbicidal activity. Some anti-inflammatory, anti-allergic and medicated transdermal patches have also been prepared with levulinic acid.

Conventionally, levulinic acid is prepared by acidic hydrolysis of furfuryl alcohol and acetyl succinate ester. The hazardous output from fuel-based technology/processes needs to be substituted by a sustainable green synthesis route to save resources and compensate energy requisites. Lignocellulose, manure, wood and brewery wastes are frequently used as sustainable bio-based feedstock applied in biorefinery approaches to synthesize value-added products (Yan et al. 2015). Levulinic acid is prepared through hydrolysis of cellulose for the synthesis of platform chemicals. Obtained cellobiose that is composed of several glucose units joined with β-glycoside bonds is treated with solid acid catalyst CP-SO_3H at 170 °C for 10 h in a stainless steel container to get levulinic acid (Zuo et al. 2014). It can also be prepared using a batch reactor containing corn starch, sulfuric acid, water under preset pressure and temperature. The concentration and amount of sulfuric acid affect the productivity of levulinic acid (Cha and Hanna 2002).

The market of levulinic acid is segregated based on its applications such as pharmaceutical, cosmeceuticals, agriculture and food industry. Geographically, the provinces of North America, Asia Pacific, Middle East, Africa and Europe are involved in the marketing of levulinic acid. The global demand estimated at 2606.2 tons in the year 2013 is expected to rise 3820 by the year 2020 with a growth rate of 4.8%. Segetis, DuPont and Biofine are the major industries involved in the preparation of levulinic acid derivatives, namely methyltetrahydrofuran and aminolevulenic acid.

14 Emerging Molecules

In addition to the above discussed biopreveliged molecules, there are a few emerging bioprivileged molecules such as muconic acid, 5-hydroxymethylfurfural and triacetic acid lactone. Muconic acid is formed during catabolism and detoxification of aromatic chemicals in some microbes. It is biotransformed into several petrochemical equivalent chemicals such as adipic acid, caprolactam and terephthalic acid. On partial hydrogenation of muconic acid, 3-hexenedioic acid is obtained that cannot be prepared via petrochemical route (Rorrer et al. 2016).

Another molecule, 5-hydroxymethylfurfuralis obtained from dehydration of fructose can be transformed into furanic acid and its derivative, 2,5-dimethylfuran. These molecules exhibit antifriction/anti-wear properties and are used as lubricant additive in gasoline. 2,5-bis (hydroxy methyl)

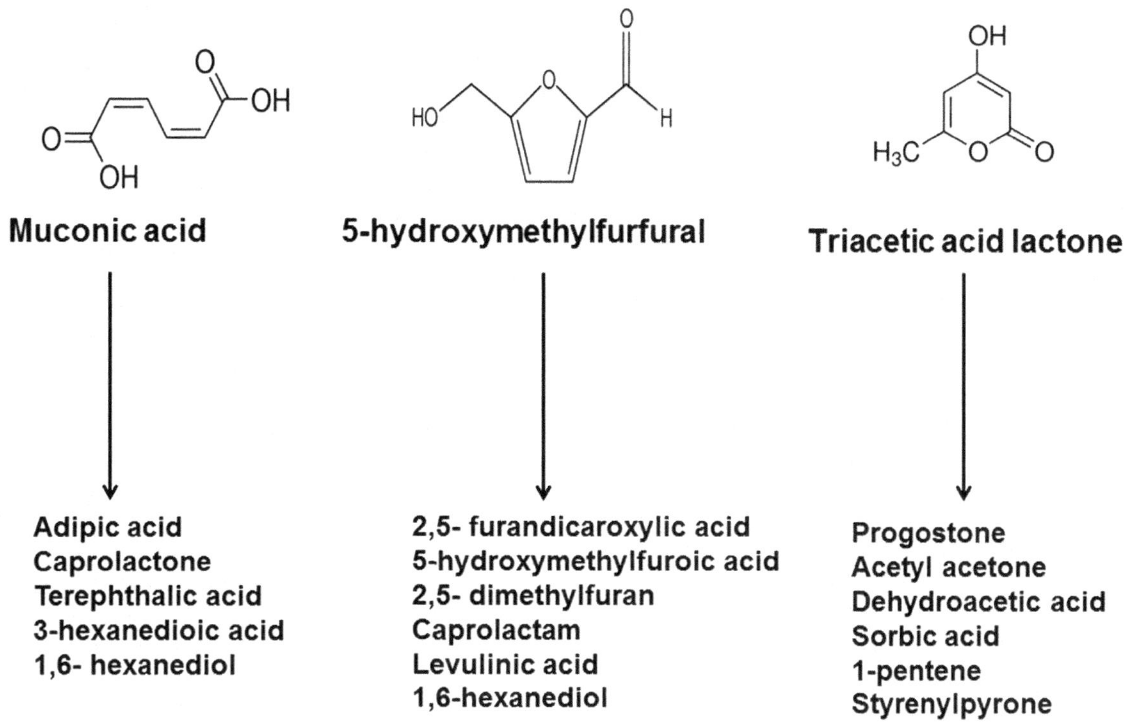

Fig. 33 Applications of muconic acid, HMF and TAL

furan is utilized for the synthesis of stiff polyurethane foam (Dunlop et al. 1982). Lastly, triacetic lactone (TAL) is found in plants as secondary metabolites and is commercially prepared through bioengineered efforts. TAL derived progestone and dehydroacetic acid finds their utility in health care owing to the antimicrobial action. Other value-added molecules styrenylpyrones and sorbic acid used as anti-obesity and food preservative, respectively, are prepared from TAL as precursor (Xie et al. 2006). Comprehensive applications of these emerging biopriviledged molecules are summarized in Fig. 33.

A summary the green synthetic modes and applications of the above described biopriviledged molecules has been comprehensively tabulated in Table 5.

Table 5 A summary of green synthesis route and potential applications of bioprivileged molecules

Bioprivileged molecule	Green synthesis route	Potential applications
Succinic acid	Bioprocessing route that involves bacterial cells (*Escherichia coli*) with enzymes and cofactors that are safe, economical and eco-friendly	Preparation of several precursors, surfactants, antifoams, detergents, chelators and synthesis of anti-convulsant therapeutics (ethosuximide, methsuximide and phensuximide, etc.)
Fumaric acid	Aerobic and anaerobic fermentation utilizing species of *Mucor, Aspergillus* and *Cirinella* with neutralizing agent ($CaCO_3$) under extreme conditions (pH 1, 160 °C)	For synthesis of dermatological, anti-carcinogenic and anti-inflammatory drugs Precursor for L-aspartic acid and malic acid Acidulant, flavoring agent, coating agent and preservative
Malic acid	Bioprocessed reductive and glyoxalate pathways under acidic condition in the presence of eukaryotic cultures, i.e., *Aspergillus* sp., *Candida* sp., *Penicillium* sp. and *Ustilago* sp.	Formulation of mouthwash, dental tablets, topical creams and effervescent powders It is used as chelating agent and antioxidant that delays oxidation of vegetable oils

(continued)

Table 5 (continued)

Bioprivileged molecule	Green synthesis route	Potential applications
Furan 2,5-dicarboxylic acid (FDCA)	The biocatalytic route of biomass (lignocellulose) in the presence of genetically engineered *P. putida* S12	Manufacturing of polyesters and plastics Suitable replacement for synthetic terephthalic acid Its film and fibers are used as packaging material. Fungicide and corrosion inhibitor
3-hydroxy propionic acid (3-HPA)	Acetyl CoA dependent and independent glycerol metabolic processes that involve acid resistant microorganism in the bioconversion of biomass into 3-HPA	As a precursor of variety of value-added chemicals such as acrylic acid, acrylonitrile, malonic acid, 1, 3-propanediol Used in coating, masonry, adhesive, as water treatment chemical
Glucaric acid	Corn stover route employing homogenous and heterogenous oxidation of cellular enzyme treated biomass	Formation of phosphate-free detergents, anti-corrosive additives, biodegradable cleaner, adhesive, coating material and various therapeutic molecules
Glycerol	Glycerol and biodiesel both are concomitantly prepared from triglyceride transesterification of oil containing feed stocks	Humectant, viscosity modifier, adhesive, plasticizer, icing agent, lubricant and cosolvent for the preparation of toothpaste, mouthwash, shaving creams, skincare products
Aspartic acid	Catalytic biotransformation of glucose-based feedstock in the presence of *Pseudomonas fluorescence, E. coli, B. subtilis, Proteus vulgaris* and *Aerobacter aerogenes,* etc.	Formation of paints, oilfield chemicals, superabsorbent polymers and sweetener aspartame, muscle strengthening protein. As a feedstock (magnesium aspartate and calcium aspartate)
Itaconic acid	Enzyme catalyzed bioconversion process utilizing *Aspergillus itaconicus* and *A. terreus* strains	Synthesis of styrene-butadiene rubber, chelant dispersant agents, superabsorbent polymers, methyl methacrylate and latex
3-hydroxy butyrolactone	Metabolically engineered technique involving *E. coli* bacteria that hydrolyses glucose-based feedstock	Active precursor of atorvastatin calcium, lovastatin, rosuvastatin, linezolid, ezetimibe Synthesis of carbapenems, oxazolidinones and L-carnitine
Sorbitol	Fermentation of lignocellulose in the presence of yeast (*Saccharomyces cerevisiae*) or fungi (*Candida boidinii*) under anaerobic conditions	Plasticizer, humectant, diluent, emulsifying agent, stabilizing agent, nonionic surfactant Precursor for vitamin C Production of ethylene glycol, propylene glycol, glycerol, sorbose
Xylitol	Biotechnological approach involving enzymatic conversion of biowastes in presence of *Gluconobacter oxydans, Mycobacterium smegmatis* and *Enterobacter liquefaciens*	Diabetic sweetener, dietary substance, pharmaceutical ingredient, skin care (cleanser), tooth remineralization, etc.
Glutamic acid	Biomass bioconversion in presence of gram-positive *Corynebacterium glutamicum* under ambient conditions (pH 7.0–8.0, 30–40 °C and stirring time 40–48 h)	Monosodium glutamate food additive, flavoring agent in beverages and soft drinks Applied as hair restorer and added in anti-wrinkles products
Levulinic acid	Catalytic conversion of Cellulose based biomass followed by fermentation	Resins, animal feed, coating materials, plasticizer, antifreeze agent, fuel additive, biodegradable herbicidal, etc

15 Conclusion

Natural resources are being consumed extensively that is impacting the limited non-renewable assets, presenting critical environmental issues and elevated greenhouse gas emissions. To overcome these issues, sustainable biotechnological processes based on green synthesis are being extensively explored for production of bioprivileged molecules. The bioprivileged molecules have the potential to resolve several waste management problems, i.e., development of disposable plastics and biodegradable pesticides. They also enrich bioeconomy through the conversion of biomass and waste stream into value-added products and bioenergy. Selection of appropriate bio-based feedstock, processing technique, microorganism, ambient parameters (temperature, catalyst and enzymes), for a particular molecule is quite challenging as it has an intense impact on the cost of the product. Moreover, channelization of biomass for ready to use is a lengthy process and tedious compared to the petrochemical route. Additionally, competition for land, environmental pressure, policy inconsistency, large financing requirements, climatic change agreements, energy obligations and food security are experienced obstacles for industries. Apart from cited issues and hurdles, there is an intense need for uplifting bio-based sustainable resources through favorable policy design, sufficient feedstock supply chain, wider business support and access to downstream markets both for novel and drop-in replacement molecules.

References

Alger BE, Teyler TJ (1976) Long-term and short term plasticity in the CA1, CA3, and dentate regions of the rat hippocampal slice. Brain Res 110:463–480

Anderson EM, KatahiraR RM et al (2016) Reductive catalytic fractionation of corn stover lignin. ACS Sustain Chem Engg. 4 (12):6940–6950

Ayoub M, Abdullah AZ (2012) Critical review on the current scenario and significance of crude glycerol resulting from biodiesel industry towards more sustainable renewable energy industry. Renew Sustain Energy Rev 16:2671–2686

Bagnato G, Iulianelli A, Sanna A (2017) Glycerol production and transformation: a critical review with particular emphasis on glycerol reforming reaction for producing hydrogen in conventional and membrane reactors. Membrane 7(2):17. https://doi.org/10.3390/membranes7020017

Besson M, Gallezot P, Pinel C (2014) Conversion of biomassinto chemicals over metal catalysts. Chem Rev 114:1827–1870.

Bin D, Wang H, Li J et al (2014) Controllable oxidation of glucose to gluconic acid and glucaric acid using an electrocatalytic reactor. ElectrochimActa 130:170–178

Biorationals: ecological pest management database (2018). Auxigrowp. Attra.ncat.org/https://www.emeraldbio.com. Accessed 13 Aug 2018

Bioprivileged molecules: a new paradigm for biobased chemical development. Accessed 13 Sept 2018

Boussie TR, Dias EL, Fresco ZM (2013) Production of adipic acid and derivatives fromcarbohydrate-containing materials. US8501989B22013, 6 Aug 2013

Bozell JJ, Petersen GR (2010) Technology development for the production of biobased products from biorefinery carbohydrates. Green Chem 12:539–554

Brent HS, Keeling PL (2017) Bioprivileged molecules: creating value from biomass. Green Chem 19(14):3177–3185

Brittain HG (2001) Malic acid. Analytical profiles of drug substances and excipients. Elsevier, pp 153–195

Burt BA (2006) The use of sorbitol and xylitol sweetened chewing gum in caries control. J Am Dent Assoc 137:190–196

Cha JY, Hanna MA (2002) Levulinic acid production based on extrusion and pressurized batch reaction. Ind Crops Prod 16 (2):109–118

Chen X, Wang X, Yao S (2013) Hydrogenolysis of biomass-derived sorbitol to glycols and glycerol over Ni-MgO catalysts. Catal Commun 39:86–89

Chung YS, Kim MD, Lee WJ (2002) Stable expression of xylose reductase gene enhances xylitol production in recombinant *Saccharomyces cerevisiae*. Enzyme Microb Technol 30:809–816

Das RK, Brar SK, Verma M (2016) Recent advances in the biomedical applications of fumaric acid and its ester derivatives: The multifaceted alternative therapeutics. Pharmacol Rep 68:404–414

Deis RC, Kearsley MW (2007) Sorbitol and Mannitol. sweeteners and sugar alternatives in food technology. 2nd edn. Wiley Blackwell. https://doi.org/10.1002/9780470996003

DengY MY, Zhang X (2016) Metabolic engineering of a laboratory-evolved Thermobifida fusca muC strain for malic acid production on cellulose and minimal treated lignocellulose. Biotechnol Prog 32:14–20

Diamond GM, Murphy V, Boussie TR (2014) Application of high throughput experimentation to the production of commodity chemicals from renewable feedstocks. Modern Applications of High Throughput R&D in Heterogeneous Catalysis Rennovia Inc., Menlo Park, California, pp 88–309

Dishisha T, Pereyra LP, Pyo SH et al (2014) Flux analysis of the *Lactobacillus reuteri* propanediol-utilization pathway for production of 3-hydroxypropionaldehyde, 3-hydroxy propionic acid and 1,3-propanediol from glycerol. Microb Cell Fact 13:76. https://doi.org/10.1186/1475-2859-13-76

Dunlop WR, Pentz WJ (1982) Low fire hazard rigid urethane insulation foam, polyol mixtures used in the manufacture thereof, and method for manufacture thereof. US Patent, US4318999A

Fachri BA, Abdilla R, Bovenkamp H (2015) Experimental and kinetic modeling studies on the sulphuric acid catalyzed conversion of d-fructose to 5-hydroxymethylfurfural and levulinic acid in water. ACS Sustain Chem Eng 3:3024–3034

Flickinger MC (2010). Encyclopedia of industrial biotechnology: bioprocess, bioseparation, and cell technology. Wiley, New York, pp 215–225. ISBN 978-0-471-79930-6

Furandicarboxylic Acid (FDCA) Market Size, Share & Trend Analysis Report by Application (PET, Polyamides, Polycarbonates, Plasticizers, Polyester Polyols), By Region, And Segment Forecast, 2012–2020. Accessed Apr 2015

3-Hydroxypropionic Acid Market Size, Share & Trends Analysis Report by Application, Regional Outlook, Competitive Strategies, And Segment Forecasts, 2019 To 2025. Market Research Report. Report ID: GVR1293

Girisuta B (2007) Levulinic acid from lignocellulose biomass. Dissertation, University of Groningen

Global Succinic Acid Market Analysis & Trends 2013–2017—Industry Forecast to 2025: $1.76 Billion Growth Opportunities/Investment Opportunities—Research and Markets. https://ceo.ca/@newswire/global-succinic-acid-market-analysis-trends-2013-2017. Accessed 6 Mar 2017

Goldberg I, Stefan RJ, Pines O (2006) Organic acids: old metabolites, new themes. J Chem Tech Biotechnol 81:1601–1611

Hollingsworth RI, Wang G (2000) Towards a carbohydrate-based chemistry: progress in the development of general purpose chiral synthons from carbohydrates. Chem Rev 100:4267–4282

Jiménez-Quero A, Pollet E, Zhao M et al (2017) Fungal fermentation of lignocellulosic biomass for itaconic and fumaric acid production. J Microbiol Biotechnol 27:1–8

Ji H, Wang B, Zhang X (2015) Synthesis of levulinic acid based polyol ester and its influence on tribological behavior as a potential lubricant. RSC Adv 5:100443–100451

Jung WS, Kang JH, Chu HS et al (2014) Elevated production of 3-hydroxypropionic acid by metabolic engineering of the glycerol metabolism in *Escherichia coli*. Metab Eng 23:116–122

Keithly L, Ferris WG, Cullen DM et al (2005) Industry research on the use and effects of levulinic acid: a case study in cigarette additives. Nicotine Tob Res 7(5):761–771

Kenar JA (2007) Glycerol as a platform chemical: Sweet opportunities on the horizon. Lipid Technol 19:249–253

Klement T, Buchs J (2013) Itaconic acid—a biotechnological process in change. Bioresour Technol 135:422–431

Kumar V, Ashok S, Park S (2013) Recent advances in biological production of 3-hydroxypropionic acid. Biotechnol Adv 31:945–961

Lee CK, Hong J (1988) Membrane reactor coupled with electrophoresis for enzymatic production of aspartic acid. Biotechnol Bioengg 32 (5):647–654

Lewkowski J (2001) Synthesis, chemistry and applications of 5-hydroxymethyl-furfural and its derivatives. Arch Org Chem 1:17–54

Liu M, Zhou Y, Zhang Y et al (2013) Preparation and structural analysis of chitosan films with and without sorbitol. Food Hydrocoll 33:186–191

Lucia LA, Argyropoulos DS, Adamopoulos L, Gaspar AR (2006) Chemicals and energy from biomass. Can J Chem 84:960–970

Lu C, Ford E (2018) Antiplasticizing behaviors of glucarate and lignin bio-based derivatives on the properties of gel-spun poly (vinyl alcohol) fibers. Macromol Mater Eng 3(4):1700523

Magnuson J, Lasure LL (2004) Organic acid production by filamentous fungi. In: Lang J, Lang L (eds) Advances in fungal biotechnology for industry, agriculture, and medicine. Kluwer Academic, New York, pp 307–340

Milsom PE, Meers JL (1985) Gluconic and itaconic acids. Comprehensive biotechnology: the principles, applications, and regulations of biotechnology in industry, agriculture, and medicine. Angewandte Chemie, Elsevier, p 588

Moon TS, Yoon S-H, Lanza AM (2009) Production of glucaric acid from a synthetic pathway in recombinant *Escherichia coli.* Appl Environ Microbiol 75(3):589–595

Muizzuddin N, Ingrassia M, Marenus KD (2013) Effect of seasonal and geographical differences on skin and effect of treatment with an osmoprotectant: sorbitol. J Cosmet Sci 64:165–174

Nhuan PN, Susanne K, Stefan S (2017) Succinic acid: technology development and commercialization. Fermentation 3(26). https://doi.org/10.3390/fermentation3020026

Ojamo H, Penttila M, Heikkila H (2009) Method for the production of xylitol. U.S. Patent 7,482,144 B2, 27 Jan 2009

Okabe M, Lies D, Kanamasa S et al (2009) Biotechnological production of itaconic acid and its biosynthesis in *Aspergillus terreus*. Appl Microbiol Biotechnol 84:597–606

Park YM, Chun JP, Rho KR et al (2004) Process for preparing optically pure (S)-3-hydroxy-γ-butyrolactone. US patent no 6,713,290, 30 Mar 2004

Pomogailo AD, Kestelman VN, Dzhardimalieva GI (2010) Monomeric and polymeric carboxylic acids. Macromolecular metal carboxylates and their nanocomposites. Springer, Berlin, pp 7–25

Reeds PT, Burrin DG, Stoll B et al (2000) Intestinal glutamate metabolism. J Nutr 130(4):978S-982S

Rivertop Renewables and DTI exceed nameplate capacity of first glucarate production facility. 12 Rivertop Press Release. Accessed 29 Mar 2016

Rorrer NA, Dorgan JR, Vardon DR (2016) Renewable unsaturated polyesters from muconic acid. ACS Sustain Chem Eng 4:6867–6876

Saguir FM, Rivero LV, Vaquero MJ et al (2018) Malic acid fermentation: influence and applications in winemaking. Nova Science Publisher, New York, p 203

Saha BC (2003) Hemicellulose bioconversion. J Ind Microbiol Biotechnol 30:279–291

Saxena RK, Saran S, Isar J (2017) Production and applications of succinic acid. Current developments in biotechnology and bioengineering production, isolation, purification of industrial production. Elsevier, Amsterdam, pp 601–630

Sefcovicova J, Filip J, Tomcik P et al (2011) A biopolymer-based carbon nanotube interface integrated with a redox shuttle and a D-sorbitol dehydrogenase for robust monitoring of D-sorbitol. MicrochimActa 175:21–30

Shwide-Slavin C, Swift C, Ross T (2012) Nonnutritive sweeteners: where are we today? Diabetes Spectrum 25:104–110

Silveira M, Jonas R (2002) The biotechnological production of sorbitol. Appl Microbiol Biot 59:400–408

Smith TN, Hash K, Davey C-L et al (2012) Modifications in the nitric acid oxidation of D-glucose. Carbohydr Res 350:6–13

Suzuki S, Sugiyama M, Mihara Y (2002) Novel enzymatic method for the production of xylitol from d-arabitol by Gluconobacter oxydans. Biosci Biotechnol Biochem 66:2614–2620

Johnson Matthey Process Technologies and Rennovia Announce On Time Start-up of Mini-Plant for Bio-Based Glucaric Acid Production Using Jointly Developed Technology. Rennovia press release. Accessed 16 July 2015

Tomic SL, Micic MM, Dobic SN, Suljovrujic EH (2010) Smart poly (2-hydroxyethyl methacrylate/itaconic acid) hydrogels for biomedical applications. Radiat Phys Chem 79:643–649

Vlakh E, Anayan A, Zashikhina N (2016) Preparation, characterization, and biological evaluation of poly(glutamic acid)-b-polyphenylalanine polymersomes. Polymers 8(6):212. https://doi.org/10.3390/polym8060212

Walaszek Z (1990) Potential use of D-glucaric acid derivatives in cancer prevention. Cancer Lett 54(1–2):1–8

Werpy T, Petersen G (2004) Results of screening for potential candidates from sugars and synthesis gas. In: Top value-added chemicals from biomass, vol 1. US Department of Energy, Biomass, Washington DC, pp 1–76

Xiaowei Z, Zachary J, Brentzel GA et al (2019) Computational framework for the identification of bioprivileged molecules. ACS Sustain Chem Eng 7(2):2414–2428

Xie D, Shao Z, Achkar J et al (2006) Microbial synthesis of triacetic acid lactone. Biotechnol Bioeng 93:727–736

Yang ST (2007) Bioprocessing—from biotechnology to biorefinery. In: Yang ST (ed) Bioprocessing for value-added products from renewable resources—new technologies and applications. Elsevier, New York, pp 1–24

Yang F, Hanna MA, Sun R (2012) Value-added uses for crude glycerol–a byproduct of biodiesel production. Biotechnol Biofuels 5:13. https://doi.org/10.1186/1754-6834-5-13

Yan K, Jarvis C, Gu J (2015) Production and catalytic transformation of levulinic acid: a platform for speciality chemicals and fuels. Sust Energ Rev 51:986–997

Zoltaszek R, Hanausek M, Kilianska ZM et al (2008) The biological role of D-glucaric acid and its derivatives: potential use in medicine. PostepyHig Med Dosw 5(62):451–462

Zuo Y, Zhang Y, Yao F (2014) Catalytic conversion of cellulose into levulinic acid by a sulfonatedchloromethyl polystyrene solid acid catalyst. Chem Cat Chem 6:753–757

Membrane Reactors for Green Synthesis

Hamidreza Bagheri, Ali Mohebbi, and Hadis Eghbali

Abstract

The chemical industries have been identified as the most important source of environmental contamination. Two critical processes in a chemical industry are separation and reaction. Membrane reactor [MR] field has been concentrated on novel membrane materials to be combined in a compressed structure. This kind of reactors can perform two roles: reaction and separation media. Subsequently, they are very useful to adjust concentrations of products and reactants. Membrane reactors, due to synergistically performing many operations, containing separation and reaction in one unit, are highly desirable in green synthesis and environmental protection. In the past three decades, there has been wide investigations in the green synthesis processes because of sustainable development, industrial safety and environmental worries, which makes green synthesis a promising substitute to the conventional synthesis methods. Promoting the efficiency of these green processes leads to the protection of the environment and personal safe. We reviewed several existing studies relevant to MRs for green synthesis applications. The MR can be applied in the synthesis of different components like ammonia, liquid fuels, methanol through reverse water gas shift reaction, enzyme synthesis using supercritical–ionic liquid system and reactions based on photocatalytic. Several parameters like operating temperature, pressure, feed flow rate in both side of MR, which affect the MR efficiency, are investigated. The MRs application for the synthesis green fuel including hydrogen and biofuels is reviewed. Then, the biocatalyst membrane reactor is introduced as a more effective reactor due to compactness and higher conversion, biocatalyst stability and activity. Finally, we focused on applications of photocatalyst MRs in wastewater treatment.

Keywords

Biofuel • Biomembrane • Green synthesis • Membrane reactors • Methanol • Photocatalyst

1 Introduction

A membrane reactor [MR] is a physical apparatus, which consists of a chemical process with a separation process to remove products or add reactants (Oliveira et al. 2020). Chemical reactors based on membranes are known as *membrane reactor*. The reaction selectivity with respect to an objective compound is increased appropriately, which leads to adjust the local reactants concentration (Colli et al. 2019; Espinosa et al. 2018; Chen et al. 2020). A unit of membrane separation is applied particularly for regulation of the product and reactant concentrations using coupling the chemical reactor unit with membrane separation (Oliveira et al. 2020; Venezia et al. 2020; Nunes et al. 2019). In the last two decades, this technique has attracted universal research and leads to the development of many chemical process (Heyse et al. 2019; Nagy 2018; Hedayati et al. 2016). Indeed, a membrane plays two different roles: (1) as a separator and (2) as a reactor (Kisszekelyi et al. 2019). The MRs including two pipes that reaction occurs in inner tube (occupied using catalyst particle) and the permeated products are removed using inert gas flowing in outer side (Colli et al. 2019; Ibrahim et al. 2020; Briceño et al. 2013). There are two structures of the membrane reactor: (1) a membrane separator and a reactor apparatus, which are connected in series state and (2) a MR combines, a membrane separator and a reactor in a single unit (Nagy 2018; Itoh et al. 2020; Al-Juaied et al. 2001). Integrated membrane reactor system is shown in Fig. 1.

H. Bagheri · A. Mohebbi (✉)
Department of Chemical Engineering, Faculty of Engineering, Shahid Bahonar University of Kerman, Kerman, Iran
e-mail: amohebbi@uk.ac.ir

H. Eghbali
Department of Chemical Engineering, Faculty of Engineering, Vali-E-Asr University of Rafsanjan, Rafsanjan, Iran

Inamuddin et al. (eds.), *Advances in Green Synthesis*, Advances in Science, Technology & Innovation,
https://doi.org/10.1007/978-3-030-67884-5_7

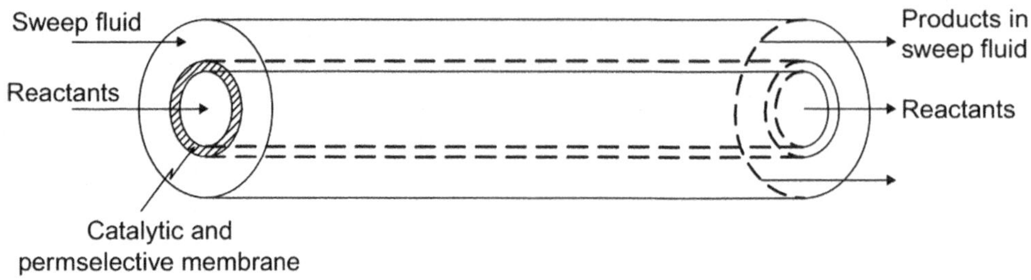

Fig. 1 Integrated membrane reactor system. Reprinted with permission of Nagy (2018)

Membrane-based processes are mostly using dense, porous and thin perm-selective layers provided using metal, inorganic and organic materials (Kisszekelyi et al. 2019; Bagheri et al. 2019; Avila et al. 2014). The choice of a dense or a porous film and the type of material applied for making a membrane depend on the preferred operating temperature, separation process and driving force and furthermore, the material selection depends on the preferred selectivity and presence and on mechanical and thermal stability desires (Espinosa et al. 2018; Avila et al. 2014; Meng et al. 2015). There are several membrane reactor structures at laboratory scale, which concentrate on the distribution of product and reactant to ameliorate selectivity–conversion efficiencies (Chen et al. 2020; Heyse et al. 2019; Murmura et al. 2017; Pomier et al. 2007). There are six membrane reactor concepts based on several problems, which can be solved by membranes in the reactor (Fig. 2). All of these reactor structures are used for improving the chemical reaction yield and selectivity (Saw et al. 2018; Herrero and Ibanez 2015). Consequently, it is very important to understand the relation between selectivity, temperature and local concentrations behavior (Murmura et al. 2017).

A membrane can be applied for several applications: catalyst support, reactant distribution and dosing, catalyst retention, reactants selective extraction and separation (Sherbo et al. 2019; Sheikhi-Kouhsar et al. 2015). The laboratory scale investigation and application of several types of membrane reactors as a promising unit operation started

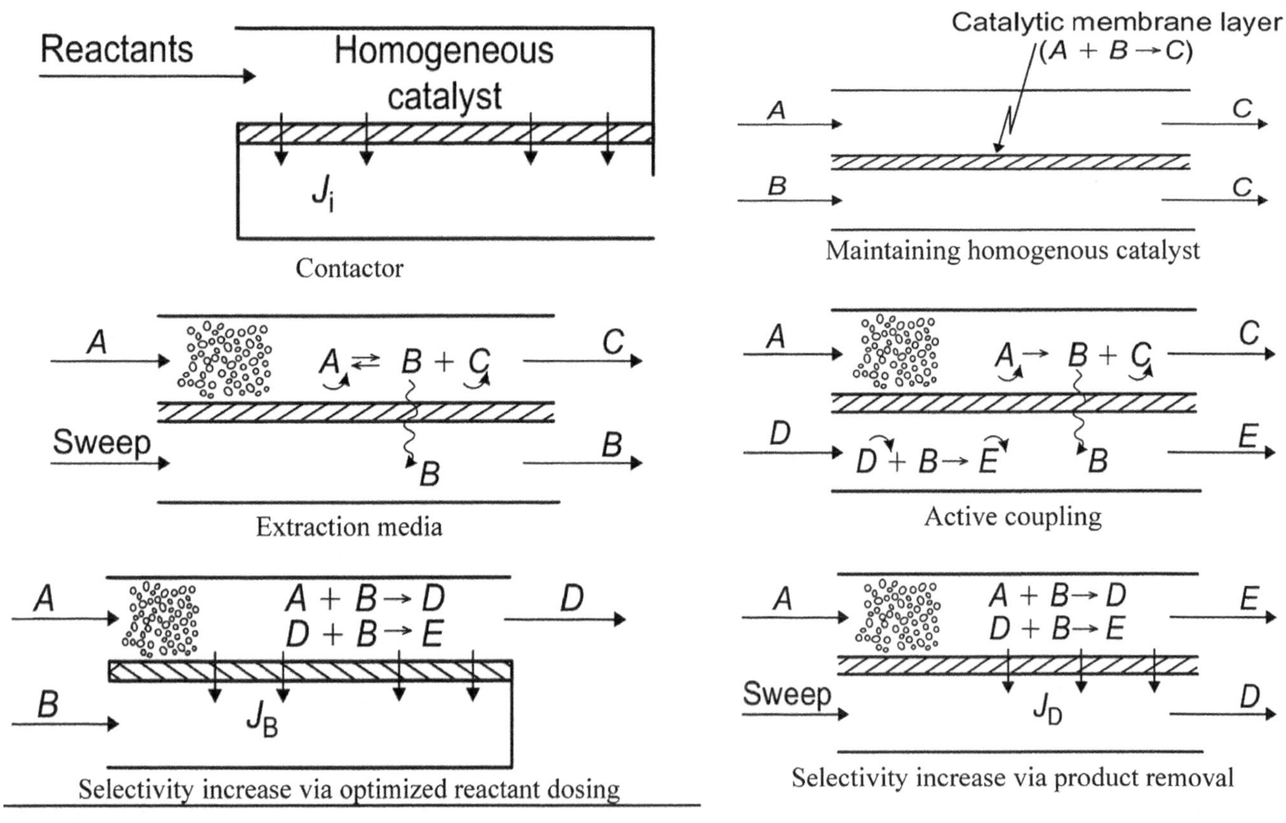

Fig. 2 Different concepts of membrane reactors. Modified after Briceño et al. (2013)

Table 1 Polymeric materials applied in polymeric membrane

Material	Advantage	Disadvantage
Cellulose acetate	Low cost	Narrow pH range Biologically active
Polyacrylonitrile	Low cost, applied for ultrafiltration membranes	Chemically resistant is less than PVDF[a]
Polyethersulfone and polysulfone	Cl_2 tolerance Sensible price	Fragile material needs support and flow inside to outside
PVDF	High Cl_2 tolerance Simple cleaning chemicals	Cannot retain pH > 10
Polypropylene	Low cost High power of hydrogen range tolerance	No Cl_2 tolerance Costly cleaning chemicals needed

Modified after Najafpour (2015)
[a]Polyvinylidene fluoride

in the 1980s (Nagy 2018; Zhang et al. 2020; Veismoradi et al. 2019). There are many applications for MR. For instance, a MR can be applied in wastewater treatment, reaction combined with pervaporation, dosing, organic matter removal (like chemical oxygen demand and biological oxygen demand processes), removal of organic micro-contaminants and selective removal (e.g., H_2 from gas steam reforming) (Wang et al. 2020; Zhao et al. 2020; Bagheri et al. 2019; Wu and Ghoniem 2019; Cannilla et al. 2018; Kumar et al. 2017; Boyd et al. 2005).

Usually, membranes are made of ceramic or organic materials. The production cost in polymeric membranes is low; however, these membranes are at degradation and fouling risk due to the variation in their pore size (Constantinou et al. 2019; Maji and Chakraborty 2019; Chakraborty and Mazzanti 2020; Chen et al. 2009). In contrariwise, some properties like high durability and quality have made membranes that made of ceramic materials, a promising choice for membrane producers. However, membranes based on ceramic materials are costly and they are impossible for industrial applications. Subsequently, most producers apply membranes that made of polymeric materials for bioreactors (Kumari et al. 2020; Najafpour 2015; Dixon 1999). The materials typically applied in polymeric membrane are given in Table 1.

Membranes are made in different modules like spiral, frame and plate and hollow fiber (Liu et al. 2020; Li et al. 2019). The hollow fibers are extruded in long fibers and collected in the modules. In this type, driving force is difference of pressure through the MR (Su et al. 2020; Lyagin et al. 2010). Flow pattern over the membrane reactors is dead-end or cross-flow. In dead-end method, feed is injected to the MR vertically and feed applied to the MR may rejected as a waste or pass through the MR. In cross-flow method, feed moves parallel to the MR surface (Najafpour 2015; Tran et al. 2013; Vankelecom 2002; Yuan et al. 2020).

Green chemistry is defined as chemical processes design and products that decrease the employ of hazardous materials. Green chemistry is described as sustainable chemistry, chemistry that is benign using pollution prevention and design at the molecular level (Pu et al. 2007). This developing field distinguishes that through the design phase of any chemical process, product and synthesis, minimized danger must be observed as an efficiency criterion (Lee et al. 2020; Oosterhout et al. 2018). This new chemistry branch looks for an alternative a new solvent with less dangerous to the environment and human health to protect the ecosystem from hazardous conventional solvents. Supercritical fluids [SCF], ionic liquids [ILs] and water are known as green solvent (Low et al. 2020; Mohammadzadeh et al. 2020; Bagheri et al. 2019). Many advantages are repeated about ILs and SCFs as green solvents to develop new processes. When a fluid is subjected to a pressure and a temperature higher than its critical points, the fluid is said to be "supercritical'. In a supercritical region, the fluid exhibits particular properties and has an intermediate behavior between that of a liquid and a gas (Kölsch et al. 2002). Two advantageous fluids, which are applied as SCF, are water and carbon dioxide [CO_2]. CO_2 has moderate critical temperature and pressure, low-cost, availability and non-toxicity (Bu et al. 2017; Farsi and Jahanmiri 2014). However, supercritical water is highly corrosive. Because of unique properties of SCFs, they are attractive as environmentally to replace for organic solvents in material processing and chemical reactions (Bagheri et al. 2018; Capello et al. 2007). ILs are kind of room temperature liquid salts that is stable on air and moisture. ILs as a new class of solvent are unique topic from three decades ago, and the number of published documents has grown rapidly (Bagheri and Mohebbi 2017; Akin et al. 2014). These liquids have significant properties like high chemical and thermal stability, negligible vapor pressure, good ionic and electrical conductivity, non-flammability, wide electrochemical range and low melting point (Low et al. 2020; Wang and Chiu 2008). ILs are low melting point organic salts, which leads to be liquids phase at ambient temperature. The thermal decomposition temperature of IL is in order of 474 K, and their boiling points are not detectable (Chambreau et al. 2012). ILs can be composed of a large number of anions and cations. The ILs properties depend on the size and nature of both their anion and cation parts. Consequently, selecting of favorite anions and cations, many ILs are designed for various applications (Lapeña et al. 2019). ILs have other scientific properties like outstanding thermal stability, organic and inorganic compounds, satisfactory dissolution properties with water, wide electrochemical windows, tunable viscosity, high ionic conductivity and highly polar and no coordinating (Bagheri and Ghader 2017; Wahidin et al. 2016).

Green synthesis has been considered as one of the promising method for synthesis of nanoparticles due to their biocompatibility, low toxicity and eco-friendly nature. Green synthesis focuses on chemical reactions that use environmentally benign reaction media and are conducted in benign reagents (Nalawade et al. 2006). This method has attracted more consideration over the last two decades, because of the need for the environmentally harmless reaction development, while attaining high yield and specificity (Zha et al. 2019; Wolfson et al. 2007). Indeed, green synthesis is a developing field in chemical area and provides environmental and economic advantages as an alternative to physical and chemical methods (Cue and Zhang 2009). In this method, nontoxic harmless reagents that are biosafety and eco-friendly are applied. Green synthesis is needed to avoid the dangerous and unwanted by-products or production via the build-up of eco-friendly, sustainable and reliable synthesis methods (Pollet et al. 2014). Applying natural resources and desirable solvent systems (like organic systems) is required to attain this purpose (Rao et al. 2017).

This communication gives a review of the green synthesis reactions using MR. The main aim of the present chapter is to describe the application of MRs in green synthesis processes. Subsequently, the most practicable and industrial synthesis based on MR is discussed, and various green synthesis processes are investigated.

2 Chemical Reaction Enzymatic MR Using Supercritical CO_2-IL

The enzymes applying in conventional solvents have significantly increased enzymes technological utilization, like the probability of performing reactions synthetic using hydrolytic enzymes and enhancement of the organic components solubility (Nalawade et al. 2006; Zha et al. 2019; Wolfson et al. 2007; Cue and Zhang 2009; Pollet et al. 2014). However, volatile organic solvents (VOSs) have a harmful effect on the human health and environment and it is subsequently necessary to extend novel alternatives for the synthesis of enzyme catalyzed in environmentally reaction medium (Antonia et al. 2007). SCFs like CO_2 have lately indicated as a new alternative to VOSs. The solvent power of SC-CO_2 can be adjusted using varying either the temperature or pressure (Bagheri et al. 2018, 2019). For the mentioned reasons, SC-CO_2 is regularly designated as a *green solvent designer*. Applying SC-CO_2 as a new media for enzymatic reactions has attracted significant attention. However, SC-CO_2 has an adverse effect on enzyme activity due to the chemical modification of the free amino groups, local pH changes caused by CO_2, or conformational changes produced during the pressurization/depressurization steps, making it necessary to develop new enzyme stabilization strategies (Antonia et al. 2007).

ILs have newly appeared as exciting non-aqueous reaction media for reactions based on enzyme catalyzed and they show a unique ability to stabilization of free enzymes. Consequently, the efficiency of ILs to preserve enzymes versus very harsh SC-CO_2 conditions has been described (Brennecke and Maginn 2001). ILs are made from an organic cation, like tetraalkylammonium, dialkylimidazolium salts and an inorganic anion (Bagheri and Mohebbi 2017; Antonia et al. 2007). Indeed, ILs have received much attention as alternative to conventional organic solvents, because of negligible vapor pressure and thermal and chemical stability. SC-CO_2/IL two-phase systems have indicated good behavior to perform green bio-catalytic processes in nonaqueous media (Kamat et al. 1995). The success of these two-phase systems is related to high solubility of SC-CO_2 in IL phase; however, IL solubility in SC-CO_2 phase is non-detectable. Indeed, high diffusivity of SC-CO_2 (Fig. 3) leads to a decrease ILs viscosity, which improves the mass transfer between two phases (Antonia et al. 2007).

The application of MRs establishes an effort to integrate catalytic conversion, product concentration and recovery and separation of catalyst in a single process (Antonia et al. 2007). Antonia et al. (2007) investigated mass transfer and chemical reactions, which occur in MRs for ester synthesizing in SC-CO_2/IL two-phase systems in an effort to develop the green enzymatic design in the mentioned systems (Antonia et al. 2007). They studied the butyl propionate synthesis of vinyl propionate, catalyzed using *Candida Antarctica* lipase B as reaction media and was carried out in IL, C_6H_{14}/IL and SC-CO_2/IL. Moreover, the IL/C_6H_6 compounds partition coefficients included in the transesterification reaction were calculated. They used various ILs (i.e., [Omim][PF_6], [Bdimim][PF_6], [C_4mim][PF_6] and [C_4mim][TFSI] ([NTf_2] = [TFSI]) to study selectivity, mass transfer and activity of *Candida Antarctica* lipase B [CLAB] for the suggested biotransformation (Antonia et al. 2007).

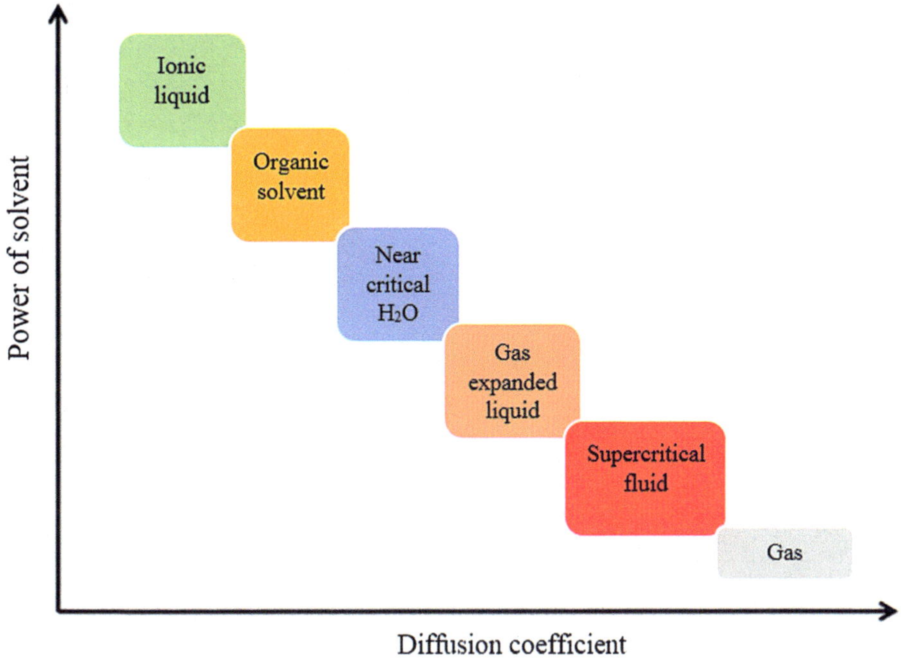

Fig. 3 Transport properties vs. solvent power for different solvents. Modified after Herrero and Ibánez (2015).

2.1 Ionic Liquid Media Effect on Free CLAB

Reaction catalysts based on lipase are affected by the reaction media used. The influence of four ILs ([Omim][PF_6], [Bdmim][PF_6], [C_4mim][PF_6] and [C_4mim][TFSI]) on the selectivity and activity of CALB for the butyl propionate synthesis using transesterification of vinyl butyrate and 1-butanol at 324 K was investigated (Antonia et al. 2007). The reaction of enzymatic was further more performed in C_6H_{14}, for synthesis of catalyzed lipase ester in nonaqueous media, in similar conditions to compare the efficiency of applied ILs as reaction environment. Figure 4 indicates the synthetic activity and selectivity using the enzyme in C_6H_{14} and in the various ILs (Antonia et al. 2007). Referring to Fig. 4, the activity of the enzyme in all of the mentioned above ILs was higher than that in C_6H_{14}, indicating the greater neoteric solvents suitability for the suggested reaction. Subsequently, the activity of enzyme order detected in ILs was: [C_4mim][PF_6] < [Bdmim][PF_6] < [C_4mim][TFSI] < [Omim][PF_6]. Originally, for the ILs with the same anion, the hydrophobicity increased by increasing the alkyl group length on the cation (Antonia et al. 2007; Ropel et al. 2005). According to Fig. 4, the synthetic activity was gradually increased with increasing ILs hydrophobicity for the similar anion (Antonia et al. 2007; Persson and Bornscheuer 2003). Indeed, difference in selectivity between the mentioned ILs is specific abilities to decrease water activity in the enzyme environment.

Subsequently, an increase in the IL hydrophobicity leads to increase water molecules, which can play as nucleophile acceptors in the transesterification reaction and consequently lead to a loss in selectivity (Antonia et al. 2007).

2.2 Butyl Propionate Synthesis Using Active Membranes SC-CO_2 and SC-CO_2/IL

The butyl propionate synthesis performance, from 1-butanol and vinyl propionate catalyzed using CLAB immobilized on dynamic membranes at 324 K and 8 MPa in SC-CO_2 and SC-CO_2/IL two-phase systems, was investigated (Antonia et al. 2007). Four ILs, i.e., [Omim][PF_6], [Bdimim][PF_6], [C_4mim][PF_6] and [C_4mim][TFSI] were used to investigate the effect of various anions and cations on selectivity and activity of immobilized CLAB. Figure 5 indicates the immobilized CLAB selectivity and synthetic activity on membranes that made of ceramic materials in SC-CO_2 media and in the four above-mentioned SC-CO_2/IL two-phase systems (Antonia et al. 2007; Mori et al. 2005).

Referring to Fig. 5, the immobilized lipase synthetic activity in SC-CO_2/IL two-phase systems is lower compared to SC-CO_2 mentioned alone. Mori et al. (2005) observed similar results in C_6H_{14}/IL two-phase systems. They revealed enzymatic membranes, which are provided with simple free CALB adsorption on surface, were more reactive compared to membranes provided by ILs. For the mentioned SC-CO_2/IL two-phase systems, the activity of synthetic increased in the subsequent order: [Bdimim][PF_6] < [C_4mim][PF_6] < [C_4mim][TFSI] < [Omim][PF_6] that was in contract by the activity order observed using CLAB in homogeneous IL systems ([C_4mim][PF_6] < [Bdmim][PF_6] < [C_4mim][TFSI] < [Omim][PF_6]) except for [Bdimim][PF_6] and [C_4mim][PF_6] (Antonia et al. 2007).

The substrate transport mechanism includes three steps (see Fig. 6). The first step is substrates diffusion through the

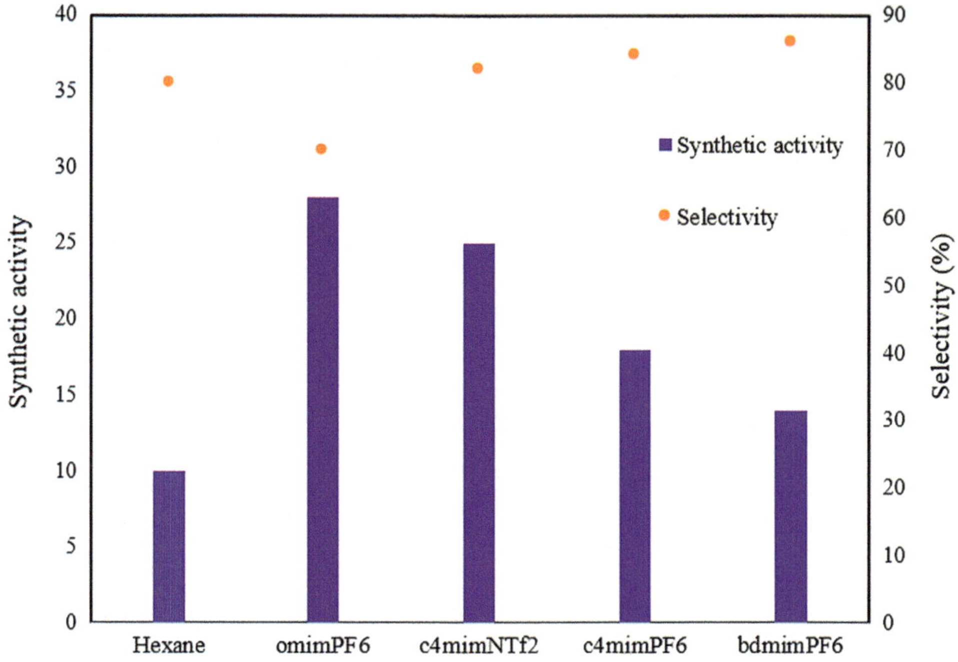

Fig. 4 Selectivity and synthetic activity variations of free *Candida Antarctica* lipase B for butyl propionate synthesis in C_6H_{14} and in various ILs at T = 323 K. Modified after Antonia et al. (2007)

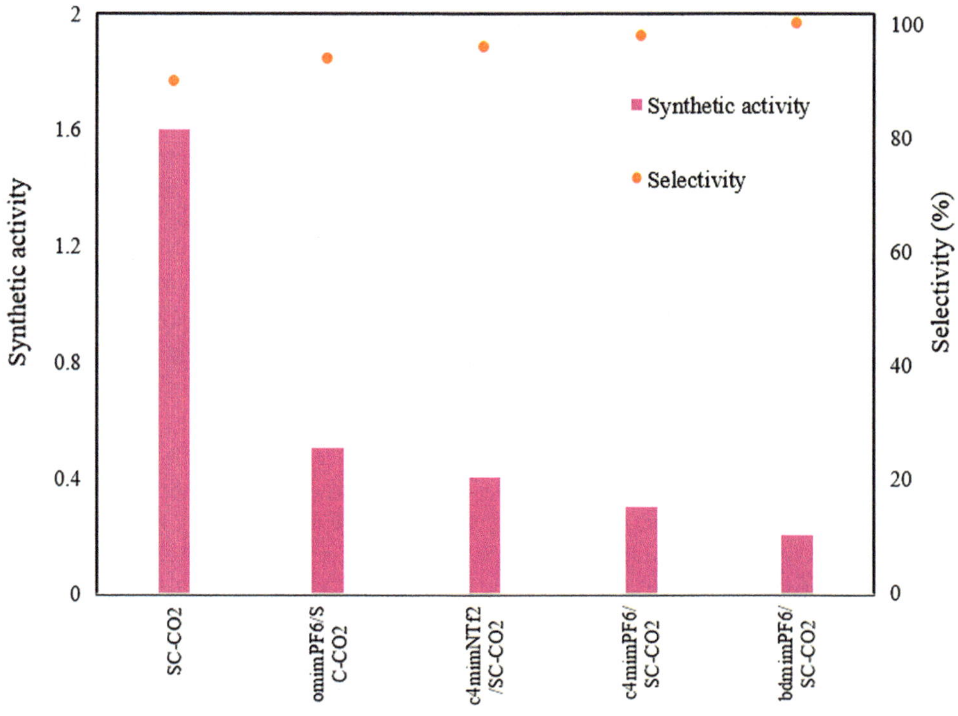

Fig. 5 Selectivity and synthetic activity variations of immobilized *Candida Antarctica* lipase B in four different IL/SC-CO_2 two-phase systems and SC-CO_2 media at T = 323 K and P = 80 bar. Modified after Antonia et al. (2007)

diffusion layer of the SC-CO_2 phase to the interface of SC-CO_2/IL. The second step is the substrates partitioning between the IL and SC-CO_2 phase, and final step is diffusion in the IL phase into the immobilized enzyme (Antonia et al. 2007). The products and substrates partition coefficients of the transesterification reaction between the ILs and C_6H_{14} in the second step were obtained and given in Table 2. C_6H_{14} selected as organic solvent reference due to C_6H_{14} solvent ability is similar to SC-CO_2 in moderate operating conditions (Antonia et al. 2007).

Referring to Table 2, the order observed in the substrates partition coefficients was: [C_4mim][TFSI] > [Omim][PF_6] > [C_4mim][PF_6] > [Bdimim][PF_6]. Consequently, the mixtures with high IL/ C_6H_{14} partition coefficients were detected in ILs (Antonia et al. 2007). This order could describe the variation in the primary reaction rate order

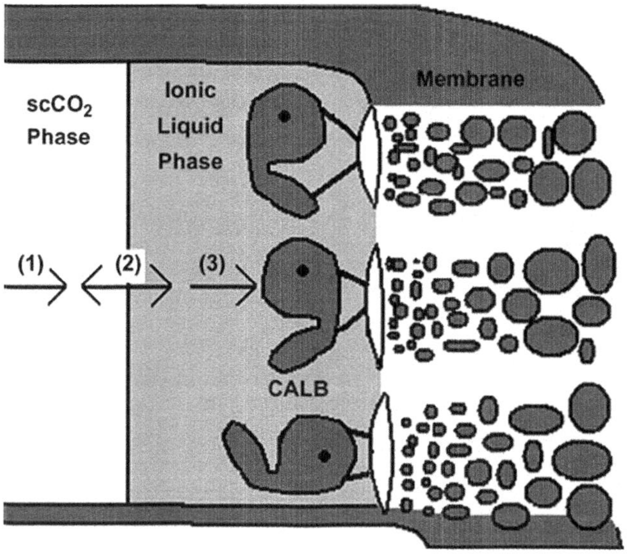

Fig. 6 Mechanism of substrate transport from SC-CO_2 phase to the immobilized enzyme in a IL/SC-CO_2 two-phase system. Reprinted with permission of Antonia et al. (2007)

between [C_4mim][PF_6] and [Bdimim][PF_6] in the SC-CO_2/IL two-phase systems with respect to IL applied as homogeneous media. Then, the higher partition coefficient for [C_4mim][PF_6] indicates more substrates absorption in this IL and subsequently an easier interaction between the enzyme catalytic core and the substrates (Antonia et al. 2007; Lozano et al. 2001). In addition, the partition coefficient indicates the reaction substrates are more strongly detected in [C_4mim][TFSI] compared to [Omim][PF_6]. This fact leads to having the lower change in the activity values between [Omim][PF_6] and [C_4mim][TFSI] in the two-phase system (see Fig. 5) with respect to in the homogeneous media (see Fig. 4).

2.3 Butyl Propionate Synthesis Using Active Membranes in Hexane/IL

The capability of free CLAB to catalyze butyl propionate synthesis of 1-butanol vinyl and propionate was investigated in hexane/IL two-phase systems at 324 K in sequence to better find out of reaction in SC-CO_2/IL two-phase systems (Antonia et al. 2007). It is required to indicate that CLAB plays in the ILs phase and which C_6H_{14} was selected as second phase due to C_6H_{14} solvent ability is comparable to supercritical carbon dioxide solvent ability in mild operating conditions. The selectivity exhibited and synthetic activity using CALB in C_6H_{14} and in various hexane/IL two-phase systems are shown in Fig. 7. As observed in Fig. 7, the synthetic activity indicated the used lipase in hexane/IL two-phase systems was lower compared to C_6H_{14} mentioned alone (Antonia et al. 2007). Subsequently, applying hexane/IL two-phase systems leads to significant limitations in mass transfer.

It has furthermore been detected that an increment in the synthetic activity in above-mentioned C_6H_{14}/ionic liquid two-phase systems pursue the identical order as the synthetic activity in SC-CO_2/IL two-phase system. Subsequently, the obtained results in hexane/IL two-phase systems confirm those in SC-CO_2/IL two-phase systems and emphasize the presence of IL affects the mass transfer and enzyme activity in IL/SC-CO_2 two-phase systems (Antonia et al. 2007). In hexane/IL two-phase systems, the selectivity was higher than when C_6H_{14} was alone (see Fig. 7). The similar behavior was indicated with the immobilized enzyme in SC-CO_2/IL two-phase systems, which SC-CO_2 was applied as reaction media.

3 Mixed Ionic Electronic MR

Separation and reaction are two fundamental processes in chemicals industries. Many difficulties, like high cost, energy consumption, huge occupancy region are in the chemical industries (Schulz et al. 2011). Subsequently, it is immediate and imperative to extend novel processes to overcome mentioned difficulties. MR has the capability of coupling separation and reaction processes in a unit operation that is important for chemical processes shortening (Maina et al. 2019). Mixed ionic electronic conducting O_2 penetrable MR is a kind of compressed membranes of ceramic material, which can conduct oxygen ions and electrons using 100% selectivity of O_2 permeation (Li et al. 2016). The partial pressure gradient of oxygen is oxygen permeation driving force across the oxygen-permeable membrane. These kinds of MRs have a well capability of coupling oxygen-related reaction and separation process (Li et al. 2019).

Table 2 IL-hexane partition coefficient values of components in four ILs: [Bdimim][PF_6], [C_4mim][PF_6], [Omim][PF_6] and [C_4mim][TFSI]

Component	[Bdimim][PF_6]	[C_4mim][PF_6]	[Omim][PF_6]	[C_4mim][TFSI]
Propionic acid	1.31	2.66	2.43	6.77
Butyl propionate	0.03	0.15	0.49	0.51
1-Butanol	0.68	2.46	3.63	5.21
Vinyl propionate	0.22	0.71	1.15	1.10

Modified after Antonia et al. (2007)

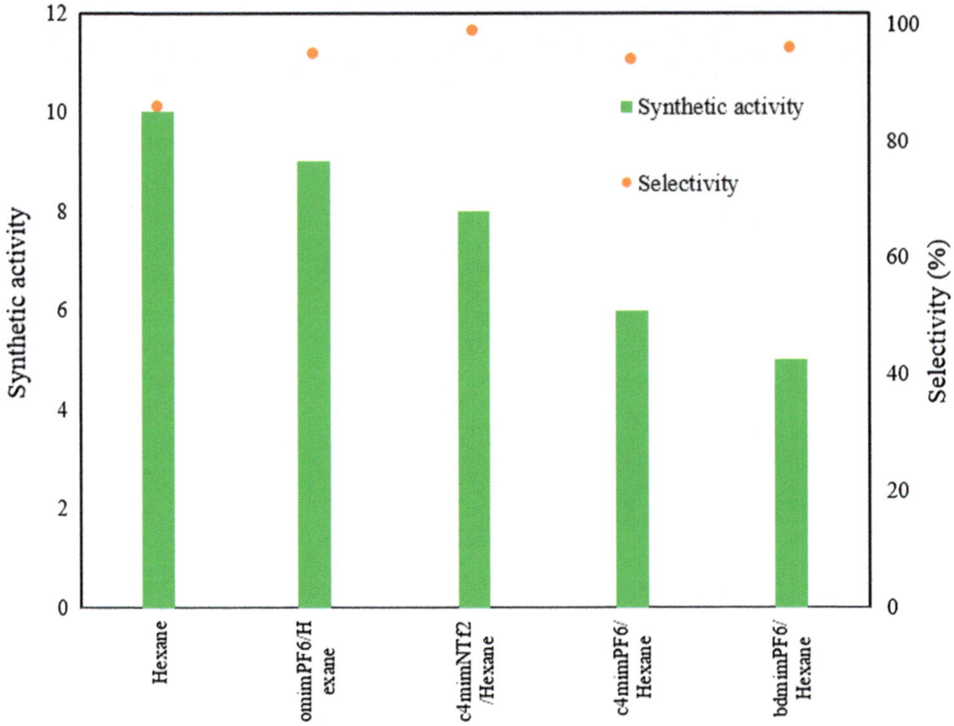

Fig. 7 Selectivity and synthetic activity variations of free *Candida Antarctica* lipase B in four different IL/hexane two-phase systems and SC-CO_2 media at $T = 323$ K. Modified after Antonia et al. (2007)

Balachandran et al. (1995) applied methane in a $La_{0.2}Sr_{0.8}Co_{0.2}Fe_{0.8}O_{3-\delta}$ perovskite MR and described the membrane broke after performing for an insufficient minute. Consequently, many research studies have concentrated on improving materials of membrane with high thermal and chemical stability in partial oxidation operating conditions of methane reaction (Wu et al. 2016). The membrane based on cobalt materials, like $SrCo_{0.8}Fe_{0.2}O_{3-\delta}$, has a best permeability for oxygen; however, the easy-reduction feature of Co makes them unstable in reducing ambient conditions (Wei et al. 2013). Subsequently, the perovskite membrane materials extension for partial oxidation of methane reaction was piecemeal transition from low and high cobalt containing to oxides based on free cobalt containing. The usual materials were $Ba_{0.5}Sr_{0.5}Co_{0.8}Fe_{0.2}O_{3-\delta}$ (Wang et al. 2003), $La_{1-x}Sr_xCo_{1-y}Fe_yO_{3-\delta}$ (Jin et al. 2000), $BaZr_xCo_yFe_{1-x-y}O_{3-\delta}$ (Wang et al. 2009), Ba (Ce, Zr, Y)$_x$ $Fe_{1-x}O_{3-\delta}$ (Zhu et al. 2006) and $La_{1-x}Sr_xGa_{1-y}Fe_yO_{3-\delta}$ (Ritchie et al. 2001). Typically, the main source of oxygen feed in oxygen-permeable MR for partial oxidation of MR is air. Replacing it by gaseous oxides, like H_2O (Zhu et al. 2015), NO (Jiang et al. 2009) and CO_2 (Liang et al. 2017), makes the MR more efficient because of the additional products achievement.

Li et al. (2019) investigated an asymmetric $Sm_{0.15}Ce_{0.85}O_{1.925}$ (75 wt%) with $Sm_{0.6}Sr_{0.4}Al_{0.3}Fe_{0.7}O_{3-\delta}$ (25 wt%) dual-phase mixed ionic electronic conducting O_2 permeable MR to produce NH_3 synthesis gas (i.e., $N_2/H_2 = 0.33$) and liquid fuels synthesis gas (i.e. $CO/H_2 = 0.5$). They investigated the effects of CH_4 concentration, CH_4 flow rate, steam flowrate and temperature on the performance of the MR. The schematic of the reactor is illustrated in Fig. 8. Catalyst based on Ruthenium material was used to catalyze the reactions in sides I and II. High efficiency was attained using the two-phase MR and the membrane material had a good structure stability (Li et al. 2019).

3.1 Methane Flow Rate and Concentration Effects on Side II of Membrane

Li et al. (2019) studied the influence of methane concentration on side II of membrane on the MR performance (including carbon dioxide selectivity and methane conversion). Figure 9 indicates effects of methane flow rate and concentration on carbon monoxide selectivity and methane conversion. The methane conversion (x_{CH_4}) and carbon monoxide selectivity (S_{CO}) are calculated using the following equations (Li et al. 2019):

$$x_{CH_4} = \left(1 - \frac{C_{CH_4}}{C_{CH_4} + C_{CO} + C_{CO_2}}\right) \times 100 \quad (1)$$

$$S_{CO} = \left(\frac{C_{CO}}{C_{CO} + C_{CO_2}}\right) \times 100 \quad (2)$$

where C is the component concentration. As can be seen in Fig. 9a, methane conversion increases slowly as the concentration of methane increases. However, the carbon monoxide selectivity decreases slightly with an increase in

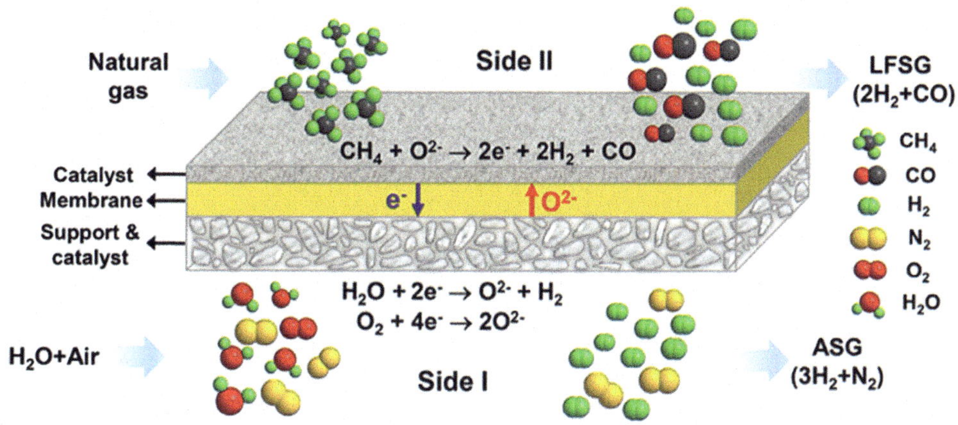

Fig. 8 Illustration of the asymmetric oxygen-permeable membrane reactor. Reprinted with permission of Li et al. (2019)

concentration of methane. Subsequently, a higher concentration of methane leads to better MR efficiency (Li et al. 2019). The methane flow rate unlike the methane concentration, has significant effect on methane conversion and carbon monoxide selectivity (Fig. 9b). As observed in Fig. 9b, methane conversion and CO selectivity have opposite behavior with the increase of methane flow rate.

3.2 Steam Flow Effect on Side I of Membrane

The influence of steam flow on side I of membrane on methane conversion and carbon monoxide selectivity, and particularly the MR energy consumption for NH_3 synthesis gas and liquid fuels synthesis coproduction, is complicated (Li et al. 2019). Consequently, the efficiency of the asymmetric $Sm_{0.15}Ce_{0.85}O_{1.925}/Sm_{0.6}Sr_{0.4}Al_{0.3}Fe_{0.7}O_{3-\delta}$ MR with several steam flow rates was investigated. The results indicated, an increase in the H_2/N_2 ratio decreases the H_2/N_2 production rate on MR side I. However, this ratio has negligible effect on the liquid fuels synthesis gas. This phenomenon indicates that varying the N_2/H_2mixed gas online requests to regulate the air feed rate on MR side I (Li et al. 2019).

3.3 Temperature Effect

The temperature effect on efficiency of the asymmetric $Sm_{0.6}Sr_{0.4}Al_{0.3}Fe_{0.7}O_{3-\delta}/Sm_{0.15}Ce_{0.85}O_{1.925}$ MR is investigated in this section.

To test the maximum MR ability at various temperatures and achieve best quality of synthesis gas, the selectivity of CO was adjusted 91% and the methane flow rate regulated using temperature. The CO selectivity is optimal parameter for MR at various temperatures. The results are indicated in Fig. 10. As one can see, the optimal specific flow rate of methane is increased with increasing in temperature. Indeed, particular flow rate is applied to indicate the MR area capacity (Li et al. 2019).

The conversion of steam on MR side I is a key factor to calculate the MR consumption of energy. High conversion of steam is proportional to low consumption of energy for the energy reduction for the heating and the unconverted steam vaporization. As shown in Fig. 11, due to the limited O_2 permeability of the O_2 permeable membrane, the steam conversion on membrane side I decreases quickly with the increase of H_2O/CH_4 molar ratio. When the H_2O/CH_4 molar ratio is reduced to 1.43, the corresponding steam conversion is high up to 57%, which is 1.2×10^5 times of the steam equilibrium conversion of water splitting reaction at 900 °C. The corresponding energy saving compared to the industrial processes is expected as high as 66%, which is close to the theoretically highest energy saving of 70% (Li et al. 2019).

4 Green Synthesis of Methanol in a Membrane Reactor

One of the greatest hazards to environment and humanity is global warming. Carbon dioxide accumulation in the atmosphere is the main source of global warming (Samimi et al. 2019; Rafiee et al. 2018). Methanol [MeOH] is great interest for the carbon dioxide conversion with hydrogen. MeOH is used in the energy industries and petro-chemical for energy or chemical application, hydrogen fuel cell and vehicle fuel (Samimi et al. 2019). H_2 must be produced through a renewable energy (like biomass, geothermal, solar, wind and water splitting). Two processes of direct and indirect for synthesis of MeOH are applied using carbon dioxide conversion (Fig. 12). In direct method, H_2 and CO_2 are directly converted into MeOH. At first step, CO_2 hydrogenation leads to syngas produced based on reverse water gas shift [RWGS] reaction, after that syngas as a raw material is used to produce MeOH. The reverse water gas shift is described as (Rafiee et al. 2018):

$$CO_2 + H_2 \leftrightarrow CO + H_2O \tag{3}$$

Indirect conversion process of carbon dioxide leads to produce green methanol, due to the feedstock includes

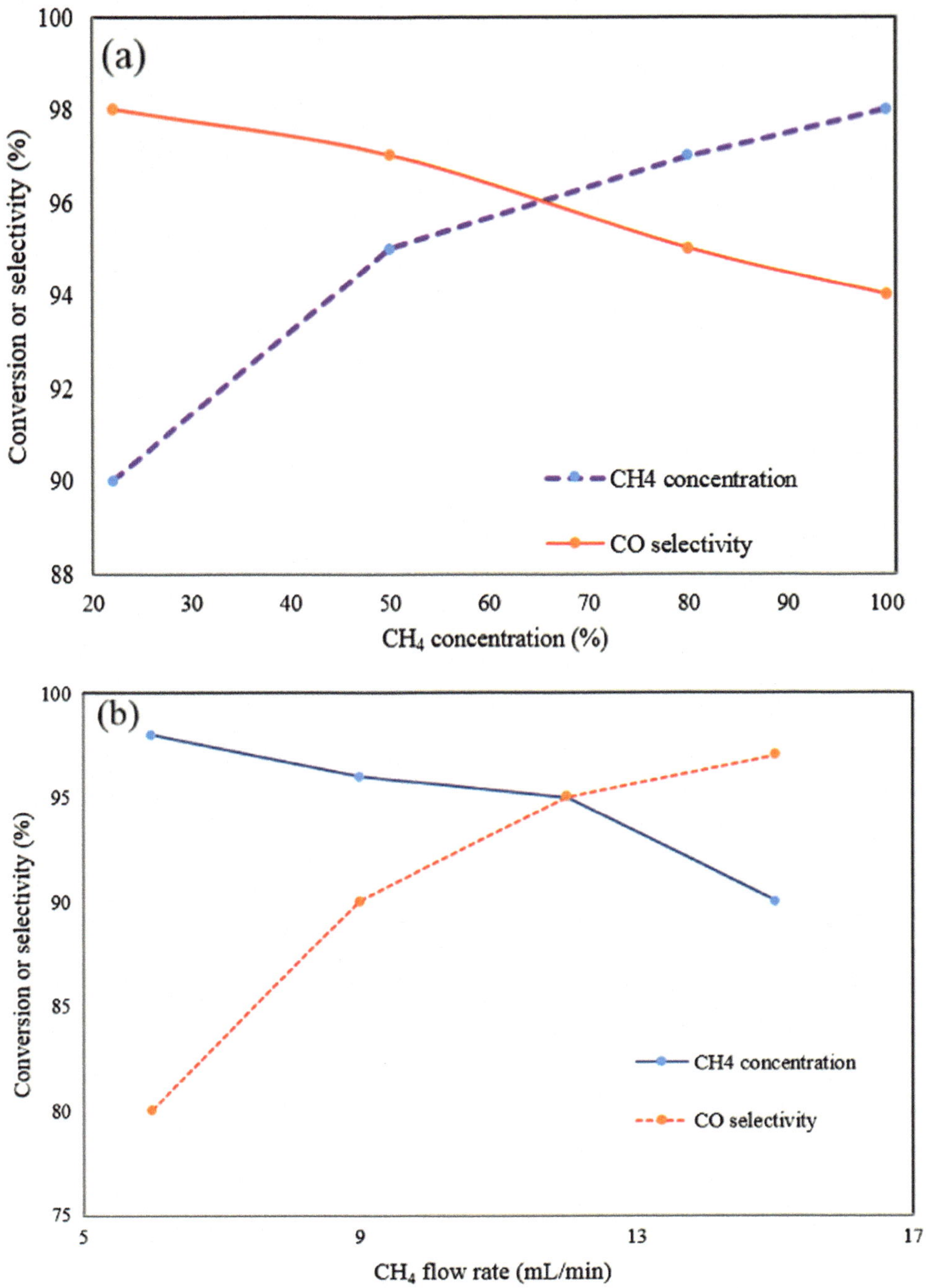

Fig. 9 Influences of **(a)** methane concentration and **(b)** methane flow rate on methane conversion and carbon monoxide selectivity at $T = 1174$ K. Modified after of Li et al. (2019)

carbon dioxide and renewable hydrogen (Samimi and Rahimpour 2018; Jadhav et al. 2014). In this method, water perm-selective membrane is used in the RWGS (Vooradi et al. 2018). Figure 12 indicates a schematic diagram of both methods. In the first method, the process contains of a MeOH synthesis and RWGS reactor (Fig. 12a). The feedstock (hydrogen and carbon dioxide) is shared in two separated streams: The mainstream is forwarded to the RWGS reactor and another one is required to regulate the produced syngas composition. Hydrogen and carbon dioxide are converted to H_2O and CO by the RWGS reaction (Samimi et al. 2019). The products of reaction that contain water and syngas are transported to a condenser, and remove H_2O from the stream (water is toxin for $Cu/ZnO/Al_2O_3$ catalyst). Finally, the syngas is fed to MeOH synthesis reactor. The procedure of operation in the second method is originally the same as the first method, except in which a hydroxyl soda lite membrane was used for H_2O removal in the reverse water gas shift reactor (Samimi et al. 2019). An illustration of RWGS MR is given in Fig. 13.

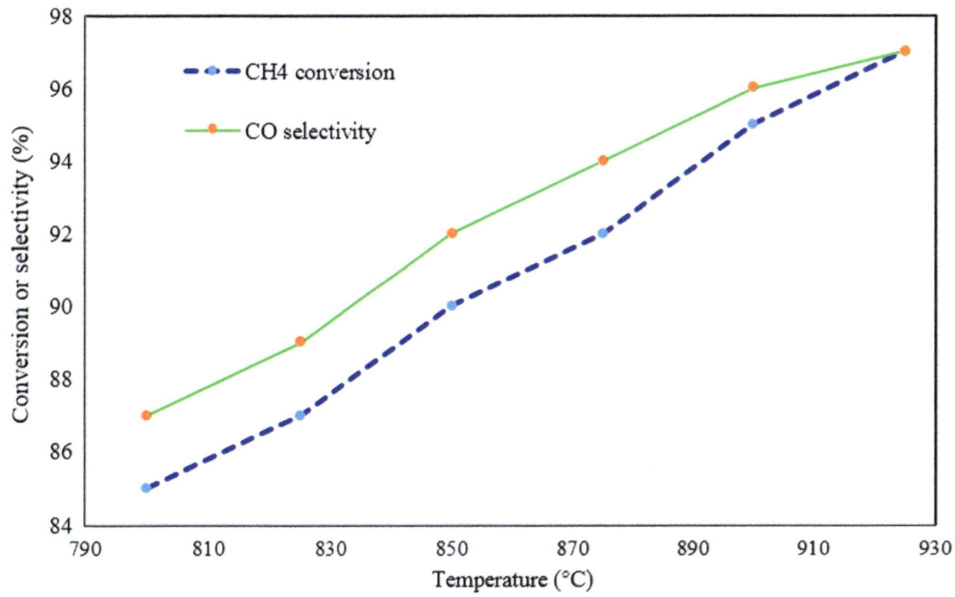

Fig. 10 Temperature influence on carbon monoxide selectivity and methane conversion. The steam flow rate value was 1.75 mL/s. Modified after of Li et al. (2019)

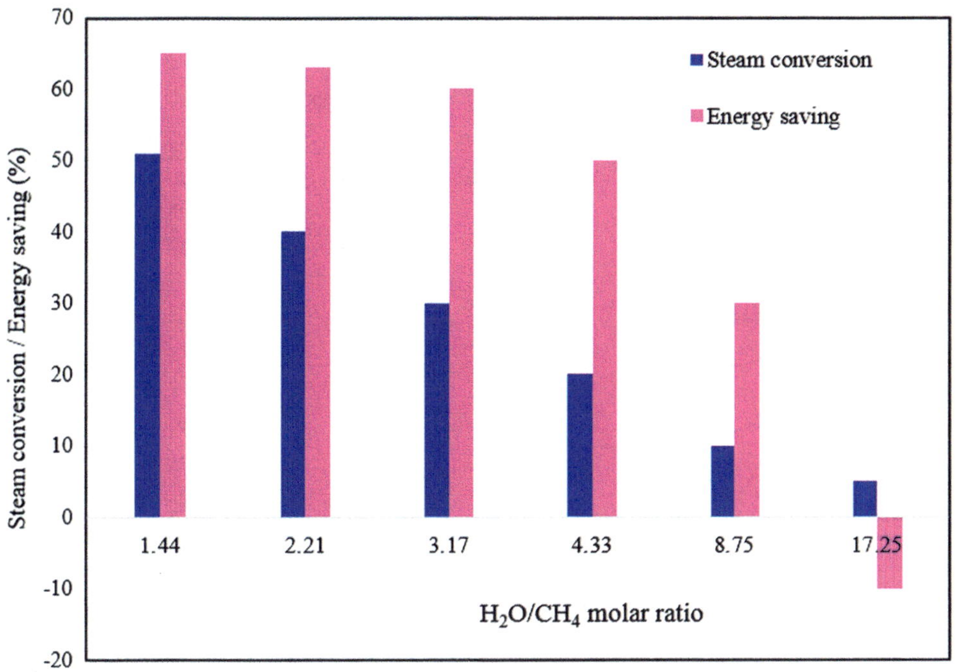

Fig. 11 H_2O/CH_4 molar ratio effects at 1174 K. Comparison between energy saving and steam conversion and industrial processes. Modified after of Li et al. (2019)

Samimi et al. (2019) applied $Fe_2O_3/Cr_2O_3/CuO$ catalyst in RWGS MR for production of MeOH. The used membrane is made of a zeolite material that can separate water in an excellent manner. They compared the two mentioned methods for the production of MeOH. The results indicated the inlet pressure and temperature have significant effect on RWGS MR performance.

At a confident CO_2/H_2 ratio and feed flow rate, the CO_2 conversion and consequently CO yield are proportional to pressure and temperature of feed for both methods. Therefore, changing the operating conditions of RWGS reactor (pressure and temperature) leads to increase produced syngas composition (Samimi et al. 2019). Because of using water perm-selective membrane to separate water, the CO_2 conversion is dependent on P_{inlet} and it enhances with increasing in P_{inlet}. Indeed, water separation driving force is increased with increasing pressure difference between reaction and permeation sides. Subsequently, water removing leads to shift the RWGS reaction (Eq. 3) into more CO production and CO_2 consumption.

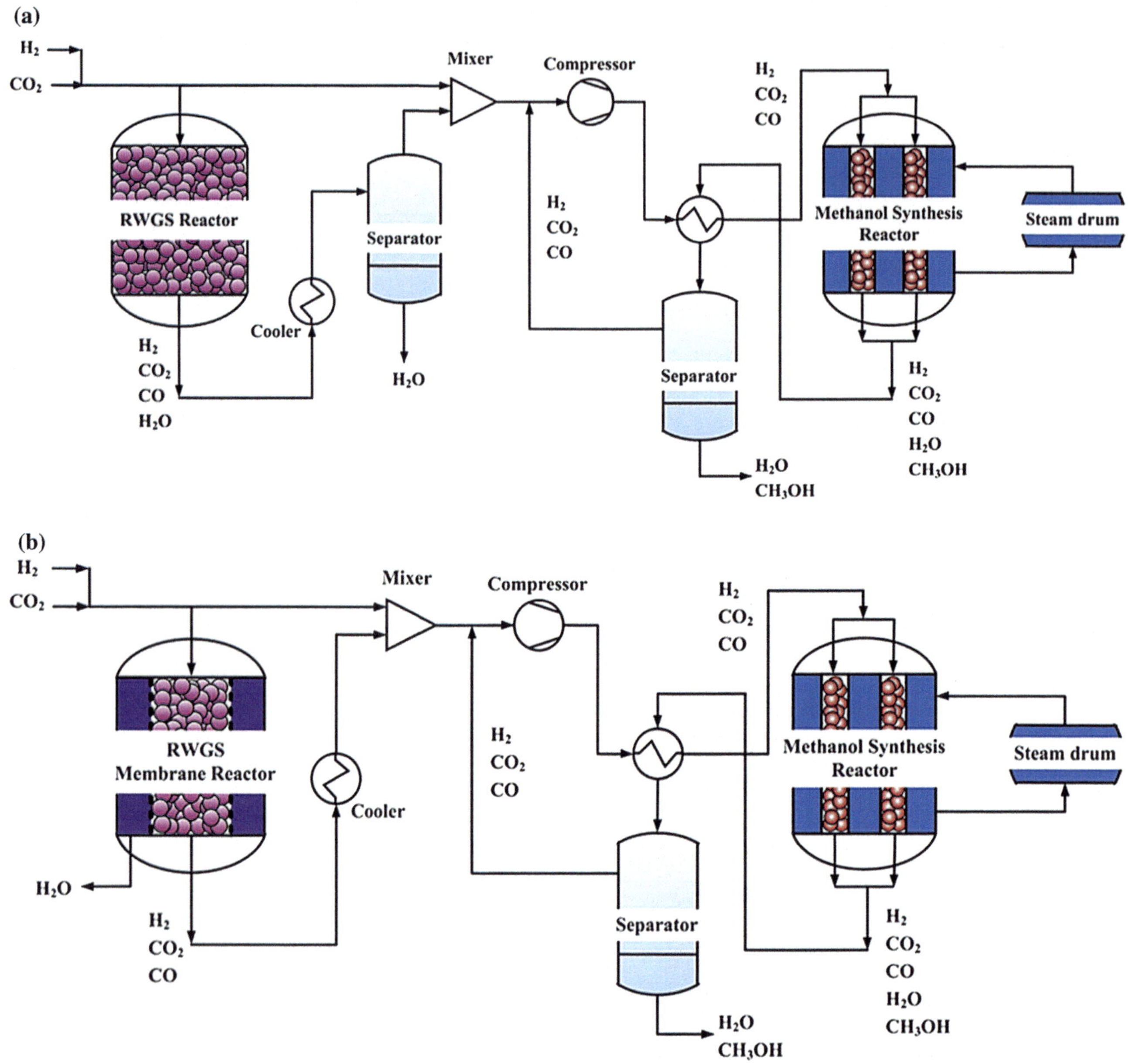

Fig. 12 Illustration of the indirect conversion of CO_2 process, **(a)** method 1 (process involves a reverse water gas shift and a reactor of methanol synthesis) and **(b)** method 2 (same as method 1 except using H-SOD (hydroxy sodalite) membrane in the reverse water gas shift reactor). Reprinted with. permission of Samimi et al. (2019)

They furthermore compared the results of H_2 and CO_2 conversions between RWGS reactor and RWGS membrane reactor (Fig. 14). Hydrogen and carbon dioxide as the feed (RWGS reaction) are consumed to produce carbon monoxide and water. In the RWGS reactor, the H_2 and CO_2 conversions are increased through the length of reactor and the constant trend is observed after reaching to the equilibrium state.

However, in the reverse water gas shift membrane reactor, H_2O eliminating through the members causes transferring thermodynamic equilibrium into consumption of H_2 and CO_2 and consequently, more conversions are obtained with respect to the reverse water gas shift reactor (Samimi et al. 2019). Figure 15 compares the results of both methods at the same operating conditions. MeOH is produced 13042 and 13584 kg/h, while production rate of H_2O is 5167 and 6209 kg/h in methods 1 and 2, respectively. Method 2 can produce 542 kg of MeOH per hour more in compared to method 1 (i.e., 4.16% increase in production rate of MeOH), that is a significant amount. Consequently, in method 2, higher carbon monoxide concentration in the produced syngas is obtained.

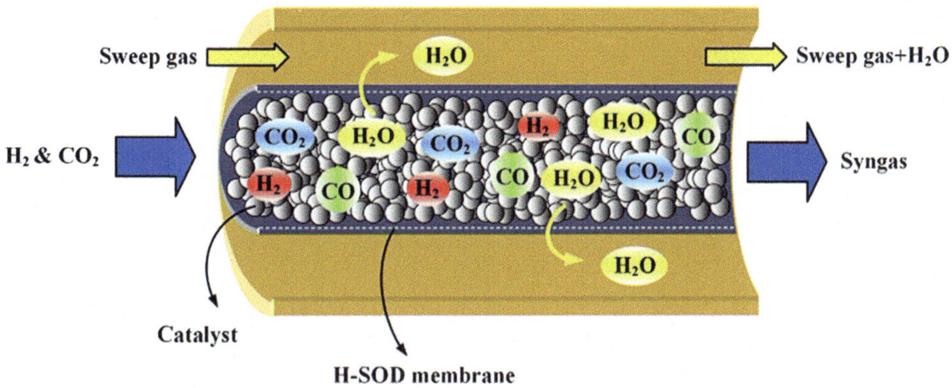

Fig. 13 Illustration of reverse water gas shift MR. Reprinted with permission of Samimi et al. (2019)

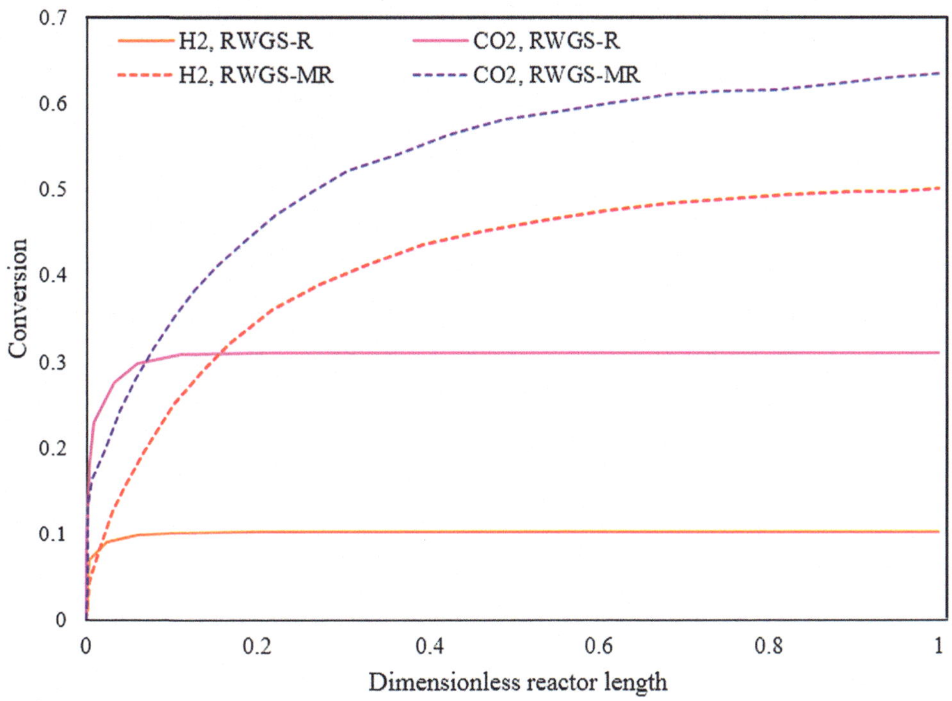

Fig. 14 Comparison between conversions of H_2 and CO_2 along reverse water gas shift reactor and reverse water gas shift membrane reactor. Modified after Samimi et al. (2019)

The results indicated that method 2 is better to method 1, because of the following reasons (Samimi et al. 2019; Samimi et al. 2018; Ribeirinha et al. 2017):

- Higher CO and CO_2 conversions were obtained in the second method.
- More appropriate composition of synthesis gas was achieved in second method.
- There is no necessity for additional water elimination from syngas in the second method.
- MeOH production in the second method is 542 kg/h more than the first method.
- Production of water was reduced 16% in method 2 than method 1.

5 Green Fuel Energy

Fossil fuel consumption produces greenhouse gases, the leading cause of global warming and climate change. Besides, natural petroleum sources are depleted by ever-increasing demand (Ardito et al. 2019). Renewable energies are the solution to climate change and mounting global energy demand. Therefore, researchers have focused on finding and developing alternative energy sources like solar energy, hydrogen energy, wind energy, bioenergy and green oxygenated fuels (Ardito et al. 2019; Chakraborty and Mazzanti 2020; Shuba and Kifle 2018; Edwards et al. 2008; Nord and Haupt 2005). These sources of power are called

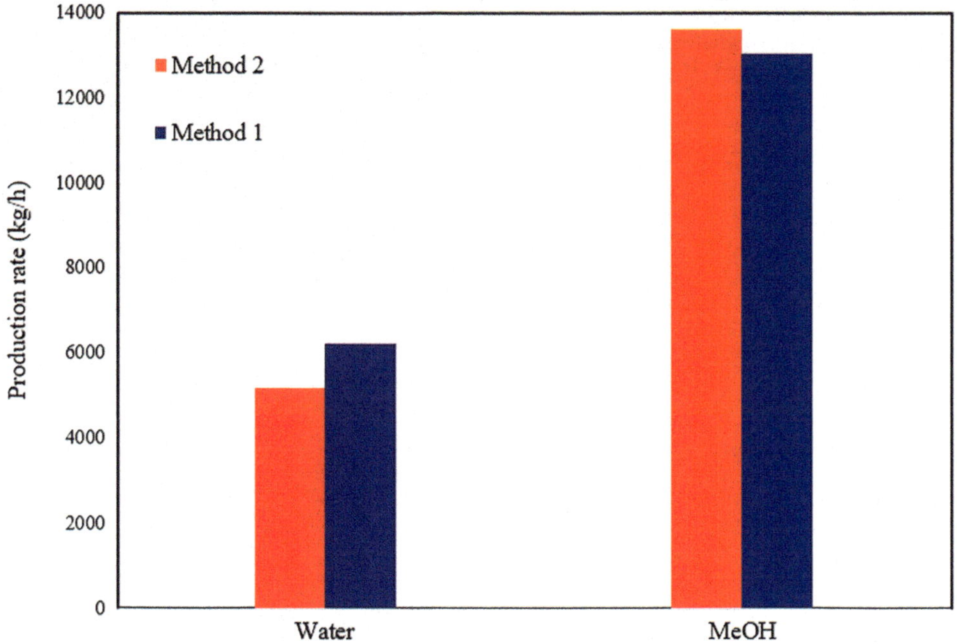

Fig. 15 Comparison between the efficiency of method 1 (process involves a RWGS and a methanol synthesis reactor) and method 2 (same as method 1 except using H-SOD membrane in the RWGS reactor) in water and MeOH production rate. Modified after Samimi et al. (2019)

clean or green energy because they can be produced from renewable sources by sustainable, clean and innovative technologies to substitute for pollutant technologies (Chakraborty and Mazzanti 2020). In this context, we reviewed some of these energies like hydrogen and biofuels, which can be produced by the membrane reactor technology in clean pathways.

5.1 Green H_2 Energy

Among the sustainable and recyclable energy sources, hydrogen is a valuable energy carrier due to clean energy sources and its high conversion efficiency. For these reasons, hydrogen has been suggested as a beneficial resource for fuel cells (Edwards et al. 2008). Conventional hydrogen gas production processes like steam reforming of methane and hydrocarbons, oxidation of fossil fuels and autothermal reforming consume a lot of energy. They require high temperatures (over 850 °C) (Armor 1999). Membrane reactors, solar energy for electrolysis of water, selective oxidation of methane, oxidative dehydrogenation, electrolysis of water using fuel cells, bio-fermentation and biomass conversion are alternative developing technologies for producing H_2. These technologies benefit from low energy consumption processes, less equipment and more environmental-friendly compared to conventional methods (Armor 1999; Hanley et al. 2018; Kapdan and Kargi 2006).

The membrane reactor produces hydrogen by applying hydrogen-selective removal by the membrane (Armor 1999; Kikuchi 2000). H_2 has a high density of energy. The temperature of H_2 rises, and it becomes flammable when the concentration increases greater than 4% (Farina et al. 2007). Using a membrane reactor, H_2 can be removed from the reaction system, which results in a decrease in operating temperature and an increase in reactant conversion (Yun and Oyama 2011). Therefore, many researchers have focused on the use of H_2 perm-selective membrane material to achieve H_2 in ultra-purity. Some examples of these materials are palladium composites, which can enhance dehydrogenation (Armor 1999; Chen et al. 2020; Jo et al. 2018). Pd composites have excellent characteristics for hydrogen purification, such as high hydrogen permeability, selectivity, significant chemical compatibility and durability (Hayakawa et al. 2019).

Ammonia has high hydrogen atom content. The conventional process of producing hydrogen from ammonia is performed at high temperature and requires independent separation unit for hydrogen purification (Jo et al. 2018). The efficient decomposition of ammonia and hydrogen purification is essential for every technology that uses ammonia as a hydrogen-carrier source. (Zhang et al. 2019). Jo et al. (2018) suggested a novel compact tubular membrane reactor in conjunction with a fuel cell that converted NH_3 to electrical energy. In this membrane reactor, reaction and separation are implemented in one step. The reactor comprised tubular Pd/Ta composite membrane and Ru/La-Al_2O_3 pellet catalysts for the decomposition of ammonia. In this reactor, NH_3 flow passes through the packed pellet catalysts, and decomposition reaction ($2NH_3 \rightarrow N_2 + 3H_2$) takes place while simultaneously the produced hydrogens permeate into Pd/Ta membrane. The membrane has high H_2 permeability and can reduce the operating temperature of NH_3 dehydrogenation (400–450 °C). It provides the in situ purification of hydrogen in the reactor.

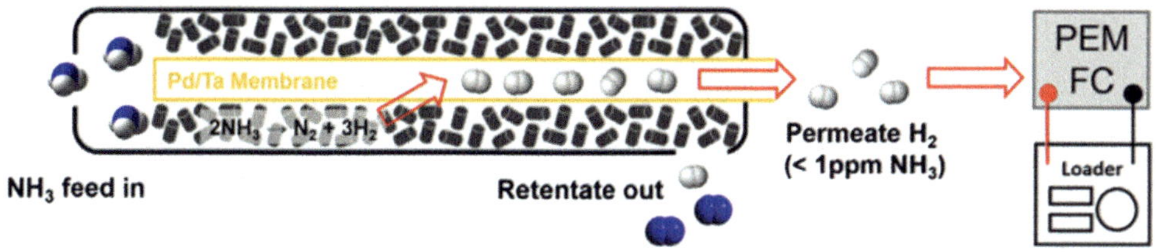

Fig. 16 Scheme of MR for sustainable H_2 production, NH_3 flow in membrane reactor and decomposition reaction ($2NH_3 \rightarrow N_2 + 3H_2$) occur by heterogeneous pellet catalysts. Hydrogen permeate into Pd/Ta membrane and directly feed proton exchange membrane fuel cell (PEM FC). Reprinted with permission of Jo et al. (2018)

Permeated hydrogen stream (over 99.9999% pure hydrogen) with a negligible amount of NH_3 (< 1 ppm) is immediately directed to the fuel cell to generate electricity (Fig. 16). The advantages of this reactor can be classified as listed below (Jo et al. 2018):

- Environmentally friendly technique.
- Reaction-separation implemented in one step.
- Ultra-pure hydrogen extraction.
- Energy consumption is low.
- No additional processing equipment is required.
- Removing other purification units for operating fuel cells.
- There are no practical limitations for related reactions concerning composite membranes.

Itoh et al. (2020) presented a tube-wall catalytic MR, which able to decrease ammonia decomposition temperature to below 400 °C and increase ammonia conversion by nearly 100% via designing of a reactor with excellent heat transfer characteristics and applying Pd membrane. In this MR, the inside of the reactor tubular shell was covered with the Ru catalyst layer and the outside shell was exposed to heat. By this design, a more uniform direct heat supply can be achieved to catalyst compared to MR with catalyst packed bed and resulted in a high ammonia conversion at lower temperatures than conventional MR (Itoh et al. 2020).

In another research (Zhang et al. 2019), a new configuration of catalytic design of MR is applied to produce hydrogen from NH_3. The catalytic MR consisted of a Ru catalyst, which is impregnated in a supported porous yttria-stabilized zirconia tube coated by a thin Pd film. In conventional membrane reactors, the catalyst particle is packed in the interior or within the annulus surrounding a tubular hydrogen-permeable membrane (Fig. 17). The traditional method suffers from two significant hydrogen transport limitations. The first limitation is effective diffusion within pellets. The second limitation is the radial diffusion of hydrogen through the bed and the support to the membrane surface before being swept in retentive flow. By the new design, it is possible to overcome transport limitations and make operating temperature as low as 400 °C and a purity degree of H_2 99.7% (Zhang et al. 2019).

Chen et al. (2016) utilized Pd/Ag ceramic membrane reactor to synthesize valuable biomaterial (amides and ureas) and hydrogen. Urea derivatives are important bioactive compounds. They are used in various areas like pharmaceuticals, therapeutics, chemical dyes and agricultural pesticides. Conventional methods for the synthesis of urea suffer from a significant amount of toxic phosgene, phosgene derivatives and CO. Another vital precursor is an amide, an essential functional group in proteins, peptides and many synthetic polymers. The standard amide production methods are complicated and produce a large amount of waste (Chen et al. 2016). Amides and urea are produced by catalytic dehydrogenative coupling of amines and volatile alcohols. The Pd-Ag/ceramic membrane has catalytic and selective features. Hydrogen permeates into the MR and is removed from the reaction system. The selective removal of produced hydrogen could monitor the process by overcoming chemical equilibrium and avoiding pressure build-up in the reactor. Produced hydrogen could be stored and used in a fuel cell (Chen et al. 2016).

5.2 Biofuel Energy

Biofuels energy is a kind of renewable energy derived from living organisms and biological sources like microorganisms, algae, animal, plant and waste biomass. These resources are nontoxic, biocompatible, biodegradable and renewable. Using biofuels has advantages over fossil fuels such as lower harmful emission (lower greenhouse gases) and less carbon (Shuba and Kifle 2018). Among biofuel sources, algae are one of the best sustainable feedstocks (Anto et al. 2020). Waste products can feed algae, and they can degrade some harmful materials into safe ones. Besides, algae are reproduced rapidly and produce suitable biomass for biofuel. Thus, algae provides benefits for environment and green chemistry by waste treatment and biofuel production (Mata et al. 2010; Kumar et al. 2010; Hou et al. 2020). Microalgae can convert greenhouse gases like CO_2 to biomass (CO_2 fixation) (Mata et al. 2010). Membrane bioreactors provide more microalgae proliferation and CO_2 fixation than conventional reactor technologies such as

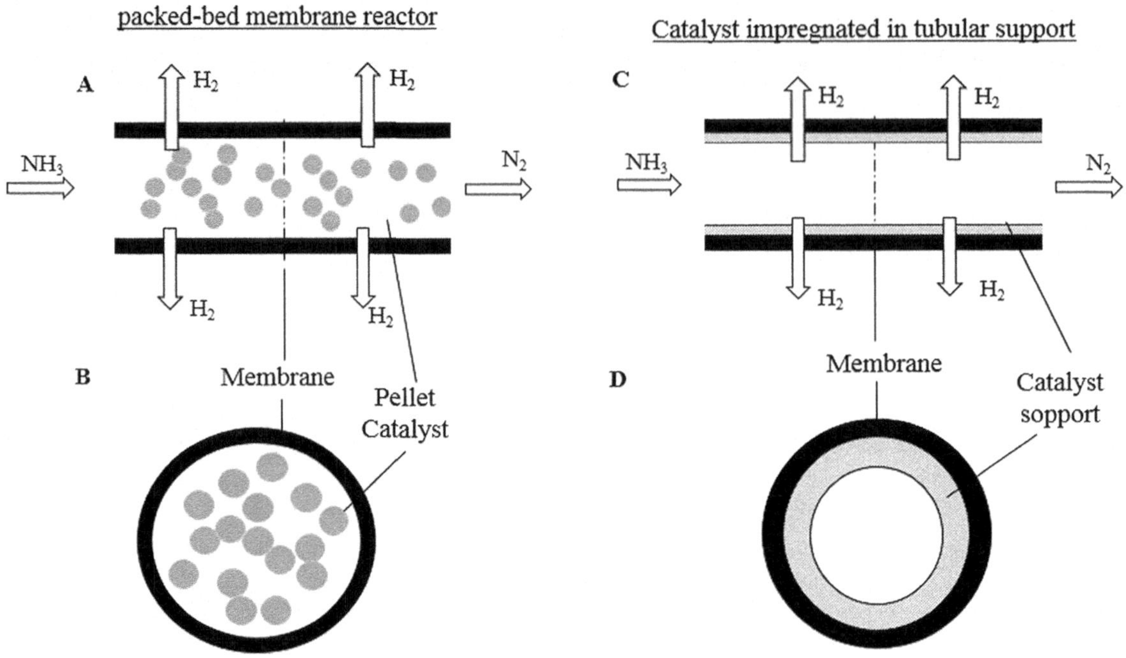

Fig. 17 (**a**) Packed-bed MR, (**b**) cross section of packed-bed MR, (**c**) catalyst membrane reactor, (**d**) cross section of catalyst MR. Modified after Zhang et al. (2019)

bubbling systems (Kumar et al. 2010). Among membrane bioreactors, the hollow fiber membrane reactors have a strong performance for microalgae cultivation (Kumar et al. 2010). Hollow fiber membrane reactors have several benefits. The first advantage is the large area per unit volume (over 30 cm^2/cm^3). Besides, they provide excellent mass diffusion, sustainable mass transfer pathways, especially when living organisms are growing, compactness, microorganism protection from turbulent hydrodynamic stresses and supporting microorganism attachment (Kumar et al. 2010; Eghbali et al. 2016). Flocculation and filamentous characteristics of algae are the greatest challenges when we use a membrane reactor for algae cultivation (Nhat et al. 2018).

Surface modification of the hollow membrane can reduce membrane fouling. Embedding nano-TiO_2 photocatalyst within the hollow membrane improves the hydrophilicity of membrane surface and reduces membrane fouling around 50% lower than that of conventional polyvinylidene fluoride hollow fiber (Hu et al. 2015). Using graphene oxide in the hollow membrane surface also improves the antifouling ability of the membrane (Wu et al. 2020). Algae hollow fiber membrane reactors can significantly treat wastewaters from chromium, ammonia-nitrogen and phosphorus and produce large amounts of biomass (Hu et al. 2015; Wu et al. 2020; Costa et al. 2019). One of the attractive sources of biofuel energy is waste biomass with plant origin (Clark 2019). Plant origin biomass produces biofuels like bioethanol, biohydrogen, biodiesel, biogas, biochar and syngas, in which valuable biochemical materials can be provided from them (Hood et al. 2013; McKendry 2002). In this field, MR can be utilized for several applications such as cell immobilization for fermentation of biomass and separation and purification of biofuel and organic extractant from substrates (Qureshi and Ezeji 2008).

5.3 Green Fuel Additive

The fuel additive, acetaldehyde dibutylacetal (1, 1-dibutoxyethane, DBE) is a derivative of acetal. Acetals are a kind of green fuel additive that can reduce the emission of particulate matter by blending with diesel and increase diesel cetane number. Pereira et al. (2012) performed a single-process numerical study on a moving bed membrane reactor integrated with pervaporation membrane reactors. Their results showed that the hybrid system could be a very efficient and promising process than the moving bed membrane reactor alone. It has more productivities and fewer adsorbent consumptions for the same purity and conversion criteria.

6 Biocatalyst Membrane Reactors

Enzymes as biocatalyst have a distinct advantage in green chemistry; in their presence, the reactions all perform in mild conditions (around ambient temperature and pressure). They consume lower energy, form minimum by-product and have higher specificity (enzymes catalyze only a single type of reaction and often they work only on one or a few substrate compounds) compared to chemical catalysts (Ugur Nigiz and Durmaz 2016; Sheldon 2005). Besides, some enzymes can biodegrade hazardous environmental pollutant compounds. For example, oxidoreductases are used to biodegrade dye, pharmaceutical and phenolic compounds (Zdarta et al. 2019). Enzymatic reactors combine two processes: enzymatic reaction (degradation or conversion) and membrane separation. Both procedures can be carried out at the same time or in a sequence (Su et al. 2020; Brunetti et al. 2018). Referring to the biocatalyst state, the biocatalyst MRs are classified into two categories: free or suspended biocatalyst membrane reactor and immobilized biocatalyst membrane reactor. In the suspended enzyme membrane reactor, the biocatalyst is free and suspended in the reaction mixture, and the membrane acts as separator (Prenosil and Hediger 1988).

The immobilized biocatalyst membrane reactors are divided into three categories (Ugur Nigiz and Durmaz 2016):

- Enzymatic batch reactor: pieces of heterogeneous biocatalyst membrane (biocatalyst is immobilized or embedded in the membrane) is placed in a conventional batch reactor.
- Biocatalyst membrane reactor: biocatalyst membrane performs both reaction and separation processes.
- Pervaporation biocatalyst membrane reactor (PVBCMR): vacuum pressure is utilized in the reactor. The biocatalyst membrane has a non-porous structure. Therefore, the evaporative separation takes place in the biocatalyst membrane.

Depending on reaction conditions, a free or immobilized enzymatic membrane reactor can be applied. Immobilization form has more benefits, such as improving the thermal and chemical stability of enzymes (Prenosil and Hediger 1988; Bilal et al. 2018; Jesionowski et al. 2014).

Pervaporation biocatalyst membrane reactor is preferred when it comes to immobilized biocatalyst membrane reactors. The energy consumption, operation time, capital and operation cost are lower. Secondly, the product has a higher purity level. Thirdly, in situ product extraction has a high conversion of the substrate. Finally, producing green organic solvent (Ugur Nigiz and Durmaz 2016) and removing additional separation are other benefits. PVBCMR can be utilized to produce 'green organic solvent.' Green organic solvents are critical in green chemistry. They should be relatively nonhazardous, nontoxic, noninflammable and noncorrosive. The emissions of the green organic solvent are completely low and environmentally safe (Sheldon 2005). The ethyl lactate is 'green solvent' because it has high boiling temperature, low vapor pressure and relatively nontoxic nature. In this research, PVBCMR was implemented to synthesize ethyl lactate from ethanol and lactic acid in mild operating conditions. The membrane of PVBCMR consisted of two layers, including the lipase biocatalyst layer and sodium alginate separation layer. The advantages of this PVBCMR compared to the batch reactor were: 1) higher acid conversion (two times higher), and 2) biocatalyst membrane has excellent stability and activity (Ugur Nigiz and Durmaz 2016). The critical challenges in enzymatic membrane reactors are low stability and efficiency of enzyme catalysts. Many studies focused on developing enzyme technology by protein and genetic engineering, immobilization techniques to reduce enzyme limitations (Madhavan et al. 2017).

The membrane of the enzymatic membrane reactor can be composed of magnetic nanoparticles and porous polymer. Magnetic nanoparticles can act as an enzyme carrier for enzyme immobilization into the membrane and nanofillers to form organic–inorganic membrane. This type of membrane has the capability of magnetic reversibility. The reversible magnetic force improves the dispersion of the enzymes over the membrane surface, allowing retention of the enzyme by a large pore, i.e. high-flux membrane. As a result, enzyme recovery becomes very convenient (Jiang et al. 2017; Gebreyohannes et al. 2015).

7 Photocatalytic Membrane Reactors

Photocatalytic membrane reactors (PMRs) combine photocatalysts with the membrane separation process. These types of reactors are mostly used for the purification of water and wastewater treatment. Photocatalyst in the presence of light can potentially decompose the organic and toxic pollutants into less toxic or harmless inorganic molecules (de Oliveira et al. 2020; Riaz and Park 2019). The most critical operating parameters in the design of PMR include the loading of photocatalysts, the initial concentration of reactant, light wavelength, light intensity, pH, temperature, flow rate, oxidants and ions (de Oliveira et al. 2020; Riaz and Park 2019). Several PMR configurations have been proposed. They can be classified into two main categories: immobilized photocatalyst system and suspended photocatalyst system. Due to the free photocatalytic active surface, suspended photocatalyst reactors have higher photocatalytic activity than the immobilized photocatalytic reactor. However, the photocatalyst must be separated from the reaction medium after detoxification (de Oliveira et al. 2020; Riaz and Park 2019).

Among photocatalysts, TiO_2 nanoparticle is frequently used in PMR, because TiO_2 has unique properties like high chemical stability, less toxicity, low cost and high surface to volume ratio and consequently more active site for the

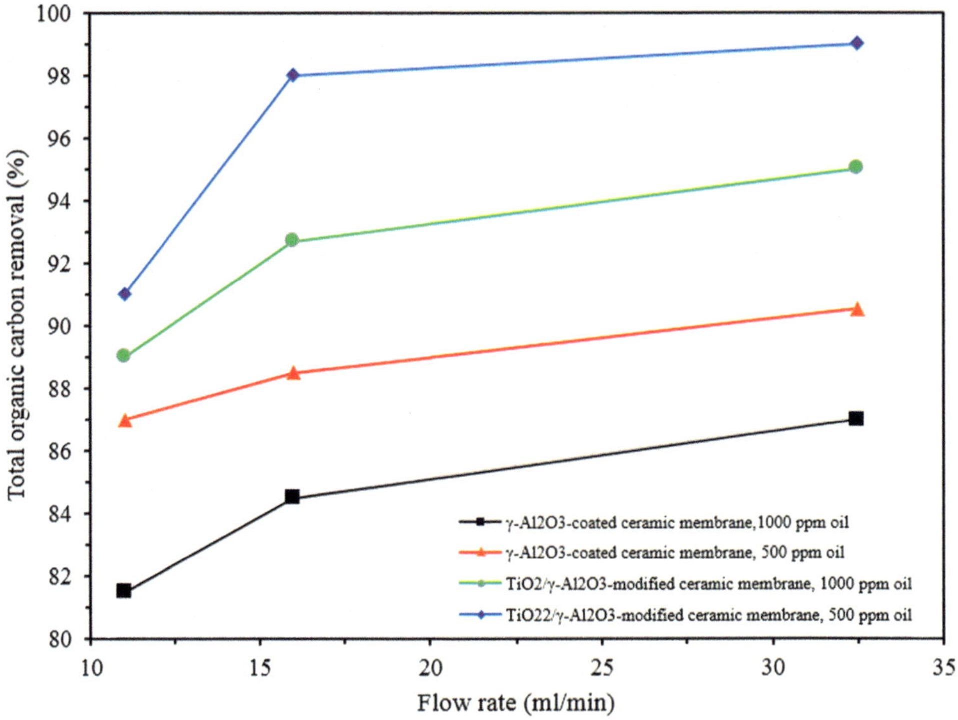

Fig. 18 Effect of γ-Al_2O_3 membrane with and without TiO_2 photocatalysts in two oil concentrations (500 and 1000 ppm) and different flow rates on TOC removal. Modified after Golshenas et al. (2020)

reaction (Riaz and Park 2019). de Oliveira et al. (2020) presented a coupling of PMR and membrane bioreactor (MBR) for oil refinery wastewater treatment. TiO_2 nano-photocatalyst and recycled osmosis membrane were employed in PMR. TiO_2 nanoparticles were synthesized with a green method by microwave radiation and utilized as a suspended nanoparticles catalyst in the PMR reaction mixture. This PMR can degrade the chemical oxygen demand (COD) of the organic matter in the MBR permeate by up to 60%. Compared to the system operated without photocatalyst, the membrane resistance attributed to fouling was 7.3 times lower and the membrane has good stability (de Oliveira et al. 2020). In another research, TiO_2 nano-photocatalysts were immobilized on the surface of γ-Al_2O_3 ceramic membrane, and they were utilized in PMR for the treatment of synthetic oily wastewater. The results showed high permeate flux and very high total organic carbon (TOC) removal efficiency (Fig. 18) (Golshenas et al. 2020).

Another photocatalyst in the PMR process is ZnO nanoparticle. It can act as a strong adsorber and improve the filtration property of the membrane. Desa et al. (2019) applied a nano-composite of ZnO capped with polyethylene glycol (ZnO-PEG) as photocatalyst in aid of the polypiperazine amide membrane to solve dye wastewater degradation problem. The textile industry discharges annually about 5000 tonnes of toxic dye components into the environment. The implementation results of this PMR under the optimum operational conditions (pH = 11, 0.10 g/L ZnO-PEG loading and 75% dilution of industrial textile wastewater) showed a complete degradation of the color and turbidity of the wastewater as well as membrane fouling reduction (Desa et al. 2019). Tungsten oxide (WO_3) is another photocatalyst that is suitable for textile dyeing wastewater treatment (Sathya et al. 2020). WO_3 has been known as a harmless, stable and low-cost photocatalyst, which reduces energy consumption due to its performance under solar irradiation (Dong et al. 2017). Pure WO_3 and its combination with 1% graphene oxide separately were utilized in a photocatalytic hollow membrane bioreactor. The reactor includes a photocatalytic and polyethersulphone hollow membrane bioreactor units, which is implemented on a pilot scale. The results showed that graphene oxide enhances the photocatalytic property of tungsten oxide. Therefore, the combination of tungsten oxide with graphene oxide improves the efficacy of dyes degradation in comparison with pure tungsten oxide. When the hollow membrane bioreactor was integrated with the photocatalyst unit, the efficiency of removal color and chemical oxygen demand increased to 45% and 28%, respectively (Sathya et al. 2020).

8 Conclusions

This chapter discusses the preference of membrane reactors for green synthesis. As evidenced throughout the context, membrane reactors employed separation and reaction processes in one unit render several advantages. Low energy consumption, high conversion, easy process control, mild operation conditions, low wastes production and being environmentally safe are few benefits to name. Membrane reactors have been improved based on individual applications. Thus, the essential designing parameters are different in each case. In membrane reactors for green H_2 synthesis, hydrogen perm-selectivity of the membrane is the most crucial parameter. A sufficient amount of biofuel is extracted from a biomembrane reactor with excellent mass transfer properties like hollow fiber membrane reactors. Types of photocatalyst, enzyme engineering and immobilization for biocatalyst membrane reactors are critical factors in MR. The common challenge in all of the membrane reactors is the fouling of membranes. The characterization and experimental results demonstrate that the efficiency of a membrane reactor is higher than that of conventional reactors.

References

Akin I, Zor E, Bingol H, Ersoz M (2014) Green synthesis of reduced graphene oxide/polyaniline composite and its application for salt rejection by polysulfone-based composite membranes. J Phys Chem B 118(21):5707–5716. https://doi.org/10.1021/jp5025894

Al-Juaied MA, Lafarga D, Varma A (2001) Ethylene epoxidation in a catalytic packed-bed membrane reactor: experiments and model. Chem Eng Sci 56(2):395–402. https://doi.org/10.1016/S0009-2509(00)00235-9

Anto S, Mukherjee SS, Muthappa R, Mathimani T, Deviram G, Kumar SS, Verma TN, Pugazhendhi A (2020) Algae as green energy reserve: technological outlook on biofuel production. Chemosphere 1(242):125079. https://doi.org/10.1016/j.chemosphere.2019.125079

Antonia P, Hernández-Fernández FJ, Gómez D, Rubio M, Tomás-Alonso F, Víllora G (2007) Understanding the chemical reaction and mass-transfer phenomena in a recirculating enzymatic membrane reactor for green ester synthesis in ionic liquid/supercritical carbon dioxide biphasic systems. J Supercrit Fluids 43(2):303–309. https://doi.org/10.1016/j.supflu.2007.06.003

Ardito L, Petruzzelli AM, Ghisetti C (2019) The impact of public research on the technological development of industry in the green energy field. Technol Forecast Soc Chang 1(144):25–35. https://doi.org/10.1016/j.techfore.2019.04.007

Armor JN (1999) The multiple roles for catalysis in the production of H_2. Appl Catal A 176(2):159–176. https://doi.org/10.1016/S0926-860X(98)00244-0

Avila AM, Yu Z, Fazli S, Sawada JA, Kuznicki SM (2014) Hydrogen-selective natural mordenite in a membrane reactor for ethane dehydrogenation. Microporous Mesoporous Mater 15(190):301–308. https://doi.org/10.1016/j.micromeso.2014.02.024

Bagheri H, Ghader S (2017) Correlating ionic liquids density over wide range of temperature and pressure by volume shift concept. J Molecul Liquid 236:172–83. https://doi.org/10.1016/j.molliq.2017.03.101

Bagheri H, Mohebbi A (2017) Prediction of critical temperature, critical pressure and acentric factor of some ionic liquids using Patel-Teja equation of state based on genetic algorithm. Korean J Chem Eng 34(10):2686–2702. https://doi.org/10.1007/s11814-017-0166-2

Bagheri H, Mansoori GA, Hashemipour H (2018) A novel approach to predict drugs solubility in supercritical solvents for RESS process using various cubic EoS-mixing rule. J Mol Liq 1(261):174–188. https://doi.org/10.1016/j.molliq.2018.03.081

Bagheri H, Ghader S, Hatami N (2019a) Solubility of ibuprofen in conventional solvents and supercritical CO_2: evaluation of ideal and non-ideal models. Chemistry 13(1):1. https://doi.org/10.23939/chcht13.01.001

Bagheri H, Hashemipour H, Ghader S (2019b) Population balance modeling: application in nanoparticle formation through rapid expansion of supercritical solution. Comput Particle Mech 6(4):721–737. https://doi.org/10.1007/s40571-019-00257-w

Bagheri H, Hashemipour H, Mirzaie M (2019c) Investigation on hydrodynamic and formation of nano particle by RESS process: the numerical study. J Mol Liq 1(281):490–505. https://doi.org/10.1016/j.molliq.2019.02.108

Balachandran U, Dusek JT, Mieville RL, Poeppel RB, Kleefisch MS, Pei S, Kobylinski TP, Udovich CA, Bose AC (1995) Dense ceramic membranes for partial oxidation of methane to syngas. Appl Catal A 133(1):19–29. https://doi.org/10.1016/0926-860X(95)00159-X

Bilal M, Rasheed T, Zhao Y, Iqbal HM, Cui J (2018) "Smart" chemistry and its application in peroxidase immobilization using different support materials. Int J Biol Macromol 1(119):278–290. https://doi.org/10.1016/j.ijbiomac.2018.07.134

Boyd T, Grace J, Lim CJ, Adris AE (2005) Hydrogen from an internally circulating fluidized bed membrane reactor. Int J Chem React Eng 3(1). https://doi.org/10.2202/1542-6580.1241

Brennecke JF, Maginn EJ (2001) Ionic liquids: innovative fluids for chemical processing. AIChE J 47(11):2384–2389. https://doi.org/10.1002/aic.690471102

Briceño K, Basile A, Tong J, Haraya K (2013) Carbon-based membranes for membrane reactors. In: Handbook of membrane reactors. Woodhead Publishing, pp 370–400. https://doi.org/10.1533/9780857097330.2.370

Brunetti A, Zito PF, Giorno L, Drioli E, Barbieri G (2018) Membrane reactors for low temperature applications: an overview. Chem Eng Process-Process Intensif 1(124):282–307. https://doi.org/10.1016/j.cep.2017.05.002

Bu T, Wu L, Liu X, Yang X, Zhou P, Yu X, Qin T, Shi J, Wang S, Li S, Ku Z (2017) Synergic interface optimization with green solvent engineering in mixed perovskite solar cells. Adv Energy Mater 7(20):1700576. https://doi.org/10.1002/aenm.201700576

Cannilla C, Bonura G, Costa F, Frusteri F (2018) Biofuels production by esterification of oleic acid with ethanol using a membrane assisted reactor in vapor permeation configuration. Appl Catal A 25(566):121–129. https://doi.org/10.1016/j.apcata.2018.08.014

Capello C, Fischer U, Hungerbühler K (2007) What is a green solvent? A comprehensive framework for the environmental assessment of solvents. Green Chem 9(9):927–934. https://doi.org/10.1039/B617536H

Chakraborty SK, Mazzanti M (2020) Energy intensity and green energy innovation: checking heterogeneous country effects in the OECD. Struct Change Econ Dyn 1(52):328–343. https://doi.org/10.1016/j.strueco.2019.12.002

Chambreau SD, Boatz JA, Vaghjiani GL, Koh C, Kostko O, Golan A, Leone SR (2012) Thermal decomposition mechanism of 1-ethyl-3-methylimidazolium bromide ionic liquid. J Phys Chem A 116(24):5867–5876. https://doi.org/10.1021/jp209389d

Chen P, Chen X, Chen X, An Z, Kita H (2009) Pervaporation separation and catalysis activity of novel zirconium silicalite-1 zeolite membrane. Chin J Chem 27(9):1692–1696. https://doi.org/10.1002/cjoc.200990284

Chen T, Zeng G, Lai Z, Huang KW (2016) Environmentally benign synthesis of amides and ureas via catalytic dehydrogenation coupling of volatile alcohols and amines in a Pd-Ag membrane reactor. J Membr Sci 1(515):212–218. https://doi.org/10.1016/j.memsci.2016.05.042

Chen T, Wang Z, Liu L, Pati S, Wai MH, Kawi S (2020a) Coupling CO_2 separation with catalytic reverse water-gas shift reaction via

ceramic-carbonate dual-phase membrane reactor. Chem Eng J 1 (379):122182. https://doi.org/10.1016/j.cej.2019.122182
Chen WH, Escalante J, Chi YH, Lin YL (2020b) Hydrogen permeation enhancement in a Pd membrane tube system under various vacuum degrees. Int J Hydrogen Energy 45(12):7401–7411. https://doi.org/10.1016/j.ijhydene.2019.04.118
Clark JH (2019) Green biorefinery technologies based on waste biomass. Green Chem 21(6):1168–1170. https://doi.org/10.1039/C9GC90021G
Colli GG, Alves JA, Martínez OM, Barreto GF (2019) Application of a catalytic membrane reactor to the selective hydrogenation of 1-Butyne. Chem Eng Process—Process Intensif 1(142):107518. https://doi.org/10.1016/j.cep.2019.04.018
Constantinou A, Wu G, Venezia B, Ellis P, Kuhn S, Gavriilidis A (2019) Aerobic oxidation of benzyl alcohol in a continuous catalytic membrane reactor. Top Catal 62(17–20):1126–1131. https://doi.org/10.1007/s11244-018-1060-9
Costa IG, Terra NM, Cardoso VL, Batista FR, Reis MH (2019) Photoreduction of chromium (VI) in microstructured ceramic hollow fibers impregnated with titanium dioxide and coated with green algae Chlorella vulgaris. J Hazard Mater 5(379):120837. https://doi.org/10.1016/j.jhazmat.2019.120837
Cue BW, Zhang J (2009) Green process chemistry in the pharmaceutical industry. Green Chem Lett Rev 2(4):193–211. https://doi.org/10.1080/17518250903258150
de Oliveira CP, Viana MM, Silva GR, Lima LS, de Paula EC, Amaral MC (2020) Potential use of green TiO_2 and recycled membrane in a photocatalytic membrane reactor for oil refinery wastewater polishing. J Clean Prod 1(257):120526. https://doi.org/10.1016/j.jclepro.2020.120526
Desa AL, Hairom NH, Ng LY, Ng CY, Ahmad MK, Mohammad AW (2019) Industrial textile wastewater treatment via membrane photocatalytic reactor (MPR) in the presence of ZnO-PEG nanoparticles and tight ultrafiltration. J Water Process Eng 1 (31):100872. https://doi.org/10.1016/j.jwpe.2019.100872
Dixon AG (1999) Innovations in catalytic inorganic membrane reactors. In: Catalysis, pp 40–92. https://doi.org/10.1039/9781847553263-00040
Dong P, Hou G, Xi X, Shao R, Dong F (2017) WO3-based photocatalysts: morphology control, activity enhancement and multifunctional applications. Environ Scie Nano 4(3):539–557. https://doi.org/10.1039/c6en00478d
Edwards PP, Kuznetsov VL, David WI, Brandon NP (2008) Hydrogen and fuel cells: towards a sustainable energy future. Energy Policy 36 (12):4356–4362. https://doi.org/10.1016/j.enpol.2008.09.036
Eghbali H, Nava MM, Mohebbi-Kalhori D, Raimondi MT (2016) Hollow fiber bioreactor technology for tissue engineering applications. Int J Artif Organs 39(1):1–5. https://doi.org/10.5301/ijao.5000466
Espinosa RB, Rafieian D, Postma RS, Lammertink RG, Lefferts L (2018) Egg-shell membrane reactors for nitrite hydrogenation: Manipulating kinetics and selectivity. Appl Catal B 1(224):276–282. https://doi.org/10.1016/j.apcatb.2017.10.058
Farina L, Boido E, Carrau F, Dellacassa E (2007) Determination of volatile phenols in red wines by dispersive liquid–liquid microextraction and gas chromatography–mass spectrometry detection. J Chromatogr A 1157(1–2):46–50. https://doi.org/10.1016/j.chroma.2007.05.006
Farsi M, Jahanmiri A (2014) Dynamic modeling and operability analysis of a dual-membrane fixed bed reactor to produce methanol considering catalyst deactivation. J Ind Eng Chem 20(5):2927–2933. https://doi.org/10.1016/j.jiec.2013.11.030
Gebreyohannes AY, Bilad MR, Verbiest T, Courtin CM, Dornez E, Giorno L, Curcio E, Vankelecom IF (2015) Nanoscale tuning of enzyme localization for enhanced reactor performance in a novel magnetic-responsive biocatalytic membrane reactor. J Membr Sci 1 (487):209–220. https://doi.org/10.1016/j.memsci.2015.03.069
Golshenas A, Sadeghian Z, Ashrafizadeh SN (2020) Performance evaluation of a ceramic-based photocatalytic membrane reactor for treatment of oily wastewater. J Water Process Eng 1(36):101186. https://doi.org/10.1016/j.jwpe.2020.101186
Hanley ES, Deane JP, Gallachóir BÓ (2018) The role of hydrogen in low carbon energy futures—a review of existing perspectives. Renew Sustain Energy Rev 1(82):3027–3045. https://doi.org/10.1016/j.rser.2017.10.034
Hayakawa Y, Miura T, Shizuya K, Wakazono S, Tokunaga K, Kambara S (2019) Hydrogen production system combined with a catalytic reactor and a plasma membrane reactor from ammonia. Int J Hydrogen Energy 44(20):9987–9993. https://doi.org/10.1016/j.ijhydene.2018.12.141
Hedayati A, Le Corre O, Lacarrière B, Llorca J (2016) Dynamic simulation of pure hydrogen production via ethanol steam reforming in a catalytic membrane reactor. Energy 15(117):316–324. https://doi.org/10.1016/j.energy.2016.06.042
Herrero M, Ibanez E (2015) Green processes and sustainability: an overview on the extraction of high added-value products from seaweeds and microalgae. J Supercrit Fluids 1(96):211–216. https://doi.org/10.1016/j.supflu.2014.09.006
Heyse A, Plikat C, Ansorge-Schumacher M, Drews A (2019) Continuous two-phase biocatalysis using water-in-oil Pickering emulsions in a membrane reactor: Evaluation of different nanoparticles. Catal Today 1(331):60–67. https://doi.org/10.1016/j.cattod.2017.11.032
Hood EE, Teoh K, Devaiah SP, Requesens VD (2013) Biomass crops for biofuels and bio-based products. Sustain Food Prod 250–79. https://doi.org/10.1007/978-1-4614-5797-8
Hou Q, Yang Z, Chen S, Pei H (2020) Using an anaerobic digestion tank as the anodic chamber of an algae-assisted microbial fuel cell to improve energy production from food waste. Water Res 1 (170):115305. https://doi.org/10.1016/j.watres.2019.115305
Hu W, Yin J, Deng B, Hu Z (2015) Application of nano TiO_2 modified hollow fiber membranes in algal membrane bioreactors for high-density algae cultivation and wastewater polishing. Biores Technol 1(193):135–141. https://doi.org/10.1016/j.biortech.2015.06.070
Ibrahim AF, Dandamudi KP, Deng S, Lin YS (2020) Pyrolysis of hydrothermal liquefaction algal biochar for hydrogen production in a membrane reactor. Fuel 1(265):116935. https://doi.org/10.1016/j.fuel.2019.116935
Itoh N, Kikuchi Y, Furusawa T, Sato T (2020) Tube-wall catalytic membrane reactor for hydrogen production by low-temperature ammonia decomposition. Int J Hydrogen Energy. https://doi.org/10.1016/j.ijhydene.2020.03.162
Jadhav SG, Vaidya PD, Bhanage BM, Joshi JB (2014) Catalytic carbon dioxide hydrogenation to methanol: a review of recent studies. Chem Eng Res Des 92(11):2557–2567. https://doi.org/10.1016/j.cherd.2014.03.005
Jesionowski T, Zdarta J, Krajewska B (2014) Enzyme immobilization by adsorption: a review. Adsorption 20(5–6):801–821. https://doi.org/10.1007/s10450-014-9623-y
Jiang H, Xing L, Czuprat O, Wang H, Schirrmeister S, Schiestel T, Caro J (2009) Highly effective NO decomposition by in situ removal of inhibitor oxygen using an oxygen transporting membrane. Chem Commun 44:6738–6740. https://doi.org/10.1039/B912269A
Jiang JF, Qiao J, Mu XY, Moon MH, Qi L (2017) Fabrication of enzyme reactor utilizing magnetic porous polymer membrane for screening D-Amino acid oxidase inhibitors. Talanta 1(165):251–257. https://doi.org/10.1016/j.talanta.2016.12.055

Jin W, Li S, Huang P, Xu N, Shi J, Lin YS (2000) Tubular lanthanum cobaltite perovskite-type membrane reactors for partial oxidation of methane to syngas. J Membr Sci 166(1):13–22. https://doi.org/10.1016/S0376-7388(99)00245-8

Jo YS, Cha J, Lee CH, Jeong H, Yoon CW, Nam SW, Han J (2018) A viable membrane reactor option for sustainable hydrogen production from ammonia. J Power Sources 1(400):518–526. https://doi.org/10.1016/j.jpowsour.2018.08.010

Kamat S, Critchley G, Beckman EJ, Russell AJ (1995) Biocatalytic synthesis of acrylates in organic solvents and supercritical fluids: III. Does carbon dioxide covalently modify enzymes? Biotechnol Bioeng 46(6):610–20. https://doi.org/10.1002/bit.260460614

Kapdan IK, Kargi F (2006) Bio-hydrogen production from waste materials. Enzyme Microbial Technol 38(5):569–582. https://doi.org/10.1016/j.enzmictec.2005.09.015

Kikuchi E (2000) Membrane reactor application to hydrogen production. Catal Today 56(1–3):97–101. https://doi.org/10.1016/S0920-5861(99)00256-4

Kisszekelyi P, Alammar A, Kupai J, Huszthy P, Barabas J, Holtzl T, Szente L, Bawn C, Adams R, Szekely G (2019) Asymmetric synthesis with cinchona-decorated cyclodextrin in a continuous-flow membrane reactor. J Catal 1(371):255–261. https://doi.org/10.1016/j.jcat.2019.01.041

Kölsch P, Smejkal Q, Noack M, Schäfer R, Caro J (2002) Partial oxidation of propane to acrolein in a membrane reactor–Experimental data and computer simulation. Catal Commun 3(10):465–470. https://doi.org/10.1016/S1566-7367(02)00177-2

Kumar A, Ergas S, Yuan X, Sahu A, Zhang Q, Dewulf J, Malcata FX, Van Langenhove H (2010) Enhanced CO_2 fixation and biofuel production via microalgae: recent developments and future directions. Trends Biotechnol 28(7):371–380. https://doi.org/10.1016/j.tibtech.2010.04.004

Kumar S, Kumar B, Kumar S, Jilani S (2017) Comparative modeling study of catalytic membrane reactor configurations for syngas production by CO_2 reforming of methane. J CO_2 Utilization 20:336–46. https://doi.org/10.1016/j.jcou.2017.06.004

Kumari P, Bahadur N, Dumée LF (2020) Photo-catalytic membrane reactors for the remediation of persistent organic pollutants—a review. Sep Purif Technol 2(230):115878. https://doi.org/10.1016/j.seppur.2019.115878

Lapeña D, Lomba L, Artal M, Lafuente C, Giner B (2019) The NADES glyceline as a potential green solvent: a comprehensive study of its thermophysical properties and effect of water inclusion. J Chem Thermodyn 1(128):164–172. https://doi.org/10.1016/j.jct.2018.07.031

Lee J, Kim EY, Chang BJ, Han M, Lee PS, Moon SY (2020) Mixed-matrix membrane reactors for the destruction of toxic chemicals. J Membr Sci 2:118112. https://doi.org/10.1016/j.memsci.2020.118112

Li W, Zhu X, Chen S, Yang W (2016) Integration of nine steps into one membrane reactor to produce synthesis gases for ammonia and liquid fuel. Angew Chem Int Ed 55(30):8566–8570. https://doi.org/10.1002/anie.201602207

Li W, Cao Z, Li H, Zhu X, Yang W (2019) Asymmetric dual-phase MIEC membrane reactor for energy-efficient coproduction of two kinds of synthesis gases. Int J Hydrogen Energy 44(8):4218–4227. https://doi.org/10.1016/j.ijhydene.2018.12.182

Liang W, Cao Z, He G, Caro J, Jiang H (2017) Oxygen transport membrane for thermochemical conversion of water and carbon dioxide into synthesis gas. ACS Sustain Chem Eng 5(10):8657–8662. https://doi.org/10.1021/acssuschemeng.7b01305

Liu Y, Liu F, Ding N, Hu X, Shen C, Li F, Huang M, Wang Z, Sand W, Wang CC (2020) Recent advances on electroactive CNT-based membranes for environmental applications: the perfect match of electrochemistry and membrane separation. Chin Chem Lett. https://doi.org/10.1016/j.cclet.2020.03.011

Low FW, Samsudin NA, Yusoff Y, Tan XY, Lai CW, Amin N, Tiong SK (2020) Hydrolytic cleavage of glycosidic bonds for cellulose nanoparticles (CNPs) production by $BmimHSO_4$ ionic liquid catalyst. Thermochim Acta 1(684):178484. https://doi.org/10.1016/j.tca.2019.178484

Lozano P, De Diego T, Carrie D, Vaultier M, Iborra JL (2001) Over-stabilization of candida antarctica lipase B by ionic liquids in ester synthesis. Biotech Lett 23(18):1529–1533. https://doi.org/10.1023/A:1011697609756

Lyagin E, Drews A, Bhattacharya S, Ansorge-Schumacher MB, Kraume M (2010) Continuous screening system for inhibited enzyme catalysis: a membrane reactor approach. Biotechnol J 5 (8):813–821. https://doi.org/10.1002/biot.201000130

Madhavan A, Sindhu R, Binod P, Sukumaran RK, Pandey A (2017) Strategies for design of improved biocatalysts for industrial applications. Biores Technol 1(245):1304–1313. https://doi.org/10.1016/j.biortech.2017.05.031

Maina JW, Pozo-Gonzalo C, Schütz JA, Wang J, Dumée LF (2019) Tuning CO_2 conversion product selectivity of metal organic frameworks derived hybrid carbon photoelectrocatalytic reactors. Carbon 1(148):80–90. https://doi.org/10.1016/j.carbon.2019.03.043

Maji NC, Chakraborty J (2019) Gram-scale green synthesis of copper nanowire powder for nanofluid applications. ACS Sustain Chem Eng 7(14):12376–12388. https://doi.org/10.1021/acssuschemeng.9b01814

Mata TM, Martins AA, Caetano NS (2010) Microalgae for biodiesel production and other applications: a review. Renew Sustain Energy Rev 14(1):217–232. https://doi.org/10.1016/j.biortech.2017.05.031

McKendry P (2002) Energy production from biomass (part 1): overview of biomass. Biores Technol 83(1):37–46. https://doi.org/10.1016/S0960-8524(01)00118-3

Meng L, Yu X, Niimi T, Nagasawa H, Kanezashi M, Yoshioka T, Tsuru T (2015) Methylcyclohexane dehydrogenation for hydrogen production via a bimodal catalytic membrane reactor. AIChE J 61 (5):1628–1638. https://doi.org/10.1002/aic.14764

Mohammadzadeh M, Bagheri H, Ghader S (2020) Study on extraction and separation of Ni and Zn using [bmim][PF_6] IL as selective extractant from nitric acid solution obtained from zinc plant residue leaching. Arab J Chem 13(6):5821–5831. https://doi.org/10.1016/j.arabjc.2020.04.019

Mori M, Garcia RG, Belleville MP, Paolucci-Jeanjean D, Sanchez J, Lozano P, Vaultier M, Rios G (2005) A new way to conduct enzymatic synthesis in an active membrane using ionic liquids as catalyst support. Catal Today 104(2–4):313–317. https://doi.org/10.1016/j.cattod.2005.03.039

Murmura MA, Cerbelli S, Annesini MC (2017) Modelling and optimization of hydrogen yield in membrane steam reforming reactors. Can J Chem Eng 95(9):1676–1682. https://doi.org/10.1002/cjce.22787

Nagy E (2018) Basic equations of mass transport through a membrane layer. Elsevier. https://doi.org/10.1016/B978-0-12-416025-5.00008-9

Najafpour G (2015) Biochemical engineering and biotechnology. Elsevier. https://doi.org/10.1016/B978-0-444-63357-6.00016-X

Nalawade SP, Picchioni F, Janssen LP (2006) Supercritical carbon dioxide as a green solvent for processing polymer melts: processing aspects and applications. Prog Polym Sci 31(1):19–43. https://doi.org/10.1016/j.progpolymsci.2005.08.002

Nhat PV, Ngo HH, Guo WS, Chang SW, Nguyen DD, Nguyen PD, Bui XT, Zhang XB, Guo JB (2018) Can algae-based technologies be an affordable green process for biofuel production and wastewater remediation? Biores Technol 1(256):491–501. https://doi.org/10.1016/j.biortech.2018.02.031

Nord KE, Haupt D (2005) Reducing the emission of particles from a diesel engine by adding an oxygenate to the fuel. Environ Sci Technol 39(16):6260–6265. https://doi.org/10.1021/es048085h

Nunes SP, Culfaz-Emecen PZ, Ramon GZ, Visser T, Koops GH, Jin W, Ulbricht M (2019) Thinking the future of membranes: Perspectives for advanced and new membrane materials and manufacturing processes. J Membr Sci 23:117761. https://doi.org/10.1016/j.memsci.2019.117761

Oliveira BR, Sanches S, Huertas RM, Crespo MB, Pereira VJ (2020) Treatment of a real water matrix inoculated with aspergillus fumigatus using a photocatalytic membrane reactor. J Membr Sci 15(598):117788. https://doi.org/10.1016/j.memsci.2019.117788

Oosterhout SD, Savikhin V, Burgers MA, Zhang J, Zhang Y, Marder SR, Bazan GC, Toney MF (2018) Absence of mixed phase in organic photovoltaic active layers facilitates use of green solvent processing. J Phys Chem C 122(20):11136–11144. https://doi.org/10.1021/acs.jpcc.8b01600

Pereira CS, Silva VM, Rodrigues AE (2012) Green fuel production using the permsmbr technology. Ind Eng Chem Res 51(26):8928–8938. https://doi.org/10.1021/ie201951d

Persson M, Bornscheuer UT (2003) Increased stability of an esterase from Bacillus stearothermophilus in ionic liquids as compared to organic solvents. J Mol Catal B Enzym 22(1–2):21–27. https://doi.org/10.1016/S1381-1177(02)00294-1

Pollet P, Davey EA, Ureña-Benavides EE, Eckert CA, Liotta CL (2014) Solvents for sustainable chemical processes. Green Chem 16 (3):1034–1055. https://doi.org/10.1039/C3GC42302F

Pomier E, Delebecque N, Paolucci-Jeanjean D, Pina M, Sarrade S, Rios GM (2007) Effect of working conditions on vegetable oil transformation in an enzymatic reactor combining membrane and supercritical CO_2. J Supercrit Fluids 41(3):380–385. https://doi.org/10.1016/j.supflu.2006.12.010

Prenosil JE, Hediger T (1988) Performance of membrane fixed biocatalyst reactors. I: Membrane reactor systems and modelling. Biotechnol Bioeng 31(9):913–21. https://doi.org/10.1002/bit.260310904

Pu Y, Jiang N, Ragauskas AJ (2007) Ionic liquid as a green solvent for lignin. J Wood Chem Technol 27(1):23–33. https://doi.org/10.1080/02773810701282330

Qureshi N, Ezeji TC (2008) Butanol, 'a superior biofuel'production from agricultural residues (renewable biomass): recent progress in technology. Biofuels Bioprod Biorefin: Innov Sustain Econ 2 (4):319–330. https://doi.org/10.2202/1542-6580.1450

Rafiee A, Khalilpour KR, Milani D, Panahi M (2018) Trends in CO_2 conversion and utilization: a review from process systems perspective. J Environ Chem Eng 6(5):5771–5794. https://doi.org/10.1016/j.jece.2018.08.065

Rao VS, Krishna TV, Mohan TM, Rao PM (2017) Physicochemical properties of green solvent 1-ethyl-3-methylimidazolium tetrafluoroborate with aniline from T = (293.15 to 323.15) K at atmospheric pressure. J Chem Thermodyn 104:150–61. https://doi.org/10.1016/j.jct.2016.09.028

Riaz S, Park SJ (2019) An overview of TiO_2-based photocatalytic membrane reactors for water and wastewater treatments. J Ind Eng Chem. https://doi.org/10.1016/j.jiec.2019.12.021

Ribeirinha P, Abdollahzadeh M, Boaventura M, Mendes A (2017) H_2 production with low carbon content via MSR in packed bed membrane reactors for high-temperature polymeric electrolyte membrane fuel cell. Appl Energy 15(188):409–419. https://doi.org/10.1016/j.apenergy.2016.12.015

Ritchie JT, Richardson JT, Luss D (2001) Ceramic membrane reactor for synthesis gas production. AIChE J 47(9):2092–2101. https://doi.org/10.1002/aic.690470919

Ropel L, Belvèze LS, Aki SN, Stadtherr MA, Brennecke JF (2005) Octanol-water partition coefficients of imidazolium-based ionic liquids. Green Chem 7(2):83–90. https://doi.org/10.1039/B410891D

Samimi F, Rahimpour MR (2018) Direct methanol fuel cell. In: Methanol. Elsevier, pp 381–397. https://doi.org/10.1016/b978-0-444-63903-5.00014-5

Samimi F, Karimipourfard D, Rahimpour MR (2018) Green methanol synthesis process from carbon dioxide via reverse water gas shift reaction in a membrane reactor. Chem Eng Res Des 1(140):44–67. https://doi.org/10.1016/j.cherd.2018.10.001

Samimi F, Hamedi N, Rahimpour MR (2019) Green methanol production process from indirect CO_2 conversion: RWGS reactor versus RWGS membrane reactor. J Environ Chem Eng 7 (1):102813. https://doi.org/10.1016/j.jece.2018.102813

Sathya U, Keerthi P, Nithya M, Balasubramanian N (2020) Development of photochemical integrated submerged membrane bioreactor for textile dyeing wastewater treatment. Environ Geochem Health 25:1–2. https://doi.org/10.1007/s10653-020-00570-x

Saw SZ, Nandong J, Ghosh UK (2018) Optimization of steady-state and dynamic performances of water-gas shift reaction in membrane reactor. Chem Eng Res Des 1(134):36–51. https://doi.org/10.1016/j.cherd.2018.03.045

Schulz M, Kriegel R, Kämpfer A (2011) Assessment of CO_2 stability and oxygen flux of oxygen permeable membranes. J Membr Sci 378 (1–2):10–17. https://doi.org/10.1016/j.memsci.2011.02.037

Sheikhi-Kouhsar M, Bagheri H, Raeissi S (2015) Modeling of ionic liquid + polar solvent mixture molar volumes using a generalized volume translation on the Peng-Robinson equation of state. Fluid Phase Equilib 15(395):51–57. https://doi.org/10.1016/j.fluid.2015.03.005

Sheldon RA (2005) Green solvents for sustainable organic synthesis: state of the art. Green Chem 7(5):267–278. https://doi.org/10.1039/B418069K

Sherbo RS, Kurimoto A, Brown CM, Berlinguette CP (2019) Efficient electrocatalytic hydrogenation with a palladium membrane reactor. J Am Chem Soc 141(19):7815–7821. https://doi.org/10.1021/jacs.9b01442

Shuba ES, Kifle D (2018) Microalgae to biofuels:'Promising'alternative and renewable energy, review. Renew Sustain Energy Rev 1 (81):743–755. https://doi.org/10.1016/j.rser.2017.08.042

Su Z, Luo J, Li X, Pinelo M (2020) Enzyme Membrane Reactors for production of oligosaccharides: a review on the interdependence between enzyme reaction and membrane separation. Sep Purif Technol 3:116840. https://doi.org/10.1016/j.seppur.2020.116840

Tran AT, Jullok N, Meesschaert B, Pinoy L, Van der Bruggen B (2013) Pellet reactor pretreatment: a feasible method to reduce scaling in bipolar membrane electrodialysis. J Colloid Interface Sci 1 (401):107–115. https://doi.org/10.1016/j.jcis.2013.03.036

Ugur Nigiz F, Durmaz Hilmioglu N (2016) Green solvent synthesis from biomass based source by biocatalytic membrane reactor. Int J Energy Res 40(1):71–80. https://doi.org/10.1002/er.3319

Vankelecom IF (2002) Polymeric membranes in catalytic reactors. Chem Rev 102(10):3779–3810. https://doi.org/10.1021/cr0103468

Veismoradi A, Mousavi SM, Taherian M (2019) Decolorization of dye solutions by tyrosinase in enzymatic membrane reactors. J Chem Technol Biotechnol 94(11):3559–3568. https://doi.org/10.1002/jctb.6158

Venezia B, Panariello L, Biri D, Shin J, Damilos S, Radhakrishnan AN, Blackman C, Gavriilidis A (2020) Catalytic teflon AF-2400 membrane reactor with adsorbed ex situ synthesized Pd-based nanoparticles for nitrobenzene hydrogenation. Catal Today. https://doi.org/10.1016/j.cattod.2020.03.062

Vooradi R, Bertran MO, Frauzem R, Anne SB, Gani R (2018) Sustainable chemical processing and energy-carbon dioxide management: review of challenges and opportunities. Chem Eng Res Des 1(131):440–464. https://doi.org/10.1016/j.cherd.2017.12.019

Wahidin S, Idris A, Shaleh SR (2016) Ionic liquid as a promising biobased green solvent in combination with microwave irradiation for direct biodiesel production. Biores Technol 1(206):150–154. https://doi.org/10.1016/j.biortech.2016.01.084

Wang JS, Chiu K (2008) Extraction of chromated copper arsenate from wood wastes using green solvent supercritical carbon dioxide. J Hazard Mater 158(2–3):384–391. https://doi.org/10.1016/j.jhazmat.2008.01.112

Wang H, Cong Y, Yang W (2003) Investigation on the partial oxidation of methane to syngas in a tubular $Ba_{0.5}Sr_{0.5}Co_{0.8}Fe_{0.2}O_{3-\delta}$ membrane reactor. Catalysis Today 82(1–4):157–66.https://doi.org/10.1016/s0920-5861(03)00228-1

Wang H, Feldhoff A, Caro J, Schiestel T, Werth S (2009) Oxygen selective ceramic hollow fiber membranes for partial oxidation of methane. AIChE J 55(10):2657–2664. https://doi.org/10.1002/aic.11856

Wang M, Zhang Y, Yu G, Zhao J, Chen X, Yan F, Li J, Yin Z, He B (2020) Monolayer porphyrin assembled SPSf/PES membrane reactor for degradation of dyes under visible light irradiation coupling with continuous filtration. J Taiwan Inst Chem Eng. https://doi.org/10.1016/j.jtice.2020.02.013

Wei Y, Yang W, Caro J, Wang H (2013) Dense ceramic oxygen permeable membranes and catalytic membrane reactors. Chem Eng J 15(220):185–203. https://doi.org/10.1016/j.cej.2013.01.048

Wolfson A, Dlugy C, Shotland Y (2007) Glycerol as a green solvent for high product yields and selectivities. Environ Chem Lett 5(2):67–71. https://doi.org/10.1007/s10311-006-0080-z

Wu XY, Ghoniem AF (2019) Mixed ionic-electronic conducting (MIEC) membranes for thermochemical reduction of CO_2: a review. Prog Energy Combust Sci 1(74):1–30. https://doi.org/10.1016/j.pecs.2019.04.003

Wu XY, Ghoniem AF, Uddi M (2016) Enhancing co-production of H2 and syngas via water splitting and POM on surface-modified oxygen permeable membranes. AIChE J 62(12):4427–4435. https://doi.org/10.1002/aic.15518

Wu W, Zhang X, Qin L, Li X, Meng Q, Shen C, Zhang G (2020) Enhanced MPBR with polyvinylpyrrolidone-graphene oxide/PVDF hollow fiber membrane for efficient ammonia nitrogen wastewater treatment and high-density Chlorella cultivation. Chem Eng J 1 (379):122368. https://doi.org/10.1016/j.cej.2019.122368

Yuan F, Cui L, Ding P, Jing W (2020) Chlorine-free emission disposal of spent acid etchant in a three-compartment ceramic membrane reactor. Chin J Chem Eng 28(1):271–278. https://doi.org/10.1016/j.cjche.2019.04.026

Yun S, Oyama ST (2011) Correlations in palladium membranes for hydrogen separation: a review. J Membr Sci 375(1–2):28–45. https://doi.org/10.1016/j.memsci.2011.03.057

Zdarta J, Meyer AS, Jesionowski T, Pinelo M (2019) Multi-faceted strategy based on enzyme immobilization with reactant adsorption and membrane technology for biocatalytic removal of pollutants: a critical review. Biotechnol Adv. https://doi.org/10.1016/j.biotechadv.2019.05.007

Zha X, Han S, Wang W, Jiao Z (2019) Experimental measurement and correlation of solubility of ethosuximide in supercritical carbon dioxide. J Chem Thermodyn 1(131):104–110. https://doi.org/10.1016/j.jct.2018.10.032

Zhang Z, Liguori S, Fuerst TF, Way JD, Wolden CA (2019) Efficient ammonia decomposition in a catalytic membrane reactor to enable hydrogen storage and utilization. ACS Sustain Chem Eng 7 (6):5975–5985. https://doi.org/10.1021/acssuschemeng.8b06065

Zhang S, Zhao Y, Yang K, Liu W, Xu Y, Liang P, Zhang X, Huang X (2020) Versatile zero valent iron applied in anaerobic membrane reactor for treating municipal wastewater: performances and mechanisms. Chem Eng J 15(382):123000. https://doi.org/10.1016/j.cej.2019.123000

Zhao Z, Liu J, Jiang J (2020) Dipeptide membranes for CO_2 separation: a molecular simulation study. Fluid Phase Equilib 20:112570. https://doi.org/10.1016/j.fluid.2020.112570

Zhu X, Wang H, Cong Y, Yang W (2006) Partial oxidation of methane to syngas in $BaCe_{0.15}Fe_{0.85}\ O_{3-\delta}$ membrane reactors. Catalysis Letters 111(3–4):179–85. https://doi.org/10.1007/s10562-006-0145-4

Zhu J, Guo S, Liu G, Liu Z, Zhang Z, Jin W (2015) A robust mixed-conducting multichannel hollow fiber membrane reactor. AIChE J 61(8):2592–2599. https://doi.org/10.1002/aic.14835

Application of Membrane in Reaction Engineering for Green Synthesis

Ahmad Mukhtar, Sidra Saqib, Sami Ullah, Muhammad Sagir, M. B. Tahir, Abid Mahmood, Abdullah G. Al-Sehemi, Muhammad Ali Assiri, Muhammad Ibrahim, and Syed Ejaz Hussain Mehdi

Abstract

Catalytic membrane reactors (CMRs) are predictable to develop a green and maintainable skill in chemical engineering which synergistically conducts separations and reacts. The use of a ceramic membrane is commonly known in CMR as it enables all temperature and chemical reaction and separation to be carried out in severe environments. Within this segment, a detailed description and review of various applications adopt specific definitions and principles for membrane reactor operation. Such programs are classified according to the items addressed. The essay explores and assesses from the present viewpoint the ability of the different definitions. Because production rates are precise diverse and in general at actual poor technology readiness levels (TRL), they have not joined a list with admiration for price effectiveness, operation quality, and applicability. To facilitate more oriented research and growth, however, it is important to demonstrate a single or limited number of membrane reactors on a market-applicable size and expense.

Keywords

Membrane • Reactors • Green synthesis • Hydrogen • Syngas • Ammonia

1 Introduction

The petrochemical and biochemical industries are important actors in the world economy, and chemical processes are crucial to both reactions and separations. However, particularly energy-intensive methods are the most modern methods of industrial separation, such as distillation. To address environmental and energy problems, the enhancement in efficacy in reactions and separations is thus becoming important. Membrane division is a large-scale and effective alternative splitting strategy, successful both in energy costs and in total costs (Dixon 2003; Marcano and Tsotsis 2002; Thursfield et al. 2012; Dong et al. 2011). The assimilation of reaction (mostly catalytic reaction) with the split-up of desirable goods has drawn substantial interest from science and engineering investigators. The idea of CMRs is the membrane-dependent synthesis of separations with catalytic reactions (Mushtaq et al. 2014a; Sagir et al. 2014a).

The systematic elimination of the drug, solidification (such as the solidified catalyst), delivery of a catalyst, and supporting catalysts (in some situations, the mucosa itself serves as a catalyst) are the most symbolic features of CMRs. The CMR not only blends membrane split-up with a catalytic procedure; rather, the two processes establish a combination that integrally connects them into one package. Climate, sustainable chemistry, and chemical engineering can be accomplished by CMR implementation with lower

A. Mukhtar
Department of Chemical Engineering, Universiti Teknologi PETRONAS, Bandar, 32610 Seri Iskandar, Perak, Malaysia

S. Saqib
Department of Chemical Engineering, COMSATS University Islamabad, Lahore Campus, Lahore, 54000, Pakistan

S. Ullah · A. G. Al-Sehemi · M. A. Assiri
Department of Chemistry, College of Science, King Khalid University, P. O. Box 9004 Abha, 61413, Saudi Arabia

M. Sagir (✉)
Department of Chemical Engineering, Khwaja fareed university of engineering and information technology, Rahim Yar khan, 50700, Pakistan
e-mail: m.sagir@uog.edu.pk

M. B. Tahir
Department of Physics, Khwaja fareed university of engineering and information technology, Rahim Yar khan, 50700, Pakistan

A. Mahmood · M. Ibrahim
School of Environmental Sciences and Engineering, Government College University, Faisalabad, 38000, Pakistan

S. E. H. Mehdi
Institute of Soil and Environmental Sciences, University of Agriculture Faisalabad, Faisalabad, 38040, Pakistan

Inamuddin et al. (eds.), *Advances in Green Synthesis*, Advances in Science, Technology & Innovation,
https://doi.org/10.1007/978-3-030-67884-5_8

energy usage, lower emissions, and improved efficiency. Since most catalytic reactions occur in harsh environments, including elevated temperatures, extreme pressure, and corrosive gasses or solutions, most CMRs occur utilizing inorganic membranes (both simple solutions and acidic ones). Such inorganic membranes are usually pottery membranes (e.g., metal oxides) and have an apparent advantage compared to the polymeric membranes in terms of chemistry and thermal resilience, tolerance to foulage, mechanical force, and life span. Such advantages have given a broad application in CMRs to inorganic membranes, such as ceramic membranes. The thick ceramic membrane, a form of a gas separation membrane, is used to support one form of CMR. Mixed ionic electronic membranes of perovskite form are among the most rigid ceramic membranes tested (Dixon 2003; Marcano and Tsotsis 2002; Thursfield et al. 2012; Dong et al. 2011). Such a membrane has a general configuration of ABO_3, where A representing a lanthanide, an earth-alkaline part, or a combination of both, however, B is usually an agent of transformation. The characteristics of the membranes of perovskite are closely related to and composed of the cations of A- and B-site. Such membranes have ionic oxygen, and electrical conductivities concurrently and technically have a 100% selectivity of oxygen at high temperatures (normally more than 700 °C). The higher oxygen flux (perm selectivity) and the higher catalytic ability of these membranes reflect the most attractive qualities. Many big catalytic gas-phase processes can be carried out on perovskite-type membranes in CMRs on perovskite-type membranes such as natural gas extraction, hydrogen processing, and absorption of greenhouse gases (Sagir et al. 2014b, 2016).

These CMRs have, therefore, often been investigated in the last ten years. For instance, Dixon (2003), Bouwmeester (2003), Yang et al. (2005), Marcano and Tsotsis (2002), Thursfield et al. (2012), Liu et al. (2006), Wei et al. (2013), and Dong et al. (2011) were among the examples of such reviews. Furthermore, numerous outstanding reports of mixed membrane materials have been published. The authoritative references to content theory and fundamental research are a chapter published by Bouwmeester and Gellings (1997) in 1997 and a study article by Sunarso et al. (2008) in 2008. A permeableclay membrane is used for the other type of CMRs, especially in heterogeneous cataloging procedures. A membrane with suitable size will efficiently isolate the catalyst after the reaction slurry in the existence of deferred ultrafine or neuro-sized catalyst. The isolation and delivery of the catalyst may be accomplished concurrently with the correct reactor layout design, which may improve the catalytic reaction as far as selectivity and performance is concerned (Talebian et al. 2015; Shahzad et al. 2018).

Difficult conditions (high temperatures and atmospheres which can reduce or are corrosive) need adequate material stability, and reaction efficiency is the difficulties intricate in the production and activity of these membrane containers. In demand to maximize energy and price performance, specific resources with long-term durability, like a catalyst, must, therefore, be produced (Mushtaq et al. 2014b). The measurable analysis of membrane container power is highly difficult for several criteria that may be added to determine, particularly if a product mixture, such as syngas, is required. Throughout the literature, there was no standardized system developed. Nonetheless, selectivity (for the commodity requested), (1) alteration ratio (of the educt), and (2) yield for the ranking membrane reactors are of interest. The selectivity quantifies the sum of the substance that is required, plus undesired by-products for all products. The alteration rate is proportional to the comparative reactant number, and the yield determines how much of the substance you choose to produce that you may define as selectivity of conversion times. The faradaic quality that represents the dependent of conveyed electrons contributing to the response is often used. In the current-supported modes, this is particularly helpful in measuring the output of a process and in measuring the electricity demand (Azam et al. 2014; Mushtaq et al. 2014c; Sagir et al. 2014c).

We chose to categorize the different processes according to the goods we needed. Therefore, we have selected many chemicals that can be synthesized in clay membrane containers that lead to ions (Dong et al. 2011; Czuprat and Jiang 2011; Hashim et al. 2011) and concentrate on.

2 Applications of Membrane Reactors in Reaction Engineering

2.1 Syngas Production

Syngas is a combination of H_2 and CO formed synthetically. It may be more refined into NH_3 through a Haber–Bosch cycle or transformed into man-made fuels by Fischer–Tropsch. This is the most valuable intermediary commodity in the chemical industrial sector. Syngas is also a process in turning regular gas or coal into liquid fuel, typically formed from restricted oil resources (Wilhelm et al. 2001), in the application of gas to liquid goods. The specific H2/CO ratios of DME (dimethyl ether), 2/1 for liquid fuel production, and 3/1 for more processing of the CO for ammonia (Aasberg-Petersen et al. 2011) are required depending on the final product. Ceramic membranes had the benefit of being extremely thermochemical in content, oxygen supply regulation, and overall oxidation, and coke forming decreased. Incomplete oxidation, steam restructuring, reversing the water–gas shift, and dry methane reforming will create syngas. Specific hydrocarbons, degradation, water, and CO_2 comprise the primary causes. Strong materials like wood or biomass (Åberg et al. 2015) and coke furnace gas are potential feedstocks (Razzaq et al. 2013). The reactions just need to be triggered thermally. Heat may be generated by certain additives, including industrial waste power or solar thermal energy through a renewable process (Agrafiotis et al. 2014). Electrical power input is not required. Various methods of heat reforming in which methane and heat are normally transferred to catalytic reactors are traditional methods to manufacture syngas on an industrial basis. The processes and necessary catalysts are outlined in Aasberg-Petersen et al. (2011). Catalyst deactivation and coke forming (Iaquaniello et al. 2015) reflect much of the problems in traditional process paths. Also explored in combination fixed and membrane processes are the possibilities for price savings even in well-recognized large-scale manufacturing (Usachev et al. 2011).

2.2 Hydrogen Production

H_2 is a potential upcoming energy transporter and is now a significant chemical raw material, for instance in ammonia synthesis and liquid fuel processing. For stationary and mobile fuel cell systems, it can be turned into energy very quickly and directly burning without toxic pollution for potential gas turbine projects. H_2 is a chief constituent of power-to-gas designs that store changing lunar or wind power in a chemical system, generating mainly through electrolysis a gas-based powertransporter (Gahleitner 2013; Yilmaz et al. 2015). Methane and other hydrocarbons are the conventional methods of processing of hydrogen. Hydrogen, CO_2, and CO (water–gas change reaction, autothermal reforming) must be isolated during this phase to produce an extremely pure substance, so this raises the difficulty and expense of the method.

This classic approach relies mainly on fossil fuels (Lu and Xie 2016), which generate large emissions of CO_2. Important CO_2 control may be the way to rising the total pollution by utilizing fossil fuels (Voldsund et al. 2016). Storage space and user preferences for the future are still required. In this context, the utilization of biogas as a crude material is an enticing alternate (Hajjaji et al. 2016). Suleman et al. (2016) have demonstrated that sustainable hydrogen processing has a far lower environmental effect than utilizing conventional fuels. The development of hydrogen employing mixed conductive ceramic membrane reactors is accomplished by three reactions: the separation of energy, the reaction to vapor–gas changes, and autothermic change. As the membrane material is fundamentally aggressive, no electric power input is needed. The thermal activation of the lattice dispersal of the charged species allows transportation through the membrane suitable at reaction temperature, as already described (Sagir et al. 2014d; Ullah et al. 2015; Mushtaq et al. 2015; Talebian et al. 2018).

Varied oxygen particle electron conductors just as varied proton–electron conductors might be utilized for the creation of hydrogen in a clay layer reactor. The 100% hydrogen selectivity of thick permeable ceramic hydrogen membranes is a benefit. Consequently, further purification steps should be dispensed for the following operations. In the (Sammells and Mundschau 2006; Fontaine et al. 2008; Kreuer 2003; Ivanova et al. 2013, 2016) work, there are suitable membrane materials with perovskite and fluoric structures. Their stability, particularly in acidic environments, i.e., should be given special attention. While several perovskite materials are carbonated, thermal as well as chemical stability materials have been recorded at temperatures over 1000 °C (Ivanova et al. 2013; Sagir et al. 2018; Ullah et al. 2019a). Dual-phase technologies may also be used to further improve the durability and efficiency of products. These materials may be cer-met or cer-cer and thus incorporate a strictly ionic with a virtuously electronic step of conduction (Ullah et al. 2019b; Tahir et al. 2019; Ramasamy et al. 2016; Rosensteel et al. 2016; Rebollo et al. 2015). In this study, the emphasis is not on existing hydrogen processing methods (e.g., electrical electrolysis), while they provide exciting potentials such as a methane reformer or an incomplete hydrocarbon oxidizing (Athanassiou et al. 2007). For more methods, the reader is referred to Lucas-Consuegra et al. (2014), Liu et al. (2002), Zhu et al. (2016).

2.3 CO_2 Thermal Decomposition

A possible approach to carbon dioxide recovery and utilization is called the thermal decay of CO into O_2. Nevertheless, the thermodynamic equilibrium prevents the decomposition of carbon dioxide. High-intensity energy sources including an extremely high temperature (>1727 °C) are needed in an implementation container to achieve a good conversion. Incorporating the TDCD and POM responses in a thick CMR suggests a substantial improvement in carbon dioxide used for the production of POM reaction oxygen (Jin et al. 2008, 2006; Zhang et al. 2007, 2009, 2014). Within the sight of an upheld palladium(Pd) catalyst, the TDCD responses happen toward one side of the layer, and methane interacts with oxygen (the oxygen that penetrates from the TDCD) on the other end, through a maintained nickel catalyst (Sagir and Talebian 2020; Sagir et al. 2020). The selectivity of carbon monoxide and CO_2 oxidation at 900 °C was 100%, while 15.8%, respectively (Jin et al. 2008). The decomposition of carbon dioxide is favorable to the improved oxygen permeation density under environmental conditions. The flow of oxygen in the cycle will typically be facilitated by raising the membrane width (i.e., where concentrations are calculated by the diffusion of bulk). Zhang et al. (2009) rendered coupling reactions with a smaller thickness of a thin tubular SCFA membrane that permitted the flux of oxygen to increase above the disk-like layer. At 950 °C, the transfer of CO_2 exceeded around 17%, which is better than the transfer accomplished with the same operating temperature by utilizing a disk-like membrane. POMs can enhance the driving energy and facilitate CO_2 decay from the opposite end of the membrane to the TDCD. In this situation, however, the membrane was probably in a somewhat more complicated setting, because on one side CO_2/CO was exposed while on the other hand CH_4/CO/H_2 was exposed. As stated in the earlier segment, a balance is required in a membrane reactor among high oxygen penetrability and adequate chemical constancy. Thus, the TDCD and POM link responses were formulated as a composite system of three layers (porous/dense/porous). The decrease resistance, carbon dioxide tolerance, and strong penetrability functions were divided into the layers of, respectively, SBFM, LSM-YSZ, and SCFNb. This architecture is important as each layer has its unique purpose and leads synergistically to enhancing stabilization and conversion. The current reactor obtained a carbon reduction of 20.58% at 900 °C and can be run continuously for over 500 h (Zhang et al. 2014).

2.4 Higher Hydrocarbon Production

In the form of higher hydrocarbons (also termed C_{2+} hydrocarbons or literature C_2 hydrocarbons), molecules that are essential intermediate products in chemistry are describable as chains composed of binary or other carbon atoms (e.g., aromatics, olefins, and alkanes). Higher hydrocarbons, such as polymer synthesis (Karakaya and Kee 2016), can often be used in liquid fuels or employed as simple chemicals. Methane can be converted into developed hydrocarbons by Fischer–Tropsch synthesis, but direct alteration will be desirable without the intermediary stage of syngas processing (Alvarez-Galvan et al. 2011). There is, however, no clear conversion process currently necessary to perform in Fischer–Tropsch. Ethane and ethylene are the primary components of the active conversion (Stoukides 2006). There is also an enticing mixture of aromatic compounds.

In general, layer procedures are very encouraging, particularly for energy-efficient separation tasks, for usage in the petrochemical industry (Ravanchi et al. 2009). The ceramic membrane reactors can be equipped for all combined oxygen ion, electron, and proton or neutron conductors. The reaction temperature of catalysts (e.g. 800 °C for direct methane conversion) may be reduced (Karakaya and Kee 2016). The combination of methane, hydrogenation, dehydrogenation, and dehydro-aromatization can be achieved to synthesize greater hydrocarbons using mixed-conducting ceramic membranes (Kirchen et al. 2013). Methane (Wood 2015), oxygen, and alkanes such as ethane are the primary ingredients for the reactions. Often, by adding an external power supply to the membrane (Morejudo et al. 2016), it can increase the efficiency of the reactors. The cutting-edge technology for extremely energy-intensive hydrocarbon processing is the thermal or catalytical splashing of crude petroleum feedstocks (Sadrameli 2015; Fakhroleslam and Sadrameli 2019). Certain potential drugs, such as dimethyl ether (DME) (Farsi et al. 2016; Azizi et al. 2014; Saeidi et al. 2014; Torrente-Murciano et al. 2014; Atsonios et al. 2016), that are not mentioned here are also available.

2.5 Methane Production

Methane is the key natural gas portion and one of the leading energy supplies at present. In several countries worldwide, methane is used for the production of heat and energy is a well-developed distribution network. In certain industrial processes, methanol is often used as a feedstock. The development of ammonia synthesis high pure hydrogen is an example. Throughout the future, the usage of carbon dioxide and surplus wind drive as a chemical drive transporter would be an essential application (Jürgensen et al. 2014). Methane can be processed utilizing CO_2 methanation and mixed-leading clay layers. The primary pollutants are CO_2 and hydrogen from energy, such as water and biomass. Coke oven gas is an effective feedstock. By adding an external power supply on the membrane, the necessary hydrogen can be generated. Catalytical methanation reactors are most commonly run on set catalyst sheets and at temperatures of around 400 °C (Schildhauer and Biollaz 2015) utilizing biomass as a feedstock. The usage of CO_2, which can play a significant part in decreasing overall CO_2 pollution from various chemical activities, is highly essential in generating synthetic methane. Certain potential goods may be produced with CO_2 and H_2 in addition to methane. Saeidi et al. (2014) explore this subject.

2.6 Ammonia Production

Ammonia is one of the world's most widely processed compounds. For fertilizer processing, it is primarily used as an intermediary component. Pharmaceuticals and coolants are other uses. Ammonia may also be considered as a possible storage medium for hydrogen since it comprises three molecular atoms of hydrogen (Klerke et al. 2008).

The storage of ammonia is better than hydrogen oil. At the pressure of a few bar (Giddey et al. 2013), it can be liquefied. The ammonia delivery with the appropriate network is in theory necessary. This may also be used onboard vehicles as a refrigerating medium (Zamfirescu and Dincer 2008). For stationary applications or as an additional control source, ammonia may also be used as a chemical in concrete oxide fuel cells. The key advantage is that it produces no methane and can be run at the end worker without CO_2 (Afif et al. 2016). Releases of CO_2 are based on the process of ammonia production and the energy sources utilized for production (Lan et al. 2012).

The standard route of production is the Haber–Bosch method, established in the early twentieth century (Haber and Oordt 1905). It is a great-pressure method performed at approximately 200–300 bars and up to 500 °C. When utilizing membrane reactors, the reaction resistance to atmospheric conditions will be reduced. This topic has also been addressed by a variety of analysis papers, with more comprehensive details (Amar et al. 2011; Garagounis et al. 2014). Comparison with the traditional method Haber–Bosch (Kugler et al. 2014), estimates for membrane reactor usage have shown that gross energy use can be reduced by 20%. This will be an immense increase, given that the production of ammonia worldwide is about 200 million tons annually (Giddey et al. 2013). Electrochemical ammonia production may be used in the manufacture of ammonia with diverse-leading clay layers. The key pipes are N_2 and H_2, of which the water separation on the membrane will also contain hydrogen. The use of polymer membrane, for example, Nafion, too, is interesting, but will not be more included in the study. Synthesis of electrochemical ammonia is a way of processing at mild temperatures of about 500 °C, ammonia without heavy strain. To carry out electrochemical ammonia synthesis (Skodra and Stoukides 2009), unadulterated proton or O_2 particle conductors required an outside supply of force. Renewable electricity from wind or solar power systems will also be installed.

Different perovskites, for example, barium cerates and fluorites, including cerium oxide with strong proton conductivity are suitable membrane materials. In addition to the pure ionic conduction of ceramics, the potential ammonium discovery may involve varied ionic electronic clay resources or even double-stage cermet, or cer-resources. No reports covering this field are currently accessible, and only the existing method is listed. A ceramic oxygen membrane is used for the synthesis of ammonia to benefit from the active usage of wet nitrogen (Amar et al. 2014; Deibert et al. 2017) or also wet air (Lan et al. 2014) for a reaction. The idea is to use water as a supply of hydrogen. Water spreading reaction happens in theory on the side of the cathode. The residual hydrogen will bind with nitrogen to ammonia when it is absorbed by the membrane, which ensures, there is no need for significant pretreatment. Ammonia forming concentrations are typically less than for proton leading membranes on an order of magnitude. It is therefore important to produce specially formulated catalysts. However, if a mixed conductor is used, a lower permeate P_{O2} is required, which is only feasible through another process, e.g., partial methane oxidation. As the Haber–Bosch NH_3 production method is being used commercially for more than 100 years, detailed work has been carried out on catalysts to enhance the efficiency of reactors of a great scale. Nevertheless, the production of appropriate and specifically developed catalysts is still in the initial stage with regards to the electrochemical ammonia synthesis. Screening experiments have been performed to identify acceptable transition metal catalysts for alternative catalyst materials (Skulason et al. 2012; Abghoui et al. 2015; Abghoui and Skúlasson 2015). Electrochemical ammonia synthesis members are also precious metals such as Ru (Back and Jung 2016).

Future work is expected to increase the production of ammonia to a point that is consistent with the conventional ammonia synthesis method in the field of catalysts. For example, the construction of small-scale ammonia plants may be a potential niche use. This is very complicated, owing to the heavy maintenance rates when current large-scale buildings are demolished.

3 Environmental Impacts

In this situation, there is no main incentive to manufacture a drug, but the emphasis is on decomposing toxic compounds. Environmental considerations may result in the spin-off of a membrane reactor. An illustration here is the nitrogen oxides, which also arise when hydrocarbon fuels are combusted in the soil. Czuprat and Jiang (2011) also suggested an increase of the N_2O conversion from 25% to almost 100% by eliminating oxygen from $BaCo_x–Fe_yZr_zO_{3-\delta}$ (BCFZ, $x + y + z = 1$). As seen in (Kondratenko and Ovsitser 2008), it is often possible to pair them with certain oxygen-intensive reactions such as oxidative ethane dehydrogenating. Catalytic research was carried out by Konsolakis et al. (2015) to create $CuO–CeO_2$ catalysts through a variety of synthesis pathways, which is essential for that reaction. More work is, however, required to improve nitrogen oxide-reduction operating systems.

The decomposition of H_2S, which can also act as a feedstock for the manufacture of hydrogen, is another function if the amount is appropriate. The development of H_2S hydrogen originating in the Black Sea was shown by Ipsakis et al. (2015) and Kraia et al. (2016). $BaZr_{0.85}Y_{0.15}O_{3-\delta}$ membranes proton-conducting is therefore applied successfully.

4 Conclusions and Future Recommendations

There is still the substantial curiosity in this technology despite more than 20 years of work into high-temperature ceramic membrane reactors even though it has not been used in commercial applications. There has been tremendous development, for instance on syngas, but there are other challenges, for example, with the stabilization of products, scalability, and joining and screening of layer constituents in modules. However, the attention in membrane reactor technology in recent years has again increased because of the great number of science papers recorded in this review study, as well as the high performance and subsequent energy-saving potential. Present features of such processes include the usage of CO_2 and heat as well as the combining of diverse reactions. Whether as a simple chemical or as an energy transporter, hydrogen may be used. All the processes of hydrogen separation, water–gas change response, and autothermic reform are encouraging; however, concerning hydrogen flux by layer and particularly engineered catalysts, these also need to be further developed. Higher hydrocarbons including ethylene are essential chemical feedstock. Direct synthesis of the ethylene production cycle by methane coupling will allow significant energy savings and generate a huge added benefit. But, at present, the technology is still at an early stage and the cost and reliability are not ideal, it cannot compare with conventional, high-scale processes. A drop in the deposition of coke caused a significant issue by catalyst deactivation, which is the extreme benefit of dehydrogenation and methane dehydro-aromatization utilizing clay layer containers. In the production of numerous chemical composites, for example, liquid fuel, or in the Haber–Bosch NH_3 manufacture cycle, Syngas is one big intermediate component. Fractional oxidation is probably the best progressive method in the field of clay layer containers as oxygen porous layers have already been produced with high flux and adequate stability. The selection of catalysts is very simple since Ni metal is a quite effective choice.

Manufacturing and upscaling are becoming the next moves toward industrialization. CO_2 methanation incorporates two main objectives: CO_2 utilization and carbon conservation. The method will provide an alternative to water electrolysis energy storage and has the benefit of using current natural gas distribution networks. Ammonia synthesis is one of the oldest processes and one of the most significant. This is well designed and configured for this purpose already. Many processing facilities with ammonia run on an extremely big scale. However, electrochemical ammonia production has several significant benefits concerning feedstock and health materials. Water may serve as a hydrogen source without strain or carbon raw materials. At present, however, the levels of ammonia synthesis are too small to deal effectively with large plants. This can be addressed by focusing on small-scale development of ammonia. Besides, membrane reactors allow the combination of various reactions so that synergies between them can be effectively produced. This combination will affect thermal administration and the distribution of reactants positively. The bulk of literature findings are focused on studies of laboratory size. Improved performance and reliability have become and remained a significant problem for all of the works studied. An efficient process with an established film material or layer worked with long-term steadiness for a (computing) optimized method has not been effectively engineered as planned. To connect a film reactor to the current procedure with no addition, no additional output and cost–benefit will be generated. Transdisciplinary work is therefore utterly necessary. A limited number of successful methods have to be established for the introduction of the first test case, in which flexible and lucrative components, part designs, and development measures are required immediately. A great deal of advancement in material science, including joining and sealing technologies, has been illustrated in this chapter, so that the analysts have a positive perspective on this technology. However, this issue would have to be resolved through substantial efforts (and time).

Acknowledgements The authors gratefully acknowledge the Departments at their respective universities for providing state-of-the-art research facilities.

References

Aasberg-Petersen K, Dybkjær I, Ovesen C, Schjødt N, Sehested J, Thomsen S (2011) Natural gas to synthesis gas–catalysts and catalytic processes. J Nat Gas Sci Eng 3:423–459

Åberg K, Pommer L, Nordin A (2015) Syngas production by combined biomass gasification and in situ biogas reforming. Energy Fuels 29:3725–3731

Abghoui Y, Skúlasson E (2015) Transition metal nitride catalysts for electrochemical reduction of nitrogen to ammonia at ambient conditions. Procedia Comput Sci 51:1897–1906

Abghoui Y, Garden AL, Hlynsson VF, Björgvinsdóttir S, Ólafsdóttir H, Skúlason E (2015) Enabling electrochemical reduction of nitrogen to ammonia at ambient conditions through rational catalyst design. Phys Chem Chem Phys 17:4909–4918

Afif A, Radenahmad N, Cheok Q, Shams S, Kim JH, Azad AK (2016) Ammonia-fed fuel cells: a comprehensive review. Renew Sustain Energy Rev 60:822–835

Agrafiotis C, von Storch H, Roeb M, Sattler C (2014) Solar thermal reforming of methane feedstocks for hydrogen and syngas production—A review. Renew Sustain Energy Rev 29:656–682

Alvarez-Galvan M, Mota N, Ojeda M, Rojas S, Navarro R, Fierro J (2011) Direct methane conversion routes to chemicals and fuels. Catal Today 171:15–23

Amar IA, Lan R, Petit CT, Tao S (2011) Solid-state electrochemical synthesis of ammonia: a review. J Solid State Electrochem 15:1845

Amar IA, Petit CT, Mann G, Lan R, Skabara PJ, Tao S (2014) Electrochemical synthesis of ammonia from N_2 and H_2O based on $(Li,Na,K)_2CO_3$-$Ce_{0.8}Gd_{0.18}Ca_{0.02}O_{2-\delta}$ composite electrolyte and $CoFe_2O_4$ cathode. Int J Hydrogen Energy 39:4322–4330

Athanassiou C, Pekridis G, Kaklidis N, Kalimeri K, Vartzoka S, Marnellos G (2007) Hydrogen production in solid electrolyte membrane reactors (SEMRs). Int J Hydrogen Energy 32:38–54

Atsonios K, Panopoulos KD, Kakaras E (2016) Thermocatalytic CO_2 hydrogenation for methanol and ethanol production: Process improvements. Int J Hydrogen Energy 41:792–806

Azizi Z, Rezaeimanesh M, Tohidian T, Rahimpour MR (2014) Dimethyl ether: a review of technologies and production challenges. Chem Eng Process 82:150–172

Azam MR, Tan IM, Ismail L, Mushtaq M, Nadeem M, Sagir M (2014) Kinetics and equilibria of synthesized anionic surfactant onto berea sandstone. J Dispersion Sci Technol 35(2):223–230

Back S, Jung Y (2016) On the mechanism of electrochemical ammonia synthesis on the Ru catalyst. Phys Chem Chem Phys 18:9161–9166

Bouwmeester HJ (2003) Dense ceramic membranes for methane conversion. Catal Today 82:141–150

Bouwmeester H, Gellings PJ (1997) The CRC handbook of solid state electrochemistry

Czuprat O, Jiang H (2011) Catalytic membrane reactors-chemical upgrading and pollution control. Chemie Ingenieur Tecknik 83:2219–2228

de Lucas-Consuegra A, Gutiérrez-Guerra N, Caravaca A, Serrano-Ruiz J, Valverde J (2014) Coupling catalysis and electrocatalysis for hydrogen production in a solid electrolyte membrane reactor. Appl Catal A 483:25–30

Deibert W, Ivanova ME, Baumann S, Guillon O, Meulenberg WA (2017) Ion-conducting ceramic membrane reactors for high-temperature applications. J Membr Sci 543:79–97

Dixon G (2003) Recent research in catalytic inorganic membrane reactors. Int J Chem Reactor Eng 1

Dong X, Jin W, Xu N, Li K (2011) Dense ceramic catalytic membranes and membrane reactors for energy and environmental applications. Chem Commun 47:10886–10902

Fakhroleslam M, Sadrameli SM (2019) Thermal/catalytic cracking of hydrocarbons for the production of olefins; a state-of-the-art review III: process modeling and simulation. Fuel 252:553–566

Farsi M, Sani AH, Riasatian P (2016) Modeling and operability of DME production from syngas in a dual membrane reactor. Chem Eng Res Des 112:190–198

Fontaine ML, Norby T, Larring Y, Grande T, Bredesen R (2008) Oxygen and hydrogen separation membranes based on dense ceramic conductors. Membr Sci Technol 13:401–458

Gahleitner G (2013) Hydrogen from renewable electricity: an international review of power-to-gas pilot plants for stationary applications. Int J Hydrogen Energy 38:2039–2061

Garagounis I, Kyriakou V, Skodra A, Vasileiou E, Stoukides M (2014) Electrochemical synthesis of ammonia in solid electrolyte cells. Front Energy Res 2:1

Giddey S, Badwal S, Kulkarni A (2013) Review of electrochemical ammonia production technologies and materials. Int J Hydrogen Energy 38:14576–14594

Haber F, Van Oordt G (1905) Über die bildung von ammoniak den elementen. Zeitschrift für Anorganische Chemie 44:341–378

Hajjaji N, Martinez S, Trably E, Steyer J-P, Helias A (2016) Life cycle assessment of hydrogen production from biogas reforming. Int J Hydrogen Energy 41: 6064–6075

Hashim SM, Mohamed AR, Bhatia S (2011) Catalytic inorganic membrane reactors: present research and future prospects. Rev Chem Eng 27:157–178

Iaquaniello G, Salladini A, Palo E, Centi G (2015) Catalytic partial oxidation coupled with membrane purification to improve resource and energy efficiency in syngas production. Chemsuschem 8:717–725

Ipsakis D, Kraia T, Marnellos G, Ouzounidou M, Voutetakis S, Dittmeyer R et al (2015) An electrocatalytic membrane-assisted process for hydrogen production from H_2S in Black Sea: preliminary results. Int J Hydrogen Energy 40:7530–7538

Ivanova M, Ricote S, Baumann S, Meulenberg WA, Tietz F, Serra JM et al (2013) Ceramic materials for energy and environmental applications: functionalizing of properties by tailored compositions. In: Doping: properties, mechanisms and applications. Nova Science Publishers, pp 221–276

Ivanova ME, Escolástico S, Balaguer M, Palisaitis J, Sohn YJ Meulenberg WA et al (2016) Hydrogen separation through tailored dual phase membranes with nominal composition $BaCe_{0.8}Eu_{0.2}O_{3-\delta}$: $Ce_{0.8}Y_{0.2}O_{2-\delta}$ at intermediate temperatures. Sci Rep 6:34773

Jin W, Zhang C, Zhang P, Fan Y, Xu N (2006) Thermal decomposition of carbon dioxide coupled with POM in a membrane reactor. AIChE J 52:2545–2550

Jin W, Zhang C, Chang X, Fan Y, Xing W, Xu N (2008) Efficient catalytic decomposition of CO_2 to CO and O_2 over Pd/mixed-conducting oxide catalyst in an oxygen-permeable membrane reactor. Environ Sci Technol 42:3064–3068

Jürgensen L, Ehimen EA, Born J, Holm-Nielsen JB (2014) Utilization of surplus electricity from wind power for dynamic biogas upgrading: Northern Germany case study. Biomass Bioenerg 66:126–132

Karakaya C, Kee RJ (2016) Progress in the direct catalytic conversion of methane to fuels and chemicals. Prog Energy Combust Sci 55:60–97

Kirchen P, Apo DJ, Hunt A, Ghoniem AF (2013) A novel ion transport membrane reactor for fundamental investigations of oxygen permeation and oxy-combustion under reactive flow conditions. Proc Combust Inst 34:3463–3470

Klerke A, Christensen CH, Nørskov JK, Vegge T (2008) Ammonia for hydrogen storage: challenges and opportunities. J Mater Chem 18:2304–2310

Kondratenko EV, Ovsitser O (2008) Catalytic abatement of nitrous oxide coupled with selective production of hydrogen and ethylene. Angew Chem Int Ed 47:3227–3229

Konsolakis M, Carabineiro S, Papista E, Marnellos G, Tavares P, Moreira JA et al (2015) Effect of preparation method on the solid state properties and the deN_2O performance of CuO–CeO_2 oxides. Catal Sci Technol 5:3714–3727

Kraia T, Konsolakis M, Marnellos GE (2016) H_2S in Black Sea: turning an environmental threat to an opportunity for clean H_2 production via an electrochemical membrane reactor. Research progress in H_2S-PROTON project. In: MATEC Web of Conferences, 2016, p. 04002

Kreuer K-D (2003) Proton-conducting oxides. Annu Rev Mater Res 33:333–359

Kugler K, Ohs B, Scholz M, Wessling M (2014) Towards a carbon independent and CO_2-free electrochemical membrane process for NH_3 synthesis. Phys Chem Chem Phys 16:6129–6138

Lan R, Irvine JT, Tao S (2012) Ammonia and related chemicals as potential indirect hydrogen storage materials. Int J Hydrogen Energy 37:1482–1494

Lan R, Alkhazmi KA, Amar IA, Tao S (2014) Synthesis of ammonia directly from wet air at intermediate temperature. Appl Catal B 152:212–217

Liu Z, Li L, Iglesia E (2002) Catalytic pyrolysis of methane on Mo/H-ZSM5 with continuous hydrogen removal by permeation through dense oxide films. Catal Lett 82:175–180

Liu Y, Tan X, Li K (2006) Mixed conducting ceramics for catalytic membrane processing. Catal Rev 48:145–198

Lu N, Xie D (2016) Novel membrane reactor concepts for hydrogen production from hydrocarbons: a review. Int J Chem Reactor Eng 14:1–31

Marcano JGS, Tsotsis TT (2002) Catalytic membranes and membrane reactors: Wiley, Weinheim

Morejudo S, Zanón R, Escolástico S, Yuste-Tirados I, Malerød-Fjeld H, Vestre P et al (2016) Direct conversion of methane to aromatics in a catalytic co-ionic membrane reactor. Science 353:563–566

Mushtaq M, Tan IM, Ismail L, Nadeem M, Sagir M, Azam R, Hashmet R (2014a) Influence of PZC (point of zero charge) on the static adsorption of anionic surfactants on a Malaysian sandstone. J Dispersion Sci Technol 35(3):343–349

Mushtaq M, Tan IM, Ismail L, Lee SYC, Nadeem M, Sagir M (2014b) Oleate ester-derived nonionic surfactants: synthesis and cloud point behavior studies. J Dispersion Sci Technol 35(3):322–328

Mushtaq M, Tan IM, Nadeem M, Devi C, Lee SYC, Sagir M (2014c) A convenient route for the alkoxylation of biodiesel and its influence on cold flow properties. Int J Green Energy 11(3):267–279

Mushtaq M, Tan IM, Rashid U, Sagir M, Mumtaz M (2015) Effect of pH on the static adsorption of foaming surfactants on Malaysian sandstone. Arab J Geosci 8(10):8539–8548

Ramasamy M, Baumann S, Palisaitis J, Schulze-Küppers F, Balaguer M, Kim D et al (2016) Influence of microstructure and surface activation of dual-phase membrane $Ce_{0.8}Gd_{0.2}O_{2-\delta}$–$FeCo_2O_4$ on oxygen permeation. J Am Ceram Soc 99:349–355

Ravanchi MT, Kaghazchi T, Kargari A (2009) Application of membrane separation processes in petrochemical industry: a review. Desalination 235:199–244

Razzaq R, Li C, Zhang S (2013) Coke oven gas: availability, properties, purification, and utilization in China. Fuel 113:287–299

Rebollo E, Mortalò C, Escolástico S, Boldrini S, Barison S, Serra JM et al (2015) Exceptional hydrogen permeation of all-ceramic composite robust membranes based on $BaCe_{0.65}Zr_{0.20}Y_{0.15}O_{3-\delta}$ and Y-or Gd-doped ceria. Energy Environ Sci 8:3675–3686

Rosensteel WA, Ricote S, Sullivan NP (2016) Hydrogen permeation through dense $BaCe_{0.8}Y_{0.2}O_{3-\delta}$–$Ce_{0.8}Y_{0.2}O_{2-\delta}$ composite-ceramic hydrogen separation membranes. Int J Hydrogen Energy 41:2598–2606

Sadrameli S (2015) Thermal/catalytic cracking of hydrocarbons for the production of olefins: a state-of-the-art review I: thermal cracking review. Fuel 140:102–115

Saeidi S, Amin NAS, Rahimpour MR (2014) Hydrogenation of CO_2 to value-added products—A review and potential future developments. J CO_2 Util 5:66–81

Sagir M, Tan IM, Mushtaq M, Ismail L, Nadeem M, Azam MR (2014a) Synthesis of a new CO_2 philic surfactant for enhanced oil recovery applications. J dispersion Sci Technol 35(5):647–654

Sagir M, Tan IM, Mushtaq M, Ismail L, Nadeem M, Azam MR, Hashmet MR (2014b) Novel surfactant for the reduction of CO_2/brine interfacial tension. J Dispersion Sci Technol 35(3): 463–470

Sagir M, Tan IM, Mushtaq M, Nadeem M (2014c) CO_2 mobility and CO_2/brine interfacial tension reduction by using a new surfactant for EOR applications. J Dispersion Sci Technol 35(11): 1512–1519 (Taylor & Francis Group)

Sagir M, Tan IM, Mushtaq M, Talebian SH (2014d) FAWAG using CO_2 philic surfactants for CO_2 mobility control for enhanced oil recovery applications. In: SPE Saudi Arabia section technical symposium and exhibition 2014. Society of Petroleum Engineers

Sagir M, Talebian SH (2020) Screening of CO_2-philic surfactants morphology for high temperature-pressure sandstone reservoir conditions. J Petrol Sci Eng 186:106789

Sagir M, Tan IM, Mushtaq M, Pervaiz M, Tahir MS, Shahzad K (2016) CO 2 mobility control using CO_2 philic surfactant for enhanced oil recovery. J Petrol Explor Prod Technol 6(3): 401–407

Sagir M, Mushtaq M, Tahir MB, Tahir MS, Ullah S, Shahzad K, Rashid U (2018) CO_2 foam for enhanced oil recovery (EOR) applications using low adsorption surfactant structure. Arab J Geosci 11(24):789

Sagir M, Mushtaq M, Tahir MS, Tahir MB, Shaik AR (2020) CO_2 philic surfactants, switchable amine-based surfactants and wettability alteration for EOR applications. Surfactants Enhanced Oil Recovery Appl 89–102

Sammells AF, Mundschau MV (2006) Nonporous inorganic membranes: for chemical processing: Wiley

Schildhauer TJ, Biollaz S (2015) Reactors for catalytic methanation in the conversion of biomass to synthetic natural gas (SNG). CHIMIA Int J Chem 69:603–607

Shahzad K, Čuček L, Sagir M, Ali N, Rashid MI, Nazir R, Nizami AS, Al-Turaif HA, Ismail IMI (2018) An ecological feasibility study for developing sustainable street lighting system. J Clean Prod 175: 683–695

Skodra A, Stoukides M (2009) Electrocatalytic synthesis of ammonia from steam and nitrogen at atmospheric pressure. Solid State Ionics 180:1332–1336

Skulason E, Bligaard T, Gudmundsdóttir S, Studt F, Rossmeisl J, Abild-Pedersen F et al (2012) A theoretical evaluation of possible transition metal electro-catalysts for N_2 reduction. Phys Chem Chem Phys 14:1235–1245

Stoukides M (2006) Methane conversion to C 2 hydrocarbons in solid electrolyte membrane reactors. Res Chem Intermediate 32:187–204

Suleman F, Dincer I, Agelin-Chaab M (2016) Comparative impact assessment study of various hydrogen production methods in terms of emissions. Int J Hydrogen Energy 41:8364–8375

Sunarso J, Baumann S, Serra J, Meulenberg W, Liu S, Lin Y et al (2008) Mixed ionic–electronic conducting (MIEC) ceramic-based membranes for oxygen separation. J Membr Sci 320:13–41

Tahir MB, Sagir M, Abas N (2019) Enhanced photocatalytic performance of CdO-WO_3 composite for hydrogen production International. J. Hydrogen Energy 44(45):24690–24697

Talebian SH, Tan IM, Sagir M, Muhammad M (2015) Static and dynamic foam/oil interactions: potential of CO_2-philic surfactants as mobility control agents. J Petrol Sci Eng 135:118–126

Talebian SH, Sagir M, Mumtaz M (2018) An integrated property–performance analysis for CO_2-philic foam-assisted CO_2-enhanced oil recovery. Energy Fuels 32(7):7773–7785

Thursfield A, Murugan A, Franca R, Metcalfe IS (2012) Chemical looping and oxygen permeable ceramic membranes for hydrogen production—A review. Energy Environ Sci 5:7421–7459

Torrente-Murciano L, Mattia D, Jones M, Plucinski P (2014) Formation of hydrocarbons via CO_2 hydrogenation—A thermodynamic study. J CO_2 Util 6:34–39

Ullah S, Suleman H, Tahir MS, Sagir M, Muhammad S, Al-Sehemi AG, Zafar MR, Kareem FAA, Maulud AS, Bustam MA (2019a) Reactive kinetics of carbon dioxide loaded aqueous blend of 2-amino-2-ethyl-1,3-propanediol and piperazine using a pressure drop method. Int J Chem Kin 51(4):291–298

Ullah S, Bustam MA, Assiri MA, Al-Sehemi AG, Sagir M, Kareem FAA, Elkhalifah AE, Mukhtar A, Gonfa G (2019b) Synthesis, and characterization of metal-organic frameworks-177 for static and dynamic adsorption behavior of CO_2 and CH_4. Micropor Mesopor Mater 288:109569

Ullah S, Bustam MA, Ahmad F, Nadeem M, Naz MY, Sagir M, Shariff AM (2015) Synthesis and characterization of melamine formaldehyde resins for decorative paper applications. J Chin Chem Soc 62(2):182–190

Usachev NY, Kharlamov V, Belanova E, Kazakov A, Starostina T, Kanaev A (2011) Conversion of hydrocarbons to synthesis gas: Problems and prospects. Pet Chem 51:96–106

Voldsund M, Jordal K, Anantharaman R (2016) Hydrogen production with CO_2 capture. Int J Hydrogen Energy 41:4969–4992

Wei Y, Yang W, Caro J, Wang H (2013) Dense ceramic oxygen permeable membranes and catalytic membrane reactors. Chem Eng J 220:185–203

Wilhelm D, Simbeck D, Karp A, Dickenson R (2001) Syngas production for gas-to-liquids applications: technologies, issues and outlook. Fuel Process Technol 71:139–148

Wood DA (2015) Conversion of natural gas and gas liquids to methanol, other oxygenates, gasoline components, olefins and other petrochemicals: a collection of published research (2010–2015). J Nat Gas Sci Eng 100:772–779

Yang W, Wang H, Zhu X, Lin L (2005) Development and application of oxygen permeable membrane in selective oxidation of light alkanes. Top Catal 35:155–167

Yilmaz C, Güttel R, Turek T (2015) Zero-emissions power plant for chemical energy storage as well as power and heat generation. Chemie Ingenieur Tecknik 87:419–425

Zamfirescu C, Dincer I (2008) Using ammonia as a sustainable fuel. J Power Sources 185:459–465

Zhang C, Chang X, Fan Y, Jin W, Xu N (2007) Improving performance of a dense membrane reactor for thermal decomposition of CO_2 via surface modification. Ind Eng Chem Res 46:2000–2005

Zhang C, Jin W, Yang C, Xu N (2009) Decomposition of CO_2 coupled with POM in a thin tubular oxygen-permeable membrane reactor. Catal Today 148:298–302

Zhang K, Zhang G, Liu Z, Zhu J, Zhu N, Jin W (2014) Enhanced stability of membrane reactor for thermal decomposition of CO_2 via porous-dense-porous triple-layer composite membrane. J Membr Sci 471:9–15

Zhu Z, Hou J, He W, Liu W (2016) High-performance Ba $(Zr_{0.1}Ce_{0.7}Y_{0.2})O_{3-\delta}$ asymmetrical ceramic membrane with external short circuit for hydrogen separation. J Alloys Compd 660:231–234

Photo-Enzymatic Green Synthesis: The Potential of Combining Photo-Catalysis and Enzymes

Pravin D. Patil, Shamraja S. Nadar, and Deepali T. Marghade

Abstract

In nature, photoautotroph organisms are capable of converting solar energy into chemical energy. Inspired by this phenomenon, natural photosynthesis could serve as an inspiration for the generation of artificial photosynthetic platforms. In particular, connecting photo-catalysis with biocatalysis can provide an efficient biotransformation system, which is highly selective and environmentally benign. In this chapter, the photo-catalytic pathways involved have been discussed in detail. Further, the structural and catalytic mechanisms of enzymes involved in light-driven catalysis are categorized. Moreover, the chapter highlights the incorporation of nanoparticles in the photo-catalytic system as an enzyme activator to make the process efficient and cost-effective. The application of photo-biocatalysis in various biotransformations has been explained with the state-of-the-art examples in halogenations, decarboxylation, and epoxidation. The chapter mainly focuses on the enzymes associated with light-driven catalysis along with the probable mechanism involved. Several applications of enzyme-assisted photo-catalysis while exploring the correlation of factors affecting the overall process are comprehensively explored. Finally, strategies for the large-scale implementation of photobiocatalyst for the production of a variety of chemicals via biological route are briefly discussed.

P. D. Patil (✉)
Department of Basic Science and Humanities, Mukesh Patel School of Technology Management and Engineering, SVKM's NMIMS University, Mumbai, Maharashtra 400056, India

S. S. Nadar
Department of Chemical Engineering, Institute of Chemical Technology Mumbai, Matunga, Mumbai, Maharashtra, 400019, India

D. T. Marghade
Department of Applied Chemistry, Priyadarshini Institute of Engineering and Technology, Nagpur, Maharashtra 440019, India

Keywords

Biocatalysis • Photo-catalysis • Bio-photo-catalysis • Simple and cascadic systems • Photo-enzymatic reduction • Co-factor regeneration • Sustainable chemistry • Green synthesis

1 Introduction

The everlasting exigence for the development of efficient, stable, and greener catalysts possessing high selectivity for the improvement of synthetic chemical processes has led to several challenges and opportunities. In recent years, inspiring from the natural biochemical reactions, scientists have employed natural enzymes with the exquisite selectivity to synthesize the required products (Seel and Gulder 2019). It is worth mentioning that combining enzymatic and chemical catalysts is a bit complex method considering the varying nature of chemical and biological catalysts that make them naturally incompatible with each other. Typically, enzymes are not compatible with harsh reaction conditions, such as high temperatures and pressures (Gacs et al. 2019). However, by employing light, the process can be made compatible with the enzymes. A novel hybrid enzymatic/photo-catalytic approach inspired by the natural photosynthesis process has now emerged as a new challenging field to explore. The combined approach of utilizing solar energy coupled with photo-catalysis and biocatalysis can make the process greener.

The hybrid enzymatic/photo-catalytic approach has been extensively researched out for the synthesis of valued products using several pathways, including hydroxylation, epoxidation, asymmetric reduction, etc. (Lee et al. 2018). The capability of nicotinamide adenine dinucleotide phosphate (NADPH) to liberate electrons and proton makes it a useful co-factor for biocatalytic redox alterations. The Baeyer–Villiger oxidation is primarily useful for the

Inamuddin et al. (eds.), *Advances in Green Synthesis*, Advances in Science, Technology & Innovation,
https://doi.org/10.1007/978-3-030-67884-5_9

synthesis of various food additives and medicines by employing redox enzymes (or oxidoreductases). Torres Pazmiño et al. efficiently developed a novel bifunctional biocatalyst and modified Baeyer–Villiger oxidation with an efficient regeneration of coenzyme having three distinct Baeyer–Villiger monooxygenases and a phosphate dehydrogenase (Torres Pazmiño et al. 2008). This approach yielded chiral intermediates involved in the formulation of bioactive compounds. The production of methanol from CO_2 using formate dehydrogenase (FDH), formaldehyde dehydrogenase (FaDH), and alcohol dehydrogenase (ADH) with an efficient revival of a co-factor (nicotinamide adenine dinucleotide hydride (NADH)) using visible light and TiO_2-based photo-catalysts was effectually demonstrated by Aresta et al. (2014). Similarly, Van Schie et al. used flavo-monooxygenases as biocatalysts for the selective epoxidation of styrene and its derivatives (Schie et al. 2019).

The synthesis of enantiomerically pure amines has become a matter of interest as amines are imperative for the synthesis of drugs and natural products of high values. Gacs et al. developed a sophisticated biocatalytic method for the transition of amine moiety using pyridoxal-5-phosphate (PLP)-dependent ω-transaminases (ω-TAs) for the formation of active pharmaceutical ingredients (Gacs et al. 2019). Zhang and his team successfully exhibited photon-driven biocatalytic decarboxylation of lower members of carboxylic acid series applying photodecarboxylase derived from *Chlorella variabilis* NC64A (CvFAP) (Zhang et al. 2019a). Tremblay et al. investigated a hybrid enzymatic/photo-catalytic approach for photosynthesis using bacterium *Ralstonia eutropha* and water-spitting *g*-C_3N_4-catalase photo-catalyst for bioplastic production where the yield was found to be increased by 1.84-fold (Tremblay et al. 2020). The other fields, including oxidation of thioethers through enantioselective sulfoxidation, halogenations of aromatic compounds using flavoenzymes, oxidative lactonization by horse liver alcohol dehydrogenase (HLADH), oxyfunctionalizations, etc., employing to hybrid enzymatic/photo-catalytic processes are widely studied (Seel and Gulder 2019). After reviewing several cases of photo-catalysis, it can be assumed that the harboring natural enzymes with co-factors having photosensitizing properties of photo-catalysts can boost the selectivity and productivities of various high-valued compounds to the uppermost level.

2 Principle

In the photo-catalytic transformations, the ultraviolet (UV)–visible-light-induced progression of photoexcited electrons and holes in semiconductor (photo-catalyst) with filled valence and empty conductance band plays an important role in boosting the transformation (Tseng et al. 2010). In the last few decades, photooxidation transformations sprouted as sustainable atom-economic alternative methods to combat various environmental issues. The high refractive index, UV absorption, dielectric constant, photo-stability, nontoxicity, and good photo-catalytic activity shoved TiO_2 (anatase form, band gap energy = 3.2 eV) as extensively used catalyst while solving a variety of environmental problems (Khalid et al. 2015). Further, the downsides such as high band gap, surface photo-activation under UV range, petite quantum efficiency, and lower photooxidation rate with TiO_2 were tackled by doping with transition metals and carbon (Pt, Au, Ag, Cr, V, C_3N_4, Nb_2O_5, etc.) (Khalid et al. 2015; Zhou et al. 2006; Torkian and Amereh 2016).

Typically, in enzymatic transformations, the natural enzymes employed as a homogeneous catalyst can facilitate the particular process. Unlike traditional metal catalysts, enzymes do not amend the thermodynamic equilibrium point of reaction and ultimately enhance the rate and feasibility of the reaction. These biological catalysts are found to be efficient due to specific properties of proteins conferred with chemical receptive nature and electrophoretic properties. The enzymatic catalytic processes have wide applicability and been employed since primitive civilizations. Generally, these enzymes-catalyzed reactions proceed through the formation of the enzyme–substrate complex (Bhatia and Bhatia 2018). In co-factor-assisted biocatalysis, the amino acid side chains of protein (enzyme) act as a binding site for distinct functional groups present on the surface of co-factors, which assist the catalysis. The co-factors then bind with the substrate, which provokes a conformational reaction on the active site (Bhatia and Bhatia 2018). Despite several advantages, enzymatic catalysis suffers a few shortcomings of enzymatic transformations, such as limited variability of substrates, less stability, and expensive manufacturing process.

Photo-catalysts have been broadly employed in solar-induced redox chemistry in the recent past. The combined approach of light-driven catalysis and biocatalysis is extensively applied to obtain mass production of desirable products. The studies proved that the hybrid light-driven biocatalysis methods have high competence in terms of sustainability and bulk production. The redox enzymes (or oxidoreductases) were widely utilized to catalyze various oxidation–reduction reactions to produce valuable products through the photon-induced biocatalysis pathway. The generalized mechanism of this redox enzyme-catalyzed photo-biocatalysis process was explained by Lee et al. (2018). In general, electron transference expedites the redox biocatalytic processes. For instance, the catalytic activity of the enzyme (oxidoreductase) hinges on an external source for electron supply, which is utilized in the transformation of the substrate into a desirable product. The most commonly used external source or natural redox equivalent is the nicotinamide adenine dinucleotide (NAD^+) co-factor. The presence of delocalized π electrons in a conjugated system of NAD^+ is responsible for its photo-catalytic activity. NAD^+ is capable of the simultaneous liberation of two electrons and one proton for enzyme-catalyzed reduction and oxidation transformation. On photon impact, these loosely held π electrons get excited to higher energy levels, procuring a reducing power of NAD^+. The liberated electron from photo-catalyst NAD^+ activates the redox enzymes either directly or indirectly through electron mediators (Lee et al. 2018). The transfer of photo-activated electron through photosensitized chemical moiety can stimulate an ample range of redox enzymes for various biocatalytic processes, as reported in the latest research (Gacs et al. 2019; Lee et al. 2018; Kim et al. 2019; Granone et al. 2018).

3 Enzymes Involved in Light-Driven Catalysis

The photo-induced electron can be transferred via direct or indirect pathway using organic (or inorganic) photosensitizers to activate the redox enzyme to catalyze various biotransformation reactions (Schmermund et al. 2019). Mostly, the enzyme from oxidoreductases class is used for catalyzing redox transformation by interchanging electrons between enzyme active site and electron mediator(s). There are two different pathways to activate the redox enzyme, i.e., direct and indirect pathways for the photo-activation of enzymes (Schie et al. 2019). The detailed mechanism of photo-activation of the enzyme is illustrated in Fig. 1 (Lee et al. 2018). In direct photo-activation, photo-induced electron directly swaps between a photosensitizer and biocatalyst, which works as an electrochemical donor and acceptor, simultaneously. On the other hand, in some cases, electron mediator(s) is acting as an intermediate counterpart to facilitate electron transfers between the enzyme active site and photosensitizer (indirect photo-activation) (Lee et al. 2019a, b).

Generally, the FDH and NADH are acting as a coupling agent in photoelectrochemical reactions. When photosensitizers absorb ultraviolet (or visible) light, it triggers the change in the redox states from a ground level to a higher excited level. This induces the delocalization of electrons that extinguishes via oxidation of e^- donor to conduct the reactions (Höfler et al. 2018; Garrone et al. 2015). The enzymes involved in photo-catalysis can be categorized on the basis of prosthetic groups, e.g., heme-containing enzymes (peroxidases, cytochrome P450s, etc.), flavins-containing enzymes (e.g., old yellow enzymes (OYEs), Baeyer–Villiger monooxygenases (BVMOs)), and metal (iron or copper) clusters-containing enzymes (e.g., carbon monoxide dehydrogenases, nitrogenases, and hydrogenases) (Gabruk and Mysliwa-Kurdziel 2015; Hanf et al. 2012). The list of reaction components and photosensitizer is summarized in Table 1.

3.1 Heme-Containing Enzymes

In the biological catalytic system, metal ions (mostly bivalent) play a critical part in electron transfer reactions along with the activation (Shaik et al. 2011; Maurya et al. 2020a). Iron is one of the vital components in biological structures, which generally forms complex, iron protoporphyrin. It is a more versatile part of proteins that functions as electron transfer and signaling (Poulos 2014). In bio-electrochemical applications, heme (iron) can form various reduced and oxidized states during the electron transfer and catalytic reaction. However, heme can exhibit different catalytic as well as functional properties, depending on a proteinaceous environment around the heme molecules (Dawson 1988; Maurya et al. 2020b). This engenders excellent applicability for the heme-containing protein in bio-electrochemical and photo-catalytic reactions. Moreover, the potential (E0′) needed for the transformation between ferrous and ferric states can be altered depending on the protein structure (Dailey 1997; Sono et al. 1996). For example, horseradish peroxidase (HRP) exhibits E0′ of -0.27 V; on the other hand, for cytochrome *c*, it is +0.26 V. There are three heme-containing enzymes such as cytochrome P450s, peroxidases, and peroxygenases, most commonly used in various photochemical reactions (Stiborová et al. 2000).

3.1.1 Cytochrome P450

The cytochrome P450 is the most popular heme-proteins which were thoroughly studied for their electrochemical properties and applications. The family of cytochrome P450s is an illustrative cysteinatoheme-containing enzyme that plays a key role in the metabolic pathways and oxidative transformations of endogenous or exogenous molecules (Ener et al. 2010). Structurally, it is mainly made up of two catalytic domains: flavin-containing and heme-containing reductase domain (Zanger and Schwab 2013). These catalytically active domains catalyze the oxidation of co-factors (mostly, NADPH) to gain electrons, which is further transferred to the reductase domain for photo-catalytic reactions (Urlacher and Girhard 2012). In a typical electron transfer reaction cycle, the heme-containing domain transfers two electrons, which undergoes stepwise reduction. One of the electrons reduces Fe(III) heme to become a Fe(II) state, followed by conversion of molecular oxygen to dioxygen complex. Another electron produces a nucleophilic compound. These two reactions of protonation generate electrophilic and nucleophilic species, which react with the reactant at the active site (Urlacher et al. 2004; Bernhardt

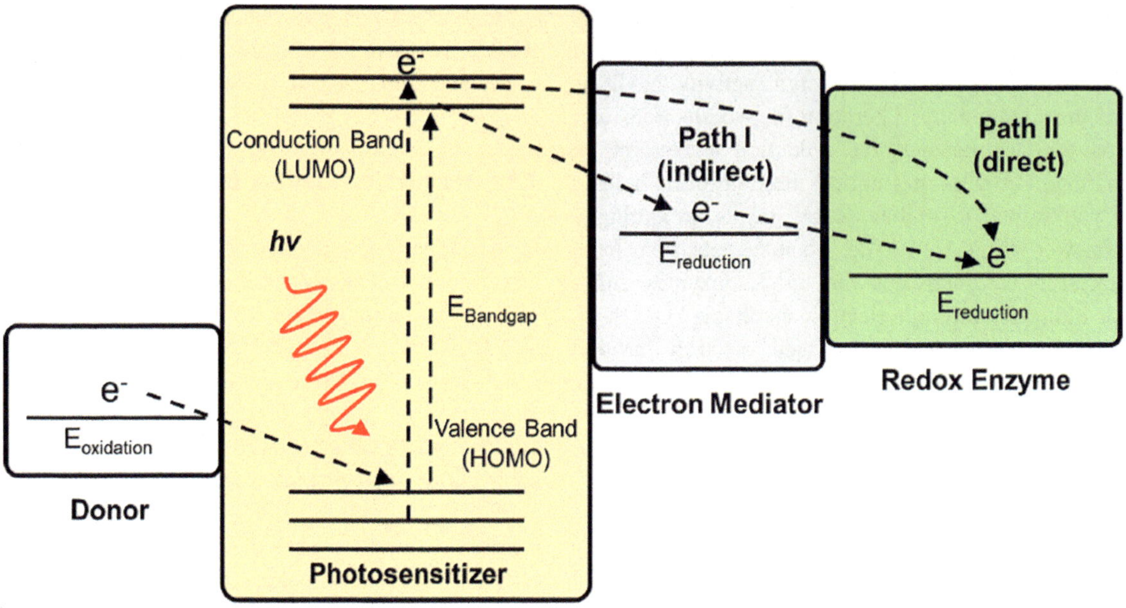

Fig. 1 Schematic illustration of direct and indirect photo-activation of redox enzymes by photo-induced electron transfer (Copyright 2018 John Wiley and Sons. All rights reserved, reprinted with permission) (Lee et al. 2018)

2006; Urlacher and Girhard 2019). The cytochrome P450 is able to catalyze an array of scientifically beneficial biotransformation reactions, such as oxidation, sulfoxidation, decarboxylation, and hydroxylation, and of a wide range of reactant such as fatty acids and antibiotics (Girvan and Munro 2016). Also, cytochrome P450 was employed in various bioremediation processes such as the biodegradation of environmentally harmful compounds such as insecticides, pesticides, and other agrochemicals (Jones et al. 2001; Kellner et al. 1997).

3.1.2 Peroxidases

Peroxidase enzyme is another class of heme-containing protein-containing Fe(III)–heme protoporphyrin that functions as a binding site (Hofrichter et al. 2010). In the catalytic cycle, it liberates two electrons, which are utilized for the reduction of hydrogen peroxide, further pursued by the oxidation of substrate molecules carried through two intermediates (Kalsoom et al. 2015). In most of the catalytic peroxide-based reduction, methylene blue is used as an activator for peroxidase under visible light. In one of the examples, peroxidase reduced oxo-ferryl π-cation and exhibited a multifold greater affinity peroxidase active site (Karmee et al. 2009). In another example, researchers conducted light-driven reactions by peroxidase immobilized onto Pt-doped magnetic films (Kamada et al. 2012). Further, peroxidase was activated by visible-light-induced excitation, which appeared via oxidation of an organic substrate (Chen et al. 2009). In most of the reactions, the photoreactions were initially supplied with peroxide followed by termination accompanied by enzyme deactivation. In some cases, the combination of semiconductor–enzyme offered control of light-driven oxidation reactions without the need for any chemicals (Paul et al. 2014). The enzyme from peroxide family was mostly used in various decolorization and remediation of wastewater containing industrial dyes such as sulfonephthalein dye, brilliant cresyl blue, methylene blue, orange G, azo, and anthraquinone dyes (Husain 2010). Thus, peroxidase-based photo-catalytic treatment will provide an efficient biotechnological process for continuous color and aromatic compound removal from various industrial effluents at large scale (Dutta et al. 2012).

3.2 Flavin-Based Enzyme

In this enzyme family, prosthetic agent flavin molecules are tightly bound to the enzyme to catalyze a wide range of biotransformation reactions through electron transfer (Joosten and Berkel 2007). Generally, the flavin group involves FAD and FMN as a prosthetic component. They can exist in three redox states: (i) oxidized, (ii) one-electron reduced, and (iii) reduced by two electrons (Walsh and Wencewicz 2013). The catalytic reduction of flavin enzymes is shown in Fig. 2. However, the properties of flavins are drastically altered by substitution and noncovalent interactions surrounding protein (Losi 2007; Chaiyen et al. 2012; Romero et al. 2018).

Table 1 A summary of redox enzymatic reactions driven by direct transfer of photo-induced electrons

Enzyme family	Enzyme	Active site/co-factor	Electron source	Photosensitizer	Applications	References
Heme-containing enzyme	Cytochrome P450	Heme	Diethyldithiocarbamate	Ru((OMe)2bpy)2PhenA	Oxidation, sulfoxidation, decarboxylation, hydroxylation	Tran et al. (2013)
	Peroxidases	Heme	Triethanolamine	Eosin Y	Decolorization and remediation of wastewater	Kamada et al. (2012)
Flavin-based enzyme	Baeyer–Villiger monooxygenases	Flavin adenine dinucleotide (FAD)	Ethylenediaminetetraacetic acid (EDTA)	FAD	Biotransformation of mono- and bicyclic ketones	Taglieber et al. (2008)
	Old yellow enzyme	Flavin mononucleotide (FMN)	EDTA, H2O, and triethanolamine	FMN and rose bengal	Trans-hydrogenation of conjugated 306 C=C double bonds	Mifsud et al. (2014a)
Metal cluster-cantered enzyme	Hydrogenases	[FeFe] and [NiFeSe] cluster	Ascorbic acid, EDTA, H2O, and triethanolamine	3-Mercaptopropionic acid capped CdS nanorod and carbon nitride–TiO_2 hybrid	Hydrogen production from water	Adam et al. (2017)

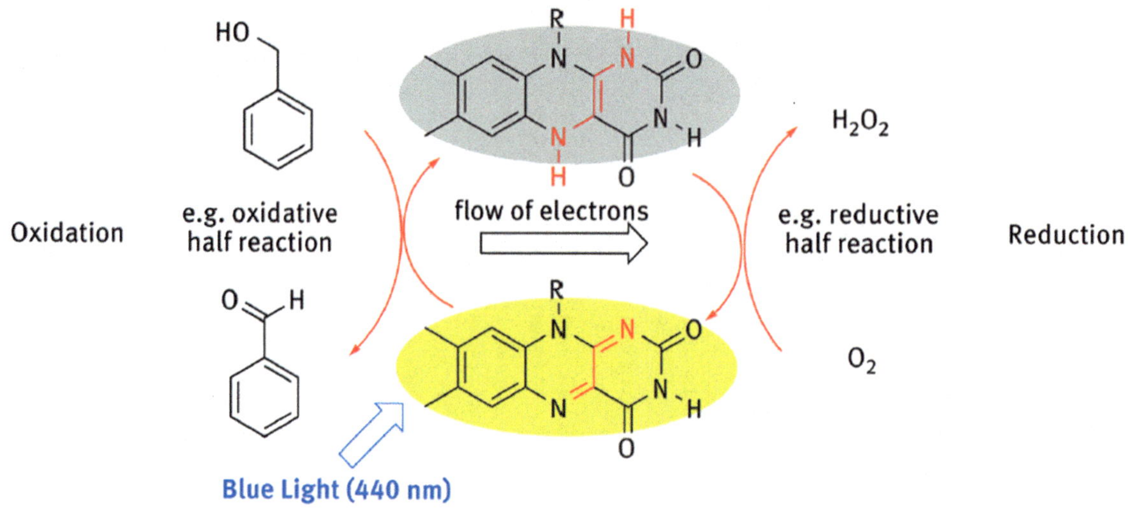

Fig. 2 Catalytic cycle of flavin redox reactions (©2018 Walter de Gruyter GmbH, Berlin/Boston, reprinted with permission) (König et al. 2019)

3.2.1 Baeyer–Villiger Monooxygenases

Flavin-containing monooxygenases signify highly attractive biocatalytic tools due to their unique way of catalyzing wide spectra of monooxygenation reactions with a remarkable selectivity (Malito et al. 2004). Baeyer–Villiger monooxygenases catalyze the insertion of an oxygen atom into a carbon–carbon bond of a carbonylic substrate in the presence of NADPH and molecular oxygen (Berkel et al. 2006). Structurally, they are made up of two-domain architecture; one domain is for FAD-binding, and other domain is for NADPH-binding. These two domains protect the active site within the cleft. Surprisingly, the active form of monooxygenase sequence motif was formed by a surface loop that connects to both domains (Huijbers et al. 2014; Leisch et al. 2011). The active site contains arginine, which was active as a stabilization agent for the peroxyflavin intermediate. Baeyer–Villiger monooxygenases are often utilized in various synthetic chemistry, typically, in oxidation reaction (Pazmino et al. 2010; Gonzalo et al. 2010). The overwhelming diversity of catalytic properties of Baeyer–Villiger monooxygenases permits access to many different classes of valuable chemicals. They were investigated mostly for the conversion of mono- and bicyclic ketones, camphor, a few aryl aliphatic ketones, and some steroids (Balke et al. 2012, 2018).

3.2.2 Old Yellow Enzymes

The enzymatic transformation of stereoselective alkene reduction is catalyzed by enzymes derived from bacteria, fungi, and plants collectively called 'old yellow enzyme' Seel and Gulder 2019). The old yellow enzyme is a single-domain enzyme containing an α/β barrel fold with the molecular weight around 45 kDa. There are two different structural barrels: bottom barrel, which is structurally made up of a β-hairpin, and top barrel containing Si-face for FMN binding and the active site Karplus et al. 1995; Williams et al. 2004). The old yellow enzyme catalyzes the asymmetric reduction of C=C bonds via Michael-type addition by reduction of flavin. Also, in nature, they play a crucial role in numerous routes such as enoyl-CoA reductase in fatty acid biosynthesis, morphine biosynthesis, and the biosynthesis of jasmonic acid (Toogood et al. 2010). One of the research teams revealed direct activation of an old yellow enzyme homolog derived from *B. subtilis* using flavins and EDTA as a photosensitizer and an electron donor, respectively. These components have the potential catalyzing ketoisophorone reduction with TOF of $1.26 \times 104\ h^{-1}$ (Pesic et al. 2017). The biocatalytic reduction of activated CC bonds was mostly carried out by OYEs in synthetic biocatalysis (Winkler et al. 2018). Also, they catalyzed the stereospecific trans-hydrogenation of conjugated C=C double bonds, creating up to two new chiral centers Rauch et al. (1868).

3.3 Metal Cluster-Centered Enzyme

In most of the redox enzymes, metal clusters are a crucial counterpart that shows significant electrochemical and catalytic properties under a biocompatible environment (Evans and Pickett 2003). Among different metal clusters, the iron–sulfur clusters are associated with various biological activities such as electron transfer, catalysis, and gene expressions. The electron transfer properties are depending upon the oxidation state switching between the (II) and (III) and portentous scaffold around the clusters (Johnson et al. 2005). Because of this unique structure, they can control potentials in the range of −500 to +300 mV, and hence they can catalyze the reaction very efficiently (Lill 2009). There are a few well-known metal-centered enzymes involved in photo-catalysis, such as nitrogenases, carbon monoxide dehydrogenases, and hydrogenases that are briefly described in the next sections (Fontecave 2006).

3.3.1 Hydrogenases

Hydrogenases have been intensively studied iron-containing metalloprotein, which can produce hydrogen gas with the reduction of the proton. They can be categorized into various types depending on the combination of the iron cluster with other metal at active sites (Lubitz et al. 2014). Generally, iron can form the clusters with Se and Ni in the form of Fe–S, Fe–Fe, and Fe–Ni (Caputo et al. 2015). In various research works, hydrogenases were activated directly by creating a bridge between organic/inorganic photosensitizers and enzymes via covalent bonding or physical adsorption (Brown et al. 2012; Sakai et al. 2013). Mostly, hydrogenases are very susceptible to molecular oxygen and undergo irreversible structural inactivation (Zhao et al. 2016). Oxygen tolerance is a crucial concern in the exploitation of hydrogenases for the biosynthesis of hydrogen gas. In order to overcome this issue, researchers came up with a platform that provides a favorable environment for the redox enzyme by avoiding contact with molecular oxygen. In another strategy, the hydrogenase was genetically modified, where one of the terminals ligated and substitutes with isosteric selenocysteine (Reisner et al. 2009). The replacement of the selenocysteine helps to demonstrate higher polarizability. The exceptional configuration permits for reducing product inhibition with enhanced overall catalytic properties, and the unavailability of photon resonance-active oxidized state (Adam et al. 2017; Parkin et al. 2008).

3.3.2 Carbon Monoxide Dehydrogenases

The nickel and iron comprising dehydrogenases derived from *Carboxydothermus hydrogenoformans* can catalyze the biotransformation of carbon monoxide and carbon dioxide (Jeoung et al. 2014; Woolerton et al. 2011). There are four different clusters, namely cluster A to cluster D. Each cluster has a different metalloprotein and different functional and catalytic properties. Every monomer comprises two types of clusters: 4Fe–4S (cubane-type, cluster B) and Ni_4Fe–4S (active site, cluster C). Additionally, 4Fe–4S (cluster D), a cluster attached to the dimer interface, covalently links both monomers (Dobbek et al. 2001). One of the research groups validated catalytic reduction of carbon dioxide to carbon monoxide by immobilizing carbon monoxide dehydrogenases, and photosensitizers on different metal oxide nanoparticles made up of zinc, titanium, etc. (Woolerton et al. 2010). During visible light excitation, photosensitizers inject electrons into the metal oxides through the conduction band. The photoexcited electrons access the cluster (D) of dehydrogenase that is delivered to the active site of the enzyme via a second cluster (B) where CO_2 to CO conversion occurred while the holes generated in photosensitizers were quenched by 2-(*N*-morpholino) ethanesulfonic acid, a sacrificial electron donor (Qin et al. 2008). Further, the researchers concluded that the morphology and chemical composition of nanomaterials drastically manipulated photon generation and transportation, which altered the total photo-biocatalytic activity (Roth et al. 2010).

4 Nanoparticle-Based Activation of Enzyme

The utmost prerequisite of photo-biocatalysis was to switch over the photo-catalytic activity from the UV range toward the visible region to make the process more economical and eco-friendly. The other prerequisite of photo-biocatalysis was to regenerate co-factors (e.g., NADH or NADPH) for the activity while making the process efficient and cost-prohibitive. This was achieved by activation of an enzyme using doped or coupled photo-catalyst nanoparticles (Lee et al. 2014). Aprile and the team noted that the photo-catalytic property of TiO_2 was not accelerated simply by doping or alteration in composition (Aprile et al. 2008). Further, it was perceived that by reducing the size of TiO_2 to nanometric scale, a highly ordered titanium was obtained with a large surface area and increased porosity. These morphological modifications upgraded the photo-catalytic characteristic by initiating quantum effects and the constraint of electrons in limited space. In this context, it was experimentally illustrated that the TiO_2 nanotube accompanying a stretched diffusion length of charge carriers unveiled the exceptional photo-catalytic activity. These observations promoted the use of nanoparticles to activate and enhance enzymatic activity. Generally, the nano-biocatalytic processes include nano-scaled photo-catalyst, oxidase reductase enzymes, mediators, and electron donors. It is largely explored by scientists for solar-induced manufacturing of commodity valued chemicals with photochemically co-factor regeneration (Patil and Yadav 2019).

The concept was further employed in stereospecific hydrogenation of ketoisophorone into (*R*)-levodione using impregnated gold nanoparticles (NPs) impregnated on TiO_2 as photo-catalyst along with natural oxidoreductase enzyme and FAD^+ as mediator and co-factor, respectively (Maciá-Agulló et al. 2015). The study signifies that coupling TiO_2 with Au nanoparticles boosted the photo-catalytic properties in visible light range. Further, the transition metal nanoparticles-doped TiO_2 photo-catalyst was extensively utilized in photo-biocatalysis. Lee et al. synthesized silica-coated nanomolecules of $NaYF_4$ doped with mixed Yb and Er or Yb and Tm and successfully employed it along with a photosensitizer named Rose Bengal (RB), redox enzymes, and nicotinamide co-factors (e.g., NADH,

NADPH), in near-infrared (NIR) light range for the synthesis of *L*-glutamate (Lee et al. 2014). The group further reported higher efficiency of RB/Si–$NaYF_4$:Yb, Er nanoparticles, than RB/Si–$NaYF_4$:Yb, Tm nanoparticles for the NIR-induced enzymatic conversion of α-ketoglutarate to *L*-glutamate. This work recommended the use of adaptable usage of NIR light in the production of high-valued chemicals using coupled photo-catalyst nanoparticles.

In textile industries, synthetic azo dyes are widely utilized that produce several toxic organic water pollutants. Azo and team developed a unique hybrid photo-catalytic enzymatic system encompassed of γ-Fe_2O_3 nanoparticles (IONPs) as a photo-catalyst with low band gap (2.2–2.3 eV), azo reductases, FMN as co-factors, NADPH, and 2-(*N*-morpholino) ethanesulfonic acid (MES) as an electron donor (Nehme et al. 2020). The high degree of photoexcited electron–hole recombination hindrance of IONPs photo-catalysts was overcome by fastening coupled catechol moiety and flavin on the surface of IONP photo-catalyst. The IONPs photo-catalysts efficiently degraded azo dye by absorbing visible light via accelerating the activity of azo reductases. Moreover, the flavin-coated iron oxide nanoparticles were found to be convenient to store at room temperature for a longer duration. These studies highlighted the development of efficient visible-light-induced photo-biocatalysis for the production of high-valued chemicals. However, the development of the economic photo-biocatalytic system needs robust co-factors (e.g., NADPH, FMN, FAD, etc.) to attain regeneration competence. Conventionally, regeneration of NADH co-factors has been triumphed using enzymatic recycling by whole-cell extracts, which is uneconomical for large-scale production. In recent years, several co-factor regeneration approaches have been explored (Hollmann et al. 2010).

Brown et al. reported the photo-induced biocatalytic conversion of aldehydes to alcohols using ADH, biohybrid complexes of CdSe quantum dots, and ferredoxin (Brown et al. 2016). The process was found to be efficient for the regeneration of NADPH co-factors. They pragmatically illustrated the proficiency of quantum dots and ferredoxin complexes to regenerate NADPH by employing *Thermoanaerobium brockii* ADH and iso-butyraldehyde. The thrust for low-cost and more eco-friendly process encouraged Choudhury et al. to develop a low-cost photo-biocatalytic system for the asymmetric reduction of prochiral ketones into chiral secondary alcohols using biomass-derived nonmetallic carbonaceous photo-catalysts (Choudhury et al. 2014). The system was comprised of ADHs biocatalyst and graphene-based nano-photo-catalyst with $Cp^*[Rh(bpy)H]^+$ as a metallic mediator and NADPH as a co-factor. This biomimetic endeavor efficiently produced chiral pharmaceutical valued 1-phenyl ethanol and their derivatives (with 64 to 74% conversion rate) along with nicotinamide co-factor (NADPH) regeneration under visible light. This study propelled researchers toward the development of more eco-friendly, cost-effective, and solar light-driven photo-catalytic/biocatalytic cascade production of high-valued chemicals.

5 Applications in Photo-Biocatalysis

Photo-catalysis can copulate either to isolated enzyme/cell lysate containing enzymes, whole-cell system, or artificial enzymes of interest, as depicted in Fig. 3. Moreover, there are multiple synthetic transformation pathways to carry out light-driven enzyme catalytic reactions as illustrated in Fig. 4.

5.1 Isolated Enzymes/Cell Lysates

Reduction of Carbonyl Functional Group

Active sites of the enzyme can facilitate a chiral environment that makes it ideally suitable for enantiospecific catalysis (Vaidya et al. 2020; Patil and Yadav 2018). Several isolated enzymes or cell lysates have been employed in the reduction of carbonyl functionalities. Commercially obtainable *Thermoanaerobium brockii* alcohol dehydrogenase (tbADH) was employed to attain the conversion of aldehydes (Brown et al. 2016). The needed reducing equivalents were regenerated (photochemically) with the help of a photo-catalyst (CdSe QD) and biohybrid complexes of ferredoxin $NADP^+$ reductase (FNR) that are specific to NADPH, from green algae (*Chlamydomonas reinhardtii*). FAD containing FNR can transfer a hydride to $NADP^+$ by corresponding natural photosynthesis where a donation of two electrons occurs. The oxidation of ascorbic acid originates electrons in the process. Similarly, chiral 1-phenylethanols were yielded via enzymatic reduction when ketones were used as substrates (Choudhury et al. 2012). Several researchers in the recent past reported comparable findings (Höfler et al. 2018; Choudhury et al. 2014). Nevertheless, the applicability of FNR as a relay enzyme is still constrained due to its limited specificity concerning phosphorylated nicotinamide co-factors.

Decarboxylation of Carboxylic Acids

Lu et al. extracted CYP102A1 (P450BM3) reductase from *Bacillus megaterium* and coupled it with a cytochrome P450 enzyme (OleTJE) to synthesize self-sufficient protein, OleT-BM3R. It carried out decarboxylation of fatty acids into linear α-olefins. The activity of engineered P450BM3's hydroxylase was noted 1000-folds higher than those of other P450 fatty acid hydroxylases. The reason for the boosted activity is possibly due to high ferricyanide and cytochrome

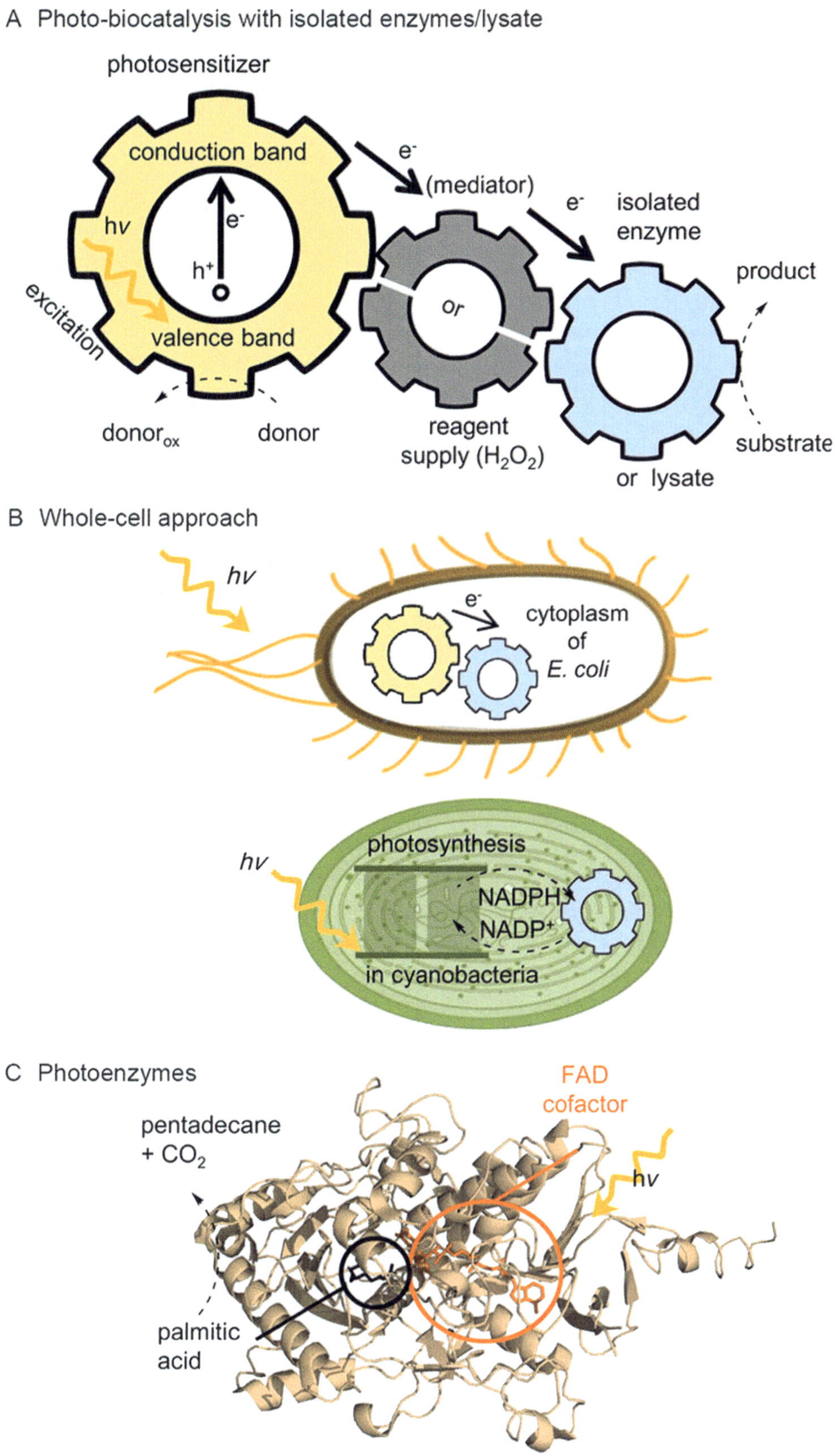

Fig. 3 **a** Principle of light-induced electron abstraction by photosensitizers and electron transfer to redox enzymes (upper pathway) or generation of the oxidant H_2O_2 in situ (lower pathway). **b** Whole-cell photo-biocatalysis. **c** Photoenzymes are light-dependent enzymes catalyzing photochemical reactions (Copyright 2019 John Wiley and Sons. All rights reserved, reprinted with permission) (Seel and Gulder 2019)

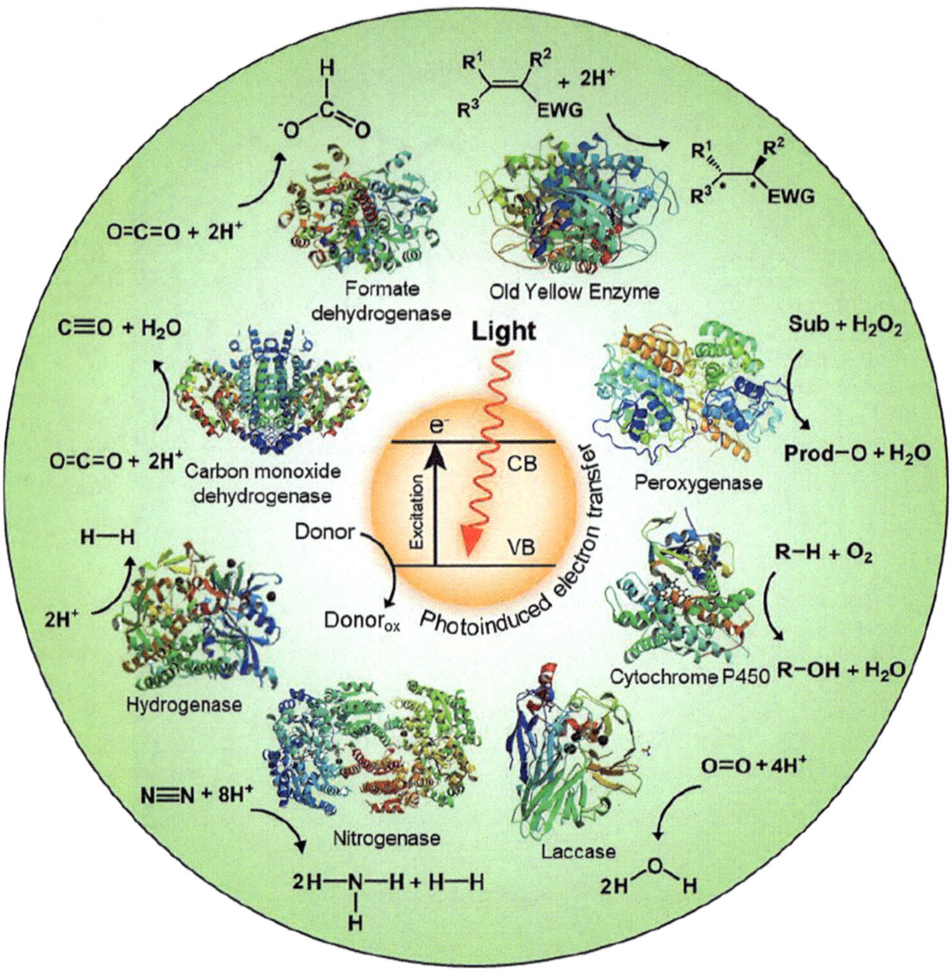

Fig. 4 Multiple synthetic transformation pathways facilitated by the combination of light as an energy source and the catalytic power of enzymes (Copyright 2018 John Wiley and Sons. All rights reserved, reprinted with permission) (Lee et al. 2018)

c reduction rates, which facilitate the rapid electron transfer in reductase (BM3R) (Lu et al. 2018). Furthermore, terminal alkenes can be generated from natural oils via decarboxylation of acids and served as chemical building blocks (feedstocks) in the production of biofuels or polymers. OleTJE enzyme originated from *Jeotgalicoccus* sp. ATCC 8456 was found to be capable of yielding terminal alkenes. The photochemical generation of H_2O_2 (in situ) via reaching the hydrogen peroxide shunt pathway of OleTJE enhanced the overall enzyme activity (Zachos et al. 2015). Recently, Zhang and the team exhumed photodecarboxylase from *Chlorella variabilis* NC64A (CvFAP) that emerged as a promising and efficient bio-photo-catalyst for the transformation of long-chain fatty acids into alkane. The thermodynamically favored catalytic cycle closed up when the flavin adduct intermediate (CH_3–FAD–) protonated to release the alkane product and regenerate the oxidized FAD (Zhang et al. 2019a).

Reduction of Alkenes

Old yellow enzyme (OYE YqjM) derived from *B. subtilis* was regenerated independently of NADPH (Taglieber et al. 2008). Cell lysate holding reaction mixtures was irradiated with white light in the presence of an electron donor (EDTA) and free FMN. It triggered a reduction of an alkene with reasonable enantioselectivity (88% ee). Another report suggested that *E. coli* cell lysate containing endogenous NemA (*N*-ethylmaleimide reductase) might be accountable for the diminished enantioselectivity on the usage of lysates (OYE recombinant) of *E. coli* (Bernard et al. 2012). In another study, a heterogeneous semiconductor (TiO_2-deposited gold NPs) was employed as a photo-catalyst to the water oxidation while coupling it to the ene-reduction using OYE homologue (TsOYE) from *Thermus scotoductus* SA-01 (Mifsud et al. 2014b). The UV light irradiation of ketoisophorone could obtain a 64% yield. Several researchers in the recent past reported similar findings (Peers et al. 2016; Grau et al. 2009; Lee et al. 2017; Kim et al. 2018). The stereo selective *trans*-hydrogenation of activated C=C bonds was carried out by Lee and his team utilizing enoate reductases as bio-photo-catalyst. They reported the transformation of xanthene dyes to the prosthetic flavin moiety mediated through NADPH-free, direct transfer of photoexcited electrons. In this study, the enantioselective reduction of 2-

methylcyclohexanone (C=C bonds) into (*R*)-2-methylcyclohexanone (80–90% yield with 99% enantiopure) was carried out in the absence of NADH under white light-emitting-diode illumination (Lee et al. 2017).

Oxidation Through C, H Activation

Several synthetically challenging transformations such as halogenations, decarboxylations, and oxyfunctionalizations can be catalyzed using oxidative enzymes. Though oxidation in synthetic chemistry is not environmentally friendly due to the extreme reaction conditions, it can naturally perform under mild conditions by employing H_2O_2 as an oxidant or molecular oxygen. The cytochrome P450 is one such enzyme that is competent in triggering selective oxidations of carbon and hydrogen (Monti et al. 2011) to yield alcohols corresponding to the aromatic and inactivated carbon atoms involved. The P450 BM3 (CYP102A1), a mutational tolerant and self-sufficient enzyme obtained from *B. megaterium*, is extensively utilized to achieve the highest catalytic hydroxylation activity (Whitehouse et al. 2012). Further, photo-enzymatically assisted conversion of fatty acid to lauric acid was yielded using a combination of deazaflavin with P450 BM3 under irradiation of visible light (100 W lamp) (Zilly et al. 2009). In another report, oxidation of ethyl group containing C, H occurred following cleavage of the hemiacetal formed in situ where cell-free extracts of P450 BM3 were employed to achieve dealkylation (Lee et al. 2013). Alternatively, hydroxylation through C, H activation can be enabled using peroxygenases that utilize H_2O_2 (Holtmann and Hollmann 2016). The photo-induced H_2O_2 generation approaches coupled with the peroxygenase from *Agrocybe aegerita* (AaeUPO) were studied by Hollmann and the team. The efficiency of the hydroxylation of various alkylbenzenes and alkanes upsurges when photosensitizer FMN and sacrificial electron donor EDTA incorporated in the process (Zhang et al. 2018). In this regard, a few more related reports were investigated (Dong et al. 2018; Churakova et al. 2011; Willot et al. 2019).

Epoxidation of Alkenes and Styrene

Peroxygenase enzymes are also capable of catalyzing epoxidations. Epoxides can be yielded using light-driven H_2O_2 generation in the presence of styrene as substrates (Churakova et al. 2011). It was found that the double bond conformation highly influences the activity/selectivity of epoxidations through enzymatic assistance. In another report, $FADH_2$ prosthetic groups can be reduced with diffusible flavo-monooxygenase from VLB120 strain of *Pseudomonas* sp. and can be favorably regenerated, employing direct excitation of photochemicals using a white light bulb (Schie et al. 2019). In recent decade, two-component, diffusible flavin monooxygenase (2CDFMO) enzyme is considered as pliable biocatalysts for selective epoxidation. Van Schie et al. successfully employed 2CDFMOs as bio-photo-catalysts for selective epoxidation of styrene and its derivatives using reduced form of flavin adenine dinucleotide ($FADH_2$) regenerated by a NADPH-dependent reductase. The accelerated reaction rates and the enantioselectivity of 2CDFMOs bio-photo-catalysts reported 95–99% formation of styrene epoxides (Schie et al. 2019).

Oxidative Lactonizations

In most of the Baeyer–Villiger oxidation research works, BVMOs were employed to transform aldehydes into the corresponding formyl esters. An inventive BVMO from *Dietzia* species is used as catalyst in the formation of profen aldehyde (Bisagni et al. 2014). The phenylacetone monooxygenase (PAMO) derived from *Thermobifida fusca* coupled with FAD and electron donor EDTA boosted the yield enantiomer-differentiating Baeyer–Villiger oxidation up to 93% and enantioselectivity to 92–97% ee (Holtmann and Hollmann 2016). In another biocatalytic approach of Baeyer–Villiger oxidation, lactones can be accessed by employing HLADH that converts α,ω-diols to monoaldehydes in a two-step oxidation reaction. These intermediates maintaining equilibrium with the cyclic hemiacetals can further be converted into lactones where photoexcited flavin is employed to regenerate NAD^+. Further, H_2O_2 was trapped by adding a catalase into the reaction mixture, which is a by-product formed in stoichiometric amounts. Similarly, an efficient photoexcitation of flavin can be achieved using blue light (LED sources) (Schroeder et al. 2018). Employing EDTA as electron reservoirs instead of NADPH in photo-induced Baeyer–Villiger (BV) reaction reduced the complexity of the regeneration system (Hollmann et al. 2007).

Oxidation of Thioethers

Photo-biocatalysis has been employed in enantioselective sulfoxidation reactions. A robust enzyme named chloroperoxidase (CPO) originated from *Caldariomyces fumago* can be used for the transformation where oxidative stress of CPO is minimized via photochemical generation of H_2O_2 based on flavin (Perez et al. 2009). It was noted that the enantioselectivity of the process was highly influenced by the electron donor, where EDTA yielded enantiopure (>99% ee) product (Churakova et al. 2013). Moreover, a surfactant-stabilized (two-liquid-phase system) could further improve the photo-biocatalytic sulfoxidation system.

Halogenations of Aromatic Compounds

Halogen substituents are found to be essential for functionalization, considering their ability to enhance bioactivity of organic compounds. Halogenases (flavin-dependent) can transfer regioselective oxidative halogen to organic compounds where chemical synthesis is constrained. However, wider adoption is limited due to the demand for co-factor regeneration. In a recent report, co-factors such as FAD was efficiently regenerated using a direct transfer of photo-induced electrons (Schroeder et al. 2018). Another class in the category, vanadium-dependent haloperoxidases (VHPOs), a halogenase enzyme that relies on H_2O_2 as an oxidant, was reported for the conversion of brominated products from thymol (Dong et al. 2018; Sabuzi et al. 2015). Further, a new robotic photo-biocatalytic method is implemented by Seel and the team for the chlorination and bromination of hetero-aromatic compounds using vanadium-dependent haloperoxidases derived from *Acaryochloris marina* (AmVHPO) and *Curvularia inaequalis* in combination with FMN. Besides, the application of tertiary amines containing buffers as an external sacrificial electron donor and conducting the experiment under with blue LEDs hugely increased excitation of the heterogeneous photo-catalyst (Seel et al. 2018).

Photo-Biocatalytic Cascade Systems

Synergistic catalysis can drastically enhance the efficiency of photo-biocatalysis, considering that the approach offers access to corresponding reactivity fields while reaching far beyond the regeneration of co-factors. Photo-biocatalytic cascades are mainly categorized into three types as follows; (a) sequential reactions comprising minimum two distinguished but independent steps where the addition of reagents or solvent exchange can take place, (b) concomitant reactions that are being carried out concurrently in the same medium while facilitating minimum two irreversible levels, and (c) collective systems facilitating minimum of two simultaneous reactions in the same medium (often comprising an initial reversible followed by an irreversible step) (Litman et al. 2018). Recent advancements in photo-biocatalytic cascades have enabled the direct conversion of several types of harmful, unstable, and difficult to isolate intermediates. A group of researchers demonstrated the photothiolation of ketones (**α**, **β**-unsaturated) with a subsequent (asymmetric) reduction where a light-triggered substrate was functionalized while employing an enzymatic transformation (Lauder et al. 2018). Similarly, a reaction sequence of photo-biocatalytic cascade inscribed the deviating access to a mixture of diverse carboxyl and carbonyl derivatives and amines derived from a simple source of alkanes (Zhang et al. 2019b). A few more related reports were investigated in the recent past (Turner 2018; Yang et al. 2018). Gacs and his team successfully carried out one-pot sequential photooxidation and reductive amination of racemic alcohols into enantiomeric amines. In this cascade system, acetophenone was synthesized in the first step through oxidation of rac-1-phenylethanol interceded by sodium anthraquinone-2-sulfonate (SAS) and heterogeneous graphitic carbon nitride (g-C_3N_4) as photo-catalysts. The study highlighted that the photooxidation of aromatic substrates facilitated much faster reaction rate with a boosted yield. The reductive amination of intermediate ketone was carried out using transaminase cascade comprised of *Aspergillus terreus* (ATωTA), *Bacillus megaterium* (BMωTA), *Chromobacterium violaceum* (CVωTA), *Pseudomonas fluorescens* (PFωTA), and *Vibrio fluvialis* (VFωTA). The use of a cascade system effectively transforms non-chiral substrates into prochiral products in the high enantiomeric form Gacs et al. (2019).

Whole-Cell Systems

Whole-cell systems can easily outperform isolated and purified enzymes-assisted biocatalytic processes due to their enhanced operational stability, additionally protected by cell environment. Moreover, whole-cell systems facilitate direct regeneration and harness of co-factors by redox-active enzymes. Several phototrophic organisms offer the natural capability to regenerate co-factors using solar power. Some endogenous enzymes originated from wild-type cyanobacteria have been investigated for catalyzing multiple reactions, including asymmetric reductions of carbonyl groups (Nakamura et al. 2000) and functionalization of monoterpenes (Balcerzak et al. 2014). A few more related reports were investigated in the recent past (Górak and Zymańczyk-Duda 2015; Yamanaka et al. 2015; Bartsch et al. 2015). Similarly, a group of researchers observed that eosin Y could penetrate into the cytoplasm of *E. coli* having heme domains of enzymes (P450). Under the illumination of visible light, white light-emitting plates could directly transfer photo-induced electrons (Park et al. 2015).

Artificial (Metallo) Enzymes

The accelerated development in protein engineering expedites the extension of the naturally available catalysts (Nadar et al. 2020). One such approach has been intriguing scientific minds where a blend of chemo-biocatalysis occurs via acquainting unnatural, entirely chemically derived catalytic entities into chiral protein frameworks. Several photosensitizers incorporated with artificial photoenzymes have been investigated in the recent past. Cheruzel et al. attached a photosensitizer (**Ru**(**II**)) to a non-native residue of cysteine, typically positioned near the heme co-factor, facilitating electron transfer between the photoexcited Ru and Fe atoms (Ener et al. 2010). It was observed that the hybrid enzyme was photochemically regenerated while not being dependent on NADPH. In a similar report, the application of hybrid catalysts was further confirmed by implementing them in a photo-catalytic cascade reaction system (Sosa et al. 2018). They coupled an earlier described process for photoredox trifluoromethylation (Beatty et al. 2015) with the enzymatic C, H oxidation. In this study, several hybrid enzymes were examined, all featuring the Ru(II) complex 18c as a photosensitizer component attached to the non-native single cysteine L407C (Lam et al. 2016). Further, for the catalytic alteration of drugs (e.g., simvastatin, lovastatin, omeprazole, etc.) and steroids (e.g., 17β-estradiol), an eosin Y (EY) attached to the photoactivated heme domain of P450 can be employed to facilitate photo-induced electron transfer (Park et al. 2015). The inorganic co-factor iron–sulfur (Fe–S) clusters are a crucial factor for redox enzymes for improved ligand-binding and catalytic properties under ambient conditions. For the direct photo-activation of hydrogenases, a fusion of [FeFe]- or [NiFe]-hydrogenase into PS-I in the presence of alkenedithiols creates a junction between inorganic photosensitizers and hydrogenases (Lubner et al. 2011). In this regard, a few more related reports were investigated in the recent past (Tran et al. 2011, 2013; Kato et al. 2014).

6 Summary and Future Scope

Biotransformation is anticipated to meet the growing demand for fuels and chemicals by sustainable and greener routes. The combination of redox photo-enzymatic in the presence of light can expedite the transformation of photon energy value-added chemicals, which helps to make the synthetic system more eco-friendly. In most of the research, artificial photo-catalysis has been restricted to the production of hydrogen gas and the mitigation of carbon dioxide. A variety of redox oxidoreductase enzymes is available for photo-catalysis; however, a few of them have been evaluated for redox conversion in the presence of light, which hints at enormous prospects for upcoming research work. Also, there is a need to understand the mechanism behind the photo-catalysis to perform various biotransformation processes. There are still certain challenges and major issues in photo-enzymatic catalysis, which must be addressed before wider adoption. In most of the photo-catalytic reactions, the co-factors are required to maintain efficient activities of enzymes. The major hurdles in the use of co-factors are the physical separation and regeneration of co-factor after the reaction. Firstly, photosensitizers are no stable as they are certainly undergoing rapid oxidation and degradation. Secondly, as the photo-catalytic enzymes are homogenous in nature, they are very difficult to separate from the reaction mixture. To overcome these problems, co-factors and enzymes can be immobilization on various matrices, such as magnetic nanoparticles, carbon dots, and other nanomaterials. In another strategy, the role of co-factors and mediators can be eliminated by direct activation of oxidoreductase enzymes in the presence of light. However, the direct photo-activation of enzyme suffers from low turnover activities due to poor electron transfer efficiencies. The efficiency of light-driven biocatalysis can be improved by rapid and easy transfer of photo-induced electrons between photosensitizer and enzyme. Hence, the mechanism and basic understanding of photo-catalytic reactions are needed in order to improve efficiency. Another major hurdle is the stability of enzyme during the photo-catalytic reactions. Recent advances in protein engineering and molecular biological evolution helped to increase stability (in terms of chemical, mechanical, and storage stability) or specificity (regio-, chemo- and stereoselectivity) of biocatalysts and thus have helped to overcome certain obstacles in biocatalysis. Also, various researchers developed a variety of novel immobilization strategies such as the incorporation of nanomaterials, biopolysaccharides, functionalized polymers, and co-immobilization of multiple biocatalysts that can make a robust biocatalytic process which is amenable to gram scale and can quickly implement in large-scale operations. However, to make the process industrially viable, there are certain requirements that need to be met, such as high turnover numbers (TONs), high concentration, and high conversion/yield.

References

Adam D, Bösche L, Castañeda-Losada L, Winkler M, Apfel U, Happe T (2017) Sunlight-dependent hydrogen production by photosensitizer/hydrogenase systems. ChemSusChem 10:894–902

Aprile C, Corma A, Garcia H (2008) Enhancement of the photocatalytic activity of TiO_2 through spatial structuring and particle size control: from subnanometric to submillimetric length scale. Phys Chem Chem Phys 10:769–783. https://doi.org/10.1039/b712168g

Aresta M, Dibenedetto A, Baran T, Angelini A, Łabuz P, Macyk W (2014) An integrated photocatalytic/enzymatic system for the reduction of CO_2 to methanol in bioglycerol-water. Beilstein J Org Chem 10:2556–2565. https://doi.org/10.3762/bjoc.10.267

Balcerzak L, Lipok J, Strub D, Lochyński S (2014) Biotransformations of monoterpenes by photoautotrophic micro-organisms. J Appl Microbiol 117:1523–1536. https://doi.org/10.1111/jam.12632

Balke K, Kadow M, Mallin H, Saß S, Bornscheuer UT (2012) Discovery, application and protein engineering of Baeyer-Villiger monooxygenases for organic synthesis. Org Biomol Chem 10:6249–6265

Balke K, Beier A, Bornscheuer UT (2018) Hot spots for the protein engineering of Baeyer-Villiger monooxygenases. Biotechnol Adv 36:247–263

Bartsch M, Gassmeyer SK, Köninger K, Igarashi K, Liauw P, Dyczmons-Nowaczyk N, Miyamoto K, Nowaczyk MM, Kourist R (2015) Photosynthetic production of enantioselective biocatalysts. Microb Cell Fact 14:53. https://doi.org/10.1186/s12934-015-0233-5

Beatty JW, Douglas JJ, Cole KP, Stephenson CRJ (2015) A scalable and operationally simple radical trifluoromethylation. Nat Commun 6:1–6. https://doi.org/10.1038/ncomms8919

Bernard J, van Heerden E, Arends IWCE, Opperman DJ, Hollmann F (2012) Chemoenzymatic reduction of conjugated C=C double bonds. ChemCatChem 4:196–199. https://doi.org/10.1002/cctc.201100312

Bernhardt R (2006) Cytochromes P450 as versatile biocatalysts. J Biotechnol 124:128–145

Bhatia S, Bhatia S (2018) Introduction to enzymes and their applications. In: Introduction to pharmaceutical biotechnology, vol 2. https://doi.org/10.1088/978-0-7503-1302-5ch1

Bisagni S, Summers B, Kara S, Hatti-Kaul R, Grogan G, Mamo G, Hollmann F (2014) Exploring the substrate specificity and enantioselectivity of a baeyer-villiger monooxygenase from dietzia sp. D5: oxidation of sulfides and aldehydes. Top Catal 57:366–375. https://doi.org/10.1007/s11244-013-0192-1

Brown KA, Wilker MB, Boehm M, Dukovic G, King PW (2012) Characterization of photochemical processes for H_2 production by CdS nanorod–[FeFe] hydrogenase complexes. J Am Chem Soc 134:5627–5636

Brown KA, Wilker MB, Boehm M, Hamby H, Dukovic G, King PW (2016) Photocatalytic regeneration of nicotinamide cofactors by quantum dot-enzyme biohybrid complexes. ACS Catal 6:2201–2204. https://doi.org/10.1021/acscatal.5b02850

Caputo CA, Wang L, Beranek R, Reisner E (2015) Carbon nitride–TiO_2 hybrid modified with hydrogenase for visible light driven hydrogen production. Chem Sci 6:5690–5694

Chaiyen P, Fraaije MW, Mattevi A (2012) The enigmatic reaction of flavins with oxygen. Trends Biochem Sci 37:373–380

Chen Y-H, Chen L-L, Shang N-C (2009) Photocatalytic degradation of dimethyl phthalate in an aqueous solution with Pt-doped TiO_2-coated magnetic PMMA microspheres. J Hazard Mater 172:20–29

Choudhury S, Baeg J-O, Park N-J, Yadav RK (2012) A photocatalyst/enzyme couple that uses solar energy in the asymmetric reduction of acetophenones. Angew Chem Int Ed 51:11624–11628. https://doi.org/10.1002/anie.201206019

Choudhury S, Baeg JO, Park NJ, Yadav RK (2014) A solar light-driven, eco-friendly protocol for highly enantioselective synthesis of chiral alcohols via photocatalytic/biocatalytic cascades. Green Chem 16:4389–4400. https://doi.org/10.1039/c4gc00885e

Churakova E, Kluge M, Ullrich R, Arends I, Hofrichter M, Hollmann F (2011) Specific photobiocatalytic oxyfunctionalization reactions. Angew Chem 123:10904–10907. https://doi.org/10.1002/ange.201105308

Churakova E, Arends IWCE, Hollmann F (2013) Increasing the productivity of peroxidase-catalyzed oxyfunctionalization: a case study on the potential of two-liquid-phase systems. ChemCatChem 5:565–568. https://doi.org/10.1002/cctc.201200490

Dailey HA (1997) Enzymes of heme biosynthesis. JBIC J Biol Inorg Chem 2:411–417

Dawson JH (1988) Probing structure-function relations in heme-containing oxygenases and peroxidases. Science 240 (80–):433–439

de Gonzalo G, Mihovilovic MD, Fraaije MW (2010) Recent developments in the application of Baeyer-Villiger monooxygenases as biocatalysts. ChemBioChem 11:2208–2231

Dobbek H, Svetlitchnyi V, Gremer L, Huber R, Meyer O (2001) Crystal structure of a carbon monoxide dehydrogenase reveals a [Ni–4Fe–5S] cluster. Science 293(80–):1281–1285

Dong JJ, Fernández-Fueyo E, Hollmann F, Paul CE, Pesic M, Schmidt S, Wang Y, Younes S, Zhang W (2018) Biocatalytic oxidation reactions: a chemist's perspective. Angew Chem Int Ed 57:9238–9261. https://doi.org/10.1002/anie.201800343

Dutta AK, Maji SK, Srivastava DN, Mondal A, Biswas P, Paul P, Adhikary B (2012) Synthesis of FeS and FeSe nanoparticles from a single source precursor: a study of their photocatalytic activity, peroxidase-like behavior, and electrochemical sensing of H_2O_2. ACS Appl Mater Interfaces 4:1919–1927

Ener ME, Lee YT, Winkler JR, Gray HB, Cheruzela L (2010) Photooxidation of cytochrome P450-BM3. Proc Natl Acad Sci U S A 107:18783–18786. https://doi.org/10.1073/pnas.1012381107

Evans DJ, Pickett CJ (2003) Chemistry and the hydrogenases. Chem Soc Rev 32:268–275

Fontecave M (2006) Iron-sulfur clusters: ever-expanding roles. Nat Chem Biol 2:171–174

Gabruk M, Mysliwa-Kurdziel B (2015) Light-dependent protochlorophyllide oxidoreductase: phylogeny, regulation, and catalytic properties. Biochemistry 54:5255–5262

Gacs J, Zhang W, Knaus T, Mutti FG, Arends IWCE, Hollmann F (2019) A photo-enzymatic cascade to transform racemic alcohols into enantiomerically pure amines. Catalysts 9:1–11. https://doi.org/10.3390/catal9040305

Garrone A, Archipowa N, Zipfel PF, Hermann G, Dietzek B (2015) Plant protochlorophyllide oxidoreductases A and B catalytic efficiency and initial reaction steps. J Biol Chem 290:28530–28539

Girvan HM, Munro AW (2016) Applications of microbial cytochrome P450 enzymes in biotechnology and synthetic biology. Curr Opin Chem Biol 31:136–145

Górak M, Zymańczyk-Duda E (2015) Application of cyanobacteria for chiral phosphonate synthesis. Green Chem 17:4570–4578. https://doi.org/10.1039/c5gc01195g

Granone LI, Sieland F, Zheng N, Dillert R, Bahnemann DW (2018) Photocatalytic conversion of biomass into valuable products: a meaningful approach? Green Chem 20:1169–1192. https://doi.org/10.1039/c7gc03522e

Grau MM, Van Der Toorn JC, Otten LG, Macheroux P, Taglieber A, Zilly FE, Arends IWCE, Hollmann F (2009) Photoenzymatic reduction of C=C double bonds. Adv Synth Catal 351:3279–3286. https://doi.org/10.1002/adsc.200900560

Hanf R, Fey S, Schmitt M, Hermann G, Dietzek B, Popp J (2012) Catalytic efficiency of a photoenzyme—an adaptation to natural light conditions. ChemPhysChem 13:2013–2015

Höfler GT, Fernández-Fueyo E, Pesic M, Younes SH, Choi E, Kim YH, Urlacher VB, Arends IWCE, Hollmann F (2018) A photoenzymatic NADH regeneration system. ChemBioChem 19:2344–2347. https://doi.org/10.1002/cbic.201800530

Hofrichter M, Ullrich R, Pecyna MJ, Liers C, Lundell T (2010) New and classic families of secreted fungal heme peroxidases. Appl Microbiol Biotechnol 87:871–897

Hollmann F, Taglieber A, Schulz F, Reetz MT (2007) A light-driven stereoselective biocatalytic oxidation. Angew Chem Int Ed 46:2903–2906. https://doi.org/10.1002/anie.200605169

Hollmann F, Arends IWCE, Buehler K (2010) Biocatalytic redox reactions for organic synthesis: nonconventional regeneration methods. ChemCatChem 2:762–782. https://doi.org/10.1002/cctc.201000069

Holtmann D, Hollmann F (2016) The oxygen dilemma: a severe challenge for the application of monooxygenases? ChemBioChem 17:1391–1398. https://doi.org/10.1002/cbic.201600176

Huijbers MME, Montersino S, Westphal AH, Tischler D, van Berkel WJH (2014) Flavin dependent monooxygenases. Arch Biochem Biophys 544:2–17

Husain Q (2010) Peroxidase mediated decolorization and remediation of wastewater containing industrial dyes: a review. Rev Environ Sci Bio/Technol 9:117–140

Jeoung J-H, Fesseler J, Goetzl S, Dobbek H (2014) Carbon monoxide. Toxic gas and fuel for anaerobes and aerobes: carbon monoxide dehydrogenases. In: The metal-driven biogeochemistry of gaseous compounds in the environment. Springer, pp 37–69

Johnson DC, Dean DR, Smith AD, Johnson MK (2005) Structure, function, and formation of biological iron-sulfur clusters. Annu Rev Biochem 74:247–281

Jones JP, O'Hare EJ, Wong L (2001) Oxidation of polychlorinated benzenes by genetically engineered CYP101 (cytochrome P450cam). Eur J Biochem 268:1460–1467

Joosten V, van Berkel WJH (2007) Flavoenzymes. Curr Opin Chem Biol 11:195–202

Kalsoom U, Bhatti HN, Asgher M (2015) Characterization of plant peroxidases and their potential for degradation of dyes: a review. Appl Biochem Biotechnol 176:1529–1550

Kamada K, Moriyasu A, Soh N (2012) Visible-light-driven enzymatic reaction of peroxidase adsorbed on doped hematite thin films. J Phys Chem C 116:20694–20699

Karmee SK, Roosen C, Kohlmann C, Lütz S, Greiner L, Leitner W (2009) Chemo-enzymatic cascade oxidation in supercritical carbon dioxide/water biphasic media. Green Chem 11:1052–1055

Karplus PA, Fox KM, Massey V (1995) Structure-function relations for old yellow enzyme. FASEB J 9:1518–1526

Kato M, Nguyen D, Gonzalez M, Cortez A, Mullen SE, Cheruzel LE (2014) Regio- and stereoselective hydroxylation of 10-undecenoic acid with a light-driven P450 BM3 biocatalyst yielding a valuable synthon for natural product synthesis. Bioorganic Med Chem 22:5687–5691. https://doi.org/10.1016/j.bmc.2014.05.046

Kellner DG, Maves SA, Sligar SG (1997) Engineering cytochrome P450s for bioremediation. Curr Opin Biotechnol 8:274–278

Khalid NR, Ahmed E, Rasheed A, Ahmad M, Ramzan M, Shakoor A, Elahi A, Abbas SM, Hussain R, Niaz NA (2015) Co-doping effect of carbon and yttrium on photocatalytic activity of tio 2 nanoparticles for methyl orange degradation. J Ovonic Res 11:107–112

Kim J, Lee SH, Tieves F, Choi DS, Hollmann F, Paul CE, Park CB (2018) Biocatalytic C=C bond reduction through carbon nanodot-sensitized regeneration of NADH analogues. Angew Chem 130:14021–14024. https://doi.org/10.1002/ange.201804409

Kim J, Lee SH, Tieves F, Paul CE, Hollmann F, Park CB (2019) Nicotinamide adenine dinucleotide as a photocatalyst. Sci Adv 5. https://doi.org/10.1126/sciadv.aax0501

König B, Kümmel S, Svobodová E, Cibulka R (2019) Flavin photocatalysis. Phys Sci Rev 3:1–17. https://doi.org/10.1515/psr-2017-0168

Lam Q, Cortez A, Nguyen TT, Kato M, Cheruzel L (2016) Chromogenic nitrophenolate-based substrates for light-driven hybrid P450 BM3 enzyme assay. J Inorg Biochem 158:86–91. https://doi.org/10.1016/j.jinorgbio.2015.12.005

Lauder K, Toscani A, Qi Y, Lim J, Charnock SJ, Korah K, Castagnolo D (2018) Photo-biocatalytic one-pot cascades for the enantioselective synthesis of 1,3-mercaptoalkanol volatile sulfur compounds. Angew Chem 130:5905–5909. https://doi.org/10.1002/ange.201802135

Lee SH, Kwon Y-C, Kim D-M, Park CB (2013) Cytochrome P450-catalyzed *O*-dealkylation coupled with photochemical NADPH regeneration. Biotechnol Bioeng 110:383–390. https://doi.org/10.1002/bit.24729

Lee JS, Nam DH, Kuk SK, Park CB (2014) Near-infrared-light-driven artificial photosynthesis by nanobiocatalytic assemblies. Chem A Eur J 20:3584–3588. https://doi.org/10.1002/chem.201400136

Lee SH, Choi DS, Pesic M, Lee YW, Paul CE, Hollmann F, Park CB (2017) Cofactor-free, direct photoactivation of enoate reductases for the asymmetric reduction of C=C bonds. Angew Chem Int Ed 56:8681–8685. https://doi.org/10.1002/anie.201702461

Lee SH, Choi DS, Kuk SK, Park CB (2018) Photobiocatalysis: activating redox enzymes by direct or indirect transfer of photoinduced electrons. Angew Chem Int Ed 57:7958–7985. https://doi.org/10.1002/anie.201710070

Lee C-Y, Zou J, Bullock J, Wallace GG (2019) Emerging approach in semiconductor photocatalysis: towards 3D architectures for efficient solar fuels generation in semi-artificial photosynthetic systems. J Photochem Photobiol C Photochem Rev 39:142–160

Lee BI, Chung YJ, Park CB (2019) Photosensitizing materials and platforms for light-triggered modulation of Alzheimer's β-amyloid self-assembly. Biomaterials 190:121–132

Leisch H, Morley K, Lau PCK (2011) Baeyer−Villiger monooxygenases: more than just green chemistry. Chem Rev 111:4165–4222

Lill R (2009) Function and biogenesis of iron–sulphur proteins. Nature 460:831

Litman ZC, Wang Y, Zhao H, Hartwig JF (2018) Cooperative asymmetric reactions combining photocatalysis and enzymatic catalysis. Nature 560:355–359. https://doi.org/10.1038/s41586-018-0413-7

Losi A (2007) Flavin-based blue-light photosensors: a photobiophysics update. Photochem Photobiol 83:1283–1300

Lu C, Shen F, Wang S, Wang Y, Liu J, Bai WJ, Wang X (2018) An engineered self-sufficient biocatalyst enables scalable production of linear α-olefins from carboxylic acids. ACS Catal 8:5794–5798. https://doi.org/10.1021/acscatal.8b01313

Lubitz W, Ogata H, Rüdiger O, Reijerse E (2014) Hydrogenases. Chem Rev 114:4081–4148

Lubner CE, Applegate AM, Knörzer P, Ganago A, Bryant DA, Happe T, Golbeck JH (2011) Solar hydrogen-producing bionanodevice outperforms natural photosynthesis. Proc Natl Acad Sci U S A 108:20988–20991. https://doi.org/10.1073/pnas.1114660108

Maciá-Agulló JA, Corma A, Garcia H (2015) Photobiocatalysis: the power of combining photocatalysis and enzymes. Chem A Eur J 21:10940–10959. https://doi.org/10.1002/chem.201406437

Malito E, Alfieri A, Fraaije MW, Mattevi A (2004) Crystal structure of a Baeyer-Villiger monooxygenase. Proc Natl Acad Sci U S A 101:13157–13162. https://doi.org/10.1073/pnas.0404538101

Maurya SS, Nadar SS, Rathod VK (2020a) A rapid self-assembled hybrid bio-microflowers of alpha-amylase with enhanced activity. J Biotechnol 317:27–33. https://doi.org/10.1016/j.jbiotec.2020.04.010

Maurya SS, Nadar SS, Rathod VK (2020b) Dual activity of laccase-lysine hybrid organic–inorganic nanoflowers for dye decolourization. Environ Technol Innov 19:100798. https://doi.org/10.1016/j.eti.2020.100798

Mifsud M, Gargiulo S, Iborra S, Arends IWCE, Hollmann F, Corma A (2014a) Photobiocatalytic chemistry of oxidoreductases using water as the electron donor. Nat Commun 5. https://doi.org/10.1038/ncomms4145

Mifsud M, Gargiulo S, Iborra S, Arends IWCE, Hollmann F, Corma A (2014b) Photobiocatalytic chemistry of oxidoreductases using water as the electron donor. Nat Commun 5:1–6. https://doi.org/10.1038/ncomms4145

Monti D, Ottolina G, Carrea G, Riva S (2011) Redox reactions catalyzed by isolated enzymes. Chem Rev 111:4111–4140. https://doi.org/10.1021/cr100334x

Nadar SS, Vaidya L, Rathod VK (2020) Enzyme embedded metal organic framework (enzyme–MOF): De novo approaches for immobilization. Int J Biol Macromol 149:861–876. https://doi.org/10.1016/j.ijbiomac.2020.01.240

Nakamura K, Yamanaka R, Tohi K, Hamada H (2000) Cyanobacterium-catalyzed asymmetric reduction of ketones. Tetrahedron Lett 41:6799–6802. https://doi.org/10.1016/S0040-4039(00)01132-1

Nehme SI, Crocker L, Fruk L (2020) Flavin-conjugated iron oxide nanoparticles as enzyme-inspired photocatalysts for azo dye degradation. Catalysts 10:324. https://doi.org/10.3390/catal10030324

Park JH, Lee SH, Cha GS, Choi DS, Nam DH, Lee JH, Lee J-K, Yun C-H, Jeong KJ, Park CB (2015) Cofactor-free light-driven whole-cell cytochrome P450 catalysis. Angew Chem 127:983–987. https://doi.org/10.1002/ange.201410059

Parkin A, Goldet G, Cavazza C, Fontecilla-Camps JC, Armstrong FA (2008) The difference a Se makes? Oxygen-tolerant hydrogen production by the [NiFeSe]-hydrogenase from Desulfomicrobium baculatum. J Am Chem Soc 130:13410–13416

Patil PD, Yadav GD (2018) Rapid in situ encapsulation of laccase into metal-organic framework support (ZIF-8) under biocompatible conditions. Chem Sel 3:4669–4675. https://doi.org/10.1002/slct.201702852

Patil PD, Yadav GD (2019) Exploring the untapped potential of solar pretreatment for deconstruction of recalcitrant Kraft lignin in fungal biotransformation. Clean Technol Environ Policy 21:579–590. https://doi.org/10.1007/s10098-018-1656-6

Paul CE, Churakova E, Maurits E, Girhard M, Urlacher VB, Hollmann F (2014) In situ formation of H_2O_2 for P450 peroxygenases. Bioorg Med Chem 22:5692–5696

Pazmino DET, Dudek HM, Fraaije MW (2010) Baeyer-Villiger monooxygenases: recent advances and future challenges. Curr Opin Chem Biol 14:138–144

Peers MK, Toogood HS, Heyes DJ, Mansell D, Coe BJ, Scrutton NS (2016) Light-driven biocatalytic reduction of α, β-unsaturated compounds by ene reductases employing transition metal complexes as photosensitizers. Catal Sci Technol 6:169–177. https://doi.org/10.1039/c5cy01642h

Perez DI, Grau MM, Arends IWCE, Hollmann F (2009) Visible light-driven and chloroperoxidase-catalyzed oxygenation reactions. Chem Commun 44:6848–6850. https://doi.org/10.1039/b915078a

Pesic M, Fernández-Fueyo E, Hollmann F (2017) Characterization of the old yellow enzyme homolog from Bacillus subtilis (YqjM). Chem Select 2:3866–3871

Poulos TL (2014) Heme enzyme structure and function. Chem Rev 114:3919–3962

Qin P, Zhu H, Edvinsson T, Boschloo G, Hagfeldt A, Sun L (2008) Design of an organic chromophore for *p*-type dye-sensitized solar cells. J Am Chem Soc 130:8570–8571

Rauch MCR, Huijbers MME, Pabst M, Paul CE, Pešić M, Arends I, Hollmann F (2020) Photochemical regeneration of flavoenzymes—an old yellow enzyme case-study. Biochim Biophys Acta (BBA)-Proteins Proteomics 1868:140303

Reisner E, Powell DJ, Cavazza C, Fontecilla-Camps JC, Armstrong FA (2009) Visible light-driven H_2 production by hydrogenases attached to dye-sensitized TiO_2 nanoparticles. J Am Chem Soc 131:18457–18466

Romero E, Gómez Castellanos JR, Gadda G, Fraaije MW, Mattevi A (2018) Same substrate, many reactions: oxygen activation in flavoenzymes. Chem Rev 118:1742–1769

Roth LE, Nguyen JC, Tezcan FA (2010) ATP-and iron−protein-independent activation of nitrogenase catalysis by light. J Am Chem Soc 132:13672–13674

Sabuzi F, Churakova E, Galloni P, Wever R, Hollmann F, Floris B, Conte V (2015) Thymol bromination—a comparison between enzymatic and chemical catalysis. Eur J Inorg Chem 2015:3519–3525. https://doi.org/10.1002/ejic.201500086

Sakai T, Mersch D, Reisner E (2013) Photocatalytic hydrogen evolution with a hydrogenase in a mediator-free system under high levels of oxygen. Angew Chem Int Ed 52:12313–12316

Schmermund L, Jurkaš V, Özgen FF, Barone GD, Büchsenschütz HC, Winkler CK, Schmidt S, Kourist R, Kroutil W (2019) Photo-biocatalysis: biotransformations in the presence of light. ACS Catal 9:4115–4144. https://doi.org/10.1021/acscatal.9b00656

Schroeder L, Frese M, Müller C, Sewald N, Kottke T (2018) Photochemically driven biocatalysis of halogenases for the green production of chlorinated compounds. ChemCatChem 10:3336–3341. https://doi.org/10.1002/cctc.201800280

Seel CJ, Gulder T (2019) Biocatalysis fueled by light: on the versatile combination of photocatalysis and enzymes. ChemBioChem 20:1871–1897. https://doi.org/10.1002/cbic.201800806

Seel CJ, Králík A, Hacker M, Frank A, König B, Gulder T (2018) Atom-economic electron donors for photobiocatalytic halogenations. ChemCatChem 10:3960–3963. https://doi.org/10.1002/cctc.201800886

Shaik S, Munro AW, Sen S, Mowat C, Nam W, Derat E, Bugg T, Proshlyakov DA, Hausinger RP, Straganz GD (2011) Iron-containing enzymes: versatile catalysts of hydroxylation reactions in nature. Royal Society of Chemistry

Sono M, Roach MP, Coulter ED, Dawson JH (1996) Heme-containing oxygenases. Chem Rev 96:2841–2888

Sosa V, Melkie M, Sulca C, Li J, Tang L, Li J, Faris J, Foley B, Banh T, Kato M, Cheruzel LE (2018) Selective light-driven chemoenzymatic trifluoromethylation/hydroxylation of substituted arenes. ACS Catal 8:2225–2229. https://doi.org/10.1021/acscatal.7b04160

Stiborová M, Mikšanová M, Martínek V, Frei E (2000) Heme peroxidases: structure, function, mechanism and involvement in activation of carcinogens. A review. Collect Czechoslov Chem Commun 65:297–325

Taglieber A, Schulz F, Hollman F, Rusek M, Reetz MT (2008) Light-driven biocatalytic oxidation and reduction reactions: Scope and limitations. ChemBioChem 9:565–572. https://doi.org/10.1002/cbic.200700435

Toogood HS, Gardiner JM, Scrutton NS (2010) Biocatalytic reductions and chemical versatility of the old yellow enzyme family of flavoprotein oxidoreductases. ChemCatChem 2:892–914

Torkian L, Amereh E (2016) Nano sized Ni/TiO_2 @ NaX zeolite with enhanced photocatalytic activity. J Nanostruct 6:307–311. https://doi.org/10.22052/jns.2016.34328

Torres Pazmiño DE, Snajdrova R, Baas B-J, Ghobrial M, Mihovilovic MD, Fraaije MW (2008) Self-sufficient Baeyer–Villiger monooxygenases: effective coenzyme regeneration for biooxygenation by fusion engineering. Angew Chem 120:2307–2310. https://doi.org/10.1002/ange.200704630

Tran NH, Huynh N, Bui T, Nguyen Y, Huynh P, Cooper ME, Cheruzel LE (2011) Light-initiated hydroxylation of lauric acid using hybrid P450 BM3 enzymes. Chem Commun 47:11936–11938. https://doi.org/10.1039/c1cc15124j

Tran NH, Nguyen D, Dwaraknath S, Mahadevan S, Chavez G, Nguyen A, Dao T, Mullen S, Nguyen TA, Cheruzel LE (2013) An efficient light-driven P450 BM3 biocatalyst. J Am Chem Soc 135:14484–14487. https://doi.org/10.1021/ja409337v

Tremblay PL, Xu M, Chen Y, Zhang T (2020) Nonmetallic abiotic-biological hybrid photocatalyst for visible water splitting and carbon dioxide reduction. iScience 23:100784. https://doi.org/10.1016/j.isci.2019.100784

Tseng TK, Lin YS, Chen YJ, Chu H (2010) A review of photocatalysts prepared by sol–gel method for VOCs removal. Int J Mol Sci 11 (6):2336–2361. https://doi.org/10.3390/ijms11062336

Turner NJ (2018) Enzymes team up with light-activated catalysts. Nature 560:310–311. https://doi.org/10.1038/d41586-018-05933-0

Urlacher VB, Girhard M (2012) Cytochrome P450 monooxygenases: an update on perspectives for synthetic application. Trends Biotechnol 30:26–36

Urlacher VB, Girhard M (2019) Cytochrome P450 monooxygenases in biotechnology and synthetic biology. Trends Biotechnol 37(8):882

Urlacher VB, Lutz-Wahl S, Schmid RD (2004) Microbial P450 enzymes in biotechnology. Appl Microbiol Biotechnol 64:317–325

Vaidya LB, Nadar SS, Rathod VK (2020) Biological metal organic framework (bio-MOF) of glucoamylase with enhanced stability. Coll Surf B Biointerf 193:111052. https://doi.org/10.1016/j.colsurfb.2020.111052

Van Berkel WJH, Kamerbeek NM, Fraaije MW (2006) Flavoprotein monooxygenases, a diverse class of oxidative biocatalysts. J Biotechnol 124:670–689

Van Schie MMCH, Paul CE, Arends IWCE, Hollmann F (2019) Photoenzymatic epoxidation of styrenes. Chem Commun 55:1790–1792. https://doi.org/10.1039/c8cc08149b

Walsh CT, Wencewicz TA (2013) Flavoenzymes: versatile catalysts in biosynthetic pathways. Nat Prod Rep 30:175–200

Whitehouse CJC, Bell SG, Wong LL (2012) P450 BM3 (CYP102A1): connecting the dots. Chem Soc Rev 41:1218–1260. https://doi.org/10.1039/c1cs15192d

Williams RE, Rathbone DA, Scrutton NS, Bruce NC (2004) Biotransformation of explosives by the old yellow enzyme family of flavoproteins. Appl Environ Microbiol 70:3566–3574

Willot SJP, Fernández-Fueyo E, Tieves F, Pesic M, Alcalde M, Arends IWCE, Park CB, Hollmann F (2019) Expanding the spectrum of light-driven peroxygenase reactions. ACS Catal 9:890–894. https://doi.org/10.1021/acscatal.8b03752

Winkler CK, Faber K, Hall M (2018) Biocatalytic reduction of activated CC-bonds and beyond: emerging trends. Curr Opin Chem Biol 43:97–105

Woolerton TW, Sheard S, Reisner E, Pierce E, Ragsdale SW, Armstrong FA (2010) Efficient and clean photoreduction of CO_2 to CO by enzyme-modified TiO_2 nanoparticles using visible light. J Am Chem Soc 132:2132–2133

Woolerton TW, Sheard S, Pierce E, Ragsdale SW, Armstrong FA (2011) CO_2 photoreduction at enzyme-modified metal oxide nanoparticles. Energy Environ Sci 4:2393–2399

Yamanaka R, Nakamura K, Murakami M, Murakami A (2015) Selective synthesis of cinnamyl alcohol by cyanobacterial photobiocatalysts. Tetrahedron Lett 56:1089–1091. https://doi.org/10.1016/j.tetlet.2015.01.092

Yang Q, Zhao F, Zhang N, Liu M, Hu H, Zhang J, Zhou S (2018) Mild dynamic kinetic resolution of amines by coupled visible-light photoredox and enzyme catalysis. Chem Commun 54:14065–14068. https://doi.org/10.1039/c8cc07990k

Zachos I, Gaßmeyer SK, Bauer D, Sieber V, Hollmann F, Kourist R (2015) Photobiocatalytic decarboxylation for olefin synthesis. Chem Commun 51:1918–1921. https://doi.org/10.1039/c4cc07276f

Zanger UM, Schwab M (2013) Cytochrome P450 enzymes in drug metabolism: regulation of gene expression, enzyme activities, and impact of genetic variation. Pharmacol Ther 138:103–141

Zhang W, Fernández-Fueyo E, Ni Y, Van Schie M, Gacs J, Renirie R, Wever R, Mutti FG, Rother D, Alcalde M, Hollmann F (2018) Selective aerobic oxidation reactions using a combination of photocatalytic water oxidation and enzymatic oxyfunctionalizations. Nat Catal 1:55–62. https://doi.org/10.1038/s41929-017-0001-5

Zhang W, Ma M, Huijbers MME, Filonenko GA, Pidko EA, Van Schie M, De Boer S, Burek BO, Bloh JZ, Van Berkel WJH, Smith WA, Hollmann F (2019a) Hydrocarbon synthesis via photoenzymatic decarboxylation of carboxylic acids. J Am Chem Soc 141:3116–3120. https://doi.org/10.1021/jacs.8b12282

Zhang W, Fueyo EF, Hollmann F, Martin LL, Pesic M, Wardenga R, Höhne M, Schmidt S (2019b) Combining photo-organo redox- and enzyme catalysis facilitates asymmetric C–H bond functionalization. Eur J Org Chem 2019:80–84. https://doi.org/10.1002/ejoc.201801692

Zhao Y, Anderson NC, Ratzloff MW, Mulder DW, Zhu K, Turner JA, Neale NR, King PW, Branz HM (2016) Proton reduction using a hydrogenase-modified nanoporous black silicon photoelectrode. ACS Appl Mater Interfaces 8:14481–14487

Zhou J, Zhang Y, Zhao XS, Ray AK (2006) Photodegradation of benzoic acid over metal-doped TiO_2. Indu Eng Chem Res 4:3503–3511

Zilly FE, Taglieber A, Schulz F, Hollmann F, Reetz MT (2009) Deazaflavins as mediators in light-driven cytochrome P450 catalyzed hydroxylations. Chem Commun 46:7152–7154. https://doi.org/10.1039/b913863c

Biomass-Derived Carbons and Their Energy Applications

Thibeorchews Prasankumar, Sujin Jose, and Meiyazhagan Ashokkumar

Abstract

In this chapter, a few interesting findings on renewable carbons derived from various biomass precursors and their electrochemical applications, especially supercapacitors and lithium-ion batteries were discussed. Electrochemical energy storage devices are progressively crucial since it helps in a significant reduction in the use of different fossils-based resources. The use of energy storage devices in a particular system rests on the nature of the electrode materials. Among an extensive range of electrodes, biomass-based carbons have gained significant consideration as an electrode because of their variable physicochemical features, environmental concern, and commercial value. We have also discussed a few recent developments of biomass-derived carbons and some important techniques such as carbonization and activation conditions that control their property and performance of carbon electrodes. Besides, some of the parameters such as pore structure, surface property, and degree of graphitization, which regulate the electrochemical functioning of the device were discussed in detail. In the final section of this chapter, we included some energy storage applications of biomass-based carbons, and their effectiveness as an electrode were summarized. In brief, this chapter provides a fundamental understanding of several biomass-derived carbon materials and suggests the essential strategies for the fabrication of various energy storage systems.

Keywords

Biomass • Porous structure • Supercapacitors • Li-ion batteries

1 Introduction

Increased use of non-renewable fossil fuels including natural gas, oil, and coal has escalated several environmental concerns (Grey and Tarascon 2017). With swift communal growth and increased insistence for growth of miniaturized portable devices, high volumetric functional storage devices are considered essential. This effort minimizes the volume of devices generated mainly due to increasing and the crucial drive to develop inexpensive, renewable energy storage devices (Niu et al. 2013). Among them, batteries and supercapacitors have been in limelight due to their overall performances. Batteries that can display extended cyclability, conventional rate performance, and high energy density which are considered essential for the manufacture of more powerful electric vehicles and portable devices with extended life (Prasankumar et al. 2019). On the other hand, supercapacitor-based systems like hybrid electric vehicles, digital communications, etc., are recognized as promising material for modern electronics applications owing to their elevated-power density, extended cyclability, lightning charge/discharge rate, and enhanced safety (Merlet et al. 2012). Furthermore, it is well known that an electrode plays an integral role in energy storage devices; therefore, efficiency, cost-effectiveness, environmental friendliness, and sustainability of electrode need consideration during the formulation of new electrodes. So far, various electrode materials have been developed, using precursors such as carbon (Zhai et al. 2011; Karnan et al. 2016), metal oxides/hydroxides (Fan et al. 2007), and conducting polymers (Meng et al. 2017). Among them, metal oxides, hydroxides, and polymers are considered non-renewable systems since

T. Prasankumar · M. Ashokkumar (✉)
Department of Materials Science and Nano Engineering, Rice University, Houston, TX 77005, USA
e-mail: ma37@rice.edu

T. Prasankumar · S. Jose
School of Physics, Madurai Kamaraj University, Madurai, 625021, India

Inamuddin et al. (eds.), *Advances in Green Synthesis*, Advances in Science, Technology & Innovation,
https://doi.org/10.1007/978-3-030-67884-5_10

they pose an environmental risk and considered expensive. Hence, it is essential to exploit easily accessible renewable resources.

Researchers widely studied different biomasses, as a reliable renewable sources available on planet, which are used for different applications such as capture of carbon dioxide (CO_2) gases (Boyjoo et al. 2017), hydrogen storage (Blankenship and Mokaya 2017), solar cells (Wang et al. 2014a), treatment of water contaminants (Ma et al. 2017), and energy applications (Chen et al. 2018; Wang et al. 2018). The presence of exceptional specific surface area (SSA), electrical conductivity, engineered pore size, and substantial mechanical strength marks these resources as an ideal material for various applications. Additionally, the biomass-based activated carbon electrodes deliver an outstanding capacitance and remarkable energy density because of their increased surface area. Significant efforts have been carried out to derive biomass-based carbons for supercapacitor applications, including the selection of precursors, carbonization, and activation process to produce carbons in high yield with precise control over pore geometry and distribution. In general, the selection of precursors was based on elemental composition, cost, easy availability, and molecular structure. For instance, rice (Zhu et al. 2017), wheat straw (Liu et al. 2018a), pistachio (Hu et al. 2007), and catkins (Su et al. 2017) are chosen as sources owing to their easy availability and low cost. Whereas eggshell membranes (Li et al. 2012), Auricularia (Long et al. 2015), kombucha (Dai et al. 2017), and orange peel (Ranaweera et al. 2017) were selected due to their exclusive porous and hierarchical microstructures. A few others include Bacillus subtilis (Zhu et al. 2013), human hair (Qian et al. 2014), chitin (Duan et al. 2016), animal bones, skins, and fish scales (Gao et al. 2016). In this chapter, some of the recent advancements in the formulation of biomass-based activated carbons were highlighted. Also, a few parameters which influence the property of derived material and their application studies related to supercapacitors and Li-ion batteries (LIBs) have been discussed.

2 Types of Biomass Materials

To date, a great variety of biomasses has been explored for the formation of porous carbon. In the below sections, a few biomass sources which were used to derive carbon materials have been discussed briefly.

2.1 Plant-Based Carbons

The chemical composition of plants changes quantitatively from each other within the same species. Table 1 displays the chemical composition in percentage (%) of various plant-based biomasses. For instance, the palm shell contains a combination of lignin and cellulose of around 83% with a larger portion of cellulose. Another example is jute and hemp which contains cellulose of about 64 and 67% approximately. The percentage of cellulose is comparably greater when compared to other sources such as poplar leaves (22%), scots pine stem wood (40%), and switchgrass (33–46%). This approximate value gives an idea of how the chemical composition of plant species, i.e., cellulose change which results in a significant change in the elemental composition of derived carbons.

To be specific, the actual carbon yield is determined by the mass ratio of cellulose, hemicellulose, lignin, and the composition of the chemical compounds present in plants. For instance, lignin is well known for its thermal stability that plays a vital part and helps in the evolution of a higher percent of carbons/chars when compared to other plant sources. The composition of cellulose, hemicellulose, and lignin are displayed in Fig. 1

2.2 Fruit-Based Carbons

Alike plant biomass, fruit-derived carbons also vary with the portions of fruit, growth condition, and the chemical composition of the species. Some of the chemical components

Table 1 Chemical compositions of different plant-based biomass (Liu et al. 2018b)

Biomass	Moisture (%)	Lignin (%)	Cellulose (%)	Hemicellulose (%)	Extractives (%)	Reference
Rice straw	–	10–18	32–41	15–24	–	Zhu et al. (2017)
Coconut coir	13.68	46.48	21.46	12.36	8.77	Arsene et al. (2013)
Corn stover	–	18–22	37–42	20–28	–	Kumar et al. (2009)
Palm shell	–	53.4	29.7	–	–	Daud and Ali (2004)
Olive waste	–	28.0	44.8	–	–	Zanzi et al. (2002)
Scots pine stem wood	–	27.0	40.7	26.9	5.0	Raisanen and Athanassiadis (2013)
Sunflower seed hull	11.8	28.7	31.3	25.2	–	Curvetto et al. (2005)

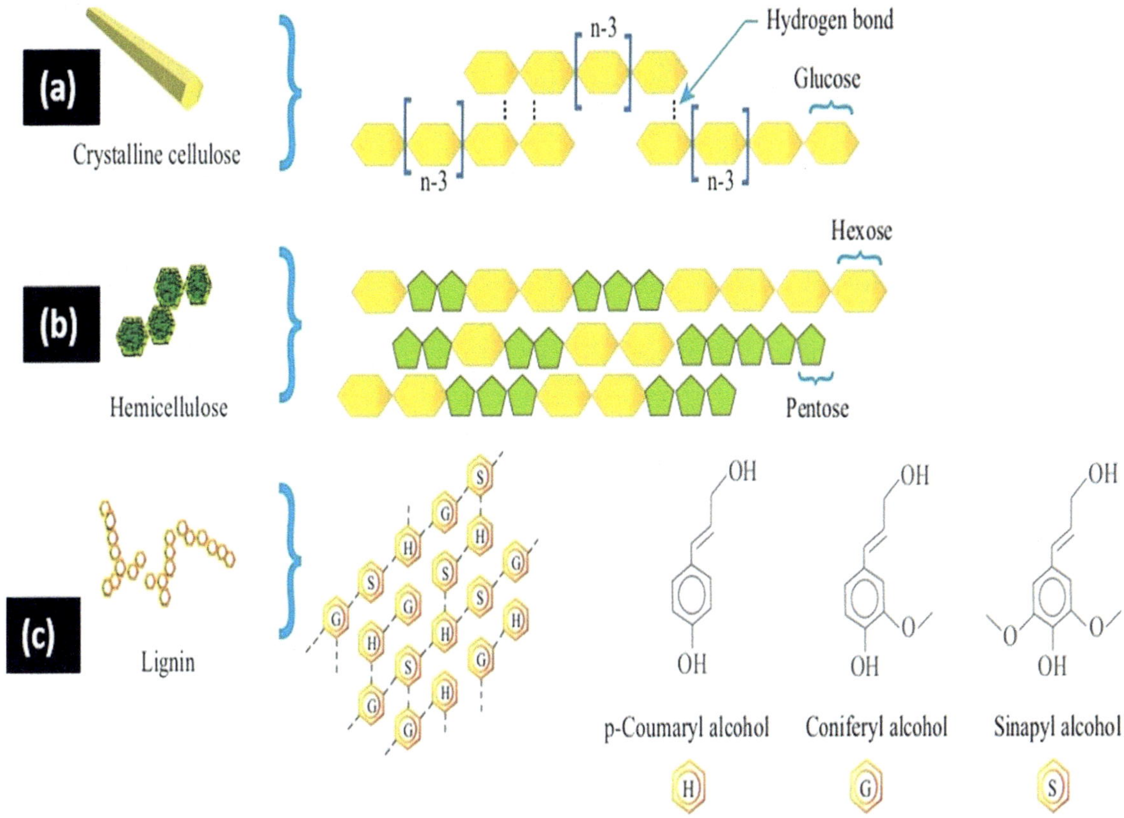

Fig. 1 Chemical configurations of **a** cellulose, **b** hemicellulose, and **c** lignin. Reproduced with permission from Springer Nature, 2015 (Biswas et al. 2015)

present in fruit-based biomass are shown in Table 2 which includes remarkable components such as crude fibers, proteins, ash, lipids, and moisture. The crude protein in pulp is found to be ~4–14% and the peel is around 2–18%, respectively. While the lipid content in pulp is reported to be ~1.4–28% and in peels, it ranges from 0.7 to 9.9%. Overall, in combination, the fruit peel and pulp were found to exhibit crude protein ~5–43% and lipids content approximately 3–28%. During heat treatment, the crude protein and lipids decompose at a reduced heat (~300 °C) because of the loss of organic contents and this limits the overall yield of derived carbons (Nawar 1969). The chemical present fruit biomasses are exhibited in Table 2.

2.3 Animal-Based Carbons

Chitin is considered a promising precursor for the formation of carbons due to chemical stability and abundance. Chitin is different from cellulose and possesses rigorous intermolecular hydrogen bonds and crosslinking networks of chitin-glycan complexes (Schwarz and Moussian 2007), which gives chitin a superior thermal strength than cellulose and thus yields a better percentage of carbon. Numerous animal sources, such as insects, mollusks, and crustaceans are used to derive chitin (Jin et al. 2019). Similar to plant and fruit-based biomass, the carbon yield was found to vary significantly according to the extraction process, precursors, and mechanical/chemical processing conditions (Kovaleva et al. 1772). The animal biomass resources such as horns, hooves, and hairs are identified as a valuable source for the formation of various carbon materials. For instance, carbonaceous flakes derived from human hair displayed notable SA, improved specific capacitance, and extended stability when used as electrodes for supercapacitors (Qian et al. 2014). Apart from chitin, animal skins are considered as an abundant source of protein, which is primarily composed of different amino acids (Chatterjee et al. 2018). A few reports suggest the formation of heteroatom and metal atom functionalized carbons using various tannery and slaughterhouse wastes (Ashokkumar and Ajayan 2020). The derived carbon materials were used for studies related to Li-ion batteries (Wei et al. 2011) and supercapacitors (Huang et al. 2011). The compositions of animal bone, skin, and fish scale are illustrated in Fig. 2.

Table 2 Chemical compositions of typical fruit-based biomass (Liu et al. 2018b)

Biomass	Fruit part	Moisture	Ash	Crude protein	Total lipids	Crude fiber
Avocado Morais et al. (2017)	Seed	67.2 ± 0.6	2.3 ± 0.4	9.6 ± 1.6	3.9 ± 0.3	10.7 ± 2.8
	Raw peel	65.7 ± 3.1	1.5 ± 0.3	6.3 ± 1.1	3.5 ± 0.7	46.9 ± 2.7
	Oven-dried peel	4.0 ± 0.1	2.0 ± 0.3	6.4 ± 0.2	4.7 ± 0.4	43.9 ± 2.1
Pineapple Morais et al. (2017)	Seed	–	–	–	–	–
	Raw peel	82.7 ± 0.7	5.0 ± 0.4	8.8 ± 0.6	1.1 ± 0.2	16.3 ± 2.5
	Oven-dried peel	8.8 ± 0.2	5.1 ± 0.1	7.3 ± 0.9	1.3 ± 0.1	15.9 ± 2.4
Banana Morais et al. (2017)	Seed	–	–	–	–	–
	Raw peel	89.8 ± 0.3	12.8 ± 0.9	9.7 ± 0.3	5.5 ± 0.1	24.2 ± 0.2
	Oven-dried peel	7.6 ± 0.2	13.4 ± 1.8	9.4 ± 0.4	6.1 ± 0.2	23.5 ± 3.8

2.4 Microorganism-Based Carbons

Microorganisms are an exciting source for the formation of carbons apart from plants, animals, and fruit-based biomass resources. Mushrooms (Fig. 3a, c) and yeasts (Fig. 3b, d) are reported as one of the potential biomasses to derive carbons. Table 3 shows the chemical compositions of some microorganisms (Abou Raya et al. 2014; Wang et al. 2014b). The microorganism-based biomass comprises of chitin, which crosslinked with glucan, acts as an essential component to derive carbons during the carbonization process (Arroyo et al. 2016). These biomasses primarily consist of cellulose, which is analogous to the carbonization behavior displayed by plant/fruit-based materials resulting in better carbon yields. The other components, i.e., fat, ash, etc., present in the crude protein leads to an immediate

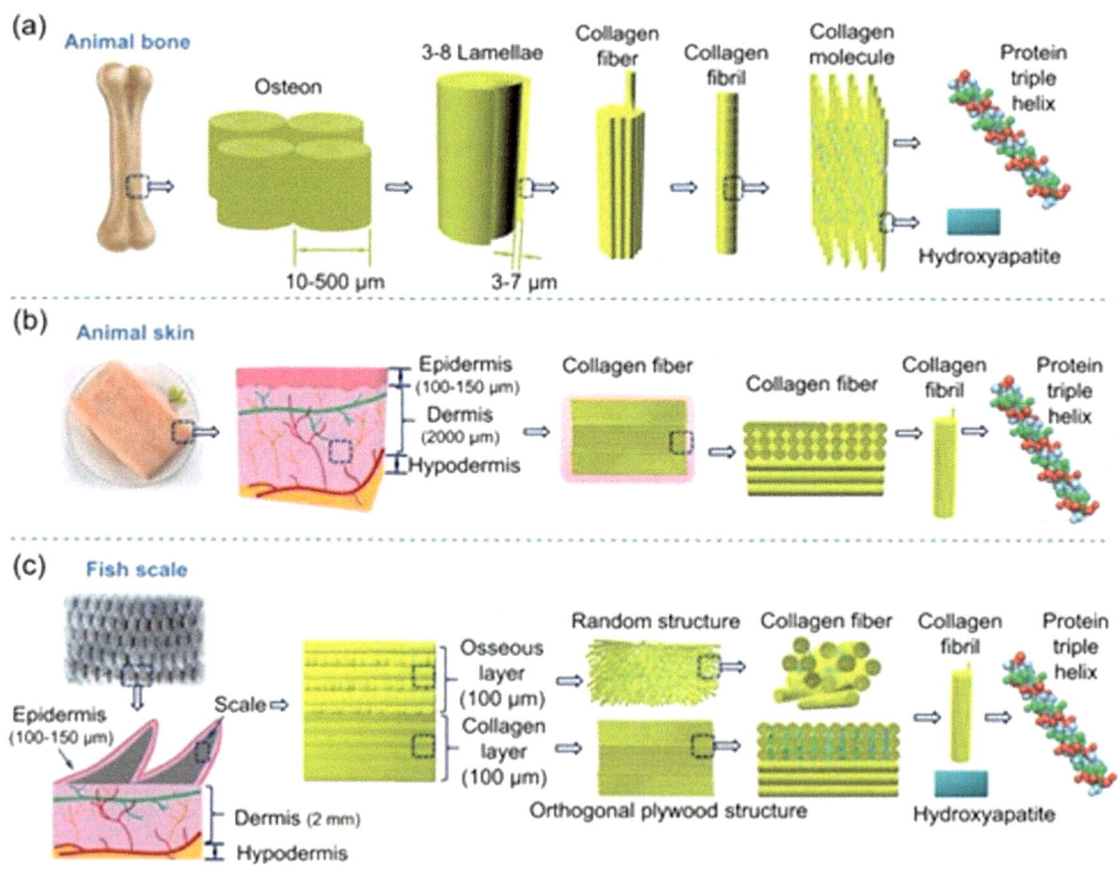

Fig. 2 Structures and compositions of **a** bone, **b** skin for livestock and poultry, and **c** scale and skin of fish. Reproduced with permission from Advanced Functional Materials, 2019 (Jin et al. 2019)

decomposition of highly volatile organics and which leads to reduced carbon content. Among various microorganism-based ingredients, mushrooms are considered as one of the chief and economical sources to derive carbons.

3 Activation of Biomass-Derived Carbons

3.1 Activation of Carbons

As discussed above, porous carbons are synthesized through different methods and most commonly activated using chemical or physical techniques. In general, the activation of carbons was carried out at an inert condition and temperatures between 600 and 800 °C. The nature and chemical composition of the precursor is one of the influential variables that control the comprehensive property of formed carbons (Williams and Reed 2006). The chemical activation includes the usage of activating agents like potassium hydroxide (KOH), phosphoric acid (H3PO_4), sodium hydroxide (NaOH), etc. It is observed that chemical activation has more advantages over the physical activation technique since it helps in the satisfactory formation of pores, with much control over their pore geometries. Moreover, this method guides toward carbons with high SA and better yields. Also, it involves a single step with a lower pyrolysis temperature, and chemical activation is more preferred for supercapacitor and other energy storage applications.

3.1.1 Chemical Activation of Carbons

Potassium hydroxide is one of the commonly used materials for the chemical activation of carbons, which involves redox reaction mechanism where the carbon compounds are scratched by the K-containing species, resulting in abundant micro/mesopores. This action boosts the metallic K group to embed toward the carbon network and enlarge the carbon frames which guides toward a substantial surge in pores in the derived carbon network. The formed porous carbons exhibit a high SSA due to the activation process. The duration of the heating, temperature, and the activation agent is considered to play an essential role in fine-tuning the porosity. Numerous carbon structures with a great variety of pore textures were synthesized using nature-based biomass precursors, including animals, food, and plant wastes, and using KOH as an activator. For example, Huang et al. pyrolyzed fish scales at 950 °C in combination with KOH (1:1 ratio). The formed fish scale-carbons displayed hierarchical lamellar carbon with ordered porosity, exhibiting numerous macropores, extraordinarily high SSA (~2200 m^2 g^{-1}), and pore volume (~2.70 cm^3 g^{-1}) (Chen et al. 2010). A few other works have demonstrated the usage of KOH to prepare a diverse variety of activated porous carbon using precursors such as bones of pig (Huang et al. 2011), sheep (Li et al. 2017), cattle (Zhang et al. 2019), fish scales (Wang et al. 2015a), and bones (Ai et al. 2017). Careful control of pyrolysis temperature and KOH dosage resulted in carbons with interconnected mesopores and macrospores.

Fig. 3 **a** Early stage of *Auricularia thailandica* on sawdust medium, Reproduced with permission from Springer Nature, 2017 (Bandara et al. 2017). **b** SEM of *Saccharomyces cerevisiae*. Reproduced with permission from Elsevier, 2003 (Sunner et al. 2003). **c** SEM image of *Psilocybe cubensis*. The arrow indicates the germ pores. Reproduced with permission from Elsevier, 2003 (Tsujikawa et al. 2003). **d** SEM of a yeast cell. Reproduced with permission from Elsevier, 2015 (Zhang et al. 2015a)

Table 3 Chemical compositions of some typical microorganisms (Liu et al. 2018b)

Microorganisms	Carbohydrates	Crude fiber	Crude protein	Crude fat	Ash	Ref.
Agaricus bisporus	42.56	13.21	33.85	2.41	7.97	Abou Raya et al. (2014)
B. aereus	34.0	17.0	26.9	2.1	8.5	Wang et al. (2014b)

Additionally, the activation agents, like H3PO$_4$ and zinc chloride ($ZnCl_2$), were also used during the process of activation to yield carbons with the better performance (Prahas et al. 2008). The SSA and pore sizes of the activated carbon procured after $ZnCl_2$ and KOH activation revealed remarkable transformations in their network. The carbon samples with higher SSA and a combination of micro/mesopores displayed better supercapacitor performance. Figure 4 demonstrates various conventional methods which were used to convert biomass into carbon materials.

3.1.2 Carbon Activation Through Physical Method

Physical activation comprises of two steps. The first step involves pyrolysis, in an inert condition to derive carbons, and the second step is gasification. The SSA and porosity of the material are improved due to the use of oxidizing gases which include air, steam, or CO_2 (El-Hendawy et al. 2001). In general, the pyrolysis procedure removes volatile resources from the organic bio-resources, while the gasification opens the closed pores due to the decomposition of tar-like products inside the pores. Besides, the active spots intensify as additional organics burn due to oxidizing agent, which results in the removal of unwanted residual impurities from the system and leads to the enhanced porous structure. Yang et al. demonstrated a one-step carbonization process using coconut shells to synthesize activated carbon under CO_2 atmosphere (Yang et al. 2010). The carbon displayed high surface area and a microporous volume of about 1667 $m^2\,g^{-1}$, and $\sim$0.87 $cm^3\,g^{-1}$, respectively, in optimized activated condition, which is greater than that of the same biomass which is carried out using steam activation process (Mi et al. 2012). The observed results exposed that SSA, pore volume improved because of pyrolysis duration, activation condition and the rate of gas flow. Compared to chemical activation, physical activation of biomass resulted in more yield and bulk density, but with an extremely reduced pore size and SSA due to a lower degree of carbon activation. However, the chemical activation method does not have influence over the SSA and electronic properties of the carbon matrix.

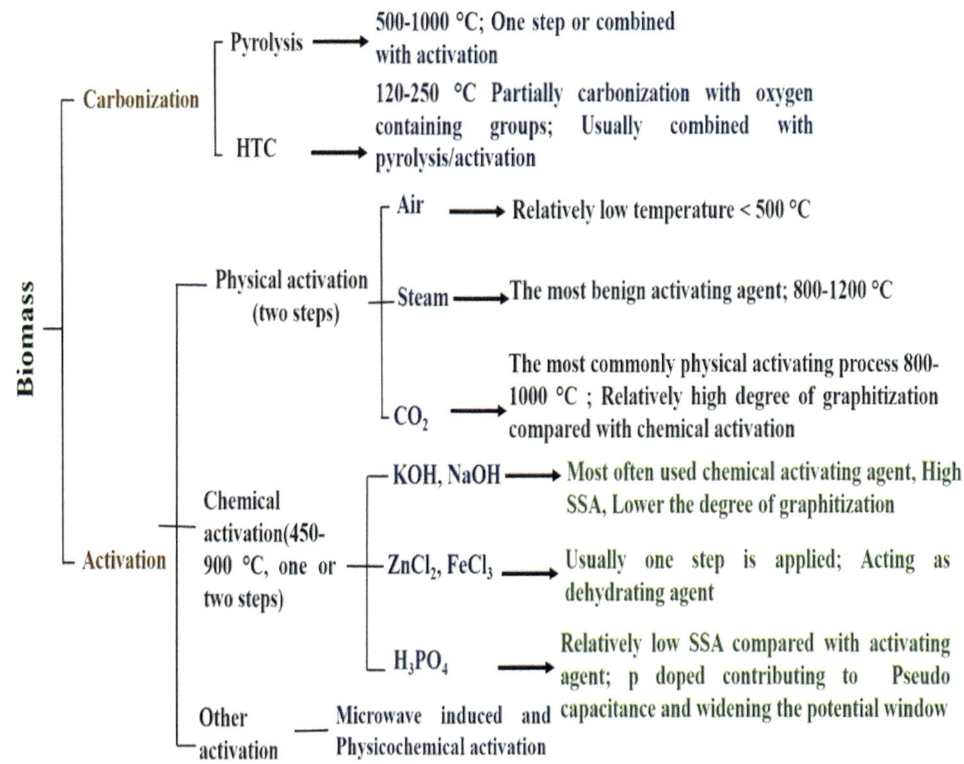

Fig. 4 Conventional methods to convert the biomass into carbon

3.1.3 Self-activation of Carbons

The self-activation process does not require additional activating reagents like physical and chemical activation techniques. It consists of two activation segments. The first part involves the utilization of gases released throughout the pyrolysis procedure to activate the transformed carbon (Bommier et al. 2015). The second part employs inorganic materials (like K^+) that were inherently loaded into biomass precursors which helps in the activation of transformed carbon (chemical self-activation) (Biswal et al. 2013). A one-step annealing process to derive high SSA activated carbons was achieved using cellulose-derived material (Bommier et al. 2015). The self-activation was accomplished due to the gases generated during the process of pyrolysis. This method efficiently uses gas released during the carbonization process to etch the carbon surfaces followed by activating them for better performance.

3.2 Pyrolysis Techniques

Pyrolysis technique has been employed to produce char and carbons (Asensio et al. 2013). It is a process that is used to breakdown the organic complexes thermochemically in the absence of an oxygen atmosphere. In general, pyrolysis can be divided into primary and secondary stages. In the preliminary stage, the temperature instable organic constituents are removed due to dehydration process, which produce bio-oil as a by-product during the condensation process. The heavyweight hydrocarbons are converted into char or gases in the second stage, which are further transformed into carbonaceous materials (Meyer et al. 2011). The pyrolysis conditions are varied from each other based on the product yields and the operating conditions (Goodman et al. 2013). For example, slow pyrolysis can be carried out at lesser temperatures (~300–400 °C), under the reduced heating degrees (~1 − 12 °C/s), and extended residence period (>5 min). Whereas the flash-pyrolysis emerges under higher temperatures (500–600 °C) (Meyer et al. 2011). For example, Huang et al. carbonized the pig bone at 850 °C, followed by acid pickling. The formed carbon exhibited a highly ordered porosity with a surface area of ~850 $m^2\ g^{-1}$ (Huang et al. 2011) and it followed development of micro and mesopore carbons. In another attempt, Redepenning and co-workers produced animal waste-derived porous carbon with a superior SSA of ~1350 $m^2\ g^{-1}$ by pyrolyzing cattle bone at ~950 °C (Goodman et al. 2013). However, at a slightly lower temperature (800 °C), he obtained hierarchical porous carbon with SSAs in between 560 to 770 $m^2\ g^{-1}$. Similarly, Lee et al. studied the comparative difference of five different agriculture wastes such as paddy straw using slow heat treatment and reaching a maximum temperature of 500 °C (Lee et al. 2013). He observed the formation of well-ordered pores with high SSA when sugar cane bagasse and wood stem were used as a source. Nevertheless, the other precursors displayed reduced porosity with a low surface area. Overall, in the pyrolysis method, factors such as residence period, temperature, the particle size of precursors, and the heating rate were found to determine the evolution of carbons. Some of the important criteria which influence the property of carbons electrodes are summarized as follows.

3.2.1 Effect of Temperature

Yang et al. heat-treated various biomass such as hickory wood, bamboo, etc., at identical temperatures and observed a reduction in char due to temperature increase (Sun et al. 2014). The carbon pyrolyzed around 600 °C exhibited a nonporous configuration, signifying the formation of ultrathin plate-like structures. However, the carbons obtained at temperatures more than 800 °C revealed ordered pores with excessive SSAs ranging between 1400 and 2500 $m^2\ g^{-1}$. It is observed that, as the temperature elevated, micropores progressively converted to mesopores, and the pore volume was boosted to ~1.829 $cm^3\ g^{-1}$ for the carbons pyrolyzed at 1100 °C. Garcia-Perez et al. reported a similar observation during the pyrolysis of Douglas fir wood at a temperature range between 350 and 600 °C (Suliman et al. 2016).

3.2.2 Effect of Residence Time

Residence time is considered as an additional variable which affects the distribution of products in different phases. As discussed, a large quantity of char yield can be obtained at prolonged residence period resulting in a better char yield since it takes more time for the re-polymerization to occur. Similarly, shorter residence time causes incomplete re-polymerization. Hou et al. studied the influence of withhold period using rapeseed stem pyrolyzed at a temperature between 150 and 800 °C and at withhold period around 10–100 min. The yield of the char decreased gradually with increases in residence time, which was due to the removal of organics. Choi et al. experimented with the yellow poplar wood under fast pyrolysis condition with the withhold duration between 1.1 and 7.7 s (Zhao et al. 2018), and he observed a surge in char produce with an escalation in residence period.

3.2.3 Heating Rate Effect

Also, the rate of heating was found to affect the yield percent to a certain degree. For instance, Mahinpey et al. examined the devolatilization of wheat straw based on the effect of heating. A growth in char production was observed ~11.5–25% as the heating frequency boosted from 5 to 20 °C/min (Mani et al. 2010). This could be due to the heat transfer effect which takes place in the inner core of the biomass. In the process of slow heating, the decomposition of the biomass results in gradual loss of volatiles organics and hence exhibits decreased char yields. This observation was plausible on account of more loss of volatiles that occur at higher heating rates which induce diminishing of pore walls to become thinner and hence decreased the SA and the pore volume remarkably.

3.2.4 Size of the Particle

The size of particle also depends on the rate of heating. It is perceived that the lesser size of the particles delivers enhanced SSA that permits extra heat to distribute inside the core of the biomass system. Thus, it enhances the pyrolytic reactions and reduces the char yield that is formed during pyrolysis. An increase in char yield is seen with particle size (Yadav and Jagadevan 2019).

3.3 Microwave-Assisted Technique

Microwave-assisted heat treatment is a rapid, simplistic, and energy-saving technique that was used for the effective conversion of biomasses into carbons. Using this technique, micropore and mesopore carbons were produced with a heat withheld duration between 6 and 30 min where the carbon aerogel becomes chemically activated during microwave irradiation (Calvo et al. 2013). This method also yields activated carbons of well-modified surface chemistry, and a remarkable decrease in the micropore volume and size was observed (Puligundla et al. 2016).

3.4 Carbonization by Hydrothermal

Hydrothermal carbonization (HTC) is an alternative technique to produce the carbon with pores for energy storage applications. In this process, a combination of water and carbon sources was subjected to heat at two different temperatures conditions such as from 150 to 300 °C for reduced-temperature HTC and between 300 and 800 °C for elevated temperature of HTC (Zhang et al. 2015b). This method reduces the amount of hydrogen and oxygen formed and leads to the formation of electrodes with an improved surface area. Jain et al. used this technique to carbonize coconut shells under various hydrothermal conditions using zinc chloride ($ZnCl_2$) and hydrogen peroxide (H_2O_2) as an activator. The mesoporous carbon exhibited a specific capacitance of around 240 F g^{-1} and cyclability of 2000 sequences. The electrode material also delivered a 7.6 Wh kg^{-1} at a higher power density of 4.5 kW kg^{-1} (Jain et al. 2015). Likewise, Wu et al. (2013) examined that the hydrothermal treatment of watermelon which leads to formation of three-dimensional (3D) carbon gels which had more porous spongy structure.

3.5 Ionothermal Carbonization

Ionothermal carbonization is a one-stage process for synthesizing porous carbons using biomass resources. In general, ionic liquids are known due to reduced melting point value, high thermal, and chemical stability. This method was used to generate carbons with enhanced SSA and the extended pore volume (Chang et al. 2015). Pampel et al. observed an extraordinary SSA of 2160 m^2 g^{-1} and a pore volume of ~ 1.70 cm^3 g^{-1} for carbons synthesized using glucose and potassium chloride (KCl). Also, it was shown that the addition of a higher amount of KCl had a significant influence in the pore size and SSA of formed carbon and this helped to attain a gravimetric capacitance of around 206 F g^{-1} (Pampel et al. 2016).

3.6 Template Method

This technique was attempted to restrict/reduce the porosity in carbons by using hard and soft templates. The carbons obtained through hard templating/nano-casting approaches were found to reproduce the morphology of templates reversely. Different carbon structures were developed using this method, some of them include hierarchical porous carbon monoliths, carbon nanosheets, ordered mesoporous carbon (OMC) matrix, and carbon spheres. The carbons derived using this method are found to be highly mesoporous. Estevez et al. investigated the double templating approach (Estevez et al. 2013) by joining silica and ice templating accompanied by physical activation. The formed carbons were found to be highly interconnected with the presence of macro, meso, and micropore characteristics.

4 Energy Storage Applications of Biomass Carbons

4.1 Supercapacitors

Supercapacitors (SCs) have gained substantial attention in energy storage applications, including hybrid automobile electric vehicles, memory back-up devices, etc., due to their rapid charging-discharging rates and extended cyclability. Based on the energy storage mechanism, supercapacitors are divided into three categories, i.e., (1) electric double layer capacitors (EDLCs), (2) pseudocapacitors, and (3) hybrid capacitors. In EDLCs, the electrostatic storage was achieved due to the separation of charge in the Helmholtz layer which happens at the interface between the surface of the electrode and the electrolyte (Gao et al. 2011). In this type, porous carbon materials such as activated carbon, mesoporous carbon, and its derivatives were used as an electrode.

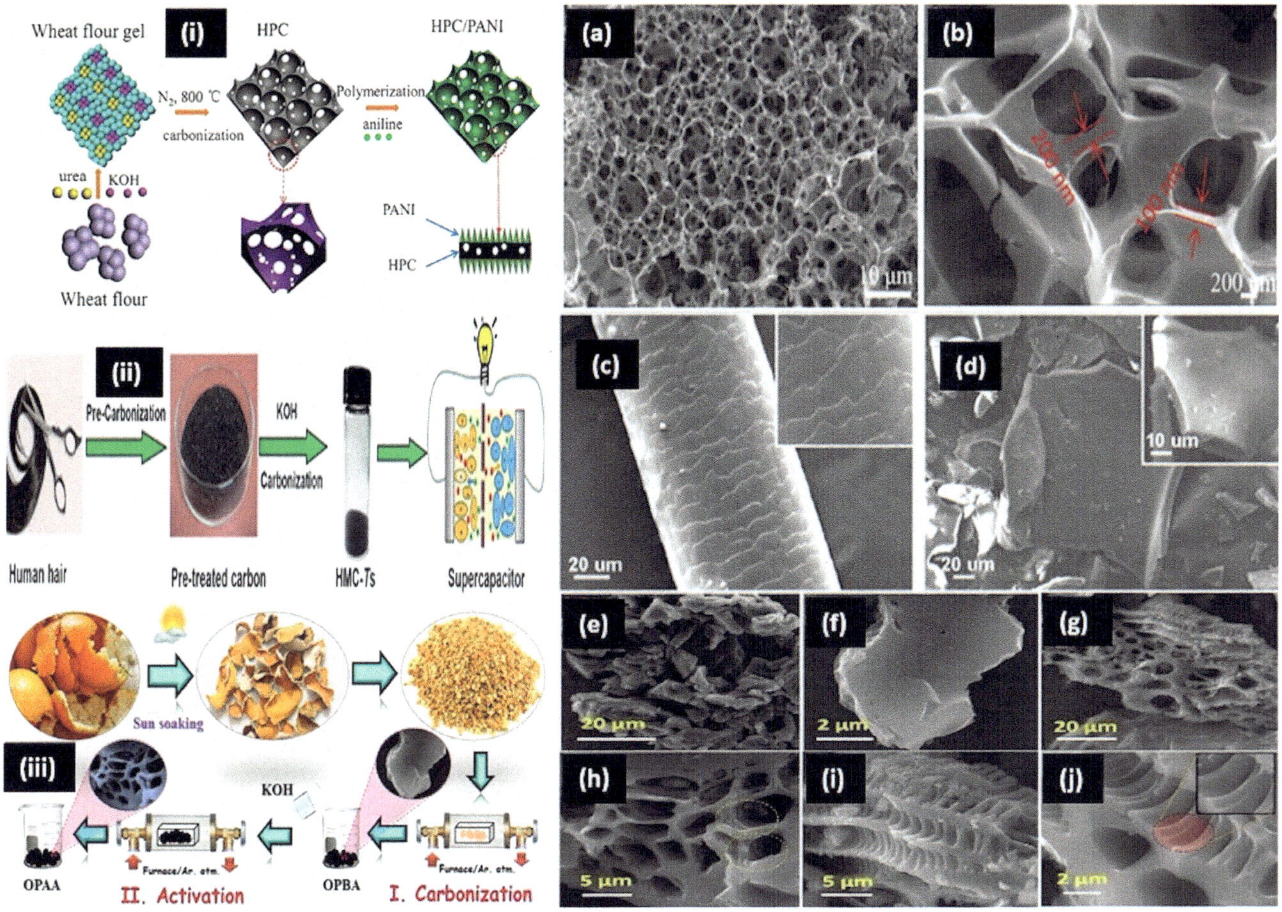

Fig. 5 (i) Schematic and SEM illustrations (**a**, **b**) of HPC/PANI composite, Reproduced with permission from Advanced Energy Materials, 2016 (Pingping et al. 2016). (ii) Flow diagram for the fabrication and SEM images (**c**, **d**) of human hair, Reproduced with permission from Energy and Environmental Science, 2014 (Qian et al. 2014). (iii) Schematic representation and SEM micrographs (**e**–**j**) of orange peel, Reproduced with permission from Energy Technology and Environmental Science, 2017 (Kaipannan et al. 2017)

These materials have high SSA, enhanced porosity, extraordinary electrical conductivity, exceptional chemical, and thermal stability. On the other hand, PCs are centered on the redox effects that happen mostly on the surface of the electrodes. Hence, it exhibits a high amount of charge storage capacity; however, reduced rate capability and cyclability were seen when compared to EDLCs. Recent studies propose biomass-derived carbons as an exciting candidate to develop a wide range of SCs electrode or substrates with effective pseudocapacitive behavior. Efforts such as tuning of pore structure, surface modifications were carried out to develop SCs with high performance. Schematic representations (Fig. 5i, ii, and iii), the SEM images of HPC/PANI composite, human hair, and orange peel are shown in Fig. 5a–j, respectively.

Yu et al. discussed the synthesis of ordered nitrogen-doped carbon (HPC)/polyaniline (PANI) nanowire arrays using the in situ polymerization method which formulates a hierarchically porous carbon structure. The prepared electrode materials delivered enhanced specific capacitance of ~383 and 1080 F g^{-1} for HPC and HPC/PANI in 1 M H_2SO_4, respectively, besides, they exposed a SSA of about 923 m^2 g^{-1}. They have also assembled the asymmetric supercapacitor, which exhibits specific capacitance (~130 F g^{-1}), and energy density (60.3 Wh kg^{-1}), with a good cycling steadiness of 91.6% capacitance following 5000 cycles (Pingping et al. 2016). Wenjing et al. demonstrated Chinese human hair fibers for the synthesis of heteroatom functionalized carbon flakes through the carbonization process. The human hairs were heat treated at 800 °C which exhibited enhanced charge storage with a specific capacitance of 340 F g^{-1} and cyclability more than 20,000 cycles (Qian et al. 2014). Similarly, Subramani et al. investigated three-dimensional (3D) non-porous carbon material obtained from orange peel wastes. The derived carbon materials were investigated for the fabrication of symmetric flexible solid-state supercapacitor (SSC) and found to display high energy density. Nano-porous carbon exhibits an extraordinary specific SSA of 2160 m^2 g^{-1} and an average pore volume of

Table 4 Supercapacitor performance and the related structural parameters of some biomass-derived carbons (Liu et al. 2018b)

Precursor	Biomass-derived carbon	Activation method	SSA ($m^2\ g^{-1}$)	Pore volume ($cm^3\ g^{-1}$)	Specific capacitance ($F\ g^{-1}$)	Cycling stability (cycles)
Poplar catkin Su et al. (2017)	Nitrogen and oxygen-doped carbon	$ZnCl_2$/C = 3:1, 800 °C, 2 h	1462.5	1.31	251	~100% 1000
Auricularia Long et al. (2015)	*Porous graphene-like carbon (PGC)*	*One-pot hydrothermal*	1103	0.54	374	99% 10,000
Human hair Qian et al. (2014)	Heteroatom-doped porous carbon flakes (HMC)	KOH/C = 2:1, 800 °C, 2 h	1306	0.90	445	98% 20,000
Orange peel Kaipannan et al. (2017)	3D nonporous carbon	KOH/C = 3:1, 600 °C, 1 h	2160	0.77	460	98% 10,000

Table 5 Comparison of cycling and rate performance of the carbon anodes

Sample	Cyclability	Reference
Sweet potato-derived carbon nanoparticles	200 cycle, 320 mAh g^{-1}	Peng et al. (2015)
Porous carbon nanofiber webs derived from bacterial cellulose	100 cycle, 914 mAh g^{-1}	Wang et al. (2015b)
Carbon fibers from bamboo chopsticks	300 cycle, 710 mAh g^{-1}	Jian et al. (2014)

0.779 cc g^{-1}. The fabricated cell delivered an enhanced specific capacitance of 460 F g^{-1}. It also exhibited remarkable electrochemical steadiness of about 98% for 10,000 cycles (Kaipannan et al. 2017). Table 4 shows the supercapacitor behavior of some biomass origin carbon electrodes.

4.2 Li/Na-Ion Batteries

In typical, graphite is studied as a prominent anode for commercial Li-ion batteries (LIBs) owing to their cost-effectiveness, electrical conductivity, extended cycling stability, and eco-friendly nature (Shen et al. 2012). But it is reported to exhibit a lesser capacity of ~370 mAh g^{-1} and a reduced rate capability since the Li^+ diffusion coefficient was at a lower value. Thus, it fails to solve the immediate requirement for the development of advanced power systems such as hybrid electric vehicles (HEVs) and unmanned aerial vehicles (UAVs). Hence, significant attempts have been put forth to derive high-performance carbon-based materials that can have an extraordinary capacity, great rate performance, and easy availability. Peng et al. investigated the possibility of deriving micro-nano-structure hard carbons using filter papers. The formed carbons were used to develop Na-ion battery and reported to have a better reversible capacity of 286 mAh g^{-1} following 100 cycles and at an applied current of 20 mA g^{-1}. The observed exceptional electrochemical functioning demonstrates that these cheap carbons could be a favorable anode for Na-ion batteries (Peng et al. 2016). Zhu et al. discussed cotton cellulose as an alternative to preparing porous carbon. The formed carbons had a SSA of 1260 $m^2\ g^{-1}$. The derived carbon was used as an anode and resulted in a reversible capacity of 793 mAh g^{-1} and current density of 0.5 A g^{-1} following 500 cycles (Chunyu and Tomohiro 2016). Some of the battery performance and the cyclability of the biomass-originated carbon anodes are tabulated in Table 5. The schematic illustration (Fig. 6a) and the SEM images of filter paper-derived carbon are displayed in Fig. 6b, c. The SEM images of cotton-derived porous carbon (Fig. 6d), the rate capabilities, cycle performance, and the galvanostatic charge–discharge profiles are demonstrated in Fig. 6e, f.

5 Conclusion

The quest for a spotless, renewable source is of high significance due to the depletion of traditional resources and to reduce the amount of CO_2 generated. Biowastes are contemplated as one of the most versatile, green, and renewable resources for the production of comprehensive variety of carbon structures. These carbon materials are reported to possess interesting characteristics suitable for several applications related to environmental and energy conversion/storage. More specifically, the consumption of various naturally available biomass resources to derive functional porous carbons with significant impact in multiple fields is considered green, economical, and sustainable. Though, the processing ability, molecular arrangement, and overall functioning of synthesized bio-derivative carbons are intensely interconnected with chemical configuration,

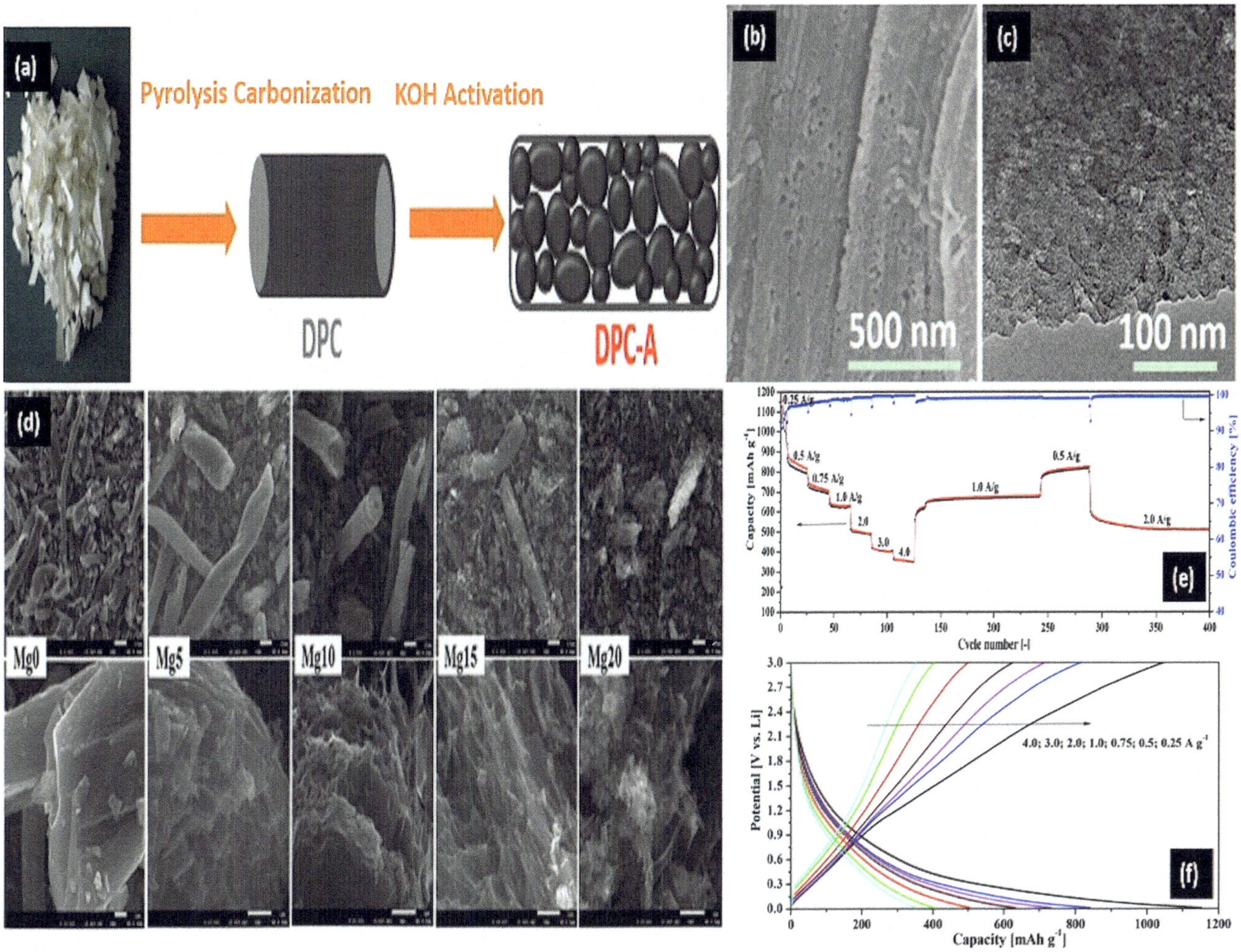

Fig. 6 Schematic illustration (**a**) and SEM images (**b**, **c**) micro-nano-carbon structure formed using filter paper, Reproduced with permission from Scientific Reports, 2016 (Peng et al. 2016). SEM images (**d**), rate capabilities and cycle performance (**e**), and galvanostatic charge–discharge profiles (**f**) of the cotton-derived porous carbon, Reproduced with permission from RSC, 2015 (Chunyu and Tomohiro 2016)

microscopic structure, including sublevel microstructural features of the precursors used for synthesis. Thus, it necessitates the sensible design, including careful selection of precursor, carbonization, and activation technique to acquire high-performance carbon or carbon composites from biomass with enhanced surface area and pore hierarchy. However, an improved process, including efficient activation and heteroatom doping, can lead to the development of high-performance carbons with tuned order and enhanced conductivity for superior electrochemical applications. In summary, some of the engineered carbons derived from cost-effectiveness, eco-friendly biomass resources, which can act as an excellent candidate for the development of various supercapacitors and Li-ion batteries with enhanced electrochemical performance, high-rate capability, and cycling stability were discussed in this chapter.

Acknowledgements TPK and SJ acknowledge the Science and Engineering Research Board (SERB), Department of Science and Technology (DST), Government of India, for financial assistance through Overseas Visiting Doctoral Fellowship Scheme (F. No. ODF/2018/000956).

References

Abou Raya MA, Shalaby MT, Hafez SA, Alshimaa MH (2014) Chemical composition and nutritional potential of some mushroom varieties cultivated in Egypt. J Food Dairy Sci 5:421–434

Ai F, Liu N, Wang W, Wang A, Wang F, Zhang H, Huang Y (2017) Heteroatoms-doped porous carbon derived from tuna bone for high performance Li-S batteries. Electrochemica Acta 258:80–89

Arroyo J, Farkas V, Sanz AB, Cabib E (2016) Strengthening the fungal cell wall through chitin–glucan cross-links: Effects on morphogenesis and cell integrity. Cell Microbiol 18:1239–1250

Arsene MA, Bilba K, Savastano H, Ghavami K (2013) Treatments of non-wood plant fibres used as reinforcement in composite materials. Mater Res 16:903–923

Asensio V, Vega FA, Andrade ML, Covelo EF (2013) Tree vegetation and waste amendments to improve the physical condition of copper mine soils. Chemosphere 90:603–610

Ashokkumar M, Ajayan PM (2020) Materials science perspective of multifunctional materials derived from collagen. Int Mater Rev. https://doi.org/10.1080/09506608.2020.1750807

Bandara AR, Karunarathna SC, Mortimer PE, Hyde KD, Khan S, Kakumyan P, Xu J (2017) First successful domestication and determination of nutritional and antioxidant properties of the red ear mushroom Auricularia thailandica (Auriculariales, Basidiomycota). Mycol Progress. 16:1029–1039

Biswal M, Banerjee A, Deo M, Ogale S (2013) From dead leaves to high energy density supercapacitors. Energy Environ Sci 6:1249–1259

Biswas R, Uellendahl H, Ahring BK (2015) Wet explosion: a universal and efficient pretreatment process for lignocellulosic biorefineries. Bioenergy Resour 8:1101–1116

Blankenship TS, Mokaya R (2017) Cigarette butt-derived carbons have ultra-high surface area and unprecedented hydrogen storage capacity. Energy Environ Sci 10:2552–2562

Bommier C, Xu R, Wang W, Wang X, Wen D, Lu J, Ji X (2015) Self-activation of cellulose: a new preparation methodology for activated carbon electrodes in electrochemical capacitors. Nano Energy 13:709–717

Boyjoo Y, Cheng Y, Zhong H, Tian H, Pan J, Pareek VK, Jiang SP, Lamonier JF, Jaroniec M, Liu J (2017) From waste coca cola to activated carbon with impressive capabilities for CO_2 adsorption and supercapacitors. Carbon 116:490–499

Calvo EG, Ferrera-Lorenzo N, Menendez JA, Arenillas A (2013) Microwave synthesis of micro-mesoporous activated carbon xerogels for high performance supercapacitors. Microporous Mesoporous Mater 168:206–212

Chang Y, Antonietti M, Fellinger TP (2015) Synthesis of nanostructured carbon through ionothermal carbonization of common organic solvents and solutions. Angewandte Chemie-Int Ed 54:5507–5512

Chatterjee K, Ashokkumar M, Gullpalli H, Gong Y, Vajtai R, Thankikaivelan P, Ajayan PM (2018) Nitrogen-rich carbon nano-onions for oxygen reduction reaction. Carbon 130:645–651

Chen W, Zhang H, Huang Y, Wang W (2010) A fish scale based hierarchical lamellar porous carbon material obtained using a natural template for high performance electrochemical capacitors. J Mater Chem 20:4773–4775

Chen JZ, Fang KL, Chen QY, Xu JL, Wong CP (2018) Integrated paper electrodes derived from cotton stalks for high-performance flexible supercapacitors. Nano Energy 53:337–344

Chunyu Z, Tomohiro A (2016) Cotton derived porous carbon via an MgO template method for high performance lithium ion battery anode. Green Chem 18:2106–2114

Curvetto NR, Figlas D, Matute RG, Delmastro S (2005) Shiitake bag cultivation on sunflower seed hulls. In: Shiitake Mushroom Growers' handbook (Chap. 4). Brighton, Mush World, pp 100–104

Dai C, Wan J, Geng W, Song S, Ma F, Shao J (2017) KOH direct treatment of kombucha and in situ activation to prepare hierarchical porous carbon for high-performance supercapacitor electrodes. J Solid State Electrochem 21:2929–2938

Daud WMA, Ali WSW (2004) Comparison on pore development of activated carbon produced from palm shell and coconut shell. Bio Resour Technol 93:63–69

Duan B, Gao X, Yao X, Fang Y, Huang L, Zhou J, Zhang L (2016) Unique elastic N-doped carbon nanofibrous microspheres with hierarchical porosity derived from renewable chitin for high rate supercapacitors. Nano Energy 27:482–491

El-Hendawy ANA, Samra SE, Girgis BS (2001) Adsorption characteristics of activated carbons obtained from corncobs. Colloids Surf Physicochem Eng Aspects 180:209–221

Estevez L, Dua R, Bhandari N, Ramanujapuram A, Wang P, Giannelis EP (2013) A facile approach for the synthesis of monolithic hierarchical porous carbons—high performance materials for amine-based CO_2 capture and supercapacitor electrode. Energy Environ Sci 6:1785–1790

Fan Z, Chen J, Cui K, Sun F, Xu Y, Kuang Y (2007) Preparation and capacitive properties of cobalt nickel-oxides/carbon nanotube composites. Electrochem Acta 52:2959–2965

Gao Z, Wang J, Li Z et al (2011) Graphene nanosheet/Ni_2^+/Al_3^+ layered double-hydroxide composite as a novel electrode for a supercapacitor. Chem Mater 23:3509–3516

Gao Z, Zhang Y, Song N, Li X (2016) Biomass-derived renewable carbon materials for electrochemical energy storage. Mater Res Lett 5:69–88

Goodman PA, Li H, Gao Y, Lu Y, Stenger-Smith J, Redepenning J (2013) Preparation and characterization of high surface area, high porosity carbon monoliths from pyrolyzed bovine bone and their performance as supercapacitor electrodes. Carbon 55:291–298

Grey C, Tarascon J (2017) Sustainability and in situ monitoring in battery development. Nat Mater 16:45–56

Hu CC, Wang CC, Wu FC, Tseng RL (2007) Characterization of pistachio shell-derived carbons activated by a combination of KOH and CO_2 for electric double-layer capacitors. Electrochemica Acta 52:2498–2505

Huang W, Zhang H, Huang Y, Wang W, Wei S (2011) Hierarchical porous carbon obtained from animal bone and evaluation in electric double-layer capacitors. Carbon 49:838–843

Jain A, Xu C, Jayaraman S, Balasubramanian R, Lee JY, Srinivasan MP (2015) Mesoporous activated carbons with enhanced porosity by optimal hydrothermal pre-treatment of biomass for supercapacitor applications. Microporous Mesoporous Mater 218:55–61

Jian J, Jianhui Z, Wei A, Zhanxi F, Xiaonan S, Chenji Z, Jinping L, Hua Z, Ting Y (2014) Evolution of disposable bamboo chopsticks into uniform carbon fibers: a smart strategy to fabricate sustainable anodes for Li-ion batteries. Energy Environ Sci 7:2670–2679

Jin N, Rong S, Mengyue L, Yongxi Z, Meiling D, Jingjun L, Zhengping Z, Yaqin H, Feng W (2019) Porous carbons derived from collagen-enriched biomass: tailored design, synthesis, and application in electrochemical energy storage and conversion. Adv Funct Mater 1905095

Kaipannan S, Nagarajan S, Manickavasakam K, Marappan S (2017) Orange peel derived activated carbon for fabrication of high-energy and high-rate supercapacitors. Chem Select 2:11384–11392

Karnan M, Subramani K, Sudhan N, Ilayaraja N, Sathish M (2016) Aloe vera derived activated high-surface-area carbon for flexible and high-energy supercapacitors. Appl Mater Interfaces 8:35191–35202

Kovaleva E, Pestov A, Stepanova D, Molochnikov L (2016) Characterization of chitin and its complexes extracted from natural raw sources. In: AIP conference proceedings, vol 1772, p 050007

Kumar R, Mago G, Balan V, Wyman CE (2009) Physical and chemical characterizations of corn stover and poplar solids resulting from leading pretreatment technologies. Bio Resour Technol 100:3948–3962

Lee Y, Park J, Ryu C, Gang KS, Yang W, Park YK, Jung J, Hyun S (2013) Comparison of bio char properties from biomass residues produced by slow pyrolysis at 500 °C. Bio Resourc Technol 148:196–201

Li Z, Zhang L, Amirkhiz BS, Tan X, Xu Z, Wang H, Olsen BC, Hol CMB, Mitlin D (2012) Carbonized chicken eggshell membranes with 3D architectures as high-performance electrode materials for supercapacitors. Adv Energy Mater 2:431–437

Li S, Xu R, Wang H, Brett DJ, Ji S, Pollet BG, Wang R (2017) Ultra-high surface area and mesoporous N-doped carbon derived from sheep bones with high electrocatalytic performance toward the oxygen reduction reaction. J Solid State Electrochem 21:2947–2954
Liu W, Mei J, Liu G, Kou Q, Yi T, Xiao S (2018a) Nitrogen-doped hierarchical porous carbon from wheat straw for supercapacitors. Sustain Chem Eng 6:11595–11605
Liu Y, Chen J, Cui B, Yin P, Zhang C (2018) Design and preparation of biomass-derived carbon materials for supercapacitors. J Carbon Res 4:53
Long C, Chen X, Jiang L, Zhi L, Fan Z (2015) Porous layer-stacking carbon derived from in-built template in biomass for high volumetric performance supercapacitors. Nano Energy 12:141–151
Ma QL, Yu YF, Sindoro M, Fane AG, Wang R, Zhang H (2017) Carbon-based functional materials derived from waste for water remediation and energy storage. Adv Mater 29:1605361
Mani T, Murugan P, Abedi J, Mahinpey N (2010) Pyrolysis of wheat straw in a thermogravimetric analyzer: effect of particle size and heating rate on devolatilization and estimation of global kinetics. Chem Eng Resour Design 88:952–958
Meng Q, Cai K, Chen Y, Chen L (2017) Research progress on conducting polymer-based supercapacitor electrode materials. Nano Energy 36:268–285
Merlet C, Rotenberg B, Madden PA, Taberna PL, Simon P, Gogotsi Y, Salanne M (2012) On the molecular origin of supercapacitance in nanoporous carbon electrodes. Nat Mater 11:306–310
Meyer S, Glaser B, Quicker P (2011) Technical, economical, and climate-related aspects of bio char production technologies: a literature review. Environ Sci Technol 45:9473–9483
Mi J, Wang XR, Fan RJ, Qu WH, Li WC (2012) Coconut-shell-based porous carbons with a tunable micro/mesopore ratio for high-performance supercapacitors. Energy Fuels 26:5321–5329
Morais DR, Rotta EM, Sargi SC, Bonafe EG, Suzuki RM, Souza NE, Matsushita M, Visentainer (2017) Mineral contents and fatty acid composition of the different parts and dried peels of tropical fruits cultivated in Brazil. J Braz Chem Soc 28:308–318
Nawar WW (1969) "Thermal degradation of lipids" A review. J Agric Food Chem 17:18–21
Niu Z, Dong H, Zhu B, Li J, Hng HH, Zhou W, Chen X, Xie SZ (2013) Highly stretchable, integrated supercapacitors based on single-walled carbon nanotube films with continuous reticulate architecture. Adv Mater 25:1058–1064
Pampel J, Denton C, Fellinger TP (2016) Glucose derived ionothermal carbons with tailor-made porosity. Carbon 107:288–296
Peng Z, Ting L, Jinzheng Z, Lifeng Z, Yi L, Jianfeng H, Shouwu G (2015) Sweet potato-derived carbon nanoparticles as anode for lithium ion battery. RSC Adv 5:40737–40741
Peng Z, Ting L, Guo S (2016) Micro-nano structure hard carbon as a high performance anode material for sodium-ion batteries. Sci Rep 6:35620
Pingping Y, Zhiming Z, Lingxia Z, Feng T, Linfeng Hand Xiaosheng F (2016) A novel sustainable flour derived hierarchical nitrogen doped porous carbon/polyaniline electrode for advanced asymmetric supercapacitors. Adv Energy Mater 6:1601111
Prahas D, Kartika Y, Indraswati N et al (2008) Activated carbon from jackfruit peel waste by H_3PO_4 chemical activation: pore structure and surface chemistry characterization. Chem Eng J 140:32–42
Prasankumar T, Vigneshwaran J, Abraham S, Jose SP (2019) 3D structures of graphene oxide and graphene analogue MoS_2 with polypyrrole for supercapacitor electrodes. Mater Lett 238:121–125
Puligundla P, Oh S, Mok C (2016) Microwave-assisted pretreatment technologies for the conversion of Lignocellulosic biomass to sugars and ethanol: a review. Carbon Let 17:1–10
Qian W, Sun F, Xu Y, Qiu L, Liu C, Wang S, Yan F (2014) Human hair-derived carbon flakes for electrochemical supercapacitors. Energy Environ Sci 7:379
Raisanen T, Athanassiadis D (2013) Basic chemical composition of the biomass components of pine, Spruce and Birch. For Refine
Ranaweera CK, Kahol PK, Ghimire M, Mishra SR, Gupta RK (2017) Orange-peel-derived carbon: designing sustainable and high-performance supercapacitor electrodes. J Carbon Res 3:25
Schwarz H, Moussian B (2007) Electron-microscopic and genetic dissection of arthropod cuticle Differentiation. Mod Res Educ Top Microsc 3:316–325
Shen L, Uchaker E, Zhang X, Cao G (2012) Hydrogenated $Li_4Ti_5O_{15}$ nanowire arrays for high rate lithium ion batteries. Adv Mater 24:6502–6506
Su X, Cheng M, Fu L, Yang J, Zheng X, Guan X (2017) Superior Super capacitive performance of hollow activated carbon nano mesh with hierarchical structure derived from poplar catkins. J Power Sour 362:27–38
Suliman W, Harsh JB, Abu-Lail NI, Fortuna AM, Dall Meyer I, Garcia-Perez M (2016) Influence of feedstock source and pyrolysis temperature on bio char bulk and surface properties. Biomass Bioenergy 84:37–48
Sun Y, Gao B, Yao Y, Fang J, Zhang M, Zhou Y, Chen H, Yang L (2014) Effects of feedstock type, production method, and pyrolysis temperature on bio char and hydro char properties. Chem Eng J 240:574–578
Sunner J, Avci R, Richards L, Groenewold G, Ingram J, Arthun M (2003) Preservation of yeast cell morphology for scanning electron microscopy using 3.28-μm IR laser irradiation. J Microbiol Methods 54:285–287
Tsujikawa K, Kanamori T, Iwata Y, Ohmae Y, Sugita R, Inoue H, Kishi T (2003) Morphological and chemical analysis of magic mushrooms in Japan. Forensic Sci Int 138:85–90
Wang L, Shi YT, Wang YX, Zhang H, Zhou HW, Wei Y, Tao SY, Ma TL (2014a) Composite catalyst of rosin carbon/Fe_3O_4: highly efficient counter electrode for dye-sensitized solar cells. Chem Commun 50:1701–1703
Wang XM, Zhang J, Wu LH, Zhao YL, Li T, Li JQ, Wang YZ (2014) A mini-review of chemical composition and nutritional value of edible wild-grown mushroom from China. Food Chem 151:279–285
Wang J, Shen L, Xu Y, Dou H, Zhang X (2015) Lamellar-structured biomass-derived phosphorus- and nitrogen-co-doped porous carbon for high-performance supercapacitors. New J Chem 39:9497–9503
Wang W, Sun Y, Liu B, Wang S, Cao M (2015) Porous carbon nanofiber webs derived from bacterial cellulose as an anode for high performance lithium ion batteries. Carbon 91:56–65
Wang C, Wang XF, Lu H, Li HL, Zhao XS (2018) Cellulose-derived hierarchical porous carbon for high-performance flexible supercapacitors. Carbon 140:139–147
Wei S, Zhang H, Huang Y, Wang W, Xia Y, Yu Z (2011) Pig bone derived hierarchical porous carbon and its enhanced cycling performance of lithium-sulfur batteries. Energy Environ Sci 4:736–740
Williams PT, Reed AR (2006) Development of activated carbon pore structure via physical and chemical activation of biomass fibre waste. Biomass Bioenerg 30:144–152
Wu XL, Wen T, Guo HL, Yang S, Wang X, Xu AW (2013) Biomass-derived sponge-like carbonaceous hydrogels and aerogels for supercapacitors. Nano 7:3589–3597
Yadav K, Jagadevan S (2019) Influence of process parameters on synthesis of biochar by pyrolysis of biomass: an alternative source of energy. Intech Open. https://doi.org/10.5772/intechopen.88204

Yang K, Peng J, Xia H, Zhang L, Srinivasakannan C, Guo S (2010) Textural characteristics of activated carbon by single step CO_2 activation from coconut shells. J Taiwan Inst Chem Eng 41:367–372

Zanzi R, Sjostrom K, Bjornbom E (2002) Rapid pyrolysis of agricultural residues at high temperature. Biomass Bioenerg 23:357–366

Zhai Y, Dou Y, Zhao D, Fulvio PF, Mayes RT, Dai SY (2011) Carbon materials for chemical capacitive energy storage. Adv Mater 23:4828–4850 (2011)

Zhang L, Wang Q, Wang BG, Yang L, Lucia A, Chen J (2015) Hydrothermal carbonization of corncob residues for hydro char production. Energy Fuels 29:872–876

Zhang J, Wang X, Li Q, Shang JK (2015) Real time, in situ observation of the photocatalytic inactivation of Saccharomyces cerevisiae cells. Mater Sci Eng C 49:75–83

Zhang X, Cui Y, Zhong Y, Wang D, Tang W, Wang X, Xia X, Gu C, Tu J (2019) Cobalt disulfide-modified cellular hierarchical porous carbon derived from bovine bone for application in high-performance lithium-sulfur batteries. J Colloid Interface Sci 551:219–226

Zhao B, O'Connor D, Zhang J, Peng T, Shen Z, Tsang DC, Hou D (2018) Effect of pyrolysis temperature, heating rate, and residence time on rapeseed stem derived bio char. Clean Prod 174:977–987

Zhu H, Yin J, Wang X, Wang H, Yang X (2013) Microorganism-derived heteroatom-doped carbon materials for oxygen reduction and supercapacitors. Adv Func Mater 23:1305–1312

Zhu L, Shen F, Smith RL, Yan L, Li L, Qi X (2017) Black liquor-derived porous carbons from rice straw for high-performance supercapacitors. Chem Eng J 316:770–777

Green Synthesis of Nanomaterials via Electrochemical Method

Aamir Ahmed and Sandeep Arya

Abstract

In nanotechnology, the application of hazardous chemicals and by-products has always been a concern. The green synthesis involves the method and ingredients which are not threatening to the human beings and the environment. The electrochemical method has the potential to be a part of the green synthesis technique. In green electrochemical synthesis, the application of non-toxic solvents such as water, ionic liquids, plant extract, and safe chemical solvents as an electrolyte is the main focus. The simple electrochemical methods along with electrodeposition are such green electrochemical techniques that have been used by the vast majority of researchers for nanomaterial synthesis. Also, there is a need for more researchers to work using this green technique so that we can take one more step toward our dream of sustainable development.

Keywords

Synthesis • Green • Electrochemical • Nanomaterials • Electrodeposition • Sustainable development • Eco-friendly • Plant extract • Nanoparticles • Nanosheets

1 Introduction

Nanotechnology in recent years has been one of the most rapidly growing concepts in science and technology. It has played a key role in the development of materials at nanoscale having unique properties in comparison with their bulk counterparts. The development of scanning tunnel microscope (STM) and atomic microscope (AM) has allowed scientists to see things at the atomic level which eventually led to the revolutions in nanotech. Most of the current researchers believe that "nanotechnology is the future." Nanotechnology has a huge potential to develop new devices, sensors, solar cells, new systems, medicines, etc., in the various fields of research and studies (Mirzaei and Darroudi 2017; Arruda et al. 2015). These nanoparticles or materials have the size in the range of nanometers (10^{-9} m) and possess catalytic properties, chemical stability, enhanced thermal conductivity, and unique optical properties; all due to their large surface area-to-volume ratio (Agarwal et al. 2017). Basically, there are two methodologies used for nanoparticle synthesis, i.e., "top-down" and "bottom-up" approach. The top-to-bottom approach makes use of various techniques such as milling, grinding, and sputtering to break down bulk material into small particles having the size in nanometers. This approach was used in ancient times and that is the reason we have some remarkable antiques like Lycurgus cup, the windows of Notre Dame Cathedral, the extraordinarily sharp and shiny Damascus swords, etc. Another method that is used for the synthesis of nanomaterials is the bottom-to-up approach. In this method, the nanoparticles are synthesized from the atomic or molecular level by the mechanism of self-assembly. It includes chemical and biological methods such as electrochemical methods, sol-gel, and flame spraying. (Mathur et al. 2018). This is the most commonly used method in modern times and is less laborious, and the particle size and shape can be controlled as well.

The chemical methods are mostly used in the synthesis of nanomaterials by most of the researchers. In this method, chemical solvents and solutes are used for the synthesis of the nanoparticles. The methods include sol-gel method, hydrothermal synthesis, chemical vapor deposition, electrochemical method, etc. The chemicals used are toxic and sometimes the nanomaterials obtained also possess toxic features. For example, cadmium (Cd) used for the preparation of cadmium sulfide (CdS) nanoparticles is very toxic,

A. Ahmed · S. Arya (✉)
Department of Physics, University of Jammu, Jammu, 180006, Jammu and Kashmir, India

Inamuddin et al. (eds.), *Advances in Green Synthesis*, Advances in Science, Technology & Innovation,
https://doi.org/10.1007/978-3-030-67884-5_11

mercury (Hg) used in the preparation of nanoparticles is also very harmful to living beings, and nickel (Ni) used in the preparation of Ni-based nanoparticles and nanowires is also very toxic and harmful. In some cases, the solvent used also possess toxicity. So, the problem with this most popularly used chemical method is the application of toxic chemicals and the toxic by-products. Also, the disposal of these products is a matter of concern. The researchers are, thus, more focused on the application of the "green synthesis" method for the production in nanotechnology.

2 Green Synthesis

Green is beautiful and striking (Rajendran et al. 2016). The term "green" is used in context to the earth; as these are the methods developed to save the earth and living beings on it from hazardous products. In products such as soaps, detergents, shoes, toothpaste, and shampoo, the application of various nanomaterials has been prominent. Nanoparticles of materials such as gold (Au) (Tsai and Thiagarajan 2010; Thiagarajan et al. 2009; Thiagarajan et al. 2011), silver (Ag) (Tsai et al. 2010; Balamurugan et al. 2009; Balamurugan and Chen 2009), and platinum (Pt) (Li et al. 2011) are widely used in these products. Moreover, the nanomaterials find their application in the medical and pharmaceutical industry as well, where they are mostly used by human beings and when disposed of finally enter the earth and its environment. It is extensively accepted today that the green synthesis involves the method and ingredients which are not threatening to the human beings and the environment. Green synthesis or green chemistry (Matlack 2001) includes the application of safer solvents, new materials, and catalysts that are more efficient in producing the nanomaterials and safe at the same time. In order to achieve this goal, the application of resources presents in nature and the perfect solvent is very important. Green synthesis is the use of techniques and methods that reduce the application and production of hazardous products that are harmful to the environment. Some of the key features of green synthesis are summarized in Fig. 1.

In nanotechnology, the application of hazardous chemicals and by-products has always been a concern. In order to stop or evade the production of these harmful products, the technique of "green synthesis" is required to achieve eco-friendly and sustainable production methods (Singh et al. 2018). This is the reason why most of the researchers are laying emphasis on the application of green synthesis techniques in nanotechnology. In the year 2005, the Noble Prize for chemistry was given to Chauvin, Grubbs, and Schrock for their work "a great step forward for green chemistry" which clearly is an indication of the development and popularity of green synthesis techniques. This was also a step to encourage more researchers toward the application of green methods in their synthesis techniques.

The green synthesis has actually two components, i.e., biological and chemical methods. The biological methods make use of microorganisms and plant extracts. Whereas in the chemical methods, use of the eco-friendly chemicals and solvents is the main concern. In some cases, the plants' extract is also used in the chemical synthesis of nanomaterials. In this section, we will try to give just a brief idea of these components of the green synthesis.

2.1 Application of Biology in Green Synthesis

In this method of green synthesis, the researchers make use of the microorganism for the preparation of nanoparticles. The mostly used organisms are fungi, yeast, plants, and bacteria. Bacteria is the microorganism species that has been used widely in the field of biotechnology for processes such as bioleaching, genetic engineering, and bioremediation (Gericke and Pinches 2006). Bacteria has always been an option for the preparation of nanomaterials due to its metal ions reducing ability (Iravani 2014). And various bacterial species such as actinomycetes and prokaryotic bacteria have been successfully applied for metal or metal oxide nanoparticle synthesis.

The metal or metal oxide nanoparticle synthesis mediated through fungi is also an efficient technique. The presence of an intracellular enzyme in various species of fungi makes them the superior biological agents in metal/metal oxide nanoparticle synthesis (Chen et al. 2009). Moreover, the nanoparticles synthesized using fungi have well-defined morphology, and in competition, to bacteria, these can produce a larger amount of nanoparticles (Mohanpuria et al. 2008). Also, they have reducing components such as proteins or enzymes on the surface of their cell as compared to other microorganisms (Narayanan and Sakthivel 2011). The enzymatic reduction on the cell wall or inside the cell of fungi results in the formation of metallic nanoparticles. Nanoparticles of metal/metal oxide such as Au, titanium (Ti), Ag, titanium oxide (TiO_2), and zinc oxide (ZnO) have been prepared using various fungi species.

Yeast is another species that has been utilized for the nanoparticle synthesis. Some 1500 species of yeast have been identified so far (Yurkov et al. 2011) and it is a single cell microorganism mostly present in a eukaryotic cell. Many research groups have reported the synthesis of nanoparticles using various yeast species. Silver and gold nanoparticles have been reported to be synthesized using yeast species.

Plants take minerals from the soil and have the ability to collect certain heavy metals in their parts. Also, the plants have the ability to reduce metal salts into nanoparticles as

Fig. 1 The diagrammatic representation of the features of Green synthesis

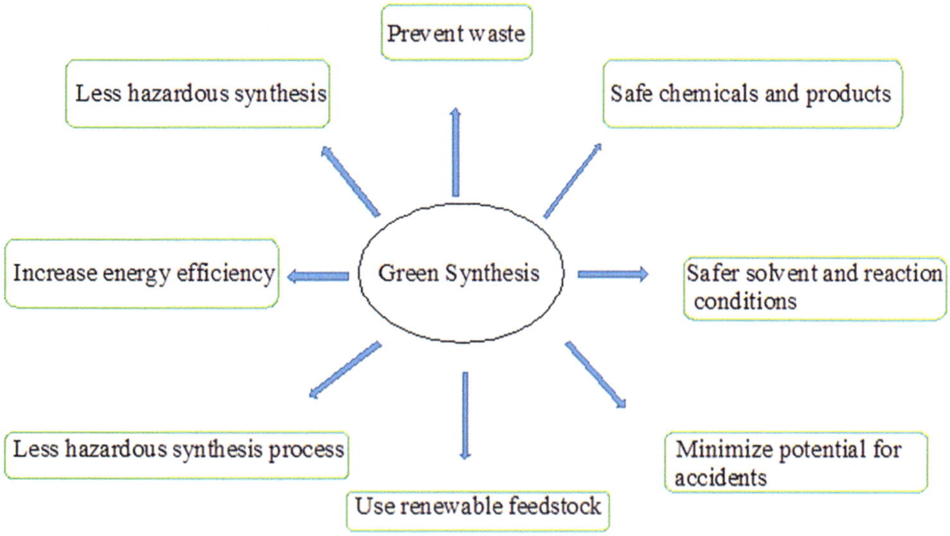

they contain biomolecules like protein, carbohydrate, and coenzyme. This is the reason that in recent times the synthesis of nanoparticles using plant extract has gained much popularity. Many researchers have synthesized metal/metal oxide nanoparticles using plant leaf extract, fruit extract, or other parts of a plant. This method using plant extract for synthesis is simple, cheap, efficient, eco-friendly, and also an alternative to various conventional methods. Using various plant species, nanoparticles can be synthesized employing the "one-pot" method. The one-pot method is a technique used in chemistry in which various chemical reactions are performed on a reactant in a single reactor or pot so that the efficiency of the chemical reaction can be enhanced. First, gold and silver nanoparticles were synthesized using various plant extracts. The reference to these research papers will be provided in the later section of this chapter. Using plants such as oat (Avena sativa), aloe vera (Aloe barbadensis miller), lemon (Citrus limon), brown mustard (Brassica juncea), alfalfa (Medicago sativa), coriander (Coriandrum sativum), neem (Azadirachta indica), tulsi (Osimum sanctum), and lemongrass (Cymbopogon flexuosus), various results have been published for synthesizing gold and silver nanoparticles (Singh et al. 2018).

2.2 Green Synthesis Based on the Application of Solvent

In a synthesis process, the role of solvent is fundamental regardless of the synthesis process we use. Water, a universal solvent, is considered to be a suitable solvent in all synthesis techniques as it can dissolve most of the substances. "The best solvent is no solvent, and if a solvent is desirable then water is ideal" was said by Sheldon regarding water as a solvent (Shanker et al. 2016). The unique fact about water is that it is easily available on earth (as 70% of the earth is water) and is also cheap. Water is also non-toxic posing, no threat to the environment. Hence, it is most widely used in the green synthesis technique. In nanotechnology, water has been used as a solvent in various published works. For example, gold and silver nanoparticles were synthesized using a bifunctional molecule and gallic acid in an aqueous medium at room temperature (Yoosaf et al. 2007). Also, using the laser ablation technique, gold nanoparticles were synthesized in an aqueous medium (Sylvestre et al. 2004). Due to the advent of technology, various other solvents have also been developed and used for synthesizing nanoparticles such as ethyl lactate (derived from processing corn), non-toxic liquids, plant extract, and ionic liquids. Thus, it can be easily understood that in the green synthesis, we either used water as a solvent or non-toxic and eco-friendly liquids as a solvent. In recent times, most of the researchers are focusing on the use of plant extracts along with water as a solvent in the nanoparticle synthesis.

3 Computational Data and Analysis

In this section of the chapter, we will look into some of the computational data that has been published in the journals regarding the trends in the synthesis of nanoparticles. The data that we will present here will be based on the green synthesis technique used in recent years and the current global trends. The first plot in Fig. 2 that we present here is about the number of papers that have been published in the reputed journals from 2003 up to 2017 using the green synthesis method for nanoparticle production. It is clear from the plot that the trend using green synthesis techniques for nanoparticle production has been growing. In the year 2003,

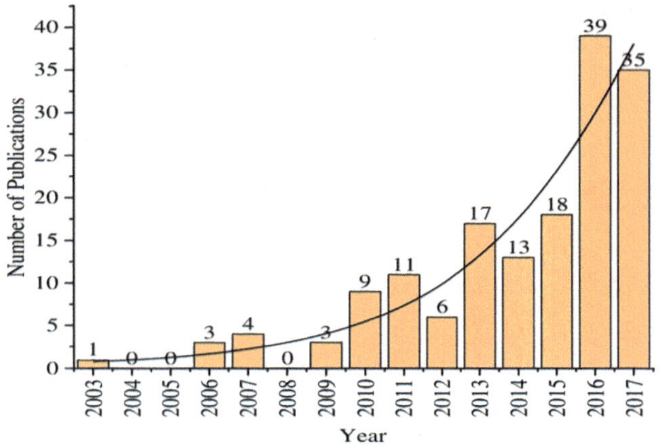

Fig. 2 Plot for the number of publications since 2003 up to 2017, Reproduced from Ribeiro et al. (2020) CC BY-NC 4.0

the number of papers published using this technique was only one, whereas up to 2017 the number of papers published increased to 35. Thus, it can be observed from the plot that the trend (green synthesis) is coming into mainstream research.

The next plot is about the citations of the research paper published using green synthesis throughout the years. And again from the plot in Fig. 3, it can be clearly observed that the popularity of the subject is increasing every year. As more are the papers cited, that means that more researchers are interested in the subject.

And the final plot that we present here is about the countries that are showing quite a growth in using green synthesis techniques for the production of nanoparticles. From the plot in Fig. 4, India has been the country with the most number of papers published in the reputed journals using the green synthesis procedure for nanoparticle synthesis. Followed by India are China, the USA, Egypt, Iran,

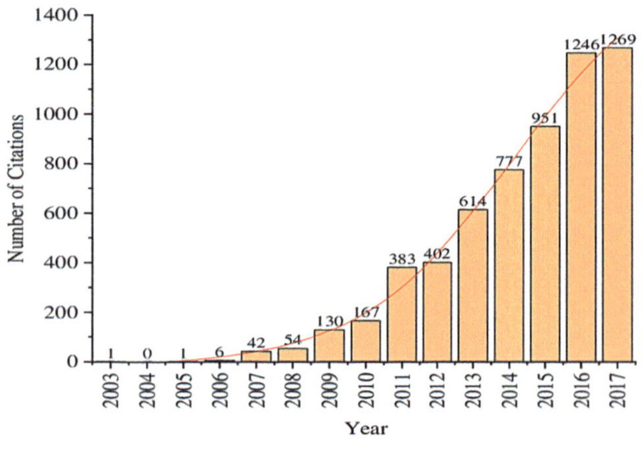

Fig. 3 Plot for the number of citations of papers based on green synthesis technique throughout the years, Reproduced from Ribeiro et al. (2020) CC BY-NC 4.0

etc. Up to 2015, India was not among the leading countries that were prominent in the field of nanoscience and nanotechnology. China, the USA, Germany, Japan, France, and South Korea were the leading countries toward the research in the nanotech field. But India showed a significant prominence in the field of nanotech using the green synthesis method. This may all be due to the availability of abundant natural resources in the country and also India has a long-standing role in sustainable development. Also, in the year 2000, investments made through a Program on Nanomaterials: Science and Devices by the Department of Science and Technology (DST) led to the serious development in India regarding nanotechnology.

The discussion made above based on some computational data reveals that there is a growing trend for the green synthesis techniques in nanotechnology. This clearly supports the use of electrochemical methods for the green synthesis of nanomaterials. In this chapter, we are focused only on the "electrochemical method for green synthesis." We will try to cover up the basic concepts of the electrochemical methods and its connection with the green technique of synthesis. In the final part of the chapter, we will also look into some of the works published in the journals where the electrochemical method was used as a green synthesis method.

4 Electrochemical Method

The electrochemical method is the process in which an electric current is passed through a medium bringing chemical changes. This process is commonly known as electrolysis and the branch of science that deals with this phenomenon is called electrochemistry. All the chemical changes are brought due to the electron, so the electron is the key factor in electrochemistry. The whole setup of the electrolysis is called an electrochemical cell. And the medium through which electron passes is known as an electrolyte. There are at least two electrodes in an electrochemical cell. The electrodes may be conducting such as liquid and solid metals, graphite, semiconductors, or can be inert such as electrodes made of carbon and platinum. An anode is an electrode connected to the positive terminal of the battery and the cathode is an electrode connected to the negative terminal of the battery. In an electrochemical reaction, the electrodes can be changed easily. Whereas the electrolyte can be aqueous or non-aqueous, solid, conducting polymer, molten salts, etc., and cannot be changed during an electrochemical reaction. The electrolyte carries the charge provided by the electrodes and this brings about the chemical changes. So, the electrochemical process can be summed in a sentence; electricity generated brings the chemical change due to the movement of electrons from one electrode to another electrode through an

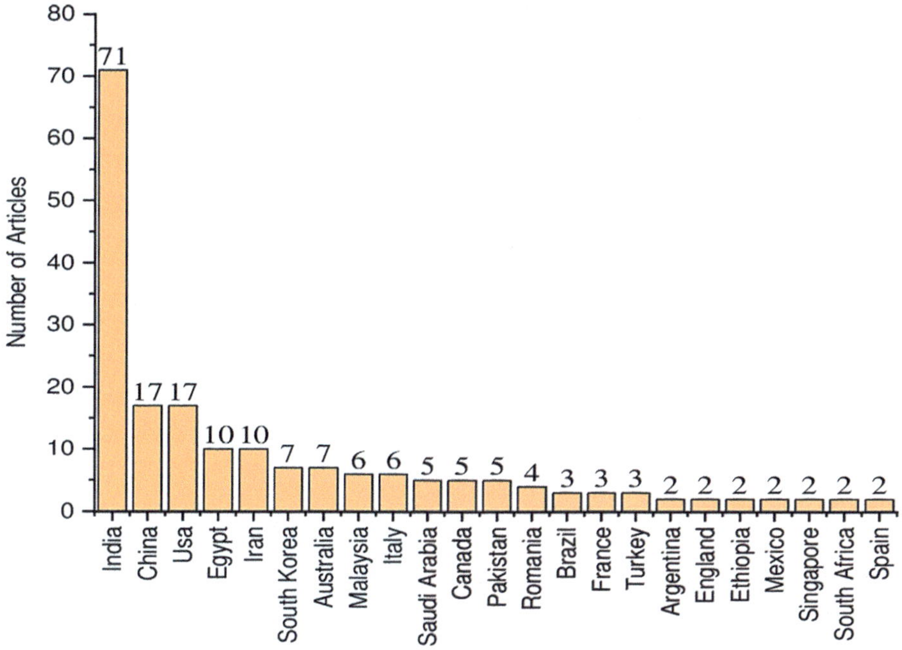

Fig. 4 The plot for the countries having the most number of papers published using green synthesis for nanoparticles throughout the years, Reproduced from Ribeiro et al. (2020) CC BY-NC 4.0

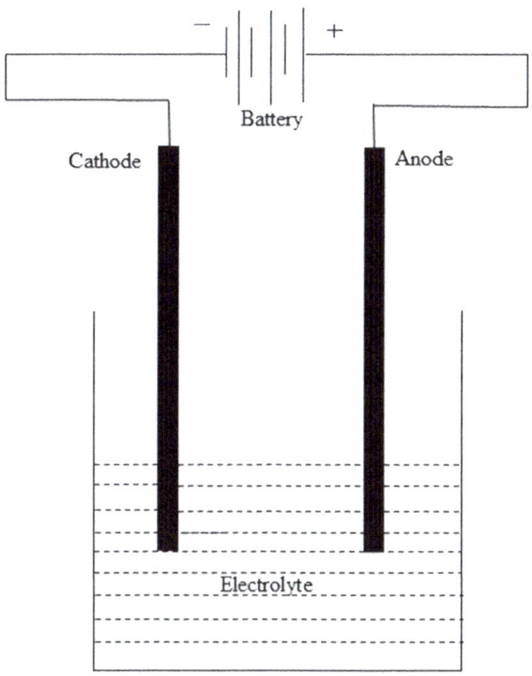

Fig. 5 An electrochemical cell

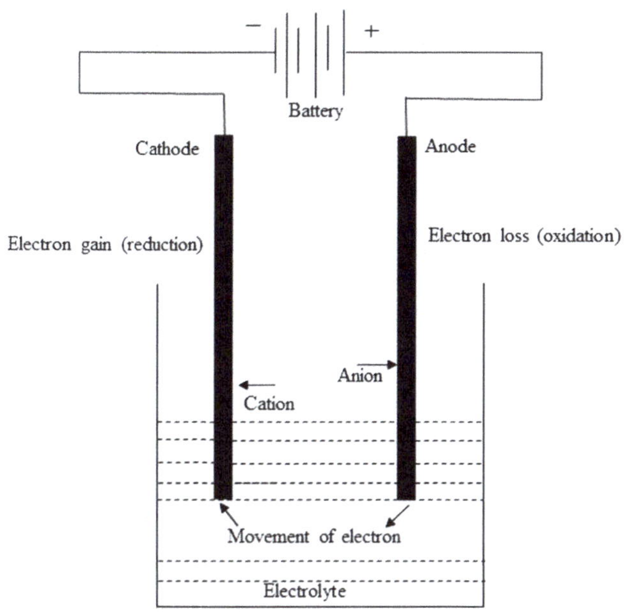

Fig. 6 Working of an electrochemical cell

electrolyte. The setup and working of an electrochemical cell are shown in Figs. 5 and 6. In an electrochemical cell, when electricity is passed through an electrolyte, cations, and anions are generated. These ions are then deposited on their respective electrodes, i.e., cation on the cathode and anion on the anode of an electrochemical cell.

The term "green electrochemistry" was most probably used by Pletcher and Weinberg for the first time in its history (Gupta et al. 2019). The main aim of explaining briefly about an electrochemical cell and its working was to show how the electrochemical method easily relates to the green synthesis. First, the electricity used to start an electrochemical reaction is environmentally friendly. Second is the choice of an electrolyte which will be used in the cell. Water is mostly used as an electrolyte in an electrochemical reaction which already discussed above is the part of green synthesis. Moreover, salts (non-toxic) dissolved in water are also used as electrolytes, thus incurring no threat to the environment.

Also, non-toxic solvents can be used as an electrolyte adding to the merits of the electrochemical method as a green synthesis technique. Third, the electrochemical reactions are usually carried out in low temperatures and non-volatile media reducing the consumption of energy, accidents (release), material failure, etc. Thus, the electrochemical method has the potential to be a part of the green synthesis technique and the above discussion explains why it is being widely used by the researchers. We would like to mention here that the research work in which water, non-toxic solvents, etc., are used for synthesizing nanoparticles by the electrochemical green method will be discussed separately in the final part of the chapter.

5 Electrodeposition Method

Electrodeposition is an electrochemical process in which a solid is deposited on the metal surface. Being easy to carry out and cheap at the same time, it acts as a basis for various industrial processes (Pasa and Munford 2006). This method is most widely used in the field of nanotechnology for the synthesis of nanomaterials, i.e., nanoparticles and nanowires. In recent years, it has been proved to be a very successful and efficient method for the synthesis of nanowires. The electrodeposition setup is similar to that of an electrochemical cell consisting of an electrolyte, electrodes, and a power source. When the current is passed through an electrolyte, the cations and anions are deposited on the respective electrodes (metal surface). The main difference between electrolysis and electrodeposition is that in electrolysis, an electric current is used to carry a chemical reaction, whereas, in electrodeposition, an electric current is used to deposit one metal on another.

In nanotechnology, the electrochemical deposition provides a direct one-step process for the development of nanostructures. And the nanostructures synthesized using the electrodeposition method have been found to be electrically active, thus possessing various applications in bio-sensing (Welch and Compton 2006; Campbell and Compton 2010), fuel cells (Maillard et al. 2005; Tian et al. 2007; Day et al. 2007), etc. Also, the electrodeposition method is cheap, selective, free from surfactants, and allows the nanoclusters to be tuned by simply changing the composition of electrolyte and parameters of deposition (Kibsgaard et al. 2012; Paunovic 2006; Staikov 2007). Materials are deposited without any expensive vacuum equipment. Originally, the electrodeposition was used for the production of metallic mirrors and surfaces resistant to corrosion. But in 1805, Luigi V. Burgnatelli who was an Italian professor, deposited Au on the metallic surface using a gold dissolved solution and a battery. And it is to him that the discovery of electrodeposition is attributed. Forty years later, John Wright from England revealed that the electroplating of silver and gold can be done using an electrolyte made of potassium cyanide. It was with this discovery, the electrodeposition became a vital industrial technique for the shelling of metallic objects in order to make them decorative and corrosion-resistant. Successively, various metals such as Ni, zinc (Zn), and tin (Sn); alloys such as brass were deposited in the next coming years. For about 100 years, the main idea behind using the electrodeposition technique was to cover the surfaces of materials using noble metals. But in the year 1940, the electronic industry revived the electrodeposition method. The gold was electrodeposited for various applications in the electronic components for the first time. The microsystems and microelectromechanical systems (MEMS) were developed using the electrodeposition of various metals, alloys, and semiconductors. For developing most of these microsystems, a simple aqueous electrolyte of metals such as Ni, copper (Cu), Au, Pt, iron (Fe), and lead (Pb) was used. Among these, nickel has been widely used in the microelectronic systems. The electrodeposited nickel and its alloys have been used in microsystem electronics for the making latch precision gears, flexure spring arms, and motors. The microelectronics added very much to the advancement of the electrodeposition technique. Various models were developed for deposition method in order to ensure that mass is transported to the electrodes, new and safer baths for electrolytes were developed, DC power supplies were developed, whole new kinetics was developed to understand the charge transfer, nucleation, and growth in the electrodeposition technique, etc. All these efforts were made in microelectronics which eventually led to the great improvement in the electrodeposition and its use for large-scale manufacturing. Since microelectronics is electronics at a very small scale; this eventually led to the utilization of electrodeposition in the nanotechnology because in nanotechnology we deal with the materials on a small scale (nanoscale).

5.1 Experimental Setup for Electrodeposition

On the industrial scale, the electrodeposition just requires a DC power supply and an electrochemical cell. In this method, the current between the electrodes is kept constant (DC supply) and is most commonly known as the galvanostatic plating system. It is simple, cheap, and easy to deposit. There is also another method for electrodeposition known as the potentiostatic method. This is the most famous and most common one used at the laboratory and industrial scale. In this method, an extra electrode known as a reference electrode is used in the electrochemical cell. With the introduction of the reference electrode, the experimental setup for electrochemical deposition changed which is shown in Fig. 7.

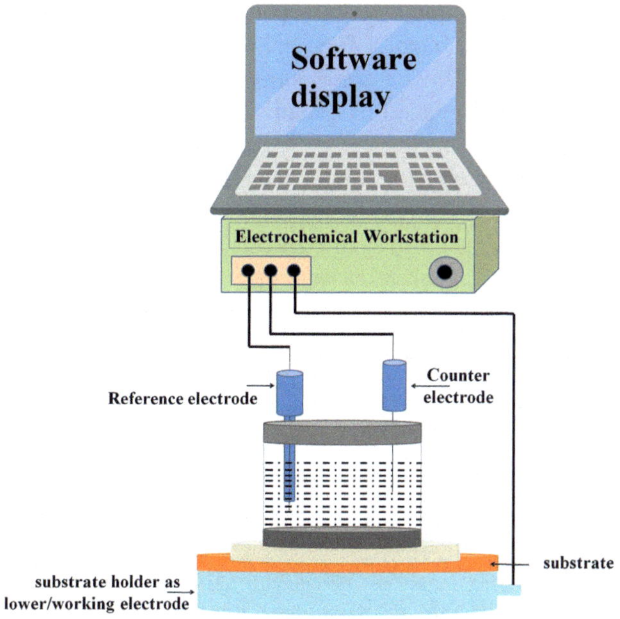

Fig. 7 Schematic representation of the setup for electrodeposition, Reproduced with permission from Gupta et al. (2019). Copyright 2019, Elsevier

There are three electrodes, namely—working electrode, reference electrode, and counter electrode. Silver, gold, platinum, inert carbon, and boron, etc., are mostly used as working electrodes. It is the working electrode where the reaction of main importance occurs in an electrochemical deposition (Kissinger and Heineman 1996; Bard et al. 2000; Zoski 2007). Whereas the counter electrode allows the passage of current to balance the current observed at the working electrode (Bard and Faulkner 2001). There is also a potentiostat that controls the potential difference between working and counter electrode, thus maintaining the potential difference across working and a reference electrode. In the modern era, with the advancement in science and technology, a more sophisticated setup has been built for performing the electrochemical deposition. The setup consists of an electrochemical workstation which has a potentiostat and a relevant software on one end, and the electrochemical setup inside Faraday cage at the other end. The electrochemical cell is designed with such a perfection that it can hold a working, reference, and counter electrode appropriately. Besides deposition, the potentiostat and the three-electrode cell can be used for various electrochemical investigations such as voltammetry, cyclic voltammetry (CV), linear sweep voltammetry (LSV), impedance, Tafel plot, and chronocoulometry.

The potentiostatic method is all due to the development and advancement of electrochemical science. In this science, the electrochemical reactions taking place at the surface of an electrode are studied carefully. After a careful investigation, the electrochemist came up with the idea of a reference electrode which can be used to measure the drop in the potential that occurs near the surface of electrodes. The development of this reference electrode was based on the assumption that there are sufficient ions in an electrolyte to evade ohmic voltage drop and the maximum voltage supplied by the battery appears near the surface of the electrodes which results in the formation of a charge region there. It is necessary to find out (measure) the voltage drop across these charge regions, as these regions control the force which is responsible for an ion to reduced state transformation. By introducing the reference electrode, the potential drop across electrodes can be measured which will help to investigate the reactions taking place at the electrodes (working or counter). The reference electrode can be made of a simple metal foil. But there must be a standard reference electrode that can be used to measure the drop in potential across electrode surface (working) in different electrolytes. The electrochemist chose a hydrogen electrode and all the electrode potentials are measured relative to this reference electrode. The standard hydrogen electrode (SHE), which is most commonly used is fabricated by passing hydrogen gas over the surface of an immersed platinum foil. The following reaction responsible for the operation of SHE takes place:

$$H_2 \leftrightarrow 2H^+ + 2e^-$$

There are various other reference electrodes used such as calomel electrode (Hg/Hg_2Cl_2) and silver/silver chloride electrode (Ag/AgCl). These are stable, strong, and are easily fabricated than SHE (Gupta et al. 2019).

The electrochemical deposition is also an electrochemical method used for the synthesis of nanomaterials. And as already discussed, by choosing the environment-friendly solvent and salts the method can be very helpful in the green synthesis technique.

6 Research Work: Using Green Electrochemical Methods for Nanomaterials Synthesis

This is the final and important section of this chapter. Because in this section, we will discuss the research that has been going on making use of the green synthesis techniques for nanomaterials; especially the electrochemical method. The section will include some research works published in the reputed journals where an electrochemical green method has been used for nanomaterial synthesis.

Dao TriThuc et al. published a paper "green synthesis of colloidal silver nanoparticles through the electrochemical method and their antibacterial activity" (Thuc et al. 2016). In the paper, an environmentally friendly and easy method was used to synthesize colloidal nanoparticles of silver using distilled water, tri-sodium citrate, a DC voltage, and silver bars. The synthesis was carried out at room temperature. The particles were spherical in shape with a size of 19.77 ± 4.3 nm. The particles were investigated for their antibacterial activity against species like Pseudomonas aeruginosa, Escherichia coli, and Staphylococcus aureus.

There were also various reports of synthesizing silver nanoparticles using plant or fruit extract. Leaf extract of Eugenia jambolana was utilized to synthesize silver nanoparticles (Firdous et al. 2017). Silver particles were synthesized using Rhynchotechum ellipticum leaves (Hazarika et al. 2014) and Hesperidin (Stephen and Seethalakshmi 2013). Application of pepper leaf extract as a capping and reducing agent for silver nanoparticles synthesis has also been reported (Mallikarjuna et al. 2014). Some other reducing agents for silver nanoparticles synthesis have also been reported such as natural rubber (Guidelli et al. 2011), cinnamon (Saliem et al. 2016), a stem derived callus of red apple (Umoren et al. 2014), lemongrass (Masurkar et al. 2011), fruit extracts of Malus domestica, coffee and tea extract (Nadagouda and Varma 2008), egg white (Lu et al. 2012), soluble starch (Vigneshwaran et al. 2006), and black tea (Begum et al. 2009). With silver nitrate as a substrate, silver nanoparticles of size 50 nm were grown using alfalfa (Medicago sativa) and mustard green (Brassica juncea) (Harris and Bali 2008). These are all the green synthesis methods for silver nanoparticles.

On a substrate of gold and copper salts, icosahedra of gold (4 nm) and copper (2 nm) were observed and investigated using M. Sativa (alfalfa) (Gardea-Torresdey et al. 2002) and Iris pseudacorus (yellow iris) (Manceau et al. 2008), respectively. Gold nanoparticles of size 20–40 nm have been reduced using rose geranium (Pelargonium graveolens) extract (Shankar et al. 2006). Also, the gold hexagons and triangles of size 50–100 nm were synthesized using neem (Azadirachta indica) extract (Shankar et al. 2004). The cubic In_2O_3 nanoparticles (5–50 nm) were produced using leaf extract of aloe vera plant (Maensiri et al. 2008).

Fauziatul Fajaroh et al. published a paper "Thermal stability of silica-coated magnetite nanoparticles prepared by an electrochemical method" (Fajaroh et al. 2013). By the oxidation of iron electrically in water, magnetite nanoparticles were synthesized and the particle size was found to be 10–30 nm. Then, the particles were coated using silica in dilute sodium silicate solution and the size of coated particles was 9–12 nm.

Jitendra Kumar Sharma et al. fabricated CuO nanoparticles using the leaf extract of Calotropis gigantean in an aqueous solution (Sharma et al. 2015). The particles synthesized were used as electro-catalytic material in the electrodes of a dye-sensitized solar cell.

Na Liu et al. published a paper "A versatile and 'green' electrochemical method for synthesis of copper and other transition metal oxide and hydroxide nanostructures" (Liu et al. 2008). The synthesis involved water as an electrolyte and no special chemicals, surfactants were utilized to produce CuO, $Cu(OH)_2$, Cu_2O nanostructures with variant morphologies. The nanostructures of ZnO, Fe_3O_4, $Ni(OH)_2$, and FeOOH were also synthesized using the same technique.

Rakesh et al. published a paper "Synthesis of Chromium(III) Oxide Nanoparticles by Electrochemical Method and Mukia Maderaspatana Plant Extract, Characterization, $KMnO_4$ Decomposition and Antibacterial Study" (Rakesh et al. 2013). The nanoparticles of chromium oxide were synthesized by reducing potassium dichromate solution with plant extract of Mukia Maderaspatana. The synthesized Cr_2O_3 nanoparticles were used as a catalyst for decomposing $KMnO_4$ and its antibacterial properties were investigated against E. coli bacteria.

A. Serrà et al. published their work in which they synthesized CoPt nanoparticles using ionic liquids via a green electrochemical method (Serrà et al. 2014). This was a unique work of its kind, easy synthesis technique, eco-friendly, and simple. In this work, the ability of ionic liquid microemulsions as microreactors was combined with the potential of electrodeposition for synthesizing nanoparticles. The microemulsions have long been established as a synthesis procedure for metal, metal oxide, and inorganic nanomaterials. The CoPt nanoparticles synthesized in this work were of size range 10–120 nm.

Lingchao Qian et al. fabricated $FePO_4$ nanoparticles for their application in $LiFePO_4/C$ cathode materials using an electrochemical method (Qian et al. 2012). In this work, amorphous and unique $FePO_4$ nanoparticles were synthesized with particle size ranging from 20 to 80 nm. The particles find their application as cathode material for $LiFePO_4/C$ cathode.

Fauziatul Fajaroh et al. published their work as "Synthesis of magnetite nanoparticles by surfactant-free electrochemical method in an aqueous system" (Fajaroh et al. 2012). In their work, they presented a simple and green electrochemical method for synthesizing magnetic nanoparticles making use of plain water as an electrolyte and iron as an anode. And various parameters that affect the development of magnetic nanoparticles were also investigated. The parameters observed were the spacing between the electrodes and the effect of OH^- ions and current density on the development of magnetite nanoparticles. The particles synthesized were of size 10–30 nm exhibiting ferromagnetic properties.

Min Zhou et al. synthesized Au-Ag alloy nanoparticles using a one-step electrochemical method (Zhoua et al. 2006). In this method, they made use of the Polyvinylpyrrolidone (PVP) along with silver and gold salt. The method was environmentally friendly, safe, and easy.

Sayed M.Ghoreishi et al. synthesized gold and silver nanoparticles using an electrochemical green method. In their synthesis, they made the use of Rosa damascene flower extract as a stabilizing and a reducing agent (Ghoreishi et al. 2011). The approach they used was cheap, simple, long time effective, and environmentally friendly. The average size of gold and silver nanoparticles synthesized was reported to be in a range of 10–30 nm.

Ping Yu et al. described a method for the green electrochemical preparation of metallic nanoparticles (Yu et al. 2010). In this work, they explained their strategy for the Pt metal nanoparticles as an example. The particles synthesized were uniform and spherical in shape with an average diameter of about 100 nm. They demonstrated how the deposition of metal nanoparticles can be made specific by simply adjusting the reducing potential of ionic liquids used as green solvents. This study opened new gates for the controlled and green metal nanoparticle synthesis requiring no reducing and capping agents, no toxic substrates, and high pressure.

Kim et al. prepared Au and Pt nanoparticles by developing a one-phase preparation technique using thiol-functionalized ionic liquids (TFILs) as stabilizers (Kim et al. 2004). Lazarus et al. prepared silver nanoparticles using an electrochemical method with both spherical and hexagonal shapes in an ionic liquid (BmimBF40) (Lazarus et al. 2012).

Aoqi Li et al. synthesized gold nanocrystals by varying the water content in deep eutectic solvents (DESs) using a green electrochemical method (Li et al. 2016). The nanocrystals synthesized were of different morphologies and were investigated for their electro-catalytic activity toward ethanol in alkaline media.

Hongcai Gao et al. synthesized PtNi nanoparticle-graphene nanocomposites using green and one-step electrochemical technique (Gao et al. 2011). They actually reduced the graphene oxide and metal precursors electrochemically. The nanocomposites obtained were successfully tested for the detection of non-enzymatic glucose.

Hui-Lin Guo et al. published their work as "A Green Approach to the Synthesis of Graphene Nanosheets" (Guo et al. 2009). The high-quality graphene sheets were synthesized by the electrochemical reduction of graphite oxide (exfoliated). The method was described to be green and material contamination free.

Haitao Li et al. synthesized fluorescent carbon nanoparticles using a green, low energy electrochemical method, and the nanoparticles were investigated for their photoluminescence properties (Li et al. 2011). The nanoparticles were prepared by one-step sodium hydroxide (NaOH)-assisted treatment of ethanol. The method described was also reported to be a future prospect in the synthesis of fluorescent carbon nanoparticles.

Su-Juan Li et al. synthesized gold-graphene nanocomposite using a green, simple, and controllable electrochemical synthesis mechanism (Li et al. 2012). The nanocomposite was reported to be sensitive toward dopamine, confirmed by its CV analysis.

Peng Miao et al. synthesized carbon nanodots using a green electrochemical synthesis technique (Miao et al. 2015). The nanodots were successfully applied for Fe^{3+} ions assay and were found to be practically applicable in real water samples as well, thus making them useful for environmental analysis.

Krishna M. Deshmukh et al. demonstrated a one-pot, green electrochemical synthesis of palladium nanoparticles using ionic liquids for the first time (Deshmukh et al. 2011). The method was reported to be surfactants and capping agent-free. The nanoparticles synthesized were investigated for their activity toward Suzuki coupling reaction of halides (un-activated).

Surjit Sahoo et al. reported a green, energy-saving electrodeposition synthesis of porous manganese-cobalt sulfide (MCS) nanosheet array on Ni-foam substrate for supercapacitor applications (Sahoo and Rout 2016). The nanosheets were described to be of potential application as supercapacitor electrodes.

Sheng Liu et al. synthesized Pt/graphene nanocomposite using a green electrochemical method (Liua et al. 2010). Their synthesis consists of a series of electrochemical processes and exhibited high and better stability and catalytic activity toward the oxidation of methanol.

R. Dehdari Vais et al. reported a one-step green electrodeposition method for the synthesis of gold nanostructures with different morphologies (Vais et al. 2016). The gold nanostructures of different morphologies were synthesized using DC potential, eco-friendly additives, and ultrasonic irradiation. The method was described to be a potential method for the preparation of other noble metal nanostructures of different morphologies.

Na Liu et al. synthesized graphene nanosheets using water and ionic liquids via a one-step electrochemical method (Liu et al. 2008). The method was reported to be green, facile, simple, and fast. Virendra V. Singh et al. developed graphene nanosheets directly from the pencil using a green electrochemical synthesis route (Singh et al. 2012). The nanosheets were then investigated for surface plasmon resonance (SPR) sensing of Salmonella typhi.

The above-mentioned work is just a few milestones of the long list that has been published using an electrochemical green method for synthesis. The main aim of citing this work was to bring it into the notice that a huge number of researchers are employing this method for the synthesis process. And there is a need for more researchers to work using this green technique so that we can take one more step toward our dream of sustainable development. With global climate change and global events leading to the degradation of the environment, it is the duty of researchers to work for the benefit of the environment. Thus, there is a global need that the synthesis in research work is done using "green" methods. The electrochemical method is one such method and can be used by researchers to achieve this global aim.

7 Conclusion

The chapter is about the use and development of green electrochemical synthesis mechanism in nanotechnology. There is a huge demand for the application of green synthesis techniques in the field of science and technology to achieve the aim of sustainable development, and nanotechnology is no exception to this. Researchers are making their efforts toward sustainable development using green methods. The electrochemical method is also one of the methods that can be used to achieve this goal. Recent developments have shown the potential that the electrochemical method possesses as a green synthesis route. This route has been followed by many researchers in nanotechnology, encouraging more researchers to apply this method in their synthesis. In the chapter, we have tried to briefly introduce this method and also cited some efficient work published using this green technique.

References

Agarwal H, Kumar SV, Rajesh Kumar S (2017) A review on green synthesis of zinc oxide nanoparticles; an eco-friendly approach. Reffit Tech 3:406–413. https://doi.org/10.1016/j.reffit.2017.03.002

Arruda SCC, Silva ALD, Galazzi RM, Azevedo RA, Arruda MAZ (2015) Nanoparticles applied to plant science: a review. Talanta 131:693–705. https://doi.org/10.1016/j.talanta.2014.08.050

Balamurugan A, Chen SM (2009) Silver nanograins incorporated PEDOT modified electrode for electro catalytic sensing of hydrogen peroxide. Electro Anal 21:1419–1423. https://doi.org/10.1002/elan.200804543

Balamurugan A, Ho KC, Chen SM (2009) One-pot synthesis of highly stable silver nanoparticles-conducting polymer nanocomposite and its catalytic application. Synth Met 159:2544–2549. https://doi.org/10.1016/j.synthmet.2009.09.004

Bard JA, Faulkner RL (2001) Electrochemical methods: fundamentals and applications. Wiley, New Jersey

Bard JA, Larry R, Faulkner (2000) Electrochemical methods: fundamentals and applications, 2nd edn. Wiley. ISBN 978-0-471-04372-0

Begum NA, Mondal S, Basu S, Laskar RA, Mandal D (2009) Biogenic synthesis of Au and Ag nanoparticles using aqueous solutions of Black Tea leaf extracts. Colloids Surf B 71:113–118. https://doi.org/10.1016/j.colsurfb.2009.01.012

Campbell FW, Compton RG (2010) The use of nanoparticles in electro analysis: an updated review. Anal Bioanal Chem 396:241–259. https://doi.org/10.1007/s00216-009-3063-7

Chen YL, Tuan HY, Tien CW, Lo WH, Liang HC, Hu YC (2009) Augmented biosynthesis of cadmium sulfide nanoparticles by genetically engineered Escherichia coli. Biotechnol Prog 25:1260–1266. https://doi.org/10.1002/btpr.199

Day TM, Unwin PR, Macpherson JV (2007) Factors controlling the electrodeposition of metal nanoparticles on pristine single walled carbon nanotubes. Nano Lett 7:51–57. https://doi.org/10.1021/nl061974d

Deshmukh KM, Qureshi ZS, Bhatte KD, Venkatesan KA, Srinivasan TG, Rao PRV, Bhanage BM (2011) One-pot electrochemical synthesis of palladium nanoparticles and their application in the Suzuki reaction. New J Chem 35:2747–2751. https://doi.org/10.1039/c1nj20638a

Fajaroh F, Setyawan H, Widiyastuti W, Winardi S (2012) Synthesis of magnetite nanoparticles by surfactant-free electrochemical method in an aqueous system. Adv Powder Technol 23:328–333. https://doi.org/10.1016/j.apt.2011.04.007

Fajaroh F, Setyawan H, Nur A, Lenggoro IW (2013) Thermal stability of silica-coated magnetite nanoparticles prepared by an electrochemical method. Adv Powder Technol 24:507–511. https://doi.org/10.1016/j.apt.2012.09.008

Firdous J, Gomathi S, Bharathi V, Shanmugapriya A, Sugunabai J, Karpagam T, Geetha S, Nachiyar S (2017) Phytochemical screening of silver nanoparticles extract of Eugenia jambolana using Fourier infrared spectroscopy. Int J Res Pharm Sci 8:383–387

Gao H, Xiao F, Ching CB, Duan H (2011) One-step electrochemical synthesis of PtNi nanoparticle-graphene nanocomposites for nonenzymatic amperometric glucose detection. ACS Appl Mater Interf 3:3049–3057. https://doi.org/10.1021/am200563f

Gardea-Torresdey JL, Parsons JG, Gomez E, Peralta-Videa J, Troiani HE, Santiago P, Yacaman MJ (2002) Formation and growth of Au nanoparticles inside live alfalfa plants. Nano Lett 10:97–401. https://doi.org/10.1021/nl015673+

Gericke M, Pinches A (2006) Microbial production of gold nanoparticles. Gold Bull 39:22–28. https://doi.org/10.1007/BF03215529

Ghoreishi SM, Behpour M, Khayatkashani M (2011) Green synthesis of silver and gold nanoparticles using Rosa damascena and its primary application in electrochemistry. Physica E 44:97–104. https://doi.org/10.1016/j.physe.2011.07.008

Guidelli EJ, Ramos AP, Zaniquelli MED, Baffa O (2011) Green synthesis of colloidal silver nanoparticles using natural rubber latex extracted from Hevea brasiliensis. Spectrochim Acta A 82:140–145. https://doi.org/10.1016/j.saa.2011.07.024

Guo HL, Wang XF, Qian QY, Wang FB, Xia XH (2009) A green approach to the synthesis of graphene nanosheets. ACS Nano 3:2653–2659. https://doi.org/10.1021/nn900227d

Gupta J, Arya S, Verma S, Singh A, Sharma A, Singh B, Prerna, Sharma R (2019) Performance of template-assisted electrodeposited Copper/Cobalt bilayered nanowires as an efficient glucose and Uric acid senor. Mater Chem Phys 238:121969

Harris AT, Bali R (2008) On the formation and extent of uptake of silver nanoparticles by live plants. J Nanoparticle Res 10:691–695. https://doi.org/10.1007/s11051-007-9288-5

Hazarika D, Phukan A, Saikia E, Chetia B (2014) Phytochemical screening and synthesis of silver nanoparticles using leaf extract of Rhynchotechum ellipticum. Int J Res Pharm Sci 6:672–674

Iravani S (2014) Bacteria in nanoparticle synthesis: current status and future prospects. Int Sch Res Not 1–18. https://doi.org/10.1155/2014/359316

Kibsgaard J, Gorlin Y, Chen Z, Jaramillo TF (2012) Meso-structured platinum thin films: active and stable electro catalysts for the oxygen reduction reaction. J Am Chem Soc 134:7758–7765. https://doi.org/10.1021/ja2120162

Kim KS, Demberelnyamba D, Lee H (2004) Size-selective synthesis of gold and platinum nanoparticles using novel thiol-functionalized ionic liquids. Langmuir 20:556–560. https://doi.org/10.1021/la0355848

Kissinger P, Heineman WR (1996) Laboratory techniques in electroanalytical chemistry, 2nd edn. Revised and expanded. CRC. ISBN 978-0-8247-9445-3

Lazarus LL, Riche CT, Malmstadt N, Brutchey RL (2012) Effect of ionic liquid impurities on the synthesis of silver nanoparticles. Langmuir 28:15987–15993. https://doi.org/10.1021/la303617f

Li Y, Wei JX, Chen SM (2011a) Platinum nanoparticles (PtNPs) with poly (tyramine)–choline oxidase (ChOx) film for the electro catalysis of choline. Int J Electrochem Sci 6:3385–3398

Li H, Ming H, Liu Y, Yu H, He X, Huang H, Pan K, Kang Z, Leeb ST (2011b) Fluorescent carbon nanoparticles: electrochemical synthesis and their pH sensitive photoluminescence properties. New J Chem 35:2666–2670. https://doi.org/10.1039/c1nj20575g

Li SJ, Deng DH, Shi Q, Liu SR (2012) Electrochemical synthesis of a graphene sheet and gold nanoparticle-based nanocomposite, and its application to amperometric sensing of dopamine. Microchim Acta 177:325–331. https://doi.org/10.1007/s00604-012-0782-9

Li A, Chen Y, Zhuo K, Wang C, Wang C, Wang J (2016) Facile and shape-controlled electrochemical synthesis of gold nanocrystals by changing water contents in deep eutectic solvents and their electro catalytic activity. RSC Adv 6:8786–8790. https://doi.org/10.1039/C5RA24499D

Liu N, Wu D, Wu H, Liu C, Luo F (2008a) A versatile and "green" electrochemical method for synthesis of copper and other transition metal oxide and hydroxide nanostructures. Mater Chem Phys 107:511–517. https://doi.org/10.1016/j.matchemphys.2007.08.026

Liu N, Luo F, Wu H, Liu Y, Zhang C, Chen J (2008b) One-step ionic-liquid-assisted electrochemical synthesis of ionic-liquid-functionalized graphene sheets directly from graphite. Adv Funct Mater 18:1518–1525. https://doi.org/10.1002/adfm.200700797

Liua S, Wanga J, Zengc J, Oua J, Li Z, Liua X, Yang S (2010) "Green" electrochemical synthesis of Pt/graphene sheet nanocomposite film and its electrocatalytic property. J Power Sources 195:4628–4633. https://doi.org/10.1016/j.jpowsour.2010.02.024

Lu R, Yang D, Cui D, Wang Z, Guo L (2012) Egg white-mediated green synthesis of silver nanoparticles with excellent biocompatibility and enhanced radiation effects on cancer cells. Int J Nanomed 7:2101. https://doi.org/10.2147/IJN.S30631

Maensiri S, Laokul P, Klinkaewnarong J, Phokha S, Promarak V, Seraphin S (2008) Indium oxide (In_2O_3) nanoparticles using Aloe vera plant extract: synthesis and optical properties. J Optoelectron Adv Mater 10:161–165

Maillard F, Schreier S, Hanzlik M (2005) Influence of particle agglomeration on the catalytic activity of carbon-supported Pt nanoparticles in CO monolayer oxidation. Phys Chem Chem Phys 7:385–393. https://doi.org/10.1039/B411377B

Mallikarjuna K, Sushma NJ, Narasimha G, Manoj L, Raju BDP (2014) Phytochemical fabrication and characterization of silver nanoparticles by using Pepper leaf broth. Arabian J Chem 7:1099–1103. https://doi.org/10.1016/j.arabjc.2012.04.001

Manceau A, Nagy KL, Marcus MA, Lanson M, Geoffroy N, Jacquet T, Kirpichtchikova T (2008) Formation of metallic copper nanoparticles at the soil–root interface. Environ Sci Technol 42:1766–1772. https://doi.org/10.1021/es072017o

Masurkar SA, Chaudhari PR, Shidore VB, Kamble SP (2011) Rapid biosynthesis of silver nanoparticles using Cymbopogan citratus (lemongrass) and its antimicrobial activity. Nano-Micro Lett 3:189–194. https://doi.org/10.1007/BF03353671

Mathur P, Jha S, Ramteke S, Jain NK (2018) Pharmaceutical aspects of silver nanoparticles. Artif Cell NanoMed B 46:115–126. https://doi.org/10.1080/21691401.2017.1414825

Matlack AS (2001) Introduction to green chemistry. Marcel Dekker Inc, New York

Miao P, Tang Y, Han K, Wang B (2015) Facile synthesis of carbon nanodots from ethanol and their application in ferric (III) ion assay. J Mater Chem A 3:15068–15073. https://doi.org/10.1039/C5TA03278D

Mirzaei H, Darroudi M (2017) Zinc oxide nanoparticles: biological synthesis and biomedical applications. Ceram Int 43:3907–3914. https://doi.org/10.1016/j.ceramint.2016.10.051

Mohanpuria P, Rana NK, Yadav SK (2008) Biosynthesis of nanoparticles: technological concepts and future applications. J Nanoparticle Res 10:507–517. https://doi.org/10.1007/s11051-007-9275-x

Nadagouda MN, Varma RS (2008) Green synthesis of silver and palladium nanoparticles at room temperature using coffee and tea extract. Green Chem 10:859–862. https://doi.org/10.1039/b804703k

Narayanan KB, Sakthivel N (2011) Synthesis and characterization of nano gold composite using Cylindrocladium floridanum and its heterogeneous catalysis in the degradation of 4-nitrophenol. J Hazard Mater 189:519–525. https://doi.org/10.1016/j.jhazmat.2011.02.069

Pasa AA, Munford ML (2006) Electrodeposition. Encycl Chem Process. doi:10.1081/E-ECHP-120037171. http://fisica.ufmt.br/escola/old-2005/material/pasa.pdf

Paunovic M (2006) Fundamentals of electrochemical deposition (Wiley-Interscience. Wiley, Hoboken, New Jersey

Qian L, Xia Y, Zhang W, Huang H, Gan Y, Zeng H, Tao X (2012) Electrochemical synthesis of mesoporous FePO4 nanoparticles for fabricating high performance $LiFePO_4$/C cathode materials. Micropor Mesopor Mat 152:128–133. https://doi.org/10.1016/j.micromeso.2011.11.048

Rajendran S, Rathish RJ, Prabha SS, Anandanc A (2016) Green electrochemistry—a versatile tool in green synthesis: an overview. Port Electrochimica Acta 5:321–342. https://doi.org/10.4152/pea.201605321

Rakesh, Ananda S, Made Gowda NM (2013) Synthesis of chromium (iii) oxide nanoparticles by electrochemical method and Mukia Maderaspatana plant extract, characterization, $KMnO_4$ decomposition and antibacterial study. MRC 2:127–135. https://doi.org/10.4236/mrc.2013.24018

Ribeiro JJ, da Silva Porto PS, Pereira RD, Muniz EP (2020) Green synthesis of nanomaterials: most cited papers and research trends. Res Soc Dev 9. https://doi.org/10.33448/rsd-v9i1.1593

Sahoo S, Rout CS (2016) Facile electrochemical synthesis of porous manganese-cobalt-sulfide based ternary transition metal sulfide nanosheets architectures for high performance energy storage applications. Electrochimica Acta 220:57–66. https://doi.org/10.1016/j.electacta.2016.10.043

Saliem AH, Ibrahim OM, Salih SI (2016) Biosynthesis of silver nanoparticles using cinnamon zeylanicum plants bark extract, |Kufa. J Vet Med Sci 7:51–63

Serrà A, Gómez E, López-Barbera JF, Nogués J, Vallés E (2014) Green electrochemical template synthesis of CoPt nanoparticles with tunable size, composition and magnetism from microemulsions using ionic liquids. ACS Nano 8:4630–4639. https://doi.org/10.1021/nn500367q

Shankar SS, Rai A, Ahmad A, Sastry M (2004) Rapid synthesis of Au, Ag, and bimetallic Au core–Ag shell nanoparticles using Neem (Azadirachta indica) leaf broth. J Colloid Interface Sci 275:496–502. https://doi.org/10.1016/j.jcis.2004.03.003

Shankar SS, Ahmad A, Sastry M (2006) Geranium leaf assisted biosynthesis of silver nanoparticles. Biotechnol Prog 19:1627–1631. https://doi.org/10.1021/bp034070w

Shanker U, Jassal V, Rani M, Kaith BS (2016) Towards green synthesis of nanoparticles: from bio-assisted sources to benign solvents: a review. Int J Environ Anal Chem 96:801–35. https://doi.org/10.1080/03067319.2016.1209663

Sharma JK, Akhtar MS, Ameen S, Srivastava P, Singh G (2015) Green synthesis of CuO nanoparticles with leaf extract of Calotropis gigantea and its dye-sensitized solar cells applications. J Alloys Compd 632:321–325. https://doi.org/10.1016/j.jallcom.2015.01.172

Singh VV, Gupta G, Batra A, Nigam AK, Boopathi M, Gutch PK, Tripathi BK, Srivastava A, Samuel M, Agarwal GS, Singh B, Vijayaraghavan R (2012) Greener electrochemical synthesis of high qualitygraphene nanosheets directly from pencil and its SPR sensing application. Adv Funct Mater 22:2352–2362. https://doi.org/10.1002/adfm.201102525

Singh J, Dutta T, Kim KH, Rawat M, Samddar P, Kumar P (2018) 'Green' synthesis of metals and their oxide nanoparticles: applications for environmentalremediation. J Nanobiotechnol 16:84. https://doi.org/10.1186/s12951-018-0408-4

Staikov G (2007) Electrocrystallization in nanotechnology (Wiley-VCH. Verlag GmbH & Co. KGaA, Weinheim

Stephen A, Seethalakshmi S (2013) Phytochemical synthesis and preliminary characterization of silver nanoparticles using hesperidin. J Nanosci 2013:5. https://doi.org/10.1155/2013/126564

Sylvestre J, Poulin S, Kabashin AV, Sacher E, Meunier M, Luong JH (2004) Surface chemistry of gold nanoparticles produced by laser ablation in aqueous media. J Phys Chem B. 108:16864–16869. https://doi.org/10.1021/jp047134

Thiagarajan S, Su BW, Chen SM (2009) Nano TiO2–Au–KI film sensor for the electro catalytic oxidation of hydrogen peroxide. Sensor Actu B 136:464–471. https://doi.org/10.1016/j.snb.2008.11.009

Thiagarajan S, Wu ZY, Chen SM (2011) Amperometric determination of sodium hypochlorite at poly MnTAPP-nano Au film modified electrode. J Electroanal Chem 661:322–328. https://doi.org/10.1016/j.jelechem.2011.08.009

Thuc DT, Huy TQ, Hoang LH, Tien BC, Van Chung P, Thuy NT, Le AT (2016) Green synthesis of colloidal silver nanoparticles through electrochemical method and their antibacterial activity. Mater Lett 181:173–177. https://doi.org/10.1016/j.matlet.2016.06.008

Tian N, Zhou ZY, Sun SG, Ding Y, Wang ZL (2007) Synthesis of tetrahexahedral platinum nanocrystals with high-index facets and high electro-oxidation activity. Science 316:732–735. https://doi.org/10.1126/science.1140484

Tsai TH, Thiagarajan S (2010) Ionic liquid assisted one step green synthesis of Au–Ag bimetallic nanoparticles. J Appl Electrochem 40:493–497. https://doi.org/10.1007/s10800-009-0020-2

Tsai TH, Thiagarajan S, Chen SM (2010) Green synthesis of silver nanoparticles using ionic liquid and application for the detection of dissolved oxygen. Electroanalysis 22:680–687. https://doi.org/10.1002/elan.200900410

Umoren S, Obot I, Gasem ZM (2014) Green synthesis and characterization of silver nanoparticles using red apple (Malus domestica) fruit extract at room temperature. J Mater Environ Sci 5:907–914

Vais RD, Sattarahmady N, Heli H (2016) Green electrodeposition of gold nanostructures by diverse size, shape, and electrochemical activity. Gold Bull 49:95–102. https://doi.org/10.1007/s13404-016-0187-3

Vigneshwaran N, Nachane R, Balasubramanya R, Varadarajan P (2006) a novel one-pot 'green' synthesis of stable silver nanoparticles using soluble starch. Carbohydr Res 341:2012–2018. https://doi.org/10.1016/j.carres.2006.04.042

Welch CW, Compton RG (2006) The use of nanoparticles in electro analysis: a review. Anal Bioanal Chem 384:601–619. https://doi.org/10.1007/s00216-005-0230-3

Yoosaf K, Ipe BI, Suresh CH, Thomas KG (2007) In situ synthesis of metal nanoparticles and selective naked-eye detection of lead ions from aqueous media. J Phys Chem C 111:12839–12847. https://doi.org/10.1021/jp073923q

Yu P, Qian Q, Wang X, Cheng H, Ohsakab T, Mao L (2010) Potential-controllable green synthesis and deposition of metal nanoparticles with electrochemical method. J Mater Chem 20:5820–5822. https://doi.org/10.1039/c0jm01293a

Yurkov AM, Kemler M, Begerow D (2011) Species accumulation curves and incidence-based species richness estimators to appraise the diversity of cultivable yeasts from beech forest soils. PLoS ONE 8. https://doi.org/10.1371/journal.pone.0023671

Zhoua M, Chena S, Zhaoa S, Maa H (2006) One-step synthesis of Au–Ag alloy nanoparticles by a convenient electrochemical method. Physica E 33:28–34. https://doi.org/10.1016/j.physe.2005.10.012

Zoski GC (2007) Handbook of electrochemistry. Elsevier, Amsterdam. ISBN 978-0-444-51958-0

Microwave-Irradiated Synthesis of Imidazo[1,2-*a*]pyridine Class of Bio-heterocycles: Green Avenues and Sustainable Developments

Ravi Kant Yadav and Sandeep Chaudhary

Abstract

A fused heterocyclic bioactive class of molecule, i.e., imidazo[1,2-*a*]pyridine abbreviated as "IP" had been comprehensively studied over the last 2–3 decades due to the presence of wide array of biological activities. Various synthetic routes have been applied for the preparation of IPs but microwave-irradiated synthesis has been recognized as an effective, fast, and high-yielding methodology. Herein, this chapter incorporates some of the most significant microwave-assisted preparation of IPs reported so far in the literature.

Keywords

Imidazo[1,2-*a*]pyridines • Microwave-assisted organic synthesis (MAOS) • N-heterocyclic compounds • 5,6-fused heterocycles

1 Introduction

The search of medicinally privileged N-containing heterocyclic structures, present as a part or as a whole in several naturally occurring natural products and other pharmaceutically important therapeutics, has been recognized as a rapidly emerging and constitutional theme of drug discovery (Pastor et al. 1995). In this context, bridge-headed N-containing heterocyclic core constitutes an important class of N-heterocyclic compounds due to its amicable existence in numerous natural products and its synthetic congeners (Pastor et al. 1995).

Among these class of bio-heterocycles, imidazo[1,2-*a*]pyridines (IPs) encompassing five-membered imidazole moiety fused with six-membered pyridine ring have been recognized as a "drug bigotry" (Kappe et al. 2009; Lu and Zhang 2004) and are endowed with a varied array of pharmaceutical properties such as anti-bacterial (Schwerkoske et al. 2005), antifungal (Masquelin et al. 2006), antiviral (DiMauro and Kennedy 2007; Rousseau et al. 2007), antiprotozoal (Tu et al. 2007), anti-inflammatory (Rousseau et al. 2007), anti-tumor (Rousseau et al. 2007), antipyretic, analgesic, peptic ulcer, anti-tuberculosis, hypnoselective, and anxioselective (Adib et al. 2010) activities (Bagdi et al. 2015). Imidazo[1,2-*a*]pyridine or its similar skeletal core is also present in several clinically used drugs/therapeutics such as Zolpidem **I** (sedative), Alpidem **II** and Saridipem **III** (both anxiolytic), Olprinone **IV** (heart failure), Zolimidine **V** (peptic ulcer), Minodronic acid **VI** (Osteoporosis), GSK812397 **VII** (anti-HIV), YM-201627 **VIII** (anti-tumor), and many other drug-like molecules (Fig. 1) (Bagdi et al. 2015). Additionally, some instances had been appealed to be antiplatelet agents, cardiotonics, acetylcholinesterase (ACE) inhibitors (Howard 1996), angiotensin II antagonists (Howard 1996), gastric H^+/K^+-ATPase inhibitors (Vagin et al. 2002), cyclin-dependent kinase inhibitors (Cai et al. 2006), β-amyloids imaging agents in Alzheimer's disease (Lockhart et al. 2005), bradykinin B-2 receptor antagonists/agonists (Heitsch 2002), 5-HT_3 antagonists (Ohta et al. 1996), and UV-induced keratinocyte apoptosis inhibitors (Enguehard-Gueiffier et al. 2005).

Apart from this, IP-containing heterocycles were also blended with fluorescent activity. Experimental studies further revealed that an increase in fluorescent activity has been observed when C-2 site of IP core is replaced with phenyl/naphthyl group (Pordel et al. 2017; Stasyuk et al. 2012; Catalán et al. 1992). Thus, the development of a fast and an efficient synthetic protocols of IP-based heterocycles becomes an attractive target for Chemist worldwide.

Microwave (MW)-assisted organic synthesis came out as a recent modification in chemical transformation and first report published in 1986 but after that more than 3500 articles have been online and counting are increasing day by

R. K. Yadav · S. Chaudhary (✉)
Laboratory of Organic and Medicinal Chemistry, Department of Chemistry, Malaviya National Institute of Technology Jaipur, Jawaharlal Nehru Marg, Jaipur, 302017, India
e-mail: schaudhary.chy@mnit.ac.in

Inamuddin et al. (eds.), *Advances in Green Synthesis*, Advances in Science, Technology & Innovation,
https://doi.org/10.1007/978-3-030-67884-5_12

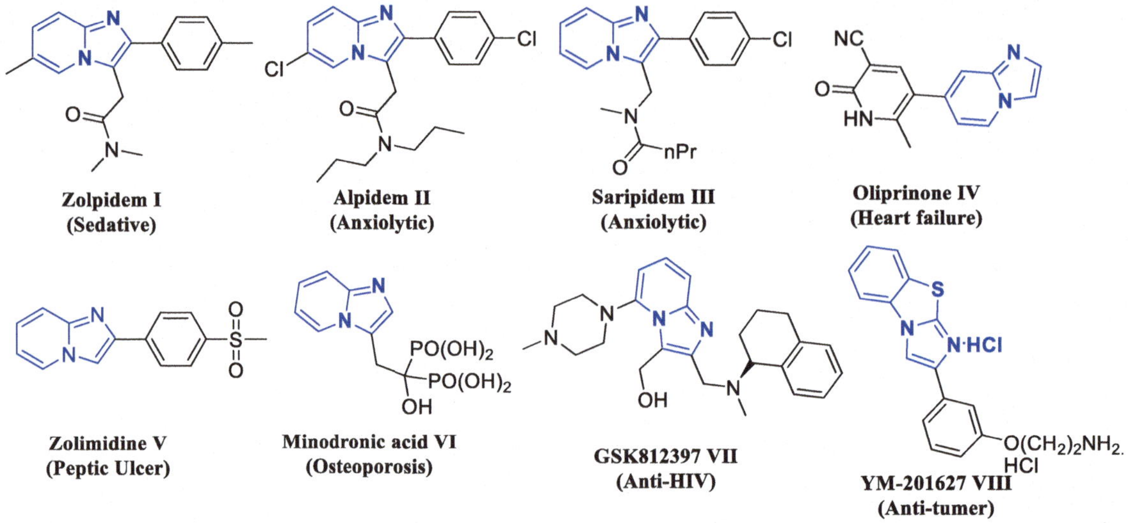

Fig. 1 Structures of some clinically used drugs based on imidazo[1,2-*a*]pyridines (IPs) and imidazo[2,1-*b*]thiazoles

day (Kappe et al. 2009). Many synthetic methodologies have been illustrated for the preparation of IPs but, in recent years, the importance of microwave-assisted organic synthesis (MAOS) has been identified as a masterstroke for its fast and efficient synthesis (Polshettiwar and Varma 2008). It has been well-established that the usage of microwave irradiation has been known to introduce the persuasive and controllable heating source for any particular type of MW-assisted organic reactions (Hoz et al. 2005). Therefore, microwave-irradiated synthesis of IPs has been extensively reconnoitered and showed better selectivity, shorter reaction time, high yields and diminished formation of by-products via suppressing side reactions (Elgemeie and Hamed 2014).

In this chapter, we accommodate most of the important historical development of microwave-irradiated synthesis of imidazo[1,2-*a*]pyridines observed during the past 2–3 decades. This chapter will also cover the developments of most efficient methodology practiced in the synthesis of imidazopyridines-based drugs/therapeutics.

Based on synthetic procedures, so far, reported in the literature, all the microwave-assisted synthesis of imidazo [1,2-*a*]pyridines have been divided into three sections.

2 Microwave-Assisted Synthesis of 2-arylimidazo[1,2-*a*]pyridines [Abbreviated as 2-Aryl-IPs].

2-Aryl-IPs (**1**), the basic skeleton of substituted 5,6-fused heterocycles, have been found in several synthetic/ naturally occurring pharmaceutically important molecules (Ramsden 1996). Therefore, microwave-irradiated preparation of 2-aryl-IPs is described year wise in a sequential manner.

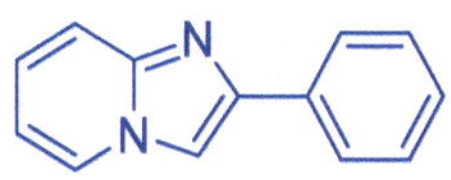

1

2.1 Synthesis of Fused Bicyclic Heteroaryl Boronates and Imidazopyridine-Quinazoline Hybrids Under MW-irradiations

Erin F. DiMauro et al. in **2006** synthesized various fused bicyclic heteroaryl boronates **4a-h** in moderate yield which were further utilized to prepare dimer of fused heterocycles via one-pot Suzuki coupling of **4a-h** with 6-bromoquinazolin-2-amine **5** which furnished imidazopyridine-quinazoline hybrids **6a-g** in moderate yield (Scheme 1) (DiMauro and Vitullo 2006).

5-(4,4,5,5-tetramethyl-1,3,2-dioxaborolan-2-yl) pyridin-2-amine **2** reacts with α-bromoketones **3a-h** in ethanol under microwave irradiation at 130 °C for 30 min furnished imidazo[1,2-*a*]pyridine-6-boronates **4a-h** as an intermediate which, then, subjected to one-pot Palladium-catalyzed Suzuki Coupling in a mixture of H2O and ethanol (1:3 ratio) as solvent at 90 °C for 30 min under microwave irradiations furnished **6a-g** in 26–54% yield range (Scheme 1) (DiMauro and Vitullo 2006).

2.2 MW-Irradiated Synthesis of IPs Using Multi-Component Strategy Under Neat Conditions

In **2010**, Mehdi Adib et al. synthesized imidazo[1,2-*a*]pyridine **10** using multi-component reaction. It initially involves the reaction of pyridine **7** with α-bromoketones **8** in situ

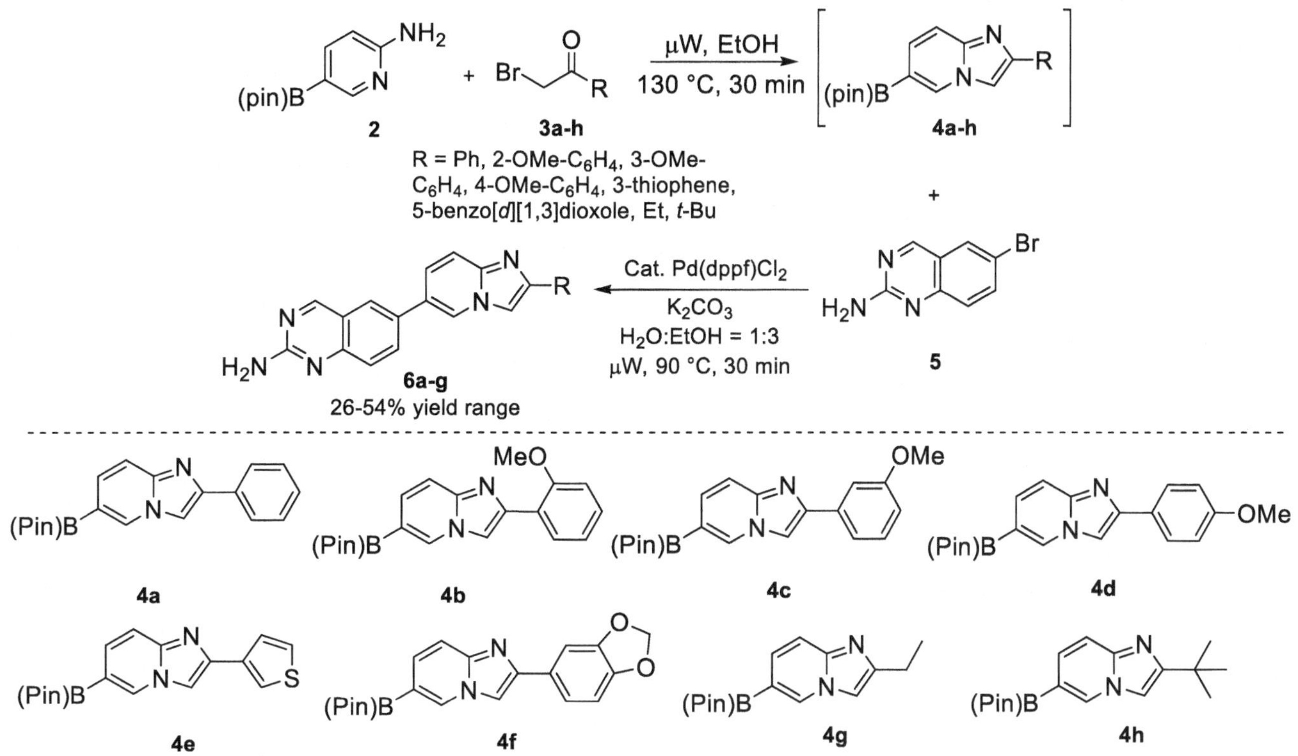

Scheme 1 Synthesis of imidazo[1,2-*a*]pyridine-6-boronates **4a-h**

which furnished *N*-phenacylpyridinium bromides as an intermediate which, then, undergone microwave-assisted nucleophilic addition of ammonium acetate **9** under solvent-free conditions at 180 °C for 4 min afforded the imidazo [1,2-*a*]pyridines **10a-j** in excellent yields using simple starting materials (Scheme 2) (Adib et al. 2010).

The reaction mechanistic pathway involved two steps. Initially, pyridine **7** undergone nucleophilic addition on the α-bromoketones **8** formed *N*-phenacyl pyridinium bromide salt. Then, in the next step, the pyridinium ion get attacked by ammonium acetate **9** (act as a source of $-NH_2$) formed adduct **I**, which then generates dihydro imidazo[1,2-*a*]pyridine intermediate **II** upon cyclization. The cyclized product **II** furnished **10** via aerial oxidation in excellent yields (Fig. 2) (Adib et al. 2010).

2.3 One-Pot, Three-Component Synthesis of 2-Phenyl-*H*-Imidazo[1,2-α]pyridine Under MW-Irradiations

A one-pot, multi-component, atom economical, benign, domino green preparation of 2-phenyl-*H*-imidazo[1,2-*a*] pyridine in excellent yields have been reported by Kourosh Motevalli et al. in the year **2012** (Scheme 3) (Motevalli et al. 2012).

This domino strategy involves the reaction with pyridine **7**, substituted phenacyl bromides **11a-h** and **12** [guanidine or urea or thiourea] at 150 °C for 3–4 min under microwave irradiation afforded the corresponding 2-phenyl-*H*-imidazo [1,2-α] pyridines **13a-h** in 75–85% yield range. The tentative mechanistic pathway occurs via the initial reaction of pyridine **7** and phenacylbromide **11a** using microwave irradiations at 100 °C for 1 min. formed N-Phenacyl pyridinium bromide **A**. Then, guanidine hydrochloride **12** was added into the same reaction mixture and further subjected to MW-irradiations with a power of 600 W at 150 °C for 2–3 min (Make: ETHOS 1600, Milestone) forms the intermediate **B** which, after the release of HBr, resulted in the formation of adduct **C**. This adduct C on further cyclization and aromatization via releasing of the formamide molecule forms the desired **13a** in 79% yield (Motevalli et al. 2012).

2.4 Microwave-Assisted Amine-Triggered Benzannulation Strategy for the Preparation of 2,8-Diaryl-6-Aminoimidazo-[1,2-*a*]pyridines

In **2012**, Shanmugam Muthusubramanian and his coworkers gave a metal-free, proficient MW-assisted, amino-triggered benzannulation approach for the formation of 2,8-diaryl-6-aminoimidazo-[1,2-*a*]pyridines (Nagaraj et al. 2012).

The reaction of aryl(4-aryl-1-(prop-2-ynyl)-1*H*-imidazol-2-yl)methanone **14a-h** with several substituted

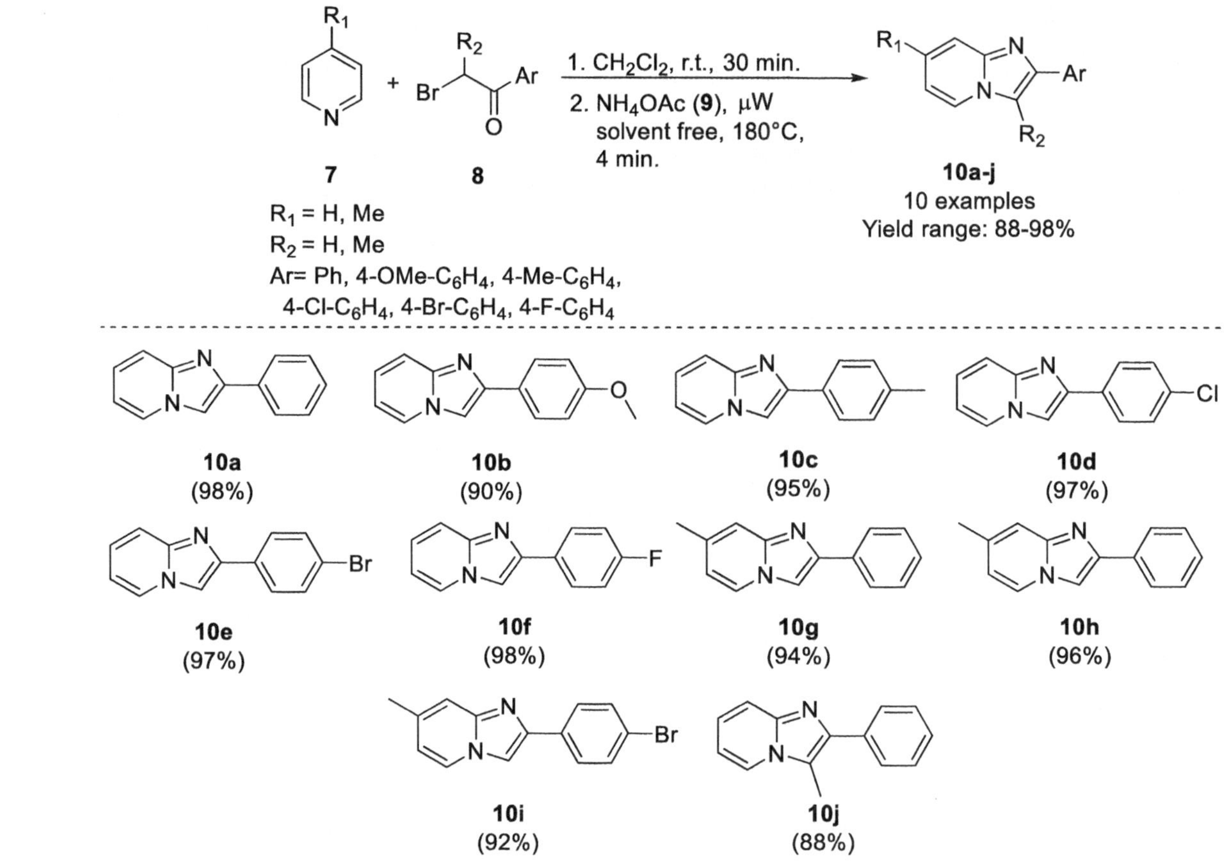

Scheme 2 MW-irradiated synthesis of imidazo[1,2-*a*]pyridines using multi-component strategy under solvent-free conditions

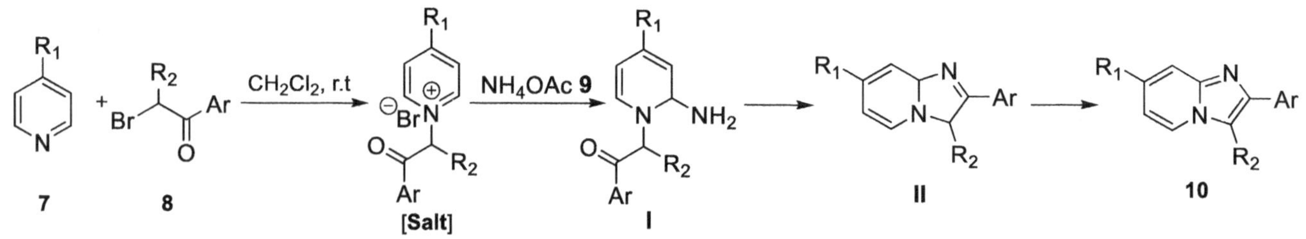

Fig. 2 Tentative mechanism for the synthesis of **10a-j**

dialkylamines **15a-h** under microwave condition at 80 °C for 10 min. using 4 Å molecular sieves in catalytic amount which furnished several substituted 2,8-diaryl-6-aminoimidazo[1,2-*a*]pyridines **16a-s** in 15–94% yield range (Scheme 4) (Nagaraj et al. 2012).

2.5 MW-Assisted $NaHCO_3$-catalyzed Synthesis of Imidazo[1,2-*a*]pyridines in PEG_{400} Media and Its Practical Application in the Synthesis of 2,3-Diaryl-IP Class of Bio-Heterocycles

Sabine Berteina-Raboin and her coworkers (**2014**) reported a straight forward $NaHCO_3$-catalyzed efficient preparation of substituted imidazo[1,2-*a*]pyridine **19a-p** via the reaction of substituted 2-aminopyridine **17a-j** with several substituted α-bromoketones **18a-g** dissolved in PEG_{400} medium under MW-irradiation conditions at 120 °C for 10 min (Scheme 5) (Heibel et al. 2014).

The above methodology was then practically applied in the one-pot synthesis of 2,3-diaryl-IP class of bio-heterocycles **20a-u** in 36–80% yield range (Scheme 7). The procedure involves initially the one-pot reaction of substituted 2-aminopyridine **17a-j** with several substituted α-bromoketones **18a-g** dissolved in PEG_{400} medium using MW-irradiation conditions at 120 °C for 10 min and then, in the same pot, aryl bromides was added in the presence of

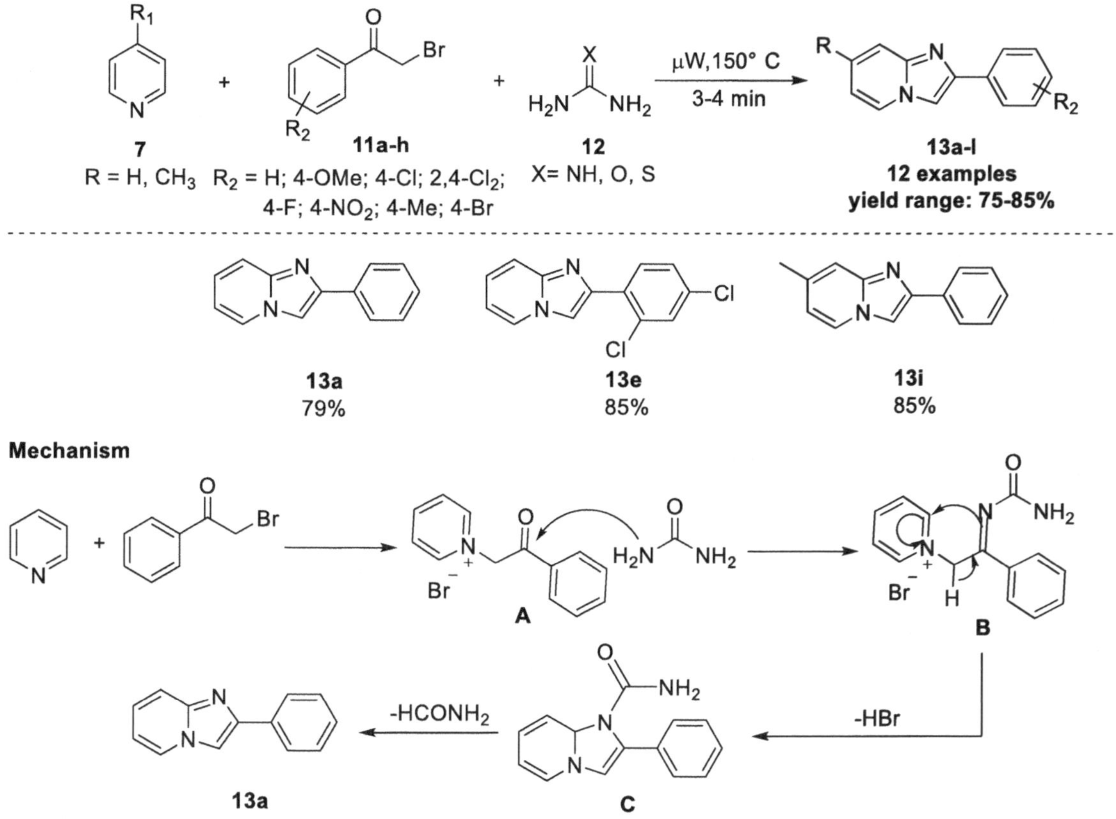

Scheme 3 Synthesis of microwave-assisted 2-phenyl-*H*-imidazo[1,2-*a*]pyridine

Palladium acetate (1 mol%), potassium acetate (4 equiv.) using MW-irradiation conditions at 220 °C for 1 h furnished 2,3-diarylimidazo[1,2-*a*]pyridine **20a-u** without the use of any ligand. PEG_{400} (polyethylene glycol) has been recognized as a suitable medium for the one-pot condensation reaction. It has also been described that functional groups containing electron-donating and electron-withdrawing groups showed tolerance under optimized reaction conditions, particularly at *para*-and *meta*-positions (Scheme 6) (Heibel et al. 2014).

2.6 MW-Irradiated, Ligand-Free, Palladium-Catalyzed, One-Pot 3-component Reaction for an Efficient Preparation of 2,3-Diarylimidazo[1,2-*a*]pyridines

Hong-yu Li and his coworkers (**2014**) reported a unique MW-irradiated, ligand-free, Palladium (Pd)-catalyzed, one-pot 3-component reaction for an efficient preparation of 2,3-diaryl-IPs in excellent yields (Scheme 7) (Wang et al. 2014a). This methodology had been described superior due to high availability of commercial grade reagents and wide substrate scope. 2-aminopyridine **21a-d**, α-bromoketones **22a-e** and aryl bromide **23a-o** were reacted in the presence of $Pd(OAc)_2$/KOAc and potassium acetate dissolved in DMF as solvent under microwave irradiation at 160 °C for an hour which furnished 2,3-diaryl-IPs analogues **24a-x** in 56–85% yield range. The one-pot mechanistic pathway proceeds with the reaction of 2-aminopyridine with α-bromoketone which furnished the imidazo[1,2-*a*]pyridine skeleton which then undergone Pd-catalyzed Suzuki product under microwave irradiations afforded the C-3 arylated product in good yields (Wang et al. 2014a).

2.7 MW-Assisted Water-PEG_{400}-mediated Synthesis of 2-Phenyl-IP via Multi-Component Reaction (MCR)

Santosh A. Jadhav et al. (2016) reported a water-PEG_{400}-mediated synthesis of 2-phenyl-IP **28** via multi-component reaction using MW-irradiations (Scheme 8) (Jadhav et al. 2016).

An efficient, simple and fast, one-pot microwave-assisted green methodology involves the multi-component reaction

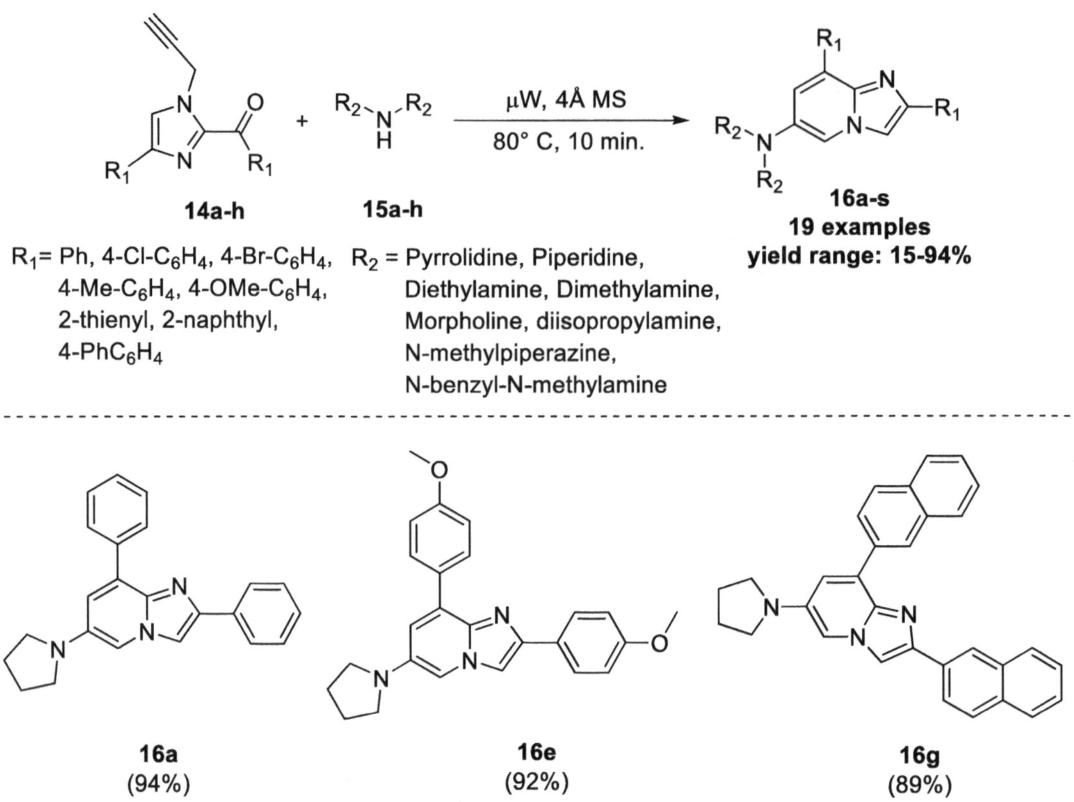

Scheme 4 Microwave-irradiated synthesis of 2,8-diaryl-6-aminoimidazo-[1,2-*a*]pyridine via amine-activated benzannulation approach

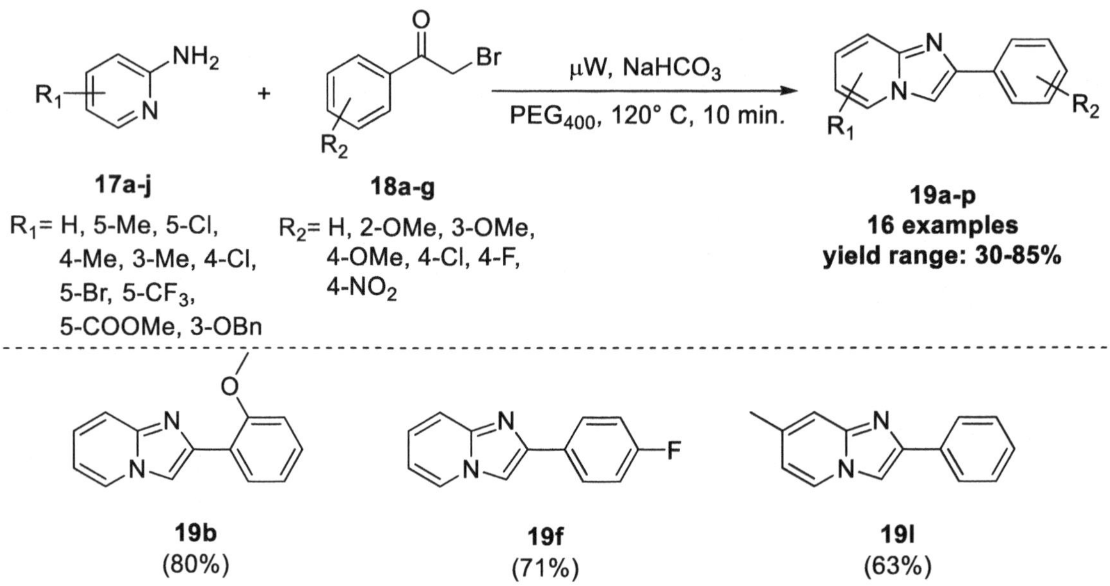

Scheme 5 MW-assisted $NaHCO_3$-catalyzed synthesis of **19a-p** in PEG_{400} medium

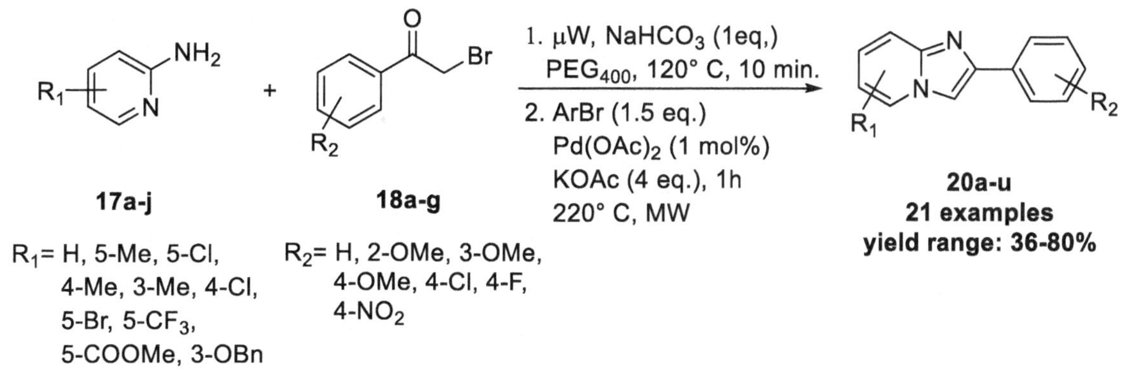

Scheme 6 Sequential one-pot synthesis of **20a-u** in PEG_{400} solvent

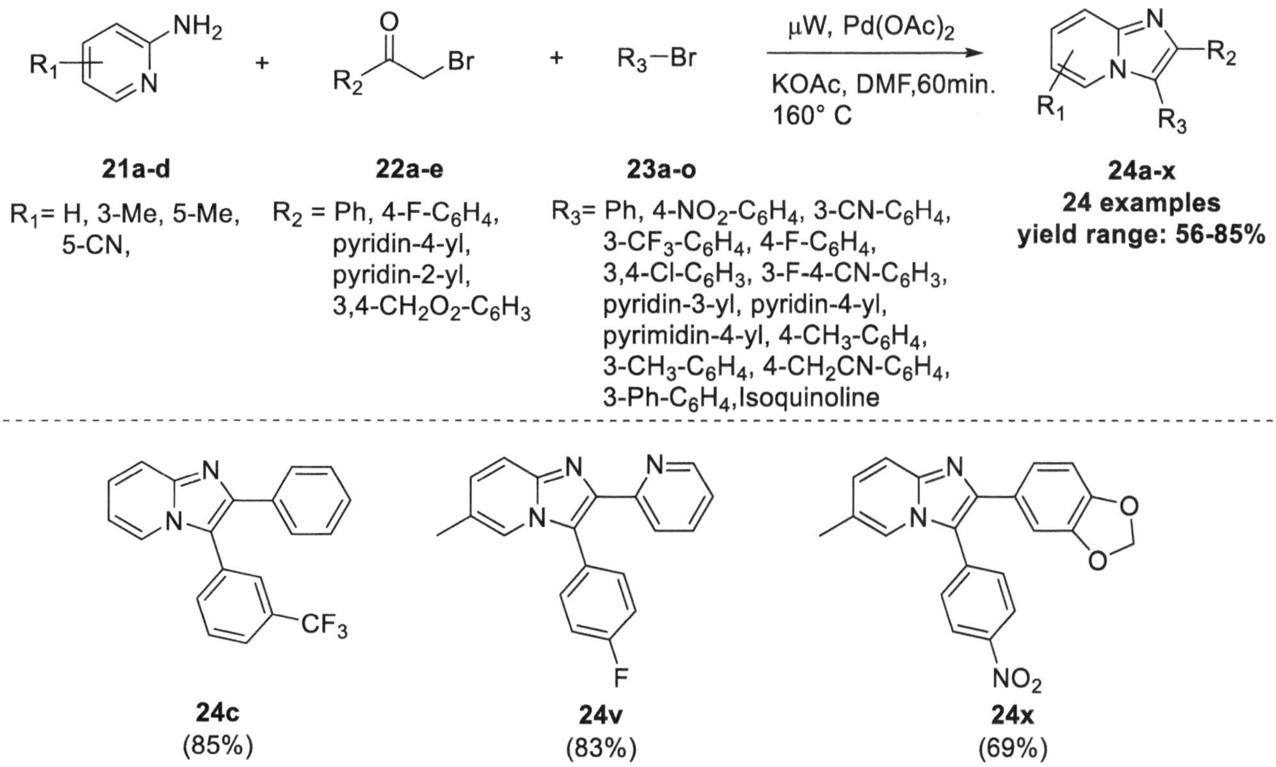

Scheme 7 Pd-catalyzed synthesis of 2,3-diarylimidazo[1,2-*a*]pyridines **24**

of succinimide **25**, substituted acetophenones **26a-f**, and substituted 2-aminopyridine **27a-d** in the stoichiometric amount of AgI and catalytic amount of iodine in the mixture of PEG_{400} and H_2O (2:1) serving as green solvent at 350 W at 95–100 °C for 7–8 min furnished substituted 2-phenylimidazo[1,2-*a*]pyridines **28a-k** in 89–98% yield range (Scheme 8). The unique feature of this methodology is the in situ generation of α-haloketones for the synthesis of targeted molecule (Jadhav et al. 2016). The probable mechanism of this reaction had been depicted as shown in Fig. 3 (Jadhav et al. 2016).

2.8 Microwave-Irradiated Synthesis of Imidazo[1,2-*a*]pyridines Under Neat, Catalyst-Free Conditions

Dulin Kong et al. in the same year **2016** gave a simple, solvent-free, catalyst-free, high-yielding greener approach toward synthesis of 2-phenylimidazo[1,2-*a*]pyridines **29a-j** (Scheme 9) (Kong et al. 2016).

Similar to previously reported synthetic procedures, substituted 2-aminopyridines **17a-d** reacts with substituted α-bromoketones **18a-d** under MW-irradiation at 65 °C for

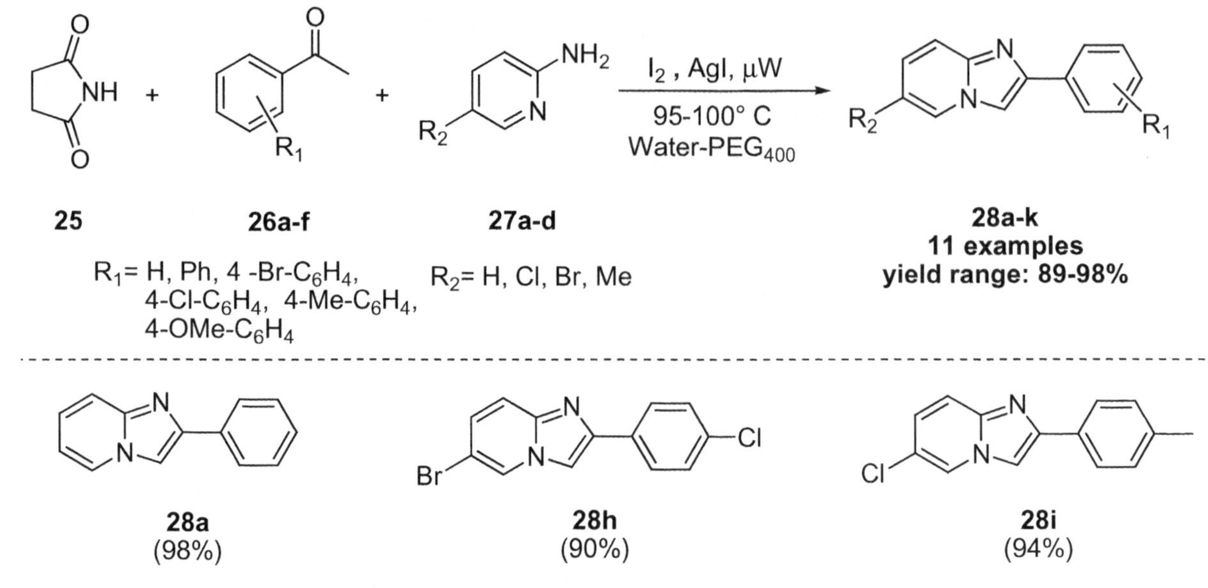

Scheme 8 AgI-catalyzed synthesis of 2-arylimidazo[1,2-*a*]pyridines

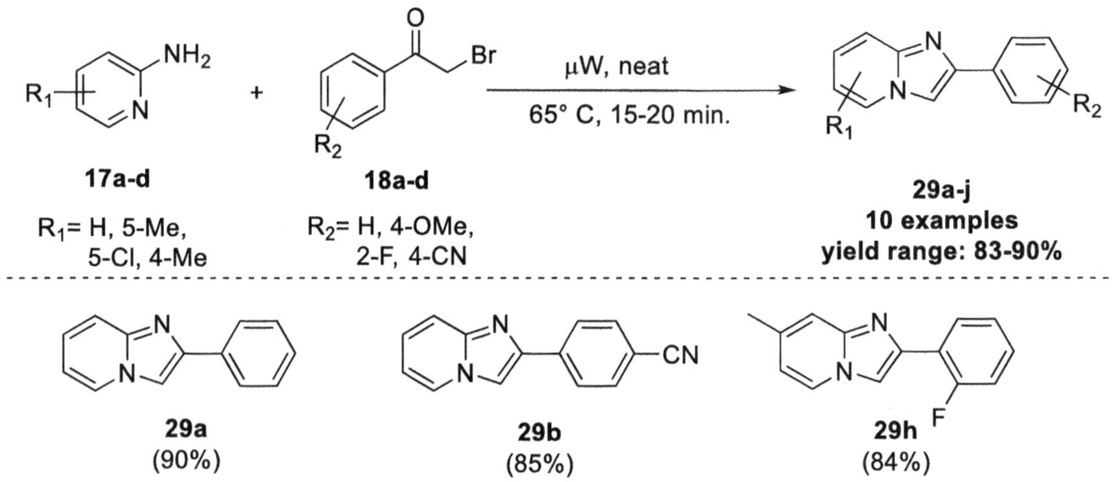

Scheme 9 Neat, catalyst-free, MW-assisted synthesis of 2-phenylimidazo[1,2-*a*]pyridines **29a-j**

15–20 min afforded 2-phenylimidazo[1,2-*a*]pyridines **29a-j** in 83–90% yield range (Scheme 9) (Kong et al. 2016).

2.9 Green Synthesis of Imidazo[1,2-*a*]pyridines in H_2O

Sheela Gopal M. and Anitha I. reported (**2016**) water-mediated synthetic protocol of 2-phenylimidazo[1,2-*a*] pyridines **30a-h** via the combination of 2-aminopyridine **17a** and α-bromoketones **18a-h** under MW conditions for 1 min. at 300 W in moderate to excellent yield (Scheme 10) (Gopal and Anitha 2016).

2.10 Microwave-Assisted Neat Synthesis of Substituted 2-Arylimidazo[1,2-a] Pyridines

Tejeswararao Dharmana and Mallika Swapna (2017) reported a MW-assisted neat method for the synthesis of substituted 2-arylimidazo[1,2-*a*]pyridines **31a-i** using the previously reported starting materials, i.e., substituted 2-aminopyridine **17a-c** and substituted α-bromoketones **18a-g** in the presence of [*1-(4-sulfonic acid) butylpyridinium hydrogen sulfate*] as green solvent at 100 °C for 30 s in 87–95% yield range (Scheme 11) (Dharmana and Swapna 2017).

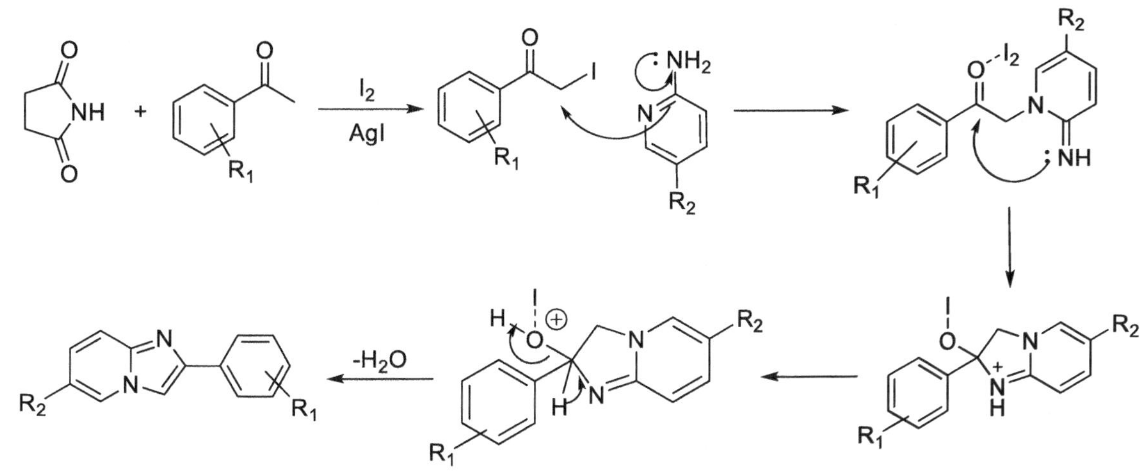

Fig. 3 Probable reaction mechanism for the synthesis of **28a-k**

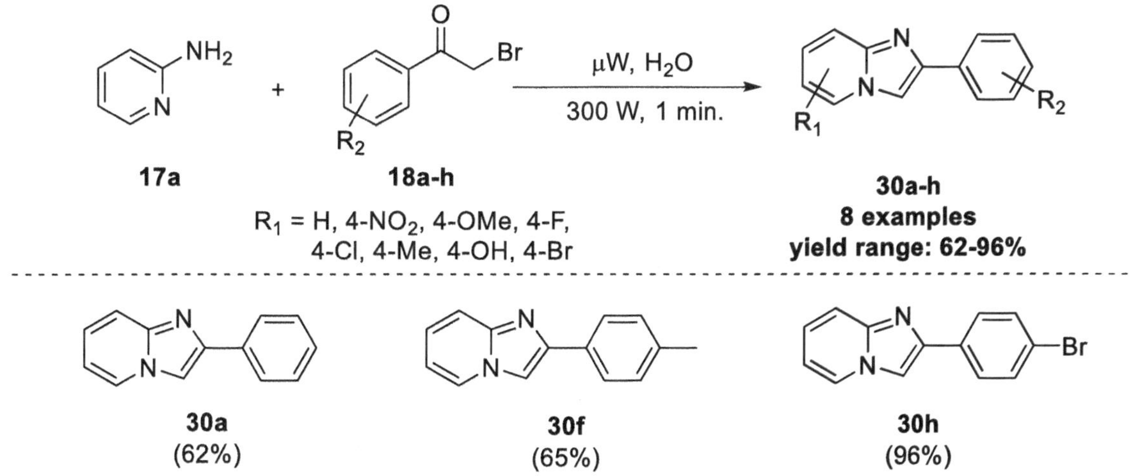

Scheme 10 Water-mediated synthesis of 2-phenylimidazo[1,2-*a*]pyridines **30a-h**

2.11 Microwave-Assisted Nano SiO_2 Neat Synthesis of Substituted 2-Arylimidazo [1,2-a]pyridines

In **2018**, Tejeswararao Dharmana et al. have synthesized the 2-phenylimidazo[1,2-*a*]pyridines **31a-i** via the fusion of substituted 2-aminopyridine **17a-c** and substituted α-bromoketones **18a-g** in catalytic assistance with nano-SiO_2 under MW-irradiations at 100° C for 30 s in 87–95% yield range (Scheme 12) (Dharmana et al. 2018).

2.12 Microwave-Assisted $NaHCO_3$-Catalyzed Synthesis of 2-phenyl-IPs

Juan C. Rodríguez et al. (**2020**) synthesized the 2-phenylimidazo[1,2-*a*]pyridines **20a-t** via the simplest route taking 2-aminopyridine **17a** and substituted α-bromoketones **18a-w** under microwave irradiation conditions. In this report, initially microwave-assisted N-bromosuccinimide (NBS)/p-toluenesulfonic acid (PTSA) catalyzed synthesis of phenacyl bromides **18a-w** were carried out via the bromination of different acetophenones in 15 min in 50–99% yield range. Then, the conjugation of these molecules **18a-w** having variety of substitution with 2-aminopyridine **17a** in the presence of sodium bicarbonate in MeOH at 80 °C for 1 min under MW-irradiations furnished imidazo[1,2-*a*]pyridine derivatives **20a-t** (20 examples) in 24–99% yield range. With substrate **18u-w** (R_1 = NHPh, 2,6-OMe and 2,4-OMe), the product **20u-w** was not formed at all under the said reaction conditions. It has also been found that the improved yields in the present methodology was observed as compared to other more tedious methodologies such as thermally and mechanically assisted routes (Scheme 13) (Rodriguez et al. 2020).

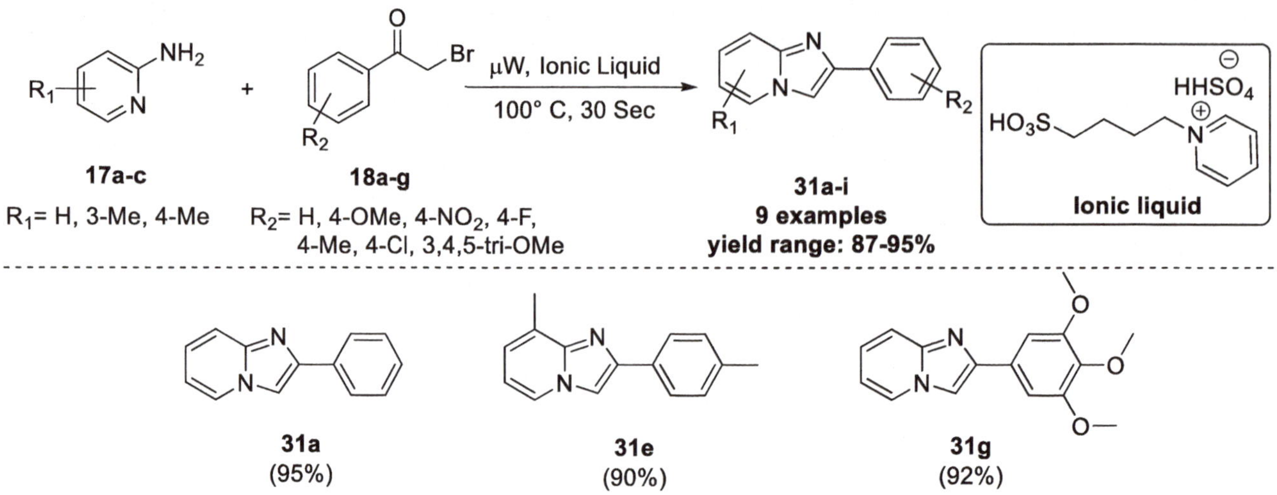

Scheme 11 Synthesis of substituted 2-arylimidazo[1,2-a]pyridines **31a-i** in ionic liquid media

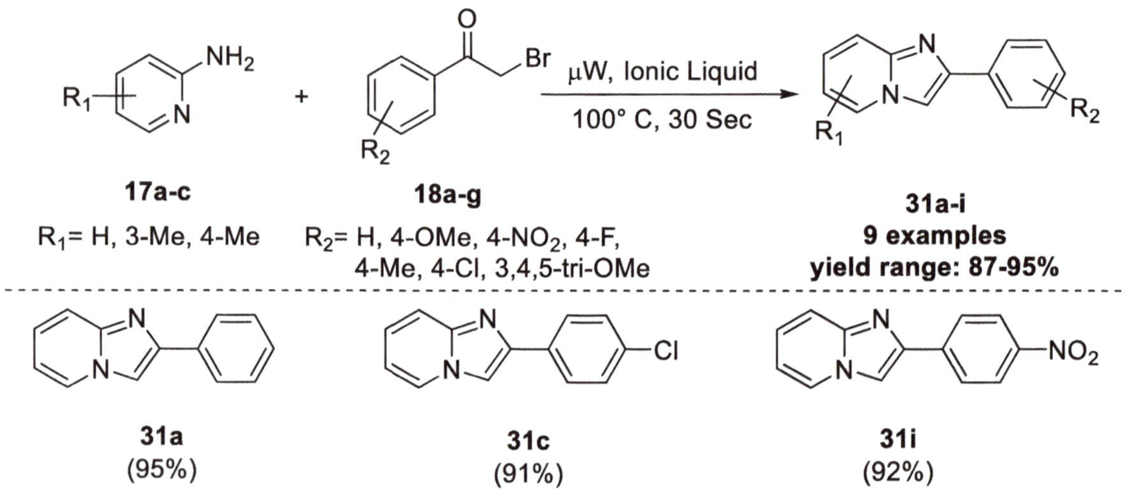

Scheme 12 Preparation of substituted 2-arylimidazo[1,2-*a*]pyridines **31a-i** using nano SiO_2 as a catalyst

3 Microwave-Assisted Synthesis of 3-amino-2-arylimidazo[1,2-*a*]pyridines [3-amino-2-aryl-IPs]

3-Amino-2-aryl-IPs (**32**), the C-3 amino substitution in the basic skeleton of substituted 5,6-fused heterocycles, have been also prepared in the laboratory via microwave-assisted reactions. Therefore, so far, the details of the synthetic protocols via microwave irradiations have been described year-wise in a sequential manner.

3.1 Microwave-Irradiated Synthesis of 3-aminoimidazo[1,2-*a*]pyridines via Fluorous Multi-component Pathway

Yimin Lu and coworkers (**2004**) described the MW-irradiated synthesis of 3-aminoimidazo[1,2-*a*]pyridines **37c** via fluorous multi-component strategy which involves Sc $(OTf)_2$-catalyzed reaction of fluorous tagged benzaldehyde **33c**, substituted 2-aminopyridine **34** and substituted iso-cyanide **35** using either under MW-irradiation at 150 °C for 10 min using DCM/MeOH (3:1) as a solvent or under thermal conditions at 80 °C for 120 min which, after purification of the reaction mixture by fluorous-solid-phase extraction (F-SPE) method afforded pure fluorous-tagged 3-aminoimidazo[1,2-*a*]pyridines **36c** in 29–103% yield range (Scheme 14c). The required fluorous benzaldehyde was prepared by the reaction of hydroxylated benzaldehyde

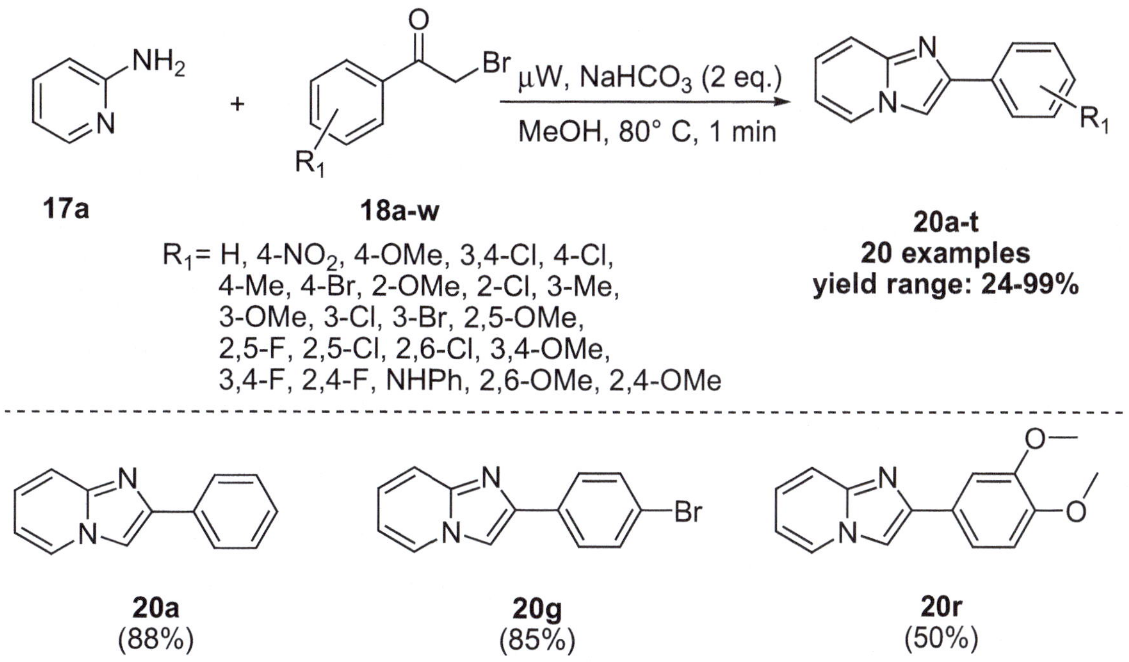

Scheme 13 $NaHCO_3$-catalyzed synthesis of substituted 2-arylimidazo[1,2-*a*]pyridines **20a-s** under MW conditions

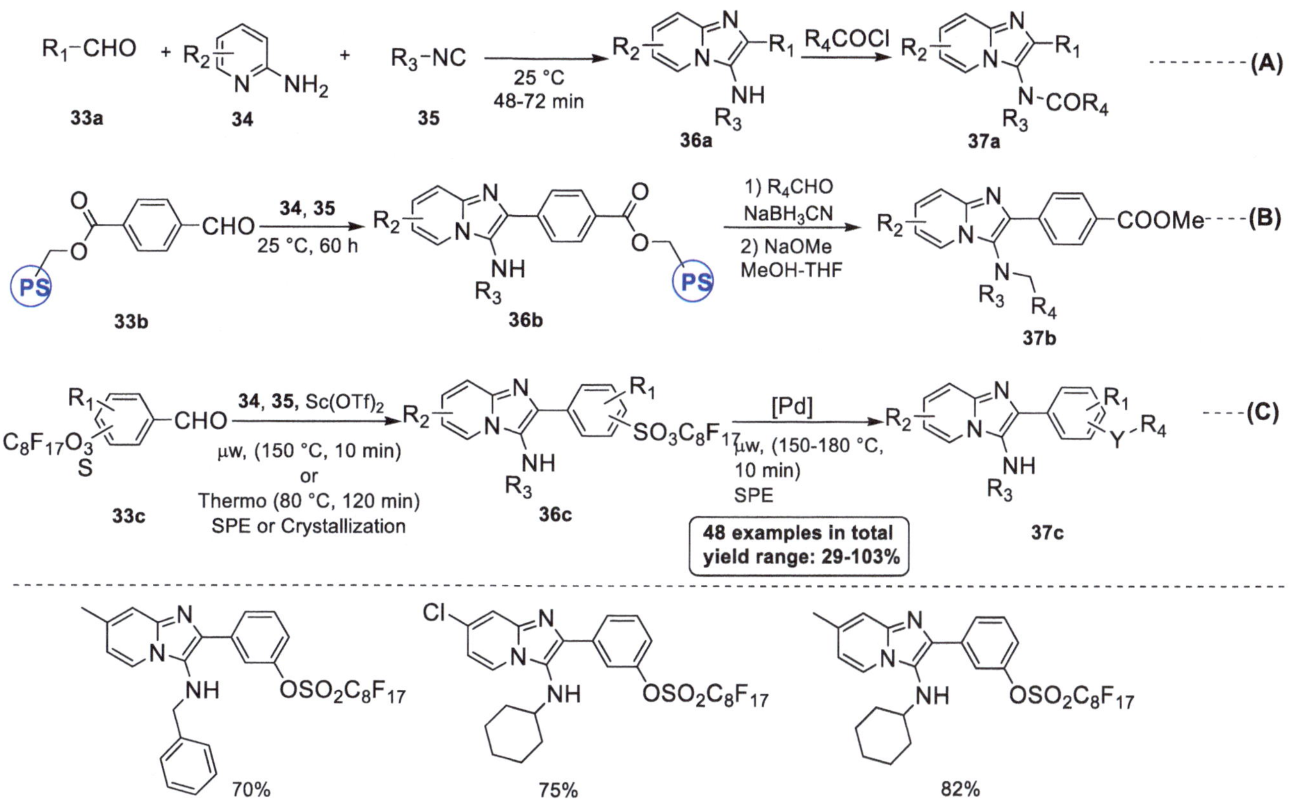

Scheme 14 Solution-phase preparation of **37a** (**A**), solid-phase synthesis of **37b** (**B**), and fluorous syntheses of **37c** (**C**)

to perfluoro-octane sulfonyl fluoride (Scheme 14c) (Lu and Zhang 2004).

A series of substituted 3-aminoimidazo[1,2-*a*]pyridine derivatives were designed to synthesize utilizing microwave as well as fluorous methodologies for up surging he speed of the reaction by decreasing the purification time in solution-phase parallel synthesis. The multi-component reactions of perfluorooctanesulfonyl-tagged benzaldehydes **33c** with substituted 2-aminopyridines **34** and substituted isocyanides **35** furnished **36c** which on subjection to Palladium-catalyzed reactions with boronic acids/thiols furnished biaryls/aryl sulfides **37c** in 29–103% yield range. It has been noticed in this strategy that the reaction mixtures had been purified either via fluorous-solid-phase extraction (F-SPE) technique or crystallization methods.

The three-component solution-phase syntheses of **37a** were delivered by Blackburn and other groups which involves the fusion of substituted aldehydes **33a**, substituted 2-aminopyridines **34**, and substituted isonitriles **35** at 25 °C for 48–72 min followed by the treatment with acid chloride R_4COCl (Scheme 14a) (Blackburn et al. 1998). This multi-component reactions (MCRs) get completed at ambient temperature, and purifications of the desired product **37a** were conducted by resin capture and release (Blackburn et al. 1998a). Another solid-phase synthesis was developed by Blackburn group using polymer-supported aldehydes **33b** in order to streamline the purification of condensation products (Scheme 14b) (Blackburn and Guan 2000). Yimin Lu and Wei Zhang reported a dual methodology incorporating both microwave irradiations and the fluorous-tagging strategy (Zhang 2004) in order to accelerate the reaction and the purification processes (Scheme 14c). Further, **36c** was subjected to Palladium-catalyzed coupling reactions which afforded **37c** without the tagging of the fluorous group and having another point of diversification ($-Y-R_4$) (Lu and Zhang 2004).

This paper also illustrates the three functions of perfluoro alkane sulfonyl tag in the fluorous approach, i.e., (1) as the protection of phenolic group before the condensation reaction; (2) as the tagging of phase for the separation of different compounds in the reaction mixture; as well as (3) acting as an activator in C–C coupling reaction. All the fused products **37a-c** installed with four variable site (R_1 to R_4) along with another diversity of Y (=nothing or S). It can be concluded that a microwave-assisted fluorous MCRs synthetic route followed by microwave-irradiated post-condensation reactions had been established for the synthesis of 3-amino-IP derivatives **37a-c**. Palladium-catalyzed coupling of the fluorous sulfonates either with boronic acids/thiols furnished biaryls and aryl sulfides **37c**, respectively, under MW-irradiations. Liquid chromatography with mass spectrometry (LC–MS) techniques or thin-layer chromatography (TLC) observation were utilized to study these conversions. The F-SPE separation technique or recrystallization methods were found fruitful in this reaction (Lu and Zhang 2004).

3.2 MW-Irradiated Synthetic Protocol for 3-aminoimidazo[1,2-*a*]pyridines via MCR Pathway

Christopher Hulme and his coworkers (**2005**) stated a one-step solution-phase MCR protocol for the direct synthesis of a series of 3-amino-IPs **39a-l** (Scheme 15) (Schwerkoske et al. 2005). It involves the $Sc(OTf)_2$-catalyzed reaction of benzaldehyde **33a-g**, substituted 2-aminopyridine **34a-d** and trimethylsilyl cyanide **38** used in place of isocyanide **35** in MeOH under MW-irradiation at 140 °C for 10 min (Schwerkoske et al. 2005).

3.3 MW-Assisted Sequential Ugi/Strecker Reactions Involving 3-Center-4-Component and 3-Center-5-Component MCR Strategy

Christopher Hulme and his coworkers (**2006**) further extended their work and prepared 3-aminoimidazo[1,2-*a*]pyridines **40**; 3-iminoaryl-IPs **41a-h** as well as imidazo[1,2-*a*]pyridyn-3-ylamino-2-acetonitrile **42a-f** via microwave-assisted sequential Ugi/Strecker reaction (Scheme 16) (Masquelin et al. 2006).

These non-isocyanide-based 3-center-4-component and 3-center-5-component MCR protocol were found to generate substrate-controlled products using polymer-bound catalytic Scandium Triflate [PS-$Sc(OTf)_3$] in methanol solvent under MW conditions at 140 °C for 5 min and 20 min furnished **41a-h** and **42a-f**, respectively. When substituted 2-aminopyridine **34a-k**, substituted aldehyde **33a-d** and trimethylsilyl cyanide (TMSCN) **38** were taken in 1:0.9:1 equivalents, respectively; it furnished pseudo-Ugi reaction product, i.e., 3-aminoimidazo[1,2-*a*]pyridines **40** in 0–10% yield range. However, when **34a-k**, **33a-d** and **38** were taken in 1:2.2:1 equivalents, respectively (increasing the amount of aldehyde from 0.9 to 2.2 equivalents); it afforded 3-iminoaryl-imidazo[1,2-*a*]pyridines **41a-h** extensively in 31–51% yield range. Sequentially, when these reagents were taken into the proportion of 1:4:3 equivalents, respectively;

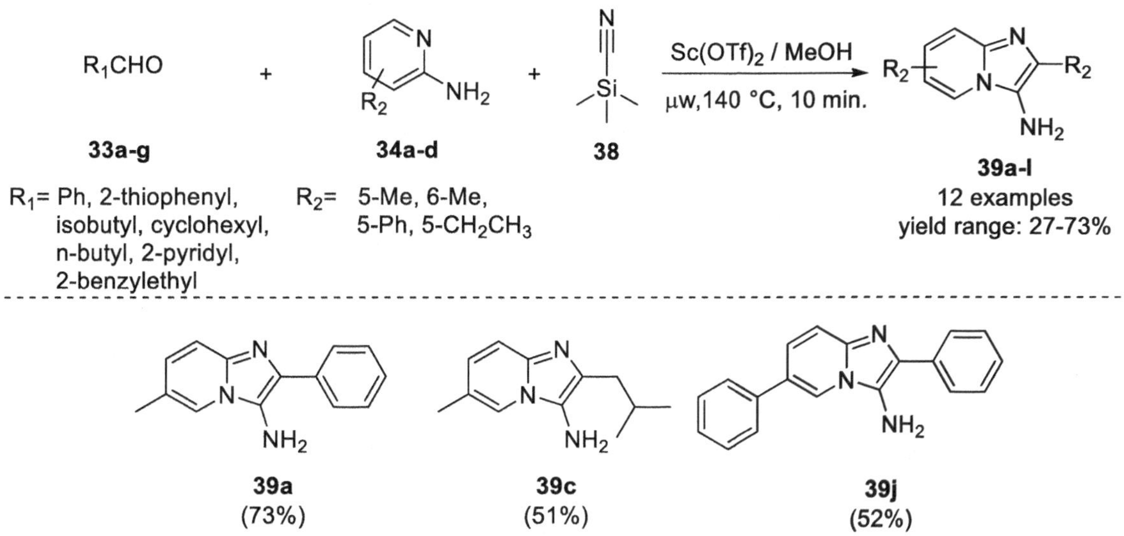

Scheme 15 One-step solution-phase MCR procedure for the direct synthesis of 3-amino-IPs **39a-l**

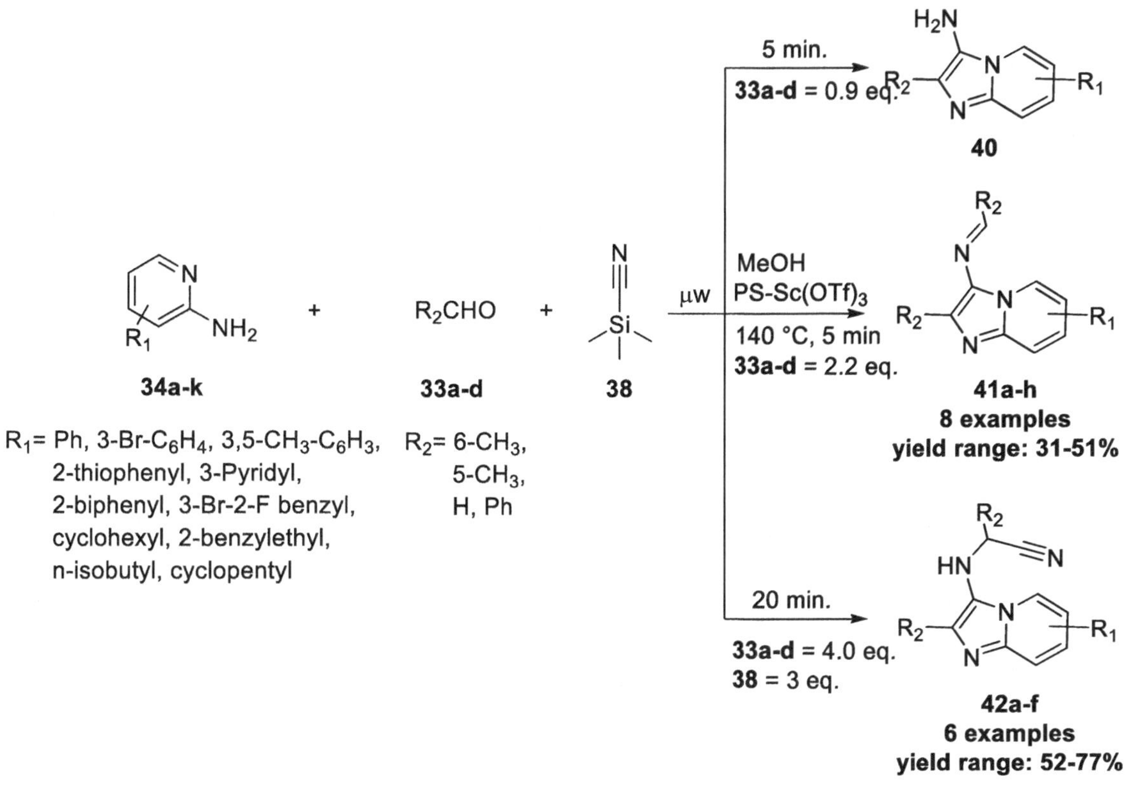

Scheme 16 MW-assisted sequential Ugi/Strecker reactions for the synthesis of 3-iminoaryl-IPs **41a-h** and imidazo[1,2-*a*] pyridyn-3-ylamino-2-acetonitrile **42a-f**

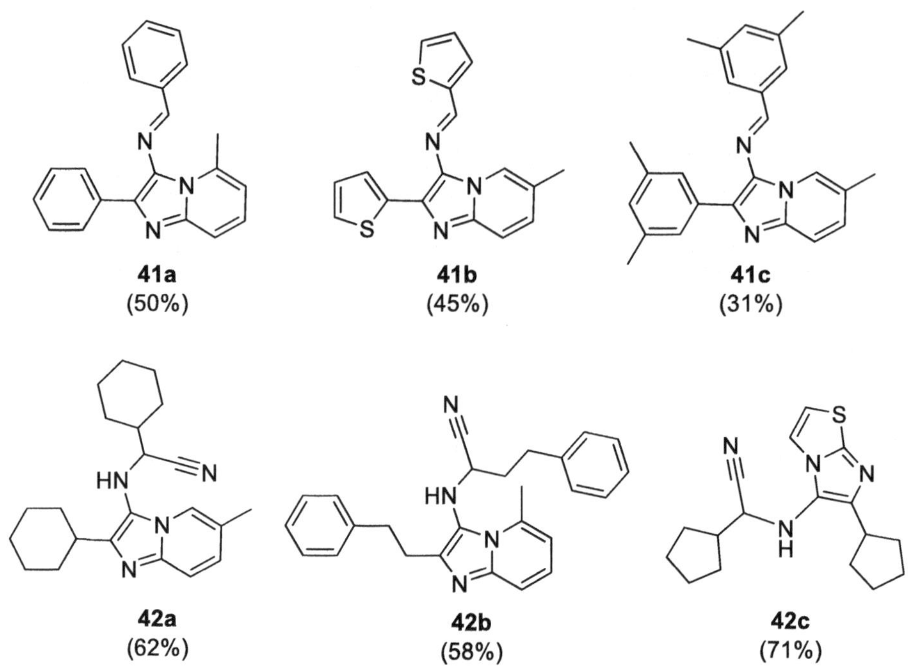

Fig. 4 Structures of few synthesized molecules **41a-c** and **42a-c**

then, it afforded imidazo[1,2-*a*]pyridyn-3-ylamino-2-acetonitrile **42a-f** in 52–77% yield range. (Scheme 16) (Masquelin et al. 2006). The structure of few synthesized compounds is shown in Fig. 4.

3.4 One-Pot, 4-component Cyclization/Suzuki Coupling Leading to the Rapid Formation of 2,6-Disubstituted-3-Amino-IPs Under Microwave Irradiations

DiMauro et al. (2007) had given a unique one-pot MW-irradiated preparation of 3-aminoimidazo[1,2-*a*] pyridines-6-boronates **44a-o** via UGI type 4-component reaction strategy utilizing the $MgCl_2$-catalyzed reaction of pinacol ester of 2-aminopyridine-5-boronic acid **43**, substituted aldehyde **33a-g**, substituted isocyanide **35a-e** in MeOH solvent under N_2 condition at 160 °C for 10 min. Furthermore, 3-aminoimidazo[1,2-*a*]pyridines-6-boronates **44a-o** were utilized as an efficient building block/ intermediate for the further Pd-catalyzed Suzuki coupling with different arylhalides **45a-d** under microwave conditions furnished 2,6-disubstituted-3-amino-IPs **46a-n** in 42–68% yield range (Scheme 17) (DiMauro and Kennedy 2007).

In this report, special emphasis had been given for the usefulness of pinacol ester of 2-aminopyridine-5-boronic acid **44a-o** as an efficient building block for the C–C bond formation. It was also mentioned that the boronate was found to be well-tolerable toward cyclization reaction catalyzed by $MgCl_2$; and simultaneously, palladium-catalyzed Suzuki reaction was also occurred smoothly in the presence of Mg^{2+} salts (DiMauro and Kennedy 2007).

3.5 $ZnCl_2$-catalyzed MCR of 3-aminoimidazo [1,2-*a*]pyridines Using MW Conditions

Amanda L. Rousseau and coworkers (**2007**) reported unique MCR protocol to synthesize 3-aminoimidazo[1,2-*a*]pyridine **47a-l** utilizing 2-aminopyridine **17a**, substituted aldehyde **33a-c** and substituted isocyanide **35c-d/35f-g** as a starting materials under microwave heating using $ZnCl_2$ and Montmorillonite clay K_{10} as a catalytic system in 1,4-dioxane as solvent at refluxing condition for 1–5 h (Scheme 18) (Rousseau et al. 2007).

3.6 Microwave-Promoted Preparation of *N*-(3-arylmethyl-2-oxo-2,3-dihydroimidazo [1,2-*a*]pyridin-3-Yl)Benzamides

Tu et al. (**2007**) described a novel methodology to synthesize fused heterocyclic core containing imidazo[1,2-*a*] pyridine-2-ones having benzyl as well as benzamido groups in habiting at C-3 position simultaneously. Thus, the reaction of 2-aminopyridine **17a** with several substituted 4-arylidene-2-phenyl-5(4*H*)-oxazolones **48a-l** in ethane-1,2-diol as a solvent under MW-irradiations at 120 °C for 4–7 min afforded *N*-(3-arylmethyl-2-oxo-2,3-dihydroimidazo[1,2-*a*]pyridin-3-yl)benzamides **49a-o** in

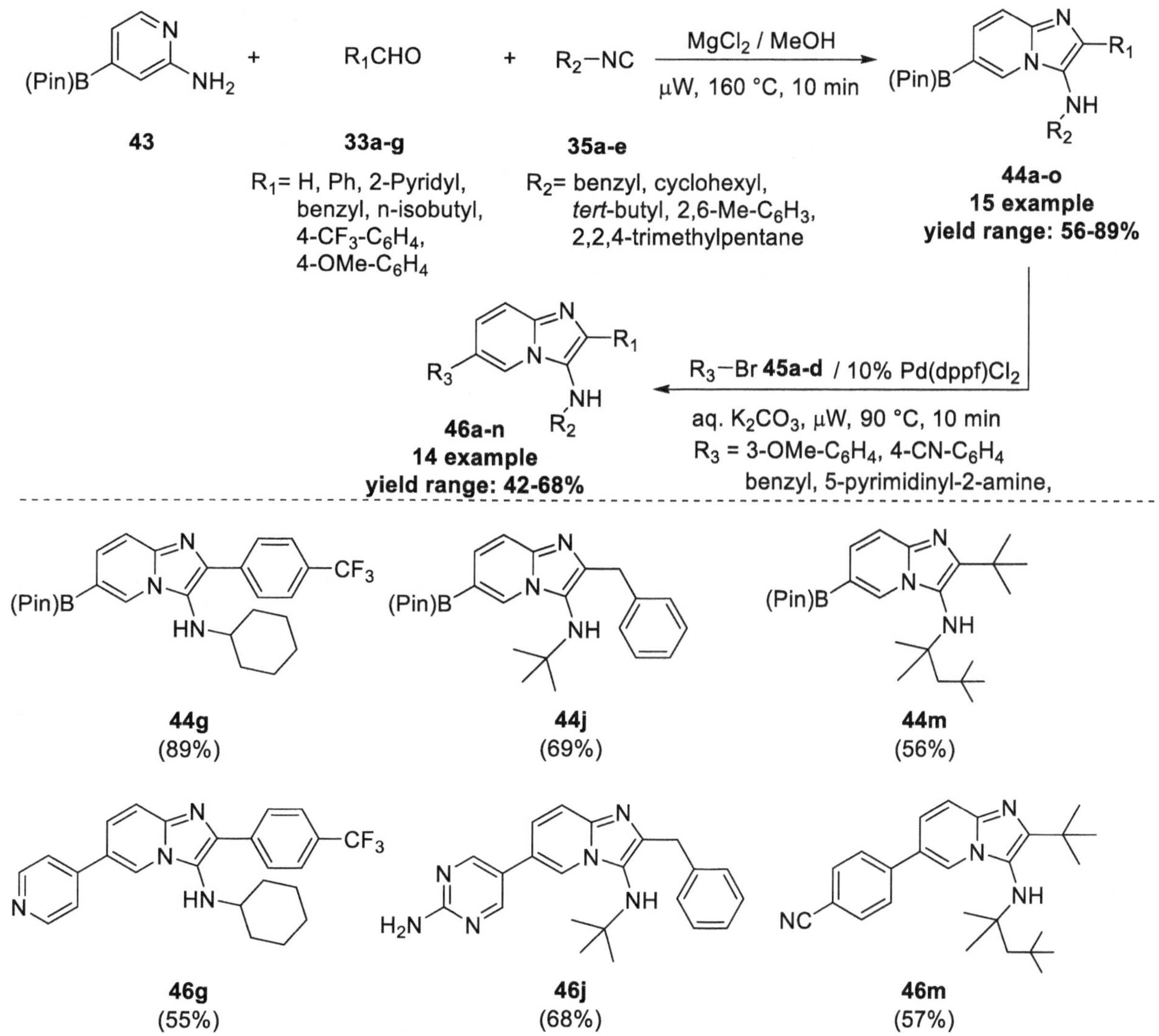

Scheme 17 Utility of **44a-o** in the Pd-catalyzed Suzuki coupling for the synthesis of **46a-n**

58–78% yield range (Scheme 19).The plausible mechanism of this reaction had been shown in Fig. 5. Under ambient conditions, the lone pair of electrons on nitrogen atom of amino group in 2-aminopyridine **17a** attacks on the carbonyl carbon atom of the oxazole **48a-l** which generated hydroxy intermediate I. The intermediate **I** through electronic rearrangement gives intermediate **II** which undergone condensation to give rise to the formation of the targeted core **49a-l** in good yields (Tu et al. 2007).

3.7 MW-Assisted Multi-component Neat Synthesis of Benzimidazolyl-Imidazo[1,2-*a*] pyridines

In **2013**, Barnali Maiti et al. have reported the greener method using multi-component and solvent-free MW-methodology for the preparation of benzimidazolyl-imidazo[1,2-*a*]pyridine **51–53** (Scheme 20) (Maiti et al. 2013). The methodology utilizes $Sc(OTf)_3$-catalyzed reaction of 2-aminopyridinyl-benzimidazole-5-carboxylate **50a-e**, substituted aldehyde **33a-i** and different isocyanides **35a-c/35h-i** under MW-irradiations at 135 °C for 5–10 min. In this reaction, three types of benzimidazolyl-imidazo[1,2-*a*]pyridines **51–53** had been formed and the synthesis of these final product were totally dependent on the linkage of 2-aminopyridine to the main core of benzimidazole to form 2-aminopyridinyl-benzimidazole-5-carboxylate **50a-e**.

Benzimidazole-linked amino pyridine **50a-e** reacts with $Sc(OTf)_3$ activated aldehyde **33a-i** afforded imine as an intermediate, which then underwent *5-exo-dig* cyclization via nucleophilic addition reaction with isonitriles **35a-c/35h-i,** respectively, to furnish imidazo[1,2-*a*]pyridine intermediate (Fig. 6). Re-aromatization of this intermediate

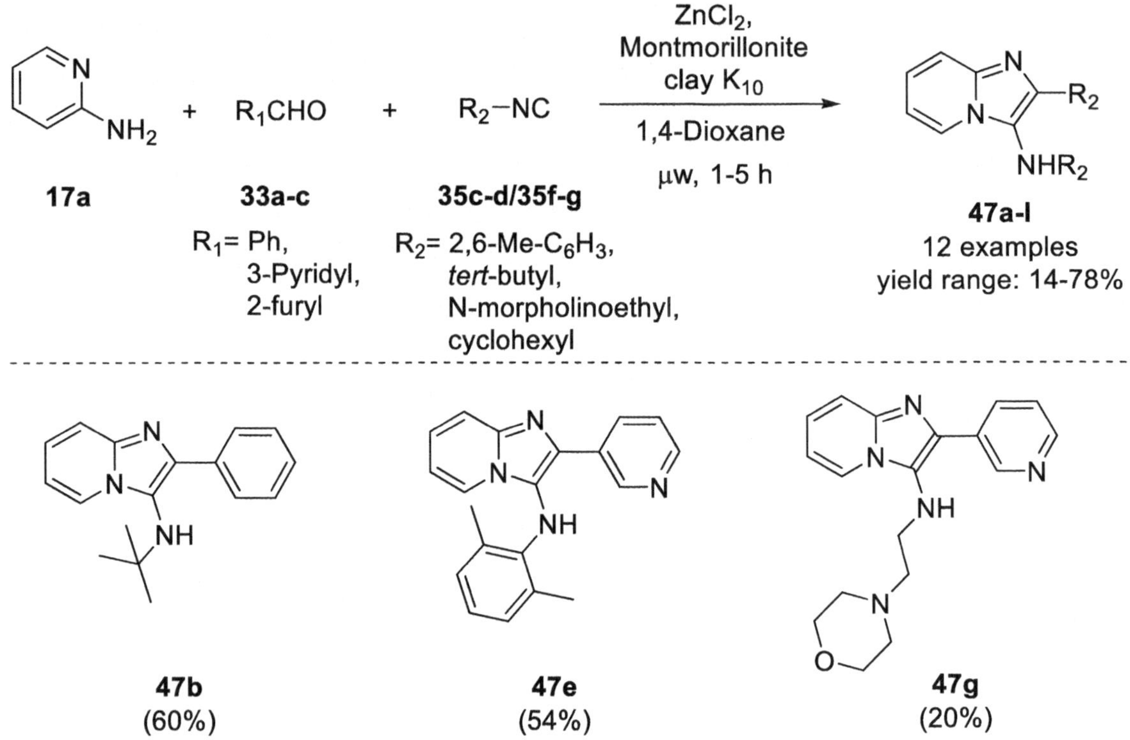

Scheme 18 MW-assisted $ZnCl_2$-catalyzed MCR protocol of **47a-o**

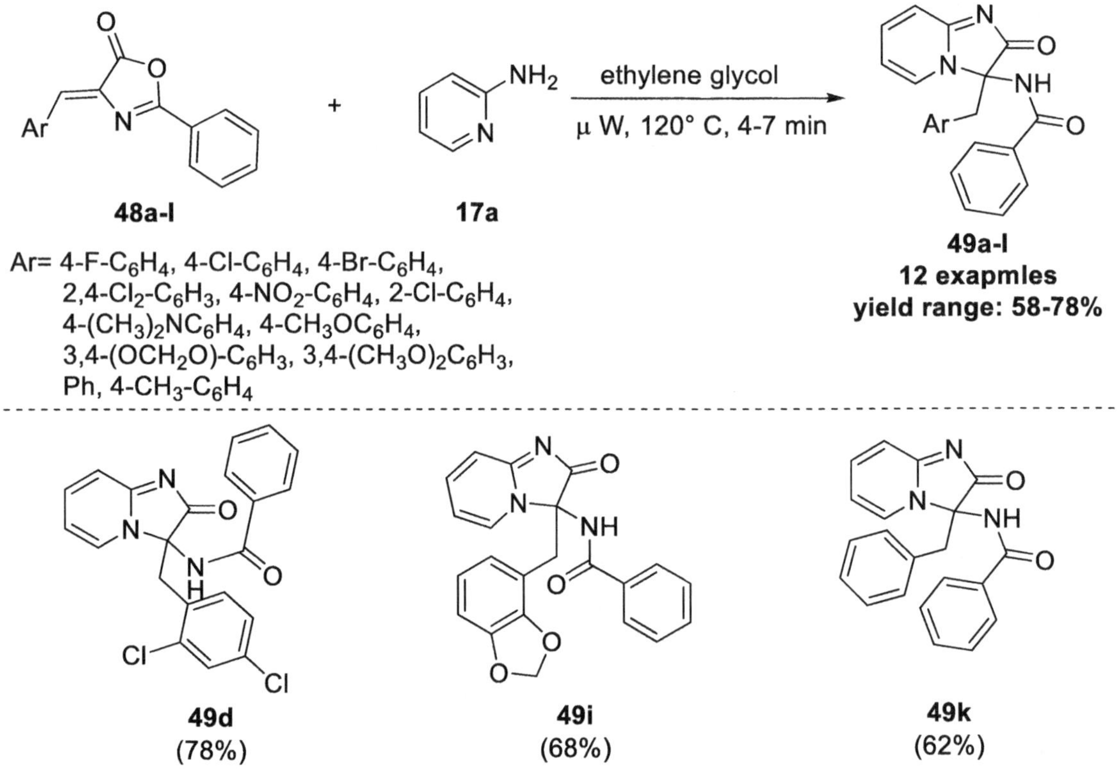

Scheme 19 Microwave-promoted synthesis of **47a-o**

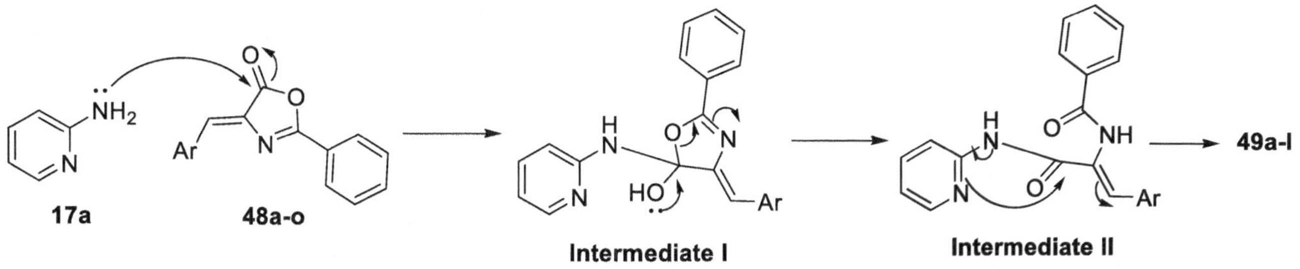

Fig. 5 Tentative mechanism

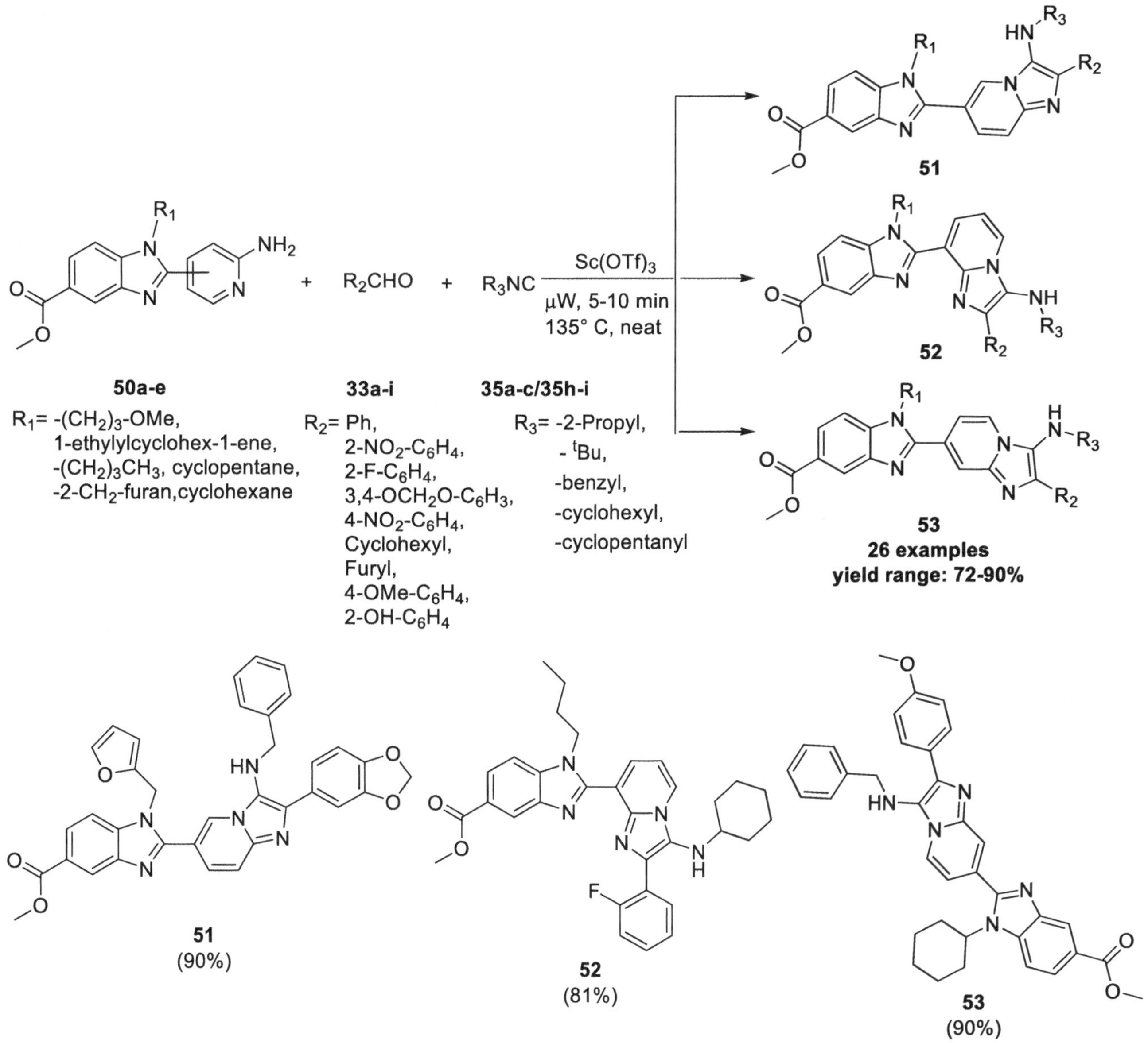

Scheme 20 MW-irradiated synthesis of **51–53**

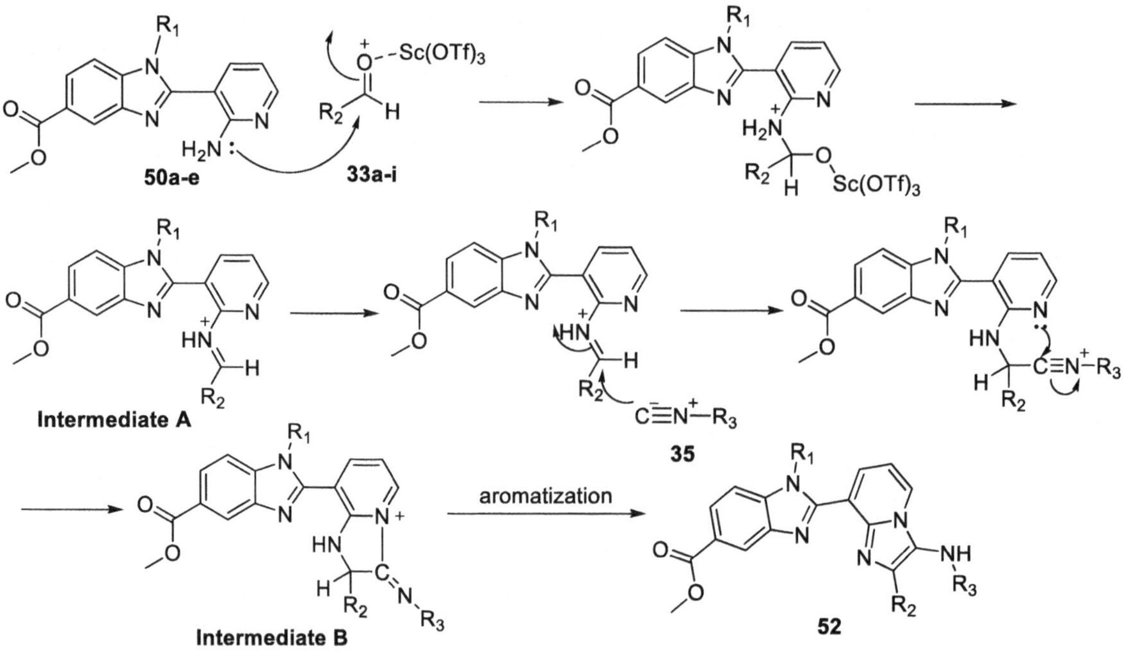

Fig. 6 Probable mechanism for the synthesis of **52**

gives the final compound **51–53** in 72–90% yield range (Maiti et al. 2013).

3.8 MW-Assisted Groebke–Blackburn–Bienaymé (GBB) Reaction for the Synthesis of Chromones-imidazo[1,2-*a*]pyridine Hybrid

In **2015**, Gámez-Montaño and coworkers had explained the synthetic methodology of chromones-imidazo[1,2-*a*]pyridine hybrid **55a-t** via GBB reaction (Scheme 21) (Kishore et al. 2015).

In this methodology, 3-formylchromone **54a-b**, 2-aminopyridine **17a-b** and substituted isonitriles **35a-e** acted as a reactants using $InCl_3$ and $ClCH_2COOH$ as a catalyst in dry MeOH under microwave conditions at 85 °C for 1 h under N_2 atmosphere furnished chromones-imidazo [1,2-*a*]pyridine hybrid **55a-t** in 70–93% yield range (Scheme 21) (Kishore et al. 2015).

3.9 Synthesis of Chromones-Imidazo[1,2-*a*] pyridine (Chromones-IP) Hybrid via MW-irradiated GBB Strategy

Gámez-Montaño and coworkers (**2019**) developed an unique eco-friendly-irradiated GBB methodology (20 mol% NH_4Cl/EtOH) for the preparation of chromones-IP hybrid **57a-e** involving the two highly important structurally core based on medicinal chemistry point of view, i.e., chromones and imidazo[1,2-*a*]pyridine. In this strategy, 3-formylchromone **56**, 2-aminopyridine **17a,** and several substituted isocyanides **35a-e** were reacted with 20 mol% NH_4Cl in EtOH under MW-irradiations at 80 °C for 15 min which afforded chromones-imidazo[1,2-*a*]pyridine hybrid **57a-e** in 21–36% yield range (Scheme 22) (Zarate-Hernandez et al. 2019).

4 MW-irradiated Miscellaneous Synthesis of Other Imidazo[1,2-*a*]pyridines (IPs)

This segment incorporates various other derivatives of IPs, particularly functionalized at C-3 position, such as pyrazinyl (**59a-m**); carboalkoxy (**62a-p**); 1,2,3,5,6,7-hexahydroimidazo[1,2-*a*]pyridine (**65a-r**); hydroxy (**68a-q**), 2,3-disubstituted IPs (**71a-t**); alkoxymethyl (**74a-u**); and formyl (**76a-o**), etc. have been discussed via microwave irradiations year-wise in a sequential manner.

4.1 MW-Promoted Synthesis of 3-pyrazinyl-IPs

In **2010**, Michael Raymond Collins and coworkers have challenged the direct Palladium-catalyzed methodology of direct arylation strategy and gave procedure for the synthesis

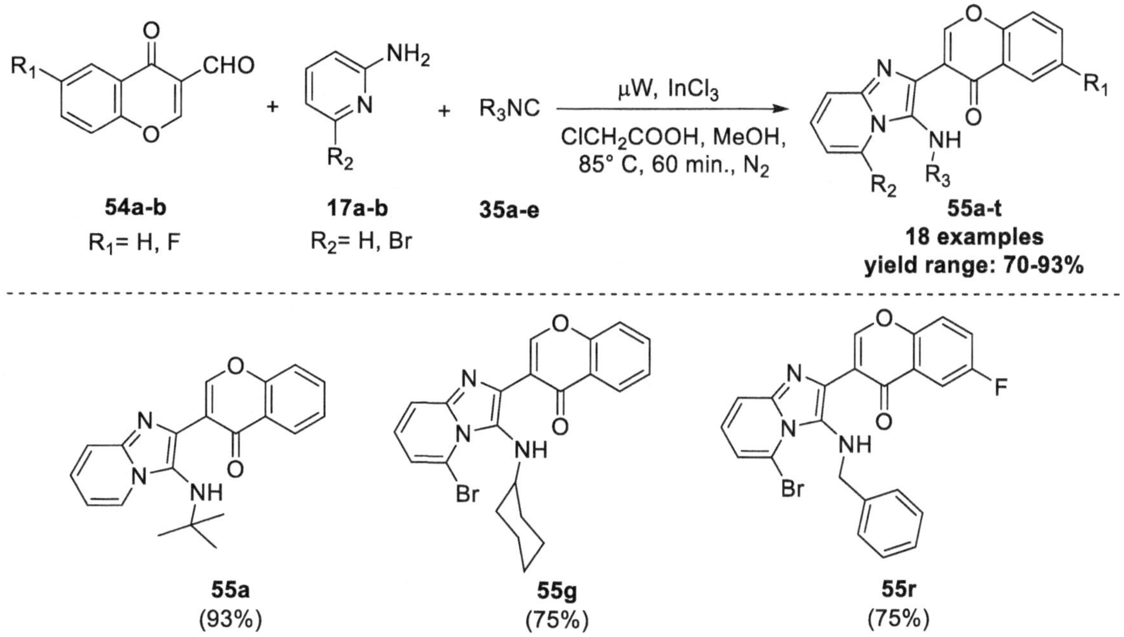

Scheme 21 MW-assisted GBB reaction: Synthesis of chromones-imidazo[1,2-*a*]pyridine hybrid **55a-t**

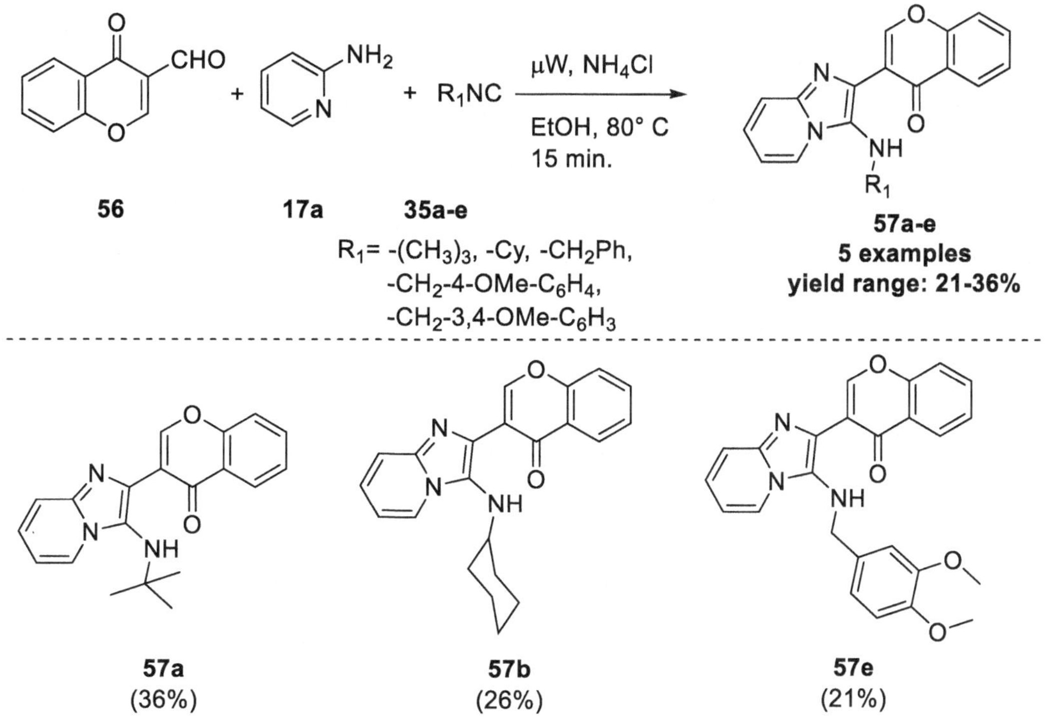

Scheme 22 Modified MW-irradiated synthesis of **57a-e**

of 3-pyrazinyl-IPs **59** comparatively in much better yields. Thus, imidazo[1,2-*a*]pyridines on Pd-catalyzed direct arylation with 2,6-dichloropyrazine **17a-b/17g** afforded 3-pyrazinyl-IPs **59a-b/59 g** in relatively lower (3–40%) yields. However, in the modified method, vinyl ether has converted into hemiacetal **58** named "2-bromo-2-(6-chloropyrazin-2-yl)-1-ethoxyethan-1-ol" which acts as a reactant along with 2-aminopyridine **17a-m** which resulted into the development of 3-pyrazinyl-imidazo[1,2-*a*]pyridines **59a-m** in 31–76% yield range. It had been observed that the

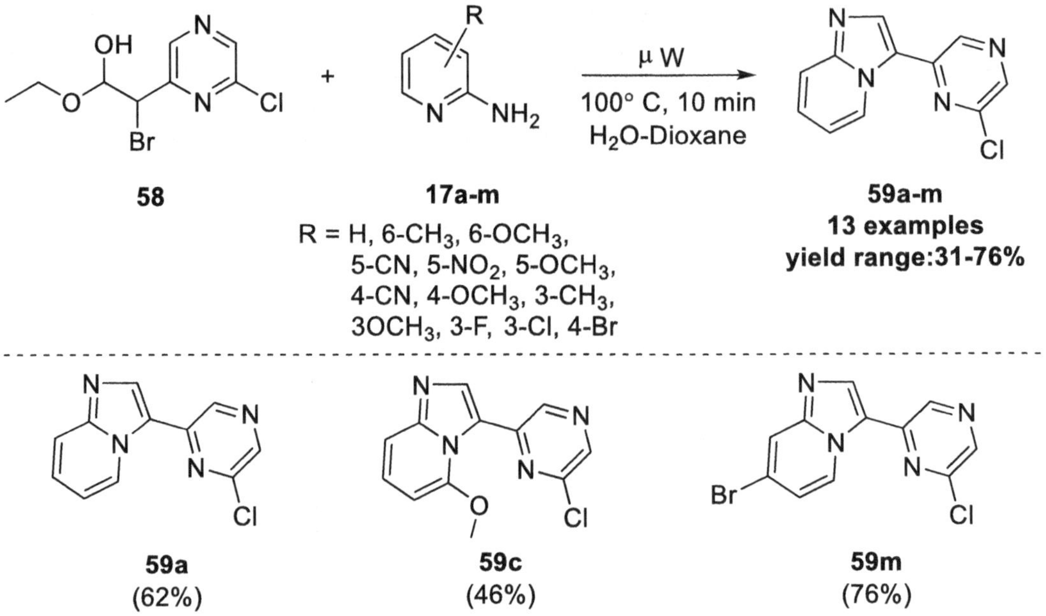

Scheme 23 Preparation of 3-pyrazinyl-imidazo[1,2-*a*]pyridines **59a-m**

C-6 substitutions (CH_3, OMe) in 2-aminopyridine significantly lower the product yield in comparison with unsubstituted or some other substitution on 2-aminopyridines (Scheme 23) (Collins et al. 2010).

4.2 Conventional and MW-assisted Fast Synthesis of Substituted IPs Under Neat Conditions

Kaushik C. Chunavala et al. in **2011** described a solvent- and catalyst-free synthetic protocol for the preparation of phenylimidazo[1,2-*a*]pyridine-3-carboxylate **62a-h** core from the reaction of 2-aminopyridine **60a-f** and *β*-keto ester **61a-c** using Al_2O_3 as a catalyst under microwave irradiation for 1 min. with time gap of 5 s after 15 s of irradiation (Scheme 24) (Chunavala et al. 2011).

A tentative mechanism for the development of products **62a-p** from **60a–f** is shown in Fig. 7. It has been assumed that the preliminary reaction of the 2-aminopyridine **60** with *α*-bromo-*β*-keto esters **61** furnished intermediate **I**. Subsequently, the more stable conjugated intermediate **II** was formed from intermediate **I** by the elimination of a water molecule. Lastly, intramolecular cyclization of intermediate **II** afforded the desired imidazo[1,2-*a*]pyridine-3-carboxylates **62** via loosing HBr molecule in 85–95% yield range (Chunavala et al. 2011).

4.3 MW-Irradiated, 3-component, Domino Reaction for the Fast Synthesis of IPs Utilizing the Application of Functionalized N,S-Ketene Acetals

In **2013**, Ming Li et al. successfully disclosed the major application of functionalized N,S-ketene acetals, i.e., ethyl 2-(3-oxo-3-arylpropanethioamido)-acetates **63a-e** to synthesis substituted 6-benzoyl-2-oxo-5-thioxo-1,2,3,5,6,7-hexahydroimidazo[1,2-*a*]pyridine-8-carbonitrile **65a-r** via DABCO-catalyzed domino annulations with malononitrile **64** and substituted aldehyde **33a-j** under microwave irradiation at 400 W at 78 °C for 5 min. The most characteristic features of this inimitable reaction were that eight different active sites were involved along with simultaneous formation of three new C–C bonds and two C–N bonds. Furthermore, two new rings were constructed with complete utilization of all reactants efficiently with the loss of H_2O and ethanol (EtOH) molecule. The microwave power significantly improves the yield of the reaction and was found directly proportional with the yield of the reaction. The employment of reaction in lower power decreased the yield and vice versa. This green cascade multi-component reactions serves to provides several diverse imidazo[1,2-*a*] pyridines derivatives (Scheme 25) (Li et al. 2013).

Mechanistically, **63** deprotonates to give its anionic synthons **A** which reacts with the intermediate **B** formed via

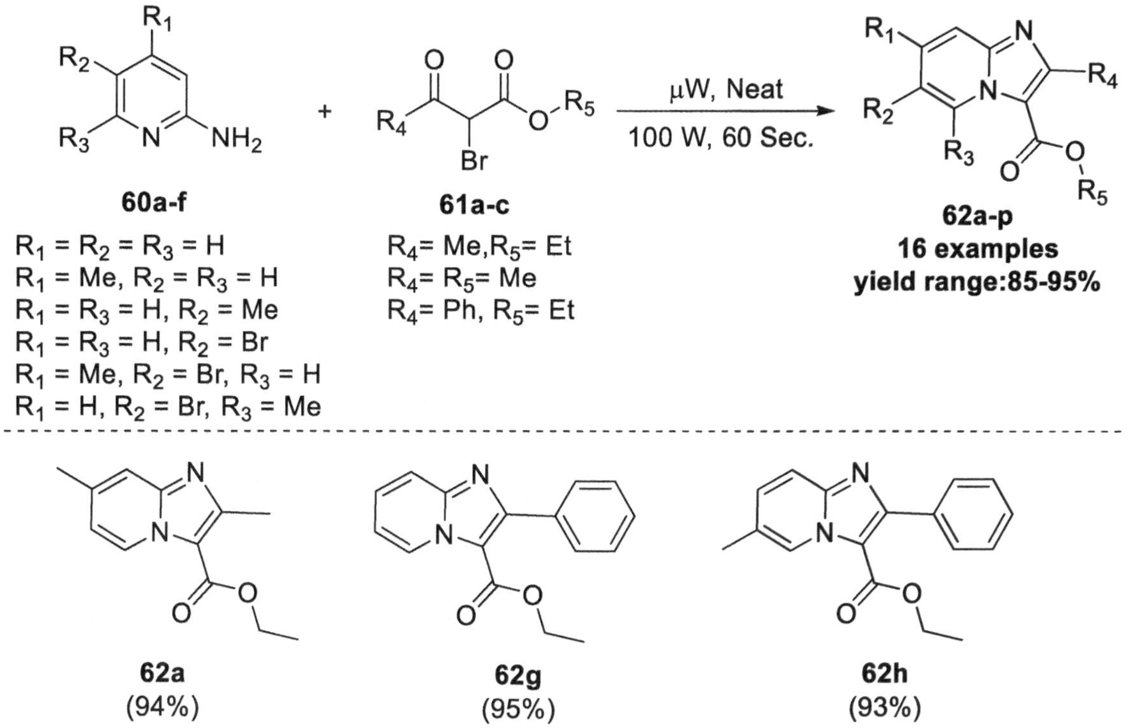

Scheme 24 Synthesis of phenylimidazo[1,2-*a*]pyridine-3-carboxylate **62a-h**

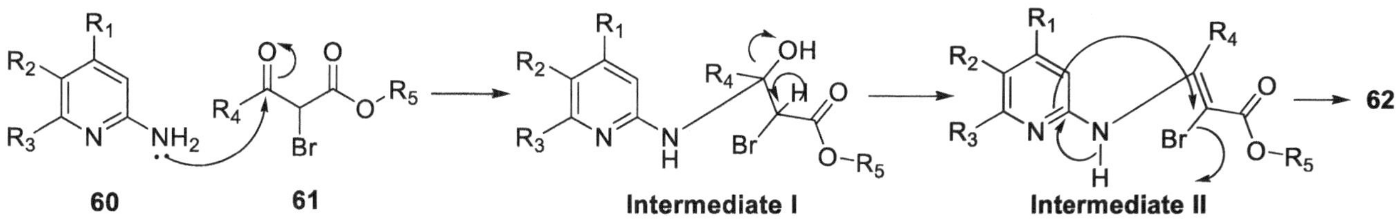

Fig. 7 Plausible mechanism for the construction of **62**

Knoevenagel condensation of malononitrile **64** with aldehydes **33**. Then after, Michael addition between **A** and **B** furnished adduct **C** which underwent an array of electronic rearrangements involving N-cyclization and imine-enamine tautomerization afforded the give final product **65** in up to 89% yield (Fig. 8) (Li et al. 2013).

4.4 A Petasis-Based Cascade Methodology to Access 3-hydroxy Functionalized 2-arylimidazo[1,2-*a*]pyridines Under Microwave Irradiation Conditions

In **2014**, Hong-yu Li and coworkers reported a novel microwave-assisted Petasis-based cascade multi-component reaction for the preparation of functionalized 2-arylimidazo [1,2-*a*]pyridin-3-ols **68a-q** (Scheme 26). Substituted 2-aminopyridine **17a-g**, substituted arylboronic acid **66a-j** and glyoxylic acid **67** were reacted in Dimethylformamide (DMF) under MW-irradiation conditions at 160 °C for 30 min. furnished substituted 2-arylimidazo[1,2-*a*] pyridin-3-ol **68a-q** in 45–85% yield range (Scheme 26) (Wang et al. 2014b). In this multi-component reaction pathway, the yield of the desired product dependent on the substitution on 2-aminopyridine ring and aryl boronic acid. It was identified that the electron-rich pyridines and boronic acids afforded higher yield than that of electron-withdrawing group containing counterparts (Wang et al. 2014b).

The tentative mechanism for the formation of **68a-q** via MW-irradiated, one-pot, Petasis-based domino MCR pathway is shown in Fig. 9. Compounds **17**, **66,** and **67** were reacted under standard Petasis reaction mechanism afforded intermediate I which, then, undergone intramolecular nucleophile cyclization which subsequently followed by dehydroxylation and aromatization afforded the thermodynamically stable **68**. Further validation of the proposed mechanism had been performed by converting intermediate **I** directly into **68** (Fig. 9) (Wang et al. 2014b).

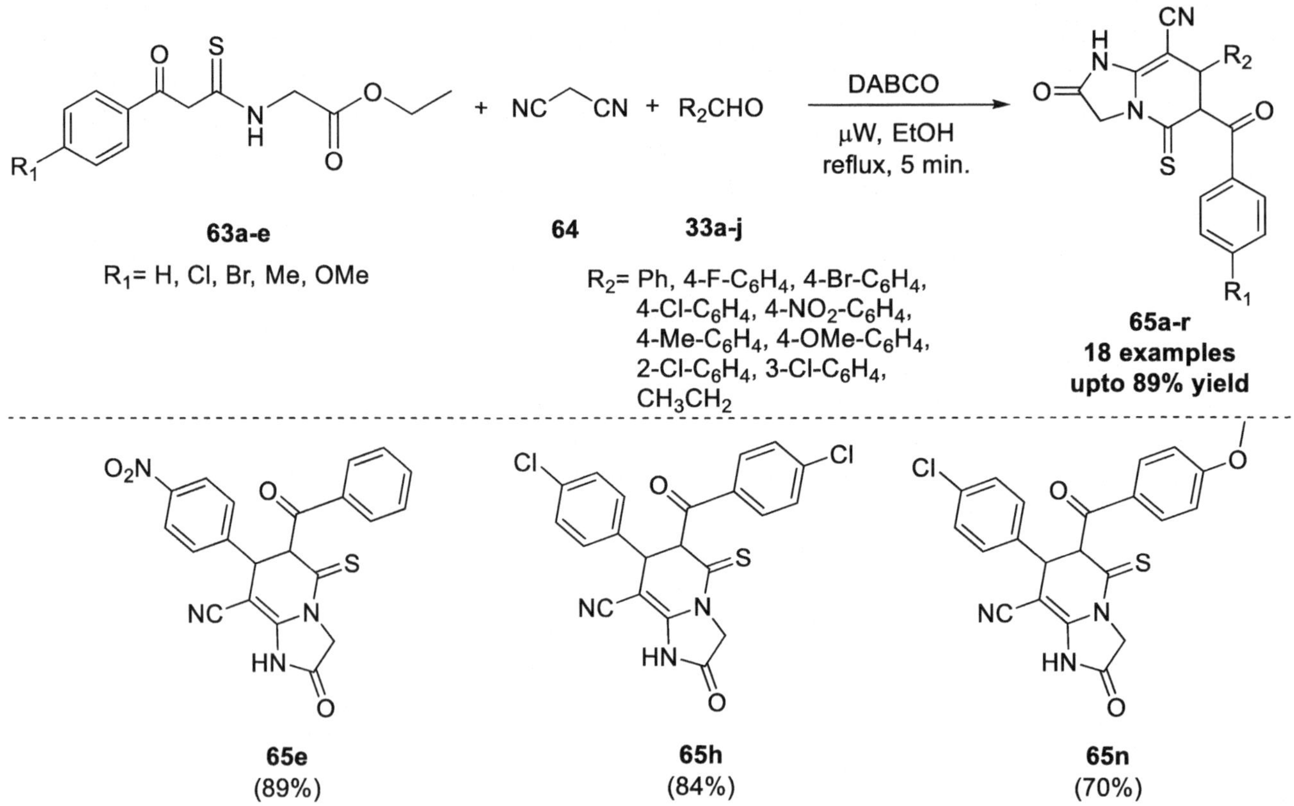

Scheme 25 MW-assisted DABCO-catalyzed domino synthesis of substituted **65a-r**

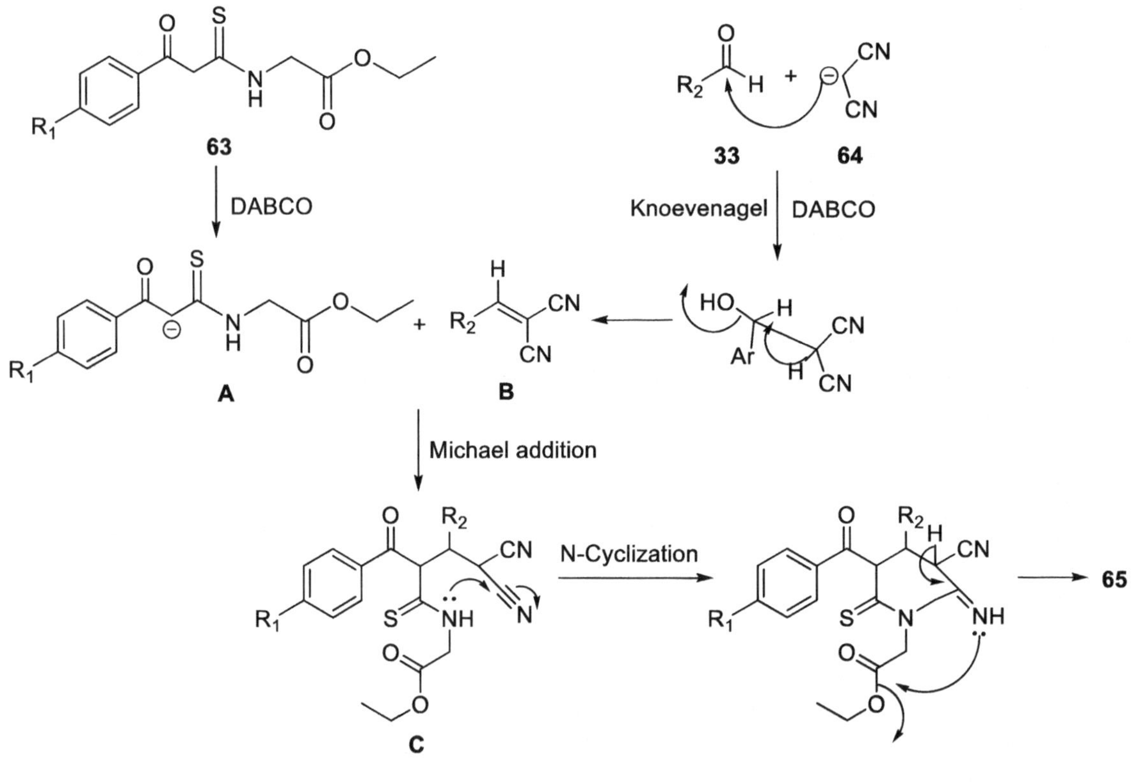

Fig. 8 Proposed mechanism for the formation of **65**

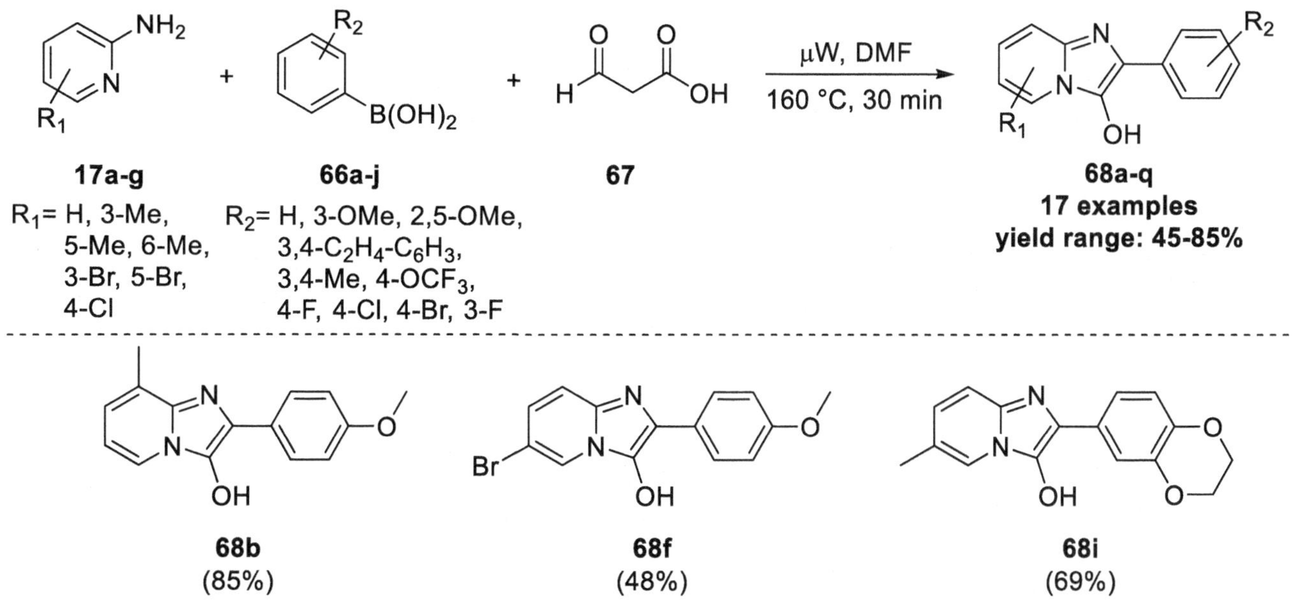

Scheme 26 Microwave-assisted Petasis-based cascade approach for the synthesis of **68a-q**

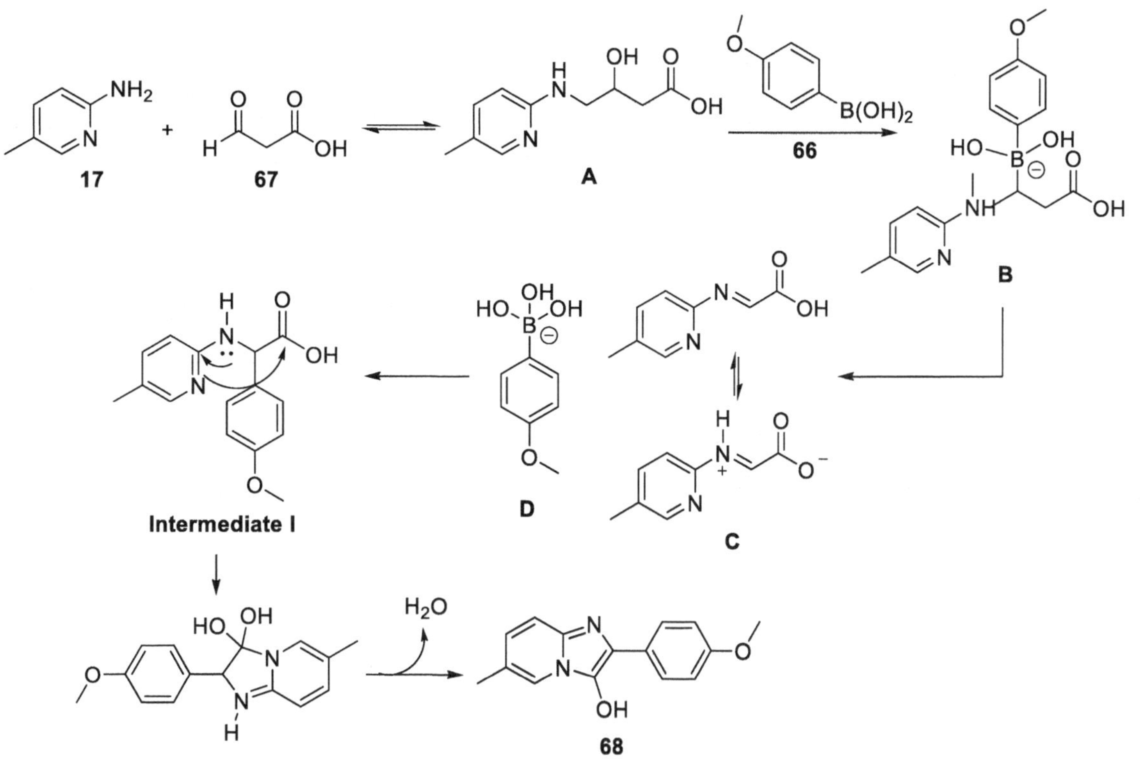

Fig. 9 Probable mechanism for the formation of **68** from **17**, **66** and **67**

4.5 Synthesis of 2,3-Disubstituted-IPs via One-Pot MCR Under MW-Irradiated Conditions

In **2015**, Shaik Karamthullam et al. developed iodine-catalyzed methodology for the preparation of 2,3-disubstituted-IPs **71a-t**. Substituted aryl-glyoxals **69a-d** (replaced the lachrymatory α-haloketones), cyclic 1, 3-dicarbonyls **70** and substituted 2-aminopyridines **17a-f** were reacted with iodine (30 mol%) in ethanol at 130 °C for 15 min. under microwave irradiations furnished 2,3-disubstituted-IPs **71a-t** in 58–84% yield range (Scheme 27) (Karamthulla et al. 2015).

Mechanistic pathway involves Knoevenagel-type condensation to form the intermediate **A**. Then, 2-aminopyridine undergone aza-Michael addition with intermediate **A** which

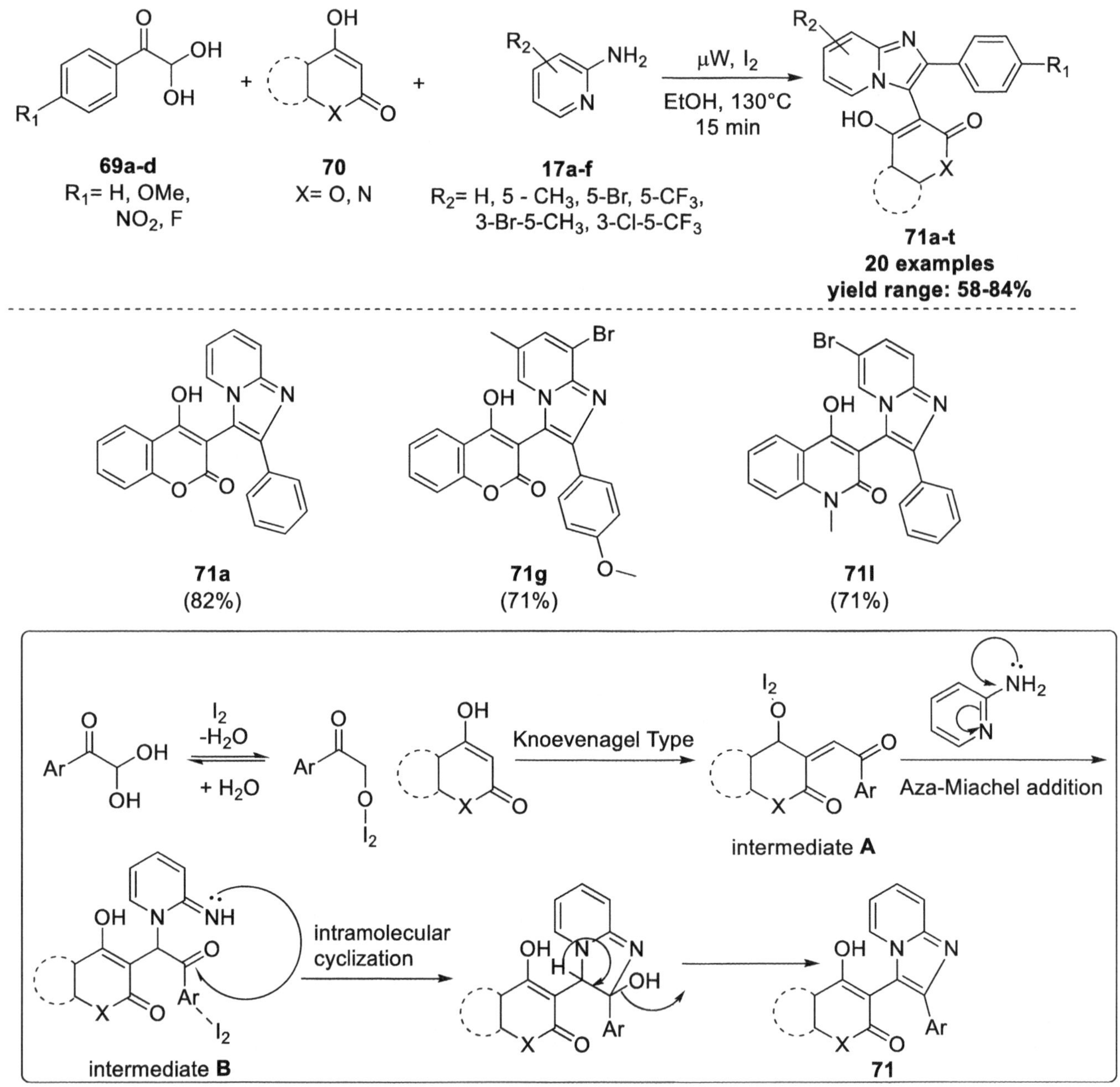

Scheme 27 Access to 2,3-substituted-IPs **71a-t** using I_2 as a catalyst

formed another intermediate **B** which subsequently undergone cyclization by the loss of H_2O molecule afforded **71** (Scheme 27) (Karamthulla et al. 2015).

4.6 MW-Irradiated TsOH-Catalyzed, Solvent-Free Synthesis of IPs via MCR Strategy

In **2015**, Zhang et al. reported a multi-component, solvent-free synthetic protocol for imidazo[1,2-*a*]pyridines unsubstituted at C-2 position using MW-irradiation conditions. In this methodology, the reaction of substituted 2-aminopyridine **17a-i**, 3-phenylpropiolaldehyde **72** and substituted alcohol **73** in the presence of TsOH under MW-irradiated and solvent-free conditions at 100 °C for 15 min. which afforded substituted 3-alkoxyalkyl-IPs **74a-u** in 75–87% yield range (Scheme 28) (Zhang and Jiang 2015).

The formation of **IP** unsubstituted at C-2 site is the characteristic feature of this methodology, making it a unique protocol. The tentative mechanism of this reaction initiates with the reaction of 2-aminopyridine **17** with aldehyde **72** which furnished imine as an intermediate **I**. Intermediate **II** is formed by the attack of alcohol to the alkyne of intermediate **I**. The cyclization and proton transfer in intermediate **II** with the help of TsOH afforded the desired product substituted 3-alkoxyalkyl-IPs **74a-u** (Fig. 10) (Zhang and Jiang 2015).

4.7 Synthesis of Substituted-3-Formyl-Imidazo [1,2-*a*]pyridines Using MW-Irradiated Conditions

In **2019**, Katarzyna M. Błazewska and coworkers reported a metal-free strategy for the synthesis of substituted 3-formyl-IPs **76a-o** via the reaction of substituted 2-aminopyridine **17a-m** and 2-bromomalonaldehyde **75** in the mixture of EtOH and H_2O as a solvents in the ratio of 1:1 under MW-irradiation conditions at 110 °C for 20 min in 23–86% yield range (Scheme 29) (Kusy et al. 2019).

The presence of the -CHO group in IPs is served as a very prominent building blocks for further functionalization. The mechanism involves the reaction of 2-aminopyridine **17a** and bromo malonaldehyde **75** formed imine **A** where was isolated and well-characterized through NMR spectroscopy and MS spectrometry. The imine **A** underwent cyclization with the removal of hydrogen bromide (HBr) molecule which furnished imidazo[1,2-*a*] pyridine-3-carbaldehydes **76a** (Fig. 11) (Kusy et al. 2019).

5 Discussion and Summary

Imidazo[1,2-*a*]pyridine-based bio-heterocycles had been considered as a very important synthon/motif in medicinal/pharmaceutical chemistry as it has been blended with a varied array of biological activities. Consequently, the

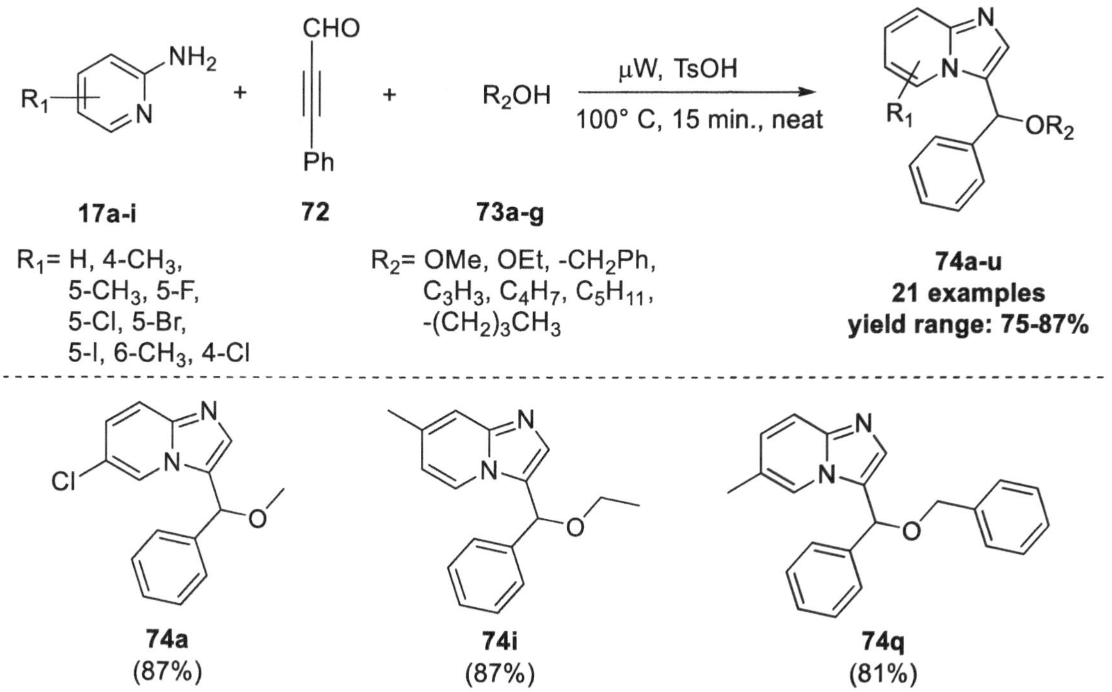

Scheme 28 Preparation of substituted 3-alkoxyalkyl-imidazo[1,2-*a*]pyridines **74a-u** under MW-assisted reaction

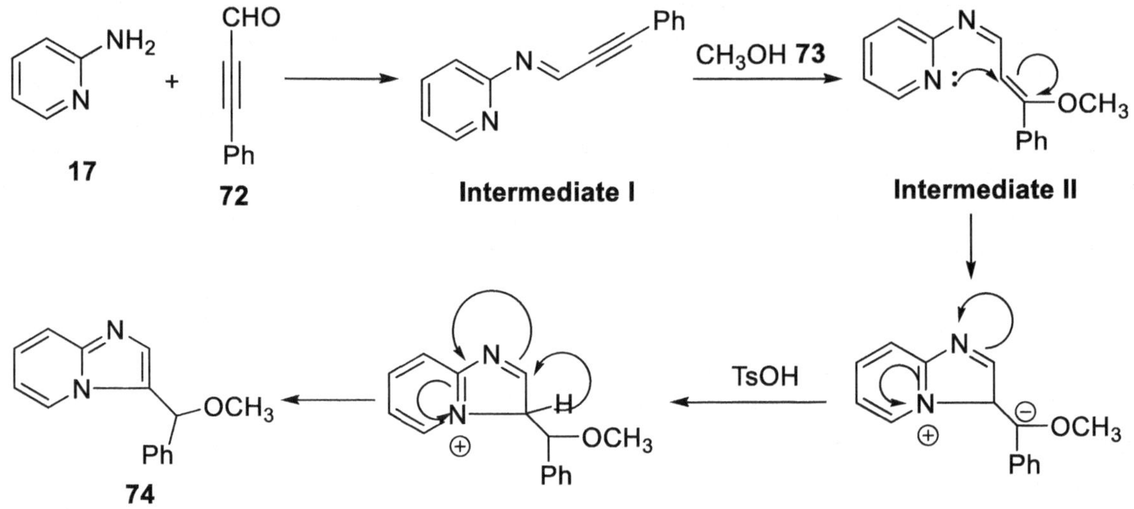

Fig. 10 Proposed mechanism for the formation of **74a-u**

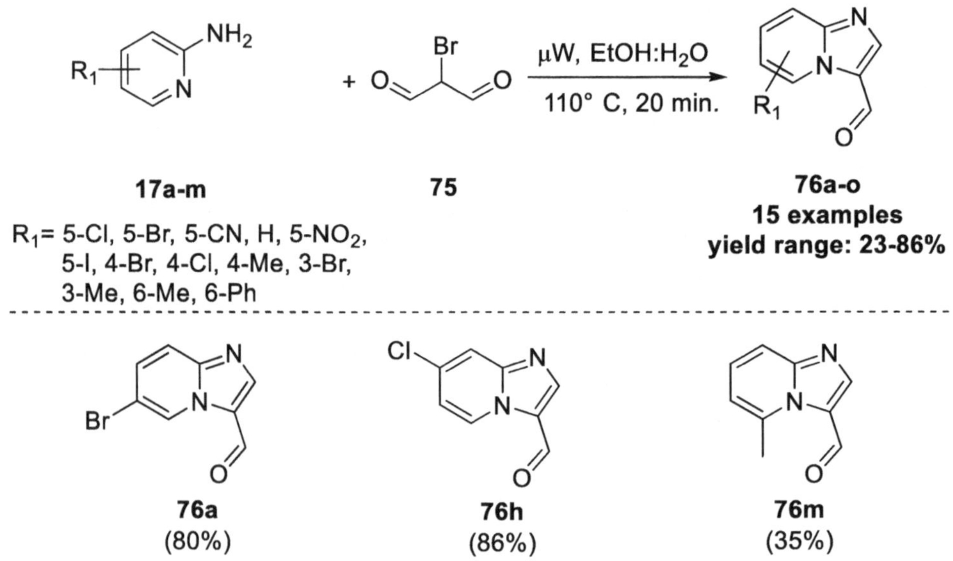

Scheme 29 Metal-free synthesis of **76a-o** under MW-irradiated conditions

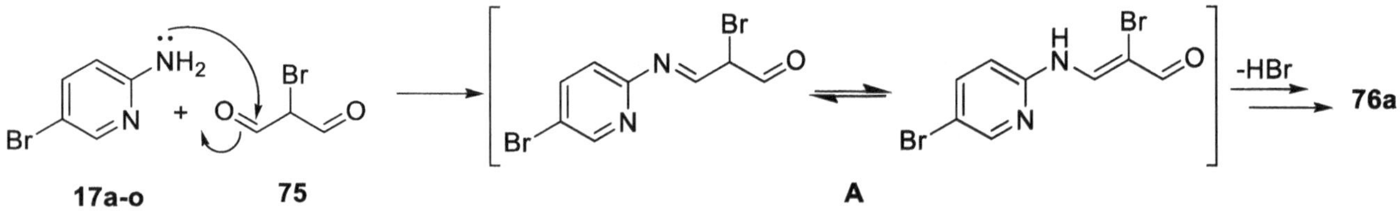

Fig. 11 Tentative mechanism for the synthesis of **76a**

imidazo[1,2-*a*]pyridines synthesis had been recognized as a decisive factor in designing core-ring of many currently marketed/leads/under clinical trial drug candidates.

Many reports had been published in the literature for the preparation of **IP** utilizing conventional heating source of energy. However, in past 2–3 decades, environmentally benign, fast, and efficient green synthesis is in focus and microwave-assisted organic synthesis (MAOS) has been an integral part of such green and sustainable developments in the field of chemical synthesis.

So, in the last 2–3 decades, several microwave-assisted synthesis of **IPs** had been established in view of generating greener approach for the sustainable synthetic routes. Several derivatives of **IPs**, for example, 2-aryl-IPs [**4a-h**, **6a-g**, **10a-j**,**13a-l**, **16a-s**, **19a-p**, **20a-u**, **24a-x**, **28a-k**, **29a-j**, **30a-h**, and **31a-i**]; various other derivatives particularly functionalized at C-3 position, for example, amino [**37a-c**, **39a-l**, **40**, **41a-h**, **42a-f**, **46a-n**, **47a-l**, **49a-l**, **51–53**, **55a-t**, **57a-e**]; pyrazinyl (**59a-m**); carboalkoxy (**62a-p**); 1,2,3,5,6,7-hexahydroimidazo[1,2-*a*]pyridine (**65a-r**); hydroxy (**68a-q**), 2,3-disubstituted-IPs (**71a-t**);alkoxymethyl (**74a-u**), and formyl (**76a-o**) have been reported to be synthesized via microwave irradiations in overall good to excellent yields.

Microwave-assisted fused bicyclic heteroaryl boronates **4a-h** had been effectively utilized to construct imidazopyridine-quinazoline hybrids **6a-g**, albeit in low to moderate yields (Scheme 1). However, high yields of 2-phenylimidazo [1,2-*a*]pyridine **10a-j** had been produced when in situ prepared *N*-phenacylpyridinium bromides (reaction of **7** with **8**)undergone MW-assisted nucleophilic addition of NH_4OAc **9** (acts as a source of nitrogen) under solvent-free conditions (Scheme 2). Another methodology afforded high yields of 2-phenylimidazo[1,2-*a*]pyridine **16a-s** (upto 94%) via metal-free amino-triggered benzannulation approach under microwave irradiations (Scheme 4). Sequentially, one-pot, 3-component synthesis (Scheme 3) and $NaHCO_3$-catalyzed synthesis in PEG_{400} media (Scheme 5) were found effective methods; AgI/I_2/Water: PEG_{400} catalytic system at 350 W (Scheme 8); catalyst- and solvent-free green synthesis (Scheme 9) and another one in H_2O or ionic liquids or nano-SiO_2 (Schemes 10, 11 and 12) were found to be the best catalytic greener synthetic approach of 2-phenyl-IPs as far as the sustainable point of view is concerned. Vast substrate scope with large variation in yields was observed for the MW-assisted synthesis of 2,8-diaryl-6-aminoimidazo[1,2-*a*]pyridines (Scheme 7).

MW-irradiation and fluorous methodologies were simultaneously engaged to accelerate the reaction and decreasing the purification time in solution-phase synthesis via the utilization of F-SPE technique (Scheme 14). MW-assisted sequential Ugi/Strecker reactions involving 3-center-4-component and 3-center-5-component MCR pathway were employed for the preparation of 3-aminoimidazo[1,2-*a*] pyridines **40**; 3-iminoaryl-imidazo[1,2-*a*]pyridines **41a-h** and imidazo[1,2-*a*]pyridyn-3-ylamino-2-acetonitrile **42a-f** using polymer-bound scandium salt [PS-Sc$(OTf)_3$] as catalyst (Scheme 16). Use of $MgCl_2$ is very useful for the synthesis of 3-aminoimidazopyridines-6-boronates with moderate substrate scope having good yields (Scheme 17). Ethylene glycol in microwave is responsible for the unique synthesis of N-(3-Arylmethyl-2-oxo-2,3-dihydroimidazo [1,2-*a*]pyridin-3-ylbenzamides keeping two heterocycles in one frame with limited substrates and moderate yield (Scheme 19). Benzimidazolyl-imidazo[1,2-*a*]pyridine and imidazo[1,2-*a*]pyridine-Chromones hybrids have also been synthesized under MW conditions using 2-aminopyridinyl-benzimidazole-5-carboxylate as a starting materials and Groebke–Blackburn–Bienaymé reaction methodology, respectively (Schemes 20 and 21). Imidazo [1,2-*a*]pyridine-chromones had also been reported to be synthesized in moderate to high yield taking $InCl_3$/ $ClCH_2COOH$ as a catalytic system in dry carbinol (MeOH) under MW-irradiation conditions (Scheme 21).

Various C-3 substituted imidazo[1,2-*a*]pyridines derivatives such as 3-pyrazinyl-imidazo[1,2-*a*]pyridines hybrid (**59a-m**); phenylimidazo[1,2-*a*]pyridine-3-carboxylates (**62a-p**); 6-benzoyl-2-oxo-5-thioxo-1,2,3,5,6,7-hexahydroimidazo[1,2-*a*]pyridine-8-carbonitriles (**65a-r**); **2**-arylimidazo[1,2-*a*]pyridin-3-ol (**68a-q**), 2,3-disubstituted imidazo [1,2-*a*]pyridines (**71a-t**);3-alkoxyalkyl-imidazo[1,2-*a*]pyridines (**74a-u**); and 3-formyl substituted imidazo[1,2-*a*]pyridines (**76a-o**), etc., had also been reported to be prepared via microwave irradiations variable from good to excellent yields.

6 Conclusion

This chapter highlights all the MW-irradiated synthetic methodologies for the imidazo[1,2-*a*]pyridine heterocycles and its derivatives during the last 2–3 decades as MW-assisted synthesis has taken its place in much higher rank than the conventional heating due to its cost effective, fast and green methodology in the preparation of IPs. Some of the reactions via microwave technique produces exceptional yields as observed in the formation of 2-phenylimidazo [1,2-*a*]pyridines ($NaHCO_3/PEG_{400}$; AgI/I_2/Water:PEG_{400}; catalyst-free neat green synthesis; H_2O/Ionic liquids/Nano SiO_2) and 3-aminoimidazo[1,2-*a*]pyridines (fluorous multi-component reactions). It has been interpreted that the imidazo[1,2-*a*]pyridines at C-2/C-3 positions had been highly explored and several medicinally important drug candidate have C-2 and/or C-3 substitutions. Hence, these two sites are very important in sustainability point of view. These microwave-assisted methodologies could offer medicinally important core in fast and efficient ways and can be utilized to generate more imidazo[1,2-*a*] pyridines analogues for fastening medicinal chemistry and drug discovery.

Declarations

Authors' Contributions

SC and RKY conceived and designed the concept. RKY carried out all the literature search. SC carried out the writing of the manuscript. Both the authors (SC and RKY) have read and approved the final version of the manuscript.

Ethics Approval

Not applicable.

Competing Interests

The authors(s) confirm that this chapter content hasve no conflicts of interest.

Acknowledgements R.K.Y. acknowledges UGC, Delhi, for the financial assistance in terms of Senior Research Fellowship (SRF).

Abbreviations

IP	Imidazo[1,2-*a*]pyridine
ACE	Acetylcholinesterase
DMF	Dimethylformamide
MCR	Multi-component reaction
NBS	*N*-Bromosuccinimide
PTSA	*p*-Toluene sulfonic acid
NMR	Nuclear magnetic resonance
MS	Mass spectrometry
DABCO	1,4-Diazabicyclo[2.2.2]octane
MW	Microwave
MAOS	Microwave-assisted organic synthesis
F-SPE	Fluorous-solid-phase extraction
TLC	Thin layer chromatography
LC-MS	Liquid chromatography mass spectrometry
GBB	Groebke–Blackburn–Bienaymé
TMSCN	Trimethylsilyl cyanide

References

Adib M, Mohamadi A, Sheikhi E, Ansari S, Bijanzadeh HR (2010) Microwave-assisted, one-pot reaction of pyridines, α-Bromoketones and ammonium acetate: an efficient and simple synthesis of imidazo [1,2-*a*]-pyridines. Synlett 11:1606–1608

Bagdi AK, Santra S, Monir K, Hajra A (2015) Synthesis of imidazo [1,2-*a*]pyridines: a decade update. Chem Commun 51:1555–1575

Blackburn C, Guan B (2000) A novel dealkylation affording 3-aminoimidazo[1,2-a]pyridines: access to new substitution patterns by solid-phase synthesis†. Tetrahedron Lett 41:1495–1500

(a) Blackburn C, Guan B, Fleming P, Shiosaki K, Tsai S (1998) Tetrahedron Lett 39:3635–3638. (b) Blackburn C (1998) Tetrahedron Lett 39:5469–5299. see also (c) Groebke K, Weber L, Fridolin M. (1998) Synlett 661–663. (d) Bienayme H, Bouzid K (1998) Angew Chem, Int Ed Engl 37:2234–2237

Cai D, Byth KF, Shapiro GI (2006) AZ703, an Imidazo[1,2-a]pyridine Inhibitor of Cyclin-Dependent Kinases 1 and 2, Induces E2F-1-Dependent Apoptosis Enhanced by Depletion of Cyclin-Dependent Kinase 9. Cancer Res 66:435–444

Catalán J, Mena E, Fabero F, Amat-Guerri F (1992) The role of the torsion of the phenyl moiety in the mechanism of stimulated ultraviolet light generation in 2-phenylbenzazoles. J Chem Phys 96:2005–2016

Chunavala KC, Joshi G, Suresh E, Adimurthy S (2011) Thermal and microwave-assisted rapid syntheses of substituted imidazo[1,2-*a*] pyridines under solvent- and catalyst-free conditions. Synthesis 4:635–641

Collins MR, Huang Q, Ornelas MA, Scales SA (2010) The synthesis of 3-pyrazinyl-imidazo[1,2-*a*]pyridines from a vinyl ether. Tetrahedron Lett 51:3528–3530

De la Hoz A, Diaz-Ortiz A, Moreno A (2005) Microwaves in organic synthesis. Thermal and non-thermal microwave effects†. Chem Soc Rev 34:164–178

Dharmana T, Swapna M (2017) Microwave-assisted-solvent free synthesis of Imidazo[1,2-*a*]pyridines. Der Pharma Chemica 9:35–38

Dharmana T, Kola KR, Bonnada NN (2018) Nano SiO_2-catalyzed synthesis of Imidazo[1,2-*a*]pyridines. ACSM 42:547–553

DiMauro EF, Kennedy JM (2007) Rapid synthesis of 3-amino-imidazopyridines by a microwave-assisted four-component coupling in one pot. J Org Chem 72:1013–1016

DiMauro EF, Vitullo JR (2006) Microwave-assisted preparation of fused bicyclic heteroaryl boronates: application in one-pot suzuki couplings. J Org Chem 71:3959–3962

Elgemeie G, Hamed M (2014) Microwave synthesis of guanine and purine analogs. Curr Microw Chem 1:155–176

Enguehard-Gueiffier C, Fauvelle F, Debouzy JC, Peinnequin A, Thery I, Dabouis V, Gueiffier A (2005) 2,3-Diarylimidazo[1,2-a] pyridines as potential inhibitors of UV-induced keratinocytes apoptosis: synthesis, pharmacological properties and interactions with model membranes and oligonucleotides by NMR. Eur J Pharm Sci 24:219–227

Gopal MS, Anitha I (2016) Green synthesis of imidazo[1,2-*a*]pyridines in aqueous medium. IOSR-JAC 9:1–5

Heibel MA, Fall Y, Scherrmann MC, Raboin SB (2014) Straightforward synthesis of various 2,3-diarylimidazo[1,2-*a*]pyridines in PEG_{400} medium through one-pot condensation and C-H Arylation. Eur J Chem 2014:4643–4650

Heitsch H (2002) Non-peptide antagonists and agonists of the bradykinin B(2) Receptor. Curr Med Chem 9: 913–928

Howard AS (1996) Comprehensive heterocyclic chemistry, vol 8, chap. 10. In: Katritzky AR, Rees CW, Scriven EVF (eds) London, Pergamon Press, pp 262–274

Jadhav SA, Shioorkar MG, Chavan OS, Sarkate AP, Shinde DB (2016) Rapid and efficient one-pot microwave-assisted synthesis of 2-phenylimidazo[1,2-a]pyridines and 2-phenylimidazo[1,2-*a*]quinoline in water–PEG-400. Synth Commun 47

Kappe CO, Dallinger D, Murphree SS (2009) Practical microwave synthesis for organic chemists: strategies, instruments, and protocols. Wiley, Weinheim

Karamthulla S, Khan MN, Choudhary LH (2015) Microwave-assisted synthesis of novel 2,3-disubstituted imidazo[1,2-*a*]pyridines *via* one-pot three component reactions. RSC Adv 5:19724–19733

Kishore KG, Basavanag UMV, Islas-Jacome A, Gamez-Montano R (2015) Synthesis of imidazo[1,2-*a*]pyridinchromones by a MW assisted Groebke- Blackburn-Bienaymé process. Tetrahedron Lett 56:155–158

Kong D, Wang X, Shi Z, Wu M, Lin Q, Wang X (2016) Solvent- and catalyst-free synthesis of imidazo[1,2-*a*]pyridines under microwave irradiation. J Chem Res 40:529–531

Kusy D, Maniukiewicz W, Blazewska K M (2019) Microwave-assisted synthesis of 3-formyl substituted imidazo[1,2-*a*]pyridines. Tetrahedron Lett 60

Li M, Li T, Zhao K, Wang M, Wen L (2013) Application of functionalized N, S-ketene acetals—microwave-assisted three-component domino reaction for rapid direct access to imidazo[1,2-a]pyridines. Chin J Chem 31:1033–1038

LockhartA YL, Judd DB, Merritt AT, Lowe PN, Morgenstern JL, Hong GZ, Gee AD, Brown JJ (2005) Evidence for the presence of three distinct binding sites for the thioflavin T class of alzheimer's disease PET imaging agents on β-amyloid peptide fibrils. Biol Chem 280:7677–7684

Lu Y, Zhang W (2004) Microwave-assisted synthesis of a 3-aminoimidazo[1,2-*a*] pyridine/pyrazine library by fluorous multicomponent reactions and subsequent cross-coupling reactions. QSAR Comb Sci 23:827–835

Maiti B, Chanda K, Selvaraju M, Tseng CC, Sun CM (2013) Multicomponent solvent-free synthesis of benzimidazolyl imidazo [1,2-*a*]-pyridine under microwave irradiation. ACS Comb Sci 15:291–297

Masquelin T, Bui H, Brickley B, Stephenson G, Schwerkoske J, Hulme C (2006) Sequential Ugi/Strecker reactions *via* microwave assisted organic synthesis: novel 3-center-4-component and 3-center-5-component multi-component reactions. Tetrahedron Lett 4:2989–2991

Motevalli K, Yaghoubi Z, Mirzazadeh R (2012) Microwave-assisted, one-pot three component synthesis of 2-phenyl H-imidazo[1,2-*a*] pyridine. J Chem 9:1047–1052

Nagaraj M, Boominathan M, Muthusubramanian S, Bhuvanesh N (2012) Microwave-assisted metal-free synthesis of 2,8-Diaryl-6-aminoimidazo-[1,2-*a*]pyridine *via* Amine-triggered benzannulation. Synlett 23:1353–1357

Ohta M, Suzuki T, Koide T, Matsuhisa A, Furuya T, Miyata K, Yanagisawa I (1996) Novel 5-hydroxytryptamine (5-HT_3) receptor antagonists. I. Synthesis and structure-activity relationships of conformationally restricted fused imidazole derivatives. Chem Pharm Bull 44:991–999

(a) Jones G (ed) (1996) Comprehensive heterocyclic chemistry. II. Volume 8: fused five-and six-membered rings with ring junction heteroatom. Oxford, Pergamon. (b) Ramsden CA (ed) (1996) Comprehensive heterocyclic chemistry. II. Volume 7: fused five-and six-membered rings without ring junction heteroatom. Oxford, Pergamon

Pastor J, Siro J, Garcia-Navio JL, Rodrigo MM, Ballesteros M, Alvarez-Builla J (1995) Synthesis of new azino fused berainid ~ o-limn salts. A new family of DNA intercalating agents. I Bioorg Med Chem Lett 5:3043–3048

Polshettiwar V, Varma RS (2008) Microwave-assisted organic synthesis and transformations using benign reaction media. Acc Chem Res 41:629–639

Pordel M, Chegini H, Ramezani S, Daee M (2017) New fluorescent heterocyclic systems from imidazo[1,2-*a*]pyridine: design, synthesis, spectral studies and quantum-chemical investigations. J Mol Struct 1129:105–112

Rodriguez JC, Maldonado RA, Ramirez-Garcia G, Cervantes ED, De la Cruz FN (2020) Microwave-assisted synthesis and luminescent activity of imidazo[1,2-*a*]pyridine derivatives. J Heterocyclic Chem 57:2279–2287

Rousseau AL, Matlaba P, Parkinson CJ (2007) Multicomponent synthesis of imidazo[1,2-*a*]pyridines using catalytic zinc chloride. Tetrahedron Lett 48:4079–4082

Schwerkoske J, Masquelin T, Perun T, Hulme C (2005) New multi-component reaction accessing 3-aminoimidazo[1,2-*a*]pyridines. Tetrahedron Lett 46:8355–8357

Stasyuk AJ, Banasiewicz M, Cyranski MK, Gryko DT (2012) Imidazo [1,2-*a*]pyridines susceptible to excited state intramolecular proton transfer: one-pot synthesis *via* an Ortoleva-King reaction. J Org Chem 77:5552–5558

Tu S, Zhang J, Jia R, Jiang B, Zhang Y, Yao C, Jiang H (2007) Synthesis of N-(3-arylmethyl-2-oxo-2,3-dihydroimidazo[1,2-*a*] pyridin-3-yl)benzamides. Chem Lett 36:222–223

Vagin O, Denevich S, Munson K, Sachs G (2002) SCH 28080, a K^+ competitive inhibitor of gastric H, K-ATPase, binds near the M5–6 luminal loop, preventing $^K+$ access to the ion binding domain. Biochemistry 41:12755–12762

Wang Y, Frett B, Li HY (2014) Efficient access to 2,3-Diarylimidazo [1,2-*a*]pyridines *via* a one-pot, ligand-free, palladium-catalyzed three-component reaction under microwave irradiation. Org Lett 16:3016–3019

Wang Y, Saha B, Li F, Frett B, Li HY (2014b) An expeditious approach to access 2-arylimidazo[1,2-*a*]pyridin-3-ol from 2-amino pyridine through a novel Petasis based cascade reaction. Tetrahedron Lett 55:1281–1284

Zarate-Hernandez C, Basavanag UMV, Gamez-Montano R (2019) Synthesis of Imidazo[1,2-*a*]pyridine-Chromones *via* microwave-assisted groebke-blackburn-bienaymé reaction[†]. In: Proceedings, vol 41, pp 1–7

Zhang W (2004) Speciality Chem Mag 24(5):30–32. Zhang W (2004) The handbook of fluorous chemistry. In: Gladysz JA, Curran DP, Horvath IT (eds) Wiley-VCH, Weinheim, pp 222–236

Zhang H, Jiang L (2015) Microwave-assisted solvent-free synthesis of imidazo[1,2-*a*]pyridines *via* a three-component reaction. Tetrahedron Lett 56:2777–2779

Green Hydrogen Synthesis Methods

Meltem Yildiz and Murat Efgan Kibar

Abstract

Hydrogen is considered as the cleanest fuel of the future due to its environmental friendliness. Since each of the hydrogen production methods is not always green, there has been a need to redesign processes to minimize waste, increase efficiency, and make it more environmentally friendly. This chapter discusses hydrogen generation techniques according to principles of green chemistry. These methods have been utilized for each twelve principles to decide whether they can correspond the deficiencies. Thermochemical, electrochemical, and biological methods for hydrogen production were investigated.

1 Introduction

Hydrogen is a glimmer of hope for its economically viable, financially promising, and energy-efficient solutions to address apprehension caused by the reduction in fossil fuel reserves. Today, the majority of the energy requirement in the world is provided by fossil fuel sources, but it is expected it will be consumed at the next 50 years. Very important and serious studies around the world have resolved hydrogen a very essential and promising energy carrier. Storage of energy accessed by different techniques such as wind turbines and solar panels can take place through hydrogen, the most common element in the universe.

Hydrogen is considered to be a great white hope fuel as it is environment-friendly. However, H_2 is not always green according to the differentiation of the production method. Every hydrogen production processes cannot provide the green chemistrys' 12 principles which are "waste prevention, atom economy, less hazardous chemical syntheses, designing safer products, safer solvents and auxiliaries, design for energy efficiency, use of renewable feedstocks, reduce derivative, catalysis, design for degradation, pollution prevention, and inherently safer chemistry for accident prevention".

Numerous methods are used for the production of hydrogen, including thermochemically, electrolytically, and biologically. Wang and coworkers (2019) classify hydrogen production sources as water and biomass. Water electrolysis, water thermolysis, photocatalytic water splitting, and thermochemical water splitting are technologies that use water for hydrogen production. Also, electrolysis, microbial and chemical methods have been classified as biomass-based hydrogen production techniques (Wang et al. 2019). In this chapter, we summarize these hydrogen production methods and discuss the connection between the green chemistry of the studies in the literature.

2 Thermochemical Processes

In thermochemical processes, water is separated into H_2 and O_2 by chemical reactions using heat. Thus, thermal energy is transformed into hydrogen energy. Thermochemical hydrogen production methods and their importance in terms of "Green Chemistry" are summarized below.

2.1 Methane Decomposition

Natural gas, containing over 95% methane, is one of the main sources for hydrogen production. Hydrogen production from methane has been focused on steam reforming, CO_2 reforming, partial oxidation, and catalytic reforming reactions. All these methods use oxygen to produce hydrogen from methane. Using catalysts decreases process temperature and increases the hydrogen amount (Kheirollahivash

M. Yildiz (✉) · M. E. Kibar
Faculty of Engineering, Department of Chemical Engineering, Kocaeli University, 41380 Kocaeli, Turkey
e-mail: myildiz@kocaeli.edu.tr

Inamuddin et al. (eds.), *Advances in Green Synthesis*, Advances in Science, Technology & Innovation,
https://doi.org/10.1007/978-3-030-67884-5_13

et al. 2019). Because of the high cleavage energy of the C–H bond of the molecules of methane, it is very difficult for methane to decompose to oxygen-free C_2 hydrocarbons and hydrogen using the conventional catalytic method (Shuanghui et al. 2013). Hydrogen production by the plasma method is thermodynamically less costly than other methods such as steam reforming (Jeremias 2018).

The fact that plasma technology is far from thermodynamic equilibrium brings a solution to this situation (Dincer 2012). For this purpose, equilibrium (thermal) plasma and non-equilibrium (non-thermal) plasma systems can be used in methane conversion to hydrogen. The methane decomposition to form hydrogen and carbon can be written as in Eq. 1 (Dincer 2007);

$$CH_4 + 75.3\text{kJ/mole} \rightarrow C_{(s)} + 2H_2 \quad (1)$$

The plasma method to produce hydrogen takes place by simultaneous decomposition of natural gas into hydrogen and solid carbon regard to an external electrical energy source. Various plasma reactors (Bespalko 2014; Qian et al. 2020; ZhiPeng et al. 2011; Mishra et al. 2004; Aleknaviciute et al. 2013; Massa et al. 2018; Khadir et al. 2017; Zhang et al. 2014; Xu 2007; Jasiński et al. 2008, 2011; Moshrefi et al. 2012; Moshrefi and Rashidi 2014) have been used in the literature.

If the hydrogen production method using plasma is evaluated considering the principles of green chemistry, it appears that it provides many benefits for converting electricity into chemical energy. This method meets the "atom economy" and "reducing derivatives" principles of green chemistry. Since the same production method produces carbon black as a by-product, it prevents the first principle of green chemistry, "waste prevention" from being fulfilled. Also, the use of high voltage as an energy source should be marked as providing the "energy efficiency". So, "using renewable feedstocks" principle is not provided (Celik and Yildiz 2017).

2.2 Dry or Steam Methane Reforming

One method to convert CO_2 and CH_4 into synthesis gas is methane dry reforming. This process has gained much importance according to its ambient interest and great use of energy (Gao et al. 2020). The reaction can be written as it can be seen in Eq. 2;

$$CH_4 + CO_2 \rightarrow 2CO + 2H_2 \Delta H^0_{298} = 248\ \text{kJ mol}^{-1} \quad (2)$$

When recent studies are examined, it can easily be seen that scientists have attracted much attention due to the environmental benefit and effective use of energy to methane dry reforming studies. The leading studies are aimed at improving the catalytic properties of the catalysts used in this reaction. After the noble metal-based catalysts that have been studied for many years, studies in recent years have turned to Ni-based catalysts for economic reasons. Although the effectiveness of different catalysts in this reaction has been studied, the expected improvements have not been possible due to the problems with the quick deactivation of the catalysts (Chen et al. 2020; Akri et al. 2017).

SMR reaction has a very important place for synthesis gas production (Lunsford 2000). It is the process of converting gas or liquid fossil fuels such as methane into synthesis gas (Eq. 3). In SMR, H_2 and CO occur as a result of steam and methane reaction that is highly endothermic (Gao et al. 2020).

$$CH_4 + H_2O \rightarrow CO + 3H_2 \Delta H^0_{298} = 206\ \text{kJ mol}^{-1} \quad (3)$$

The reactions were performed at high temperatures and pressures (Farsi and Monsouri 2016). Steam methane reforming is economical and effective method for H_2 production. To improve the efficiency of this process, different catalysts with different properties are used. In literature, noble metals such as ruthenium (Homsi et al. 2014), rhodium (Hassan et al. 2016), palladium (Fernandes and Soares 2006), iridium (Cheah et al. 2018), and platinum (Castro et al. 2018) are used as the active element on catalysts. Also, nickel-based catalysts were studied at large amounts because of its catalytic activity, lower cost, and higher availability properties (Giuliano and Gallucci 2018).

One of the most important problems in the long-term use of Ni catalysts is the formation of carbon deposits that turn into carbon filaments on the catalyst surface. Various studies have been carried out to solve this problem. The results acquired with nickel catalysts showed that attributed the performance of the alkaline catalysts to the large oxygen storing capability and the ability to increase the disintegration of active metals (Pashchenko 2019; Oliveira Rocha et al. 2019; Zeppieri et al. 2010).

Hydrogen production via the steam methane reforming method satisfies renewable feedstocks (7), derivative reduction (8), and catalysis (9) items of green chemistry principles. Unfortunately, this method produces CO as waste by this method depending on the chosen reaction conditions, and, this reaction is against the atomic economy (2). Also, using high temperature and pressure prevents this reaction to provide the 6th principle (energy efficiency) of green chemistry.

2.3 Coal Gasification

With this process, coal turns into hydrogen, CO, and CO_2 at high temperatures (>700 °C) with oxygen. The CO formed then reacts with H_2O to react to form CO_2 and H_2. The composition of gases obtained by gasification of coals depends on the reaction ability of coal, type of gases used, and applied gasification process (pressure, temperature, coal, and the flow directions of the gas, etc.).

Coal is a fuel that is abundant and low-cost, prone to hydrogen. However, concerns over the control of CO_2 released into the atmosphere raise some doubts for its future use (İrfan et al. 2011). Important reactions encountered about coal gasification reactions (Eqs. 4–8) can be listed as seen in Table 1. The thermodynamic states of the reactions are cited from Sajjad and Rasul (2015). In reaction 4, water vapor combines with carbon to form CO and H_2 gases mixture called "synthesis gases".

Except for the reactions listed in Table 1, homogeneous reactions also occur between the gas products and/or substances used as gasifiers. All of the gasification reactions begin with pyrolysis. As a result of thermal decomposition, CO_2, CH_4, saturated and unsaturated hydrocarbons are formed. They interact with each other again during gasification. The post-reaction products are CO, CO_2, and H_2.

CO is a product of this hydrogen production procedure. And also, it is a reactant of water–gas shift reaction, CO reacts with H_2O, and the product is CO_2.

Water–gas reaction occurs between water and carbon (coke or coal) at high temperatures and pressures. The products of the reaction such as CO_2 and H_2 are called synthesis gas and can be overly useful in fixing with metallic heterogeneous catalysts. The Boudouard reaction is important according to its transforming capability of the carbon content and non-combustible carbon dioxide into a usable gas-carbon monoxide property. And also, the heating value of the produced syngas increases. Partial oxidation is the burning of hydrocarbons in an oxygen environment less than necessary. In general, the efficiency of partial oxidation is 50% lower than steam reforming. The combustion of coal causes approximately 15% CO_2 formation in the flue gas. Due to the low CO_2 concentration and the presence of other gas types such as sulfur oxides, it is difficult to capture CO_2 from flue gas (İrfan et al. 2011).

Reactions at equilibrium mean that carbon monoxide is always produced and waste generation is an inevitable ending (violation of the first principle). And also, the atom economy is impossible (violation of the second principle), and the reactions take place at high temperatures and pressures (violation of the sixth principle). This method meets the principles of raw materials, derivative reduction, and catalysis.

2.4 Biomass Gasification

Biomass can be explicable as renewable organic resources such as corn stover, wheat straw, switchgrass, and organic urban wastes. Gasification is a suitable method to obtain hydrogen from these renewable resources. Hydrogen from biomass is produced mostly by pyrolysis and gasification (Cao et al. 2020).

Pyrolysis is the process of gasification of biomass without oxygen. Since there is no oxygen, there are other HCs in the gas mixture coming out after the treatment. An extra step should be used to separate these hydrocarbons from the gas mixture to obtain a clean mixture of H_2, CO, and CO_2. Also, another step required for H_2 production is carried out, and carbon monoxide is converted into carbon dioxide by steam. The direct conversion of biomass into liquid fuel by pyrolysis still has not resolved problems today (Prasertcharoensuk et al. 2019).

An oxidizing agent is used in the biomass gasification method and occurs at higher temperature (Cao et al. 2020). Biomass gasification has two main routes for hydrogen production as (i) steam gasification of biomass, (ii) supercritical water gasification of biomass.

2.4.1 Steam Gasification

This method includes the process that converts raw materials to high-temperature inflammable gases as H_2, CO, CO_2, and hydrocarbons. It facilitates the introduction of water vapor as a gasifier. Reaction temperature, biomass ratio, and gasification working conditions have a crucial effect in the hydrogen content of the synthesis gas to be produced. Steam is the gas that is accepted as the most suitable gasifier (Cao et al. 2020). The biomass vapor gasification reaction mainly involves carbon gasification reactions such as carbon gasification, water–gas shift, methane reforming, and hydrocarbon reforming. The catalysts and adsorbents to be used in all these reactions are of great importance for the purity and amount of hydrogen to be obtained.

Table 1 Important reactions encountered in coal gasification reactions (Sajjad and Rasul 2015)

Chemical synthesis	Reaction	Thermodynamic sate
Water–gas shift reaction (4)	$C + H_2O \rightarrow CO + H_2$	ΔH = +118.5 kJ/mol
Boudouard reaction (5)	$C + CO_2 \rightarrow 2CO$	ΔH = +159.9 kJ/mol
Gasification with hydrogen (6)	$C + 2H_2 \rightarrow CH_4$	ΔH = −87.5 kJ/mol
Partial oxidation (7)	$C + \frac{1}{2}O_2 \rightarrow CO$	ΔH = −123.1 kJ/mol
Combustion (8)	$C + O_2 \rightarrow CO_2$	ΔH = −406.0 kJ/mol

2.4.2 Supercritical Water Gasification

Water has incomparable physical properties over the critical conditions, resulting in better transport properties due to high diffusion abilities, low viscosities, and new reaction alternatives for hydrolysis or oxidation (Reddy et al. 2014). SCWGs' future looks bright in hydrogen production. The supercritical water gasification process uses the supercritical water properties as 374 °C and 22.1 MPa. That is, the essential feature that distinguishes SCWG relates to gasification media, namely supercritical water. Biomass components such as lignocellulose can split into smaller molecules and produce synthesis gas during supercritical water gasification (SCWG).

For biomass gasification reactions, unfortunately, waste generation cannot be prevented (1st principle of green chemistry), atom economy cannot be satisfied, and derivatives exist (8th principle of green chemistry). Also, the input energy required to observe the 6th principle of energy efficiency should not be excessive. "Use of raw materials" and "catalysts" requirements of green chemistry can be achieved by this method (7th and 9th principles of green chemistry).

2.5 Biomass-Derived Liquid Reforming

Biomass sources are converted into biomass-based fluids such as cellulosic ethanol, bio-oils, or other liquid biofuels. The process of converting biomass-based liquids into hydrogen can be produced by a few steps. Firstly H_2, CO, and CO_2 are produced from liquid fuel using a catalyst. Then, carbon monoxide formed at the previous stage is reacted with steam at high-temperature "water–gas shift reaction" to produce CO_2 (Cortright et al. 2002).

Hydrogen production with the help of liquids from biomass is like reforming of natural gas. However, fluids from biomass are made up of larger molecules with more carbon atoms than natural gas and are therefore more difficult to regenerate. Studies with different catalysts are ongoing to increase the efficiency and the selectivity in the literature.

2.6 Solar Thermochemical Hydrogen

High temperature is required to carry out reactions required for hydrogen production by thermochemical water separation processes. With this production method, chemicals can be reused in every cycle. Dias and coworkers (2018) defined five different thermochemical routes for solar hydrogen production as solar (i) thermolysis, (ii) thermochemical cycles, (iii) reforming, (iv) cracking, and (v) gasification. All of these pathways involve endothermic reactions (Dias and Mendes 2018).

Difficulties remain in the research and development of suitable reactors for the commercial development of all these methods. Reaction efficiency and durability should be increased for the thermochemical cycle.

3 Electrolytic Processes

3.1 Electrochemical

Electrolysis, a hopeful option for hydrogen production from renewable sources, uses electricity to separate water into H_2 and O_2. Units called electrolyzers are the rigs where this reaction takes place. Electrolyzers include an electrolyte to separate anode and a cathode. There are various types of electrolyzers due to the electrolyte material involved. Equation 9 shows the reaction:

$$H_2O \xrightarrow{\text{electricity}} H_{2(g)} + 1/2O_{2(g)} \tag{9}$$

The most basic electrolysis cell system is alkaline electrolysis, a method extensively used in the industry. Besides, there are different types of electrolyzers such as polymer electrolyte membrane electrolyzers, alkaline electrolyzers, and solid fuel electrolyzers. Polymer electrolyte membrane electrolyzer (PEME) splits water molecules into hydrogen and oxygen gas. PEME is a widely used technology that produces "green hydrogen". In the PEME, as hydrogen ions selectively advance to the cathode, water molecules flow from an external cycle and in the cathode, and hydrogen gas is generated. In alkaline electrolyzers, OH^- ions generate H_2 on the cathode side. It can be possible to recover and reuse the alkaline solution that is used as electrolyte. Solid oxide electrolyzer produces hydrogen with the help of water electrolysis using a solid oxide or ceramic electrolyte. Nevertheless, since it is a commercialized electrolyzer, the alkaline system was evaluated in this study in terms of green chemistry properties. The alkali electrolysis method is suitable for the first principle from the green chemistry principles as it is a non-waste method, the 2nd principle because it provides the atomic economy principle, the 7th principle because it uses water as a renewable raw material, and the 8th principle because it does not have derivatives. Unfortunately, the use of KOH does not accord principles 3 and 12. It can also be assumed that principle 10 is violated in terms of having an unknown cell material life.

3.2 Photo-Electrochemical

Hydrogen production by the photo-electrochemical method can also be realized by the electrolysis of water with the photoelectrode, an electrode coated with photocatalyst materials. In a photocatalytic reaction, a photon hits the photocatalyst to form an electron–hole pair, thereby obtaining an electric charge. A photo-electrochemical cell also consists of electrodes, such as an electrolysis cell. But one or both electrodes are photoelectrodes, and at least one of them is a semiconductor. Hydrogen production in this way is a promising solar hydrogen pathway and offers high conversion efficiency potential at low operating temperatures using thin film and/or particulate semiconductor materials at low-cost temperatures (Steinfeld 2005).

Hydrogen production via photo-electrochemical methods, many of green chemistry methods can be satisfied. This method can be accepted as the greenest hydrogen production method. There is no waste formation by photo-electrochemical method, and therefore, it provides principle 1. It is also suitable for the second principle of green chemistry because it provides the atomic economy principle. KOH is not used in this method as in alkaline electrolysis, and therefore, it is in accordance with principle 5. It fulfills principle 7 because it uses renewable resources, principle 8 due to a one-step reaction, and finally the principle 9 of green chemistry due to the use of photocatalyst.

4 Biological Processes

Hydrogen can be obtained by using microorganisms which are called biohydrogen. According to the process, hydrogen is produced from carbon sources generally renewable energy resources (Saratale et al. 2019; Wang et al. 2020). The operating conditions are attractive related to ambient temperature and atmospheric pressure (Chandrasekhar et al. 2015). Therefore, the mentioned reasons make biohydrogen production processes important. Unfortunately, biohydrogen production processes are not interested deeply for industrial-scale due to low production yields (Sinha and Pandey 2011; Sung et al. 2003). At present, hydrogen production from natural gases, heavy oil, naphtha, and coal is more favorable, and the hydrogen is produced from these raw materials comprise almost 90% of all hydrogen production (Hallenbeck et al. 2019). High demand for fossil fuels causes the depletion of the sources. Other hydrogen production methods such as by electrolysis are 4%, and only 1% of hydrogen production is available from biomass. To minimize the pressure on fossil fuel usage, the alternative production methods have to be improved and might be integrated into commercial hydrogen production systems.

Although the biohydrogen production level is low, it is still a promising production method due to its mild operating conditions. By using renewable energy sources, biohydrogen production can be categorized as (i) Photofermentation, (ii) Dark fermentation, (iii) Biophotolysis (direct and indirect), and (iv) Microbial electrolysis cell (MEC) (Xu 2007; Mohanraj et al. 2019; Nikolaidis and Poullikkas 2017). In general, according to economical experts or low levels of hydrogen production, biohydrogen production is given by the first two categories, but the other categories also have to be considered related to their improvements in literature.

4.1 Photofermentation

In the photofermentation process, biohydrogen can be produced by the decomposition of organic matters. The photosynthetic bacteria decompose the substrate with nitrogenase catalysis. Nitrogen-fixing purple non-sulfur bacteria are the most commonly preferred specie to produce biohydrogen (Ghosh et al. 2017). In photofermentation, all types of purple bacteria and green bacteria are able to convert light energy to chemical energy. The first mechanism is given by Eq. 10.

$$N_2 + 8H^+ + 8e^- + 16ATP \rightarrow 2NH_3 + H_2 + 16ADP + 16P_i \quad (10)$$

If nitrogen exists, nitrogenase enzyme catalyses the reduction of N_2–NH_3, and during the reduction reaction, 1 mol of H_2 is produced per mole of N_2. On the other hand, under limited nitrogen, the nitrogenase enzyme catalyses the reaction in a different pathway. Now, the reduction of protons is responsible for hydrogen production in Eq. 11.

$$2H^+ + 2e^- + 4ATP \rightarrow H_2 + 4ADP + 4P_i \quad (11)$$

By comparing the Eqs. (10) and (11), 4 times more biohydrogen can be produced by nitrogen-limited conditions with the same energy demand. Hydrogen production by photofermentation method meets most of the principles except the complex reaction system, which is the 8th principle of green chemistry. Working under ambient T and P, microorganisms acting as catalysts make this production method quite Green.

4.2 Dark Fermentation

Anaerobic or facultative bacteria can also produce hydrogen by using preferably simple sugars as substrates such as glucose without using light energy, means in the dark media. In dark fermentation, glucose is converted to pyruvate through glycolysis. While the pyruvate is formed, the decomposition differs according to the enzymes. If the enzyme is pyruvate-formate lyase, liquid products such as lactate, acetate, and ethanol are formed. On the other hand, ethanol, propionate, butyrate, and butanol are the end products of pyruvate ferredoxin oxidoreductase mechanism with regard to the type of the microorganism and fermentation media (Balachandar et al. 2013). The theoretical hydrogen production yield is 4 mol H_2 mol^{-1} glucose, Eq. 12.

$$C_6H_{12}O_6 + 2H_2O \rightarrow 2CH_3COOH + 2CO_2 + 4H_2 \quad (12)$$

According to Eq. 13, final product is butyric acid,

$$C_6H_{12}O_6 + 2H_2O \rightarrow 2CH_3CH_2COOH + 2CO_2 + 2H_2 \quad (13)$$

It was concluded that the concept of green chemistry fulfills the principles 6, 7, 9, and 10, respectively, since it uses biomass during hydrogen production with this method and is studied under moderate temperature and pressure conditions. However, the production of wastes such as CO_2 does not meet the principles of green chemistry 1, 2, and 8 due to the fact that it does not provide the atomic economy principle due to organic acid production and creates wastes as a result of complex biological mechanisms.

4.3 Biophotolysis

Biophotolysis is a reaction series of photochemical reactions that are triggered by absorption of light energy and continues by converting to chemical energy. Molecular hydrogen and oxygen are produced during the direct biophotolysis process by dissociation of the water molecule, Eq. 14.

$$2H_2O \xrightarrow{\text{Light energy}} 2H_2 + O_2 \quad (14)$$

Green algae and cyanobacteria produce hydrogen by biophotolysis. Related to Eq. 14, green algae generate H_2 in the reduction side of photosystem I and generate O_2 in the oxidation side of photosystem II (Hay et al. 2013). While the O_2 is produced, another limitation occurs, reversible hydrogenase is sensitive to O_2, and this phenomenon inhibits the H_2 production. To overcome this limitation, an indirect biophotolysis process can be suggested. According to indirect biophotolysis, O_2 and H_2 evolution can be separated temporally or spatially. Therefore, indirect biophotolysis arises from two degrees. The first degree, CO_2, is fixed by cyanobacteria to convert it to a carbon source, and then, hydrogen production proceeds (Deo et al. 2012; Lam and Lee 2013). The overall hydrogen production mechanism through cyanobacteria is represented by Eqs. 15 and 16 (Xiao et al. 2010).

$$6H_2O + 6CO_2 \xrightarrow{\text{Light energy}} C_6H_{12}O_6 + 6O_2 \quad (15)$$

$$C_6H_{12}O_6 + 6H_2O \xrightarrow{\text{Light energy}} 12H_2 + 6CO_2 \quad (16)$$

The biophotolysis method meets most of the green chemistry principles applicable as in the dark fermentation method. Likewise, it does not comply with the 8th principle, and it is concluded that it meets the 1, 2, 6, 7, 9, and 10th principles.

4.4 Microbial Electrolysis Cell

Microbial electrolysis cell (MEC) is a promising technology for H_2 production from degradable matters such as wastewater (Mohan et al. 2013) and organic substrates (Escapa et al. 2016; Rahman et al. 2016). Electrochemically active bacteria refer to respiring bacteria produce CO_2, electrons, and protons by oxidizing the organic substrates. The bacteria have the ability to transfer the electrons to the anode. The generated protons released to the solution. The electrons at the anode travel to the cathode. The reaction on the cathode produces hydrogen by combining the electrons and protons (Kadier et al. 2016). Electrochemical reactions are represented in Eqs. 17 and 18 (Cardeña et al. 2019; Rivera et al. 2019). Anode reaction is given for acetate oxidation.

Oxidation (anode):

$$C_2H_4O_2 + 2H_2O \rightarrow 2CO_2 + 8H^+ + 8e^- \quad (17)$$

Reduction (cathode):

$$8H^+ + 8e^- \rightarrow 4H_2 \quad (18)$$

MEC requires an external power supply to input a voltage to initiate the cathodic proton reduction. In Fig. 1, the overview of biohydrogen production is given. These cells can benefit from renewable biomass sources and clean hydrogen production from wastewater. Of course, there are some basic challenges. For example, the hydrogen production rate is low and should be increased.

There are several mechanisms for biohydrogen production, but these processes have some advantages and disadvantages. Especially, biohydrogen production from organic wastes offers an environmentally friendly process compared to thermochemical and physicochemical routes. Although biohydrogen production is an environmental method, the production yield and rate are the bottleneck of commercialization. These problems should be addressed through for the economical and technological viability otherwise

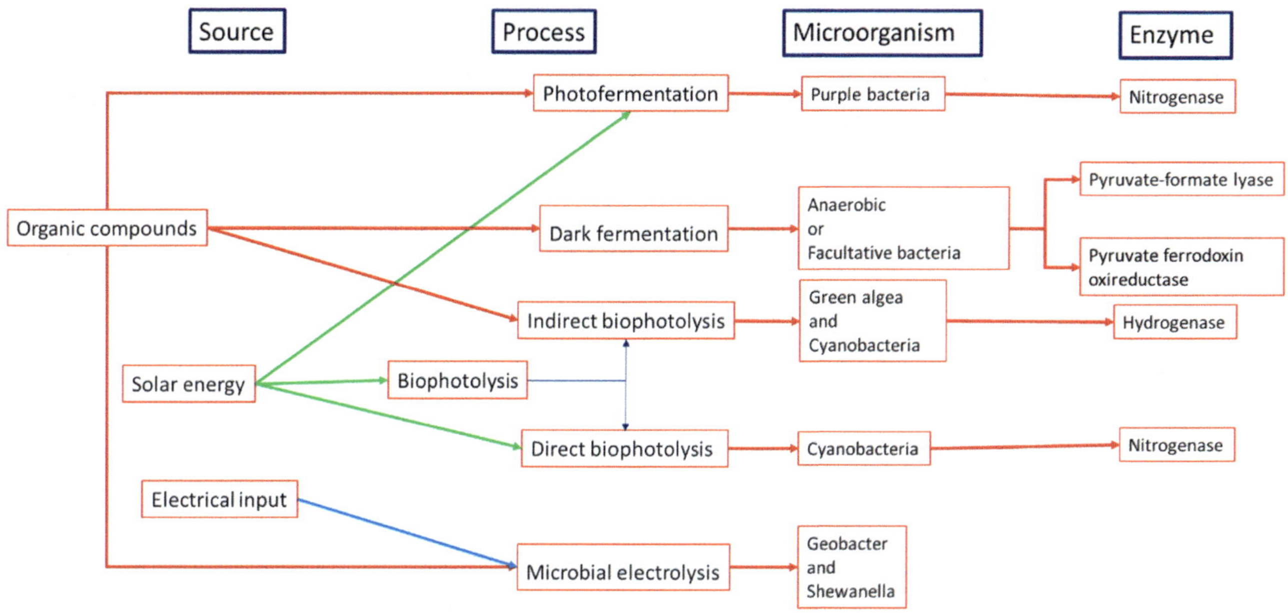

Fig. 1 Biohydrogen production routes

biohydrogen will not contribute to the hydrogen economy and sustainable growth.

5 Conclusions

Each of the hydrogen production methods is examined in this study by considering 12 principles which are accepted as the basis of green chemistry. Three main hydrogen production processes (thermochemical, electrochemical, and biological) were evaluated per each principle, and the results were discussed.

As all hydrogen production methods are compared, the results are as follows: Electrolysis is a greener option to plasma arc dissociation. When thermal methods are compared, it is concluded that the best choice for the environment is the thermal decomposition method. Biomass vapor reform or gasification methods may be an alternative since such high temperatures can only be achieved by using contradictory nuclear energy sources. It is concluded that the best alternative to the environment among the hybrid systems is photo-electrochemical water splitting. And, consequently, all biological methods are environmentally friendly because they use mimic or natural pathways.

References

Akri M, Pronier S, Chafik T, Achak O, Granger P, Simon P, Trentesaux M, Batiot-Dupeyrat C (2017) Development of nickel supported La and Ce-natural illite clay for autothermal dry reforming of methane: toward a better resistance to deactivation. Appl Catal B 205:519–531

Aleknaviciute I, Karayiannis TG, Collins MW, Xanthos C (2013) Methane decomposition under a corona discharge to generate COx-free hydrogen. Energy 59:432–439

Balachandar G, Khanna N, Das D (2013) Biohydrogen production from organic wastes by dark fermentation. Biohydrogen. ISBN 9780444595553

Bespalko S (2014) Recent advances in hydrogen generation by contact glow discharge electrolysis: review. J Int Sci Publ Mater Methods Technol 8:355–363

Cao L, Yu IKM, Xiong X, Tsang DCW, Zhang S, Clark JH, Hu C, Ng YH, Shang J, Ok YS (2020) Biorenewable hydrogen production through biomass gasification: a review and future prospects. Environ Res 186:109547

Cardeña R, Cercado B, Buitrón G (2019) Microbial electrolysis cell for biohydrogen production. Biomass, Biofuels, Biochem. ISBN 9780444642035

Castro TP, Silveira EB, Rabelo-Neto RC, Borges LEP, Noronha FB (2018) Study of the performance of Pt/Al_2O_3 and $Pt/CeO_2/Al_2O_3$ catalysts for steam reforming of toluene, methane and mixtures. Catal Today 299:251–262

Celik D, Yildiz M (2017) Investigation of hydrogen production methods in accordance with green chemistry principles. Int J Hydrogen Energy 42:23395–23401

Chandrasekhar K, Lee YJ, Lee DW (2015) Biohydrogen production: strategies to improve process efficiency through microbial routes. Int J Mol Sci 16(4):8266–8293

Cheah SK, Massin L, Aouine M, Steil MC, Fouletier J, Gélin P (2018) Methane steam reforming in water deficient conditions on $Ir/Ce_{0.9}Gd_{0.1}O_{2-x}$catalyst: metal-support interactions and catalytic activity enhancement. Appl Catal B 234:279–289

Chen S, Zaffran J, Yang B (2020) Dry reforming of methane over the cobalt catalyst: theoretical insights into the reaction kinetics and mechanism for catalyst deactivation. Appl Catal B 270:118859

Cortright RD, Davda RR, Dumesic JA (2002) Hydrogen from catalytic reforming of biomass-derived hydrocarbons in liquid water. Nature 29(418, 6901):964–967

De Oliveira Rocha K, Marques CMP, Bueno JMC (2019) Effect of Au doping of Ni/Al_2O_3 catalysts used in steam reforming of methane: mechanism, apparent activation energy, and compensation effect. Chem Eng Sci 207:844–852
Deo D, Ozgur E, Eroglu I, Gunduz U, Yucel M (2012) Photofermentative hydrogen production in outdoor conditions. In: Hydrogen energy—challenges and perspectives. Dragica Minic, IntechOpen
Di Giuliano A, Gallucci K (2018) Sorption enhanced steam methane reforming based on nickel and calcium looping: a review. Chem Eng Process Process Intensif 130:240–252
Dias P, Mendes A (2018) Hydrogen production from photoelectrochemical water splitting. In: Meyers R (ed) Encyclopedia of sustainability science and technology. Springer, New York, NY
Dincer İ (2007) Environmental and sustainability aspects of hydrogen and fuel cell systems. Int J Energy Res 31:29–55
Dincer İ (2012) Green methods for hydrogen production. Int J Hydrogen Energy 37:1954–1971
El Hassan N, Kaydouh MN, Geagea H, El Zein H, Jabbour K, Casale S, El Zakhem H, Massiani P (2016) Low temperature dry reforming of methane on rhodium and cobalt based catalysts: active phase stabilization by confinement in mesoporous SBA-15. Appl Catal A 520:114–121
Escapa A, Mateos R, Martínez EJ, Blanes J (2016) Microbial electrolysis cells: an emerging technology for wastewater treatment and energy recovery. From laboratory to pilot plant and beyond. Renew Sustain Energy Rev 55:942–956
Farsi A, Monsouri SS (2016) Influence of nanocatalyst on oxidative coupling, steam and dry reforming of methane: a short review. Arab J Chem 9:28–34
Fernandes FAN, Soares AB (2006) Methane steam reforming modeling in a palladium membrane reactor. Fuel 85:569–573
Gao N, Cheng M, Zheng Y (2020) Syngas production via combined dry and steam reforming of methane over Ni-Ce/ZSM-5 catalyst. Fuel 273:117702
Ghosh S, Dairkee UK, Chowdhury R, Bhattacharya P (2017) Hydrogen from food processing wastes via photofermentation using purple non-sulfur bacteria (PNSB)—a review. Energy Convers Manage 141:299–314
Hallenbeck PC, Lazaro CZ, Sagir E (2019) Biohydrogen. Compre Biotechnol 3:128–139
Hay JXW, Wu TY, Juan JC, Jahim JM (2013) Biohydrogen production through photo fermentation or dark fermentation using waste as a substrate: overview, economics, and future prospects of hydrogen usage. Biofuels, Bioprod Biorefining 7(3):334–352
Homsi D, Aouad S, Gennequin C, Aboukais A, Abi-Aad E (2014) A highly reactive and stable $Ru/Co_{6-x}Mg_xAl_2$ catalyst for hydrogen production via methane steam reforming. Int J Hydrogen Energy 39 (19):10101–10107
İrfan MF, Usman MR, Kusakabe K (2011) Coal gasification in CO_2 atmosphere and its kinetics since 1948: a brief review. Energy 36:12–40
Jasıński M, Dors M, Nowakowska H, Mizeraczyka J (2008) Hydrogen production via methane reforming using various microwave plasma sources. Chem Listy 102:1332–1337
Jasıński M, Dors M, Nowakowska H, Nichipor GV, Mizeraczyk J (2011) Production of hydrogen via conversion of hydrocarbons using a microwave plasma. J Phys D Appl Phys 44(19):194002
Jeremias M (2018) Hydrogen production via thermal plasma. In: EuroSciCon conference on biochemical, chemical engineering and mass spectrometry, vol 2
Kadier A, Simayi Y, Abdeshahian P, Azman NF, Chandrasekhar K, Kalil MS (2016) A comprehensive review of microbial electrolysis cells (MEC) reactor designs and configurations for sustainable hydrogen gas production. Alexandria Eng J 55(1):427–443
Khadir N, Khodja K, Belasri A (2017) "Methane conversion using a dielectric barrier discharge reactor at atmospheric pressure for hydrogen production. Plasma Sci Technol. 19:095502–095512
Kheirollahivash M, Rashidi F, Moshrefi MM (2019) Experimental study and kinetic modeling of methane decomposition in a rotating arc plasma reactor with different cross-sectional areas. Int J Hydrogen Energy 44(33):17460–17469
Lam MK, Lee KT (2013) Biohydrogen production from algae. Biohydrogen. ISBN 9780444595553
Lunsford JH (2000) Catalytic conversion of methane to more useful chemicals and fuels: a challenge for the 21st century. Catal Today 63(2–4):165–174
Massa L, Retter JE, Elliot GS, Freund JB (2018) Dielectric-barrier-discharge plasma-assisted hydrogen diffusion flame. Part 2: Modeling and comparison with experiments 191:541–557
Mishra LN, Shibata K, Ito H, Yugami N, Nishida Y (2004) Conversion of methane to hydrogen via pulsed corona discharge. J Nat Gas Chem 13:82–86
Mohan SV, Chandrasekhar K, Chiranjeevi P, Babu PS (2013) Biohydrogen production from wastewater. Biohydrogen. ISBN 9780444595553
Mohanraj S, Pandey A, Venkata Mohan S, Anbalagan K, Kodhaiyolii S, Pugalenthi V (2019) Metabolic engineering and molecular biotechnology of biohydrogen production. Biohydrogen. ISBN 9780444595553
Moshrefi MM, Rashidi F (2014) Hydrogen production from methane by DC spark discharge: effect of current and voltage. J Nat Gas Sci Eng 16:85–89
Moshrefi MM, Rashidi F, B ozorgzadeh H.R., Zekordi, S.M. (2012) Methane conversion to hydrogen and carbon black by DC-Spark discharge. Plasma Chem Plasma Process 32:1157–1168
Nikolaidis P, Poullikkas A (2017) A comparative overview of hydrogen production processes. Renew Sustain Energy Rev 67:597–611
Pashchenko D (2019) Experimental investigation of reforming and flow characteristics of a steam methane reformer filled with nickel catalyst of various shapes. Energy Convers Manage 185:465–472
Prasertcharoensuk P, Bull SJ, Phan AN (2019) Gasification of waste biomass for hydrogen production: effects of pyrolysis parameters. Renew Energy 143:112–120
Qian JX, Chen TW, Enakonda LR, Liu DB, Mignani G, Basset J, Zhou L (2020) Methane decomposition to produce CO_x-free hydrogen and nano-carbon over metal catalysts: a review. Int J Hydrogen Energy 45(15):7981–8001
Rahman SNA, Masdar MS, Rosli MI, Majlan EH, Husaini T, Kamarudin SK, Daud WRW (2016) Overview biohydrogen technologies and application in fuel cell technology. Renew Sustain Energy Rev 66:137–162
Reddy SN, Nanda S, Dalai AK, Kozinski JA (2014) Supercritical water gasification of biomass for hydrogen production. Int J Hydrogen Energy 39(13):6912–6926
Rivera I, Schröder U, Patil SA (2019) Microbial electrolysis for biohydrogen production. Microbial Electrochem Technol. ISBN 9780444640529
Sajjad M, Rasul MG (2015) Prospect of underground coal gasification in Bangladesh. Procedia Eng 105:537–548
Saratale GD, Saratale RG, Banu JR, Chang J-S (2019) Biohydrogen production from renewable biomass resources. Biohydrogen. Elsevier, ISBN 9780444595553
Shuanghui H, Baowei W, Yijun LV, Wenjuan Y (2013) Conversion of methane to C_2 hydrocarbons and hydrogen using a gliding arc reactor. Plasma Sci Technol 15(6):555–561
Sinha P, Pandey A (2011) An evaluative report and challenges for fermentative biohydrogen production. Int J Hydrogen Energy 36 (13):7460–7478

Steinfeld A (2005) Solar thermochemical production of hydrogen—a review. Sol Energy 78:603–615

Sung S, Contact P, Bazylinski DA, Raskin L (2003) Biohydrogen production from renewable organic wastes. In: Hydrogen, fuel cells, and infrastructure technologies FY 2003 Progress Report

Wang M, Wang G, Sun Z, Zhang Y, Xu D (2019) Review of renewable energy-based hydrogen production processes for sustainable energy innovation. Glob Energy Interconnection 2(5):436–443

Wang S, Zhang T, Bao M, Su H, Xu P (2020) Microbial production of hydrogen by mixed culture technologies: a review. Biotechnol J 15 (1):1–8

Xiao B, Han Y, Liu J (2010) Evaluation of biohydrogen production from glucose and protein at neutral initial pH. Int J Hydrogen Energy 35(12):6152–6160

Xu Z (2007) Biological production of hydrogen from renewable resources. In: Bioprocessing for value-added products from renewable resources. Elsevier. ISBN 9780444521149

Zeppieri M, Villa PL, Verdone N, Scarsella M, De Filippis P (2010) Kinetic of methane steam reforming reaction over nickel and rhodium-based catalysts. Appl Catal A 387:147–154

Zhang H, Changming D, Wu A, Bo Z, Yan J, Li X (2014) Rotating gliding arc assisted methane decomposition in nitrogen for hydrogen production. Int J Hydrogen Energy 39:12620–12635

ZhiPeng Z, JiMin Z, TaoHong Y, PingHui Z, WeiDong X (2011) Hydrogen production by reforming methane in a corona inducing dielectric barrier discharge and catalyst hybrid reactor. Energy Sci Technol 56(20):2162–2166

Fundamental and Principles of Green Synthesis

Mukta Sharma and Manoj Sharma

Abstract

The hazard factors attributed to chemical and physical methods used in the synthesis of organic, inorganic, hybrid, coordinated compounds have led to the emergence of alternative methods that do not pose any risk to the environment. "Green Synthesis" is an evolving science that accords benefits to the environment and to the economy. In recent years, the application of "Green Synthesis" in the production of vital components using techniques of bionanotechnology has provided benefits and alternatives to physical and chemical methods. The fundamental, principle, and concept of "Green Synthesis" is based on the twelve standard principles of "Green Chemistry". These principles involve the process of sustainability, saving consumption of energy, low toxic potential of chemical reagents and final products formed, minimum harm to the ecosystem, low risk to global warming, use of resources exploited naturally, and other agrarian wastes generated rationally. The practices and systems of "Green Synthesis" have not only been made pragmatic in the synthesis of many well-known chemical compounds including but not limiting to nanoparticles of metals and nonmetals but also to improve various other materials such as polymers, bioplastics, and aerosols. This has been achieved by adopting new routes of sustainability and using new materials. Physical methods such as ball mill, heating assisted with microwave irradiation, hydrothermal processes when used in combination with precursors of natural origin pose significant importance not only in the greener synthesis but also in the biosynthesis and solventless procedures and techniques. Non-hazardous solvents such as plant extracts, bacteria, viruses, fungi, and yeasts are also part of "Green Synthesis". The chapter highlights the fundamental and principles governing "Green Synthesis".

Keywords

Green synthesis • Green chemistry • Bionanotechnology • Sustainable chemistry

1 Introduction

During the past three decades, rampant exploitation of chemical resources and other metallic and nonmetal agents for faster productivity and low cost have led to the deleterious effect on the ecosystem. This includes utility and exploitation of chemicals to prevent corrosion which is a major challenge in different industries (automobile, pharmaceutical, food, etc.). Studies conducted by Sharma et al. (2009), Sharma and Singh (2011) have explored corrosion inhibitors under different concentrations. The chronic exploitation of chemicals and their associated deleterious effects on ecosystems have necessitated the demand of "Green Synthesis" using the concepts of "Green Chemistry". The application of this emerging science has paved the way for new paradigms to achieve sustainable approaches in the area of production of pharmaceuticals, food and beverages, solvent and paint industry, petroleum (biofuels), and paper industry. In 1991, for the first time, the term "Green Chemistry" was presented to the scientific fraternity. The term was mainly used to describe the processes and methods for reducing or eliminating substances that are detrimental to the ecosystem along with robust scientific approaches for reduction of chemical exposure to the environment and humans. A large number of green processes have been

M. Sharma (✉)
Department of Chemistry, Faculty of Engineering and Technology, Manav Rachna International Institute of Research and Studies, Faridabad, Haryana, India
e-mail: mukta.fet@mriu.edu.in

M. Sharma
Global Drug Safety Department, Win Medicare Pvt. Ltd. 1311, Modi Tower, 98 Nehru Place, New Delhi, India
e-mail: manoj.sharma@winmedicare.com

Inamuddin et al. (eds.), *Advances in Green Synthesis*, Advances in Science, Technology & Innovation,
https://doi.org/10.1007/978-3-030-67884-5_14

widely applied for energy conservation and generation, purification of water, and in the arena of pharmaceuticals.

The hazardous nature in a chemical substance is attributed due to its molecular structure. The different hazards can be categorized as:

- Biohazards (due to biological substances that cause risk to living organisms)
- Hazards due to toxicity of chemicals or drugs (e.g., mortality, cancer, toxic epidermal necrolysis)
- Hazards due to physical properties of chemicals (e.g., explosion, flammability)
- Hazards on ecosphere (e.g., variation in climatic conditions, ozone layer depletion, global warming, etc.)

The principles of "Green Synthesis" are based on the twelve key principles of "Green Chemistry", according to which if these principles are practiced there is either no impact or at least there is a less deleterious influence on living beings and on the ecosystem (Alberto and Ade 2013). Henceforth, a pragmatic approach on "Green Synthesis" needs to be adopted for more robust and effective outcomes (Ribeiro et al. 2010). Although it is difficult for the existence of an absolute "Green Synthesis", however, pragmatically the term "Greener Synthesis" is more rational as an alternative. The description of principles of green chemistry was documented by Anastas et al. (1998). Although the principles laid down are simple, valuable, and useful (Parveen et al. 2016), these lack in describing the detailed processes more particularly the scaling up of chemical processes in industrial settings and the assessment of the financial impact. Pragmatically, the processes involved in "Green Synthesis" can reduce harmful effects on the ecosystem, but they cannot solely preclude the interaction and impact of hazardous substances on the atmospheric surroundings. The principle (s) can be used for the progression and development of systems and processes that not only reduce waste formation but also are more energy efficient and utilize organic solvents that are safer. These standards can be used

- to produce less toxic substances;
- to produce low quantities of by-products and their derivatives.
- to reduce cost of the synthesis of final compounds. This is achieved by reducing precursor loss and formation of intermediary compounds during final compound synthesis;
- to lower chemical hazards by using less toxic and safe alternate chemical substances like extracts from plant origin;
- to minimize energy consumption by using ideal conditions (normal pressure and room temperature);
- to generate sustainable sources of raw material;
- to decrease the generation of waste by evading steps of reaction and additional chemicals;
- to make use of catalyst(s) for formation of products of high yield;
- to prevent contamination through study of reaction intermediates;
- to produce and use small quantities of reactants for prevention of hazards and accidents that may occur due to chemical explosions, leakages etc.

It is worthwhile to note that the green synthesis processes are inexpensive, possess cost effectiveness, and commonly generate products with enhanced quality thereby facilitating non-evasion of any regulations and laws related to the environment.

"Green Synthesis" has importance and applications in organic chemistry as the hazardous organic substances and solvents pose severe damage to the ecosystem. The utility of green techniques in the arena of organic chemistry mainly comprises activation of carbon–hydrogen bond; usage of ionic liquids as green solvents; use of water; reactions conducted using microwaves, ultraviolet waves, and ultrasound (Hang et al. 2012). The main characteristic features of green techniques in the utility and application of organic chemistry mainly include the precise choice of solvent and conduct of reactions assisted with catalyst use. While using green techniques, it is recommended that aromatic solvents containing chlorine shall not be used as these are not only volatile but also toxic and detrimental to the ozone layer. Accordingly, the approach shall be for the alternate non-aqueous, nonvolatile and polar solvents, for example ionic liquids. Other greener techniques include microwave irradiations, organic reactions in carbon dioxide supercritical medium without solvents. In several reactions where toxic benzene is used as a solvent, in such reactions the solvent toluene can act as an alternate. The other green alternative includes use of solvents that are biodegradable (Do 2016). The wide applications of green methods are further extended to synthesize drugs, synthesize medicinal compounds for the purpose of clinical development in the treatment of tropical and other diseases. The term "Green Synthesis" differs from "Biosynthesis" in the sense that while "Biosynthesis" means the usage and application of extracts from plant, microbes, algae, and fungi, the term "Green Synthesis" is mainly to address the concerns related to minimization of deleterious effects of chemicals on the living beings and ecosystem.

In the recent past, the utility of green technologies has mainly been in the synthesis of nanoparticles (metal nature) for pharmaceutical use in the form of targeted drug delivery.

This chapter provides an overview of applicability of fundamentals and principles of "Green Synthesis" in the current research activities using green technology methods.

2 Fundamentals and Principle in "Green Synthesis"

The fundamental and principles of "Green Synthesis" based on different green technology methods are as follows:

(A) **Physical Methods**
- Mechano synthesis using ball mill
- Radiation through non-ionization process (microwave irradiation)
- Synthesis mediated by ultrasound technology
- Synthesis induced by magnetic field
- Synthesis using hydro(solvo)thermal technology

(B) Chemical Methods
- Use of solvents and catalysts

(C) Biological Methods (Biogenesis)

2.1 Physical Methods

In accordance with the requirements of "Green Synthesis" although methods of "Green Chemistry" like mechanical synthesis and microwave techniques are widely used, however, some of these methods like synthesis using magnetic fields are very less common. During past decades novel cleaner synthetic methods have been developed for chemical transformations using the approaches of mechanochemistry. Microwave technology has been applied in the processes involving drying, calcination, sintering, and other chemical processes control. Microwave sintering has shown more effectiveness than conventional sintering, and hence, this technology has been applied to ceramics and sintered powder. Studies on inorganic nanoparticles and ultrafine particles using hydrothermal synthesis processes have further shown the applicability of greener techniques (Chudoba and Wojnarowicz 2018; Darr et al. 2017; Ye et al. 2018). The different physical methods used in the process of "Green Synthesis" mainly include:

2.1.1 Mechano Synthesis Using Ball Mill

Mechanical energy is required to drive chemical reactions efficiently and effectively. This can be achieved by applying mechanical energy using a ball mill. Such type of synthesis achieved using mechanical energy is also known as mechanochemical synthesis or tribochemical synthesis. The concept of mechanochemical synthesis is not only extensively used in organic chemistry for synthesis of compounds but for its applications in processes involving inorganic solid-state as well (Ranu and Stolle 2016; Margetic and Štrukil 2016). The ball mill procedure not only enhances efficiency of energy but also circumvents the involvement of toxic solvents and reagents. Owing to its simplicity and nonhazardous nature to the environment, the ball mill has been widely considered as a good and widely acceptable tool for green chemistry and green synthesis. Further, the mechanochemical synthesis reaction and microwave-irradiated processes do not involve any use of solvents and can also take place at room temperature. Other reactions that can be carried out using ball mill are those reactions which involve the usage of reducing, oxidizing agents for oxidation and reduction purpose; synthesis of polymers; synthesis of coordinate compounds; synthesis of compounds involving coupling reaction using dehydrogenative processes; formation of peptides and amino acids. There are certain reactions in which organic substrates being used are sensitive to the temperature, for example reactions which involve formation of carbon–carbon bond, carbon–nitrogen bond like during synthesis of imines, azines, oximes, imines, guanidine compounds, amine arylation, etc., carbon–oxygen bond, carbon–halogen bond, reactions involving cycloadditions, oxidations, reductions, and other similar reactions. In such types of synthesis reactions, mechanosyntheis is essentially required and is highly useful. An example of widely used mechanosynthesis equipment using ball mill technique is high energy Retsch planetary-type ball mill (ScienceDirect 2016).

2.1.2 Radiation Through Non-ionization Process (Microwave Irradiation)

The use of non-ionization radiation process for heating is the fast, effective, clean, and cheap method used as a green mode of heating in the synthesis of organic and inorganic compounds as it does not require any organic solvents to be used for heating. This is based on the principle of dipolar polarization, conduction, and is widely recognized as a conservative tool for heating in synthetic chemistry (Cravotto et al. 2017). Microwave irradiation techniques are the non-ionizing radiation techniques having no effect on the molecular structure. The electromagnetic energy generated during irradiation gets transformed into heat energy which further pushes interaction amongst compounds. Since microwave irradiation and reactants experience an altered interfacial interaction so a minimum energy is required for the purpose of heating with no need of further escalating the process (Kitchen 2013). Heating through microwave irradiation heavily relies on the dielectric constant of a substance. Accordingly substances like water, methanol, acetone get heated rapidly, whereas substances like aliphatic hydrocarbons, toluene, carbon tetrachloride have differences. Following are the characteristic features of heating through microwave irradiation.

- microwave-heated substances are hotter at interior, and the external surface is cooler;
- microwave heating is a volumetric phenomenon;
- less energy is involved in transfer of heat;
- heating process is fast and prompt due to fast transfer or conduction of heat in the form of microwave radiations;
- heating does not involve any medium requirement;

Heating through the microwave irradiation process is boon in chemical engineering as it facilitates volumetric heating efficiently and effectively thereby reducing further costs, for example, nanoparticles can be produced using different frequencies of microwave in a much shorter time as compared to conventional mode of heating (Leadbeater and McGowan 2010). The generation and transmission of microwaves involve the electrical system and source of microwaves from magnetrons. At small scale, chemical reactions have been carried out using commercially available laboratory-scale microwave reactors at high temperature and pressure. The various applications of microwave heat-mediated synthesis include synthesize of polymers, synthesis of pharmaceuticals, and many other secondary plant metabolites (Bogdal and Prociak 2007; Dworakowska et al. 2012; Kappe et al. 2012).

2.1.3 Synthesis Mediated by Ultrasound Technology

When sound waves possess frequency of more than the human auditory range varying from 20 to 100 kHz, these form ultrasound. The applications of ultrasound are mainly based on the process of cavitation which may be formed in any liquescent medium. At frequencies of 20–100 kHz, the sound waves mainly result in formation, expansion, and implosion (shrinking) of bubbles present in the liquids thereby leading to the process of cavitation. Ultrasound mediated technology is eco-friendly as it rarely causes any damage to raw/fresh materials due to heat. In a recent review conducted by Chen et al. (2019), it has been seen that ultrasound-mediated thermal technology is widely used in the prevention of contamination from microbes in food industry, pharmaceutical industry wherein it has been used as sterilizing agent, disinfectant, for inactivation of enzymes, etc. Ultrasound-mediated technology has emerged as an efficient and green technology for processing of food items that are ready to eat.

Recent advancements in ultrasound technology have been widely used in the environmental and energy applications. In an overview given by Jayaraman et al. (2020), the integration of electrochemistry with ultrasound technology (Sono-electrochemistry) provides many advantages. These advantages mainly include but are not limited to faster rate of reaction, increased activation of surfaces.

2.1.4 Synthesis Induced by Magnetic Field

The integration of chemical reduction reaction with external source of magnetic field is the basis of synthesis mediated through magnetic field. The synthesis using magnetic fields heavily relies on various parameters including but not limited to temperature of the ongoing reaction, pH of the entire reaction, magnetic field strength, surfactants presence, and other solvents. These factors exert an impact on physical, chemical, and mechanical properties of wire-like materials and alter their chemical, physical, and mechanical properties (Krajewski 2017). Currently, this has served as an alternative to traditional synthesis methods which require harmful solvents that are toxic in nature. The traditional methods can also require extra steps for synthesis and may lead to generation of unsolicited substances, residues. During synthesis of desirable substances, the additional steps required in the synthesis can be minimized using the technology of magnetic field. The progress of formation of desired substances can be affected using magnetic field-assisted technology. Although this mode of synthesis has many advantages, however, owing to its limitation related to magnetic field, it can be applied to the substances where compounds as reactants and products formed possess magnetic properties.

In a study conducted by Samadi et al. (2018), the nanoparticles of cobalt ferrite were synthesized using magnetic field-assisted technology. The methodology involved co-precipitation in a reverse mode and further analysis of magnetic and other structural characteristics using electron microscope (field emission), diffractometer, and vibrating sample magnetometer. The outcome of the study revealed that the powder with larger unit cells gets produced in the presence of an external field and not in the nonexistence of any external field.

In the study, there was an accrual in the coercivity and magnetization. No significant changes were seen in morphology and particle size.

In another study by Xia et al. (2015), the magnetic characteristics, surface, and shape of ferric oxide nanoparticles were altered using magnetic fields during thermal decomposition.

Nanoparticles of $Bi_6Fe_2Ti_3O_{18}$ (BFTO) have also been synthesized using hydrothermal methods. The methods were adopted in a large magnetic field as this can affect the growth behavioral characteristics (Liu 2020).

The compounds containing cobalt, nickel, and bismuth can facilitate the synthesis through magnetic fields as these have ferromagnetic characteristics (Wu et al. 2010; Hu and Sugawara 2009).

Nanoparticles have been generated using solvothermal procedures mediated with magnetic fields (Yang et al. 2020). This is regarded as a green synthesis method as it reduces generation of waste. It is worthwhile to mention that the magnetic field application does not produce any supplementary product.

2.1.5 Synthesis Using Hydro(Solvo)thermal Technology

The synthesis using hydro(solvo)thermal technology involves heating of solvents in closed vessels. The solvents in these sealed vessels are heated using pressure technology, for example in autoclaves (Li et al. 2015). The procedure is termed as "hydrothermal" when water is used as a solvent and "solvothermal" when organic compound is used as a solvent. A thick walled cylinder made of stainless steel is used as an autoclave.

The autoclaves work differently depending on the type of material which may include alloys of high strength, material made from quartz, glass. From the purpose of "Green Chemistry", hydrothermal reactions have been regarded as more appropriate and eco-friendly as these can be widely used to produce different materials. In solvothermal processes, organic solvents for example amine, methanol, and toluene are largely used.

For preparation of nanomaterials, the methods using technology of hydrothermal and solvothermal processes are widely used as these synthesis methods produce large quantities of high crystalline nanomaterials at a low cost. These synthesis methods can be integrated with magnetic fields and microwaves.

2.1.6 Photocatalysis

This is one of the methods of "Green Synthesis" that involve chemical reaction occurring in the presence of light and a semiconductor that enhances the rate of reaction in its presence (photocatalyst). The process of photocatalysis has applications in removal of cyanotoxins from polluted water bearing small and large permeable rocks (aquifer). Owing to its capability, the process of photocatalysis has got many applications like deodorization, fogg removal, antibacterial, etc. (Crisenza Giacomo and Melchiorre 2020).

Photocatalytic reactions are also used in the synthesis of organic compounds. The catalytic phenomenon in photoredox reactions facilitates sustainability as it fulfills principles of Green Chemistry. In a photoredox reaction, the primary energy source is obtained from light energy which is not only free of cost but is also nonhazardous to nature and highly energy efficient. Unlike the thermal activation process where high temperature or other harsh conditions are required to achieve the desired reaction process, photons provide sufficient energy to attain the desired reactivity without such conditions. The photocatalysts are light-absorbing species that are used in low quantity (catalytic reagent). When these photocatalysts reach an excited state, they stimulate the transfer of a single electron to form stable compounds, for example, a large number of heterocyclic compounds of nitrogen have been synthesized by photo-oxygenation of furan derivatives (Crisenza Giacomo and Melchiorre 2020).

2.2 Chemical Methods

2.2.1 Use of Solvents and Catalysts

For green synthesis, it is imperative to make use of materials in the form of solvents and catalysts which do not pose risk to the environment. The role of catalysts is significant in reducing environment pollution as these are known to reduce usage of organic compounds that are volatile (Menges 2018). A large number of waste compounds and other by-products can be eliminated using catalytic methods. There are two types of catalytic reactions, homogenous catalysis, and heterogeneous catalysis which may be used on the basis of type of reaction needed. A homogeneous catalytic reaction is a reaction involving liquid/liquid (single phase) phase and heterogeneous catalytic reaction is bi-phasic or multi-phasic. Homogenous catalysis saves energy by lowering down temperature conditions. Catalysts are compounds of very expensive metals, and therefore, any significant loss would be very expensive to purchase.

Like catalysts, solvents used for green synthesis must possess the characteristics of biodegradability, low toxic potential, recyclability, and water immiscibility. The commonly used solvents include polyethylene glycols, water, and ionic liquids. Glycerol is the widely used solvent as it is biodegradable, nontoxic, readily available, cheap, high boiling point, possess low vapor pressure, high polarity, has capacity to get solubilized with both organic and inorganic compounds. Other solvents include supercritical carbon dioxide, supercritical water which is widely used in synthesis of polymers. The technology of supercritical fluid is widely used in processes such as cleansing, polymerization, and extraction.

The concept of green synthesis also utilizes dry media to synthesize green compounds. The processes such as heating, microwave technique, and ultrasound reduce contamination. These are eco-friendly and can be scaled up for green synthesis reactions.

2.3 Biological Methods

The biological synthesis methods include usage of microorganisms, plants, and animals for their capabilities and capacity to synthesize nanoparticles (metallic) using "Green Synthesis" techniques. The natural sources (microorganisms, plants, and animals) have been evaluated for the "Green Synthesis" of diverse metallic nanoparticles without involving any toxic chemical and other hazardous material. In last decades, there has been significant development in the production of nanoparticles using techniques of biosynthesis (Venkateswarlu et al. 2013). Green synthesis methods based on biological methods not only make use of viruses, bacteria, and yeasts but also make use of secondary plant metabolites extracted from seed, root, leaf, and stem as these behave as reducing agents or as stabilizing agents (Yallappa et al. 2015). Several forms of algae are also used as these possess capabilities for removal of toxic metals thereby converting them into more usable forms (Patel et al. 2014).

The rate of success of green synthesis process using microbes and animals is not that attractive as compared to the method adopted using plant extracts. Research activities conducted with plants of varying taxonomic groups have elicited capabilities of these plants for capping and green synthesis (St. Angelo and Hartz 2012; Punuri et al. 2011). In addition to plants, green synthesis potential has been exhibited by biomolecules derived from animal sources such as silk, chitosan, and alginates (Aramwit et al. 2014; Hemlata et al. 2020).

Owing to easy availability, usage in treatment of human and animal malignancies, and edible use, angiosperms have been largely used for green synthesis of metallic nanoparticles (Petrovska 2012). Their use has also avoided the use of hazardous and toxic chemical substances such as citrate of trisodium and sodium borohydride. Plants detoxify higher concentration of copper which is required as a micronutrient. Plants detoxify copper by a reduction process involving reduction of copper ions into copper neutral atoms, and copper is required as micronutrient for plants. Higher concentration of copper is detoxified by plants by the reduction of cuprous ions (Cu^{2+}) into neutral atoms and successively into nanoparticles of copper (Manceau et al. 2008). From the formation of nanoparticles, it can be established that type of plant and extract of its type of part, pH of synthesis process, heavily affects quantity, size, and morphology of nanoparticles. Among pteridophytes, the pteris genus plants were studied by De Britto et al. (2012) in reference to synthesis of silver nanoparticles. The maximum antibacterial properties were shown by *Pteris biaurita* silver nanoparticles. The antibacterial properties have also been seen in nanoparticles synthesized from fern *Nephrolepi sexaltata* (Das et al. 2020).

Bryophytes also produce compounds that are biologically active and in a way similar to pteridophytes safeguard themselves from other living organisms. These active compounds help bryophytes in protection. Algae are macroscopic or microscopic, multicellular or unicellular organisms mainly present on moist surface, sea water or fresh water (Dahoumane et al. 2017). These can be categorized in different kingdoms as plantae, bacteria, protozoa, and chromista. Algae of different types have been screened for green synthesis of nanoparticles of gold (Senthilkumar et al. 2019; Rahman et al. 2020). The mechanism of action for synthesis of nanoparticles by algae involves cationic accumulation and reduction. In a study conducted by Khanna et al. (2019) on applications, characterization, and synthesis of algae-based nanoparticles, the species of saragassum (*Sargassum* spp.) and chorella (*Chorella* spp.) were widely experimented to develop nanoparticles of zinc oxide, silver, and gold. The role of algal enzymes has also been explored for biosynthesis of marine caretonids. In a study conducted by Dautermann et al. (2020), it was seen that biosynthesis of fucoxanthin was catalyzed by violaxanthinde-epoxidase (VDL). The enzyme violaxanthin de-epoxidase is an enzyme that exhibits photoprotective action and functions mostly in algae and plants.

Synthesis of metallic nanoparticles using microbes is also one of the approaches in green synthesis. In biotechnology, the interaction of metal with microbes is well recognized vide processes such as biocorrosion, bioremediation, biomineralization, and bioleaching (Joshi et al. 2014; Haferburg and Kothe 2007). For synthesis, it is imperative that metallic toxicity shall be overcome by bacteria. The metallic nanoparticles and nanocrystals exhibit the same characteristics whether synthesized using chemical processes or microbes.

3 Application of "Green Synthesis"

3.1 Synthesis of Metal Salts, Metal Complexes, and Metal Organic Frameworks

The application of "Green Synthesis" methods have been realized in synthesis of organic compounds using green processes as these adhere to the twelve rules of green chemistry (Unterlass 2016; Deshmukh and Bhanage 2018; Chen 2018). For example, microwave treatment technique has been used in obtaining high yield of benzimidazole derivatives in a short span of time. Green synthesis techniques have also resulted in development of compounds for pharmacological action, for example thalidomide, pyrimidine derivatives, and methyl nitroacetate (Liu et al. 2018; Benjamin and Hijji 2017). The characteristic features of these processes include milder conditions and high yields.

"Inorganic–organic hybrid materials" are inorganic building blocks present in an organic, polymeric, and matrix in the colloidal form. The salient features of these hybrid molecules involve the presence of considerably less amount of inorganic polymeric component than that of the organic component. Owing to features related to conductivity, magnetic properties, catalytic or redox activities, and enriched mechanical strength, these compounds can considerably be used in improving current technologies for membranes, fuel cells, and electronic devices. The principles of green chemistry which are appropriate for organic synthesis can be largely applied in the synthesis of hybrid materials also. Thus, the biodegradable and biocompatible surfactants are required to make the hybrid materials of organic and inorganic components compatible. Green methods have been analyzed separately for organic and inorganic components. Renewable resources can be used in the case of inorganic colloids, whereas organic components can be used using green routes and other renewable sources (Choudhary et al. 2013). In accordance with green chemistry principles, the essential parameters required in the synthesis of metallic organic frameworks mainly include: (i) solvent, (ii) sources of cation, and (iii) synthesis conditions such as temperature and pressure of the reactor. During synthesis of metallic organic frameworks, all precursors shall be used in maximum considering that any waste or side product may be reused. For this, oxides of metal or lesser sulfate hydroxides are widely used as metal sources thereby avoiding loss of toxic anions in the form of chlorides, nitrates or perchlorates.

3.2 Synthesis of Metallic Nanoparticles

In the synthesis of nanoparticles through green synthesis methods, capping agents are mainly used for prevention of aggregation of nanoparticles, to control shape of nanoparticles and stabilization. The examples of capping agents used in the nanochemistry mainly include long-chain hydrocarbons, polymers such as polyvinyl alcohol, polyethylene glycol, polyacrylic acid, and polystyrene. For synthesis of nanoparticles, following agents are considered as greener : (i) polysaccharides such as starch, dextran with mild capping ability, and water solubility characteristics as these allow escaping toxic solvents and further allow easy separation of nanoparticles from reaction media; (ii) biomolecules such as peptides and proteins possessing high biocompatibility; and (iii) other small molecules which act as capping agents. The natural products used during the synthesis of nanoparticles and nanomaterials can be applied as surfactants, capping agents, solvents, reactants, carriers, catalysts, and templates. The ligands are used to coat the surface of nanoparticles to stabilize them and prevent agglomeration. The ligands used in the nanoparticle synthesis include alkynyls, thiols, amines, phosphines, and carbenes. Low and large-scale synthesis of nanoparticles can be carried out using water and supercritical fluids such as carbon dioxide. Green methods such as microwave- and ultrasound-assisted synthesis are required to synthesize larger nanoparticles. The ultrasonic treatment allows formation of small nanoparticles and maintenance of the same without any further agglomeration. After synthesis and preparation, nanoparticles thus formed are separated from the reaction medium using precipitation and washing method (Gour and Jain 2019).

3.3 Synthesis of Elemental Nanoparticles of Nonmetals

The synthesis of elemental nanoparticles of nano-metals can be done using a greener route involving allotropic forms of carbon nanosize, nanodots, and nanotubes (Zahid et al. 2018). The techniques such as microwave, hydrothermal, and pyrolysis have been mainly used in the synthesis of nanoparticles. Microwave pyrolysis method has been used in a study conducted on the green synthesis of nitrogen-doped carbon dots from sesame seeds (Roshni and Divya 2017). In the study, the characterization of carbon dots was done using techniques such as fluorescence spectroscopy, visible spectroscopy, and Fourier transform infra-red spectroscopy (FTIR) techniques. The results in the study indicated carbon dots that were highly fluorescent, photostable, and soluble in aqueous solvents. The study demonstrated an environment and cost friendly, waste-recyclable synthetic method for preparation of carbon dots.

4 Nanoparticles of Metal and Nonmetal Oxide

The nanoparticles are usually the oxides of metals. The green methods employed in their formation include vapor phase oxidation method, hydrothermal, polyol method, condensation for nanoparticles of zinc oxide; co-precipitation method, thermal decomposition method for nanoparticles of magnetite; spray pyrolysis, heating through laser, hydrothermal methods for nanoparticles of indium oxide (Singh 2018; Kalpana and Rajeswari 2018). In processes including the ones where plant extracts are used, the formation of hydroxides occurs first followed by decomposition by annealing/calcination.

5 Conclusion

Green synthesis methods have necessitated the importance of alternate to chemical and toxic substance use. The methods based on the principle of green synthesis have exhibited large potential in the synthesis of nanoparticles which are used as novel agents in treatment of various ailments. For exploitation of various methods of "Green Synthesis", it is imperative to understand the various underlying principles and mechanisms involved in the process. The green processes such as microwaves, ultrasound, plant-based extracts, and biosynthesis processes involving a wide variety of microorganisms have resulted in generation of high yield of nanoparticles.

Microwave irradiation method enhances the kinetics of reaction. Reducing agents such as starch and glucose can also be further used in synthesis of nanoparticles. Nanoparticles synthesized using extracts from plants mainly use aqueous solvent rather than any chemical solvent. Ultrasound-assisted techniques facilitate nanoparticle synthesis in a controlled manner. Different varieties of microorganisms exert enzymatic reactions and act as a catalyst in the nanoparticle synthesis. Biosynthesis methods are advantageous as they are cheap, economical, have low requirements for energy, and do not require high pressure. These methods mainly make use of algae, fungi, viruses, bacteria, yeast, and extracts of plants thereby acting as an ideal source of "Green Synthesis". Biosynthesis method reduces the toxicity associated with a metal compound with the help of enzymes present in the microbes. The fundamentals and principles discussed have a wide role in reducing carbon footprints which is the need of hour in the circumstances of ozone layer depletion and to lessen foot prints of carbon.

Acknowledgements The authors acknowledge Ms. Sasha Raina, Student Department of Biology, Rutgers University, New Brunswick, USA, for proofreading the chapter.

Conflict of Interest

None.

References

Alberto C, Ade M (2013) Environmental sustainability: implications and limitations to Green Chemistry. Found Chem 16:125–147. https://doi.org/10.1007/s10698-013-9189-x

Anastas PT, Warner JC (1998) Principles of green chemistry. In: Anastas P, Warner J (eds) Green chemistry: theory and practice. Oxford University Press, Oxford, pp 29–56

Aramwit P, Bang N, Ratanavaraporn J, Ekgasit S (2014) green synthesis of silk sericin-capped silver nanoparticles and their potent anti-bacterial activity. Nanoscale Res Lett 9:79. https://doi.org/10.1186/1556-276X-9-79

Benjamin E, Hijji YM (2017) A novel green synthesis of thalidomide and analogs. J Chem 2017:6436185. https://doi.org/10.1155/2017/6436185

Bogdal D, Prociak A (2007) Microwave enhanced polymer chemistry and technology. Blackwell Publishing, Hoboken

Chen DY (2018) A personal perspective on organic synthesis: past, present, and future. Isr J Chem 58:85–93. https://doi.org/10.1002/ijch.201700113

Chen F, Zhang M, Yang C-H (2019) Application of ultrasound technology in processing of ready-to-eat fresh food: a review. Ultrason Sonochem 63:104953. https://doi.org/10.1016/j.ultsonch.2019.104953

Choudhary MK, Kataria J, Bhardwaj VK, Sharma S (2013) Nanoscale advances Green biomimetic preparation of efficient Ag–ZnO heterojunctions with excellent photocatalytic performance under solar light irradiation: a novel. Nanoscale ADV 1:1035–1044. https://doi.org/10.1039/c8na00318a

Chudoba T, Wojnarowicz J (2018) Current trends in the development of microwave reactors for the synthesis of nanomaterials in laboratories and industries: a review. Crystals 8:379. https://doi.org/10.3390/cryst8100379

Cravotto G, Carnaroglio D (2017) Microwave chemistry; De Gruyter Textbook; De Gruyter, Berlin. ISBN 978-3-11-047993-5

Crisenza Giacomo EM, Melchiorre P (2020) Chemistry glows green with photoredox catalysis. Nat Commun 11:803. https://doi.org/10.1038/s41467-019-13887-8

Dahoumane SA, Jeffryes C, Mechouet M, Agathos SN (2017) Biosynthesis of inorganic nanoparticles: a fresh look at the control of shape, size and composition. Bioengineering (Basel, Switzerland) 4(1):14. https://doi.org/10.3390/bioengineering4010014

Darr JA, Zhang J, Makwana NM, Weng X (2017) Continuous hydrothermal synthesis of inorganic nanoparticles: applications and future directions. Chem Rev 117:11125–11238. https://doi.org/10.1021/acs.chemrev.6b00417

Das G, Patra JK, Shin H (2020) Biosynthesis, and potential effect of fern mediated biocompatible silver nanoparticles by cytotoxicity, antidiabetic, antioxidant and antibacterial, studies. Mater Sci Eng C https://doi.org/10.1016/j.msec.2020.111011

Dautermann O, Lyska D, Andersen-Ranberg J, Becker M, Fröhlich--Nowoisky J, Gartmann H, Krämer LC, Mayr K, Pieper D, Rij LM, Wipf HM-L, Niyogi KK, Lohr M (2020) An algal enzyme required for biosynthesis of the most abundant marine carotenoids. Sci Adv 6 (10):eaaw9183. https://doi.org/10.1126/sciadv.aaw9183

De Britto AJ, Gracelin DHS, Kumar PBJR (2012) Biogenic silver nanoparticles by *Adiantum caudatum* and their antibacterial activity. Int J Univ Pharm Life Sci 2(4):92–98

Deshmukh DS, Bhanage BM (2018) Molecular iodine catalysed benzylic sp3 C-H bond amination for the synthesis of 2-arylquinazolines from 2-aminobenzaldehydes,2-aminobenzophenones and 2-aminobenzyl alcohols. Synlett 29:979–985. https://doi.org/10.1055/s-0037-1609200

Do J (2016) Mechanochemistry: a force of synthesis. ACS Cent Sci 3:13–19. https://doi.org/10.1021/acscentsci.6b00277

Dworakowska S, Bogdał D, Prociak A (2012) Microwave-assisted synthesis of polyols from rapeseed oil and properties of flexible polyurethane foams. Polymers 4:1462–1477

Gour A, Jain NK (2019) Advances in green synthesis of nanoparticles. Artif. Cells. Nanomed. Biotechnol. 47:844–851. https://doi.org/10.1080/21691401.2019.1577878

Haferburg G, Kothe E (2007) Microbes and metals: interactions in the environment. J Basic Microbiol 47(6):453–467. https://doi.org/10.1002/jobm.200700275

Hang W, Cue BW, Wiley J (2012) Green techniques for organic synthesis and medicinal chemistry. Wiley, Hoboken

Hemlata PRM, Pratap Singh A, Tejavath KK (2020) Biosynthesis of silver nanoparticles using Cucumis prophetarum aqueous leaf extract and their antibacterial and antiproliferative activity against cancer cell lines. ACS Omega 5(10):5520–5528. https://doi.org/10.1021/acsomega.0c00155

Hu H, Sugawara K (2009) Magnetic-field-assisted synthesis of Ni nanostructures: selective control of particle shape. Chem Phys Lett 477:184–188. https://doi.org/10.1016/j.cplett.2009.06.085

Jayaraman T, Madhavan J, Lee SJ, Choi M, Ashokkumar M, Pollet B (2020) Sonoelectrochemistry for energy and environmental applications. Ultrason Sonochem 63:104960. https://doi.org/10.1016/j.ultsonch.2020.104960

Joshi SR, Kalita D, Kumar R, Nongkhlaw M, Swer PB (2014) Metal–microbe interaction and bioremediation. In: Gupta D, Walther C (eds) Radionuclide contamination and remediation through plants. Springer, Cham

Kalpana VN, Rajeswari VD (2018) A review on Green synthesis, biomedical applications, and toxicity studies of ZnO NPs. Bioinorg Chem Appl 2018:3569758. https://doi.org/10.1155/2018/3569758

Kappe CO, Stadler A, Dallinger D (2012) Microwaves in organic and medicinal chemistry, vol 52. Wiley-VCH Verlag, Weinheim
Khanna P, Kaur A, Goyal D (2019) Algae-based metallic nanoparticles: Synthesis, characterization and applications. J Microbiol Methods 163:105656. https://doi.org/10.1016/j.mimet.2019.105656
Kitchen HJ et al (2013) Modern microwave methods in solid-state inorganic materials chemistry: from fundamentals to manufacturing. Chem Rev 114:1170–1206. https://doi.org/10.1021/cr4002353
Krajewski M (2017) Magnetic-field-induced synthesis of magnetic wire-like micro- and nanostructures. Nanoscale 9(43):16511–16545. https://doi.org/10.1039/c7nr05823c
Leadbeater NE, McGowan CB (2010) Laboratory experiments using microwave heating. CRC Press, Boca Raton
Li J, Wu Q, Wu J (2015) Synthesis of nanoparticles via solvothermal and hydrothermal methods. In: Aliofkhazraei M (eds) Handbook of nanoparticles. Springer, Berlin
Liu Q, Liu Q, Liu B, Hu T, Liu W, Yao J (2018) Green synthesis of tannin hexamethylendiamine based adsorbents for efficient removal of Cr (VI). J Hazard Mater 352:27–35. https://doi.org/10.1016/j.jhazmat.2018.02.040
Manceau A, Nagy KL, Marcus MA et al (2008) Formation of metallic copper nanoparticles at the soil-root interface. Environ Sci Technol 42(5):1766–1772. https://doi.org/10.1021/es072017o
Margetic D, Štrukil V (2016) Mechanochemical organic synthesis, 1st edn. Elsevier, Amsterdam
Menges N (2018) The role of green solvents and catalysts at the future of drug design and of synthesis. https://doi.org/10.5772/intechopen.71018
Parveen K, Banse V, Ledwani L (2016) Green synthesis of nanoparticles: their advantages and disadvantages. AIP Conf Proc 1724:20048. https://doi.org/10.1063/1.4945168
Patel V, Berthold D, Puranik P, Gantar M (2014) Screening of cyanobacteria and microalgae for their ability to synthesize silver nanoparticles with antibacterial activity. Biotechnol Rep (Amst) 5:112–119. Published 5 Dec 2014. https://doi.org/10.1016/j.btre.2014.12.001
Petrovska BB (2012) Historical review of medicinal plants' usage. Pharmacognosy Rev 6(11):1–5. https://doi.org/10.4103/0973-7847.95849
Punuri J, Das RK, Kumar A (2011) Microwave-mediated synthesis of gold nanoparticles using coconut water. Int J Green Nanotechnol 3:13–21. https://doi.org/10.1080/19430892.2011.574534
Rahman A, Kumar S, Nawaz T (2020) Biosynthesis of nanomaterials using algae. https://doi.org/10.1016/B978-0-12-817536-1.00017-5
Ranu B, Stolle A (2016) Ball milling towards Green synthesis: applications, projects, challenges. Johnson Matthey Technol Rev 60:148–150. https://doi.org/10.1595/205651316X691375
Ribeiro MGTC, Costa DA, Machado AASC, Ribeiro MGTC, Costa DA, Machado AASC (2010) Green Chemistry letters and reviews 'Green Star': a holistic Green Chemistry metric for evaluation of teaching laboratory experiments. Green Chem Lett Rev 3:149–159. https://doi.org/10.1080/17518251003623376
Roshni V, Divya O (2017) One-step microwave-assisted green synthesis of luminescent N-doped carbon dots from sesame seeds for selective sensing of Fe (III). Curr Sci 112:385–390. https://doi.org/10.18520/cs/v112/i02/385-390
Samadi M, Shokrollahi H, Zamanian A (2018) The magnetic-field-assisted synthesis of the co-ferrite nanoparticles via reverse co-precipitation and their magnetic and structural properties. Mater Chem Phys 215. https://doi.org/10.1016/j.matchemphys.2018.05.067
ScienceDirect (2016) Ball Mill. See https://www.sciencedirect.com/topics/chemistry/ball-mill. Accessed on 29 May 2020
Senthilkumar P, Surendran L, Sudhagar B et al (2019) Facile green synthesis of gold nanoparticles from marine algae *Gelidiella acerosa* and evaluation of its biological Potential. SN Appl Sci 1:284. https://doi.org/10.1007/s42452-019-0284
Sharma M, Singh G (2011) Effect of Brij35 on mild steel corrosion in acidic medium. Indian J Chem Technol 18:351–356
Sharma M, Chawla J, Singh G (2009) Cetyltrimethylammonium bromide as corrosion inhibitor for mild steel in acidic medium. Indian J Chem Technol 16:339–343
Shengman L (2020) Structural and magnetic properties of high magnetic-field-assisted hydrothermal synthesized $Bi_6Fe_2Ti_3O_{18}$ particles. Mod Phys Lett B 34(03). https://doi.org/10.1142/S0217984920500438
Singh A (2018) Zinc oxide nanoparticles: a review of their biological synthesis, antimicrobial activity, uptake, translocation and biotransformation in plants. J Mater Sci 53:185–201. https://doi.org/10.1007/s10853-017-1544-1
St. Angelo S, Hartz E (2012) Ginkgo as a green reducing agent for gold nanoparticles and nanoplatelets. Int J Green Nanotechnol 4:111–116. https://doi.org/10.1080/19430892.2012.678706
Unterlass MM (2016) Green synthesis of inorganic–organic hybrid materials: state of the art and future perspectives. Eur J Inorg Chem 2016:1135–1156. https://doi.org/10.1002/ejic.201501130
Venkateswarlu S, Yakkate S, Balaji T, Prathima B, Jyothi NV (2013). Biogenic synthesis of Fe_3O_4 magnetic nanoparticles using plantain peel extract. Mater Lett 100:241–244. https://doi.org/10.1016/j.matlet.2013.03.018
Wu M, Liu G, Li M, Dai P, Ma Y, Zhang L (2010) Magnetic field-assisted solvo thermal assembly of one-dimensional nanostructures of Ni–Co alloy nanoparticles. J Alloys Compd 491:689–693. https://doi.org/10.1016/j.jallcom.2009.10.273
Xiao W, Liu X, Hong X, Yang Y, Lv Y, Fang J, Ding J (2015) Magnetic-field-assisted synthesis of magnetite nanoparticles via thermal decomposition and their hyperthermia properties. CrystEngComm 17. https://doi.org/10.1039/C5CE00442J
Yallappa S, Manjanna J, Dhananjaya BL (2015) Phytosynthesis of stable Au, Ag and Au–Ag alloy nanoparticles using *J. sambac* leaves extract, and their enhanced antimicrobial activity in presence of organic antimicrobials. Spectrochim Acta A Mol Biomol Spectrosc 137:236–243. https://doi.org/10.1016/j.saa.2014.08.030
Yang Z, Yin Z, Li J, Zhao Z, Yu J, Ren Z, Yu G (2020) Magnetic-field-assisted solvothermal synthesis and magnetic properties of Fe-doped CeO_2 nanoparticles. J Asian Ceramic Societies. https://doi.org/10.1080/21870764.2020.1769815
Ye N, Yan T, Jiang Z, Wu W, Fang T (2018) A review: conventional and supercritical hydro/solvo thermal synthesis of ultrafine particles as cathode in lithium battery. Ceram Int 44:4521–4537. https://doi.org/10.1016/j.ceramint.2017.12.236
Zahid MU, Pervaiz E, Hussain A, Shahzad MI, Bilal M, Niazi K (2018) Synthesis of carbon nanomaterials from different pyrolysis techniques: a review. Mater Res Exp 5:052002. https://doi.org/10.1088/2053-1591/aac05b

Electrochemical Green Synthesis

N. Suresh Kumar, R. Padma Suvarna, K. Chandra Babu Naidu, H. Manjunatha, A. Ratnamala, and M. Ajay Kumar

Abstract

Nowadays, numerous green processes are being employed for purification of water, generation of energy, fabrication of various materials, etc. The main theme of the green process is eliminating or reducing hazardous properties such as flammability, toxicity, change of climate conditions and overheating through modifying the structure of the chemical substances. Further, in green synthesis, the shape and morphology of the nanoparticles can be regulated through employing plant extracts. In addition, the consistent properties can be achieved by controlling the morphology at the time of biosynthesis process which influences their optical, electrochemical and physical properties. This chapter is mainly focused on electrochemical properties of recently synthesized materials by green synthesis approach. In addition, the applications of prepared materials in supercapacitors, electrochemical hydrogen storage and the role of electrochemical green synthesis in fabrication of dopamine sensors are also discussed.

Keywords

Green synthesis • Nanoparticles • Biodegradability • Supercapacitors • Hydrogen storage

N. Suresh Kumar (✉) · R. Padma Suvarna
Department of Physics, JNTUA College of Engineering, Anantapuramu, AP 515002, India

K. Chandra Babu Naidu (✉)
Department of Physics, GITAM Deemed to be University, Bangalore, Karnataka 562163, India

H. Manjunatha · A. Ratnamala
Department of Chemistry, GITAM Deemed to be University, Bangalore, Karnataka 562163, India

M. Ajay Kumar
Department of ECE, GITAM Deemed to be University, Bangalore, Karnataka 562163, India

1 Introduction

In science and technology, one of the major research topics receiving the tremendous advancement in recent years is nanotechnology. In general, the term nanotechnology is referred to as a branch of science which deals with the study of materials in nano-range, usually between 1 and 100 nm. The materials which are in nanoscale consist of large surface area to volume ratio and show high chemical stability, improved catalytic activity, thermal conductivity, electrical conductivity, etc. (Agarwal et al. 2017) Owing to distinct physicochemical properties, the nanomaterials have great potential to introduce newfangled devices, structures and nanoplatforms with imminent bids in diverse fields of science such as pharmaceuticals, bio-engineering and dentistry (Mirzaei and Darroudi 2017; Arruda et al. 2015; Rafique et al. 2017). These superior properties gained much attention by many researchers to discover novel synthesis techniques for the preparation of nanomaterials. However, based on the precursors, overall preparation techniques are categorized into two types which are top-down and bottom-up techniques. The examples for top-down techniques are ball milling, grinding, laser ablation, sputtering, etc., in which appropriate bulk material is shattered down to fine small particles with the help of size reduction process. Similarly, the examples for bottom-up technique are electrochemical methods, chemical reduction, hydrothermal, solvothermal, sol–gel, etc., in which the nanoparticles are prepared with the help of biological and chemical reactions, where the nanoparticles are grown by self-assembly of atoms. The production of substantial amount of nanoparticles within very short time is one of the greatest advantages of the conventional synthesis techniques. But in this synthesis method, in order to maintain stability, the toxic chemicals are used which cause toxicity in the environment. In this regard, the researchers from all over the world focused on synthesizing the nanoparticles by green synthesis technique. Usually, the green synthesis can be partly accomplished by

Inamuddin et al. (eds.), *Advances in Green Synthesis*, Advances in Science, Technology & Innovation,
https://doi.org/10.1007/978-3-030-67884-5_15

replacing the toxic organic solvents with non-toxic organic solvents or water. Any reactions, chemicals and processes which are hazardous to environment and humans are completely evaded in green synthesis (Salam et al. 2014). Therefore, the green synthesis is a good choice for the preparation of nanoparticles because in green synthesis, the non-toxic and ecofriendly nanoparticles can be prepared from natural extracts. Hence, emerging green technique is important for the prospect of nanomaterials in future. In this chapter, we have focused on recent developments in electrochemical green synthesis of materials along with their applications in various fields.

2 Green Synthesis

The term green approach came into light in the year 1991, which was designed to decrease the exposure of humans and environment from hazardous and toxic substances or chemicals. Nowadays, numerous green processes are being employed for purification of water, generation of energy, fabrication of electronics, etc. The main theme of the green process is eliminating or reducing hazardous properties such as flammability, toxicity, change of climate conditions and overheating through modifying the structure of the chemical substances. So, green synthesis involves the preparation of materials without/at least reducing the harmful effect on environment and human health. Generally, green synthesis can be achieved by.

(i) utilization of microorganisms such as yeasts, bacteria, fungi, etc.,
(ii) utilization of plants of plant extracts and
(iii) utilization of templates like viruses, deoxyribonucleic acid (DNA), membranes, diatoms, etc.

Furthermore, ecofriendliness, cost effectiveness, preventing unnecessary wastes, lower hazard chemical reactions, renewable sources, biodegradability, etc., are considered as advantages of green approach over other techniques (Alberto and Ade 2013; Ribeiro et al. 2010). Owing to all these reasons, the green synthesis gained tremendous attention in material synthesis.

3 Electrochemical Green Synthesis

Nowadays, advancement in electronic technology and deeper dependence of people on compact electronic devices such as hybrid-electric vehicles, mobiles and laptop require more energy to consume the natural sources such as fossil fuels cannot satisfy these excessive energy demands. This also causes energy deficiency and environment pollution. The introduction of suitable energy storage devices becomes most necessary to fulfill the current energy requirements (Shen et al. 2016). In this concern, the electrochemical energy storage has been getting much attention for promising applications in electric vehicles and renewable energy systems from intermittent solar and wind resources. Owing to their high power and energy densities, outstanding cyclic stability, quick charge/discharge rates, environmental friendliness, etc., the supercapacitors have considerable potential for using next generation electrochemical energy storage devices (Lukatskaya et al. 2016). However, the component materials, exclusively the materials are utilized in their electrodes which can greatly influence the performance of the supercapacitors. Conventional techniques (hydrothermal, solvothermal, sol–gel, etc.) are being used to fabricate electrode materials in nano-range; nevertheless, these methods have their own disadvantages such as toxicity and external additives during which the reaction is considered as damaging to environment and health (Besner et al. 2008; Kumar and Yadav 2009). Thus, the researchers have been putting continuous efforts from last decade to attain new production techniques with low impact on environment (Paul et al. 2016).

In recent years, to overcome the drawbacks of traditional synthesis techniques, researchers focused on the fabrication of nanomaterial thru green synthesis which is ecofriendly and economical (Suvith and Philip 2014). Further, in green synthesis, the shape and morphology of the nanoparticles can be organized by employing medicinal plant extracts. In addition, the consistent properties can be achieved by controlling the morphology at the time of biosynthesis process which influences their optical, electrochemical and physical properties. In this chapter, electrochemical properties of recently synthesized materials by greener approach and their supercapacitor and electrochemical hydrogen storage applications have been discussed.

In this connection, by using the by-products of orange such as orange peel and extract of orange juice as bases for biological antioxidants (e.g., flavonoids, ascorbic acid, pectins and phenolic compounds), Abuzeid et al. (2019) prepared the nanosized α-MnO_2 materials via cost-effective green synthesis technique for the first time for supercapacitor electrode applications. They reported that the X-ray diffraction (XRD) analysis confirms the α-MnO_2 structures with some secondary phases. Transmission electron microscope (TEM) images disclosed the nanosized nature of the prepared materials. Further, the distribution of pore size and N_2 adsorption and desorption of isotherms of the prepared compounds orange peel MnO_2 (OP-MnO_2) and orange juice MnO_2 (OJ-MnO_2) exhibited the surface areas as 8.40 and 5.63 $m^2\ g^{-1}$. From the above results, the surface area of OP-MnO_2 dominates the surface area of OJ-MnO_2. In addition, the electrochemical studies revealed that at 0.5 Ag^{-1}

current density, the OP-MnO_2 compound exhibits the specific capacitance of 139 Fg^{-1}, whereas OJ-MnO_2 compound exhibits the specific capacitance of 50 Fg^{-1}. Hence, among the prepared materials, OP-MnO_2 compound can be the potential aspirant for supercapacitor electrode materials. Further, Matinise et al. (2018) demonstrated the preparation of $ZnFe_2O_4$ nanocomposites by using green synthetic route in which natural plant extract called Moringa Oleifera acted as chelating and capping agent throughout synthesis of nanocomposites. As-prepared composites exhibit good crystallinity which is confirmed by XRD studies. Electrochemical impedance spectroscopy (EIS) and cyclic voltammetry (CV) investigations revealed the electrochemical behavior of $ZnFe_2O_4$ nanocomposites. Further, from the EIS calculations, the time constant is found to be 5.2001×10^{-4} s/rad and exchange current of 6.594×10^{-4} A is obtained. Furthermore, the fabricated glassy carbon loaded $ZnFe_2O_4$ (GCE/$ZnFe_2O_4$) electrode shows the excellent voltammetric response, and electrochemical performance confirms that the prepared composites are suitable for fabricating the supercapacitor electrode materials with good electrochemical performance.

In continuing this development, for the first time, by using the garbage collected Huggies (used baby diaper), Atchudan et al. (2019), synthesized novel mesoporous nitrogen-doped carbon nano-grass (N-CNG) through carbonization at 800 °C temperature under argon atmosphere for flexible supercapacitor applications. Figure 1 exhibits the schematic representation of preparation of N-CNG from the used baby diaper (Atchudan et al. 2019). They reported that the prepared N-CNG shows grass-like morphology with large surface area of 183 m^2g^{-1} and mean pore size of 3.3 nm.

On the basis of novel morphology and porosity, N-CNG can serve as good electroactive material. Besides, in order to explore the electrochemical properties of prepared nano-grass, the N-CNG is coated on a carbon cloth (CC). Furthermore, it was found that the CV curves which are in quasi-rectangular shape reveal the considerable capacitive behavior of N-CNG. Moreover, at the current density of 0.5 Ag^{-1}, N-CNG has large specific capacitance of 81 Fg^{-1} and also shows the excellent retention of specific capacitance about 95% even after 10,000 charge–discharge cycles. In addition, with increasing current density, the specific capacitance is decreased in a very slower rate which evidenced the excellent rate capability of prepared N-CNG. All the results evidenced the N-doped carbon nano-grass can be promising for supercapacitor applications.

Anand et al. (2019) prepared zinc oxide (ZnO) nanoparticles utilizing almond gum (Prunus dulcis) via greener synthesis route. The XRD studies revealed that ZnO nanoparticles exhibit wurtzite structure with hexagonal phase. Besides, from the Fourier transform infrared spectral (FTIR) analysis, it is observed that the cutoff wavelength is 243.93 nm, and the direct band obtained is 5.17 eV. The high-resolution scanning electron microscopy (HRSEM) images show the homogeneous agglomeration in the morphology of the prepared ZnO particles with mean particle size of 25 nm. Further, the electrochemical properties of the ZnO nanoparticles are carried out by CV and EIS. From the CV curve (Anand et al. 2019), it is clear that owing to diffusion mechanism between the electrode and electrolyte, there is an increment in current density with respect to the scan rates. Also, the CV curve is increased with scan rate which results in the enhancement in capacitive performance of the ZnO nanoparticles. The EIS studies revealed the electrochemical properties of prepared materials from the Nyquist plots (Anand et al. 2019). At higher-frequency regime, straight line indicates the Warburg impedance which causes diffusion control and interfacial charge transfer process which strongly influences the electrochemical performance of prepared materials. Moreover, at high-frequency region, ZnO has low internal resistance. Hence, all the outcomes suggest that green-synthesized ZnO nanoparticles can be suitable for electrochemical capacitor applications.

In addition, through electrochemical green synthesis route, Sportelli et al. (2020) prepared ZnO nanoparticles and explored the properties of the obtained nanoparticles. They reported that the prepared materials exhibit flower and rod-like structures based on concentration ratio. The agar diffusion technique confirms the consistent antimicrobial efficiency of the prepared benzyl-hexadecyl-dimetylammoniumchloride (BAC) and poly-diallyl-(dimethylammonium)chloride (PDDA) containing ZnO materials against B. subtilis. Hence, the present approach can be considered as more efficient technique than other conventional techniques for producing elongated ZnO nanomaterials in cationic capping agents and aqueous solution. Further, Herrero-Calvillo et al. (2020) explored electrochemical properties of the gold nanoparticles prepared by green chemical technique using different concentrations of Loeselia mexicana leaf extract. They reported that the prepared nanoparticles exhibit spherical and triangular-shaped morphologies confirmed by scanning electron microscopy (SEM) analysis, and the leaf extract concentration greatly influences the morphology of the gold nanoparticles. Finally, the cyclic voltammetry analysis revealed the gold nanotriangles showing superior electrochemical response and stability over gold nanospheres. In addition, the gold nanotriangles have large specific surface area which influences the degradation efficiency of the methylene blue, rhodamine-B and gentian violet organic dyes in catalytic activity. Rashmi et al. (2020) reported the electrochemical, photocatalytic and antimicrobial properties of the silver oxide nanoparticles prepared with the help of green combustion technique utilizing Centella asiatica and tridax plant powder. Novel pentacyclic ingredients of Centella

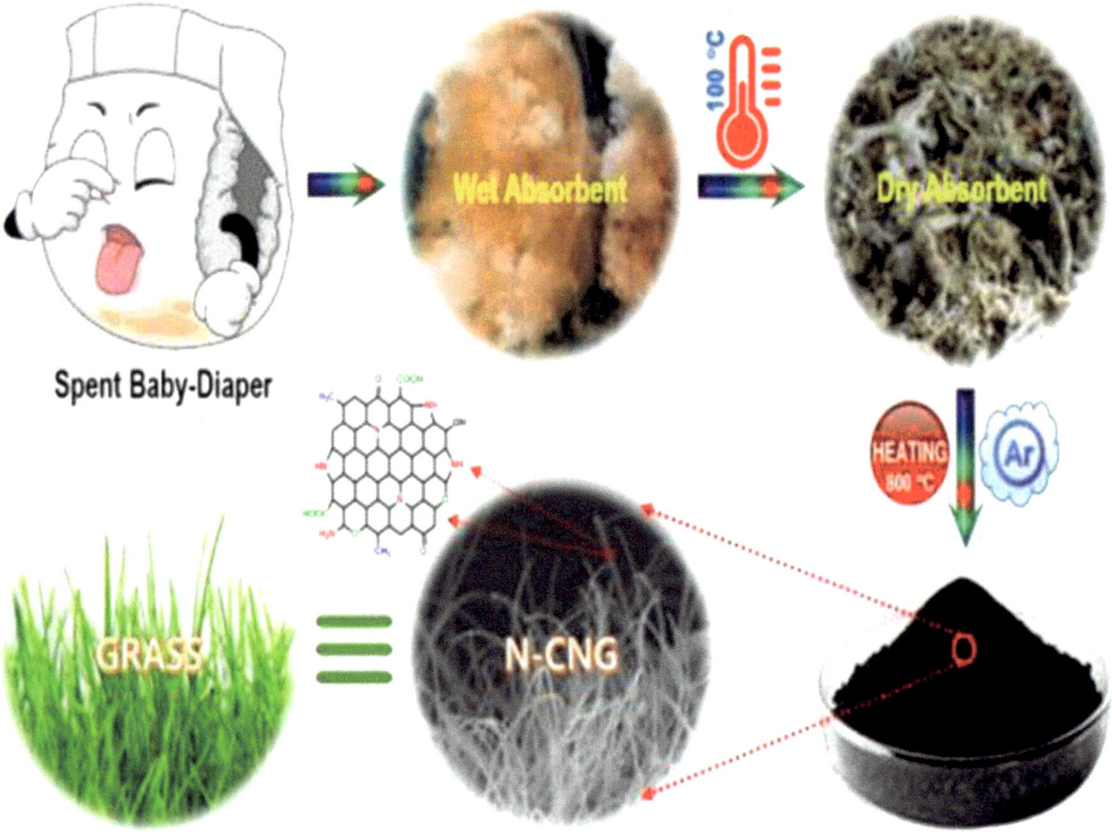

Fig. 1 Preparation of N-doped carbon nano-grass from the garbage collected baby diaper (Atchudan et al. 2019)

asiatica and triterpene constituents of tridax powder boost up the biological, electrochemical and photocatalytic activities of silver oxide nanoparticles. They reported that the prepared nanoparticles cathodic and anodic peak potentials display hysteresis in voltage range of 0.3–0.1 V which is confirmed by cyclic voltammetric analysis. Besides, in photocatalytic activity, they observed high percentage of deprivation of acid-orange-8 dye by the prepared nanoparticles. In addition, antimicrobial studies carried in contradiction of *S. epidermidis* and *S. aureus* and antifungal studies carried out against *A. fumigates* and *A. aureus* revealed that the silver oxide nanoparticles have the ability to prevent the development of several disease-causing pathogens. Therefore, silver oxide nanoparticles that are prepared via green combustion technique can serve as active electrode material in electrochemical supercapacitor, good photocatalyst and excellent antimicrobial agent. Owing to this, silver oxide nanoparticles may contribute to numerous environmentally benign industrial applications.

In designing an electrical double-layer capacitor (EDLC) with improved power and energy densities, carbon materials with optimized hierarchical pore structures play a pivotal role. In this regard, the advancement in the synthesis techniques gained significant attention for producing larger amount of hierarchically porous carbon (HPC) materials. In continuing this, Koo Kim et al. (2020) demonstrated a facile and ecofriendly technique called spray pyrolysis for synthesizing hierarchical pore-structured carbon materials. They reported that two steps were involved in preparation of HPC materials: in first step the spray solution comprising sucrose, H_2SO_4 and NaCl is aerosolized using ultrasonic nebulizer to provide NaCl-templated carbon microspheres, whereas the second step is the activation step followed by cleaning to obtain HPC. Figure 2 illustrates procedure for synthesis of NaCl-activated carbon microsphere through spray pyrolysis and post-treatment (Koo Kim et al. 2020). Further, the prepared HPC materials exhibit large specific surface area of 1704 m^2g^{-1} and pore volume of 1.81 cm^3g^{-1}. Besides, the outcomes suggest that the HPC consists of interconnected mesoporous network which offers short diffusion length leading to efficient ion transportation and low ionic resistance. Furthermore, the electrochemical studies disclosed that, at the current density of 30 Ag^{-1}, the prepared HPC materials show high specific capacitance of 102 Fg^{-1} along with outstanding cyclic retention of 92% even after 10,000 charge–discharge cycles at 10 Ag^{-1} as an electrode in organic electrolyte-based EDLC. Hence, all outcomes suggest that proposed HPC materials are promising candidates for supercapacitor as well as energy storage applications.

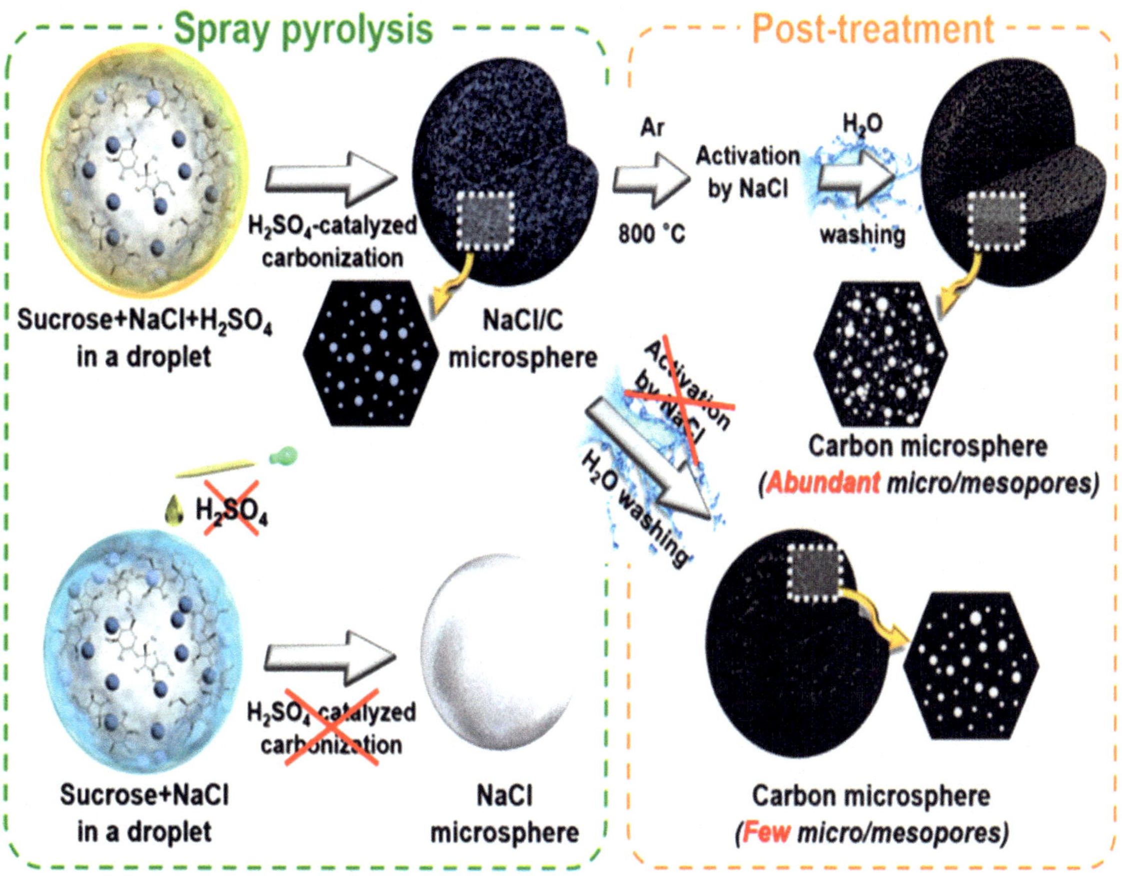

Fig. 2 Schematic representation of synthesis of NaCl-activated carbon microsphere through spray pyrolysis and post-treatment (Koo Kim et al. 2020)

In addition, Wang et al. (2019) demonstrated a facile, cost-effective and ecofriendly green synthesis technique for synthesizing TiN/C composites. Herein, the proposed green synthesis route contains two steps: Initially in the dispersion process the nitrogen and titanium sources are combined with the help of complexing agent, subsequent annealing of complexing compound precursors in nitrogen atmosphere to get TiN/C composites. Besides, annealing temperature and the amount of the complexing agent are greatly influencing the microstructure of the prepared composites. The prepared composites exhibit large specific surface area of 148 m^2g^{-1}. Further, the electrochemical studies revealed that when TiN/C composites were utilized as electrodes in supercapacitors, the composite shows large specific capacitance of 159 Fg^{-1} at a current density of 0.5 Ag^{-1} similarly 96 Fg^{-1} at a current density of 20-Ag^{-1}. The proposed simple and cost-effective strategy might be the prominent approach for production of metal nitride/carbon composites for a large scale without harming the environment.

Owing to cost effectiveness and low utility ruthenium-based materials like RuO_2 attracted much in supercapacitors. However, these ruthenium-based materials exhibit non-degradable nature. Generally, leaf extracts of Acalypha indica, Aspasaths linearis, Dictyota dichotoma, etc., have been using as capping, stabilizing and reducing agents in biogenesis of ruthenium nanoparticles. In addition, Akarkara (Anacyclus pyrethrum) is a medicinal plant; the extracts of the Akarkara can also serve as reducing agent in the biogenesis of ruthenium nanoparticles. To overcome the drawbacks of ruthenium-based materials, using extracts of Akarkara as reducing agent Nisha et al. (2020), synthesized the RuO_2 nanoparticles through green synthesis technique without adding any external catalyzing agent for reaction. Further, they explored the structural and electrochemical properties of prepared ruthenium oxide nanoparticles. From structural analysis, it is clear that the RuO_2 nanoparticles are highly crystalline with mean crystallite size of 13 nm and spherically shaped. Besides, the cyclic voltammetric studies revealed that RuO_2 nanoparticles shows high stability after repeating usage and the carbon nanosheet coated with the prepared nanoparticles exhibit high specific capacitance of 209 Fg^{-1} at scan rate of 5 Mv s^{-1}. These green-synthesized RuO_2 nanoparticles can also be employed as electrode materials in electrochemical capacitors. Bashir et al. (2020)

demonstrated the electrochemical performance of $NiFe_2O_4$-nanoparticles synthesized via green chemistry route using extracts of Persa americano seeds. The presence of metal oxide bonds is confirmed by FTIR and XRD studies disclosed the spinel structure of $NiFe_2O_4$ nanoparticles. Besides, morphological studies showed the nano-cube shapes of the prepared nanoparticles with mean size of 15–20 nm. From UV-Visible spectra, optical band gap is found to be 4.25 eV. Further, they explored the electrochemical properties of the prepared nanoparticles by CV and EIS analysis. It is clear from the CV curves (Bashir et al. 2020) that the prepared $NiFe_2O_4$ nanoparticles as electrodes exhibit excellent capability rate and electrochemical reversibility which indicates that fast diffusion for charge transfer kinetics among electrode and analyte. As well, the redox peak increases linearly with respect to scan rate which endorses the steadiness of $NiFe_2O_4$ nanoparticles on electrode. In addition, the EIS plot or Nyquist plot (Bashir et al. 2020) of GCE/$NiFe_2O_4$ electrode provides the information about capacitance and charge transfer resistance. From the Nyquist plot, it is noticed that low value of charge transfer resistance reveals quick transfer of charges between electrode and electrolyte which in turn confirms excellent conductivity of prepared nanoparticles. Hence, the regulated charge transfer kinetics, diffusion process, large electronic conductivity and improved electrochemical stability of prepared $NiFe_2O_4$ nanoparticles recommend them as prominent candidates for applications in electrochemical capacitors.

With the help of hydrothermal-assisted green synthesis technique and using Punica granatum (pomegranate) juice as a reducing agent, Das et al. (2020) prepared nickel and silver nanoparticles and explored electrochemical properties of the prepared nanoparticles by decorating them on reduced graphene oxide (rGO). They reported that the Ni/Ag@rGO electrode exhibits excellent electrochemical properties. Further, EIS studies revealed that the green-synthesized Ni/Ag@rGO electrode exhibits low electrode resistance of 30 Ω along with large electroactive area of 0.149 cm^2 and roughness factor of 0.379, which indicates the presence of large amount of active sites in prepared nanocomposites that are responsible for electron transfer process. Besides, by applying the prepared electrode materials, the square wave voltammetry (SWV) is performed to detect the ascorbic acid (AA) electrochemically, which disclosed the linear range of 4.89–90.09 μM, limit of detection (LOD) of 0.16 μM and detection sensitivity of 23,381.8 μAcm^{-2} mM^{-1}, respectively. The interference study confirms the AA detection capability of Ni/Ag@rGO electrode material. It also shows outstanding stability up to 45 days. All the results indicate that the green-synthesized Ni/Ag@rGO composite can be a prominent candidate for supercapacitor applications.

Owing to low toxicity, good electrical conductivity, high charge density, low cost, excellent environmental stability, etc., polypyrrole (PPy) gained considerable research attention toward supercapacitor electrode applications (Zhang et al. 2019). However, during repeated charge and discharge process PPy polymer backbone experiences shrinking and swelling which lead to volumetric changes and in turn causes cycling instability (Huang et al. 2016). In addition, in conventional preparation techniques PPy shows poor capacitive performance due to inhibition of diffusion of electrolyte ions. Hence, numerous techniques have seen light to improve the electrochemical performance of the PPy through modifying its morphology or establishing the composite with highly conductive carbon-based materials (Luo et al. 2018). On the other hand, PPy shows various structures in nano-level which includes nano-wires/rods/tubes, nano-brushes, nanosheets, hallow spheres, etc., which can have several advantages compared to bulk PPy. Among, PPy nanotubes provide short diffusion paths and large area for electrolyte–electrode interface. Therefore, PPy tubes received considerable attention for supercapacitor electrode applications. Massive research work has been going on PPy-based electrode materials to expand the electrochemical performance of the supercapacitors. In continuing this advancement and to overcome the drawbacks of conventional preparation techniques, Jyothibasu and Lee (2020) proposed a one-step synthetic technique called chemical oxidative polymerization to prepare PPy-based electrode materials with improved electrochemical performance. By using ecofriendly natural plant extract called curcumin (haul out from spice turmeric (Curcuma longa) and chemically it is a diarylheptanoid), they prepared ultra-long hollow PPy tubes. Further, PPy tubes (PPyT:PPyC1T1, PPyC1T2,PPyC1T4, PPyC2T2,andPPyC3T2) are synthesized under diverse circumstances and combining with functionalized carbon nanotubes(f-CNTs) to form freestanding electrodes. However, PPyC3T2/f-CNT electrode exhibits homogeneous morphology, hierarchically porous structure, outstanding electrochemical properties among all prepared electrode materials. The electrochemical studies revealed that PPyC3T2/f-CNT electrode at high mass loading of 30 $mgcm^{-2}$ and 2 $mAcm^{-2}$ current density shows large areal capacitance of 11,830.4 $mFcm^{-2}$. Furthermore, asymmetric supercapacitor fabricated with PPyC3T2/f-CNT electrode shows outstanding areal capacitance of 2732 $mFcm^{-2}$ at 2 $mAcm^{-2}$ along with excellent cyclic retention of 118% even after 12,500 cycles, and it also shows high energy density of 242.84 $\mu Whcm^{-2}$ and power density of 129.35 $mWcm^{-2}$. Hence, all these outcomes evidenced that the prepared PPyC3T2/f-CNT can serve as prominent electrode material in high-performance supercapacitors.

In addition, Zhu et al. (2020) prepared PPy-coated manganese vanadate nano ($Mn(VO_3)_2$@PPy) nanocomposite via green synthesis approach. First, they prepared $Mn(VO_3)_2$ precursors by hydrothermal technique, and then precursors are wrapped by PPy to attain $Mn(VO_3)_2$@PPy nanoflower composites. The electrochemical studies revealed that the prepared nanocomposites exhibit discharge capacity of 102.6 $mAhg^{-1}$ at 0.1 Ag^{-1} current density and specific discharge capacitance of 75.3 $mAhg^{-1}$ at 1 Ag^{-1}. It also shows 100% cyclic stability after 500 cycles. All the outcomes suggest that the prepared $Mn(VO_3)_2$@PPy nanoflower composites can serve as positive electrode materials in electrochemical capacitors.

Sahan et al. (2019) synthesized CoO@Co_3O_4@C composite via green synthesis technique by using Punica granatum extract for electrochemical applications. They reported that the prepared composite exhibits mixed phases as confirmed by XRD, and all the precursors are distributed on amorphous carbon network confirmed by SEM analysis. The electrochemical properties are studied by cyclic voltammetry and galvanostatic charge–discharge measurements which revealed that the prepared composite exhibits reversible specific capacitance of 447 $mAhg^{-1}$ after ten cycles at 100 mAg^{-1} and reversible specific capacitance of 10 $mAhg^{-1}$ after 50 cycles at 50 mAg^{-1} along with improved cyclic and rate capability. This enhanced electrochemical performance evinced that CoO@Co_3O_4@C composite can be suitable for electrochemical applications.

As we know that multiple morphologies of cobalt–nickel hybrid materials exhibit considerable electrochemical activity due to multiple valence states. By using homogeneous precipitation technique, Yang et al. (2020) prepared Co–Ni hollow microspheres by using urea as precipitation reactant for the detection of dopamine. Further, ammonia and water are the decomposition products, and also the entire process does not contain any organic solvents. Hence, the entire preparation technique is environmentally friendly. Besides, the electrochemical studies revealed the excellent electrochemical activity of the prepared nanospheres as sensor electrodes. CV curves (Yang et al. 2020) with dopamine concentrations are observed at scan rate of 50 Mv s^{-1}. However, the oxidation peak potential is observed at 0.379 V, and the peak current is found to be 17.88 A. Nevertheless, the oxidation peak is influenced by the concentration of dopamine owing to quick reaction rate of dopamine. Further, redox potential increases with respect to the dopamine concentration. All these results specify that the proposed material can be utilized as modified electrode material for the detection of dopamine through electrochemical process.

4 Applications in Electrochemical Hydrogen Storage

In recent years, the research on hydrogen storage attracted immense attention because it is an inexpensive and good energy carrier also. In fuel cells, the hydrogen in used to generate electricity without emitting any pollutants like CO_2 into the atmosphere and water as its only by-product (Fang et al. 2006; Kim et al. 2016). Numerous metal oxides are being used in hydrogen storage, which can be prepared by various chemical, physical and electrochemical approaches (Ouyang et al. 2017). Among, electrochemical hydrogen storage is commonly used to store hydrogen because in electrochemical method hydrogen adsorption occurs directly in aqueous medium at the time of electrochemical decomposition (Zhang et al. 2014). However, one of the major challenges in electrochemical method is finding appropriate materials as a source for hydrogen storage with good discharge capacity. Till now, researchers introduced various materials such as metal oxides (Butt et al. 2014), CNTs (Mohammadi et al. 2016) and alloy/graphene composites (Ouyang et al. 2014) for hydrogen storage. Owing to low surface acidity, good chemical stability and high thermal conductivity, zinc aluminate ($ZnAl_2O_4$) is considered as one of the prominent materials for hydrogen storage (Battidton et al. 2014). Till now, $ZnAl_2O_4$ is prepared by different techniques such as sol–gel, hydrothermal and thermal decomposition for applications various fields.

Recently, with the help of green synthesis technique, Gholami et al. (2018) synthesized $ZnAl_2O_4$ nanoparticles and $ZnAl_2O_4$/graphene nanocomposites for electrochemical storage applications in which green tea and olive leaf extracts are used as reagents. They reported that the prepared materials exhibit good structural and morphological properties as confirmed by XRD and SEM analysis. Further, under 1 mA current in 6 M KOH electrolyte, they tested hydrogen storage performance of prepared materials with chronopotentiometry technique and also compared their Coulombic efficiency. Figure 3 represents the flowchart of synthesis of $ZnAl_2O_4$ and $ZnAl_2O_4$/graphene nanocomposites and their cyclic performance (Gholami et al. 2018). The prepared materials exhibit the highest discharge capacity of 3100 $mAhg^{-1}$ and Coulombic efficiency of 67.5%. The outcomes of the investigation suggest that the prepared $ZnAl_2O_4$ nanoparticles and $ZnAl_2O_4$/graphene nanocomposites are prominent candidates for hydrogen storage. In addition, using the novel fuel, for the first time, Zinatloo-Ajabshir et al. (2019) prepared $Dy_2Ce_2O_7$ nanoparticles by green synthesis route for electrochemical storage applications. Besides, they prepared different $Dy_2Ce_2O_7$ nanoparticles with different structures by varying fig extract at different temperatures. All the prepared structures exhibit different hydrogen storage properties and Coulombic efficacy which is confirmed by chronopotentiometry method at potash solution. All results suggest that the $Dy_2Ce_2O_7$ nanoparticles molded with aid of fig extract at 400 °C show good electrochemical hydrogen storage properties, i.e., maximum discharge capacity observed for above-said sample is 3070 $mAhg^{-1}$ after 18 charge/discharge cycles, and it also exhibits good Coulombic efficacy. Hence, the $Dy_2Ce_2O_7$ nanoparticles originated with the assistance of fig extract at 400 °C are the good candidate for electrochemical hydrogen storage. Finally, the usage of fig extract as new and ecofriendly fuel can be important to fabricate the nanostructured $Dy_2Ce_2O_7$ nanoparticles which are efficiently capable for electrochemical hydrogen storage and can be beneficial for energy storage technology.

Ecofriendly nature is one of the major advantages of electrochemical green synthesis. In practical point of view, the electrochemical reactions generally show good functional group tolerance. In addition, the electrochemical reactions provide energy-saving option at elevated pressure and temperatures. Owing to their high-reaction efficacy, electrochemical techniques require short reaction time as compared to traditional techniques. Further, the electrochemical reaction can easily be stopped at any time by turning-off the power switch, whereas traditional methods require frequent quenching. Finally, most of the electrochemical methods such as electrochemical–synthetic approaches for 1,4-dicyanobutane and sebacic-acid have fruitfully accomplished the industrialization (Pletcher and Walsh 1990). However, despite of having many advantages, still some shortcomings exist in application point of electrochemical green synthesis. For instance, to enhance the conductivity of solution, some techniques use the hazardous electrolytes which results in a higher risk of explosion. The selection of corrosive solvents, flammable and toxic solvents may also negatively impact on safety features of entire process. The usage of these additives also leads to production of higher amount of waste (Anastas and Warner 1998; Anastas and Eghbali 2010). In order to overcome these, the researchers have to be focused on developing new ion exchange membranes, electrode materials, electrolytic cells and electrochemical mediators, also, diminishing, recycling and even avoiding the usage of supporting electrolytes. Finally, current researchers are attentive on the designing

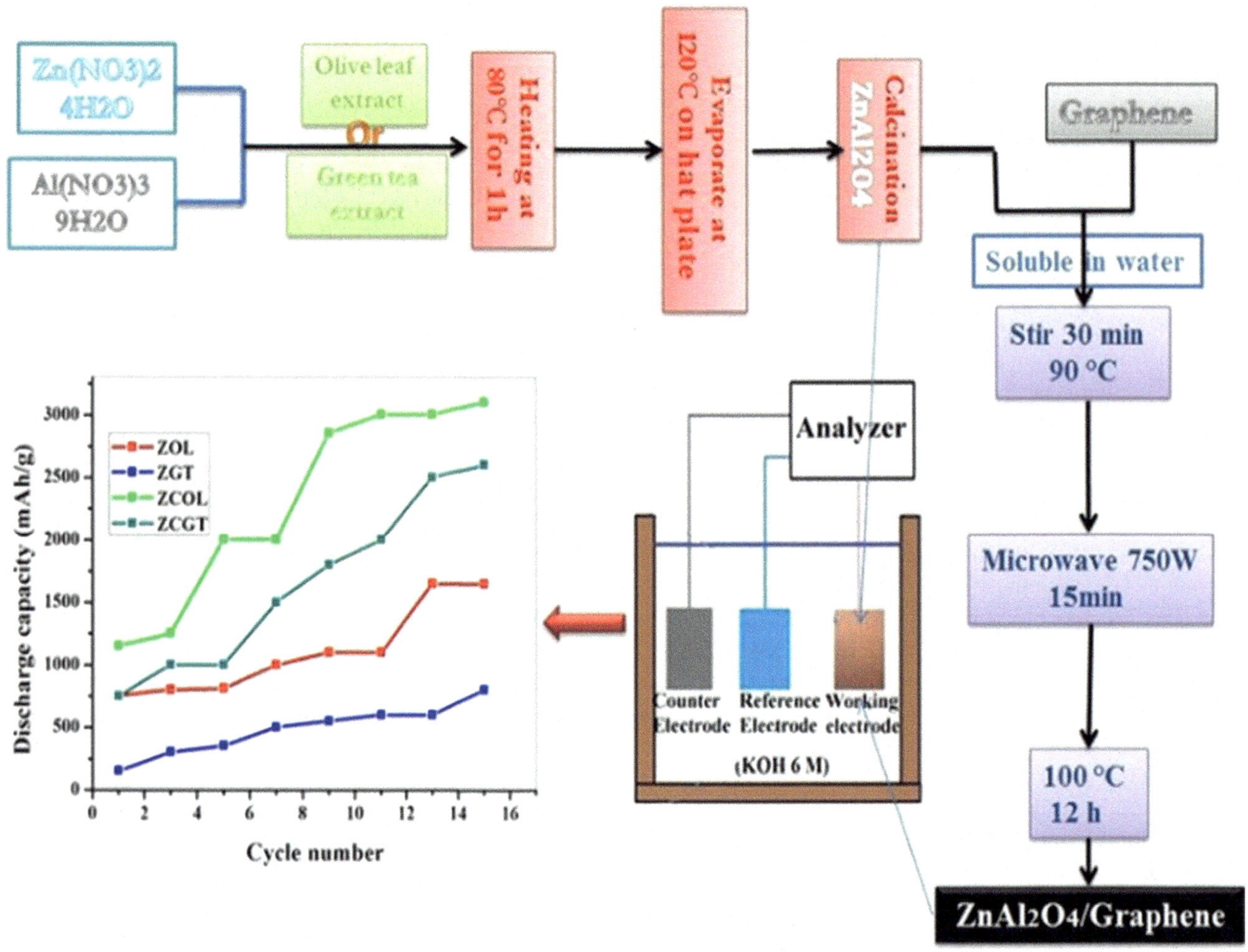

Fig. 3 Flowchart of synthesis of $ZnAl_2O_4$ and $ZnAl_2O_4$/graphene nanocomposites and their cyclic performance (Gholami et al. 2018)

and manufacturing of novel electrolyzers which can improve the efficiency and practical applicability of electrochemical green synthesis (Yuan and Lei 2020).

5 Conclusions

Electrochemical green synthesis is simple, ecofriendly, economical for the production of materials with good physical and electrochemical properties. Owing to this electrochemical green synthesis of gains, the considerable research attention in fabricating inexpensive and ecofriendly materials for diverse applications was developed. In this chapter, we have explored some of the recent materials such as α-MnO_2, $ZnFe_2O_4$, N-CNG, ZnO, RuO_2, $NiFe_2O_4$, Ni/Ag@rGO, $ZnAl_2O_4$, $Dy_2Ce_2O_7$, Co–Ni and CoO@-Co_3O_4@C have been prepared via green synthesis route, and their applications in electrochemical capacitors, hydrogen storage, detection of dopamine, antimicrobial, etc., have been discussed as well.

References

Abuzeid HM, Elsherif SA, AbdelGhany NA, Hashem AM (2019) Facile, cost-effective and eco-friendly green synthesis method of MnO2 as storage electrode materials for supercapacitors. J Energy Storage 21:156–162

Agarwal H, Kumar SV, Rajesh kumar S (2017) A review on green synthesis of zinc oxide nanoparticles—an eco-friendly approach. Resour Efficient Technol 3(4):406–413

Alberto C, Ade M (2013) Environmental sustainability: implications and limitations to Green Chemistry. Found Chem 16:125–147

Anastas P, Eghbali N (2010) Green chemistry: principles and practice. Chem Soc Rev 39:301–312

Anastas P, Warner JC (1998) (1998) Green chemistry: theory and practice. Oxford University Press, New York, NY

Arruda SCC, Silva ALD, Galazzi RM et al (2015) Nanoparticles applied to plant science: a review. Talanta 131:693–705

Atchudan R, Edison TNJI, Perumal S, Thirukumaran P, Vinodh R, Lee YR (2019) Green synthesis of nitrogen-doped carbon nanograss for supercapacitors. J Taiwan Inst Chem Eng 102:475–486

Bashir AK, Matinise N, Sackey J, Kaviyarasu K, Madiba IG, Kodseti L et al (2020) Investigation of electrochemical performance, optical and magnetic properties of NiFe2O4 nanoparticles prepared by a green chemistry method. Physica E: Low-Dimension Syst Nanostruct 119:114002

Battidton S, Rigo C, Cruz Severo E, Mazutti MA, Kuhn RC, Gundel A, Foletto EL (2014) Synthesis of zinc aluminate ($ZnAl_2O_4$) spinel and its application as photocatalyst. Mater Res 17:735–738

Besner S, Kabashin AV, Winnik FM, Meunier M (2008) Ultrafast laser based "green" synthesis of non-toxic nanoparticles in aqueous solutions. Appl Phys A 93(4):955–959

Butt Fk, Tahir M, Cao C, Idrees F, Ahmed R, Khan Zulfiqar A (2014) Synthesis of novel ZnV_2O_4 hierarchical nanospheres and their applications as electrochemical supercapacitor and hydrogen storage material. ACS Appl Mater Interfaces 6:13635–13641

Das TR, Sharma PK (2020) Hydrothermal-assisted green synthesis of Ni/Ag@rGO nanocomposite using Punica granatum juice and electrochemical detection of ascorbic acid. Microchem J 104850 (in press). doi:https://doi.org/10.1016/j.microc.2020.104850

Fang B, Zhou H, Houma I (2006) Ordered porous Carbon with tailored pore size for electrochemical hydrogen storage application. J Phys Chem B 110:4875–4880

Gholami T, Salavati-Niasari M, Sabet M (2018) Novel green synthesis of $ZnAl_2O_4$ and $ZnAl_2O_4$/graphene nanocomposite and comparison of electrochemical hydrogen storage and Coulombic efficiency. J Clean Prod 178:14–21

Herrero-Calvillo R, Santoveña-Uribe A, Esparza R, Rosas G (2020) A photocatalytic and electrochemical study of gold nanoparticles synthesized by a green approach. Mater Res Exp 7(1):015019

Huang Y, Li H, Wang Z, Zhu M, Pei Z, Xue Q et al (2016) Nanostructured polypyrrole as a flexible electrode material of supercapacitor. Nano Energy 22:422–438

Jyothibasu JP, Lee R-H (2020) Green synthesis of polypyrrole tubes using curcumin template for excellent electrochemical performance in supercapacitors. J Mater Chem A 8:3186–3202

Kim TH, Bae J, Lee TH, Hwang J, Jung JH, Kim DK, Lee JS, Kim DO, Lee YH (2016) Room-temperature hydrogen storage via two-dimensional potential well in mesoporous graphene oxide. J Nano Energy 27:402–411

Koo Kim J, Yoo Y, Chan Kang Y (2020) Scalable green synthesis of hierarchically porous carbon microspheres by spray pyrolysis for high-performance supercapacitors. Chem Eng J 382:122805

Kumar V, Yadav SK (2009) Plant-mediated synthesis of silver and gold nanoparticles and their applications. J Chem Technol Biotechnol 84 (2):151–157

Lukatskaya MR, Dunn B, Gogotsi Y (2016) Multidimensional materials and device architectures for future hybrid energy storage. Nat Commun 7:12647

Luo S, Zhao J, Zou J, He Z, Xu C, Liu F et al (2018) Self-standing polypyrrole/black phosphorus laminated film: promising electrode for flexible supercapacitor with enhanced capacitance and cycling stability. ACS Appl Mater Interfaces 10(4):3538–3548

Matinise N, Kaviyarasu K, Mongwaketsi N, Khamlich S, Kotsedi L, Mayedwa N, Maaza M (2018) Green synthesis of novel zinc iron oxide ($ZnFe_2O_4$) nanocomposite via Moringa Oleifera natural extract for electrochemical applications. Appl Surf Sci 446:66–73

Mirzaei H, Darroudi M (2017) Zinc oxide nanoparticles: biological synthesis and biomedical applications. Ceram Int 43(1):907–914

Mohammadi M, Khoshnevisan B, Varshoy Sh (2016) Electrochemical hydrogen storage in EPD made porous Ni-CNT electrode. Int J Hydrogen Energy 41:10311–10315

Nisha B, Vijayalakshmi Y, Abdul Razack S (2020) Enhanced formation of ruthenium oxide nanoparticles through green synthesis for highly efficient supercapacitor applications. Adv Powder Technol (in press). https://doi.org/10.1016/j.apt.2019.12.026.

Ouyang L, Chen W, Liu J, Felderhoff M, Wang H, Zhu M (2017) Enhancing the regeneration process of consumed NaBH4 for hydrogen storage. Adv Energy Mater 7:1700299

Ouyang Guo L, Cai W, Ye J, Hu R, Liu J, Yang L, Zhu M (2014) Facile synthesis of Ge@FLG composites by plasma assisted ball milling for lithium ion battery anodes. J Mater Chem A 2:11280–11285

Paul B, Bhuyan B, Purkayastha DD, Vadivel S, Dhar SS (2016) One-pot green synthesis of gold nanoparticles and studies of their anticoagulative and photocatalytic activities. Mater Lett 185:143–147

Pletcher D, Walsh FC (1990) Industrial electrochemistry. Springer

Rafique M, Sadaf I, Rafique MS et al (2017) A review on green synthesis of silver nanoparticles and their applications. Artif Cells Nanomed Biotechnol 45:1272–1291

Rashmi BN, Harlapur SF, Avinash B, Ravikumar CR, Nagaswarupa HP, Kumar MRA et al (2020) Facile green synthesis of silver oxide nanoparticles and their electrochemical, photocatalytic and biological studies. Inorg Chem Commun 111:107580

Ribeiro MGTC, Costa DA, Machado AASC, Ribeiro MGTC, Costa DA, Machado AASC (2010) Green Chemistry letters and reviews 'Green Star':a holistic Green Chemistry metric for evaluation of teaching laboratory experiments. Green Chem Lett Rev 3:149–159

Şahan H, Göktepe H, Yıldız S, Çaymaz C, Patat Ş (2019) A novel and green synthesis of mixed phase $CoO@Co_3O_4@C$ anode material for lithium ion batteries. Lonics 25:447–455

Salam HA, Sivaraj R, Venckatesh R (2014) Green synthesis and characterization of zinc oxide nanoparticles from Ocimum basilicum L. var. purpurascens Benth.-Lamiaceae leaf extract. Mater Lett 131:16–18

Shen LX, Du LH, Tan SZ et al (2016) Flexible electrochromic supercapacitor hybrid electrodes based on tungsten oxide films and silver nanowires. Chem Commun 52:6296–6299

Sportelli MC, Picca RA, Izzi M, Palazzo G, Gristina R, Innocenti M et al (2020) ZnO nanostructures with antibacterial properties prepared by a green electrochemical-thermal approach. Nanomaterials 10(3):473

Suvith V, Philip D (2014) Catalytic degradation of methylene blue using biosynthesized gold and silver nanoparticles, Spectrochim. Acta A 118:526–532

Theophil Anand G, Renuka D, Ramesh R, Anandaraj L, John Sundaram S, Ramalingam G et al (2019) Green synthesis of ZnO nanoparticle using prunus dulcis (Almond Gum) for antimicrobial and supercapacitor applications. Surf Interfaces 17:100376

Wang T, Li K, An S, Song C, Guo X (2019) Facile and green synthesis of TiN/C as electrode materials for supercapacitors. Appl Surf Sci 470:241–249

Yang C, Sun X, Zhang C, Liu M (2020) Green synthesis of Co-Ni hollow spheres for its electrochemical detection of dopamine. J Nanopart Res 22(3):2020

Yuan Y, Lei A (2020) Is electrosynthesis always green and advantageous compared to traditional methods? Nat Commun 11:802

Zhang M, Song Y, Guo D, Yang D, Sun X, Liu X-X (2019) Strongly coupled polypyrrole/molybdenum oxide hybrid films via electrochemical layer-by-layer assembly for pseudocapacitors. J Mater Chem A 7:9815–9821

Zhang Ch, Li J, He Ch, Liu E, Zhao N, Effect of Ni (2014) Fe and Fe-Ni alloy catalysts on the synthesis of metal contained carbon nano-onions and studies of their electrochemical hydrogen storage properties. J Energy Chem 23:324–330

Zhu H, Ma L, Jiang J et al (2020) Green synthesis of polypyrrole coated manganesee(II) vanadate nanoflower composite as cathode materials. Int J Electrochem Sci 15:371–381

Zinatloo-Ajabshir S, Salehi Z, Amiri O, Salavati-Niasari M (2019) Green synthesis, characterization and investigation of the electrochemical hydrogen storage properties of $Dy_2Ce_2O_7$ nanostructures with fig extract. Int J Hydrogen Energy 44(36):20110–20120

Enzyme-Mediated Synthesis of Heterocyclic Compounds

Deepshikha Rathore, Geetanjali, and Ram Singh

Abstract

Heterocyclic compounds are cyclic organic molecules possessing at least one atom other than carbon in the ring structure. They are a widely used class of organic compounds. The heterocyclic scaffolds represent the central framework of many biologically active molecules. The other applications, like agrochemicals, veterinary products, dyes, etc., also make them essential. Due to its high global demand, there is always a need for new and efficient methodology in synthesizing these molecules. The sustainable process with a minimal environmental impact is the need of the present day and biocatalysis supports this. Enzymes are biocatalysts and play a progressively significant role in the synthesis of heterocyclic molecules. This chapter discusses the utility of different enzymes for the synthesis of nitrogen-, oxygen- or both containing heterocyclic molecules.

Keywords

Enzyme • Biocatalysts • Heterocyclic compounds • Green synthesis • Sustainable synthesis

1 Introduction

Heterocyclic compounds are cyclic organic molecules possessing at least one atom other than carbon in the ring structure (Sabir et al. 2015). Their both properties, physical and chemical, are reliant on the presence of heteroatom(s). These compounds find their applications in almost all types of industries, including pharmaceuticals, agrochemicals, dyes, and pigments and others (Joule and Mills 2013; Rudi et al. 2005; Broughton and Watson 2004; Arunkumar 2015; Arora et al. 2012; Kozikowski 1984). Due to its increasing demand, researchers continuously keep an eye on their synthetic routes. They are always trying to develop various efficient and environmentally friendly methodologies (Busto et al. 2011; Feber 2004; Shoda et al. 2016). The conventional methods, like chemical catalysis, electrochemical, microwave-assisted, and using ionic liquids, solid and solution phase synthesis, are some of the well-known methods (Busto et al. 2011; Feber 2004; Shoda et al. 2016). Still, the development of a sustainable synthetic process is widely explored in organic synthesis. Biocatalysis has shown potential towards the synthesis of organic molecules, including heterocyclic molecules in a sustainable and environmentally friendly manner (Milner and Maguire 2012; Mane et al. 2018; Wu et al. 2019; Singh et al. 2006).

Biocatalysis is the use of bio-based catalysts or enzymes in organic synthesis (Dalal et al. 2016; Xie et al. 2013; Li et al. 2008; Xiang et al. 2013, 2014; Xue et al. 2012; Ding et al. 2015). Apart from the other advantages possessed by enzyme-catalyzed reactions, they also take care of the production of single enantiomers instead of racemic mixtures (Singh et al. 2006). The two fundamental properties of catalysts also apply to enzyme-catalyzed reactions: (i) increase in the rate of reaction; and (ii) remain non-consumed after the reaction. The focus of research revolves around the development of stereo-, regio- and chemo-selective reactions (Shoda et al. 2016; Kobayashi et al. 1997, 1996). Different classes of enzymes such as hydrolases, oxidoreductases, transferase, lyase, isomerase and ligase perform different types of reactions (Fig. 1) (Shoda et al. 2016; Singh et al. 2006; Groger and Asano 2012; Webb 1992).

In this chapter, the enzyme-catalyzed synthesis of pyrroles, indoles, phenazines, benzocarbazoles, benzimidazoles,

D. Rathore · R. Singh (✉)
Department of Applied Chemistry, Delhi Technological University, Delhi, 110042, India
e-mail: ramsingh@dtu.ac.in

Geetanjali
Department of Chemistry, Kirori Mal College, University of Delhi, Delhi, 110007, India

Inamuddin et al. (eds.), *Advances in Green Synthesis*, Advances in Science, Technology & Innovation,
https://doi.org/10.1007/978-3-030-67884-5_16

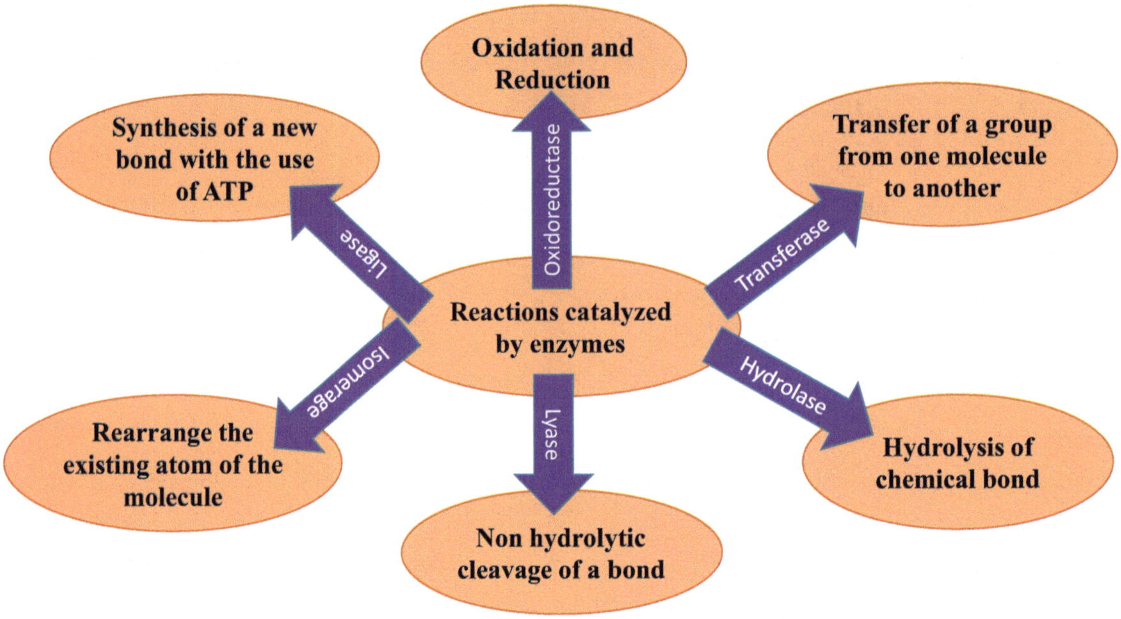

Fig. 1 Reactions catalyzed by enzymes in organic synthesis

pyrazoles, benzofurans, chromenes, dioxins, lactones and oxazolidinones have been discussed.

2 Enzyme-Mediated Synthesis of Heterocyclic Compounds

Enzyme-mediated synthesis is based on the ability of its active site to allow a particular substrate to enter into it and further get transformed into a suitable product (Yang et al. 2015; Kłossowski et al. 2013). Many name reactions like Morita–Baylis–Hillman reaction (Reetz et al. 2007), Michael addition (Zhang et al. 2017a), aldol addition (Li et al. 2008) and others (Hu et al. 2012; Wang et al. 2017) have been successfully carried out using enzyme-catalyzed method. Some of the important enzyme-mediated reactions have been discussed in this section towards the synthesis of heterocyclic compounds.

2.1 Nitrogen-Containing Heterocycles

Nitrogen heterocycles are mainly found in peptides and alkaloids, and these both are most widely spread in mostly all-natural products. The one, two and more number of nitrogen-containing heterocycles with 5-, 6- and 7-membered aromatic compounds is very important moiety in the field of medicine (Jampilek (2019); Vitaku et al. 2014; Singh and Geetanjali 2011). The synthesis of N-containing heterocyclic molecules has always been explored for sustainable processes (Veer and Singh 2019; Poonam 2019; Chauhan and Geetanjali 2000; Chauhan et al. 2003). There is always demand for the wide-structural diversity of N-heterocycles and hence exploration of synthetic protocols. They are prepared by synthetic reactions as well as via enzymatic methods. The biosynthetic pathways of naturally occurring heterocyclic compounds help in designing their enzymatic synthetic paths (Hemmerling and Hahn 2016; Junghanns et al. 1995).

2.1.1 Pyrroles

Pyrroles are five-membered ring, N-containing heterocyclic molecule which showed their utility in diverse biological activities, speciality polymeric materials, etc. (Trofimov et al. 2004; Bellina and Rossi 2006). The Paal–Knorr reaction has been used to synthesize the derivatives of N-substituted pyrrole in 60–99% yield catalyzed by α-amylase obtained from hog pancreas. This reaction was standardized using aniline (R=Ph) and 2,5-hexanedione as the starting materials (Fig. 2) (Zheng et al. 2013). The α-amylase from hog pancreas yielded N-phenyl-2,5-dimethylpyrrole in 94% yield; however, from *Aspergillus oryzae* gave the same product in 65% yield. This suggests that the sources of enzyme also play important role in product formation.

The authors have optimized the biocatalyst from eight different enzymes they tried for the reaction. The optimized enzyme, α-amylase from hog pancreas under mild reaction condition such as 50 °C temperature in methanol showed excellent activity using wide variety of primary amines (Zheng et al. 2013). The reactions showed high yields and better efficiency under mild reaction conditions.

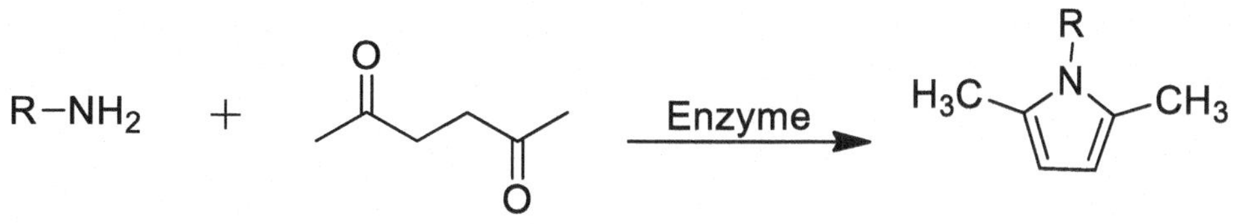

Fig. 2 Condensation of aniline and 2,5-hexanedione

2.1.2 Indoles

Spirooxindoles are indole derivatives present in many secondary metabolites and biologically important molecules (Ding et al. 2006; Galliford and Scheidt 2007). These molecules have been synthesized using enzymes as biocatalysts (Chai et al. 2011). The reaction of isatin, malononitrile and ethyl acetoacetate in the presence of lipase from porcine pancreas (PPL) in water–ethanol gave spirooxindole derivatives in 82–95% yield (Fig. 3) (Chai et al. 2011). The reaction condition with respect to reaction time, catalyst, temperature and solvent was optimized by the authors.

The use of others solvents like acetone, dichloromethane, hexane, chloroform, tetrahydrofuran, acetonitrile and dimethylformamide either did not promote the reaction or gave only moderate yield (Chai et al. 2011). Other commercially available hydrolytic enzymes were also evaluated for the reaction but PPL gave better results. The enzymes amano lipase M from *Mucor javanicus* and Amano lipase A from *Aspergillus niger* also gave the products in 82 and 84% yield, respectively. The lipase acrylic resin from *Candida antarctica* gave only trace of spirooxindoles. This was proposed that the specific spatial conformation along with the tertiary structure of PPL gave better yield of spirooxindole derivatives (Chai et al. 2011).

2.1.3 Phenazines

Phenazine is dibenzo annulated pyrazine with the formula $(C_6H_4)_2N_2$. Their derivatives are multifunctional in nature and successfully used in pharmaceuticals (Zhuo et al. 2013; Gamage et al. 2006), energy sectors (Okazaki et al. 2017; Lee et al. 2010), sensors (Pauliukaite et al. 2010), etc. Due to their broad applications, enzymatic synthesis has also been performed on this molecule synthesis (Sousa et al. 2014, 2018). Sousa et al. studied the use of Laccases, a multi-copper oxidase for the synthesis of phenazine and phenoxazinone frameworks from substituted aromatic amines (Fig. 4) (Sousa et al. 2014). This has been observed that the laccase-catalyzed reactions give only water as waste product when reactions are performed in aqueous solvent systems (Witayakran and Ragauskas 2009; Mikolasch and Schauer 2009). The reaction was performed under mild reaction condition using solvent methanol in phosphate buffer (pH 6–7) at temperature of 37 °C under aerobic conditions. The enzymatic oxidation of *meta*, *para*-disubstituted amine derivatives afforded phenazines. With the starting reagent, 1,2-diaminobenzene, 2,3-diaminophenazine was formed in 66% yield, whereas with *ortho*-aminophenol, phenoxazine formed in 83% yield (Sousa et al. 2014). The oxidation of 1-amino-2-naphthol with PPL at pH 7 gave 14H-dibenzo[a,j]phenoxazine-5,6-diol in 59% yield (Fig. 5) (Sousa et al. 2014).

This group further utilized the optimized protocol for a one-step asymmetric phenazines and phenoxazinones synthesis using spore coat protein A (CotA)-laccase enzyme as catalyst from *ortho*-substituted diamines and *ortho*-substituted hydroxylamines through aerobic oxidations (Sousa et al. 2018). Some of the other similar molecules synthesized using similar procedure with different reactants are given with their yields in Fig. 6 (Sousa et al. 2014, 2018).

2.1.4 Benzocarbazoles

Carbazole consists of two six-membered benzene rings fused on either side of a five-membered nitrogen-containing ring. This is based on indole structure where a second benzene ring is fused at the 2–3 position of indole. Carbazole derivatives possess pharmaceutical properties (Knölker and

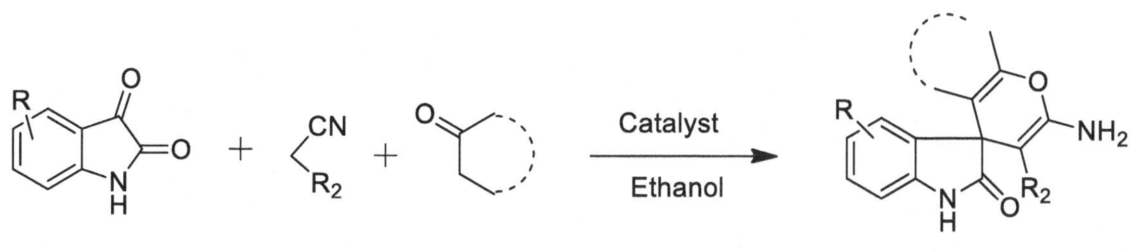

Fig. 3 Synthesis of different spirooxindole derivatives

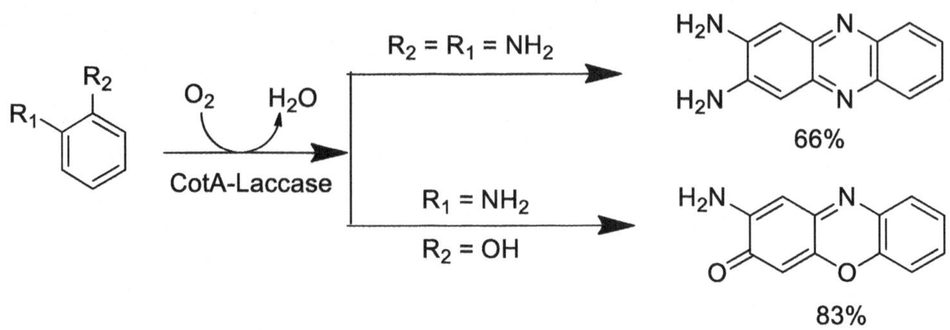

Fig. 4 Laccase-catalyzed reactions with substituted aromatic amines

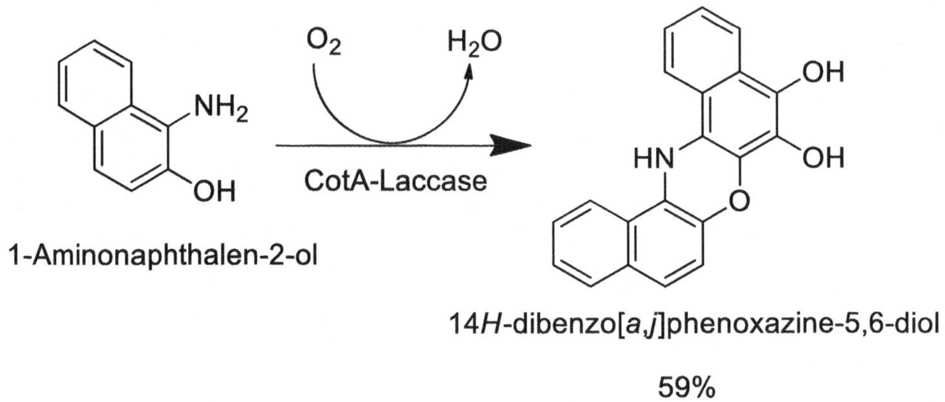

Fig. 5 Laccase-catalyzed synthesis of 14H-dibenzo[a,j] phenoxazine-5,6-diol

Reddy 2002; Pecca and Albonico 1971) and have applications in material sciences (Grazulevicius et al. 2003; Lia and Grimsdale 2010).

Sousa et al. performed the enzymatic synthesis of carbazoles using CotA laccase (Sousa et al. 2015). The oxidation of the *meta*, *para*-disubstituted arylamine 2,4-diaminophenyldiamine afforded benzocarbazole derivative in 74% yield and hence developed a clean method to construct in one-step C–C and C–N bonds (Fig. 7) (Sousa et al. 2015). The electrochemical behaviour of the target substrate plays essential role for product formation through an intramolecular oxidative coupling step (Sousa et al. 2015).

2.1.5 Benzimidazoles

Benzimidazole is a bicyclic heteroaromatic system which is composed of a benzene ring fused with an imidazole ring, i.e. five-membered ring with three number of carbon and two nitrogen is attached. It is occurring in nature as part of the vitamin B12 molecule with chemical formula $C_7H_6N_2$. This type of molecules possesses therapeutic properties including broad-spectrum anthelmintic, fungicidal or antimicrobial action (Salahuddin and Mazumder 2017).

Wang et al. in 2010 developed an efficient synthesis of bioactive compound in an ecologically and economically favourable way (Wang et al. 2010). In their study, they developed the enzyme-mediated synthesis of 2-alkylbenzimidazole. *o*-Phenylenediamine is used as primary reactant for the synthesis of the desired product (Fig. 8). In this synthesis, *o*-phenylenedimine reacted with the corresponding ester in the presence of immobilized lipase from *Mucor miehei* (MML) as a catalyst. This mixture was stirred at 50 °C for 60 h to complete the reaction and achieve the synthesis of benzimidazole derivatives (Wang et al. 2010). The reaction with other hydrolases like lipase acrylic resin from *Candida antarctica* B, Amano lipase M from *Mucor javanicus* and lipase from *Candida rugosa* gave low to poor yields. The spatial conformation of the lipases plays very important role in the product formation (Chai et al. 2011; Wang et al. 2010).

2.1.6 Pyrazoles

Pyrazoles are five-membered heterocyclic molecules possessing two nitrogen atoms in the ring. Pyrazole derivatives are biologically active compounds and also useful for other applications like dyes and luminophores (Mishra and Sasmal 2011; Stellrecht and Chen 2011). These molecules have been synthesized through enzyme-catalyzed reaction. Mane et al. used a whole cell biocatalyst, *Saccharomyces cerevisae* (Baker's yeast), with 1,3-dicarbonyl compound and hydrazines at room temperature to give the N-substituted pyrazole derivatives in 70–92% yield through oxidative cyclocondensation reaction (Fig. 9) (Mane et al. 2015). This synthetic method was found to be useful to a series of

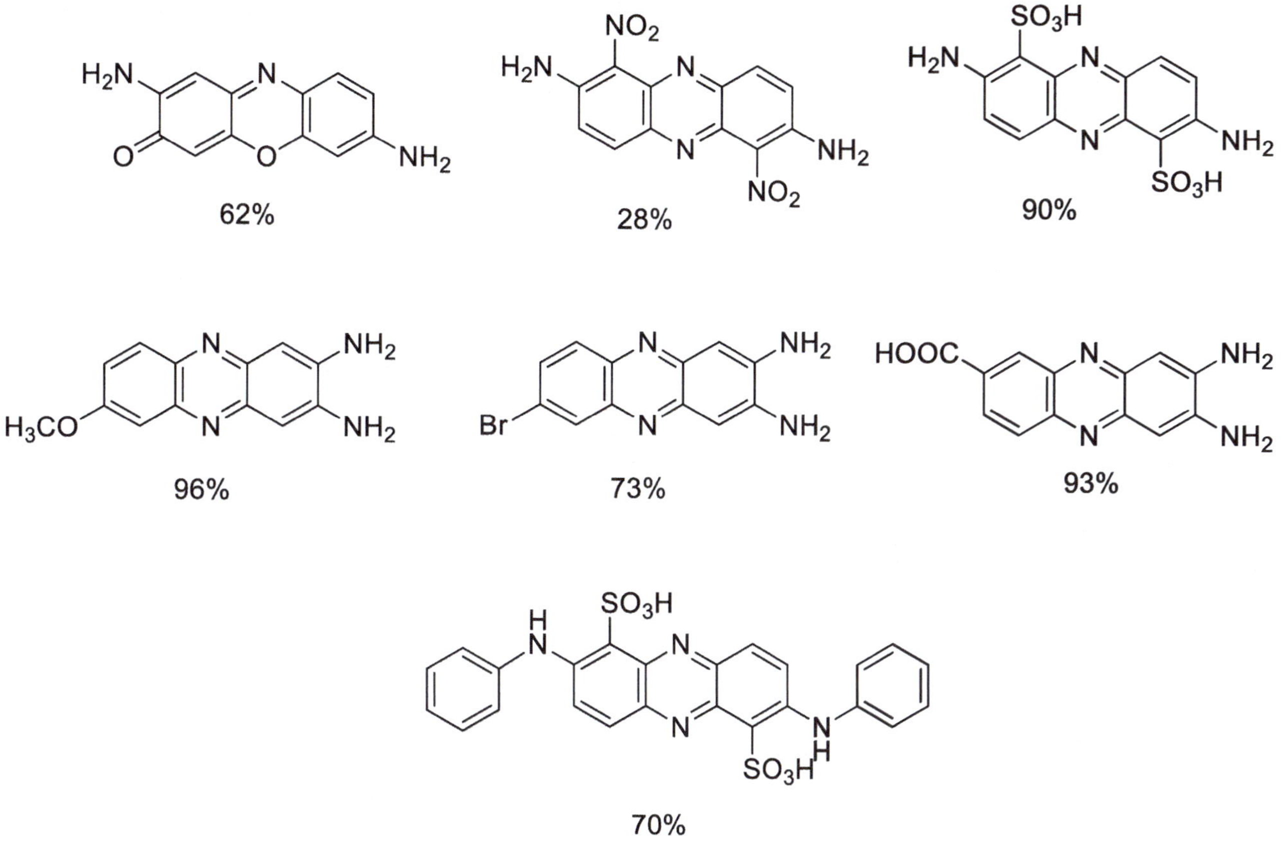

Fig. 6 Examples of laccase-catalyzed synthesis of some molecules

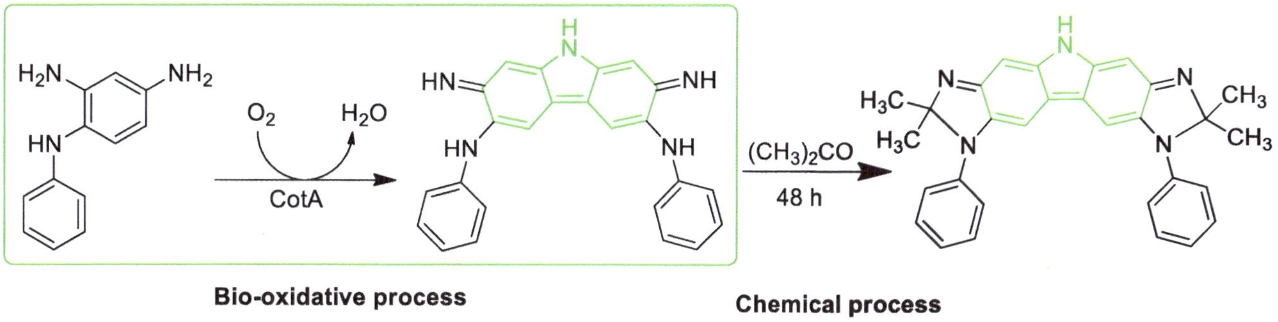

Fig. 7 Chemoenzymatic synthesis of benzocarbazole

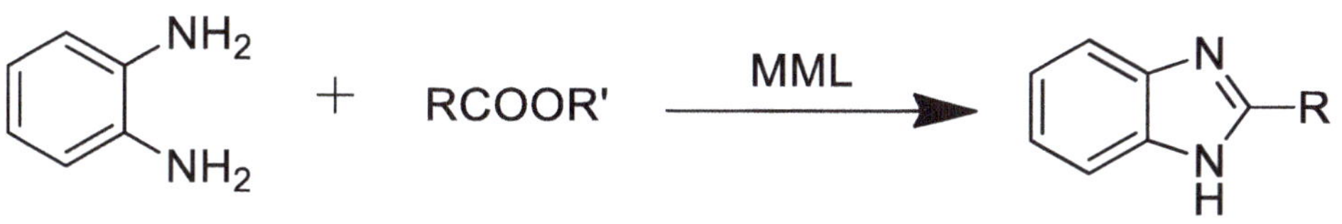

Fig. 8 Enzyme-mediated 2-alkylbenzimidazole synthesis

pyrazole derivatives where fermented Baker's yeast played role in efficient cyclocondensation 1,3-diketones with hydrazines/hydrazides. The study also suggested that the presence of the enzyme lipase in Baker's yeast accelerated the reaction leading to the desired pyrazole derivatives (Mane et al. 2015).

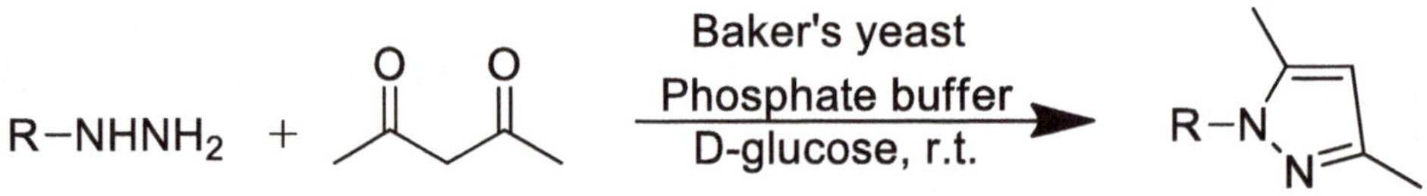

Fig. 9 Baker's yeast catalyzed synthesis of pyrazoles

Pyranopyrazoles are another important group of heterocyclic compounds. A four-component cyclocondensation reaction of hydrazine hydrate, malononitrile, ethyl acetoacetate and benzaldehyde afforded 6-amino-3-methyl-4-(3-nitrophenyl)-2,4-dihydropyrano[2,3-c]pyrazole-5-carbonitrile using lipase from fungi *Aspergillus niger* as catalyst (Fig. 10) (Bora et al. 2013). This method of synthesis was successfully utilized for different carbonyl compounds as one of the reactant yielding dihydropyrano[2,3-c]pyrazoles in 75–98% yield (Bora et al. 2013). The cyclic ketones also successfully gave spiro-substituted dihydropyrano[2,3-c] pyrazoles in 70–80% yield. The enzyme showed its utility towards wide range of substrates, reusability and mild reaction condition, i.e. room temperature and ethanol solvent. The study was also done on lipases from different sources such as *Pseudomonas cepacia*, Amano AK, *Penicillium camemberti*, *Porcine pancreas* and *Aspergillus niger* giving the yield of 65, 72, 75, 91 and 95%, respectively, for the product 6-amino-3-methyl-4-(3-nitrophenyl)-2,4-dihydropyrano[2,3-c]pyrazole-5-carbonitrile.

2.2 Oxygen-Containing Heterocycles

There is no doubt that oxygen-containing heterocycles play important role in industrial, medicinal and nutritional applications due to their diverse biological functions and natural abundance (Venkatachalam and Kumar 2019). Their synthetic methods are always being explored. There are many chemical synthetic methods to produce oxygen-containing heterocycles but due to toxicity and unfriendly approach towards environment and economy, there is demand for green synthesis. Enzyme-mediated synthesis is one of the methodologies to produce oxygen-containing heterocyclic compounds.

2.2.1 Chromenes

4H-chromene derivatives are shown their potential in various pharmaceutical activities (Zhang et al. 2018). They have been synthesized with the help of immobilized *mucor miehei* lipase (Fig. 11) through multi-component reaction using aldehydes, active methylene compounds and suitable nucleophile in 81–96% yield (Fig. 12) (Zhang et al. 2018). The immobilization of enzymes on magnetite nanoparticles (MNPs) has advantages for low mass transfer resistance, high specific surface area and easy separation from the reaction mixture in the presence of magnetic field (Vaghari et al. 2016; Hola et al. 2015). Zhang et al. used silica-coated MNPs as starting material whose surface was functionalized with 3-aminopropyltriethoxy silane to introduce amino groups. This was further treated with 2,4,6-trichloro-1,3,5-triazine (TCT) to develop support for covalent immobilization of enzyme (Zhang et al. 2018; Abbasi et al. 2016; Ranjbakhsh et al. 2012). Xu et al. also reported lipase-catalyzed synthesis of tetrahydrochromene derivatives using 1,3-dicarbonyl compound, aldehyde and malononitrile (Xu et al. 2011).

2.2.2 Dioxins

Dioxin is 6-membered heterocyclic, non-aromatic organic molecule which consists of four carbon atoms and two oxygen atoms. The molecular formula is $C_4H_4O_2$. It is also refereed to 1,4-dioxin or *p*-dioxin. There is an isomeric form which is 1,2-dioxin (o-dioxin) and is very unstable due to its peroxide nature. Agarwal et al. in 2014 proposed an enzymatic synthesis of polybrominated dioxins by using monooxygenase halogenase CPY450 Bmp7 as catalyst in

Fig. 10 Dihydropyrano[2,3-c] pyrazole synthesis

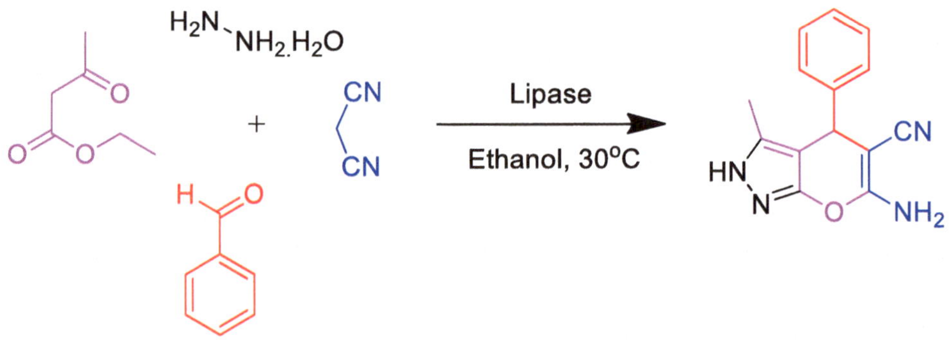

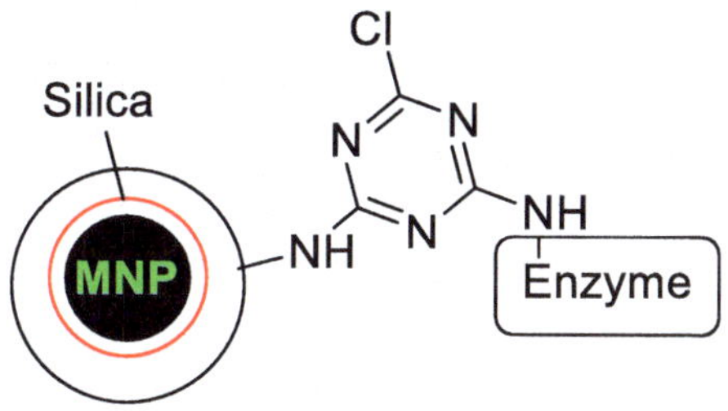

Fig. 11 Covalent immobilization of lipase enzyme

the reaction using bromocatechol as substrates (Fig. 13) (Agarwal and Moore 2014). Both bromocatechol and the electrophilic quinone, i.e. 3,5-dibromo-1,2-dibenzoquinone undergo coupling reaction in the presence of CYP450 Bmp7 enzyme as catalyst to produce dibenzo-p-dioxins. In this case, the benzoquinone provided both 1,4-dioxin oxygen atoms (Fig. 13) (Agarwal and Moore 2014). This reaction method is indicative of synthetic hetero-Diels–Alder coupling between orthoquinones and enamines leading to the formation of 1,4-benzodioxin frameworks. The whole reaction has very mild reaction conditions. The synthesis of desired dioxins can be achieved at room temperature. Excess bromine was quenched by the addition of sodium thiosulphate, and the reaction was extracted twice with the ethyl acetate. This methodology was simple, quite easy to handle and environment-friendly (Agarwal and Moore 2014).

2.2.3 Lactones

Lactones are cyclic esters and have potential applications in the field of synthetic intermediates, pharmaceutical molecules and polymers (Fischer and Pietruszka 2010). A monoclonal antibody (Fig. 14) was utilized as biocatalyst for the synthesis of -lactone (Kitazume et al. 1996). This antibody behaved as enzyme-like catalyst (abzyme) leading to the formation of carbon–carbon bond through the generation of carbanion and the internal nucleophilic attack on the carbonyl carbon to give -lactone (Fig. 15) (Kitazume et al. 1996). Drozdz et al. gave a chemoenzymatic method for the synthesis of lactone using catalyst acyltransferase from *Mycobacterium smegmatis* in high yield of 84–99% through Baeyer–Villiger (BV) oxidation method (Drożdż et al. 2016). The enzyme retained its activity even in harsh reaction condition like oxidation with 60% aq. H_2O_2 at 45 °C. The practical potential of this method was established by the use of different ketones as starting material to give their corresponding lactones (Drożdż et al. 2016).

2.2.4 Benzofuran

A five-membered ring possessing an oxygen atom fused with a benzene ring is known as benzofuran. Benzofuran derivatives have wide range of applications mainly in the field of pharmaceutical industries. They exhibited selective cytotoxicity against tumourigenic cell lines (Hayakawa et al. 2004), antiviral and antitumor and activities (Kim et al. 2006), and pharmaceutical agents (Murata et al. 2003; Murota et al. 1990). Kidwai et al. gave an enzymatic synthesis of this molecule (Kidwai et al. 2013). They studied the enzymatic oxidation of catechols/hydroquinones in aqueous solution using laccase as a catalyst and pyrazolin-5-ones as co-substrate (Fig. 16). Here, the enzyme laccase performs one-electron oxidation on catechol to quinone which undergoes 1,4-addition reaction with co-substrate to develop furan ring leading to benzofuro[2,3-c]pyrazolin-5-ones derivatives (Fig. 16) (Kidwai et al. 2013). The optimized synthetic process has been successfully extended towards the synthesis of a new series of benzofuropyrazole derivatives through the coupling of 3-methyl-1-phenyl-pyrazolin-5-one/3-methyl-pyrazolin-5-one and catechols/hydroquinones (Kidwai et al. 2013).

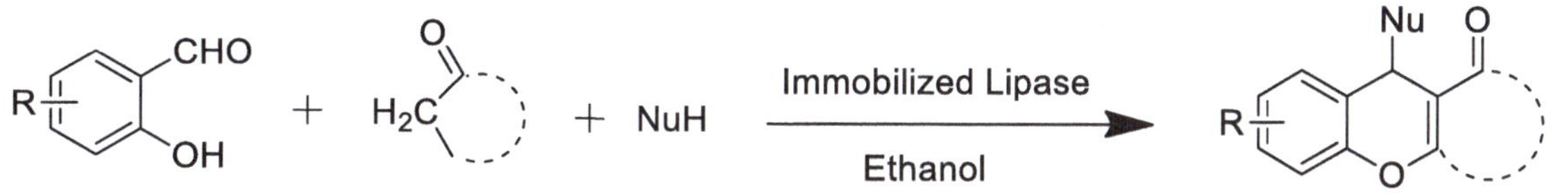

Fig. 12 Enzymatic synthesis of 4H-chromenes

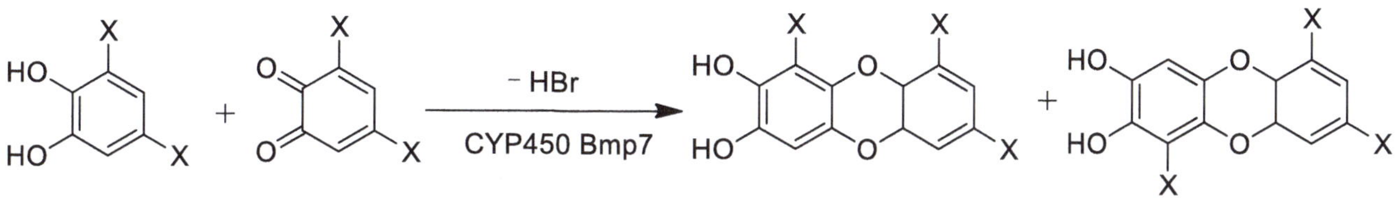

Fig. 13 Enzymatic synthesis of dioxin catalyzed by CYP450 Bmp7 (X=Br)

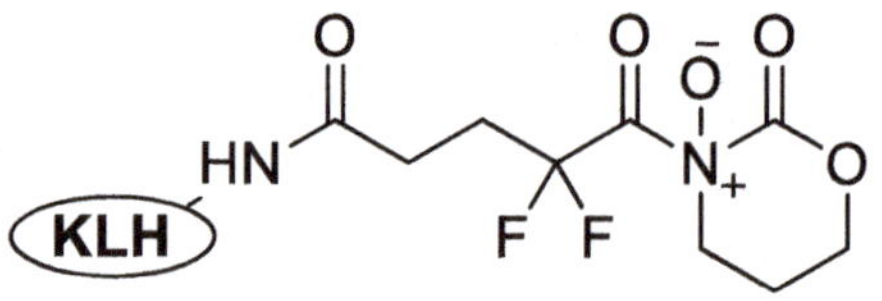

Fig. 14 Keyhole limpet haemocyanin (KLH) antibody

2.3 Nitrogen- and Oxygen-Containing Heterocycles

Heterocycles possessing nitrogen and oxygen in single ring or in a molecule are well-known moiety in medicinal chemistry. Their utility as immunomodulator, antifungal, psychotropical, antibacterial, neuro-related drugs, etc., has been established (Bhattacharya et al. 1991; Kakeya et al. 1998; Danielmeier and Steckhan 1995; Mishra et al. 2019). This section discusses the enzymatic synthesis of those heterocyclic molecules which have nitrogen and oxygen as heteroatoms.

2.3.1 Oxazolidinones

Oxazolidinones are nitrogen- and oxygen-containing five-membered heterocyclic molecules possessing varied applications (Kakeya et al. 1998; Danielmeier and Steckhan 1995). Yadav et al. studied the synthesis of 3-ethyl-1,3-oxazolidin-2-one using Novozyme 435 as catalyst from 2-aminoalcohol and dimethyl carbonate in 61–89% yield (Fig. 17) (Yadav and Pawar 2014). Among the eight immobilized lipases studied, the *Candida antarctica* lipase B (Novozyme 435) was considered as the choice of the catalyst for the reaction. The authors optimized the effect of various parameters like catalyst loading, temperature, agitation speed, solvent and mole ratio for the reaction (Yadav and Pawar 2014).

The lipase-catalyzed reaction was also utilized for the synthesis of enantioenriched oxazolidinone derivatives with excellent enantiopurities (Zhang et al. 2015). The reaction of 2-(methylamino)-1-phenylethanol and disubstituted carbonate as substrates yielded corresponding oxazolidinone in 46% yield with an absolute (S)-configuration as the major enantiomer (ee 92%) (Fig. 18) (Zhang et al. 2015). Different lipases such as from *Burkholderia* (*Pseudomonas*) *cepacia*, *Pseudomonas fluorescens* and *Candida antarctica* were studied. The immobilized *P. cepacia* gave better result and faster substrate transformation in chosen solvent *tert*-butyl methyl ether (Zhang et al. 2015). Various enzyme-mediated synthesized heterocyclic compounds have been represented in Table 1.

3 Summary and Outlook

Heterocyclic compounds are industrially essential molecules. Their history started in the eighteenth century, and since then, they developed themselves as both natural products and synthetic molecules. There is always a demand for the wide-structural diversity of heterocycles and hence

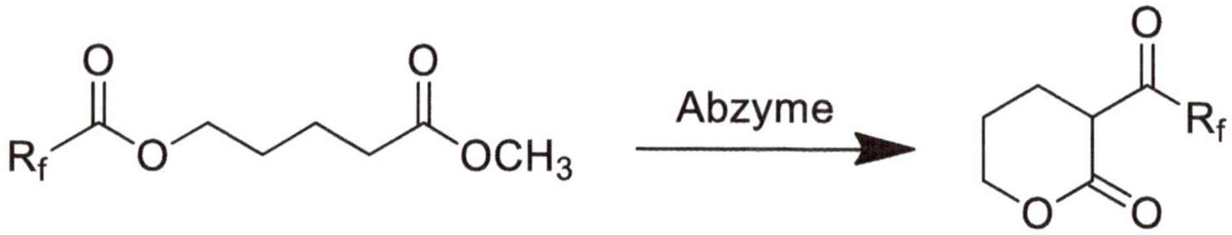

Fig. 15 δ-Lactone synthesis with abzyme catalyst

Fig. 16 Benzofuro [2,3-c] pyrazolin-5-ones synthesis

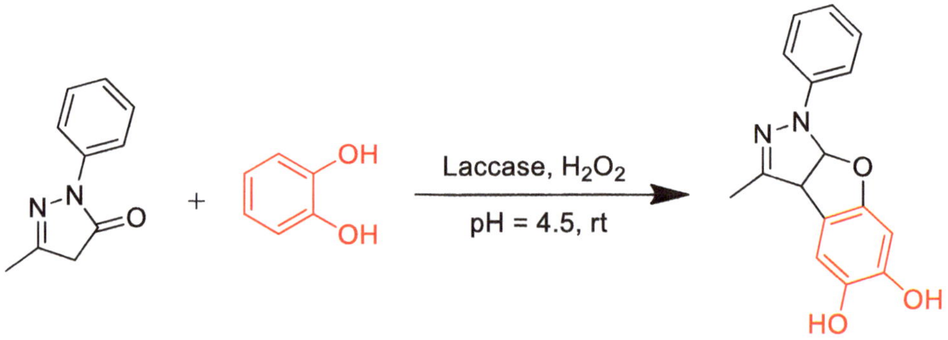

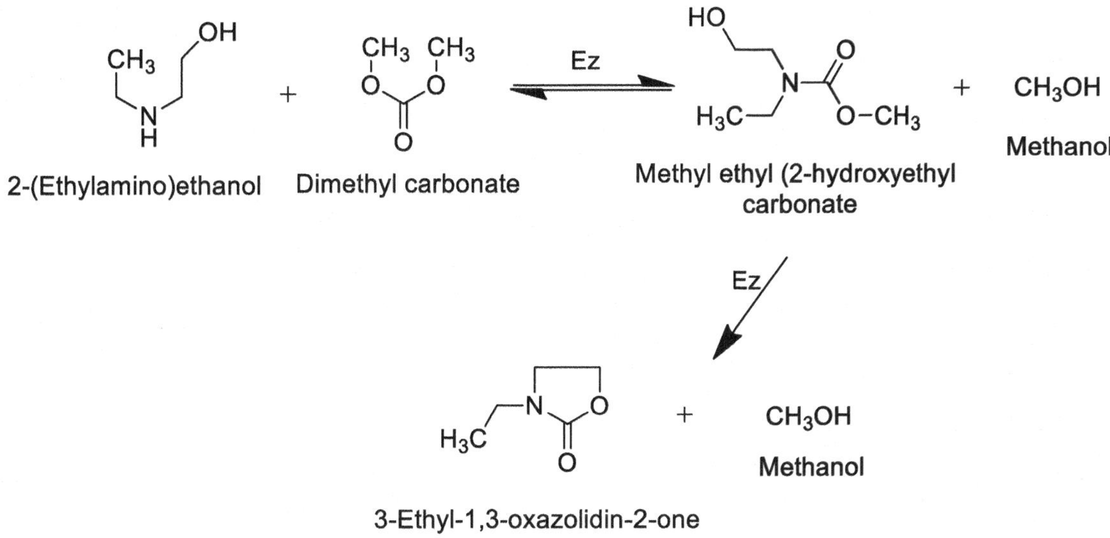

Fig. 17. 3-Ethyl-1,3-oxazolidin-2-one synthesis

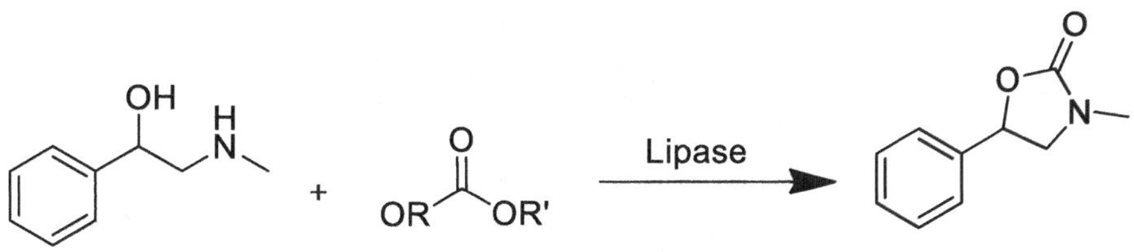

Fig. 18 Synthesis of oxazolidinone through lipase-catalyzed reaction

Table 1 Various enzyme-mediated synthesized heterocyclic compounds

S. No.	Heterocyclic compounds	Enzyme used for synthesis	References
1	*N*-Substituted pyrrole derivatives	Amylase from hog pancreas	Zheng et al. (2013)
2	Spirooxindole derivatives	Lipase from porcine pancreas	Chai et al. (2011)
3	Phenazine derivatives	Laccases	Sousa et al. 2014)
4	Phenoxazinone derivatives	Laccases	Sousa et al. (2014), Mihovilovic (2006), Bruyneel et al. (2009)
5	Carbazole derivatives	CotA laccase	Sousa et al. (2015)
6	2-Alkylbenzimidazole derivatives	Immobilized lipase from *Mucor miehei*	Wang et al. (2010)
7	Pyrazole derivative derivatives	*Saccharomyces cerevisae* (Baker's yeast)	Mishra and Sasmal (2011)
8	Pyranopyrazole derivatives	Lipase from *Aspergillus niger*	Bora et al. (2013)
9	4H-Chromene derivatives	Immobilized *mucor miehei* lipase	Zhang et al. (2018)
10	Tetrahydrochromene derivatives	Lipase	Xu et al. (2011)
11	Dioxin derivatives	Monooxygenase halogenase CPY450 Bmp7	Agarwal and Moore(2014)
12	δ-Lactone derivatives	Abzyme	Kitazume et al. (1996)
13	Lactone derivatives	Acyltransferase from *Mycobacterium smegmatis*	Drożdż et al. (2016)
14	Benzofuran derivatives	Laccase	Kidwai et al. (2013)
15	3-Ethyl-1,3-oxazolidin-2-one derivatives	Novozyme 435	Yadav and Pawar (2014)
16	Oxazolidinone derivatives	Lipase	Zhang et al. (2015)

(continued)

Table 1 (continued)

S. No.	Heterocyclic compounds	Enzyme used for synthesis	References
17	Indolyl 4H-Chromenes	Lipase	Zhang et al. (2017b)
18	Lactone	Engineered Baeyer–Villiger monooxygenase (BVMO)	Farhat (2020), Mihovilovic (2006)

exploration of synthetic protocols. Due to the environmental issues related to chemical synthesis, the researchers started looking for alternatives to chemical synthesis in accordance with green and sustainable chemistry. Out of the several modified methods, biocatalysis or using enzymes for synthesis showed potential. The enzymatic heterocyclic synthesis has contributed significantly to the structural diversity of heterocyclic compounds. This also allowed their applicability in various fields. No doubt, the chemical synthesis has also provided structural diversity, but for many asymmetric syntheses has better been performed using enzymatic methods.

This chapter discussed the enzymatic synthesis of nitrogen- and oxygen-containing or possessing both the atoms heterocyclic molecules. The synthesis of pyrroles, indoles, phenazines, benzocarbazoles, benzimidazoles, pyrazoles, benzofurans, chromenes, dioxins, lactones and oxazolidinones has been discussed. This has been observed that the enzyme-catalyzed reactions are more suitable than chemical reactions due to their high selectivity and mild reaction conditions. With the development of genetic engineering and enzyme engineering, the cost of enzymes continues to decrease. Enzyme activity and stability continue to increase. The superiority of enzymatic clean production is definitely promoting breakthrough development in this field. Enzyme-catalyzed biotransformation technology represents the future development of this synthetic industry.

References

Abbasi M, Amiri R, Bordbar AK, Ranjbakhsh E, Khosropour AR (2016) Improvement of the stability and activity of immobilized glucose oxidase on modified iron oxide magnetic nanoparticles. Appl Surf Sci 364:752–757

Agarwal V, Moore BS (2014) Enzymatic synthesis of polybrominated dioxins from the marine environment. ACS Chem Biol 9:1980–1984

Arora P, Arora V, Lamba HS, Wadhwa D (2012) Importance of heterocyclic chemistry: a review. Int J Pharma Sci Res 3:2947–2954

Arunkumar SS (2015) A review on synthetic hetrocyclic compounds in agricultural and other applications. Int J PharmTech Res 8:170–179

Bellina F, Rossi R (2006) Synthesis and biological activity of pyrrole, pyrroline and pyrrolidine derivatives with two aryl groups on adjacent positions. Tetrahedron 62:7213–7256

Bhattacharya SK, Clow A, Przyborowska A, Halket J, Glover V, Sandler M (1991) Effect of aromatic amino acids, pentylenetetrazole and yohimbine on isatin and tribulin activity in rat brain. Neurosci Lett 132:44–46

Bora PP, Bihani M, Bez G (2013) Multicomponent synthesis of dihydropyrano[2,3-c]pyrazoles catalyzed by lipase from *Aspergillus niger*. J Mol Catal B: Enzym 92:24–33

Broughton HB, Watson IA (2004) Selection of heterocycles for drug design. J Mol Graph Model 23:51–58

Bruyneel F, Payen O, Rescigno A, Tinant B, Marchand-Brynaert J (2009) Laccase-mediated synthesis of novel substituted phenoxazine chromophores featuring tuneable water solubility. Chem Eur J 15:8283–8295

Busto E, Fernandez VG, Gotor V (2011) Hydrolases in the stereoselective synthesis of N-heterocyclic amines and amino acid derivatives. Chem Rev 111:3998–4035

Chai SJ, Lai YF, Xu JC, Zheng H, Zhu Q, Zhanga PF (2011) One-pot synthesis of spirooxindole derivatives catalyzed by lipase in the presence of water. Adv Synth Catal 353:371–375

Chauhan SMS, Geetanjali, Singh R (2000) A mild and efficient synthesis of 10-substituted isoalloxazines in the presence of solid acids. Ind J Heterocycl Chem 10:157–158

Chauhan SMS, Singh R, Geetanjali (2003) Microwave assisted synthesis of 10-substituted isoalloxazines in the presence of solid acids. Synth Commun 33:1179–1184

Dalal KS, Wagh YB, Trivedi DR, Dalal DS, Chaudhari BL (2016) Bovine Serum Albumin catalyzed one pot, three-component synthesis of dihydropyrano[2,3-c]pyrazole derivatives in aqueous ethanol. RSC Adv 6:14868–14879

Danielmeier K, Steckhan E (1995) Efficient pathways to (*R*)- and (*S*)-5-hydroxymethyl-2-oxazolidinone and some derivatives. Tetrahedron Asymmetry 6:1181–1190

Ding K, Lu Y, Coleska ZN, Wang G, Qiu S, Shangary S, Gao W, Qin D, Stuckey J, Krajewsky K, Roller PP, Wang S (2006) Structure-based design of spiro-oxindoles as potent, specific small-molecule inhibitors of the MDM2−p53 interaction. J Med Chem 49:3432–3435

Ding Y, Ni X, Gu M, Li S, Huang H, Hu Y (2015) Knoevenagel condensation of aromatic aldehydes with active methylene compounds catalyzed by lipoprotein lipase. Catal Commun 64:101–104

Drożdż A, Hanefeld U, Szymańska K, Jarzębski A, Chrobok A (2016) A robust chemo-enzymatic lactone synthesis using acyltransferase from *Mycobacterium smegmatis*. Catal Commun 81:37–40

Farhat W, Biundo A, Stamm A, Malmström E, Syrén PO (2020) Lactone monomers obtained by enzyme catalysis and their use in reversible thermoresponsive networks. J Appl Polym 137:48949

Feber K (2004) Biotransformations in organic chemistry, 5th edn. Springer, Berlin, Heidelberg, p 454

Fischer T, Pietruszka J (2010) Key building blocks *via* enzyme-mediated synthesis. In: Piel J (eds) Natural products *via* enzymatic reactions. Topics in current chemistry, vol 297. Springer, Berlin

Galliford CV, Scheidt KA (2007) Natural pyrrolidinylspirooxindoles as templates for the development of medicinal agents. Angew Chem 119:8902–8912

Gamage SA, Rewcastle GW, Baguley BC, Charlton PA, Denny WA (2006) Phenazine-1-carboxamides: structure–cytotoxicity relationships

for 9-substituents and changes in the H-bonding pattern of the cationic side chain. Bioorg Med Chem 14:1160–1168
Grazulevicius JV, Strohriegl P, Pielichowski J, Pielichowski K (2003) Carbazole-containing polymers: synthesis, properties and applications. Prog Polym Sci 28:1297–1353
Groger H, Asano Y (2012) Principle of enzyme catalysis. In: Drauz K, Groger H, May O (eds) Enzyme catalysis in organic synthesis, 3rd edn. Wiley-VCH Verlag GmbH & Co KGaA, pp 1–42
Hayakawa I, Shioya R, Agatsuma T, Furukawa H, Naruto S, Sugano Y (2004) 4-Hydroxy-3-methyl-6-phenylbenzofuran-2-carboxylic acid ethyl ester derivatives as potent anti-tumor agents. Bioorg Med Chem Lett 14:455–458
Hemmerling F, Hahn F (2016) Biosynthesis of oxygen and nitrogen-containing heterocycles in polyketides. Beil J Org Chem 12:1512–1550
Hola K, Markova Z, Zoppellaro G, Tucek J, Zboril R (2015) Tailored functionalization of iron oxide nanoparticles for MRI, drug delivery, magnetic separation and immobilization of biosubstances. Biotechnol Adv 33:1162–1176
Hu W, Guan Z, Deng X, He YH (2012) Enzyme Catalytic promiscuity: the papain-catalyzed knoevenagel reaction. Biochimie 94:656–661
Jampilek J (2019) Heterocycles in medicinal chemistry. Molecules 24:3839
Joule JA, Mills K (2013) Heterocyclic chemistry at a glance: applications and occurrences of heterocycles in everyday life, 2nd edn. Wiley, New York, pp 180–194
Junghanns KT, Kneusel RE, Baumert A, Maier W, Gröger D, Matern U (1995) Molecular cloning and heterologous expression of acridone synthase from elicited *Ruta graveolens* L. cell suspension cultures. Plant Mol Biol 27:681–692
Kakeya H, Morishita M, Kobinata K, Osono M, Ishizuka M, Osada H (1998) Isolation and biological activity of a novel cytokine modulator, cytoxazone. J Antibiot 51:1126–1128
Kidwai M, Jain A, Sharma A, Kuhad RC (2013) Laccase—a natural source for the synthesis of benzofuro[2,3-c]pyrazolin-5-ones. Catal Sci Technol 3:230–234
Kim S, Salim AA, Swanson SM, Kinghorn AD (2006) Potential of cyclopenta[b]benzofurans from Aglaia species in cancer chemotherapy. Anti-Cancer Agents Med Chem 6:319–345
Kitazume T, Tsukamoto T, Murata K, Yoshimura K (1996) Preparation of heterocyclic compounds via carbon-carbon bond formation catalyzed by an antibody. J Mol Catal B: Enzym 2:27–31
Kłossowski S, Wiraszka B, Berłożecki S, Ostaszewski R (2013) Model studies on the first enzyme-catalyzed Ugi reaction. Org Lett 15:566–569
Knölker HJ, Reddy KR (2002) Isolation and synthesis of biologically active carbazole alkaloids. Chem Rev 102:4303–4428
Kobayashi S, Okamoto E, Wen X, Shoda S (1996) Chemical synthesis of native-type cellulose and its analogues via enzymatic polymerization. J Macromol Sci Part A: Pure Appl Chem 33:1375–1384
Kobayashi S, Shoda S, Wen X, Okamoto E, Kiyosada T (1997) Choroselective enzymatic polymerization for synthesis of natural polysaccharides. J Macromol Sci Part a: Pure Appl Chem 34:2135–2142
Kozikowski A (1984) Comprehensive heterocyclic chemistry. In: Katritzky AR, Rees CW (eds), vol 1. Pergamon Press Ltd., Oxford, p 567
Lee DC, Cao B, Jang K, Forster PM (2010) Self-assembly of halogen substituted phenazines. J Mater Chem 20:867–873
Lia J, Grimsdale AC (2010) Carbazole-based polymers for organic photovoltaic devices. Chem Soc Rev 39:2399–2410
Li C, Feng XW, Wang N, Zhou YL, Yu XQ (2008) Biocatalytic promiscuity: the first lipase-catalysed asymmetric aldol reaction. Green Chem 10:616–618
Mane A, Salokhe P, More P, Salunkhe R (2015) An efficient practical chemo enzymatic protocol for the synthesis of pyrazoles in aqueous medium at ambient temperature. J Mol Catal B: Enzym 121:75–81
Mane AH, Patil AD, Kamat SR, Salunkhe RS (2018) Biocatalyst mediated synthesis of tryptanthrins performed under ultrasonication. Chem Select 3:6454–6458
Mihovilovic MD (2006) Enzyme mediated Baeyer-Villiger oxidations. Curr Org Chem 10:1265–1287
Mikolasch A, Schauer F (2009) Fungal laccases as tools for the synthesis of new hybrid molecules and biomaterials. Appl Microbiol Biotechnol 82:605–624
Milner SE, Maguire AR (2012) Recent trends in whole cell and isolated enzymes in enantioselective synthesis. ARKIVOC 1:321–382
Mishra N, Sasmal D (2011) Development of selective and reversible pyrazoline based MAO-B inhibitors: virtual screening, synthesis and biological evaluation. Bioorg Med Chem Lett 21:1969–1973
Mishra D, Fatima A, Singh R, Munjal NS, Mehta V, Malairaman U (2019) Design, synthesis and evaluation of coumarin-phenylthiazole conjugates as cholinesterase inhibitors. Chem Biol Lett 6:23–30
Murata K, Kumagai H, Kawashima T, Tamitsu K, Irie M, Nakajima H, Suzu S, Shibuya M, Kamihira S, Nosaka T, Asano S, Kitamura T (2003) Selective cytotoxic mechanism of GTP-14564, a novel tyrosine kinase inhibitor in leukemia cells expressing a constitutively active Fms-like tyrosine kinase 3 (FLT3). J Biol Chem 278:32892–32898
Murota S, Morita I, Suda N (1990) The control of vascular endothelial cell injury. Ann N Y Acad Sci 598:182–187
Okazaki M, Takeda Y, Data P, Pander P, Higginbotham H, Monkman AP, Minakata S (2017) Thermally activated delayed fluorescent phenothiazine–dibenzo[*a*, j]phenazine–phenothiazine triads exhibiting tricolor-changing mechanochromic luminescence. Chem Sci 8:2677–2686
Pauliukaite R, Ghica ME, Barsan MM, Brett CMA (2010) Phenazines and polyphenazines in electrochemical sensors and biosensors. Anal Lett 43:1588–1608
Pecca JG, Albonico SM (1971) Synthetic trypanocides. 2. Substituted 5,6-dihydro[c]benzocarbazoles. J Med Chem 14:448–449
Poonam, Singh R (2019) Facile one-pot synthesis of highly functionalized pyrazoles using alumina-silica-supported MnO_2 as recyclable catalyst in water. Res Chem Intermed 45:4531–4542
Ranjbakhsh E, Bordbar AK, Abbasi M, Khosropour AR, Shams E (2012) Enhancement of stability and catalytic activity of immobilized lipase on silica-coated modified magnetite nanoparticles. Chem Eng J 179:272–276
Reetz MT, Mondière R, Carballeira JD (2007) Enzyme promiscuity: first protein-catalyzed Morita-Baylis-Hillman reaction. Tetrahedron Lett 48:1679–1681
Rudi A, Erez, Y, Benayahu Y, Kashman Y (2005) Omriolide A and B; two new rearranged spongian diterpenes from the marine sponge *Dictyodendrilla aff. Retiara*. Tetrahedron Lett 46:8613–8616
Sabir S, Alhazza MI, Ibrahim AA (2015) A review on heterocyclic moiety and their applications. Catal Sustain Energy 2:99–115
Salahuddin SM, Mazumder A (2017) Benzimidazoles: a biologically active compounds. Arab J Chem 10:S157–S173
Shoda S, Uyama H, Kadokawa J, Kimura S, Kobayashi S (2016) Enzymes as green catalyst for precision macromolecular synthesis. Chem Rev 116:2307–2413
Singh R, Geetanjali, Singh V (2011) Exploring alkaloids as inhibitors of selected enzyme. Asian J Chem 23:483–490
Singh R, Sharma R, Tewary N, Geetanjali, Rawat DS (2006) Nitrilase and its application as a green catalyst. Chem Biodiv 3:1279–1287
Sousa AC, Oliveira MC, Martins LO, Robalo MP (2014) Towards the rational biosynthesis of substituted phenazines and phenoxazinones by laccases. Green Chem 16:4127–4136

Sousa AC, Fátima M, Piedade MM, Martins LO, Robalo MP (2015) An enzymatic route to a benzocarbazole framework using bacterial CotA laccase. Green Chem 17:1429–1433

Sousa AC, Oliveira MC, Martins LO, Robalo MP (2018) A sustainable synthesis of asymmetric phenazines and phenoxazinones mediated by CotA-Laccase. Adv Synth Catal 360:575–583

Stellrecht CM, Chen LS (2011) Transcription inhibition as a therapeutic target for cancer. Cancers 3:4170–4190

Trofimov BA, Sobenina LN, Demenev AP, Mikhaleva AI (2004) C-vinylpyrroles as pyrrole building blocks. Chem Rev 104:2481–2506

Vaghari H, Jafarizadeh-Malmiri H, Mohammadlou M, Berenjian A, Anarjan N, Jafari N, Nasiri S (2016) Application of magnetic nanoparticles in smart enzyme immobilization. Biotechnol Lett 38:223–233

Veer B, Singh R (2019) Facile synthesis of 2-arylimidazo[1,2-a] pyridines catalyzed by DBU in aqueous ethanol. P Roy Soc A-Math Phy 475:1–12

Venkatachalam H, Kumar NVA (2019) Heterocycles: synthesis and biological activities: the oxygen-containing fused heterocyclic compounds. Nandeshwarappa BP, Sadashiv SO (eds) IntechOpen

Vitaku E, Smith D T, Njardarson J T (2014) Analysis of the structural diversity, substitution patterns, and frequency of nitrogen heterocycles among U.S. FDA approved pharmaceuticals. J Med Chem 57:10257–10274

Wang L, Li C, Wang N, Li K, Chen X, Yu XQ (2010) Enzyme-mediated domino synthesis of 2-alkylbenzimidazoles in solvent-free system: a green route to heterocyclic compound. J Mol Catal B: Enzym 67:16–20

Wang Z, Chen X, Wang C, Zhang L, Li FX, Zhang WA, Chen P, Wang L (2017) A mild and efficient dakin reaction mediated by lipase. Green Chem Lett Rev 10:269–273

Webb E C (1992) Enzyme nomenclature 1992, recommendation of the nomenclature committee of the international union of biochemistry and molecular biology on the nomenclature and classification of enzymes. International Union of biochemistry and molecular biology, 6th edn. Academic Press, San Diego, p 863

Witayakran S, Ragauskas AJ (2009) Synthetic applications of Laccase in green chemistry. Adv Synth Catal 351:1187–1209

Wu J, Wang X, Wang Q, Lou Z, Li S, Zhu Y, Qin L, Wei H (2019) Nanomaterials with enzyme-like characteristics (nanozymes): next generation artificial enzymes (II). Chem Soc Rev 48:1004–1076

Xiang Z, Liu Z, Chen X, Wu Q, Lin X (2013) Biocatalysts for cascade reaction: porcine pancreas lipase (PPL)-catalyzed synthesis of bis (indolyl)alkanes. Amino Acids 45:937–945

Xiang Z, Liang Y, Chen X, Wu Q, Lin X (2014) D-aminoacylase-initiated cascade aldol condentation/robinson annulation for the synthesis of substituted cyclohex-2-enones from simple aldehydes and acetones. Amino Acids 46:1929–1937

Xie ZB, Wang N, Jiang GF, Yu XQ (2013) Biocatalytic asymmetric aldol reaction in buffer solution. Tetrahedron Lett 54:945–948

Xue Y, Li L, He Y, Guan Z (2012) Protease-catalsed direct asymmetric mannich reaction in organic solvent. Sci Rep 2:761

Xu JC, Li WM, Zheng H, Lai YF, Zhang PF (2011) One-pot synthesis of tetrahydrochromene derivatives catalyzed by lipase. Tetrahedron 67:9582–9587

Yadav GD, Pawar SV (2014) Novelty of immobilized enzymatic synthesis of 3-ethyl-1,3-oxazolidin-2-one using 2-aminoalcohol and dimethylcarbonate: mechanism and kinetic modeling of consecutive reactions. J Mol Catal B: Enzym 109:62–69

Yang FJ, Wang HR, Jiang LY, Yue H, Zhang H, Wang Z, Wang L (2015) A green and one-pot synthesis of Benzo[g]chromene Derivatives through a multi-component reaction catalyzed by lipase. RSC Adv 5:5213–5216

Zhang Y, Ren Y, Ramström O (2015) Synthesis of chiral oxazolidinone derivatives through lipase-catalyzed kinetic resolution. J Mol Catal B: Enzym 122:29–34

Zhang MJ, Li R, He YH, Guan Z (2017a) Pepsin-catalyzed vinylogous Michael addition of deconjugated butenolides and maleimides in water. Catal Commun 98:85–89

Zhang WA, Zhao ZY, Wang Z, Guo C, Wang CY, Zhao R, Wang L (2017b) Lipase-catalyzed synthesis of indolyl 4H-chromenes via a multicomponent reaction in ionic liquid. Catalysts 7:185

Zhang W, Chen P, Zhao Z, Wang L, Wang S, Tang Y, Wang B, Wang Z, Zhuang H (2018) Synthesis of functionalized 4H-chromenes catalyzed by lipase immobilized on magnetic nanoparticles. Green Chem Lett Rev 11:246–253

Zheng H, Shi Q, Du K, Mei Y, Zhang P (2013) A novel enzyme-catalyzed synthesis of N-substituted pyrrole derivatives. Mol Divers 17:245–250

Zhuo ST, Li CY, Hu MH, Chen SB, Yao PF, Huang SL, Ou TM, Tan JH, An LK, Li D, Gu LQ, Huang ZS (2013) Synthesis and biological evaluation of benzo[*a*]phenazine derivatives as a dual inhibitor of topoisomerase I and II. Org Biomol Chem 11:3989–4005

Solid-State Green Synthesis of Different Nanoparticles

Madhuri Hembram, Rashmirekha Tripathy, Jagannath Panda, Tejaswini Sahoo, Saraswati Soren, Deepak Senapati, J. R. Sahu, C. K. Rath, Alok Kumar Panda, and Rojalin Sahu

Abstract

In material science, "green synthesis" has achieved more attention as an environmentally sustainable, trustable and eco-friendly way for the large-scale synthesis of nanomaterial containing metal and metal oxides. Metal nanoparticles like silver and gold are synthesized by utilizing plant extracts. Further, metal oxides of Zn, ZnO and copper (CuO) are also synthesized by using plant metabolites. It is an essential tool for decreasing the detrimental effects and has great contribution in various applications such as drug delivery, dentistry, X-ray imaging and agricultural engineering and in many more fields. Synthesis of metal nanoparticles and metal oxide nanoparticles by following a greener route is summarized in this chapter. This chapter also throws light on the various applications of metal and metal oxide nanoparticles in different fields.

Keywords

Green synthesis • Nanoparticles • Metal oxide nanoparticles • Eco-friendly • Microorganism

1 Introduction

Nanotechnology or nanoscience is the field that majorly constitutes the synthesis of particles which ranges from 1 to 100 nm. These ranges of nanoparticles (NPs) are used in an array of areas such as in physical science, chemical science, pharmaceutical industries, molecular biology and material sciences (Heiligtag and Niederberger 2013; De et al. 2008). In the past several decades, various NPs involving silver, gold, copper and copper oxide, zinc oxide nanoparticles, etc., have been synthesized among which silver NPs (AgNPs) have found a variety of different applications. Different techniques can be applied to control the size as well as shapes of the nanoparticles (Abid et al. 2002; Itakura et al. 1995; Pol et al. 2002; Stiger et al. 1999; Harfenist et al. 1996; Komarneni et al. 2002; Liz-Marzán and Lado-Touriño 1996; Petit et al. 1993; Heath et al. 1997). Among these various techniques, the most important method is solid-state synthesis for synthesizing NPs of desired shape and size. This method of synthesis is mainly based on the grinding technique. But in the case of inorganic and organic substrates, this method involves the use of solvent. The use of a solvent in this method is quite limited, and this limited exposure of solvent leads to the fabrication of materials with unique structure and composition which is beneficial for piezoelectric substances (Iravani et al. 2014). As the synthesis of NPs by solid-state involves minimum use of solvents, it is considered as a green synthetic approach (Roy and Barik 2010). The characteristics of the NPs generally depend on their shape, size, composition, and its crystalline form. The two other methods such as top-down and bottom-up approaches have been implemented to synthesize nanomaterials having a suitable shape, size and functionalities, which are illustrated in Fig. 1. Metallic nanoparticles are especially useful due to its different catalytic, electrical as well as optical characteristics. The colloidal nanoparticle of silver is one of the extensively used nanomaterials, and application of it in various fields generally depends on its desired shape and size (Kim 2007; Polte 2015; El Khoury et al. 2015; Al-Namil et al. 2019). Silver nanoparticles are used in numerous fields such as biotechnology, medicine and catalysis. For example, silver nanoparticles are synthesized with *Acacia nilotica* pod and modified glassy carbon

M. Hembram · R. Tripathy · J. Panda · T. Sahoo · S. Soren · D. Senapati · J. R. Sahu · C. K. Rath · A. K. Panda (✉) · R. Sahu (✉)
School of Applied Sciences, Kalinga Institute of Industrial Technology, Deemed to be University, Bhubaneswar, 751024, India
e-mail: alok.pandafch@kiit.ac.in

R. Sahu
e-mail: rsahufch@kiit.ac.in

Inamuddin et al. (eds.), *Advances in Green Synthesis*, Advances in Science, Technology & Innovation,
https://doi.org/10.1007/978-3-030-67884-5_17

electrode, when undergoes reduction with benzyl chloride exhibits more catalytic properties as compared to the use of glassy carbon and metallic silver electrode only. The combination of silver nanoparticles with the extract of *Gloriosa superba* promotes the degeneration of the dye methylene blue. The production of nanoparticles of Ag using the most nonhazardous and safe reagent curcumin, is considered as a solid-state green synthetic approach. The formation of nanoparticles by the usage of curcumin is a magnificent way for the green synthetic method. The shape and size of the silver nanoparticles are affected by the change in the temperature during the synthesis. Hence, temperature plays a very important role in the formation of the nanoparticles. The preparation of curcumin conjugated silver nanoparticles has been used as a catalyst to reduce p-nitrophenol.

To enhance the physical and chemical properties of nanosized metal particles, further investigations have been employed to dominate its structure, shape and size. The particle size at nanometers enables them for many applications such as in biological and different optoelectronic devices, mainly due to its efficient optical as well as electronic characters (Alivisatos 1996; Coe et al. 2002; Bruchez et al. 1998). There are numerous chemical and physical synthesis techniques by which silver nanoparticles can be synthesized. These are laser ablation, dispersion, electrochemical, chemical reduction and photochemical reduction. In a few of these techniques, some reagents are required as a surface passivator to avoid the accumulation of nanoparticles. To synthesize large amounts of nanoparticles, organic passivators like thiourea, mercaptoacetate and thiophenol are used which lead to environmental contamination due to their toxicity (Bae et al. 2002; Smetana et al. 2005; Liu and Lin 2004; Sandmann et al. 2000; Yu 2007; Tan et al. 2002; Petit et al. 1993; Vorobyova et al. 1999; Mallick et al. 2005; Pattabi and Uchil 2000; Ravindran et al. 1999). As the majority of the techniques used to synthesize nanoparticles involve a variety of toxic reagents and solvents, the formation of nanoparticles by the biosynthetic process is obtaining significance in the present scenario as it involves greener and eco-friendly ways of synthesizing nanoparticles. To illustrate this, a large amount of inorganic substance has been biosynthesized such as the synthesis of silver nanoparticles by microorganism (Mandal et al. 2006; Basavaraja et al. 2008; Vigneshwaran et al. 2006, 2007; Shahverdi et al. 2007a, b). The problems regarding the preparation of nanoparticles and its standardization in a greener way can be resolved by adopting biosynthetic processes. Among the various biosynthetic processes, the methods involving different microorganisms specifically prokaryotic bacteria have gained great interest. Klaus et al. have illustrated the synthesis of silver-based nanoparticles at the cell poles in the microorganism *Pseudomonas stutzeri* AG259 (Klaus et al. 1999). Eukaryotic organisms like *Verticillium* sp. have been used to synthesize the metal nanoparticles which is reported by Sastry et al.

They reported about the easier processing and management of biomass by replacing bacteria with fungi (Pum and Sleytr 1999; Sleytr et al. 1999). Some microorganisms that produce inorganic nanomaterials such as S-layer bacteria (Lovley et al. 1987; Philipse and Maas 2002) and magnetotactic bacteria (Philipse and Maas 2002; Dickson 1999) are the best

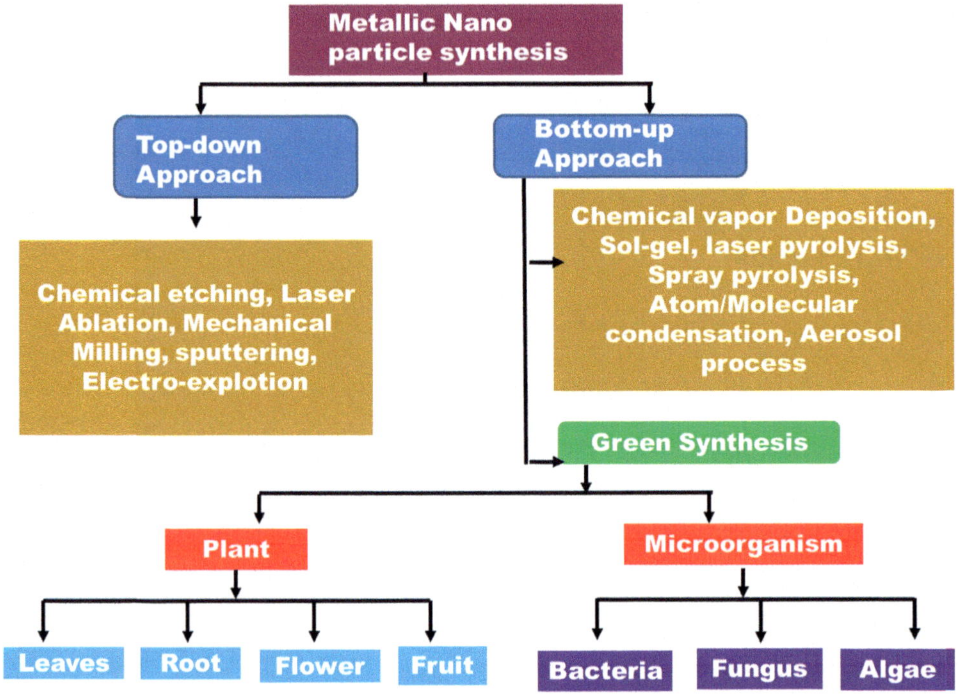

Fig. 1 Different approach for synthesis of nano particles from different sources

examples of synthesizing nanoparticles in an environmentally friendly way. The microscopic organisms like actinomycetes, fungi and bacteria have been investigated in the production of metal nanoparticles like Ag NPs. The usage of whole parts of the plant in the synthesis of nanoparticles is a new and promising frontier in the synthesis of nanoparticles in greener ways. Although the gold nanoparticles (AuNPs) are biocompatible, still it may have detrimental effects in medical operation due to the absorbance of some hazardous chemical substances on the surface of AuNPs during its preparation by using chemical synthesis process. To avoid these hazardous chemical substances and to make the nanoparticle more biocompatible, synthesizing the NPs with plant extracts or microorganisms is a lucrative alternative. Production of metal nanoparticles by utilizing plant derivatives is more beneficial than other environmentally favorable biological methods, and it removes the complicated methods of preserving and maintaining the microorganism cell cultures. Synthesis of gold and silver nanoparticles by using plant extracts and their applications is reported by Jose-Yacaman and co-workers (Dickson 1999; Gardea-Torresdey et al. 2002; Gardea-Torresdey et al. 2003). The above methods which utilize natural ways are considered as green synthetic methods of synthesis for metal nanoparticles.

Nowadays, the preparation of green Ag nanoparticles has been carried out by utilizing various natural products such as neem (*Azadirachta indica*) leaf broth, aloe vera plant extract, green tea (*Camellia sinensis*), leguminous shrub (*Sesbania drummondii*), starch, lemongrass leaves extract (Shankar et al. 2004a, b; Chandran et al. 2006; Vilchis-Nestor et al. 2008; Sharma et al. 2007; Vigneshwaran et al. 2006; Bakar et al. 2007), etc. From an economic point of view, the *Jatropha curcas* plant is most essential as biodiesel is taken out from its seed on a large scale. Nevertheless, the plant *Jatropha latex* has some ethnic medical treatment such as blood coagulating and wound healing. Apart from this, it is also corrosive and irascible to skin. From the widespread literature review, the latex of *J. curcas* contains the major constituents of curcacycline A (a cyclic octapeptide), curcacycline B (a cyclic nonapeptide) and curcain (an enzyme). After analyzing the function of peptides during the preparation of nanoparticles, the researchers hypothesized to develop a "green" route for nanoparticle preparation by utilizing *J. latex* owing to its reducing as well as capping properties. The acquired metal nanoparticles were analyzed by various advanced techniques such as UV-visible spectroscopy, X-ray diffraction (XRD) and high-resolution transmission electron microscopy (HRTEM) study. Other important metal/metal oxide nanoparticles are Cu/CuO nanoparticles and are used in several applications such as solar energy conversion, high-temperature superconductors, batteries, antimicrobials, gas sensors and so on. For many centuries, people have been using Cu metal and its complexes for different purposes such as fungicides, algaecides, for water purification as well as antifouling and antibacterial agents (Perelshtein et al. 2009). The Cu nanoparticles were synthesized by using plant materials like stem latex of *Euphorbia nivulia* and magnolia leaf extract. The excellent or magnificent antimicrobial performance against *Escherichia coli* cells is the main function of these nanoparticles. This nanoparticle is also treated as nontoxic in aqueous conditions and hence may exhibit anti-cancer activity (Ghorbani et al. 2015). Copper (Cu) and copper oxide (CuO) nanoparticles are considered as prospective antimicrobial agents against various infectious organisms like *Pseudomonas aeruginosa*, *Bacillus subtilis*, *E. coil*, *Staphylococcus aureus*, *Vibria cholera* and *Syphillis typhus*. The role of nontoxic gum arabic in the synthesis of nanoparticles was explored by the scientists Kattumuri and co-workers (2007). They used it as a natural hydrocolloid and nonpoisonous for the synthesis of biocompatible gold nanoparticles, which was then used in diagnostic and medicinal applications. The natural gum acts as a nontoxic carrier for in vivo applications of NPs. There are various methods that have been used for the preparation of copper oxides (CuO) nanoparticles such as electrochemical, sonochemical, sol-gel, solvothermal synthesis, solid-state reaction, quick precipitation and microwave irradiation (Carnes and Klabunde 2003; Vijaya Kumar et al. 2001; Xu et al. 1999; Hong et al. 2002). The synthesized copper nanoparticles have been characterized by various instruments like Tem, SEM, UV-Vis and FT-IR spectrophotometer. From the analysis, it has been observed that the copper oxide nanoparticles have some antibacterial activity. Recently, it was found that copper oxide (CuO) nanoparticles are used in the gas sensor. Thus, in the face of the availability of such a wide array of methods to synthesize NPs, it is challenging to find beneficial, favorable, nontoxic and natural products to synthesize metal nanoparticles in an aqueous medium. The nanoparticles formed by the plant are more stable and have diverse shapes and sizes. Apart from this, the rate of production of NPs is faster in presence of the plant product as compared to the microorganisms (Iravani et al. 2014).

The main principle of green chemistry is the biosynthesis of metal/metal oxide by utilizing eco-friendly techniques without the use of severe and various toxic reducing agents like ethylene glycol, hydrazine hydrate, dimethylformamide, sodium borohydride and expensive chemicals. Most of the reducing agents are connected with different natural toxicity and biological toxicity. Currently, the removal of waste and the managing of renewable processes by the utilization of the basic rule of green chemistry are most necessary to execute advanced eco-friendly processes for the synthesis of different nanoparticles with controlled size and shapes (Raveendran et al. 2003; Rao et al. 2002). Among various polysaccharides, hydrocolloids having high-molecular-weight macromolecules such as agar-agar, alginate, guar, starches, carrageenan, gums (karaya, tragacanth and arabic), the most essential, economical and easily available are the gums. They have various applications like viscosifiers, food emulsifiers, sweeteners, thickeners, and the drug discharge changer in pharmaceutical applications (Rana et al. 2011). Generally, India produces gum karaya (GK) extensively and exports it, with the main importer being the European countries. The different research groups studied the occurrence, production, structure, and both physical and chemical properties, as well as the food and nonfood utilization of gums karaya. The GK has a little acetylated polysaccharide and has a derivative structure with a molecular mass of 16.0×10^9 Da and is associated under replacement of Rhamnogalacturonoglycan (pectic)-kind tree gums. It consists of neutral sugars (galactose and rhamnose) and acidic sugars (galacturonic acids and glucuronic acid) about 60% and 40%, respectively. In the synthesis of nanoparticles, the gum acts as a capping agent as well as a reducing agent (Laha et al. 2019). The field of biosynthesis of various metal nanoparticles by plants is still a budding field with a lot of scope of research. The preparation of metal nanoparticles by utilizing plant extracts, exudates, deactivated plant tissues and different parts of the plant is a current alternative method for the preparation of eco-friendly nanoparticles, which is a profitable as well as environmentally accepted (Huang et al. 2007). The bioreduction of metal nanoparticles by a combination of biomolecules found in plant extracts (e.g., polysaccharides, enzymes, organic acids like citrates, amino acids, proteins, and vitamins) is naturally favorable, still chemically complex. These metal nanoparticles act as antibacterial agents against the human pathogenic bacteria generally *S. aureus* and *E. coli*. Many industries apply color pigment for their products, and analysis of these coloring substances has shown that the color gets discharged in water which causes environmental pollution. These dye contaminants are chemically stable. Hence, technologies like hydrogen peroxide oxidation and UV radiation are not able to completely degrade these color pigments. In recent times, there is increasing use of photocatalytic processes in the successful degradation of these color pigments (Sankar et al. 2014). The metal nanoparticles are the best for degradation of color pigment, and it can be synthesized by two methods like physical and chemical methods. Figure 1 shows the different approach for the synthesis of nanoparticles from different sources.

2 Green Synthesis

Green chemistry employs a set of principles in chemical reactions that decreases the production of unpredictable and dangerous substances as a byproduct of the chemical reaction. The green synthesis is easy, efficient and environmentally friendly and is preferred over the conventional techniques which can eliminate the use of wastes such as nontoxic or hazardous chemical, a renewable resource, improve atom efficiencies and catalyst usage in place of reagents. Among the various methods, there is an increase in the interest and demand of green synthesis methods. Green synthesis is a prominent region in the bio-nanotechnology field and gives environmental and economic benefits even as a preference to physical and chemical methods. The physical methods involve diffusion flame synthesis of nanoparticles, thermal evaporate, spray pyrolysis, ultra-thin films, lithographic techniques, layer-by-layer growth, etc. The chemical methods include chemical vapor deposition and chemical solution deposition. The main purpose of green synthesis methods is to implement the processes which are sustainable. The various green synthesis methods involving plant extracts or microorganism are used as reducing agents for the synthesis of nanoparticles (Luechinger et al. 2010). The key merits of the green synthesis method are shown in Fig. 2.

2.1 Green Approach for Synthesis of NPs

Previously, researchers generally use conventional processes, but now they have demonstrated well-known green methods which are the most useful for the formation of metal nanoparticles with fewer disadvantages, less expensive and can be easily characterized. Physical and chemical methods of nanoparticle synthesis or combination have various stresses on the atmosphere by virtue of their harmful metabolic activity. The best useful and beneficial procedure is plant-based metal nanoparticle synthesis. The metal salts amalgam reacts with the extract of the plant, and the reaction completes in a few minutes at room temperatures (Abdelghany et al. 2018). This method has increased much interest among metals like transition metal nanoparticles of silver and gold. Further metal oxides such as zinc oxides and copper oxide nanoparticles are also synthesized by this method. The metal nanoparticles synthesized by this method are most protective as compared with alternative metallic nanoparticles. The formation of nanoparticles by green methods could be increased effectively. The synthesis of nanoparticles by normal conventional methods involves the usage of excess chemicals, which is detrimental and poisonous for the environment and also creates undesirable and unfavorable results on human health. The green synthesis process is more important due to its possibilities for decreasing the generation of many harmful byproducts in the synthesis of nanoparticles (Hussain et al. 2016). Therefore, the plant extracts, amino acids and vitamins are enormously used in the current times due to their more utilization in green synthetic processes.

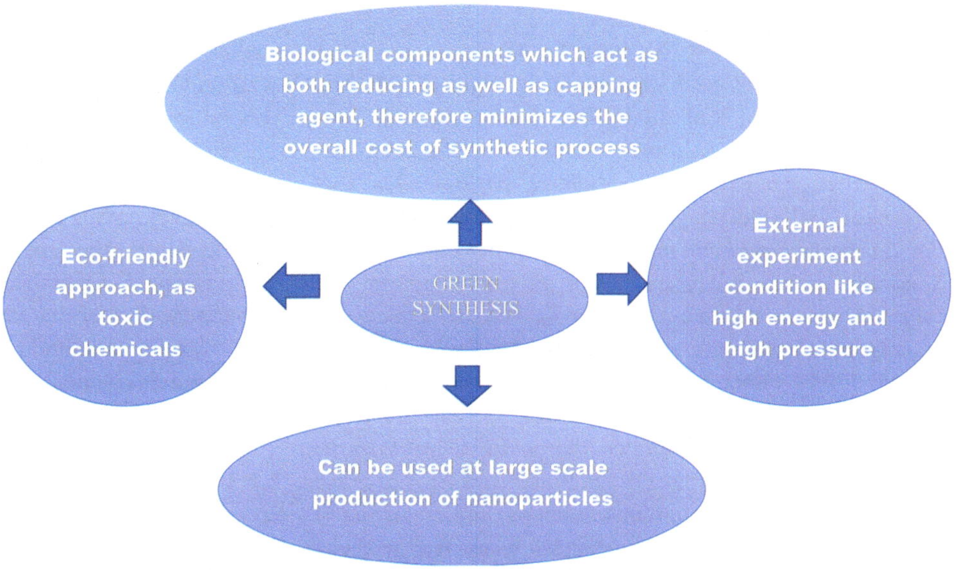

Fig. 2 Key merits of green synthesis methods

2.1.1 From Vitamins

The preparation of silver as well as palladium nanoparticles for the synthesis of nanowires, nanospheres, as well as nanorods through the use of vitamin B_2 which acts as both capping and reducing agents has been reported in recent times. The nanorods and nanowires are synthesized by the utilization of vitamin B_2, which acts as a reducing agent. This is an excellent perspective in the nanotechnology field which recommends the usage of crucial agents for the improvement of this area, for instance, their influence on distinct tumor cells (Nadagouda and Varma 2006). In the synthesis of various nanoparticles, chitosan is used as a stabilizing agent. Ascorbic acid is used as a reducing as well as capping agents in many nanoparticle syntheses. During the synthesis of nanoparticles, chitosan plays an important role and the nanoparticle concentration is directly preoperational to the concentration of chitosan used (Zain et al. 2014). The expansions of nanoparticles of their reliable and homogeneous size are carried out by applying ascorbic acid which acts as capping and reducing materials. Nicotinamide adenine dinucleotide (NAD) which is manufactured from plants acts as a reducing agent during the glycolysis process. This method has been found to be the most profitable in the development of silver nanoparticles (Anjum et al. 2016).

2.1.2 From Enzymes

Enzymes have been utilized for the preparation of metal nanoparticles. Purified and specific enzymes take part in the pure synthesis of metal nanoparticles. The growth of pure nanoparticles depends on the substrate nature of the enzyme. The polymeric-layered enzyme having more layers has strong electrostatic interactions which is a potential molecule for the preparation of bi-metallic metal nanoparticles following a greener route (Smuleac et al. 2011). Extracellular amylase enzymes have been used for the synthesis of gold nanoparticles which reduces the AuCl to make the reaction (Manivasagan et al. 2015). Gold nanoparticles have been synthesized using sulphate reduced enzyme which is collected from *E. coli* by ion-exchange chromatography. These synthesized nanoparticles have good antifungal activity against human pathogenic fungi (Gholami-Shabani et al. 2015). Agri waste is loaded in different biological compounds, and the use of these biomolecules acts as a reducing agent for the synthesis of nanoparticles. The agri waste used for the synthesis of nanoparticles is *Cocos nucifera coir*, corn cob, fruit seeds and peels, wheat and rice bran, palm oil, etc. (Adelere and Lateef 2016). In another work, beet juice has been applied for the synthesis of NPs. In this process, decreasing the amount of beet juice increases the particle size of the NPs. This process enhances the catalytic activity of NPs. Green tea extracts are also used for the synthesis of bimetallic NPs of Fe/Pd. The green tea extracts act as a reducing agent as well as the capping agent for the synthesis (Smuleac et al. 2011). Redox enzymes coated on the AuNPs can perform as an electron transmitter between biocatalyst and electrode. These NPs can be used in different sensor applications.

2.2 Biobased Methods

Among various methods, the production of nanoparticles by biobased methods is highly reliable, secure, well characterized than the chemical methods. These chemical methods are not generally eco-friendly and less secure.

2.2.1 Bacteria

The development of AgNPs from the supernatants of the cell-free culture has been carried out by Shivaji and their co-workers (Shivaji et al. 2011). They have worked on psychrophilic bacteria *Pseudomonas antarctica*, *Pseudomonas proteolytica*, *Pseudomonas meridiana*, *Arthrobacter kerguelensis*, *Arthrobacter gangotriensis*, *Bacillus indicus* and *Bacillus cecembensis* and made stable NPs in a dark place for 8 months. Simon Silver has given an explanation of the bacterial resistance in the culture media and also reduces the toxicity in the solution media by employing a particular gene (Silver 2003). Other type of bacteria such as *P. stutzeri* AG259 have been used for the successful synthesis of Ag and Au nanocrystals. The bacteria *Geobacter* sp. *Magnetospirillum magnetotacticum* intake the iron and convert it to ferric oxide. These ferric ions are then reduced, and magnetite is formed inside the magnetosome vesicles, where ferritin keeps these nanoparticles solubilized. These nanoparticles produced have high purity and are mono-dispersed in nature. Similarly, there are many thermophilic bacteria that serve as a versatile process for the preparation of Ag and Au nanoparticles, which are often environmentally friendly and it also reduces the downstream processing cost. In one report it is shown that with the aid of the bacteria *Shewanellao neidensis*, gold nanoparticles with size as low as 2 nm have been synthesized. It is also interesting to note that by using various bacterial strains, the size of nanoparticles can be modulated from around 2 nm on the lower side to around 400 nm on the higher side. This shows that various bacterial strains can be exploited to synthesize nanoparticles of various shapes and sizes (Menon et al. 2017).

2.2.2 Yeasts and Fungi

Silver oxide is synthesized into a nano- and properly scattered form by the metabolic action of a microorganism, *Fusarium oxysporum*. The process is controlled by the release of nitrate reductase which gives stable AgNPs (Ahmed et al. 2018). Platinum NPs have been synthesized with the help of *Alternaria alternata*. The NPs are collected from the culture filtrate of the species and characterized by using different techniques. From the different investigations, it is known that the NPs having the particle size of 2–30 nm, and different shapes like spherical and triangular can be synthesized. From the FTIR spectra, it shows O–H stretching and C–H stretching bands due to the presence of different organic residues present in the solution like amide protein bonds (Sarkar and Acharya 2017). Selenium NPs have been synthesized by using fungi *Candida albicans*. The fungi *C. albicans* with the chitosan is used as a reductive agent in the synthesis of the selenium NPs (Lara et al. 2018). *Trichoderma viride* has been applied for the synthesis of AgNPs from silver nitrate solution by extracellular biosynthesis method (Fayaz et al. 2010). A stable and smaller size AgNPs have been developed by using *F. oxysporum* with size 5–15 nm (Pantidos and Horsfall 2014). This biological method of synthesis is considered as environmentally benign, nontoxic and clean. Various microorganisms produce intracellular or extracellular metal nanoparticles having different shapes, sizes and efficiency. In order to control the size of silver nanoparticles, *F. oxysporum* has opted. Optimization parametrically displayed the very minute size of the particle on treating *F. oxysporum* at 50 °C with silver nitrate. Antibacterial property against *S. aureus* and *E. coli* displayed the highest zone of inhibition of 1.6 and 2 mm, respectively, with these silver nanoparticles.

2.2.3 Algae

Several algae have been used for the biorecovery of metal from the liquid sources. Different algae like cyanobacteria and eukaryotic green have been utilized for the recovery of metal (Bakir et al. 2018). Uma Suganya et al. have reported that the use of algae protein from *Spirulina platensis* in green synthesis of AuNPs. The property of a high surface-to-volume ratio and small size of gold nanoparticles enhance the antibacterial properties on interaction with microorganisms. Nanomaterials change the bacterial cell membrane permeability, thereby forming gaps which reduce the respiratory chain enzymes activity finally leading to the death of a cell. Gold nanoparticles functionalized with drug like doxorubicin have been reported to be effective in the delivery and transport of drug into the human body. Gold nanoparticles functionalized with *S. platensis* protein act as an effective antibacterial system for gram positive bacteria. *S. platensis* protein of blue green algae of freshwater is preferred for research owing to its characteristics like anticancer, antiviral and antioxidant properties. It is reported that *S. platensis* aqueous extracts exhibit inhibitory activity on Langerhan cells, peripheral blood mononuclear cells and viral replication of human T cells. Owing to the development of resistance toward multiple drugs in bacteria, there is a need for the development of new antibacterial systems against gram negative and positive bacteria. In case of gram negative bacteria, there is a presence of a thin cell wall which is more prone to antibacterial activities of nanomaterial in comparison to the gram positive bacteria. Opposite to this, gram positive bacteria has a thick peptidoglycan layer of cell wall like a mesh, which displays high resistance toward different antibacterial substances. *S. aureus* and *B. subtilis* (example of gram positive bacteria) cause different types of pathogenic health problems that need to be controlled by stronger inhibitory systems. The mechanism of the formation of the AuNPs is to control the size by the reduction of silver particles of chloroauric acid to Au^0 with the aid of algae protein (Suganya et al. 2015).

3 Metal Oxide Nanoparticles Synthesized from Green Synthesis

3.1 Copper Oxide (CuO)

The copper oxide (CuO) nanoparticles are formed by the use of a colloidal heat combination process. The integrated copper oxide was disinfected and dehydrated to gain the different sizes and shapes of the metal oxide nanoparticles (Padil and Černík 2013). Nanoparticle-based sensor for the detection of nitrate ions was reported by Manoj and co-workers. By using the substrate carboxymethyl cellulose (CMC), they informed that this method produces extremely sturdy and impressionable nanoparticles of copper (Manoj et al. 2018). The highly biocompatible and stable copper nanocomposite is used for electrochemical sensing. Production of copper nanoparticles which are stabilized by carboxymethyl cellulose in presence of aqueous media is reported. For surface properties identification of produced copper nanoparticles, studies like X-ray photoelectron spectroscopic and X-ray diffraction are performed. These copper nanoparticles are subjected to dispersion with carbon nanotubes which are multi-walled, and the final dispersion is placed on a carbon glassy electrode for obtaining modified copper multi-walled glass carbon electrode. The glassy carbon electrode after modification exhibits a better oxidation peak for nitrite oxidation. This altered electrode shows great reproducibility and selectivity and is applied successfully for determining real nitrite samples.

3.2 Zinc Oxide (ZnO)

The zinc oxide nanoparticle has been synthesized by the use of *Cassia auriculate* blossom extract which produced nanoparticles in the size range of about 110–280 nm (Ramesh et al. 2014). These different nanoparticles like ZnO and Ag/ZnO nanoparticles that are acquired by green synthesis have been used in clinical antimicrobial wound-healing bandages (Khatami et al. 2018). Coffee and *Prosophis fracta* are used for cost effective and environmentally benign production of zinc oxide and silver nanoparticles, where zinc oxide and silver nanoparticles are characterized physicochemically by SEM, X-ray diffraction, and UV-visible spectroscopy. Zinc oxide and silver nanoparticles synthesized by the green method consists of an average size of 26 and 16 nm, respectively. Zinc oxide and silver nanoparticles and their mixture show a minimum inhibitory concentration against *P. aeruginosa* and *Acinetobacter baumannii* cultures. Mixed silver–zinc oxide and zinc oxide and silver nanoparticles were impregnated with cotton wound bandages, and it has been observed that both the nanoparticles exhibit antibacterial property in bandages. This type of antibacterial bandages displayed effectiveness in covering and treatment of wounds which are sensitive to infection like burn or diabetic wounds.

3.3 Silver and Gold

The two important nanoparticles like silver and gold nanoparticles are considered as the most useful in various fields such as catalysis, medicine, sensing, optoelectronics, etc. Francis et al. reported the microwave-assisted synthesis of silver and gold nanoparticles utilizing the extract of leaves *M. glabrata*. Various metal nanoparticles can be synthesized by using this plant material from the respective sources of the metal salts. These nanoparticles are used for the purification of water due to their enormous antimicrobial activity preventing pathogenic microscopic organisms such as *P. aeruginosa*, *Bacillus pumilus*, *E. coli*, *Penicillium chrysogenum*, *Aspergillus niger* and *S. aureus* (Francis et al. 2017). Synthesis of silver nanoparticles is synthesized in a more economical way by using microorganisms, *Bacilus subtilus* 10,833 and *Bacilus amylococus* 1853 (Ghiuță et al. 2018). The general problem of this method is the consumption of more time, impurity and lack of reproducibility to some extent. The combination of gold nanoparticles with three different microorganisms such as yeast, fungus and bacteria has been analyzed by Shen et al. (2015). They have reported that the size of nanoparticles with different microorganisms used is 22.2, 9.5 and 18.8 nm, respectively. They also observed that fungus exhibits better results relative to other microorganisms. Green synthesis of CeO_2 nanoparticle has been synthesized by using the aerial extract of *Prosopis farcta* (Miri and Sarani 2018).

3.4 Cerium Oxide (CeO_2)

CeO_2 nanoparticle has been used as an antioxidant for the treatment of obesity (Pourkhalili et al. 2011). The CeO_2 nanoparticles not only possess fast electron transfer kinetics but also act as a classic co-immobilization catalyst for various enzymes like glucose oxidase, horseradish peroxidase and cholesterol oxidase (Njagi and Kagwanja 2011). Cerium oxide nanoparticle has been synthesized using *G. superba* leaf which increases the antibacterial activity of the nanoparticles. The shape of the synthesized nanoparticles is spherical, and the size is about 5 nm (Arumugam et al. 2015). Sarani and Miri have studied cerium oxide nanoparticles which are synthesized biologically. According to their study, *P. farcta* ethereal parts aqueous extract can be utilized for cerium oxide nanoparticle biosynthesis and provides round-shaped NPs with 30 nm size. This plant consists of a phenolic functional group which acts as capping and reducing agents. The particles produced by biosynthesis are subjected to characterization through FTIR, Raman, EDX, FESEM, TEM, PXRD, and UV-Vis. There is a characteristic peak of CeO_2 nanoparticles at 317 nm confirming the spherical shape. TEM and FESEM showed that the size of the nanoparticles is 30 nm and uniform and EDX reported that only oxygen and cerium existed in the sample which is biosynthesized (Miri and Sarani 2018). The range of concentrations between 0 and 800 Âμg per mL of cerium oxide nanoparticles displayed nontoxicity, and hence, it can be considered that this category of nanoparticles is effective for applications in sectors like drug delivery and medicine.

4 Applications of Green Nanotechnology

In current years, there is spurring attention from many researchers to expand green nanotechnology as well as to increase its publication regularly in science. Application of nanoparticles synthesized by following a greener route possess distinct effects when compared with nanoparticles synthesized by other routes. Nanoparticles synthesized following the greener route are very useful in the field of pharmaceutical.

4.1 Application of Nanoparticles in Agriculture

Novel nanosized materials derived from materials obtained from plant sources have established one more platform for these advanced classes of materials. Pesticides with nanodimension and fertilizers interweave with coatings with nanosize and nanoherbicides. In conventional agriculture, the number of agrochemicals which reach the active site of the crops is extremely low in number. This low number turnout to the plants mainly happens due to chemical leaching, hydrolysis, photolysis and microbial ingestion. Nanoparticles help in delivering the agrochemicals in the field to the plants in a controlled manner and avoid the major factors which lead to the low number turnout of these chemicals to the plants. It is seen that graphene oxide film encapsulated with potassium nitrate leads to prolonged release of the fertilizers in the field (Shang et al. 2019). Similarly, conjugating nanoparticles with the agrochemicals protect these chemicals from degradation and increases their effectiveness. Nanoparticles carrying foreign DNA as delivery vectors are used to modify target genes in many plants (Shang et al. 2019). In an interesting research work, silicon dioxide nanoparticles carrying DNA sequences have been used to modify gene targets in corn and tobacco plants. Similarly, the CRISPR/Cas9 system has been used in a phenomenally successful manner to edit genomes in plants but has a low delivery efficiency (Shang et al. 2019). In this regard, nanoparticles increase the efficiency of the system by enhancing the specificity of the CRISPR/Cas9 system (Shang et al. 2019; Rossi et al. 2019; Wang et al. 2019; Zheng et al. 2005; Iqbal et al. 2019; Hojjat and Kamyab 2017; Wang et al. 2018; Sturikova et al. 2018)

4.2 Dentistry

The silver nanoparticles (Ag–NP) are used in the bandage and dental appliances. Connecting silver nanoparticles into orthodontic gum can enhance the chemical bonding in the cement which is used for orthodontic application (Burduşel et al. 2018). For centuries, silver has been utilized for oral health care for tooth restoration and dental amalgams and attracted researchers throughout the world during the nineteenth century. Other fields for silver nanoparticles applications include implantology, endodontic dentistry, restorative dentistry and dental prostheses. Silver in the form of nanosilver diamine fluoride has potential effects in behaving as prophylaxis, but it also has few side effects like staining of teeth. The reduction of silver nanoparticles size leads to an increase in the contact surface considerably and hence silver antimicrobial activity improves, and utilization of silver nanoparticles reduces teeth black staining which generally results due to silver diamine fluoride applications. This resin can be used for denturing or filling base materials for restorative dentistry (Hamouda 2012). Hence, for improvement of their antimicrobial and physico-mechanical characteristics, incorporation of acrylic resin denture base material with silver nanoparticles has been developed. The oral environment is an active system that generally gets filled with many pathogenic microorganisms, and hence, implant and dental materials have a high risk if they get contaminated. In this context, nanosystems based on silver incorporated with dental composite, orthodontic cement, and adhesive resin display potential results. Additionally, silver nanoparticles also have challenging and attractive dental applications like utilization as a biocide or biostatic coatings for traditional dental implants based on titanium. Even though silver nanoparticles exhibit effective properties for dental applications, there is a certain area where its application is restricted owing to its toxicity in certain biological environments (Anu 2016; Argueta-Figueroa et al. 2014). Hence, prior to silver nanoparticles applications in dentistry, there must be vigorous research for biofunctional performance and physicochemical features optimal compromise.

4.3 X-Ray Imaging

The gold nanoparticles are used as the primary application on the point of X-ray as it demonstrates a nontoxicity, high X-ray retention coefficient, surface functionalization for colloidal dependability and simplicity of engineered control (Jakhmola et al. 2012; Hainfeld et al. 2004; Cho 2005). In the field of X-ray imaging, alternative emerging solutions are being developed. Gold nanoparticles exhibit very high contrasting properties. Apart from these various kind inorganic nanoparticles are gadolinium, bismuth, platinum, and silver (Dadashi et al. 2018). In comparison to iodinated substance, the inorganic nanoparticles include a higher absorption coefficient of X-ray producing higher contrast thereby reducing the amount of dosage and substantial control during the process of synthesis. The control of bioconjugation chemistry, growth and nucleation of inorganic materials are simple and capable of adapting to specified targeting systems. The contrast agent role is to enhance the internal body structure visibility by modifying the contrast between various tissues. The contrast between tissues takes place because of X-ray various attenuation during their movement to the detector through the body. For fulfilling this requirement, the X-ray contrast system should be different from X-ray attenuation from the surrounding organ and tissue. X-ray absorption by any kind of material is dependent on the mass absorption coefficient of contrast medium elements as well the energy of X-ray, sample thickness, concentration, etc. (Boisselier and Astruc 2009). Various applications of nanoparticles are shown in Fig. 3.

4.4 Drug Delivery

Due to the comprehensive appearance of gold nanoparticles such as nontoxicity, physicochemical properties, controlled disparity, biocompatibility, remarkable optical and viable flexibility, it performs as an imperative nanocarrier in the drug delivery process. Viruses are mostly used in the delivery of genes to the desired targets inside the cells, but they raise many safety concerns such as cytotoxicity and immune responses. To avoid these complications, synthetic DNA delivery systems majorly based on nanotechnology are very good alternatives. In this prospect, gold nanoparticles are the major active candidates in gene delivery. It is shown that gold nanoparticles capped with quaternary ammonium ions bind strongly to the DNA and hence protect them from nucleases. Nanoparticles grafted with nucleic acid strands such as siRNA are used for delivery inside the cells. It is also shown that RNA modified with polymeric nanoparticles acts as a delivery system and can silence particular genes in HuH cells. In another work, nanoparticles conjugated with fluorophores when incubated with cells exhibited the efficient uptake of the nanoparticles which takes place by the process of endocytosis, and it happens due to the adsorption of the serum protein on the surface of the particles. Similarly, nanoparticles are also used in the delivery of a variety of therapeutic proteins such as insulin. Nanoparticles stabilized by chitosan help in the absorption of insulin and the trans-mucosal delivery of insulin. Therefore, taking all these into account nanoparticles are emerging as a promising delivery system for many drugs and genes (DeLouise 2012; Ghosh et al. 2008; Han et al. 2007a, b, c; Kim et al. 2009; Duncan

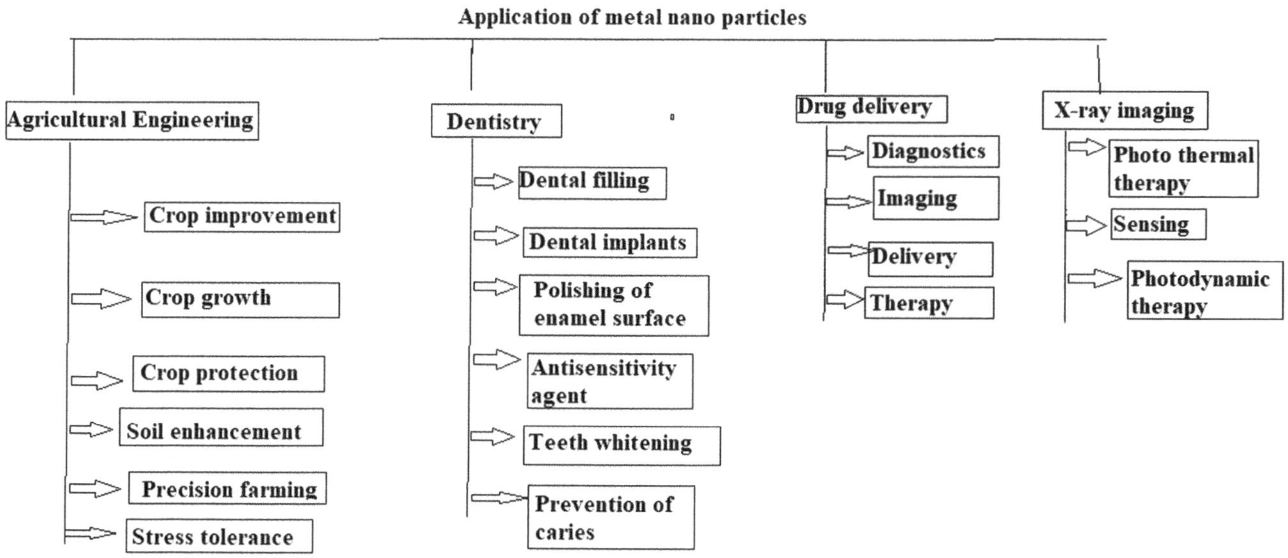

Fig. 3 Applications of nano particles

et al. 2010; Trono et al. 2011; Brown et al. 2010; Dhar et al. 2008; Paciotti et al. 2006; Khan et al. 2014).

5 Conclusion

The solid-state green synthesis of different metal/metal oxide nanoparticles has been increasing more interest in various research fields such as chemical, physical, biomedicine, material science, biosensing, dentistry, pharmaceuticals and many others. Recently, people are very conscious about the advantages of metal nanoparticle synthesis by utilizing the plant extract, and it has also high efficiency as reducing, stabilizing and capping agents. The production of nanoparticles by using natural resources is nontoxic, low cost, sustainable, eco-friendly, protection to human health, and there is no chemical effect for medical and biological applications. The synthesized nanoparticles by virtue of the green route are more useful and reliable in comparison with physicochemical methods.

References

(a) Abdelghany TM, Al-Rajhi AM, Al Abboud MA, Alawlaqi MM, Magdah AG, Helmy EA, Mabrouk AS (2018) Recent advances in green synthesis of silver nanoparticles and their applications: about future directions. A review. BioNanoSci 8(1):5–16. (b) Makarov VV, Love AJ, Sinitsyna OV, Makarova SS, Yaminsky IV, Taliansky ME, Kalinina NO (2014) "Green" nanotechnologies: synthesis of metal nanoparticles using plants. Acta Naturae (англоязычная версия) 6(1):20

Abid JP, Wark AW, Brevet PF, Girault HH (2002) Preparation of silver nanoparticles in solution from a silver salt by laser irradiation. Chem Commun 7:792–793

Adelere IA, Lateef A (2016) A novel approach to the green synthesis of metallic nanoparticles: the use of agro-wastes, enzymes, and pigments. Nanotech Rev 5(6):567–587

Ahmed AA, Hamzah H, Maaroof M (2018) Analyzing formation of silver nanoparticles from the filamentous fungus *Fusarium oxysporum* and their antimicrobial activity. Turkish J Biol 42(1):54–62

Alivisatos AP (1996) Semiconductor clusters, nanocrystals, and quantum dots. Science 271(5251):933–937

Al-Namil DS, El Khoury E, Patra D (2019) Solid-state green synthesis of Ag NPs: higher temperature harvests larger Ag NPs but smaller size has better catalytic reduction reaction. Sci Rep 9(1):1–9

Anjum S, Abbasi BH, Shinwari ZK (2016) Plant-mediated green synthesis of silver nanoparticles for biomedical applications: challenges and opportunities. Pak J Bot 48(4):1731–1760

Anu K (2016) Wet biochemical synthesis of copper oxide nanoparticles coated on titanium dental implants. Int J Adv Res Sci Eng Technol 3:1191–1194

Argueta-Figueroa L, Morales-Luckie RA, Scougall-Vilchis RJ, Olea-Mejía OF (2014) Synthesis, characterization and antibacterial activity of copper, nickel and bimetallic Cu–Ni nanoparticles for potential use in dental materials. Prog Nat Sci Mater Int 24(4):321–328

Arumugam A, Karthikeyan C, Hameed ASH, Gopinath K, Gowri S, Karthika V (2015) Synthesis of cerium oxide nanoparticles using *Gloriosa superba* L. leaf extract and their structural, optical and antibacterial properties. Mater Sci Eng, C 49:408–415

Bae CH, Nam SH, Park SM (2002) Formation of silver nanoparticles by laser ablation of a silver target in NaCl solution. Appl Surf Sci 197:628–634

Bakar NA, Ismail J, Bakar MA (2007) Synthesis and characterization of silver nanoparticles in natural rubber. Mater Chem Phys 104(2–3):276–283

Bakir EM, Younis NS, Mohamed ME, El Semary NA (2018) Cyanobacteria as nanogold factories: chemical and anti-myocardial infarction properties of gold nanoparticles synthesized by *Lyngbya majuscula*. Mar Drugs 16(6):217

Basavaraja S, Balaji SD, Lagashetty A, Rajasab AH, Venkataraman A (2008) Extracellular biosynthesis of silver nanoparticles using the fungus *Fusarium semitectum*. Mater Res Bull 43(5):1164–1170

Boisselier E, Astruc D (2009) Gold nanoparticles in nanomedicine: preparations, imaging, diagnostics, therapies and toxicity. Chem Soc Rev 38(6):1759–1782

Brown SD, Nativo P, Smith JA, Stirling D, Edwards PR, Venugopal B, Flint DJ, Plumb JA, Graham D, Wheate NJ (2010) Gold nanoparticles for the improved anticancer drug delivery of the active component of oxaliplatin. J Am Chem Soc 132(13):4678–4684

Bruchez M, Moronne M, Gin P, Weiss S, Alivisatos AP (1998) Semiconductor nanocrystals as fluorescent biological labels. Science 281(5385):2013–2016

Burduşel AC, Gherasim O, Grumezescu AM, Mogoantă L, Ficai A, Andronescu E (2018) Biomedical applications of silver nanoparticles: an up-to-date overview. Nanomaterials 8(9):681

Carnes CL, Klabunde KJ (2003) The catalytic methanol synthesis over nanoparticle metal oxide catalysts. J Mol Catal A: Chem 194(1–2):227–236

Chandran SP, Chaudhary M, Pasricha R, Ahmad A, Sastry M (2006) Synthesis of gold nanotriangles and silver nanoparticles using Aloevera plant extract. Biotechnol Prog 22(2):577–583

Cho SH (2005) Estimation of tumour dose enhancement due to gold nanoparticles during typical radiation treatments: a preliminary Monte Carlo study. Phys Med Biol 50(15):N163

Coe S, Woo WK, Bawendi M, Bulović V (2002) Electroluminescence from single monolayers of nanocrystals in molecular organic devices. Nature 420(6917):800–803

Dadashi S, Poursalehi R, Delavari H (2018) Optical and structural properties of oxidation resistant colloidal bismuth/gold nanocomposite: an efficient nanoparticles based contrast agent for X-ray computed tomography. J Mol Liq 254:12–19

De M, Ghosh PS, Rotello VM (2008) Applications of nanoparticles in biology. Adv Mater 20(22):4225–4241

DeLouise LA (2012) Applications of nanotechnology in dermatology. J Invest Dermatol 132(3):964–975

Dhar S, Reddy EM, Shiras A, Pokharkar V, Prasad BEE (2008) Natural gum reduced/stabilized gold nanoparticles for drug delivery formulations. Chem Eur J 14(33):10244–10250

Dickson DP (1999) Nanostructured magnetism in living systems. J Magn Magn Mater 203(1–3):46–49

Duncan B, Kim C, Rotello VM (2010) Gold nanoparticle platforms as drug and biomacromolecule delivery systems. J Controlled Release 148(1):122–127

El Khoury E, Abiad M, Kassaify ZG, Patra D (2015) Green synthesis of curcumin conjugated nano silver for the applications in nucleic acid sensing and anti-bacterial activity. Colloids Surf, B 127:274–280

Fayaz M, Tiwary CS, Kalaichelvan PT, Venkatesan R (2010) Blue orange light emission from biogenic synthesized silver nanoparticles using *Trichoderma viride*. Colloids Surf, B 75(1):175–178

Francis S, Joseph S, Koshy EP, Mathew B (2017) Green synthesis and characterization of gold and silver nanoparticles using *Mussaenda glabrata* leaf extract and their environmental applications to dye degradation. Environ Sci Pollut Res 24(21):17347–17357

Gardea-Torresdey JL, Parsons JG, Gomez E, Peralta-Videa J, Troiani HE, Santiago P, Yacaman MJ (2002) Formation and growth of Au nanoparticles inside live alfalfa plants. Nano Lett 2 (4):397–401

Gardea-Torresdey JL, Gomez E, Peralta-Videa JR, Parsons JG, Troiani H, Jose-Yacaman M (2003) Alfalfa sprouts: a natural source for the synthesis of silver nanoparticles. Langmuir 19(4):1357–1361

Ghiuţă I, Cristea D, Croitoru C, Kost J, Wenkert R, Vyrides I, Munteanu D (2018) Characterization and antimicrobial activity of silver nanoparticles, biosynthesized using Bacillus species. Appl Surf Sci 438:66–73

Gholami-Shabani M, Shams-Ghahfarokhi M, Gholami-Shabani Z, Akbarzadeh A, Riazi G, Ajdari S, Razzaghi-Abyaneh M (2015) Enzymatic synthesis of gold nanoparticles using sulfite reductase purified from *Escherichia coli*: a green eco-friendly approach. Process Biochem 50(7):1076–1085

Ghorbani HR, Mehr FP, Poor AK (2015) Extracellular synthesis of copper nanoparticles using culture supernatants of *Salmonella typhimurium*. Orient J Chem 31(1):527–529

Ghosh P, Han G, De M, Kim CK, Rotello VM (2008) Gold nanoparticles in delivery applications. Adv Drug Deliv Rev 60 (11):1307–1315

Hainfeld JF, Slatkin DN, Smilowitz HM (2004) The use of gold nanoparticles to enhance radiotherapy in mice. Phys Med Biol 49 (18):N309

Hamouda IM (2012) Current perspectives of nanoparticles in medical and dental biomaterials. J Biomed Res 26(3):143–151

Han G, Ghosh P, De M, Rotello VM (2007a) Drug and gene delivery using gold nanoparticles. NanoBiotechnology 3(1):40–45

Han G, Ghosh P, Rotello VM (2007b) Functionalized gold nanoparticles for drug delivery. Nanomedicine 2(1):113–123

Han G, Ghosh P, Rotello VM (2007c) Functionalized gold nanoparticles for drug delivery. Nano Med 2(1):113–123

Harfenist SA, Wang ZL, Alvarez MM, Vezmar I, Whetten RL (1996) Highly oriented molecular Ag nanocrystal arrays. J Phys Chem 100 (33):13904–13910

Heath JR, Knobler CM, Leff DV (1997) Pressure/temperature phase diagrams and super lattices of organically functionalized metal nanocrystal monolayers: the influence of particle size, size distribution, and surface passivant. J Phys Chem B 101(2):189–197

Heiligtag FJ, Niederberger M (2013) The fascinating world of nanoparticle research. Mater Today 16(7–8):262–271

Hojjat SS, Kamyab M (2017) The effect of silver nanoparticle on Fenugreek seed germination under salinity levels. Rus Agric Sci 43 (1):61–65

Hong ZS, Cao Y, Deng JF (2002) A convenient alcohothermal approach for low temperature synthesis of CuO nanoparticles. Mater Lett 52(1–2):34–38

Huang J, Li Q, Sun D, Lu Y, Su Y, Yang X, Hong J (2007) Biosynthesis of silver and gold nanoparticles by novel sundried *Cinnamomum camphora* leaf. Nanotechnology 18(10):105104

Hussain I, Singh NB, Singh A, Singh H, Singh SC (2016) Green synthesis of nanoparticles and its potential application. Biotech Lett 38(4):545–560

Iqbal M, Raja NI, Hussain M, Ejaz M, Yasmeen F (2019) Effect of silver nanoparticles on growth of wheat under heat stress. Iranian J Sci Technol Trans A: Sci 43(2):387–395

Iravani S, Korbekandi H, Mirmohammadi SV, Zolfaghari B (2014) Synthesis of silver nanoparticles: chemical, physical and biological methods. Res Pharm Sci 9(6):385

Itakura T, Torigoe K, Esumi K (1995) Preparation and characterization of ultrafine metal particles in ethanol by UV irradiation using a photo initiator. Langmuir 11(10):4129–4134

Jakhmola A, Anton N, Vandamme TF (2012) Inorganic nanoparticles based contrast agents for X-ray computed tomography. Adv Healthc Mater 1(4):413–431

Kattumuri V, Katti K, Bhaskaran S, Boote EJ, Casteel SW, Fent GM, Katti KV (2007) Gum arabic as a phytochemical construct for the stabilization of gold nanoparticles: in vivo pharmacokinetics and X-ray-contrast-imaging studies. Small 3(2):333–341

Khan AK, Rashid R, Murtaza G, Zahra A (2014) Gold nanoparticles: synthesis and applications in drug delivery. Tropical J Pharm Res 13(7):1169–1177

Khatami M, Varma RS, Zafarnia N, Yaghoobi H, Sarani M, Kumar VG (2018) Applications of green synthesized Ag, ZnO and Ag/ZnO nanoparticles for making clinical antimicrobial wound-healing bandages. Sustain Chem Pharm 10:9–15

Kim JS (2007) Reduction of silver nitrate in ethanol by poly (N-vinylpyrrolidone). J Ind Eng Chem 13(4):566–570

Kim CK, Ghosh P, Rotello VM (2009) Multimodal drug delivery using gold nanoparticles. Nanoscale 1(1):61–67

Klaus T, Joerger R, Olsson E, Granqvist CG (1999) Silver-based crystalline nanoparticles, microbially fabricated. Proc Natl Acad Sci 96(24):13611–13614

Komarneni S, Li D, Newalkar B, Katsuki H, Bhalla AS (2002) Microwave-polyol process for Pt and Ag nanoparticles. Langmuir 18(15):5959–5962

(a) Laha B, Goswami R, Maiti S, Sen KK (2019) Smart karaya-locust bean gum hydrogel particles for the treatment of hypertension: optimization by factorial design and pre-clinical evaluation. Carbohyd Polym 210:274–288. (b) Aspinall GO, Khondo L, Williams BA (1987) The hex-5-enose degradation: cleavage of glycosiduronic acid linkages in modified methylated Sterculia gums. Can J Chem 65(9):2069–2076. (c) Krishnappa PB, Badalamoole V (2019) Karaya gum-graft-poly (2-(dimethylamino) ethyl methacrylate) gel: an efficient adsorbent for removal of ionic dyes from water. Int J Biol Macromol 122:997–1007

Lara HH, Guisbiers G, Mendoza J, Mimun LC, Vincent BA, Lopez-Ribot JL, Nash KL (2018) Synergistic antifungal effect of chitosan-stabilized selenium nanoparticles synthesized by pulsed laser ablation in liquids against *Candida albicans* biofilms. Int J Nanomed 13:2697

Liu YC, Lin LH (2004) New pathway for the synthesis of ultrafine silver nanoparticles from bulk silver substrates in aqueous solutions by sono electrochemical methods. Electrochem Commun 6 (11):1163–1168

Liz-Marzán LM, Lado-Touriño I (1996) Reduction and stabilization of silver nanoparticles in ethanol by nonionic surfactants. Langmuir 12 (15):3585–3589

Lovley DR, Stolz JF, Nord GL, Phillips EJ (1987) Anaerobic production of magnetite by a dissimilatory iron-reducing microorganism. Nature 330(6145):252–254

Luechinger NA, Grass RN, Athanassiou EK, Stark WJ (2010) Bottom-up fabrication of metal/metal nanocomposites from nanoparticles of immiscible metals. Chem Mater 22(1):155–160

Mallick K, Witcomb MJ, Scurrell MS (2005) Self-assembly of silver nanoparticles in a polymer solvent: formation of a nano chain through nanoscale soldering. Mater Chem Phys 90(2–3):221–224

Mandal D, Bolander ME, Mukhopadhyay D, Sarkar G, Mukherjee P (2006) The use of microorganisms for the formation of metal nanoparticles and their application. Appl Microbiol Biotechnol 69 (5):485–492

Manivasagan P, Venkatesan J, Kang KH, Sivakumar K, Park SJ, Kim SK (2015) Production of α-amylase for the biosynthesis of gold nanoparticles using *Streptomyces* sp. MBRC-82. Int J Biol Macromol 72:71–78

Manoj D, Saravanan R, Santhanalakshmi J, Agarwal S, Gupta VK, Boukherroub R (2018) Towards green synthesis of monodisperse

Cu nanoparticles: an efficient and high sensitive electrochemical nitrite sensor. Sens Actuators B: Chem 266:873–882
Menon S, Rajeshkumar S, Kumar V (2017) A review on biogenic synthesis of gold nanoparticles, characterization, and its applications. Res Eff Technol 3(4):516–527
Miri A, Sarani M (2018) Biosynthesis, characterization and cytotoxic activity of CeO_2 nanoparticles. Ceram Int 44(11):12642–12647
(a) Nadagouda MN, Varma RS (2006) Green and controlled synthesis of gold and platinum nanomaterials using vitamin B2: density-assisted self-assembly of nanospheres, wires and rods. Green Chem 8(6):516–518. (b) Nadagouda MN, Varma RS (2008) Green synthesis of Ag and Pd nanospheres, nanowires, and nanorods using vitamin: catalytic polymerisation of aniline and pyrrole. J Nanomater
Njagi JI, Kagwanja SM (2011) The interface in biosensing: improving selectivity and sensitivity. In: Interfaces and interphases in analytical chemistry. American Chemical Society, pp 225–247
Paciotti GF, Kingston DG, Tamarkin L (2006) Colloidal gold nanoparticles: a novel nanoparticle platform for developing multifunctional tumor-targeted drug delivery vectors. Drug Dev Res 67 (1):47–54
Padil VVT, Černík M (2013) Green synthesis of copper oxide nanoparticles using gum karaya as a biotemplate and their antibacterial application. Int J Nanomed 8:889
Pantidos N, Horsfall LE (2014) Biological synthesis of metallic nanoparticles by bacteria, fungi and plants. J Nanomed Nanotechnol 5(5):1
Pattabi M, Uchil J (2000) Synthesis of cadmium sulphide nanoparticles. Sol Energy Mater Sol Cells 63(4):309–314
Perelshtein I, Applerot G, Perkas N, Wehrschuetz-Sigl E, Hasmann A, Gübitz G, Gedanken A (2009) CuO–cotton nanocomposite: formation, morphology, and antibacterial activity. Surf Coat Technol 204 (1–2):54–57
Petit C, Lixon P, Pileni MP (1993) In situ synthesis of silver nanocluster in AOT reverses micelles. J Phys Chem 97(49):12974–12983
Philipse AP, Maas D (2002) Magnetic colloids from magnetotactic bacteria: chain formation and colloidal stability. Langmuir 18 (25):9977–9984
Pol VG, Srivastava DN, Palchik O, Palchik V, Slifkin MA, Weiss AM, Gedanken A (2002) Sonochemical deposition of silver nanoparticles on silica spheres. Langmuir 18(8):3352–3357
Polte J (2015) Fundamental growth principles of colloidal metal nanoparticles–a new perspective. CrystEngComm 17(36):6809–6830
Pourkhalili N, Hosseini A, Nili-Ahmadabadi A, Hassani S, Pakzad M, Baeeri M, Mohammadirad A, Abdollahi M (2011) Biochemical and cellular evidence of the benefit of a combination of cerium oxide nanoparticles and selenium to diabetic rats. World J Diab 2(11):204
Pum D, Sleytr UB (1999) The application of bacterial S-layers in molecular nanotechnology. Trends Biotechnol 17(1):8–12
Ramesh P, Rajendran A, Meenakshisundaram M (2014) Green syntheis of zinc oxide nanoparticles using flower extract cassia auriculata. J Nanosci Nanotechnol 2(1):41–45
Rana V, Rai P, Tiwary AK, Singh RS, Kennedy JF, Knill CJ (2011) Modified gums: approaches and applications in drug delivery. Carbohyd Polym 83(3):1031–1047
Rao CNR, Kulkarni GU, Thomas PJ, Edwards PP (2002) Size-dependent chemistry: properties of nanocrystals. Chem Eur J 8(1):28–35
Raveendran P, Fu J, Wallen SL (2003) Completely "green" synthesis and stabilization of metal nanoparticles. J Am Chem Soc 125 (46):13940–13941
Ravindran TR, Arora AK, Balamurugan B, Mehta BR (1999) Inhomogeneous broadening in the photoluminescence spectrum of CdS nanoparticles. Nanostruct Mater 11(5):603–609
Rossi L, Fedenia LN, Sharifan H, Ma X, Lombardini L (2019) Effects of foliar application of zinc sulfate and zinc nanoparticles in coffee (*Coffea arabica* L.) plants. Plant Physiol Biochem 135:160–166
Roy N, Barik A (2010) Green synthesis of silver nanoparticles from the unexploited weed resources. Int J Nanotechnol 4:95
Sandmann G, Dietz H, Plieth W (2000) Preparation of silver nanoparticles on ITO surfaces by a double-pulse method. J Electroanal Chem 491(1–2):78–86
Sankar R, Manikandan P, Malarvizhi V, Fathima T, Shivashangari KS, Ravikumar V (2014) Green synthesis of colloidal copper oxide nanoparticles using *Carica papaya* and its application in photocatalytic dye degradation. Spectrochim Acta Part A Mol Biomol Spectrosc 121:746–750
Sarkar J, Acharya K (2017) *Alternaria alternata* culture filtrate mediated bioreduction of chloroplatinate to platinum nanoparticles. Inorganic Nano-Metal Chem 47(3):365–369
Shahverdi AR, Minaeian S, Shahverdi HR, Jamalifar H, Nohi AA (2007a) Rapid synthesis of silver nanoparticles using culture supernatants of Enterobacteria: a novel biological approach. Process Biochem 42(5):919–923
Shahverdi AR, Fakhimi A, Shahverdi HR, Minaian S (2007b) Synthesis and effect of silver nanoparticles on the antibacterial activity of different antibiotics against *Staphylococcus aureus* and *Escherichia coli*. Nanomed Nanotechnol Biol Med 3(2):168–171
Shang Y, Hasan M, Ahammed J, Li M, Yin H, Zhou J (2019) Applications of nanotechnology in plant growth G and crop protection: a review. Molecules 24(14):2558
Shankar SS, Rai A, Ahmad A, Sastry M (2004a) Rapid synthesis of Au, Ag, and bimetallic Au core–Ag shell nanoparticles using Neem (*Azadirachta indica*) leaf broth. J Colloid Interface Sci 275(2):496–502
Shankar SS, Rai A, Ankamwar B, Singh A, Ahmad A, Sastry M (2004b) Biological synthesis of triangular gold nanoprisms. Nat Mater 3(7):482–488
Sharma NC, Sahi SV, Nath S, Parsons JG, Gardea-Torresde JL, Pal T (2007) Synthesis of plant-mediated gold nanoparticles and catalytic role of biomatrix-embedded nanomaterials. Environ Sci Technol 41 (14):5137–5142
Shivaji S, Madhu S, Singh S (2011) Extracellular synthesis of antibacterial silver nanoparticles using psychrophilic bacteria. Process Biochem 46(9):1800–1807
Silver S (2003) Bacterial silver resistance: molecular biology and uses and misuses of silver compounds. FEMS Microbiol Rev 27(2–3):341–353
Sleytr UB, Messner P, Pum D, Sára M (1999) Crystalline bacterial cell surface layers (S layers): from supramolecular cell structure to biomimetics and nanotechnology. Angew Chem Int Ed 38(8):1034–1054
Smetana AB, Klabunde KJ, Sorensen CM (2005) Synthesis of spherical silver nanoparticles by digestive ripening, stabilization with various agents, and their 3-D and 2-D super lattice formation. J Colloid Interface Sci 284(2):521–526
Smuleac V, Varma R, Baruwati B, Sikdar S, Bhattacharyya D (2011a) Nanostructured membranes for enzyme catalysis and green synthesis of nanoparticles. Chemsuschem 4(12):1773–1777
Smuleac V, Varma R, Sikdar S, Bhattacharyya D (2011b) Green synthesis of Fe and Fe/Pd bimetallic nanoparticles in membranes for reductive degradation of chlorinated organics. J Membr Sci 379(1–2):131–137
Stiger RM, Gorer S, Craft B, Penner RM (1999) Investigations of electrochemical silver nanocrystal growth on hydrogen-terminated silicon (100). Langmuir 15(3):790–798
Sturikova H, Krystofova O, Huska D, Adam V (2018) Zinc, zinc nanoparticles and plants. J Hazard Mater 349:101–110
Suganya KU, Govindaraju K, Kumar VG, Dhas TS, Karthick V, Singaravelu G, Elanchezhiyan M (2015) Blue green alga mediated

synthesis of gold nanoparticles and its antibacterial efficacy against gram positive organisms. Mater Sci Eng, C 47:351–356

Tan Y, Wang Y, Jiang L, Zhu D (2002) Thiosalicylic acid-functionalized silver nanoparticles synthesized in one-phase system. J Colloid Interface Sci 249(2):336–345

Trono JD, Mizuno K, Yusa N, Matsukawa T, Yokoyama K, Uesaka M (2011) Size, concentration and incubation time dependence of gold nanoparticle uptake into pancreas cancer cells and its future application to X-ray drug delivery system. J Rad Res 52(1):103–109

Vigneshwaran N, Kathe AA, Varadarajan PV, Nachane RP, Balasubramanya RH (2006a) Biomimetics of silver nanoparticles by white rot fungus, *Phaenerochaete chrysosporium*. Colloids Surf, B 53 (1):55–59

Vigneshwaran N, Nachane RP, Balasubramanya RH, Varadarajan PV (2006b) A novel one-pot 'green' synthesis of stable silver nanoparticles using soluble starch. Carbohyd Res 341(12):2012–2018

Vigneshwaran N, Ashtaputre NM, Varadarajan PV, Nachane RP, Paralikar KM, Balasubramanya RH (2007) Biological synthesis of silver nanoparticles using the fungus *Aspergillus flavus*. Mater Lett 61(6):1413–1418

Vijaya Kumar R, Elgamiel R, Diamant Y, Gedanken A, Norwig J (2001) Sonochemical preparation and characterization of nanocrystalline copper oxide embedded in poly (vinyl alcohol) and its effect on crystal growth of copper oxide. Langmuir 17(5):1406–1410

Vilchis-Nestor AR, Sánchez-Mendieta V, Camacho-López MA, Gómez-Espinosa RM, Camacho-López MA, Arenas-Alatorre JA (2008) Solventless synthesis and optical properties of Au and Ag nanoparticles using *Camellia sinensis* extract. Mater Lett 62(17–18):3103–3105

Vorobyova SA, Lesnikovich AI, Sobal NS (1999) Preparation of silver nanoparticles by interphase reduction. Colloids Surf, A 152(3):375–379

Wang X, Sun W, Zhang S, Sharifan H, Ma X (2018) Elucidating the effects of cerium oxide nanoparticles and zinc oxide nanoparticles on arsenic uptake and speciation in rice (*Oryza sativa*) in a hydroponic system. Environ Sci Technol 52(17):10040–10047

Wang Y, Lin Y, Xu Y, Yin Y, Guo H, Du W (2019) Divergence in response of lettuce (var. ramosa Hort.) to copper oxide nanoparticles/microparticles as potential agricultural fertilizer. Environ Pollut Bioavailability 31(1):80–84

Xu JF, Ji W, Shen ZX, Tang SH, Ye XR, Jia DZ, Xin XQ (1999) Preparation and characterization of CuO nanocrystals. J Solid State Chem 147(2):516–519

Yu DG (2007) Formation of colloidal silver nanoparticles stabilized by Na^{+}–poly (γ-glutamic acid)–silver nitrate complex via chemical reduction process. Colloids Surf, B 59(2):171–178

Zain NM, Stapley AGF, Shama G (2014) Green synthesis of silver and copper nanoparticles using ascorbic acid and chitosan for antimicrobial applications. Carbohyd Polym 112:195–202

Zheng L, Hong F, Lu S, Liu C (2005) Effect of nano-TiO_2 on strength of naturally aged seeds and growth of spinach. Biol Trace Elem Res 104(1):83–91